AF342512

MOSFET Modeling with SPICE

Prentice Hall Series in Innovative Technology

Dennis R. Allison, David J. Farber, and Bruce D. Shriver   *Series Advisors*

Bhasker	*A VHDL Primer*
Bhasker	*VHDL Syntax*
Blachman	*Mathematica: A Practical Approach*
Chan and Mourad	*Digital Design Using Field Programmable Gate Arrays*
El-Rewini, Lewis, and Ali	*Task Scheduling in Parallel and Distributed Systems*
Henricson and Nyquist	*Industrial Strength C+−*
Foty	*MOSFET Modeling with SPICE: Principles and Practice*
Jenkins	*Designing with FPGAs and CPLDs*
Johnson	*Superscalar Microprocessor Design*
Kane and Heinrich	*MIPS RISC Architecture, Second Edition*
Kehoe and Mixon	*Children and the Internet: A Zen Guide for Parents and Educators*
Kehoe	*Zen and the Art of the Internet, Fourth Edition*
Lawson	*Parallel Processing in Industrial Real-Time Applications*
Nelson, ed.	*Systems Programming with Modula-3*
Rose	*The Internet Message: Closing the Book with Electronic Mail*
Rose	*The Little Black Book: A Practical Perspective on OSI Directory Services*
Rose	*The Open Book: A Practical Perspective on OSI*
Rose	*The Simple Book: An Introduction to Management of TCP/IP-Based Internets*
Rose	*The Simple Book: An Introduction to Networking Management*
Slater	*Microprocessor-Based Design*
SPARC International, Inc.	*The SPARC Architecture Manual, Version 9*
Steinmetz and Nahrstedt	*Multimedia: Computing, Communications & Applications*
Wirfs-Brock, Wilkerson, and Weiner	*Designing Object-Oriented Software*

MOSFET MODELING WITH SPICE
Principles and Practice

Daniel P. Foty

Gilgamesh Associates, Inc.

To join a Prentice Hall PTR mailing list, point to:
http://www.prenhall.com/register

Prentice Hall PTR, Upper Saddle River, NJ 07458
http://www.prenhall.com

Library of Congress Cataloging-in-Publication Data

Foty, D. (Daniel)
    MOSFET modeling with SPICE: principles and practice / Daniel P.
Foty.
      p.   cm.—(Prentice Hall innovative technology series)
    Includes bibliographical references and index.
    ISBN 0-13-227935-5
    1. Metal oxide semiconductor field-effect transistors--Computer
simulation.   2. SPICE (Computer file) I. Title. II. Series.
TK7871.95.F37   1997
621.3815'284--dc21   96-49414
                                       CIP

Acquisitions editor: **Karen Gettman**
Editorial/production supervision
  and interior design: **bookworks**
Manufacturing manager: **Alexis Heydt**
Marketing manager: **Miles Williams**
Cover design: **Design Source**
Cover design director: **Jerry Volta**
Composition: **PREPARÉ Inc.**

 © 1997 by Prentice Hall PTR
Prentice-Hall, Inc.
A Simon & Schuster Company
Upper Saddle River, New Jersey 07458

The publisher offers discounts on this book when ordered
in bulk quantities. For more information, contact:

Corporate Sales Department
PTR Prentice Hall
One Lake St.
Upper Saddle River, NJ 07458

Phone: 800-382-3419
FAX: 201-236-7141
E-mail: corpsales@prenhall.com

ISBN 0-13-227935-5

Prentice-Hall International (UK) Limited, *London*
Prentice-Hall of Australia Pty. Limited, *Sydney*
Prentice-Hall Canada Inc., *Toronto*
Prentice-Hall Hispanoamericana, S.A., *Mexico*
Prentice-Hall of India Private Limited, *New Delhi*
Prentice-Hall of Japan, *Tokyo*
Simon & Schuster Asia Pts. Ltd., *Singapore*
Editora Prentice-Hall do Brasil, Ltda., *Rio de Janeiro*

Contents

8   BSIM   *213*

9   HSPICE Level 28   *277*

10   BSIM2   *318*

11   BSIM3   *379*

Preface

In recent years, the international integrated circuit (IC) industry has enjoyed unprecedented growth. Interestingly, supply is now creating its own demand, as integrated circuits appear in a wider and more diverse set of applications. For a variety of reasons, complementary metal-oxide semiconductor (CMOS) technology has emerged as the leading silicon choice, and clearly dominates the present marketplace.

This growth has spurred the continuing evolution of computer-aided design (CAD) technology for integrated circuit simulation. The SPICE circuit simulator long ago emerged as the de facto industry standard in this field, and is used today in several different forms. To simulate a CMOS integrated circuit, SPICE must make use of element models of the field-effect transistor (FET); these FET models provide a description of how a transistor will behave in the designed circuit. Thus, the FET models serve as the connection between the designed circuit and the underlying fabrication technology.

At this time, several trends are transforming FET model quality into a major issue. First, there has been rapid growth in low-power CMOS technology; this imposes added requirements for model accuracy and robustness. Second, there has been a surge of interest in analog and mixed signal IC design; this design field places more stringent requirements on the finer details of FET modeling, and forces careful evaluation of topics which, in the past, could be safely neglected. Finally, a major trend is the emergence of the fabless design company; such a firm carries out integrated circuit design and then contracts the fabrication to a silicon foundry. In this business, circuit designers and their fabrication technology are separated both geographically and philosophically. Under these circumstances, the FET models serve as the critical communication vehicle between the circuit designer and the foundry.

It therefore behooves the circuit designer to become an educated model consumer of SPICE FET models. This text attempts to provide a single source reference book on the FET models used in mainstream versions of SPICE. A designer must work with the existing modeling infrastructure, and make the best use of currently available simulation technology. This book carefully describes the FET models and various related issues. Chapter 1 introduces the basic reasons for the growth of the IC industry and the concomitant importance of SPICE simulation, while Chapter 2 describes the basic structure used to model FETs in SPICE. Chapter 3 reviews some basic concepts in semiconductor physics and small geometry effects in FETs; Chapter 4 compares these basic analytical results with more accurate numerical simulations, to illustrate the approximations used to derive the analytical expressions. Chapters 5–10 discuss the FET models currently in common use (Level 1, Level 2, Level 3, BSIM, HSPICE Level 28, and BSIM2), evaluating each model's strengths, weaknesses, behavior, and applicability for particular types of circuit design. It is carefully noted that the model equations represent an upper limit of what is possible from a particular type of model, while good parameter extraction is required to reach that limit. Chapters 11 and 12 address, respectively, two new models (BSIM3 and MOS Model 9) which are now being introduced. Chapter 13 discusses gate capacitance modeling and charge conservation; Chapter 14 discusses methods of modeling statistical process variations; and Chapter 15 considers methods of correlating the SPICE FET models and circuit-level results. Chapter 16 considers several candidate FET models, which may find use in SPICE in the near future; these models also serve to identify important trends in FET modeling. Chapter 17 briefly examines historical trends, the present situation, and future directions in analytical FET modeling.

The author hopes that this text will serve as a single reference source for circuit designers and model building engineers in this field. The discussion is intended to be complete and comprehensive, while providing a full set of bibliographic citations for detailed analysis.

This book was made possible by assistance from a number of individuals and organizations. I am particularly indebted to Paul Findley, Morgan Smith, Ali Rezvani, and their colleagues at VLSI Technology, Inc., San Jose, California, for providing the data which were used in the parameter extraction sections of this book, and to Hamed Emami, David Perejda, and their colleagues at Meta-Software, Inc., Campbell, California, for providing the copy of HSPICE which was used for all the parameter extraction and circuit simulation detailed in this text. I also wish to thank Prof. Gennady Gildenblat of Pennsylvania State University, University Park, Pennsylvania, Howard Russell of Opal Engineering, San Jose, California, and Jeff Deutsch of Deutsch Technology Research, Los Altos, California, for reading the manuscript and providing numerous helpful comments and suggestions. I also owe a debt to many colleagues with whom I continue to discuss FET modeling issues, including Peter Bendix, Christian Enz, Matthias Bucher, Seamus Power, Jared Zerbe, John Cooley, Jeff Deutsch, Mohammed Ismail, Richard Kaul, Ann Spratt, Alex Sinar, Steve Michael, Robert Taft, Narain Arora, and Al Kordesch. Finally, I wish to thank Karen Gettman of Prentice-Hall for her patience as the editor for this project, and for seeing the effort through to its completion. Thanks are also due to Karen Fortgang for production assistance and to Donna Mulder for copy editing.

Daniel Foty
Fletcher, Vermont

MOSFET Modeling with SPICE

1

SPICE Modeling and the Dominance of CMOS Technology

1.1 Introduction

In recent years, the international integrated circuit (IC) industry has undergone explosive growth. Electronics is now the dominant international industry, with an annual growth rate which is much higher than other types of business. At this time, integrated circuits are appearing in a larger and more diverse range of product systems.

During the 1990s, much of this growth has been in complementary metal-oxide semiconductor (CMOS) technology. CMOS technology has proven to be ideal for the expansion of integrated circuit usage outside the traditional "computation" role, and has been able to meet the more stringent cost constraints inherent in these more diverse mainstream applications. The reasons for the success of CMOS will be considered below.

At the same time, this growth has fueled a vast expansion of circuit modeling and simulation. As more ICs are designed, there is clearly a corresponding need for more circuit simulation. In addition, increasingly clever circuit designs are creating a demand for more accurate and precise modeling capabilities. As part of this expansion, a set of standards has emerged; this is due to both the common educational background of most engineers and the growing need for collaboration among various design groups.

For reasons to be discussed more fully in Chapter 2, the SPICE circuit simulator, in its many varieties, has emerged as the lingua franca standard for circuit simulation. Many fine texts are available to aid the circuit designer in the use of SPICE; however, this is not the focus of this particular text.

For CMOS circuit simulation, the FET element models that are available in SPICE must be invoked (that is, these FET element model subroutines must be called by the circuit main program). For good simulation results, the ability to make effective use of

1

these element models is essential; the designer wishes to have an FET model available which produces close agreement with the actual fabrication results. With accurate FET models, the basic goals are to reduce the number of design-and-fab iterations required to achieve proper fabricated circuit results, while providing the best possible yield results when design success is achieved.

A new factor on the IC scene is the explosive growth of the *fabless* segment of the industry. A fabless firm has no dedicated fabrication facility of its own; instead, the firm specializes in design, and then contracts fabrication to a silicon *foundry*, which specializes in manufacturing. This division of the business, away from a vertically integrated model, puts added requirements on modeling and simulation, as the ability of a design firm and a foundry to communicate properly becomes essential for success.

Therefore, before considering the details of analytical FET modeling and SPICE, the present state of the industry should be briefly sketched.

1.2   The Dominance of CMOS Technology

During the 1980s and 1990s, CMOS emerged as the dominant technology for general-purpose integrated circuit applications [1]. This trend continues, as CMOS has crowded out less competitive technologies, such as bipolar and NMOS, while relegating more exotic technologies (e.g., GaAs MESFETs and HEMTs) to niche applications.

CMOS circuits have come to the technology forefront primarily due to their low passive power consumption. Aside from leakage currents, power in digital circuits is dissipated only during switching events; when the circuit is in a stable (nonswitching state), virtually no power is consumed. This is a major improvement over the situation in bipolar and NMOS circuits, where considerable power is dissipated in the quiescent state.

CMOS technology has also advanced due to the availability of simple scaling laws [2]; these scaling laws have served as an outline for the reduction of FET dimensions. By combining these scaling laws with the inherently low power dissipation of CMOS, extremely dense integration has been made possible; millions of FETs are now found in many ICs.

Another significant feature of CMOS is its stability of operation; unlike bipolar and NMOS technologies, small deviations in device characteristics do not perturb the circuit operating point. Consequently, CMOS has proven to be a robust manufacturing technology that permits large quantities and varieties of ICs to be fabricated with high yield.

Finally, CMOS technology is amazingly flexible for a wide variety of applications. For example, the portfolio of CMOS applications includes (but is certainly not limited to) digital logic, static RAMs, dynamic RAMs, signal processing, and a variety of analog functions.

As a result, CMOS technology has reached the enviable position where supply creates its own demand. The wide applicability of CMOS has given rise to a broad, high-quality infrastructure for development and fabrication that allows advances to be made widely available at a reasonable cost. This is evidenced by the proliferation of CMOS ICs in a growing variety of applications that would not have been considered less than a decade ago. This proliferation is particularly evident in areas outside of the traditional computation field, where growth has been particularly rapid, with amazing diversity. It is reasonable to postulate that in the near future, "If it doesn't rot, it will have silicon in it."

The need to serve this diversity is adding new demands to CMOS technology. In the traditional computation role, the main goal was the increase of clock (switching) speeds in digital logic. While still a strong factor, this trend is being joined by the need for increased integrated logic functionality; that is, designers now integrate functions (which previously were distributed across several ICs and then combined at the card level) onto a single IC (or a much smaller number of ICs). As a result, some de facto technology trends have emerged.

First, simply reducing the size of various structures continually improves device and circuit performance, and smaller structures consume less power. Second, since smaller structures take up less space, and continual improvements in photolithography equipment allow for physically larger ICs, the number of FETs that can be placed in a single IC has grown enormously, now approaching tens of millions. Finally, CMOS fabrication is very flexible, and a single basic process is easily employed for a wide variety of integrated circuits. Combined with improvements in computer-aided design (CAD) systems, this is leading to mixed functions on the same chip. Most large ICs now contain both digital and analog functions, integrated into the same piece of silicon.

1.3   The Growth of the Fabless Design Industry

While CMOS technology has proliferated, a new and very interesting phenomenon has appeared on the technology landscape—the fabless design company. As the name suggests, these firms do not have their own fabrication facilities; instead, they specialize in design, and contract out final silicon manufacturing to a silicon foundry, which eschews design, and specializes in wafer fabrication. This is a very important trend, which must be examined closely.

Until recently, most semiconductor technology companies were vertically integrated, containing development, design, and manufacturing in a single entity. This model, common in the older smokestack industries, allowed for economies of scale, and for the concentration of the large amounts of capital necessary to maintain the entire business. Unfortunately, in the IC industry, many large firms demonstrated an inability to make effective use of this strength, and instead slid toward mediocrity in all facets of the business.

In recent years, a new model of specialization and excellence has burst onto the scene. In this model, the key feature is the choice of a small number of activities; it is then possible to become very good at this short list of activities, and thus maintain focus, agility, and quickness. As a somewhat surprising consequence, a new business ecology has developed, with well-defined "spheres of influence" [3] for the various specialized firms. Interfirm interactions are actually enhanced, since cooperation is more easily carried out among companies that do not in fact compete directly with each other.

In this form, the integrated circuit industry is now undergoing a significant amount of fragmentation. Due to the growth of CMOS technology described in the previous section, manufacturing capacity is no longer at a premium. Instead, the key issues in CMOS manufacturing are low-cost and high-volume production; the centralization of these activities lowers the cost for all users. In addition, this new focus has caused silicon manufacturing to become very sensitive to the "cost of business"; this accounts for the shift of silicon fabrication in the United States to geographic areas that offer lower costs (e.g., away from California and toward, e.g., Texas, Arizona, and New Mexico). Along this line of thought, it is important to note that the largest growth of silicon capacity is occurring in Asia, a region long known for its aptitude for low-cost, high-volume manufacturing. While the techniques for CMOS manufacturing are now well known, the business is extremely capital intensive, with a very high cost of entry; the cost of a brand new, state-of-the-art wafer fabrication facility now runs in the neighborhood of $1.2 billion.

In contrast, CMOS circuit design is very knowledge intensive and requires a very high skill level. Factors such as creativity, innovation, and a rapid response to the marketplace dominate this wing of the business. Intellectual property is very important here, as the "value-add" of design is very high.

Of particular note is the drive to condense functions that used to be served at the board or system level directly into silicon. To serve such specific needs generally requires specific solutions. This has led to the growth of the application-specific integrated circuit (ASIC) business, where an IC (or group of ICs) is designed specifically for the application in question. The creation of a specific (rather than general) solution allows for better overall performance, and for the creation of unique features to address the particular application. At present, ASICs represent about 10% of the integrated circuit business [4], with rapid growth projected and in progress. A small, highly skilled design team is the key for ASIC success; this is in contrast to the very large design teams that are assembled for large-scale ICs, such as mainstream microprocessors.

When compared with the requirements for semiconductor manufacturing, the cost of entry for design is rather low, and the design team can be quite small. This had led to the creation of a number of small fabless companies. Here the firm can concentrate on the detailed design of a small number of profitable ICs, while not having to deal with the cost or difficulties of final fabrication. Fabless companies rely on an extreme degree of flexibility, in both the types of ICs they design, and in ensuring that their designs are portable among a number of silicon foundries. A fabless company must offer products that require specialization and innovation, rather than the easy availability of volume manufacturing.

Today the most innovative developments in the integrated circuit industry are emerging from the many small fabless companies that have sprung up (and which continue to appear on the scene). At present, fabless companies represent about 3% of semiconductor revenue; however, fabless revenue is growing at twice the rate of the revenue growth of the IC industry as a whole [5].

The development of the fabless company is a major event on the integrated circuit industry landscape. However, this new form of business puts an additional burden on the development of analytical FET models for circuit simulation. In the situation of design firm and foundry supplier, the circuit design group and the silicon fabrication group are likely to have no formal connection. The FET element models thus become the only method of communication between the circuit designers and the foundry. It thus behooves a circuit designer to become an "educated model consumer," and take a strong interest in the FET element models that are being used to design circuits [6].

1.4   The Present Modeling and Simulation Infrastructure

In the integrated circuit industry, there are a number of proprietary circuit simulation programs, as well as a few that are not proprietary but which are infrequently used. Many of these simulators are based on SPICE, but with a large number of added features. However, with the basic SPICE structure available, a new user can quickly use the basic capabilities of the simulator, while picking up the more detailed proprietary features at a later time.

The situation in analytical FET models is quite similar. There are a number of proprietary FET models in use, along with a large number of models that have been published in the technical literature; however, these models find little general use.

In this text, the focus will be on mainstream FET element models and circuit simulators, which find widespread use throughout the integrated circuit industry. These are the FET models that are available in mainstream versions of SPICE, such as University of California/Berkeley SPICE, HSPICE, and PSPICE.

In addition, small fabless design firms, which are proliferating rapidly, do not have the resources to maintain large in-house efforts in circuit simulation and FET element model development. Instead, generally available circuit simulators and FET element models must be used as is. In such a situation, it is imperative that ways be found to use the available models to the greatest effect. This type of effort is the focus of this text.

BIBLIOGRAPHY

1. D. Foty and E. Nowak, "MOSFET Technology for Low Voltage/Low Power Applications," *IEEE Micro*, June 1994, pp. 68–77.

2. R. Dennard et al., "Design of Ion Implanted MOSFETs with Very Small Physical Dimensions," *IEEE J. Sol. St. Circ.*, vol SC-9, pp. 256–268 (1974).

3. B. Caven, *The Punic Wars*, Barnes & Noble Books, 1980.

4. J. Schroeter, *Surviving the ASIC Experience*, Prentice-Hall, 1992.

5. M. Phelps, "Chips' Fabulous Future," *Upside*, August 1995, p. 16.

6. Y. Tsividis and K. Suyama, "MOSFET Modeling for Analog Circuit CAD: Problems and Prospects," *IEEE J. Sol. St. Circ.* vol. SC-29, pp. 210–216 (1994).

2
SPICE Modeling
and the Formalism
of Model Building

2.1 Introduction

As Abraham Lincoln noted, "We cannot escape history" [1]. Any discussion of SPICE and its FET models must begin by considering the historical development of the SPICE circuit simulator. This evolution has imposed a structure on circuit simulation and FET modeling which has grown in complexity over time. By sketching an outline for these developments, an unintended but clearly discernible order can be identified. The core of the SPICE circuit simulator has been improved but in basic structure has essentially remained unchanged; however, increasingly complex FET models have been required to keep up with the pace of improvements in CMOS technology. In addition, when the demands of industrial circuit design are included, new requirements are imposed on circuit simulation; in particular, the need to properly account for the behavior of the device characteristics has led to model *binning*, and methods of dealing with process variability have been developed. Finally, the formulation of the model equations implies that certain types of data must be made available for parameter extraction; this requires that certain test structures be devised if that task is to be properly performed.

Here the basic infrastructure that has been assembled for the simulation of CMOS circuits in SPICE is briefly reviewed. The history and methods of SPICE and its FET models are discussed, along with the various adaptations which have been introduced for the industrial circuit design environment. This will serve as the basis for the more detailed discussions of succeeding chapters.

2.2   The SPICE Circuit Simulator

During the 1960s, two complementary trends combined to provide both the requirements and abilities to simulate electronic circuits. As circuits grew ever larger and more complex, it became desirable to simulate their behavior on an electronic computer, rather than through the traditional breadboard evaluation. This was particularly true as the integrated circuit appeared on the scene; with hundreds and then thousands of transistors, the task of setting up a circuit on a breadboard was becoming complicated to the point of impossibility. At the same time, as computational power was increasing, it became possible to create a simulation program that would provide accurate answers in a reasonable amount of time.

In the best engineering tradition, growing complexity led to a clever solution. Although this approach is taken for granted now, at the time of its introduction, the concept was somewhat revolutionary. Rather than employing an iterative succession of breadboarding and testing of designs, a circuit would be designed, evaluated, and redesigned using only computer simulation.

The SPICE circuit simulator was the result of this new approach to circuit design [2, 3]. The acronym SPICE is assembled from **S**imulation **P**rogram with **I**ntegrated **C**ircuit **E**mphasis, reflecting its origin as a new method of dealing with the monumental complexity inherent in the design of integrated circuits. The original SPICE core was developed at the University of California/Berkeley in the late 1960s, as a successor to a first attempt at a computer-aided circuit simulator with the somewhat egregious name of CANCER (**C**omputer **A**nalysis of **N**onlinear **C**ircuits, **E**xcluding **R**adiation) [4].

In SPICE, any circuit is treated in a node/element fashion; the circuit is depicted as a collection of various elements (e.g., resistors, capacitors, etc.) connected at numbered nodes. The circuit is described in its entirety by the number of nodes and elements it contains, and by the nodes to which each particular element is connected. In a circuit with n nodes, the SPICE representation can be thought of as an $n \times n$ matrix; if two nodes are connected by some element, the corresponding matrix element will be nonzero. By defining one or more state variables (such as the voltage) at each node, and fixing the state variables at external nodes (such as the power supply, ground, etc.), the matrix can be solved to determine the values of the state variables at the various internal circuit nodes. To reach a solution, models specific to any particular circuit element are required. In this form, the circuit can be thought of as the main program, while the element models are subroutines that are called when the behavior of an element must be evaluated.

In addition to SPICE, there are a number of circuit simulation programs which employ the same basic approach. However, SPICE has become the dominant circuit simulator for two reasons. First, the entire SPICE package, including the program code itself, was (and continues to be) made available by the University of California/Berkeley for a nominal fee. Thus, it was possible to purchase ready-to-run SPICE for circuit simulation, or to acquire the core program and customize it for the user's own purposes. Second, SPICE became available in the right place at the right time. Its availability coincided with the nearby growth of the integrated circuit industry, and was thus easily adopted as the circuit simulation standard.

Because of this widespread adoption, SPICE has become the lingua franca for circuit simulation in the integrated circuit industry. The circuit designer may encounter SPICE in any one of three different groups. The first incarnation is the basic SPICE that is still maintained and updated by the University of California/Berkeley. In addition, there are also enhanced versions of SPICE available with numerous additional features, developed with the industrial circuit designer in mind. The most common such package is HSPICE [5] from Meta-Software, which is used extensively on UNIX workstations in the integrated circuit industry. Another package is PSPICE [6], which was developed for personal computers; PSPICE is also available in an inexpensive student edition, and is frequently used in university integrated circuit classes. Finally, many firms maintain their own internally modified and continually developing version of SPICE, based on the original Berkeley core. In all these versions of SPICE, the basic method of describing and simulating a circuit is the same; in general, the circuit description can be entered into any of these versions of SPICE for simulation. The differences exist beneath the surface, in the mathematical methods used for circuit simulation, and in the element models which are called during circuit simulation.

Therefore, to describe the use of SPICE to simulate integrated circuits, the problem can be separated into three distinct levels. The first level involves the basic methods of using SPICE to simulate various circuits. This subject is well covered in a number of texts and is not described here. At the second level, there is a variety of issues revolving around the mathematical and numerical methods involved in simulating circuits, where the focus is on ensuring the convergence of the simulation while making efficient use of computing resources, so that a useful simulation result may be obtained in a reasonable time. This subject is not discussed in detail here; it is only considered for a few specific issues where problems at the circuit simulation level are intertwined with the main focus of the discussion. Finally, there is a third level, which comprises the element models; this level is the focus of this text.

The discussion in the remainder of this text will focus on one specific part of this third level; the element models used to describe MOSFETs. Due to the present overwhelming dominance of CMOS technology, the vast majority of industrial use of SPICE concerns the simulation of FET-based circuits. In addition, FET technology is evolving rapidly, which provides impetus for evaluating how available models describe that technology, while also demonstrating what requirements must be imposed on new FET models. To make effective use of the FET models available in SPICE, the user must understand the construction and implementation of those models.

2.3 Modeling FETs in SPICE

As noted above, FET technology has evolved considerably over the past 30 years. Thus, the methods employed to model FET elements in SPICE have followed an historical path, paralleling that evolution. The roots of that history may be found in the older, simpler FET technology of the 1960s. The current-voltage and capacitance-voltage characteristics of the FET were originally described by a small set of rather simple equations, with a few parameters to match the model to the specific silicon technology in question. These parameters either represented particular and quantifiable physical values, or had strong physical meaning. The values of those parameters which had physical meaning were provided directly from process information, while those that could not be quantified directly in that manner were determined from electrical data by very simple parameter extraction, using a linear least squares method.

This approach set the rules for the description of FETs as model elements to be called during a SPICE simulation. A SPICE FET model consists of a tabular set of model parameters which describes that FET technology, while the general equations describing the device characteristics are implemented in SPICE. The provided parameter set is plugged into the device equations, and the equations are solved to describe the specific device characteristics during circuit simulation.

As FET technology has evolved, this situation has become a good deal more complex. As various effects due to small-device geometries and/or high field strengths have become more important, more equations have been introduced to describe the device behavior, and those equations have become more complicated. Not surprisingly, this has also led to an increasing number of model parameters. This has had the effect of increasingly shifting the burden of modeling the FET from the the model equation formulation to the process of extracting the model parameters. These more complicated model equations have forced the use of nonlinear least squares curve-fitting techniques, which are numerical and iterative; thus, rather than merely being able to use a hand calculator for parameter extraction, more sophisticated mathematical techniques and software are required. As a result, the analytical formulation of a particular FET model represents an upper limit of how well that model is able to describe any particular FET technology. Well-executed parameter extraction is required to come as close as possible to that limit; in more complex models, parameter extraction is at least as important a consideration as the model formulation.

With the continuing development of the FET models which are used in SPICE, the relationship between the model parameters and the process technology they describe has become increasingly murky. In a completely ideal situation, a set of purely physical process parameters, comprised of easily measurable quantities, is compiled, and the FET model is complete. Reality has never quite been able to reach this level of simplicity, and some extraction of a few parameters from measured electrical data has always been necessary. As a consequence, any physical basis for these extracted parameters is weakened. For the discussions in the remainder of this text, it will be useful to divide the model parameters into two categories. *Physical* parameters are treated as having direct and quantifiable physical meaning; an example is the gate oxide thickness. *Electrical* parameters are defined from parameter extraction, and thus may actually have little direct physical meaning. Some parameters which, in name, would seem to be physical parameters may in fact be electrical parameters; an example is the junction depth, which is a physical parameter in the Level 1 model but, without a name change, becomes an electrical parameter in the Level 2 and Level 3 models. As the basic model formulation has evolved into more complex forms, the number of electrical parameters has become very large. For example, in the Level 1 model, there are three physical parameters and three electrical parameters, while in the BSIM2 model, there are five physical parameters and 35 electrical parameters. When a large number of electrical parameters must be determined, parameter extraction grows to assume a very large degree of importance.

Concurrent with this evolution in FET technology, new analytical model formulations for describing modern FETs are continually appearing in the technical literature. Regardless of their quality or potential usefulness, few of these models are adopted for common use in circuit simulation. To become widely used, a particular model requires a commonly used circuit simulation *vehicle* (a commonly used version of SPICE) to facilitate its adoption. Due to the availability of SPICE as such a vehicle, the lion's share of the most commonly used FET models was developed at the University of California/Berkeley; these are the most generally encountered models for use in circuit design. The only mainstream exception is the HSPICE Level 28 model, which uses the HSPICE circuit simulator as the necessary vehicle for acceptance. Although the model formulation is largely proprietary, this model is frequently employed in circuit simulation. It is possible that the recognition of the need for a vehicle (other than simple publication) for model adoption will soon cause new models from other sources to find general use; MOS Model 9, from Philips Laboratories, is a new model which is being made widely available to facilitate its potential adoption. Finally, many firms have developed their own FET models for internal use; however, these models remain strictly proprietary, so they cannot be examined and will not find their way into common use.

With this backdrop, the models that *are* commonly employed can be divided into three historical generations. The first-generation models represent the oldest efforts, and come closest to the ideal of describing the FET from very simple, physically based parameters. This generation consists of the Level 1, Level 2, and Level 3 models. The second-generation models introduce a very large number of empirical electrical parameters, clearly shifting the focus to the circuit design user. Extensive mathematical conditioning is introduced to improve the robustness and convergence behavior of the model when used in circuit simulation, and a new approach to describing the geometry dependence, involving geometry-dependent parameters, is introduced. Due to their highly empirical nature, successful use of these models requires a tremendous amount of parameter extraction effort. This generation of models is composed of BSIM (sometimes referred to as BSIM1), HSPICE Level 28, and BSIM2. Both HSPICE Level 28 and BSIM2 were improved so that the derivatives of the current equations are continuous; this is required for analog circuit design. The development of the third generation of SPICE FET models is currently under way, and is being shaped by two major forces. The first is the need for models suitable for analog design, as reflected in the widespread use of HSPICE Level 28 and BSIM2. The second force is a desire to reduce the number of model parameters from that found in the second-generation models, and return more physical meaning to those parameters. As a result, these newer models contain one current equation to describe all regions of device operation; this guarantees the continuity of the current equation and its derivatives. At present, this situation is in a state of flux, or perhaps chaos. The University of California/Berkeley has developed two versions of its BSIM3 model (versions 2 and 3), which bear little resemblance to the two forebearers which share the same name (BSIM and BSIM2). However, in contrast to the situation in the past, models from other sources are receiving consideration for widespread use. As noted above, MOS Model 9 from Philips Electronics [7] is being made available, and is being included in a number of mainstream circuit simulators. MOS Model 9 is somewhat unique in that it has all of the third-generation characteristics but maintains a second-generation approach to describing the geometry dependence. Among possible new candidate models are the PCIM model from Digital Equipment Corporation [8], and the EKV model from the Swiss Federal Institute of Technology in Lausanne [9]. At this time, it is uncertain if any of these candidate models will find widespread use, or if other potential candidates will appear.

Some final points on the FET models should be made. The SPICE FET models have been developed for *circuit simulation* rather than *process characterization*. This is a very important distinction; with the growth of the application-specific integrated circuit (ASIC) industry, the SPICE FET models serve as the vehicle for communication between the ASIC design firm and the silicon fabrication foundry. To proceed successfully, the designer requires a set of stable, reliable, and sufficiently accurate models for circuit design and development. As silicon process technology evolves, predictable performance goals should be attainable, and the FET models should be very stable. The FET models should not chase the process technology; if any chasing is necessary, it should be in the other direction, as the process technology should conform to the model predictions (or be modified to do so). Continually changing FET model parameter sets indicate that the process is unstable, and the circuit designer should use extreme caution before investing time and effort in a design.

In addition, in an ideal world, all the model parameters would have direct physical meaning, allowing a circuit designer to examine directly the effect of process changes on circuit behavior by merely changing one or more parameters. However, modern FET models are largely empirical, preventing such a simple evaluation. Furthermore, attempts to use circuit simulation results to suggest process modifications are very risky, as the complexity of process technology renders it vulnerable to nonlinear behavior; a very small attempted change can have larger repercussions that may be difficult to control. Instead, the role of process technology development is to set goals for FET behavior, to develop processes to meet those goals, and to make that stable process technology available at a low cost to circuit designers. SPICE FET models are constructed to describe this stable process technology and are used to communicate with the circuit designer.

2.4   Model Binning

In the situation of an ideal model, for any given FET process technology, one set of model parameters (both process parameters and electrical parameters) would be valid over the entire geometry range. The model equations and this set of model parameters would accurately describe the FET characteristics for all possible values of the channel length and the channel width.

In general, however, while good model results can be obtained for carefully chosen target geometries, the simulated characteristics vary quite widely from the actual device

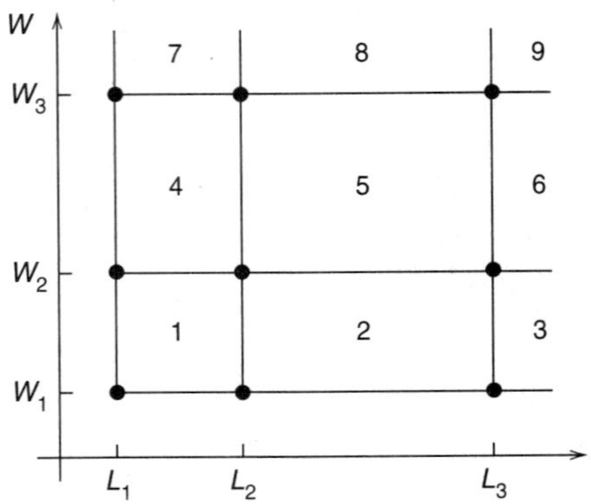

Figure 2.1   Model binning, in which the model is divided into subregions (or bins). A submodel is created for each subregion and is valid only in that region.

characteristics when the channel length and width are varied. This problem is handled in a clumsy and after-the-fact (but necessary) approach known as model *binning*, in which the channel length and width geometry space is divided into smaller subregions, as shown in Figure 2.1. Based on this division, submodels are created, each one of which is deemed to be valid in one particular subregion of geometry space. This method is commonly used and can be implemented with any SPICE FET model. However, the practice of binning introduces a set of problems that is all its own. Since the details of binning are specific to each particular SPICE FET model, a discussion of this approach and its side effects will accompany the description of each FET model.

If binning is used, the modeling situation becomes more complicated. The main burden is carried during parameter extraction and model construction, where the bin sizes and scopes must be selected, and separate models are constructed for each bin. The circuit designer must now ensure that the proper submodel for each given length and width is called for each FET in the circuit. In Berkeley SPICE and in PSPICE, this must be done by hand in each FET call in the circuit netlist. HSPICE includes an automatic model selector; the bins and their geometric limits of validity are tabulated, and when an FET model is called, the length and width of the FET in the circuit netlist are used to select the appropriate submodel.

2.5   Accounting for Process Variations

In an ideal world, absolutely identical FETs would be fabricated every time, on every die and on every wafer. In reality, even when a process is stable and well controlled, there will be systematic variations around the statistical center of that process. Since these variations will affect the device characteristics and circuit behavior, they must be described by any SPICE FET model.

While accounting for this variability may seem to be a nuisance, it is a very important part of the information that is communicated to the circuit designer by the FET model. At a descriptive level, the circuit designer must know what range of distribution

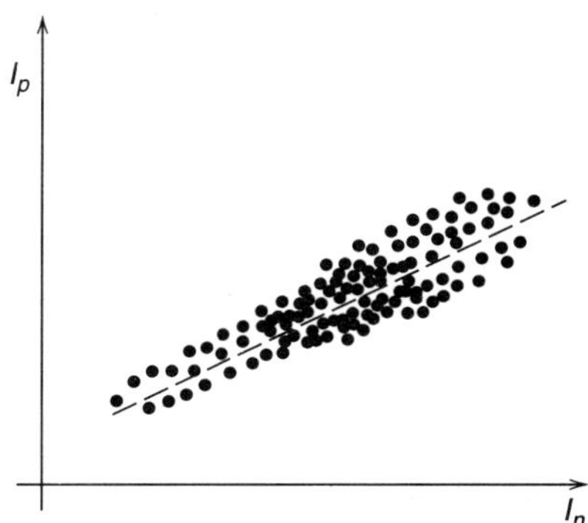

Figure 2.2   A plot of n-channel current versus p-channel current, showing the typical and 3σ slow and fast points along the line. By attempting to push the design center toward the fast corner of the process, the sorted yield of high-frequency parts can be increased, at a cost of a decrease in the overall yield.

the process will extend over, so that it can be ensured that the circuit will work over the entire expected random variation of that process. In addition, the description of that range of process variation can be used as a design variable itself, as will be discussed later.

In digital circuit design, the model must have the capability of describing how process variations influence the final circuit speed. This is demonstrated somewhat crudely in Figure 2.2, where the n-channel current is plotted against the p-channel current. The typical values of the two currents, noted in the plot, represent the center of a typical design. Lower current values define the "slow" variation of the process; here, the model description is used to ensure that circuit switching speeds will still be fast enough to meet the design goals and specifications. Higher current values define the "fast" variation of the process; in this case, the designer must ensure that no waveforms switch too quickly for the intended signal to be properly received by a succeeding circuit. If this variability is well described, it can even be put to use in certain design situations. Instead of targeting the design toward the typical center of the process, the design can be intentionally pushed up toward the fast end of the process. This practice is often used to increase the sorted yield of high-frequency integrated circuits, at a cost of reducing the overall yield, as the highest end of the distribution will move beyond the point of ensured circuit functionality.

The requirements of analog circuit design are somewhat different, where accuracy is generally more important than speed. The modeling of systematic process variations is used to describe the statistical distribution of currents and conductances in filters and signal processing circuits. Here the ability to model the variations in frequency response and bandpass width typically are paramount concerns. In addition, there is a more stringent requirement to understand how the process varies within individual wafers and dies. An example of a circuit that is vulnerable to within-die process variations is the current mirror circuit shown in Figure 2.3. If the two FETs are absolutely identical, the current mirror will mirror perfectly. However, process variations between these two FETs will determine how different the two current values can become.

If all the parameters in the FET model were physically based process parameters, it would be a straightforward matter to compile the measured process variations, and

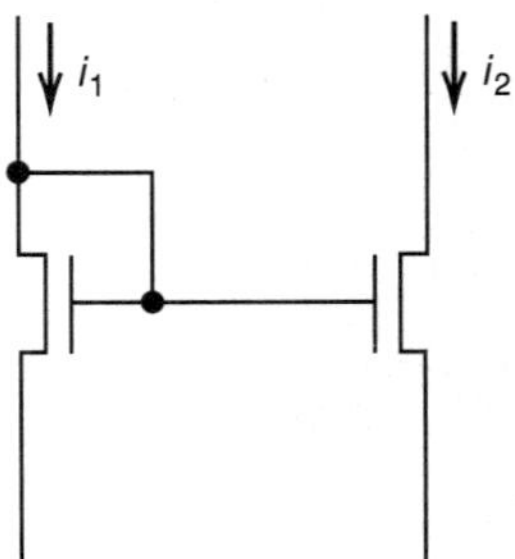

Figure 2.3   A current mirror circuit, consisting of two nFETs.

then use that compilation to describe the expected variability in the device characteristics and circuit behavior. However, with the lion's share of model parameters being electrical parameters, the situation is very complex. Thus, this subject is discussed in further detail in Chapter 14.

2.6   The Design of Test Patterns for Model Building

A final and somewhat neglected issue is the requirement that certain types of data be available if proper parameter extraction and model construction are to be carried out. As the various element and FET models are developed in the other chapters of this text, it will be implicitly shown what data are required, which suggest what types of structures should be available to produce those data. For convenience, the basic test pattern requirements will be tabulated here.

The overall data requirements tend to vary from FET model to FET model. However, it is likely that for any given process technology, more than one type of FET model will be required to meet various (and often conflicting) circuit design requirements. Thus, the best approach involves designing test patterns that can be made to serve all the possible models which might have to be developed for a particular CMOS process technology.

For the SPICE FET models, the most important structures are the test FETs which are used to describe the current-voltage characteristics. For both physical and historical reasons, all the FET models require the use of a long, wide base device, and some short and narrow devices to describe small geometry effects. To cover all the possible FET models, it is best to have a width/length matrix of FETs, as shown in Figure 2.4. A matrix of this sort is required for parameter extraction for the most sophisticated FET models, and will easily meet the requirements of the less complex models. Usually, a minimum of four channel lengths and three channel widths is chosen. The total number of devices included is chosen at the discretion of the test pattern designer, often based on how rapidly the device characteristics vary with geometry for the particular process technology in question.

In addition to FETs, a number of large structures are included. The large size is used to allow the measurement of currents and capacitances which would have prohibitively

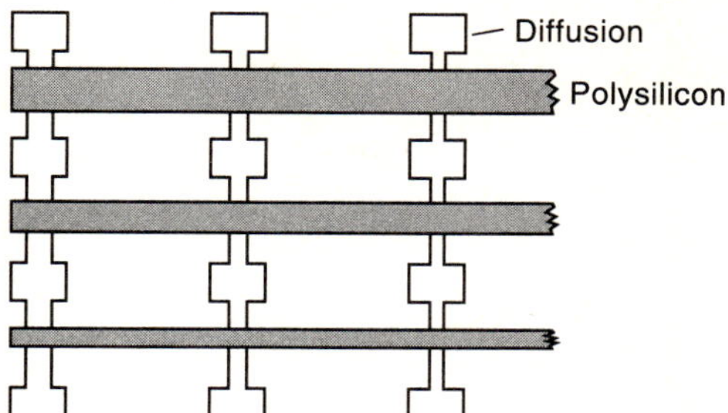

Figure 2.4 The layout of a matrix of nine FETs, a combination of three channel lengths and three channel widths.

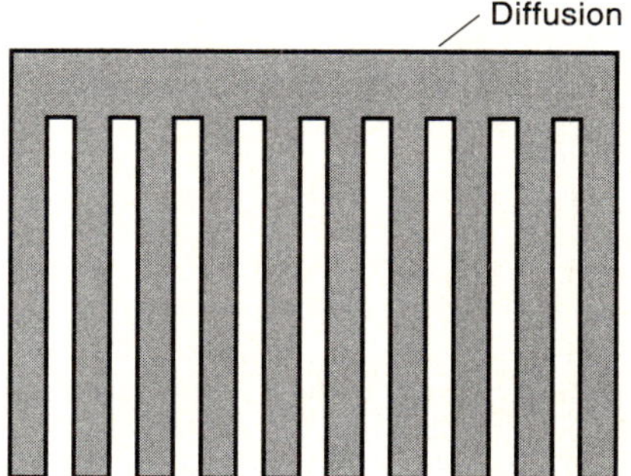

Figure 2.5 A diffusion comb, used to determine perimeter current and capacitance.

small values in FET-size structures. A large-area thin oxide capacitor is always included, and is used to determine the gate oxide thickness by capacitance-voltage measurements. Such structures are generally included in any test pattern, and are used to routinely monitor the gate oxide quality. A number of diffused diodes are also included. A large-area diffused diode must be available, along with a large isolation perimeter-length diode, in the form of a comb (Figure 2.5) or a serpentine. These structures are used to build the current-voltage and capacitance-voltage diode models, while the complementary structures serve to allow the separation of area and perimeter effects; this is described in detail in Chapter 3. A large perimeter length of gate material, as a comb (Figure 2.6) or serpentine over a diffusion is also required; this structure is used to determine the edge capacitance on the side of a diffusion abutting the FET gate, and is also used to extract the zero bias gate-diffusion capacitance.

All of these structures must occur twice; once for NMOS parameter extraction, and once for PMOS parameter extraction.

Finally, some test circuits should be included for correlation of the constructed model with circuit behavior. For digital circuits, one or more ring oscillators (Figure 2.7) are commonly employed. Since a ring oscillator produces simple data (a ringing frequency) rather than a waveform, it is a very straightforward structure for use in comparing model and hardware results. The only major design issue is ensuring that a

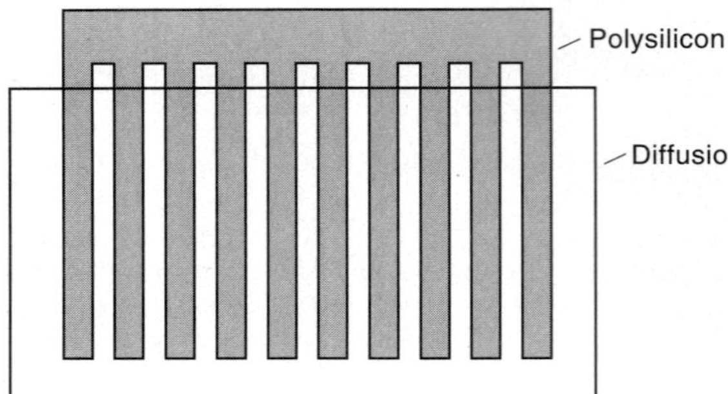

Figure 2.6 A polysilicon comb over a diffusion. This structure contains a large polysilicon perimeter over the diffusion edge, which mimics the FET gate edge over the source and drain diffusions. This structure is used to measure gate-bounded diffusion perimeter current and capacitance, and the gate-diffusion overlap capacitance.

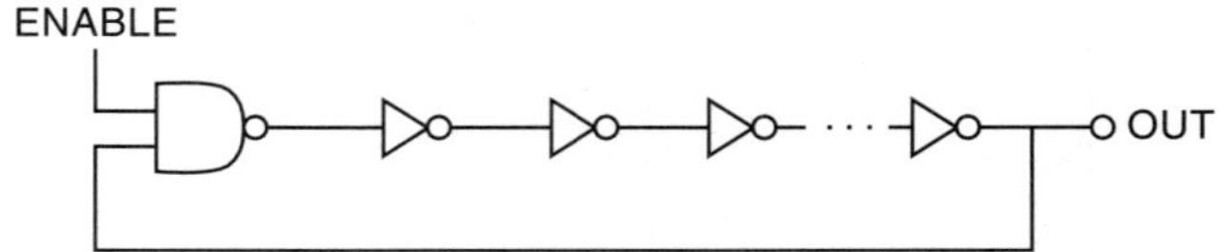

Figure 2.7 A simplified schematic of a ring oscillator circuit, in which a signal propagates around the ring; the delay time for the propagation of the signal is used to determine the switching delay per inverter stage. The NAND gate serves as an inverter which is enabled or disabled by the voltage on the Enable pin; switching Enable from Low to High (ground to V_{dd}) turns on the ring oscillator.

sufficient number of inverters is used, so that each inverter will have the opportunity to switch to either the power supply voltage or ground before the next signal pulse comes around the ring again; the use of more than 100 inverters is common. Note that care must be used to ensure the use of an **odd** number of stages, or the ring oscillator will not ring. It is also common practice to include CMOS inverters and two-input NAND gates to compare model and hardware output waveforms. Additional circuits (such as current mirrors) should be included to meet the particular circuit design needs of each specific situation.

BIBLIOGRAPHY

1. A. Lincoln, Annual Message to Congress, December 1, 1862.

2. L. Nagel and D. Pederson, "Simulation Program with Integrated Circuit Emphasis," University of California/Berkeley, Electronics Research Laboratory Memorandum No. UCB/ERL M352 (1973).

3. L. Nagel, "SPICE2: A Computer Program to Simulate Semiconductor Circuits," University of California/Berkeley, Electronics Research Laboratory Memorandum No. UCB/ERL M520 (1975).

4. L. Nagel and R. Rohrer, "Computer Analysis of Nonlinear Circuits, Excluding Radiation (CANCER)," *IEEE J. Sol. St. Circ.* vol. SC-6, pp. 166–182 (1971).

5. *HSPICE User's Manual*, Meta-Software, Inc., Campbell, California, 1993.

6. *PSPICE User's Guide*, MicroSim, Inc., Irvine, California, 1992.

7. R. Velghe, D. Klaassen, and F. Klaassen, "MOS Model 9," Philip Electronics N.V., Unclassified Report NL-UR 003/94 (1994).

8. N. Arora, R. Rios, C. Huang, and K. Raol, "PCIM: A Physically Based Continuous Short Channel IGFET Model for Circuit Simulation," *IEEE Trans. Elec. Dev.* vol. ED-41, pp. 988–997 (1994).

9. C. Enz, F. Krummenacher, and E. Vittoz, "An Analytical MOS Transistor Model Valid in All Regions of Operation and Dedicated to Low Voltage and Low Current Applications," *Analog Int. Circ. and Sig. Processing* vol. 8, pp. 83–114 (1995).

3

The Semiconductor Physics
of MOS Structures

3.1 Introduction

In this chapter, the basic physics of two-, three-, and four-terminal MOS structures is reviewed. This will provide the foundation for the discussions which follow in later chapters, as the basic analytical expressions will be developed here. The models used in SPICE for the capacitance and current of the p-n junction diode have been developed independently of the FET current models, and are fully evaluated in this chapter.

For the four-terminal MOS structure, the basic differential equation that is used as the starting point for the derivation of the various drain current models is developed. The origins of the gate capacitance models are also considered, and a brief synopsis of the various high field and small geometry effects encountered in modern FETs is also given. The physical cause of any particular effect does not change, whereas a variety of evolving methods is used in modeling these phenomena. Thus, the device physics will be considered here, while specific modeling methods are discussed in the individual chapters on the various FET models. The intent is to introduce the required background material without delving into the deepest details of semiconductor physics.

The approaches described here lead to analytical models, which require various approximations to allow analytical solutions. In Chapter 4, these analytical models are compared with more accurate numerical solutions to demonstrate the differences between the two methods, and to indicate where caution must be taken in employing analytical models.

3.2 The p-n Junction Diode

The simplest semiconductor device is the p-n junction diode, shown in Figure 3.1. The analysis begins [1, 2] by assuming that the junction depletion regions extend to depths

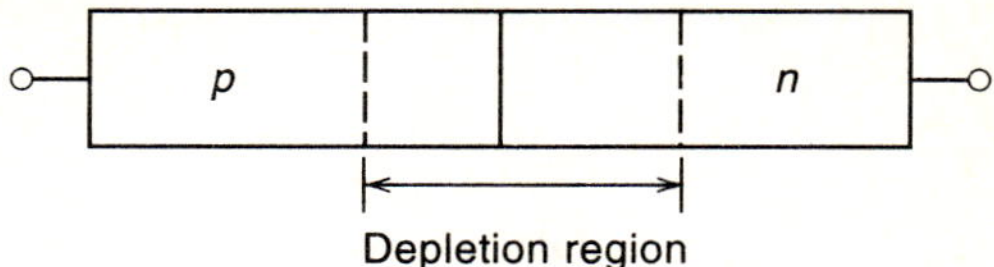

Figure 3.1 A simple abrupt p-n junction, showing the depletion region near the metallurgical junction.

of W_{no} and W_{po} from the junction, and that they end abruptly (the *delta depletion approximation*). It is then a simple matter to solve the one-dimensional Poisson equation:

$$\frac{d^2\psi(x)}{dx^2} = -\frac{\rho}{\epsilon_{Si}} = -\frac{q}{\epsilon_{Si}}(p - n + N_D - N_A),\qquad(3.1)$$

where $\psi(x)$ is the potential along the device; complete impurity ionization is assumed here. Since

$$\frac{d\psi(x)}{dx} = -E_x,\qquad(3.2)$$

(3.1) can be rewritten in Gauss' law form:

$$\frac{dE_x}{dx} = \frac{q}{\epsilon_{Si}}(p - n + N_D - N_A).\qquad(3.3)$$

It is reasonable to assume that the field has swept the depletion region free of carriers:

$$(N_D - N_A) \gg (p - n),\qquad(3.4)$$

and (3.3) becomes

$$\frac{dE_x}{dx} = \frac{q}{\epsilon_{Si}}(N_D - N_A).\qquad(3.5)$$

Since the junction shown in Figure 3.1 is abrupt, (3.5) can be split into two expressions. To the left of the junction, $N_D = 0$, and (3.5) becomes

$$\frac{dE_x}{dx} = -\frac{q}{\epsilon_{Si}}N_A.\qquad(3.6)$$

To the right of the junction, $N_A = 0$, and (3.5) becomes

$$\frac{dE_x}{dx} = -\frac{q}{\epsilon_{Si}}N_D.\qquad(3.7)$$

It is then easy to integrate (3.6) and (3.7),

$$\int_0^{E_x(x)} dE_x = -\frac{q}{\epsilon_{Si}} N_A \int_{-W_{po}}^{x} dx, \tag{3.8}$$

$$\int_0^{E_x(x)} dE_x = \frac{q}{\epsilon_{Si}} N_D \int_{W_{no}}^{x} dx, \tag{3.9}$$

yielding, respectively,

$$E_x(x) = -\frac{q}{\epsilon_{Si}} N_A \left(x + W_{po}\right), \tag{3.10}$$

and

$$E_x(x) = -\frac{q}{\epsilon_{Si}} N_D \left(W_{no} - x\right). \tag{3.11}$$

Setting $x = 0$ in (3.10) and (3.11) produces two expressions for the field at the junction:

$$E_j = -\frac{q N_A W_{po}}{\epsilon_{Si}}, \tag{3.12}$$

$$E_j = -\frac{q N_D W_{no}}{\epsilon_{Si}}. \tag{3.13}$$

These results can be related to the junction potential via the definition

$$E_x(x) = -\frac{dV_x}{dx}, \tag{3.14}$$

or

$$dV_x = -E_x(x)dx. \tag{3.15}$$

Since there are two different regions, the integration is once again split into two parts:

$$\int_{\phi_p}^{\phi_j} dV_x = -\int_{-W_{po}}^{0} E_x(x)dx, \tag{3.16}$$

$$\int_{\phi_j}^{\phi_n} dV_x = -\int_0^{-W_{no}} E_x(x)dx, \tag{3.17}$$

where ϕ_j is the junction potential and ϕ_n and ϕ_p are, respectively, the Fermi potentials in the bulk of the n and p regions. Substituting (3.10) into (3.16) and (3.11) into (3.17), the integrals become

$$\int_{\phi_p}^{\phi_j} dV_x = \frac{qN_A}{\epsilon_{Si}} - \int_{-W_{po}}^{0} (x + W_{po}) \, dx,$$
(3.18)

$$\int_{\phi_j}^{\phi_n} dV_x = \frac{qN_D}{\epsilon_{Si}} - \int_{0}^{-W_{no}} (W_{no} - x) \, dx.$$
(3.19)

Integrating the left sides of (3.18) and (3.19) and rearranging the right sides leads to

$$(\phi_j - \phi_p) = \frac{qN_A}{\epsilon_{Si}} \left[\int_{-W_{po}}^{0} x \cdot dx + W_{po} \int_{-W_{po}}^{0} dx \right],$$
(3.20)

$$(\phi_n - \phi_j) = \frac{qN_D}{\epsilon_{Si}} \left[W_{no} \int_{0}^{-W_{po}} dx - \int_{0}^{-W_{po}} x \cdot dx \right].$$
(3.21)

Integrating (3.20) and (3.21) produces

$$(\phi_j - \phi_p) = \frac{qN_A}{\epsilon_{Si}} \left[-\frac{W_{po}^2}{2} + W_{po}^2 \right] = \frac{qN_A W_{po}^2}{2\epsilon_{Si}},$$
(3.22)

$$(\phi_n - \phi_j) = \frac{qN_D}{\epsilon_{Si}} \left[W_{no}^2 - \frac{W_{no}^2}{2} \right] = \frac{qN_D W_{no}^2}{2\epsilon_{Si}}.$$
(3.23)

Adding (3.22) and (3.23) leads to

$$(\phi_n - \phi_p) = \frac{q}{2\epsilon_{Si}} \left(N_A W_{po}^2 + N_D W_{no}^2 \right).$$
(3.24)

However, $(\phi_n - \phi_p)$ is the junction potential V_{bi}, so (3.24) becomes

$$V_{bi} = \frac{q}{2\epsilon_{Si}} \left(N_A W_{po}^2 + N_D W_{no}^2 \right).$$
(3.25)

Expressions are now required for W_{po} and W_{no}. Recall the two expressions for the field at the junction, (3.12) and (3.13). Setting (3.12) equal to (3.13) yields

$$N_A W_{po} = N_D W_{no}.$$
(3.26)

Solving (3.25) and (3.26) together produces, after some algebraic manipulation, expressions for the depletion region depths on either side of the junction:

$$W_{po} = \left[\frac{2\epsilon_{Si}}{q} V_{bi} \frac{N_D}{N_A(N_D + N_A)} \right]^{\frac{1}{2}},$$
(3.27)

$$W_{no} = \left[\frac{2\epsilon_{Si}}{q} V_{bi} \frac{N_A}{N_D(N_D + N_A)} \right]^{\frac{1}{2}}.$$
(3.28)

The total depletion width W is found by summing (3.27) and (3.28). After algebraic manipulation,

$$W = W_{po} + W_{no} = \left[\frac{2\epsilon_{Si}}{q} V_{bi} \left(\frac{N_A + N_D}{N_A N_D} \right) \right]^{\frac{1}{2}} . \tag{3.29}$$

If there is a voltage across the junction, V_{bi} in (3.25) is replaced by $V_{bi} + V_R$, where V_R is the magnitude of the applied reverse bias voltage. In this case, (3.29) becomes

$$W = W_{po} + W_{no} = \left[\frac{2\epsilon_{Si}}{q} (V_{bi} + V_R) \left(\frac{N_A + N_D}{N_A N_D} \right) \right]^{\frac{1}{2}} . \tag{3.30}$$

For a forward bias, V_R is negative, and (3.30) is valid as long as $(-V_R) < V_{bi}$.

In FETs, junctions are either n$^+$/p or p$^+$/n, and a depletion region forms only on the more lightly doped side of the junction. In an n$^+$/p junction, $N_D \gg N_A$, and (3.30) becomes

$$W = \left[\frac{2\epsilon_{Si}}{q N_A} (V_{bi} + V_R) \right]^{\frac{1}{2}} . \tag{3.31}$$

For a p$^+$/n junction, $N_A \gg N_D$, and (3.30) becomes

$$W = \left[\frac{2\epsilon_{Si}}{q N_D} (V_{bi} + V_R) \right]^{\frac{1}{2}} . \tag{3.32}$$

Note that (3.31) and (3.32) are identical in form; this allows the same junction capacitance model to be used for both n$^+$/p and p$^+$/n junctions. This form is universal; however, the derivation was carried out assuming an abrupt junction, which leads to the exponent of $\frac{1}{2}$. It can be shown [1] that in the case of a linearly graded doping profile across the junction, (3.31) becomes

$$W = \left[\frac{2\epsilon_{Si}}{q N_A} (V_{bi} + V_R) \right]^{\frac{1}{3}} , \tag{3.33}$$

with a similar modification to (3.32). Since real junctions fall somewhere between the two extremes, it is best to modify (3.31) (and, similarly, (3.32)) to

$$W = \left[\frac{2\epsilon_{Si}}{q N_A} (V_{bi} + V_R) \right]^{m} , \tag{3.34}$$

where m is generally between $\frac{1}{3}$ and $\frac{1}{2}$. Since a certain amount of curve fitting occurs with real data, m is occasionally found to be outside of this range.

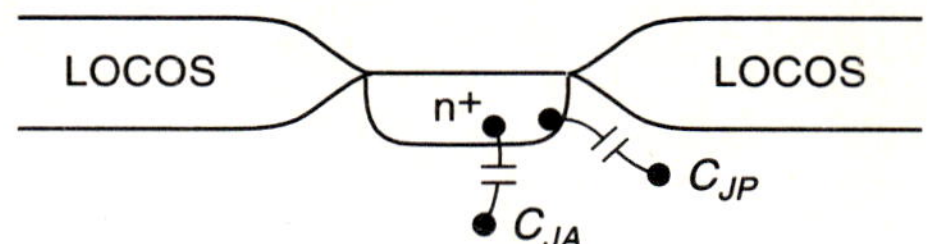

Figure 3.2 A schematic of a diffusion surrounded by isolation, showing how the total capacitance is comprised of area and perimeter components.

3.2.1   The Junction Capacitance Model

The junction capacitance is modeled in a very simple manner. The depletion region is taken to be a dielectric between two capacitor plates, in which case the capacitance is

$$C_j = \frac{\epsilon_{Si}}{W}. \tag{3.35}$$

Since, as described earlier, the form of (3.31) and (3.32) is identical, the derivation presented here will be for an n^+/p junction. In this case, using (3.34), (3.35) becomes

$$C_j = \epsilon_{Si} \left[\frac{q N_A}{2 \epsilon_{Si}(V_{bi} + V_R)} \right]^m = \left[\frac{q \epsilon_{Si} N_A}{2(V_{bi} + V_R)} \right]^m \tag{3.36}$$

For zero bias ($V_R = 0$), C_{jo} can be defined as

$$C_{jo} = \left[\frac{q \epsilon_{Si} N_A}{2 V_{bi}} \right]^m. \tag{3.37}$$

By further algebraic manipulation, (3.36) can be rewritten as

$$C_j = \frac{C_{jo}}{\left(1 + \frac{V_R}{V_{bi}}\right)^m} = C_{jo} \left(1 + \frac{V_R}{V_{bi}}\right)^{-m}. \tag{3.38}$$

This is the junction capacitance model that is used in SPICE [3].

The model is implemented in two parts. Consider the junction depicted in Figure 3.2. The total junction capacitance C_J is found from

$$C_J = C_{JA} A + C_{JP} P, \tag{3.39}$$

where C_{JA} is the area junction capacitance per unit area, A is the total junction area, C_{JP} is the perimeter junction capacitance per unit length, and P is the total junction perimeter. A general model is built for C_{JA} and C_{JP}, while A and P are specific to a particular junction structure. Using the convention of bold-facing parameters, the area junction capacitance is written from (3.38) as

$$C_{JA} = \mathbf{CJ} \left(1 + \frac{V_R}{\mathbf{PB}}\right)^{-\mathbf{MJ}}. \tag{3.40}$$

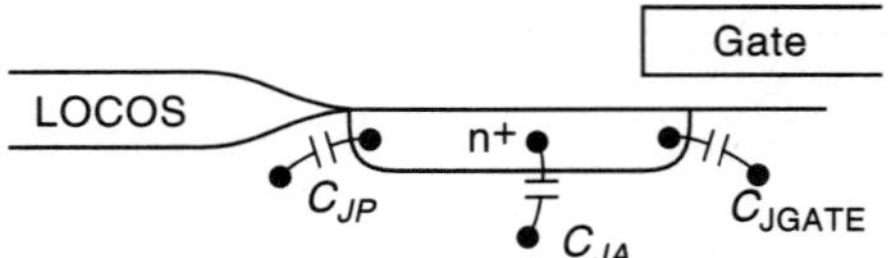

Figure 3.3 A source or drain diffusion of an FET. Three sides of the diffusion are bounded by isolation, while the fourth side is bounded by the polysilicon gate.

The perimeter junction capacitance is expressed as

$$C_{JP} = \mathbf{CJSW}\left(1 + \frac{V_R}{\mathbf{PBSW}}\right)^{-\mathbf{MJSW}}. \tag{3.41}$$

The term **PBSW** is written in that form for notational clarity; it appears in this form in PSPICE. In Berkeley SPICE, there is no separate built-in sidewall potential, and the area potential **PB** is used instead of **PBSW** in (3.41). In HSPICE, the built-in sidewall potential is defined as **PHP**; **PHP** replaces **PBSW** in (3.39). In all cases, if the perimeter bulk potential (**PBSW** or **PHP**) is available but not specified, it defaults to **PB**.

HSPICE [4] also makes available a perimeter capacitance expression for a diffusion surrounded by polysilicon gate rather than isolation. As shown in Figure 3.3, an FET source or drain diffusion is bordered on three sides by isolation and on one side by the gate. The perimeter capacitance can be different along the gate edge. HSPICE introduces another expression, based on (3.41), to account for this behavior:

$$C_{JP} = \mathbf{CJGATE}\left(1 + \frac{V_R}{\mathbf{PHP}}\right)^{-\mathbf{MJSW}}. \tag{3.42}$$

The parameter **CJGATE** represents the zero bias capacitance, while the other two parameters (**PHP** and **MJSW**) are taken directly from the model for the isolation-bounded perimeter capacitance.

Parameter extraction using (3.40) and (3.41) is fairly simple. As (3.39) indicates, a measurement using a junction produces data for C_J as a function of voltage. However, (3.39) contains two unknowns, C_{JA} and C_{JP}. Thus, two measurements on two junctions, with different areas and perimeters, are required to compute C_{JA} and C_{JP} as functions of the applied bias. These data are then used to find the parameters of (3.40) and (3.41).

An example is shown in Figures 3.4 to 3.7. Figure 3.4 contains a measurement of C_J for $A = 120{,}000$ μm^2 and $P = 1{,}400$ μm. Figure 3.5 contains an identical measurement for a junction with $A = 61{,}584$ μm^2 and $P = 30{,}800$ μm. This allows separate data for C_{JA} and C_{JP} to be found. It is then a simple matter to find the parameters by a nonlinear least squares fit, as shown in Figures 3.6 and 3.7. For any particular diode, it is possible to extract both the built-in potential (e.g., **PB**) and the exponent (e.g., **MJ**) directly from this nonlinear fitting; however, an unrealistically low value of the built-in potential may occur. While this will give good results for the temperature at which the extraction is performed, the simple temperature dependence of the built-in potential can cause a great deal of trouble. Instead, it is best to determine the built-in potential externally; a simple

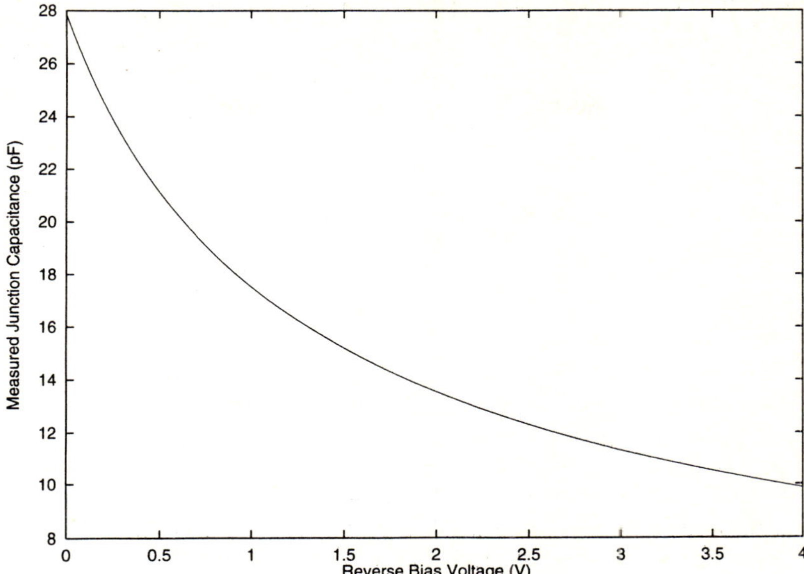

Figure 3.4 Junction capacitance data from a diffusion with $A = 120,000$ μm^2 and $P = 1,400$ μm.

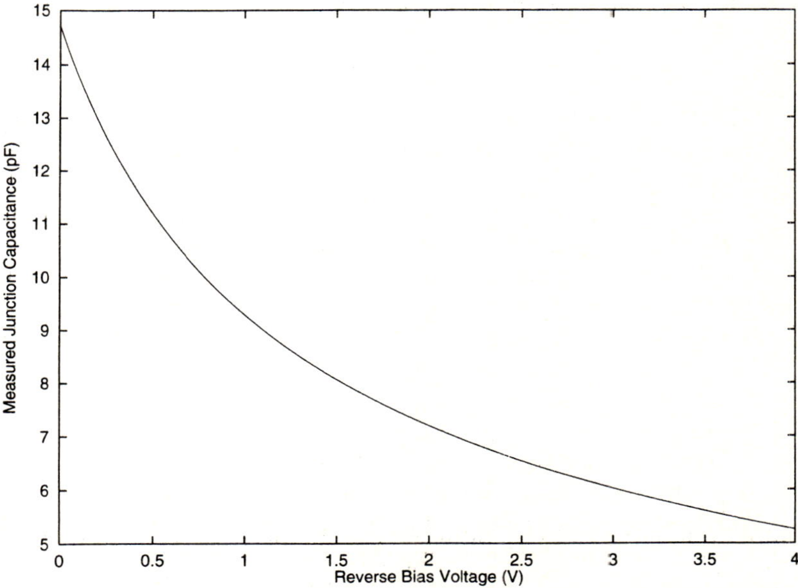

Figure 3.5 Junction capacitance data from a diffusion with $A = 61,584$ μm^2 and $P = 30,800$ μm.

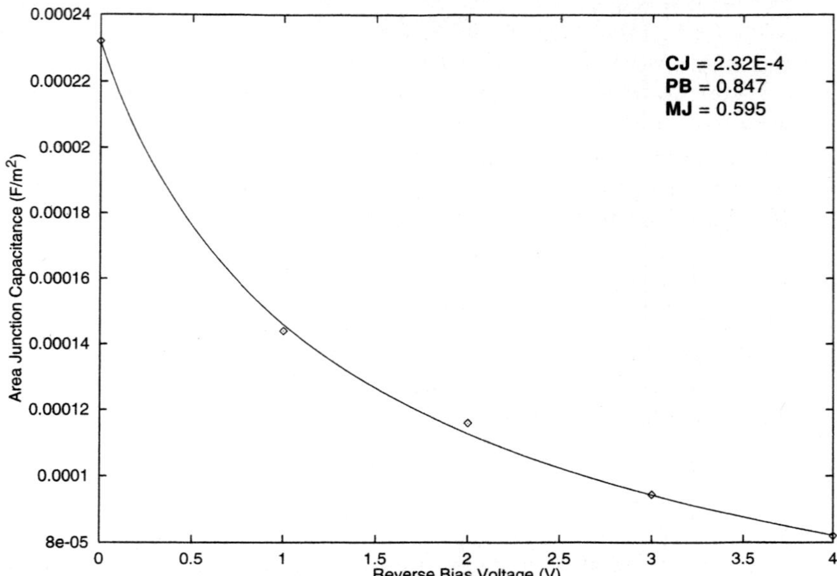

Figure 3.6 Extraction of the parameters **CJ**, **PB**, and **MJ** after the data have been processed to separate the area and perimeter contributions.

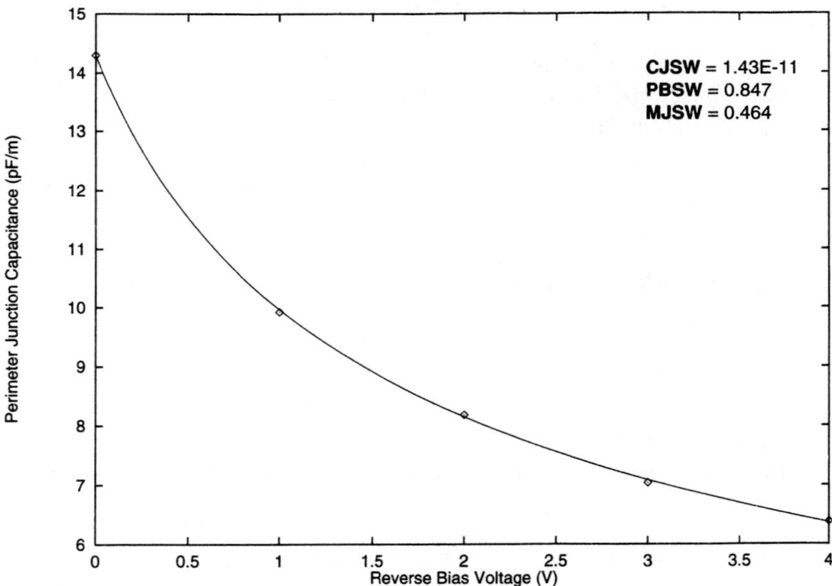

Figure 3.7 Extraction of the parameters **CJSW**, **PBSW**, and **MJSW**.

method is described below when the forward bias current is considered (see Figure 3.11). With the built-in potential thus defined, the exponent is extracted by nonlinear fitting; this is the method that was employed here.

If the HSPICE model for the polysilicon-bounded diffusion is to be used, the extraction is done after the one just described is completed. Two capacitor structures, similar to those detailed above, are required to determine a value for **CJGATE**. Only this zero bias parameter is extracted, since, as shown in (3.42), the other two parameters (**PHP** and **MJSW**) are taken directly from the model for **CJSW**.

Temperature Dependence

The temperature dependence of the junction capacitance is modeled very simply. For illustrative purposes, the area component of an n^+/p junction will be considered here.

Using (3.35), the relationship between the junction capacitance at different temperatures can easily be seen to be

$$\frac{C_{T1}}{C_{T2}} = \frac{W_{T2}}{W_{T1}}, \tag{3.43}$$

where the subscripts indicate, respectively, the junction capacitance and depletion region widths at two different temperatures, $T1$ and $T2$. Using (3.36), (3.43) is transformed to

$$\frac{C_{T1}}{C_{T2}} = \left[\frac{V_{bi,T2} + V_R}{V_{bi,T1} + V_R}\right]^m, \tag{3.44}$$

where it is assumed that m is a function only of the junction doping profile [1] and not of the temperature. For purposes here, consider the zero bias situation ($V_R = 0$), in which case (3.44) becomes

$$\frac{C_{o,T1}}{C_{o,T2}} = \left[\frac{V_{bi,T2}}{V_{bi,T1}}\right]^m, \tag{3.45}$$

This expression can be solved for the zero bias capacitance at $T2$:

$$C_{o,T2} = C_{o,T1}\left[\frac{V_{bi,T2}}{V_{bi,T1}}\right]^{-m}. \tag{3.46}$$

Now let $T1$ be the temperature at which the parameters were extracted. The $T1$ terms in (3.46) are thus the model parameters described earlier, so (3.46) is rewritten as

$$C_{o,T2} = \mathbf{CJ}\left[\frac{V_{bi,T2}}{\mathbf{PB}}\right]^{-\mathbf{MJ}}. \tag{3.47}$$

An expression is now required for $V_{bi,T2}$. From first principles, $V_{bi,T2}$ could be computed solely from the value of the bulk doping:

$$V_{bi,T2} = \frac{E_{g,T2}}{2} + \frac{k_b \cdot T2}{q} ln\left(\frac{N_A}{n_{i,T2}}\right), \tag{3.48}$$

where $E_{g,T2}$ is the band gap at the temperature $T2$. However, **PB** (the parameter representing $V_{bi,T1}$) is not computed from first principles, but is an extracted parameter. Some method of finding $V_{bi,T2}$ from **PB** must be found. Equation (3.48) can be recast in a more generic form as

$$V_{bi,T2} = \frac{E_{g,T2}}{2} + \phi_{p,T2}. \tag{3.49}$$

Using the simple expression for the bulk Fermi potential

$$\phi_p = \frac{k_b T}{q} ln\left(\frac{N_A}{n_i}\right), \tag{3.50}$$

an expression relating the bulk Fermi potential at the temperatures $T1$ and $T2$ can be found:

$$\frac{\phi_{p,T2}}{\phi_{p,T1}} = \frac{\phi_{p,T2}}{\mathbf{PB}} = \frac{T2}{T1} \cdot ln\left(\frac{N_A}{n_{i,T2}} - \frac{N_A}{n_{i,T1}}\right). \tag{3.51}$$

This expression can be solved for $\phi_{p,T2}$:

$$\phi_{p,T2} = \mathbf{PB} \cdot \frac{T2}{T1} \cdot ln\left(\frac{N_A}{n_{i,T2}} - \frac{N_A}{n_{i,T1}}\right). \tag{3.52}$$

Equation (3.52) is then substituted into (3.49) to yield a final expression for $V_{bi,T2}$:

$$V_{bi,T2} = \frac{E_{g,T2}}{2} + \mathbf{PB} \cdot \frac{T2}{T1} \cdot ln\left(\frac{N_A}{n_{i,T2}} - \frac{N_A}{n_{i,T1}}\right). \tag{3.53}$$

Equation (3.53) requires some sort of expression for $E_g(T)$ and $n_i(T)$; various suggested offerings are plentiful in the literature (see, for example, [5, 6]).

Using (3.47) and (3.53), the junction capacitance at an arbitrary temperature $T2$ is computed from

$$C_{j,T2} = C_{o,T2}\left[1 + \frac{V_R}{V_{bi,T2}}\right]^{-\mathbf{MJ}}. \tag{3.54}$$

The discussion presented here can easily be applied to a p$^+$/n junction, and to the perimeter capacitances.

Some simulators, such as HSPICE [4], make available more complex methods of computing the temperature dependence of the junction capacitance.

3.2.2 The Current-Voltage Model

The diode current-voltage model [1, 2] begins with the one-dimensional, drift-diffusion current density equation:

$$J_{x,n} = q n \mu_n E_x + q D_n \frac{dn(x)}{dx},\qquad(3.55)$$

where $J_{x,n}$ is the cross-sectional electron current density and D_n is the diffusion constant for electrons. The first term to the right of the equals sign represents drift transport, while the second term represents diffusion transport. A similar equation could also be written down for holes. Outside the depletion region, there is no field, so carriers cannot move by drift transport. Inside the depletion region, there are no free carriers. Carriers can enter the depletion region; they enter by diffusion, and are then swept across the depletion region by the field. Since the diffusion of carriers into the depletion region is a much slower process than the drift of carriers across that depletion region, diffusion transport is the rate limiting step; diffusion transport will thus determine the current behavior of the diode. This observation simplifies (3.55) to

$$J_{x,n} = q D_n \frac{dn(x)}{dx}.\qquad(3.56)$$

To find an expression for the diode current, the carrier density in the junction must be evaluated.

First, consider a p-n junction with no applied bias. In the bulk on either side of the junction, it can be shown [1] that the free carrier concentrations are

$$n_n = n_i e^{q\phi_n/k_b T},\qquad(3.57)$$

$$p_n = n_i e^{-q\phi_n/k_b T},\qquad(3.58)$$

$$p_p = n_i e^{-q\phi_p/k_b T},\qquad(3.59)$$

$$n_p = n_i e^{q\phi_p/k_b T},\qquad(3.60)$$

where n and p represent the concentration of, respectively, free electrons and holes, and the subscripts n and p denote that the expression represents a free carrier concentration in, respectively, the n or p doped region; n_i is the intrinsic carrier concentration in silicon. In the discussion which follows, only the hole density will be considered. Combining (3.58) and (3.59), the ratio of the hole density in the p-bulk to the hole density in the n-bulk is

$$\frac{p_p}{p_n} = e^{q V_{bi}/k_b T}.\qquad(3.61)$$

The situation is depicted in Figure 3.8, where the differences in the carrier densities in the n-bulk and p-bulk must be connected continuously in the depletion region. Carriers

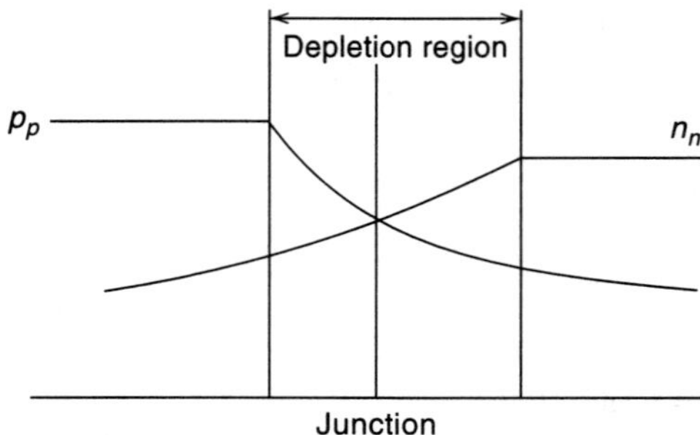

Figure 3.8 Carrier densities in a p-n junction with no bias voltage applied.

diffuse into the depletion region and recombine; the process of carrier entrance into the depletion region is balanced by recombination.

Now consider the same junction when a bias is applied. It is relatively straightforward to show that (3.61) can be rewritten as a ratio of the hole concentrations at the edges of the depletion region on each side:

$$\frac{p(-x_{po})}{p(x_{no})} = e^{q(V_{bi}-V)/k_bT}. \tag{3.62}$$

The concentration $p(-x_{po})$, being in the p region, is approximately p_p. Thus, (3.62) can be rewritten as

$$\frac{p_p}{p(x_{no})} = e^{q(V_{bi}-V)/k_bT}. \tag{3.63}$$

Using p_p to manipulate (3.61) and (3.63) leads to an expression for $p(x_{no})$:

$$p(x_{no}) = p_n e^{qV/k_bT}. \tag{3.64}$$

When the bias V applied, the extent of the depletion region changes; the hole concentration at x_{no} is not equal to p_n, as (3.64) shows. The difference can be found using (3.64):

$$\Delta p_n = p(x_{no}) - p_n = p_n e^{qV/k_bT} - p_n = p_n \left(e^{qV/k_bT} - 1\right). \tag{3.65}$$

The situation is depicted in Figure 3.9. As holes are injected into the n-region, they recombine with electrons, and their concentration decays exponentially with a characteristic length L_p. It is also convenient for what follows to redefine the x-coordinate with $x = 0$ at x_{no}. The excess hole concentration as a function of x is

$$\delta p(x) = \Delta p_n e^{-x/L_p} = p_n \left(e^{qV/k_bT} - 1\right) e^{-x/L_p}. \tag{3.66}$$

The change in $\delta p(x)$ as a function of x will determine the current density, using (3.56):

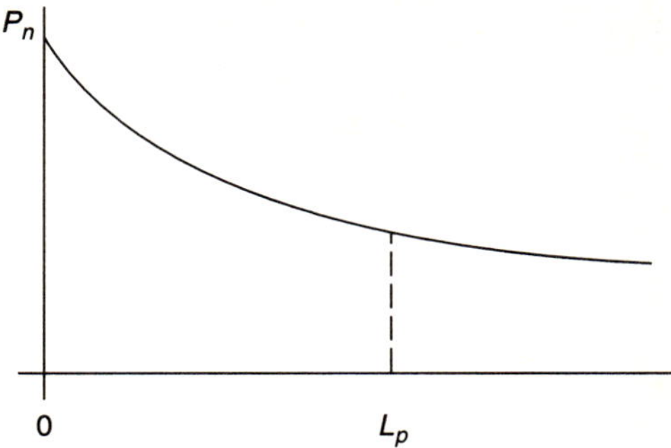

Figure 3.9 The computed hole density on the n-doped side of the junction.

$$J_p = -q D_p \frac{dp(x)}{dx} = -q D_p \frac{d(\delta p(x))}{dx}. \tag{3.67}$$

Substituting (3.66) into (3.67) leads to

$$J_p = -q D_p \frac{d}{dx}(\delta p(x)) = -q D_p p_n \left(e^{qV/k_bT} - 1\right) \frac{d}{dx} e^{-x/L_p}$$

$$= \frac{-q D_p}{L_p} p_n \left(e^{qV/k_bT} - 1\right) e^{-x/L_p}. \tag{3.68}$$

Since in the absence of sources (generation) or sinks (recombination) of free carriers, J_p is constant for all x, (3.68) can be evaluated at $x = 0$, becoming

$$J_p = \frac{q D_p}{L_p} p_n \left(e^{qV/k_bT} - 1\right). \tag{3.69}$$

By a similar derivation

$$J_n = \frac{q D_n}{L_n} n_p \left(e^{qV/k_bT} - 1\right). \tag{3.70}$$

The total current through the junction is found by adding (3.69) and (3.70):

$$J = J_n + J_p = \left(\frac{q D_p}{L_p} p_n + \frac{q D_n}{L_n} n_p\right)\left(e^{qV/k_bT} - 1\right) = J_o \left(e^{qV/k_bT} - 1\right). \tag{3.71}$$

This is the ideal diode equation derived by Shockley [7], and is the diode model that is implemented in SPICE. The current-voltage characteristic is shown in Figure 3.10. The term J_o is known as the reverse saturation current density, as in the ideal situation described by (3.71) the current takes on a constant value of J_o with increasing reverse bias. The ideal diode equation is, in practice, valid only for small forward bias voltages

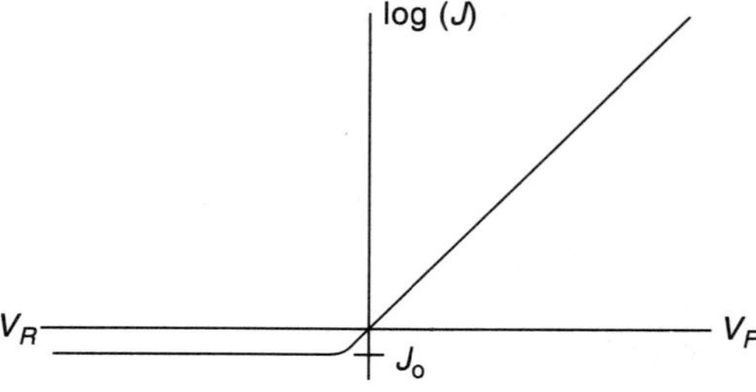

Figure 3.10   The diode current-voltage characteristic as described by the Shockley diode equation.

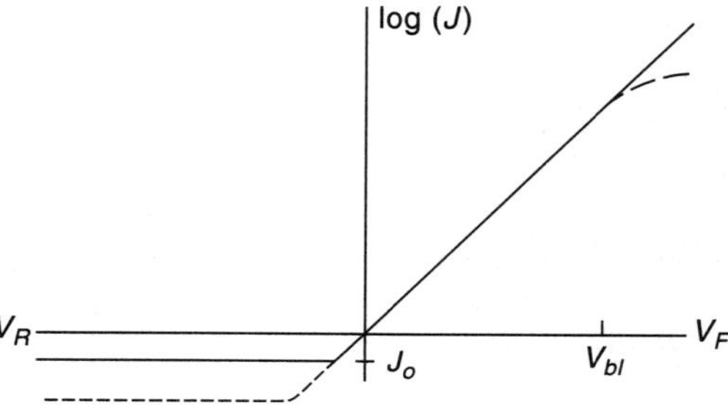

Figure 3.11   A comparison of the ideal Shockley diode equation (solid line) with data from a typical diode (dashed line). As shown, this deviation from the ideal characteristic can be used to extract the built-in junction potential.

[1]; as shown in Figure 3.11, for larger forward biases ($V_f > V_{bi}$, with, in general, $V_{bi} \sim 0.7\text{V}$), the built-in diode potential barrier is no longer important; instead, the material resistivity is the determining factor, and the current departs from the ideal characteristic. As shown in Figure 3.11, this deviation from the ideal characteristic can be used to extract the built-in potential of the junction; this information is used in the reverse bias junction capacitance model described above. In addition, the reverse bias current is rarely (if ever) described by J_o. Instead, as also shown in Figure 3.11, a larger reverse bias current, due to generation from amphoteric mid-gap states (Shockley-Read-Hall generation) [8, 9] occurs. Due to this generation, the reverse bias current is several orders of magnitude larger than J_o. However, in all but a few special situations (such as DRAM cell design), the reverse bias current should be too small to be of concern during circuit design; if this is not the case, there are problems with the process technology that should be resolved before circuit design proceeds. A fuller description of diode reverse bias leakage current may be found in [10].

More formally, (3.71) is usually implemented as

$$J = J_o \left(e^{qV/nk_bT} - 1\right), \tag{3.72}$$

where n is an *ideality factor* added to the model. During the 1950s, when the Shockley diode model was developed, process technology was considerably less clean than it is today. This led to a high concentration of amphoteric mid-gap states (the same as those just described in the preceding paragraph [8, 9]), which act as generation states. However,

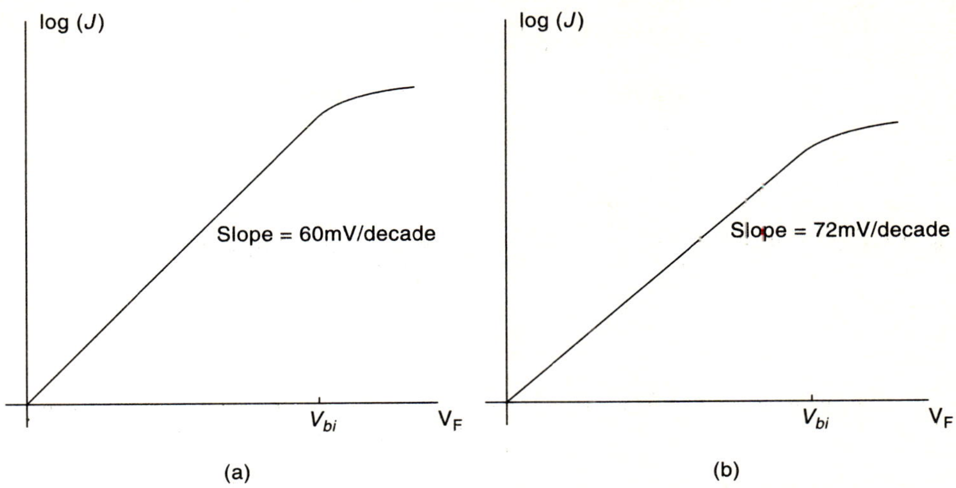

Figure 3.12   Log-linear diode forward bias characteristics; (a) a "clean" diode, with a slope of 60 mV/decade; (b) a questionable diode, with a slope greater than 60 mV/decade.

with a large influx of carriers into the depletion region under forward bias, these states act as recombination centers, reducing the total number of carriers that successfully transit the junction and appear as terminal current. Under the ideal case of (3.71), a semi-log plot of J versus V should show a slope of 60 mV/decade, as shown in Figure 3.12a. However, in a "dirty" junction (Figure 3.12b), the slope is greater than 60 mV/decade, due to recombination; this behavior is described empirically by including n in the model equation. In the ideal case, $n = 1$ and does not affect (3.72).

In modern process technology, the density of generation-recombination states has been greatly reduced. Values have become low enough that n should equal 1; the ideality factor should be superfluous. If a model is provided with $n > 1$, it should be treated as a warning sign that something is wrong with the fabrication process technology, and caution is advised. In silicided diffusions (which are now commonplace), a value of $n > 1$ often indicates the presence of a Schottky conduction path in the junction [11].

Examination of J_o indicates that it is a composite parameter composed of several different physically based variables. The bulk carrier concentrations (p_n and n_p) could be determined from knowledge of the value of the substrate doping. However, nonuniform substrate doping, common in modern FETs, makes it difficult to stipulate a single precise value of the bulk carrier concentrations. The other variables, such as D_n and L_n, should have physically meaningful values, stipulated for various doping concentrations. However, determining precise values for these variables is not simple either. Instead, the entire matter is simplified by treating J_o as a model parameter to be determined by extraction, as will be described below.

The implementation of the diode current model is identical to that of the diode capacitance model; it is divided into an area component and a perimeter component. The area component is written as

$$J_{area} = \mathbf{JS}\left(e^{qV/k_bT} - 1\right),\tag{3.73}$$

where J_{area} is in A/m^2. The perimeter component is

$$J_{area} = \mathbf{JSW}\left(e^{qV/k_bT} - 1\right),\tag{3.74}$$

where J_{perim} is in A/m. (Note the discussion above on why the ideality factor n should no longer be included in (3.73) and (3.74)).

As in the case of the diode capacitance model, (3.73) and (3.74) are used in combination with the specific area and total perimeter of a particular diode to calculate the total current through the junction:

$$I_{diode} = J_{area}A + J_{perim}P,\tag{3.75}$$

where A and P are, respectively, the diode area and perimeter. Thus, as in the diode capacitance case, two diodes, with different areas and perimeters, are required to determine the parameters $\mathbf{JS}$ and $\mathbf{JSW}$. An example is shown in Figures 3.13 to 3.16 for the same structures used earlier to determine the junction capacitance model parameters. Figure 3.13 shows a current-voltage characteristic for a structure with $A = 120,000$ μm^2 and $P = 1,400$ μm, while Figure 3.14 contains the same characteristic for a structure with $A = 61,584$ μm^2 and $P = 30,800$ μm. By solving (3.75) for each voltage point on the two characteristics (two equations and two unknowns), separate values for J_{area} and J_{perim} as functions of voltage are found. The parameters $\mathbf{JS}$ and $\mathbf{JSW}$ are then determined by a nonlinear least squares fit, as shown in Figures 3.15 and 3.16.

Temperature Dependence

Although the form of (3.73) and (3.74) explicitly includes the temperature, those equations contain no expression for the temperature dependence of the reverse saturation current parameters $\mathbf{JS}$ and $\mathbf{JSW}$. The simplest method of including temperature dependence is to note that the forward bias current increases approximately exponentially with temperature. With $T1$ once again defined as the temperature at which parameter extraction was performed (see the discussion above on the temperature dependence of the junction capacitance model), the area current density at another temperature $T2$ can be written as

$$J_{area}(T2) = J_{area}(T1) \cdot e^{Q},\tag{3.76}$$

where Q is a parameter which describes the temperature dependence. Using (3.73) to expand (3.76) leads to

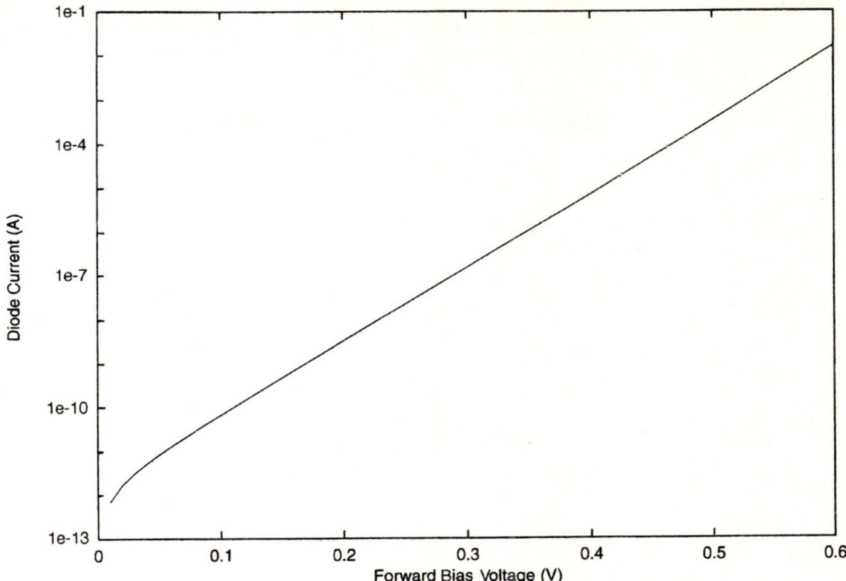

Figure 3.13 Forward bias diode data from a junction with $A = 120,000$ μm^2 and $P = 1,400$ μm.

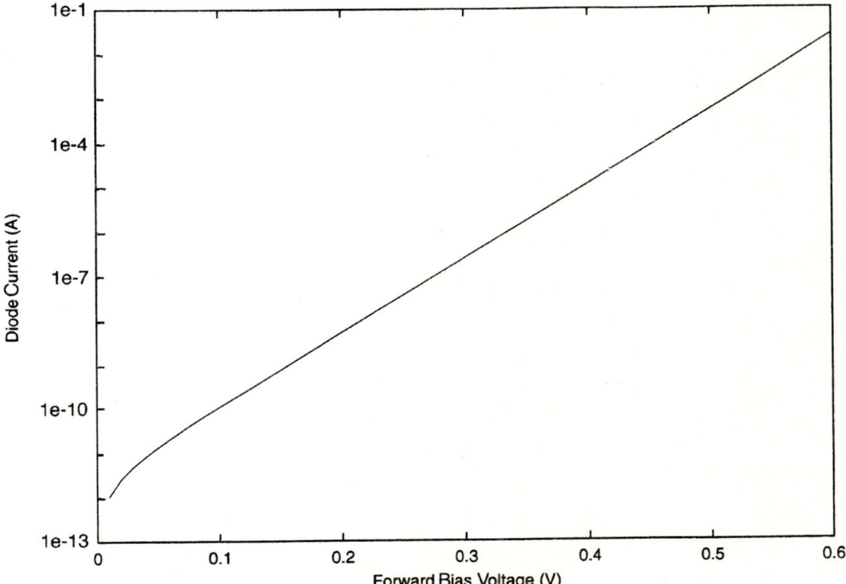

Figure 3.14 Forward bias diode data from a junction with $A = 61,584$ μm^2 and $P = 30,800$ μm.

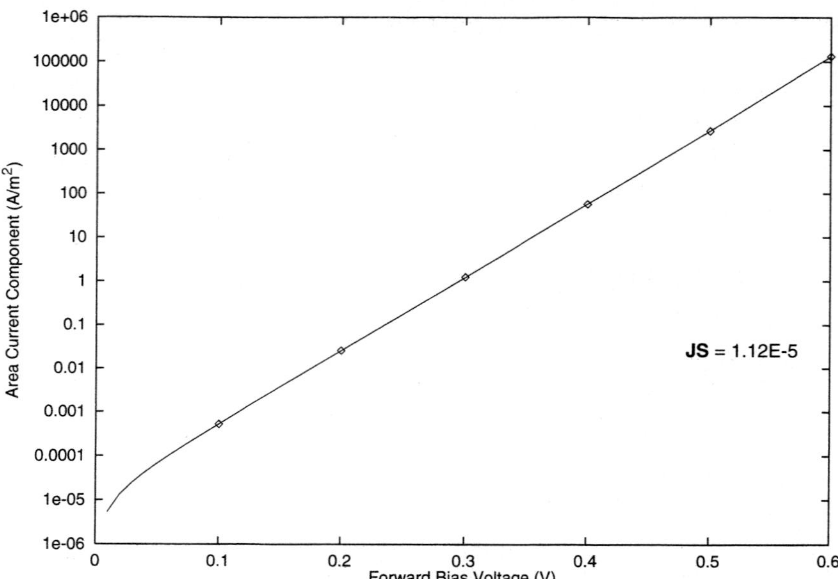

Figure 3.15   Extraction of the parameter **JS** after the data have been processed to separate the area and perimeter components.

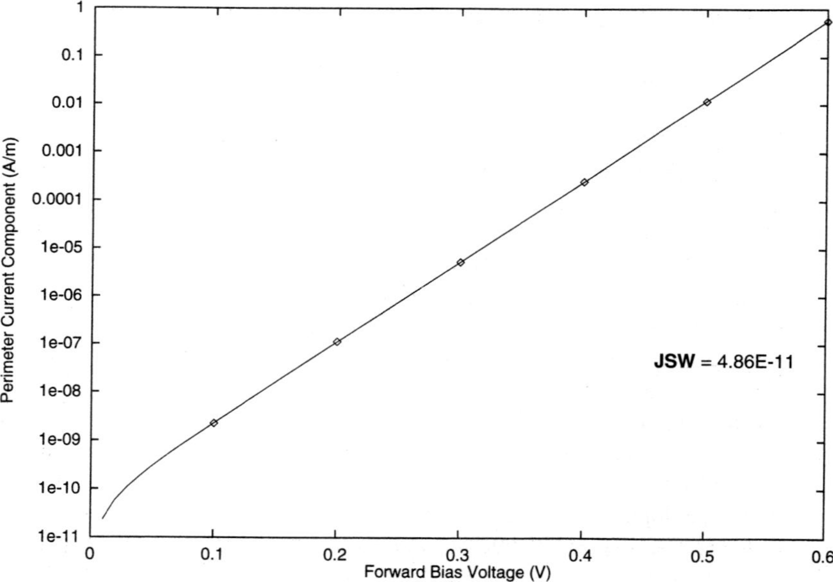

Figure 3.16   Extraction of the parameter **JSW**.

$$J_{area}(T2) = \mathbf{JS}\left(e^{qV/k_b\cdot(T1)} - 1\right) e^{Q}. \tag{3.77}$$

Since $J_{area}(T2)$, $\mathbf{JS}$, and $T1$ are all known, (3.77) can be solved for Q. A similar expression can be derived for the perimeter current.

This is the general form that expressions for the temperature dependence take. Different versions of Berkeley SPICE [12] use a variety of specific schemes [13]. HSPICE [4] provides a number of more complex models to choose from.

3.2.3   The Complete Junction Diode Model

For most applications encountered in digital and analog circuit design, the junction diode model developed here is sufficient; it contains area and perimeter terms to describe the reverse bias capacitance characteristics, and area and perimeter terms for the forward bias current characteristics for $V_f < V_{bi}$. As described above, there is no model for reverse bias leakage due to Shockley-Read-Hall generation, nor are there models for any of the more involved generation/leakage mechanisms; however, a detailed description of this current should not be necessary.

A wide variety of more complicated models is available to describe the behavior of the junction diode. Models exist to account for the space charge limited current when $V_f > V_{bi}$, a variety of high-frequency phenomena (such as the diffusion capacitance under forward bias), and $\frac{1}{f}$ noise. However, as these aspects of diode modeling are less frequently required and are well documented elsewhere [14], they will not be considered here.

3.3   The Two-Terminal MOS Capacitor Structure

The simple MOS capacitor [2, 15] is shown schematically in Figure 3.17, with a bias of V_{gb} applied between the gate and the substrate. The relevant energy band structures are shown in Figure 3.18. In the accumulation condition ($V_{gs} < V_{fb}$) (Figure 3.18a), the gate bias attracts carriers from the substrate to the surface. At the flatband voltage ($V_{gs} = V_{fb}$), the gate bias exactly balances the workfunction difference between the gate material and the substrate. The field is zero throughout the substrate, as reflected by the flat bands. For a further increase in the gate bias, the sign of the surface potential ϕ_s will be opposite that of the Fermi potential ϕ_f. This is shown for the depletion condition (Figure 3.18c), where $V_{gs} > V_{fb}$ and $\phi_s < \phi_f$. A depletion region forms at the silicon surface. In weak inversion (Figure 3.18d), $\phi_f < \phi_s < 2\phi_f$; at the surface, there is a free electron layer with a density less than the doping concentration (i.e., $n < N_A$). In strong inversion (Figure 3.18e), $\phi_s \geq 2\phi_f$; the surface electron density now exceeds the substrate doping concentration ($n > N_A$).

In the depletion and inversion conditions, the surface potential ϕ_s reflects the potential drop due to the applied gate bias. The total potential drop V_{gb} is accounted for by the workfunction difference between the gate and the silicon substrate, and any charge layers in the silicon. This can be written as

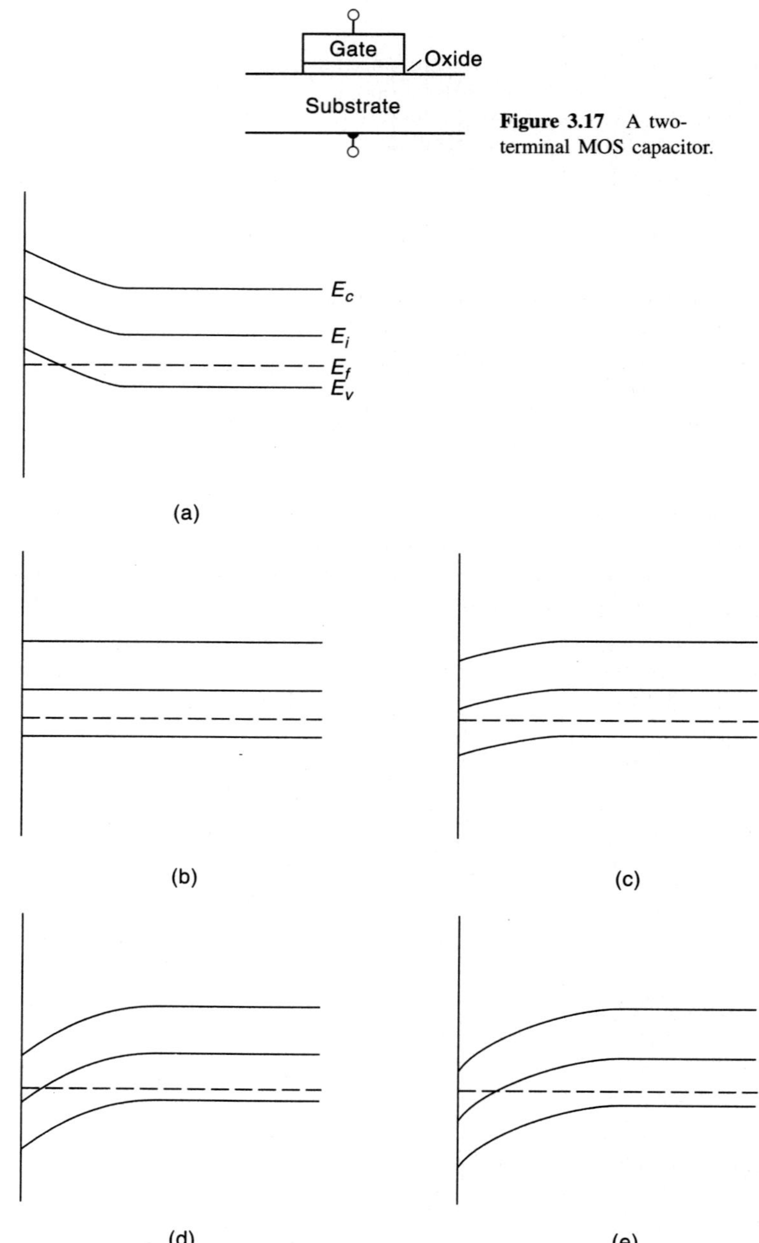

Figure 3.17 A two-terminal MOS capacitor.

Figure 3.18 The band structure of a two-terminal MOS capacitor; (a) accumulation; (b) flatband; (c) depletion; (d) weak inversion; (e) strong inversion.

$$V_{gb} = V_{fb} + \phi_s - \frac{Q_s}{C_{ox}},\tag{3.78}$$

where V_{fb} is the flatband voltage, and Q_s is the total surface charge density induced in the substrate by the gate bias. The surface charge can be separated into two components—the inversion charge, composed of free carriers at the interface, and depletion charge due to fixed ionized impurities:

$$Q_s = Q_{inv} + Q_{depl}\tag{3.79}$$

The depletion charge will be examined first.

Assuming complete impurity ionization and a uniform uncompensated doping density N_A, Gauss' law in one dimension for the depletion charge is

$$\frac{dE}{dx} = \frac{\rho}{\epsilon_{Si}} = -\frac{qN_A}{\epsilon_{Si}}.\tag{3.80}$$

The simplifying assumption can be made that the effect of the applied gate bias terminates abruptly at $x = x_D$ (the delta depletion approximation). Below the depletion region edge ($x > x_D$), the bands are flat, implying a field strength of zero. At the same time, for free carriers, $p \gg n$, and $p = N_A$.

For $x < x_D$, it be will assumed that $p = 0$, as the applied field has swept all the free holes into the substrate. This leaves a space charge region of fixed, ionized impurities. Equation (3.80) is integrated from the point where the field vanishes (x_D) to the surface:

$$\int_0^{E_s} dE = -\frac{qN_A}{\epsilon_{Si}} \int_{x_D}^0 dx.\tag{3.81}$$

This yields an expression for the surface field:

$$E_s = \frac{qN_A x_D}{\epsilon_{Si}}.\tag{3.82}$$

Also, the potential V is defined as

$$E = -\frac{dV}{dx},\tag{3.83}$$

which allows (3.82) to be rewritten as

$$dV = -E_s dx = -\frac{qN_A x_D}{\epsilon_{Si}}.\tag{3.84}$$

This expression is integrated in a fashion similar to (3.81):

$$\int_0^{\phi_s} dV = -\frac{qN_A}{\epsilon_{Si}} \int_{x_D}^0 x_D dx, \tag{3.85}$$

which leads to

$$\phi_s = \frac{qN_A x_D^2}{2\epsilon_{Si}}. \tag{3.86}$$

Equation (3.86) can be rearranged to yield an expression for x_D:

$$x_D = \left(\frac{2\epsilon_{Si}\phi_s}{qN_A}\right)^{\frac{1}{2}}. \tag{3.87}$$

The total depletion charge is then

$$Q_{depl} = qN_A x_D = (2\epsilon_{Si}qN_A\phi_s)^{\frac{1}{2}}. \tag{3.88}$$

Next the inversion charge is considered. To begin its computation, (3.78) is rewritten as

$$Q_s = -C_{ox}(V_{gb} - V_{fb} - \phi_s). \tag{3.89}$$

At the onset of strong inversion, $\phi_s = 2\phi_f$, so (3.89) becomes

$$Q_s = -C_{ox}(V_{gb} - V_{fb} - 2\phi_f). \tag{3.90}$$

(This substitution can be put to general use under strong inversion, as ϕ_s increases very little above $2\phi_f$ for further increases in the gate bias.) From (3.79),

$$Q_{inv} = Q_s - Q_{depl}. \tag{3.91}$$

Now, (3.90) is substituted for Q_s and (3.88) is substituted for Q_{depl}, leading to

$$Q_{inv} = -C_{ox}(V_{gb} - V_{fb} - 2\phi_f) - \left(2\epsilon_{Si}qN_A(2\phi_f)\right)^{\frac{1}{2}}, \tag{3.92}$$

where the strong inversion condition $\phi_s = 2\phi_f$ has been inserted into (3.88). It is convenient to rewrite (3.92) as

$$Q_{inv} = -C_{ox}\left\{(V_{gb} - V_{fb} - 2\phi_f) - \frac{1}{C_{ox}}\left[2\epsilon_{Si}qN_A(2\phi_f)\right]^{\frac{1}{2}}\right\}. \tag{3.93}$$

Defining

$$\gamma = \frac{(2\epsilon_{Si}qN_A)^{\frac{1}{2}}}{C_{ox}}, \tag{3.94}$$

(3.93) becomes

$$Q_{inv} = -C_{ox}\left[V_{gb} - V_{fb} - 2\phi_f - \gamma \cdot (2\phi_f)^{\frac{1}{2}}\right]. \tag{3.95}$$

From (3.95), it is convenient to define the threshold voltage V_t as the gate bias V_{gb} corresponding to the condition $\phi_s = 2\phi_f$, as described by (3.95):

$$V_t = V_{fb} + 2\phi_f + \gamma \cdot (2\phi_f)^{\frac{1}{2}}. \tag{3.96}$$

This allows (3.95) to be rewritten as

$$Q_{inv} = -C_{ox}(V_{gb} - V_t). \tag{3.97}$$

It will be seen later that this operational definition of V_t is fairly satisfactory for the description of most facets of FET behavior. However, examination of (3.97) indicates that when $V_{gb} = V_t$, $Q_{inv} = 0$. Thus, this approach neglects inversion charge in the weak inversion regime, when $\phi_f < \phi_s < 2\phi_f$. This has consequences that will be discussed later.

The Flatband Voltage

The implementation of the flatband voltage V_{fb} in SPICE is an important matter. In the first generation models, V_{fb} is computed from an essentially first principles approach. For a standard MOS system, the flatband voltage is defined as [1]

$$V_{fb} = \phi_{gs} - \frac{1}{C_{ox}}\left(Q_f + Q_m + Q_{it}\right), \tag{3.98}$$

where Q_f is the fixed interface charge density, Q_m is the mobile charge density in the gate oxide, and Q_{it} is the interface state density. The term ϕ_{gs} is the gate to substrate workfunction difference, to be defined below. In modern MOS technology, Q_f, Q_m, and Q_{it} are all negligible and can be disregarded. This simplifies (3.98) to

$$V_{fb} = \phi_{gs} = \chi_g - \chi_s, \tag{3.99}$$

where χ_g and χ_s are, respectively, the gate and substrate workfunctions (the energy required to remove an electron from the surface of the gate material or the substrate).

This is best demonstrated graphically, as shown in Figure 3.19. Figure 3.19a shows a band diagram for an n^+ polysilicon gate over a p^- substrate. From Figure 3.19a, it can be seen that

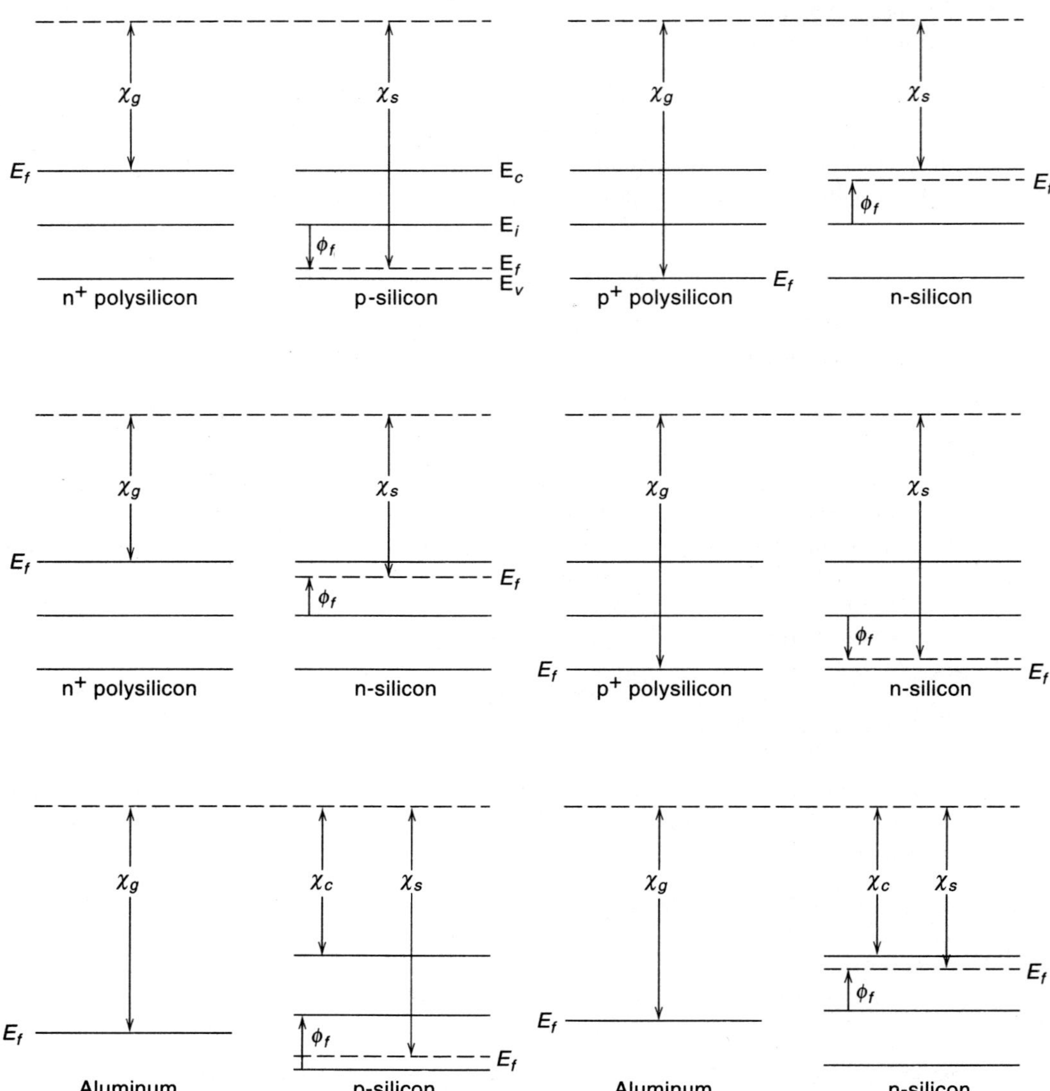

Figure 3.19 Band diagrams showing how the flatband voltage is computed for various types of MOS capacitors; (a) n$^+$ polysilicon gate over a p-substrate; (b) p$^+$ polysilicon gate over an n-substrate; (c) n$^+$ polysilicon gate over an n-substrate; (d) p$^+$ polysilicon gate over a p-substrate; (e) aluminum gate over a p-substrate; (f) aluminum gate over an n-substrate.

$$V_{fb} = \phi_{gs} = \chi_g - \chi_s = -\left(\frac{E_g}{2} + \phi_f\right) = -\frac{E_g}{2} - \phi_f. \qquad (3.100)$$

The band diagram for a p$^+$ polysilicon gate over an n$^-$ substrate is shown in Figure 3.19b. Here it can be seen that

$$V_{fb} = E_g - \left(\frac{E_g}{2} - \phi_f\right) = \frac{E_g}{2} + \phi_f. \qquad (3.101)$$

Next consider an n$^+$ polysilicon gate over an n$^-$ substrate, as in Figure 3.19c. Here

$$V_{fb} = -\left(\frac{E_g}{2} - \phi_f\right) = -\frac{E_g}{2} + \phi_f. \qquad (3.102)$$

Finally, consider a p$^+$ polysilicon gate over a p$^-$ substrate (Figure 3.19d), where

$$V_{fb} = E_g - \left(\frac{E_g}{2} + \phi_f\right) = \frac{E_g}{2} - \phi_f. \qquad (3.103)$$

Note that these four structures described by (3.100)–(3.103) and Figures 3.19a–3.19d correspond to, respectively, an n$^+$-gate NMOS, a p$^+$-gate PMOS, an n$^+$-gate PMOS, and a p$^+$-gate NMOS. While the first three of these systems are common, the fourth (p$^+$-gate NMOS) is rarely encountered.

These various structures are handled in the first-generation SPICE models with the parameter **TPG**, which represents the type of gate material in the MOS system. If the gate material is of the opposite polarity of the substrate doping (Figures 3.19a and 3.19b), **TPG** $= 1$. If the gate material is of the same polarity of the substrate doping (Figures 3.19c and 3.19d), **TPG** $= -1$. The two possible values of **TPG** combined with the system being either NMOS or PMOS allows the appropriate choice among (3.100)–(3.103) to be made. This is the approach that is implemented in the first-generation FET models.

The first-generation models also have available **TPG** $= 0$, which corresponds to an aluminum gate. For an aluminum gate over a p$^-$ substrate (NMOS, Figure 3.19e),

$$V_{fb} = \chi_g - \left[\chi_c + \left(\frac{E_g}{2} + \phi_f\right)\right], \qquad (3.104)$$

while for an aluminum gate over an n$^-$ substrate (PMOS, Figure 3.19f),

$$V_{fb} = \chi_g - \left[\chi_c + \left(\frac{E_g}{2} - \phi_f\right)\right]. \qquad (3.105)$$

The gate and substrate are now composed of different materials, so the system is a little more complicated than the cases involving polysilicon gates. To solve (3.104) or

(3.105), the workfunction from the substrate conduction band (χ_c) must be known. Since aluminum used to be a commonly employed gate material, solutions to (3.104) and (3.105) have been known for some time and can be looked up either in tables or graphs [15]. However, aluminum is now obsolete as a gate material due to its inability to serve in any self-aligned source/drain process. It is unlikely to be encountered in a model development or usage situation.

The second-generation FET models simplify the situation greatly by making the flatband voltage an extracted model parameter. As implemented, a portion of the reduction of the threshold voltage in short channel devices is assigned to the flatband voltage, as described in Chapters 8–10. This is an empirical occurrence, as the flatband voltage is actually defined from a two-dimensional MOS capacitor structure. This introduction of geometry dependence into the flatband voltage represents a change in the surface potential rather than a change in the actual flatband voltage.

The third-generation models adopt a different approach. The flatband voltage is either included as a provided process parameter, or it is computed internally using equations different from (3.100)–(3.103). This is described in Chapter 11.

3.4   The Three-Terminal MOS Structure

The discussion now expands to include the three-terminal MOS structure, as depicted in Figure 3.20. The third terminal is the source diffusion, which forms a p-n junction with the substrate. The voltage between the various terminals must obey

$$V_{gb} = V_{gs} + V_{sb}.   \tag{3.106}$$

It has become conventional to use V_{bs}, and since $V_{bs} = -V_{sb}$, (3.106) is rewritten as

$$V_{gb} = V_{gs} - V_{bs}.   \tag{3.107}$$

This relationship will become important, as it is standard practice in FETs to define the source as the location of the reference potential for all the other terminals (in contrast to the two-terminal case, where the substrate body is defined as the reference location).

The band diagram for the situation is depicted in Figure 3.21. Here, at the onset of strong inversion, $\phi_s = 2\phi_f - V_{bs}$. Thus, from (3.88), the total depletion charge is

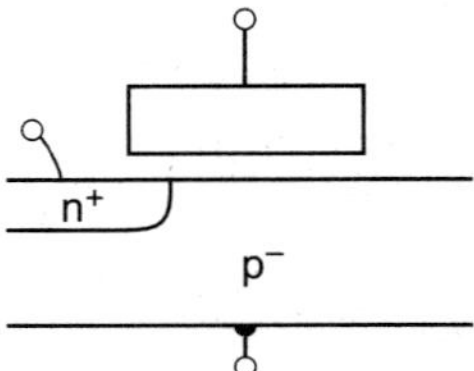

Figure 3.20   A three-terminal MOS capacitor structure.

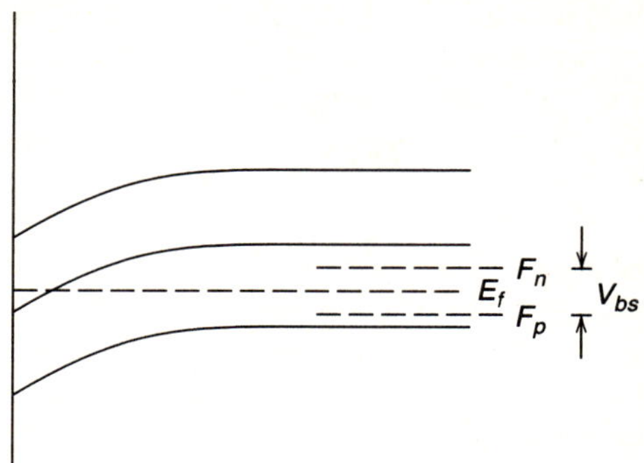

Figure 3.21   The band structure of a three-terminal MOS capacitor.

$$Q_{depl} = \left[2\epsilon_{Si} q N_A (2\phi_f - V_{bs}) \right]^{\frac{1}{2}}, \tag{3.108}$$

while, from (3.89), the total surface charge is

$$Q_s = -C_{ox} \left[V_{gb} - V_{fb} - (2\phi_f - V_{bs}) \right]. \tag{3.109}$$

Using (3.91), the inversion charge now becomes

$$Q_{inv} = -C_{ox} \left\{ \left[V_{gb} - V_{fb} - (2\phi_f - V_{bs}) \right] - \frac{1}{C_{ox}} \left[2\epsilon_{Si} q N_A (2\phi_f - V_{bs}) \right]^{\frac{1}{2}} \right\}, \tag{3.110}$$

or

$$Q_{inv} = -C_{ox} \left\{ \left[V_{gb} - V_{fb} - (2\phi_f - V_{bs}) \right] - \gamma \cdot (2\phi_f - V_{bs})^{\frac{1}{2}} \right\}. \tag{3.111}$$

Substituting the expression (3.107) to account for the change in the reference potential location (from the substrate contact in the two-terminal case to the source diffusion in the three-terminal case) produces the final expression

$$Q_{inv} = -C_{ox} \left\{ (V_{gs} - V_{fb} - 2\phi_f) - \gamma \cdot (2\phi_f - V_{bs})^{\frac{1}{2}} \right\}. \tag{3.112}$$

In a fashion similar to that of (3.96), the threshold voltage can be defined as

$$V_t = V_{fb} + 2\phi_f + \gamma \cdot (2\phi_f - V_{bs})^{\frac{1}{2}}, \tag{3.113}$$

which allows (3.112) to be rewritten as

$$Q_{inv} = -C_{ox}(V_{gs} - V_t). \tag{3.114}$$

Note that (3.114) is essentially identical to (3.107). Thus, the behavior of the three-terminal structure is essentially the same as that of the two-terminal structure, with the additional effect of the third-terminal voltage taken as a method of modifying the threshold voltage.

3.5   The Four-Terminal MOS Structure

The four-terminal MOS structure is shown in Figure 3.22. As in the three-terminal structure, the source electrode is designated as the voltage reference point. Thus, the voltages of concern in this structure will be V_{gs}, V_{ds}, and V_{bs}. In contrast to the two- and three-terminal structures, there is now a two-terminal current path (source to drain) modulated by a third terminal (the gate). This requires a discussion of the current behavior, along with consideration of the capacitance between the gate terminal and the source and drain terminals.

3.5.1   The Drain Current Behavior

Consider a piece of material conducting an electric current [16]. The current density in that material is defined as

$$j = -qnv, \tag{3.115}$$

where n is the conducting (free) charge density and v is the carrier velocity. The Ohm's law approximation (which is derived in Chapter 4) for the velocity/field relationship is

$$v = \mu E, \tag{3.116}$$

which introduces the proportionality constant μ, the mobility. This allows (3.115) to be rewritten as

$$j = -qn\mu E. \tag{3.117}$$

The simple Drude theory of conduction [17, 18] (originally derived for metals, but generally applicable to conduction in any material) finds that

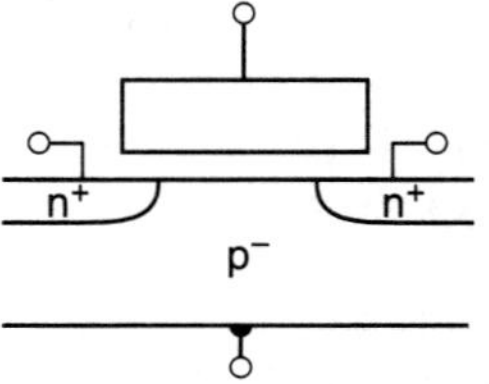

Figure 3.22   The four-terminal MOS structure.

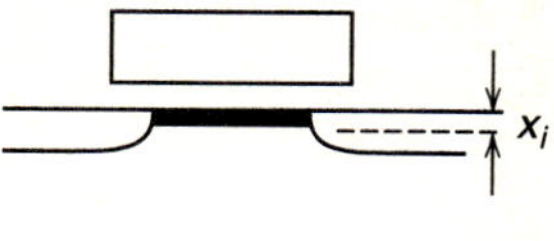

Figure 3.23 The simplified approach to describing the conducting channel of the MOSFET, where the inversion layer is treated as uniform for $x < x_i$ and nonexistent for $x > x_i$.

$$j = \sigma E, \tag{3.118}$$

where σ is the material conductivity. Thus,

$$\sigma = -qn\mu, \tag{3.119}$$

which leads to an expression for the material resistivity:

$$\rho = \frac{1}{\sigma} = -\frac{1}{qn\mu}. \tag{3.120}$$

While the resistivity is a material property, to compute a current it will be necessary to find an expression for the resistance. The approximate structure is shown in Figure 3.23, where W is the FET width, y is the distance between the source and drain diffusions, and x_i is the depth of the conducting layer. It will be assumed here for simplicity that for $x < x_i$, the charge density is uniform with a value of n, and for $x > x_i$, no conducting charge is present. The resistance of this piece of material is

$$R = \frac{\rho y}{xW} = -\frac{y}{xWqn\mu}. \tag{3.121}$$

It is also reasonable to assume that $W \gg x$ and $y \gg x$, and treat the conducting layer as a vanishingly thin two-dimensional charge layer; in this manner, the (areal) inversion charge is defined as

$$Q_{inv}(y) = xqn, \tag{3.122}$$

where now Q_{inv} is a function of position (y) along the channel. Equation (3.121) is thus rewritten as

$$R = -\frac{y}{\mu W Q_{inv}(y)}. \tag{3.123}$$

The lateral field in this structure is provided by the drain to source voltage V_{ds}. Using Ohm's law, the expression for the current is

$$I_{ds} = \frac{V_{ds}}{R} = -\frac{\mu W Q_{inv}(y) V_{ds}}{y}. \tag{3.124}$$

The voltage drop across a small part of the channel dy is

$$dV_{ds} = I_{ds}dR. \tag{3.125}$$

The differential element dR is found by taking the derivative of (3.123):

$$\frac{dR}{dy} = -\frac{1}{\mu W Q_{inv}(y)}. \tag{3.126}$$

Thus,

$$dV_{ds} = -\frac{I_{ds}}{\mu W Q_{inv}(y)}dy. \tag{3.127}$$

To evaluate the drain-source current I_{ds} using (3.127), an expression for $Q_{inv}(y)$ must be found. This necessitates that the various charge components be examined, while also considering the dependence of the charge on the specific position y along the channel. Therefore, (3.78) is rewritten here as

$$V_{gb} = V_{fb} - \phi_s(y) - \frac{Q_s(y)}{C_{ox}}, \tag{3.128}$$

while (3.79) is recast as

$$Q_s(y) = Q_{inv}(y) + Q_{depl}(y). \tag{3.129}$$

The depletion charge, from (3.88), is

$$Q_{depl}(y) = (2\epsilon_{Si}qN_A\phi_s(y))^{\frac{1}{2}}. \tag{3.130}$$

However, here $\phi_s(y)$ is

$$\phi_s(y) = V(y) + 2\phi_f - V_{bs}, \tag{3.131}$$

where $V(y)$ is the voltage at any point y along the channel due to the drain-source voltage V_{ds}. The depletion charge is thus

$$Q_{depl}(y) = \left[2\epsilon_{Si}qN_A\left(V(y) + 2\phi_f - V_{bs}\right)\right]^{\frac{1}{2}}. \tag{3.132}$$

Next note that (3.89) becomes

$$Q_s(y) = -C_{ox}\left(V_{gb} - V_{fb} - \phi_s(y)\right). \tag{3.133}$$

Using (3.131), (3.133) becomes

$$Q_s(y) = -C_{ox}(V_{gb} - V_{fb} - V(y) - 2\phi_f + V_{bs}). \tag{3.134}$$

Following (3.91), (3.129) is rearranged to

$$Q_{inv}(y) = Q_s(y) - Q_{depl}(y), \tag{3.135}$$

leading to an expression for $Q_{inv}(y)$:

$$Q_{inv}(y) = -C_{ox}\left[V_{gb} - V_{fb} - V(y) - 2\phi_f + V_{bs}\right]$$

$$- \left[2\epsilon_{Si}qN_A\left(V(y) + 2\phi_f - V_{bs}\right)\right]^{\frac{1}{2}}. \tag{3.136}$$

Using (3.107) to change from V_{gb} to V_{gs}, and inserting the expression (3.94) for γ produces

$$Q_{inv}(y) = -C_{ox}\left[V_{gs} - V_{fb} - V(y) - 2\phi_f\right] - \gamma \cdot \left[V(y) + 2\phi_f - V_{bs}\right]^{\frac{1}{2}}. \tag{3.137}$$

(Note for reference that the substrate-source voltage V_{bs} no longer appears in the first term of the right-hand side of the equation.) Using (3.137) in (3.127) will provide the basis for the discussion of the drain current in subsequent chapters.

Drain Current Saturation

Consider an FET with a fixed gate bias ($V_{gs} > V_t$) and an increasing drain bias. As shown in Figure 3.24, for a low drain bias, the drain current increases linearly with the drain bias (hence the moniker of *linear region*). At larger drain biases, the current departs from the linear characteristic. At some voltage, known as the saturation voltage V_{dsat}, the drain current no longer increases with increasing drain bias.

This behavior is briefly described in Figure 3.25. For low drain biases (Figure 3.25a), there is an inversion layer connecting the source and drain, which behaves as an ideal ohmic resistor. For larger drain biases (Figure 3.25b), the drain field prevents the

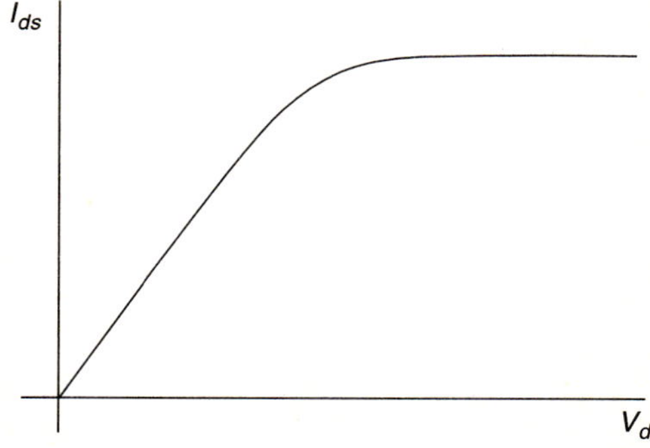

Figure 3.24   FET drain current for a fixed gate bias with $V_{gs} > V_t$, showing drain current saturation for $V_{ds} > V_{dsat}$.

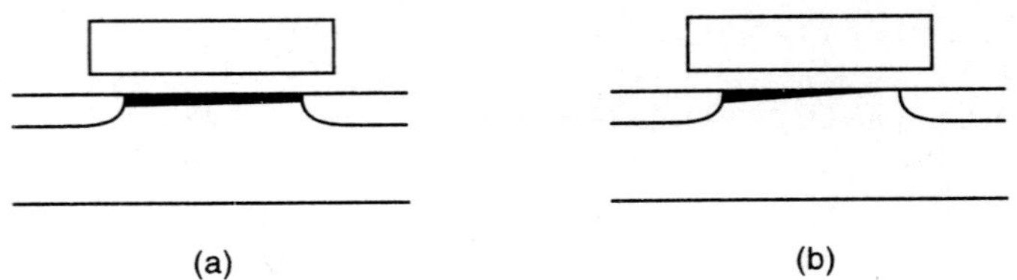

(a)   (b)

Figure 3.25   FET inversion layer for $V_{gs} > V_t$; (a) $V_{ds} < V_{dsat}$; the inversion layer contacts the drain diffusion; (b) $V_{ds} > V_{dsat}$; the inversion layer no longer contacts the drain diffusion.

inversion layer from reaching the drain diffusion; the channel pinches off, and the drain current is no longer affected by the drain bias.

The Subthreshold Behavior

Recall that it was noted that (3.97) reduced to $Q_{inv} = 0$ when $V_{gs} < V_t$, which implies that $I_{ds} = 0$ in this voltage regime. However, this is a consequence of the simplicity of the analysis, in which an expression for the total surface charge Q_s can be written down, a straightforward expression for the depletion charge Q_{depl} derived, and the inversion charge Q_{inv} taken as the difference between the two. This approach allowed the inversion charge to be computed without considering the energy band structure of the MOS capacitor. To evaluate the current behavior of the FET for $V_{gs} < V_t$, a different approach must be taken [14].

For $\phi_s < \phi_f$, the inversion charge density is less than the intrinsic silicon charge density n_i and can be neglected, while for $\phi_s > 2\phi_f$, the analysis of the previous section applies. Thus, the region of interest is $\phi_f < \phi_s < 2\phi_f$. Consider bulk p-doped silicon. Assuming complete impurity ionization, the Fermi potential in the material can be written as

$$\phi_f = k_b T \cdot ln\left(\frac{N_A}{n_i}\right) = k_b T \cdot ln\left(\frac{p}{n_i}\right), \tag{3.138}$$

which can be solved for the free hole density:

$$p = n_i e^{\phi_f / k_b T}. \tag{3.139}$$

Since detailed balance requires that in equilibrium or a steady state

$$n_i^2 = np, \tag{3.140}$$

the free electron density is

$$n = n_i e^{-\phi_f / k_b T}. \tag{3.141}$$

Now introduce the effect of band bending due to the applied gate bias. Since the surface potential described in Figure 3.18 serves to increase the surface electron concentration, (3.141) is amended to

$$n_s = n_i e^{(\phi_s - \phi_f)/k_b T},\tag{3.142}$$

where n_s denotes that (3.142) describes the electron concentration at the silicon surface. When $\phi_s = \phi_f$, $n_s = n_i$, which defines the onset of inversion of the p-type silicon surface. When $\phi_s = 2\phi_f$,

$$n_s = n_i e^{\phi_f/k_b T};\tag{3.143}$$

that is, the surface electron concentration equals the hole concentration in the bulk p-type silicon substrate, which defines the onset of strong inversion.

A sketch of the behavior of the subthreshold current can now be developed. Recall (3.78), relating the surface potential and the applied gate bias. Rearranging (3.78) to solve for the surface potential produces

$$\phi_s = V_{gb} - V_{fb} + \frac{Q_s}{C_{ox}}.\tag{3.144}$$

As just described, the surface electron concentration is less than the bulk hole concentration, which implies that $Q_{inv} \ll Q_{depl}$; this allows Q_s in (3.144) to be replaced with Q_{depl}. Equation (3.107) can also be employed to change from the MOS capacitor expression V_{gb} to the FET expression V_{gs}:

$$V_{gb} = V_{gs} - V_{bs}.\tag{3.145}$$

Equation (3.144) is thus recast as

$$\phi_s = V_{gs} - V_{fb} - V_{bs} + \frac{Q_{depl}}{C_{ox}}.\tag{3.146}$$

The areal inversion layer charge density is computed by integrating the three-dimensional electron charge density:

$$Q_{inv} = q \int_{-\infty}^{0} n(x)dx.\tag{3.147}$$

Finally, recalling (3.124), the drain current is expressed as

$$I_{ds} = -\frac{\mu W Q_{inv} V_{ds}}{y}.\tag{3.148}$$

Examining (3.146), (3.147), and (3.148), it can be seen that in the most general form,

$$I_{ds} \propto e^{V_{gs}}. \tag{3.149}$$

Equation (3.149) describes, in the most general terms, the subthreshold current, a drain current that changes exponentially with V_{gs}. More detailed analyses will be delayed until consideration of the subthreshold current portions of the specific SPICE FET models.

Discontinuity at the Subthreshold-Superthreshold Transition

A more physically rigorous analysis [14, 19] of the drain current models indicates that when $V_{gs} > V_t$, current transport is dominated by drift due to the lateral electric field, while for $V_{gs} < V_t$, current transport is dominated by diffusion of carriers over the channel potential barrier. Due to the approximations used in deriving the expressions, these analytical models show discontinuities in Q_{inv} and I_{ds} at the transition between the subthreshold region and the superthreshold region. As will be discussed in Chapter 4, these discontinuities are not found in a more detailed numerical solution. Analytical models cope with these discontinuities by introducing some form of empirical curve-fitting expression, as will be discussed in the individual SPICE FET model chapters.

3.5.2   The Gate Capacitance Behavior

In addition to the source-substrate and drain-substrate capacitances, the FET contains capacitances between the gate and the rest of the device. These capacitances are strongly bias dependent, and in most cases the gate capacitances are the largest capacitive load introduced into a circuit by an FET.

The situation is diagrammed in Figure 3.26. In the zero bias condition (Figure 3.26a), there are small capacitances between the gate and the other three terminals. The situation becomes more interesting when an inversion layer is present. In the linear region ($V_{gs} > V_t$, $V_{ds} < V_{dsat}$), as shown in Figure 3.26b, the inversion layer is electrically connected to the source and drain diffusions. The inversion layer screens any charge in the bulk from the gate, greatly reducing or eliminating this component of the capacitance. However, the capacitance between the gate and either the source or drain is now in fact the capacitance between the gate and the inversion layer. This is the dominant intrinsic capacitance in an FET. In saturation ($V_{gs} > V_t$, $V_{ds} < V_{dsat}$), as shown in Figure 3.26c, the inversion layer is no longer electrically connected to the drain diffusion; thus, the gate

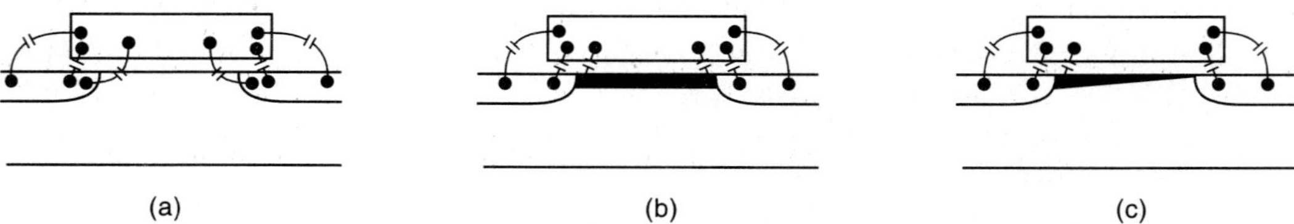

(a)   (b)   (c)

Figure 3.26   The gate capacitance of the FET structure; (a) no biases applied; (b) $V_{gs} > V_t$, $V_{ds} < V_{dsat}$; (c) $V_{gs} > V_t$, $V_{ds} > V_{dsat}$.

to drain capacitance becomes small again. However, the inversion layer is still connected to the source diffusion, so the total gate capacitance retains its large value.

The approach to modeling the gate capacitance is conceptually simple but difficult in practice. The charge on at least the gate, and preferably on all the other terminals, must be found as a function of the applied biases; these charges and their variations with the applied biases are used to compute the gate capacitance values. Since the computation of terminal charges is intertwined with the formulation of the drain current model, the specific node charge models implemented in SPICE will be considered as the individual current models are developed. The issues associated with converting those node charge expressions into descriptions of the gate capacitance are discussed in Chapter 13.

Conservation of Charge

An additional requirement imposed on a gate capacitance model is that charge must be conserved. For any time and bias, as guaranteed by the principle of charge neutrality, the total charge in the FET must sum to zero:

$$Q_{GATE} + Q_{DEPL} + Q_{INV} = 0. \qquad (3.150)$$

(The capitalized subscripts indicate the *total* charge, rather than areal charge density.) Forcing (3.150) to be true is a necessary but not sufficient condition for assuring that charge is conserved.

As described in Chapter 2, the original implementations of SPICE used the voltage as the state variable at each circuit node, but not the charge. In this case, the node charges are not computed directly from the FET current model, but are instead computed from the node voltages. It is easily shown [20, 21] that this can lead to nonconservation of charge. The problem is solved by making charge a state variable at each circuit node.

Thus, the computation of the gate capacitance involves two separate issues. First, using (3.150), the charge at each device terminal is computed. This computation of charge is carried out using the FET model; specific details will be given with each particular model derivation. Second, the computed charges are used to calculate the gate capacitance. This computation, and whether or not it conserves charge, is really an issue of implementation in the SPICE circuit simulator (rather than a concern of the individual FET models). A complete discussion of this problem is presented in Chapter 13.

3.5.3 Small Geometry Effects

The basic and original analytical MOSFET models were introduced during the mid-1960s [22, 23]. These models are relatively simple, and were developed for what are referred to today as "long, wide" FETs; they form the underlying foundation for all the analytical FET models available in SPICE.

As device geometries have decreased (channel length and width, and oxide thickness), device structures have become more complicated (e g., nonuniform substrate doping), and internal electric fields have become stronger (due to the reduction in feature sizes proceeding more rapidly than the reduction of the power supply voltage [24]). Thus, a number of new features have appeared in FET behavior. New models, extending beyond the original models, are continually being introduced to account for these additional effects.

This is the basis for all the FET current models that have been developed and introduced into SPICE. The original long channel FET model is treated as a base, and the small geometry effects are added as "corrections." Over time, decreasing device geometries have spawned a growing list of such effects; this list will be considered in a simple form here. The objective is to quantitatively describe the physical origin of each effect, and to postpone a detailed discussion of specific descriptions until the individual SPICE FET models are developed. Often, the same effect is modeled differently in different FET models. On the other hand, some effects discussed here are not yet included in any of the SPICE FET models; their description is included for completeness, and in anticipation of their addition to future SPICE FET models.

The descriptions to be given here are qualitative, with quantitative discussion postponed until the inclusion of each effect in a particular SPICE FET model is considered; the intent is to introduce each small geometry effect in a simple and clear manner. Exhaustive mathematical descriptions of these effects may be found in two excellent textbooks [14, 25].

Depletion Region Overlap

The basic FET structure combines the depletion regions of two p-n junctions (source-substrate and drain-substrate) with the depletion region of a MOS capacitor. In a large FET, these depletion regions can be described by the equations for the corresponding two-terminal structures. In a smaller FET, the situation is as shown in Figure 3.27. The shaded regions show where the junction depletion regions overlap the gate depletion region. Thus, the total depletion charge is no longer simply the sum from the three individual regions; instead, it is reduced from that value. Since, according to (3.79),

$$Q_s = Q_{inv} + Q_{depl},\qquad(3.151)$$

and as Q_s merely images the gate charge, a decrease in Q_{depl} from its expected value implies an increase in Q_{inv}. The net result is that Q_{inv} will be larger than that predicted by the theory of the simple two-dimensional MOS capacitor; this must be accounted for in the FET model.

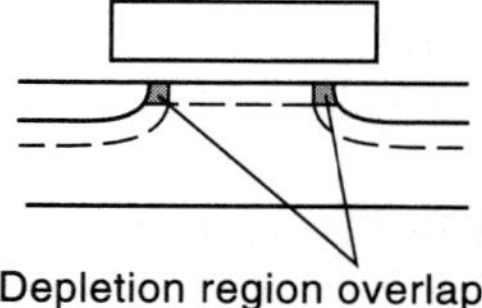

Figure 3.27 FET depletion regions. The gate-induced depletion region overlaps the source and drain depletion regions.

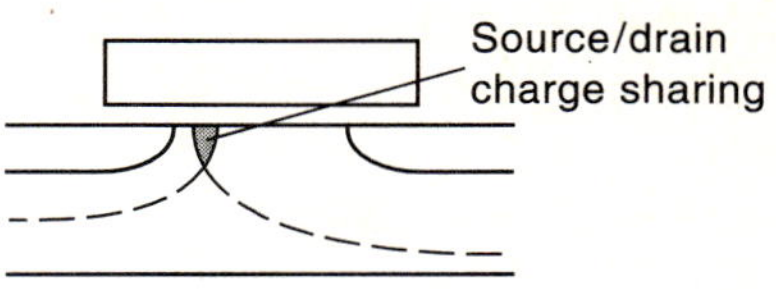

Figure 3.28   FET depletion regions, showing charge sharing between the source and drain.

Source-Drain Charge Sharing

As the channel length decreases further, the situation just described becomes even more complicated. As shown in Figure 3.28a, the source and drain depletion regions can overlap. This is often referred to as *charge sharing between the source and the drain*. This further decreases the depletion charge Q_{depl} from the value predicted by the two-terminal MOS capacitor model, leading to a concomitant increase in the inversion charge Q_{inv} above its expected value. In addition, the application of a drain bias will expand the drain junction depletion region (Figure 3.28b), further increasing the extent of charge sharing.

Depletion Region Spreading Outside the Channel Width

Consider the cutaway view (across the width of the channel) of an FET shown in Figure 3.29. In a very wide FET (Figure 3.29a), the gate-induced depletion region is treated as not extending beyond the gate edge; since the FET width is large, this approximation is good, and the depletion region is well described by the simple MOS capacitor model. The situation is different in a narrow device, defined by an isolation oxide "groove," as shown in Figure 3.29b. Here the depletion region spreading outside the channel region is significant and cannot be ignored. In this case, the depletion charge Q_{depl} is larger than

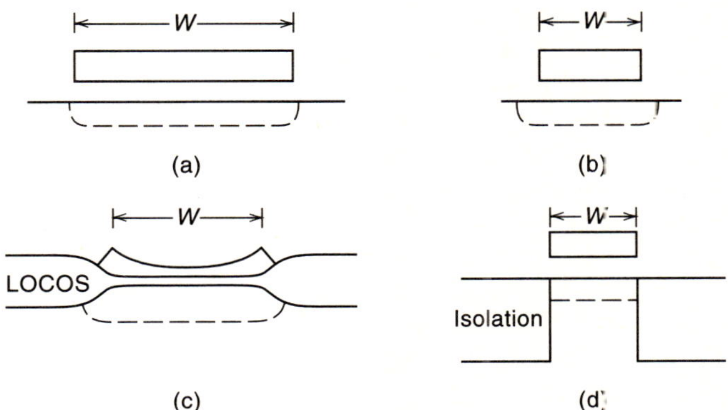

Figure 3.29   The gate-induced depletion region across the width of the FET; (a) very wide device; (b) simple narrow device, showing how the depletion region spreads outside the defined channel; (c) depletion region spreading in a LOCOS-bounded device; (d) an isolation-trench-bounded structure, in which the isolation prevents the depletion region from spreading outside the defined channel width.

that predicted by the two-terminal MOS capacitor model, so the inversion charge Q_{inv} is decreased from the simply predicted value.

The same situation holds in the more complex LOCOS isolation structure. As shown in Figure 3.29c, LOCOS complicates the shape of the extended depletion region, but the basic result is the same. Interestingly, more recent (and more complex) oxide-filled trench isolation structures [26, 27] eliminate this problem. As shown in Figure 3.29d, the presence of the isolation trench prevents the depletion region from spreading outside the width of the defined channel; assuming that the doping is relatively uniform across the device "pedestal," the depletion region is once again well described by the simple MOS capacitor model.

Mobility Reduction Due to the Gate Field

The mobility, μ, was introduced in (3.116) as a method of relating the carrier velocity to the lateral field:

$$v = \mu E. \tag{3.152}$$

This relationship is simple and approximate, but actually describes carrier conduction in a bulk semiconductor material under low fields. This approach is adequate for large FETs, and (3.152) has been used effectively with μ as a constant.

However, as gate oxides have become thinner and the vertical field at the silicon surface has increased, the mobility no longer remains constant. Instead, as shown in Figure 3.30, it decreases with increasing gate field. Although the physical basis of this mobility reduction is still a matter of some debate [28–30], it must be included in the FET models.

Lateral Mobility Reduction and Velocity Saturation

In addition to the mobility reduction due to the gate field, the simple Ohm's law relationship (3.152) does not hold for larger values of the lateral field. Instead, as shown in Figure 3.31, the carrier velocity departs from its linear relationship with the lateral field and becomes constant; this phenomenon is known as *carrier velocity saturation*, and the constant maximum velocity is referred to as the *saturation velocity*.

At the most detailed level, since the lateral field varies along the channel, the mobility will also vary along the channel. However, this makes the situation too complicated for description by an analytical model. Instead, velocity saturation and the lateral mobility reduction are linked to the drain current saturation described by the saturation voltage V_{dsat}. As detailed earlier, for a constant gate bias V_{gs} ($V_{gs} > V_t$), the drain current I_{ds} increases linearly with the applied drain bias until V_{ds} reaches the saturation voltage V_{dsat}. For $V_{ds} > V_{dsat}$, the drain current remains essentially constant. However, in short channel FETs, due to the high lateral field, the carriers reach the saturation velocity before the expected value of V_{dsat}. As shown in Figure 3.32, the drain current thus saturates at a lower value.

In short channel FET models, this behavior is accounted for in the following way. Carrier velocity saturation is used to describe the reduction of the mobility by the high

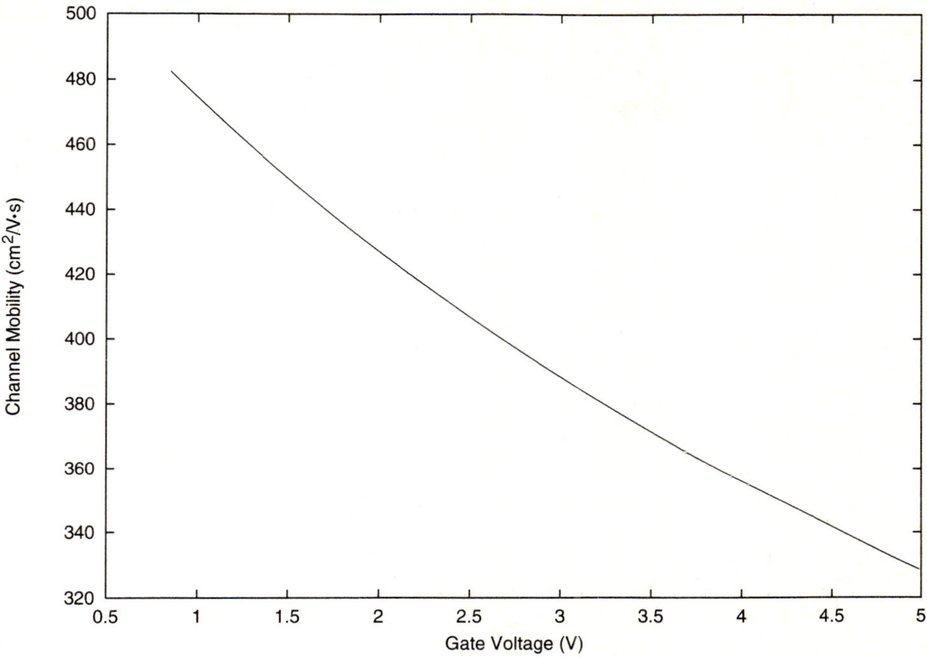

Figure 3.30   The channel electron mobility versus gate voltage of a thin oxide FET.

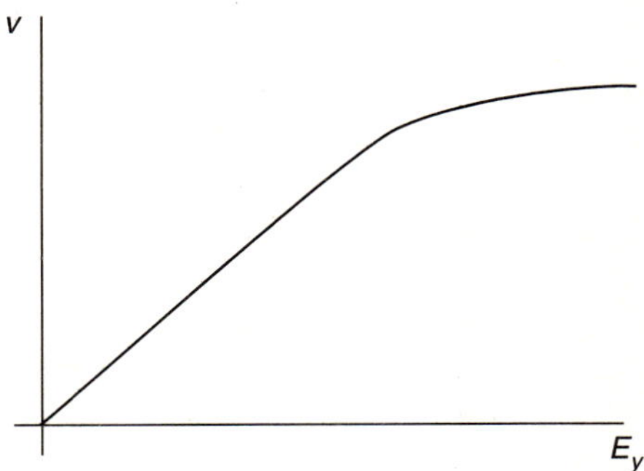

Figure 3.31   The velocity of electrons in silicon versus the applied field. For low fields, the velocity increases linearly with the field (Ohm's law regime), while for higher fields the velocity no longer increases.

lateral field. This mobility result is then used to compute a reduced value of the saturation voltage V_{dsat}, which is used to describe the drain current behavior.

Interestingly, *drain current saturation* and *carrier velocity saturation* originally described two separate FET phenomena, the former being an inherent characteristic of any FET, while the latter is a high field phenomenon in a bulk material. However, it turns

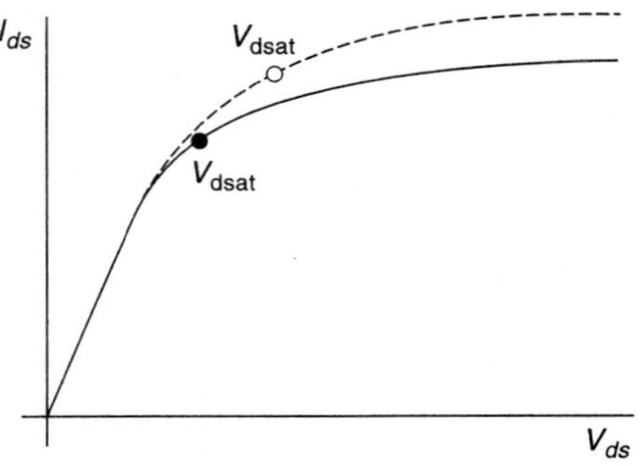

Figure 3.32   Drain current characteristics of an FET with a fixed gate bias ($V_{gs} > V_t$). The dashed line shows the expected behavior (if carrier velocity saturation is ignored), while the solid line shows the effect of velocity saturation. Note how carrier velocity saturation reduces the saturation voltage V_{dsat}.

out that the two phenomena are closely related, and in short channel devices the current saturation characteristics are determined by carrier velocity saturation.

Channel Length Modulation

As described earlier in this chapter and just above, for a constant gate bias with $V_{gs} > V_t$ and $V_{ds} > V_{dsat}$, the drain current no longer increases with increasing drain bias. In short channel devices, this is not exactly the case. As shown in Figure 3.33a, when $V_{ds} = V_{dsat}$, the channel pinches off and the inversion layer no longer reaches the drain. When $V_{ds} > V_{dsat}$, as shown in Figure 3.33b, the pinch-off point moves away from the drain junction and toward the source. This phenomenon is negligible in long channel devices, but becomes more important in short channel FETs. Since the lateral voltage at the pinch-off point is always V_{dsat}, if the pinch-off point moves toward the source, the lateral field along the channel (between the source and the pinch-off point) increases above its value when $V_{ds} = V_{dsat}$. This increased field in turn increases the drain current. This is illustrated in Figure 3.34 for a short channel device; the drain current increases slightly for increasing drain bias when $V_{ds} > V_{dsat}$.

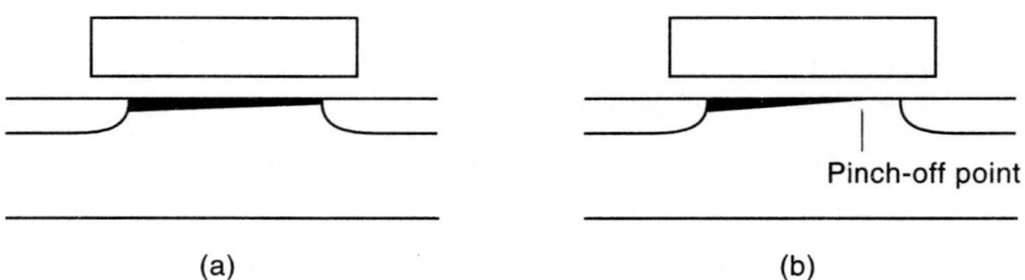

Figure 3.33   Channel length modulation in an FET; (a) $V_{ds} = V_{dsat}$; (b) $V_{ds} > V_{dsat}$; the pinch-off point has moved away from the drain junction and toward the source.

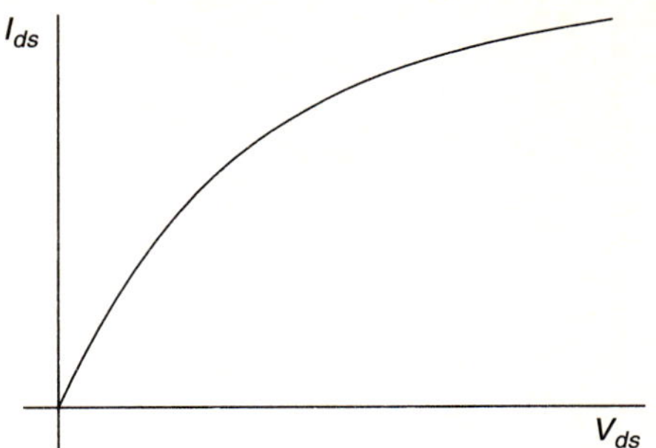

Figure 3.34 FET drain current characteristic for a fixed gate bias ($V_{gs} > V_t$). Due to channel length modulation, the drain current increases slightly as V_{ds} increases above V_{dsat}.

Series Resistance

In a long channel FET, the resistance of the channel region is by far the largest resistance in the source-drain current path. In a short channel device, this is no longer the case, as shown in Figure 3.35; the channel resistance has become small enough for the resistance of the source and drain regions to become important. A portion of the applied drain-source bias V_{ds} is in fact dropped across the source and drain resistances, decreasing the drain current from its expected value.

Drain-Induced Barrier Lowering (DIBL)

Consider the potential profile along the channel axis of a long channel FET with no biases applied, as depicted in Figure 3.36a. There is a barrier between the diffusions and the channel region equal to the built-in zero bias junction potential between the diffusions

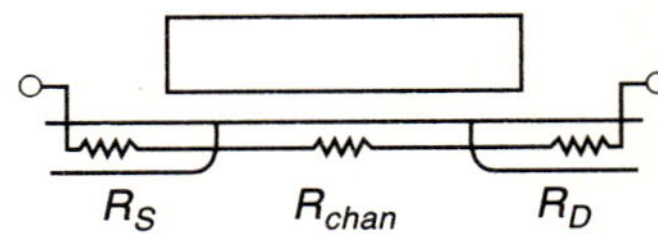

Figure 3.35 Parasitic source-drain series resistance. These parasitic resistances become important in short channel devices, in which the channel resistance has become small.

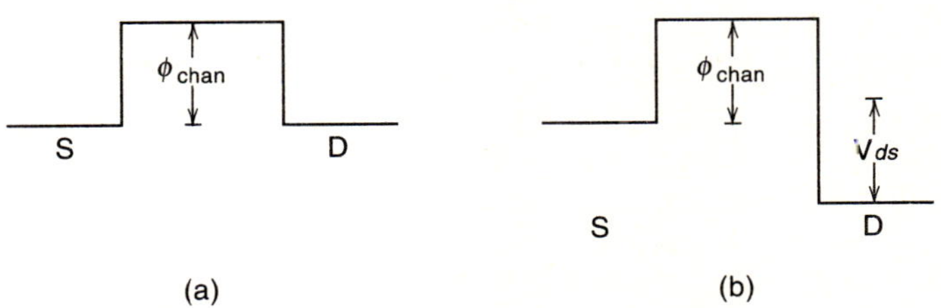

Figure 3.36 Channel potential profile of a long channel device; (a) $V_{ds} = 0$; (b) $V_{ds} > 0$.

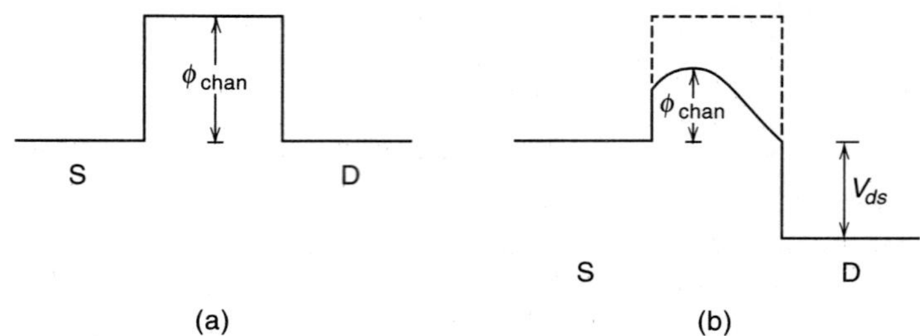

Figure 3.37 Channel potential profile of a short channel device; (a) $V_{ds} = 0$; (b) $V_{ds} > 0$; the drain bias lowers the potential barrier at the source end of the channel.

and the bulk silicon. When a drain bias is applied (Figure 3.36b), only the immediate region of the drain is affected, and the channel potential profile is basically unaffected.

Now consider a short channel device. As shown in Figure 3.37a, with no biases applied, the potential profile is the same as that in the long channel device. However, with a drain bias applied (Figure 3.37b), the channel potential profile is affected. The drain bias now changes the potential profile along the entire channel, lowering the barrier at the source-bulk junction; this is referred to as *drain-induced barrier lowering*, or DIBL. For a given drain bias, this allows carriers to traverse the channel at a lower gate bias than would otherwise be expected. It is most common to note the effect on the threshold voltage, defined by (3.96):

$$V_t = V_{fb} + 2\phi_f + \gamma \cdot (2\phi_f)^{\frac{1}{2}}. \tag{3.153}$$

Since carriers are able to traverse the channel for lower than expected gate biases, DIBL is usually included as a modification to the threshold voltage expression (3.153), as it serves to decrease the threshold voltage.

Punchthrough

When drain-induced barrier lowering occurs in a short channel FET, the channel potential barrier is decreased, but still exists; although a drain current will flow at a lower gate bias than would otherwise be expected, the gate is still able to control the current.

Recall the above discussion of charge sharing between the source and the drain, in which the drain depletion region expands with increasing drain bias to partially overlap the source depletion region. A more detailed analysis [31] shows that the extent of the source and drain depletion regions in the lateral (channel) direction is in fact restricted by the presence of the gate. As shown in Figure 3.38a, under zero bias, the lateral depletion depth is less than the vertical depletion depth. The same situation holds when a drain bias is applied (Figure 3.38b). However, note that away from the silicon surface, the depletion region is larger. As the drain bias increases, this "bulge" in the depletion region can reach the source depletion region (Figure 3.38c). When this occurs, the channel

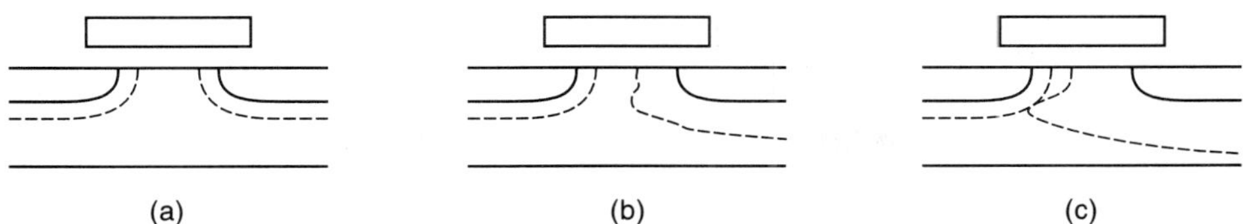

(a)   (b)   (c)

Figure 3.38 Source-drain depletion regions in a short channel FET; (a) $V_{ds} = 0$; (b) $V_{ds} > 0$; note how the lateral depletion region is larger away from the surface; (c) $V_{ds} \gg 0$; below the surface, the drain depletion region has contacted the source depletion region, and a punchthrough current flows.

potential barrier no longer exists in the "contact" region; a large drain current which is not controlled by the gate begins to flow. This phenomenon is known as punchthrough, and the current which ensues is referred to as the punchthrough current.

Punchthrough is a failure mechanism which limits the minimum channel length of a design. The technology design rules should keep the designer from encountering this problem; its presence in a design indicates a problem with the silicon process technology.

Hot Carrier Effects

During the 1970s, simple scaling rules for MOSFET technology were developed [32]. These rules stipulated that the power supply voltage should be reduced in proportion to the reduction of FET feature sizes (the channel length and width, the oxide thickness, etc.). However, a continually changing power supply voltage is very difficult to accommodate in designs; it is much simpler to maintain a constant supply voltage for compatibility.

Since constant voltage scaling was chosen over constant field scaling [24], a number of interesting problems have manifested themselves in FET operation. With feature sizes decreasing more rapidly than the power supply voltage, field strengths in the FET channel increased. This is particularly notable at the drain end of a short channel device, where the applied bias V_{dd} is dropped over a smaller channel length. Under the simple model described earlier, all carriers have the same velocity, reaching a maximum at the saturation velocity v_{SAT}. The carrier energy is then simply computed from

$$U = \frac{1}{2}m^* v_{SAT}{}^2, \tag{3.154}$$

where m^* is the carrier effective mass. However, as will be described in Chapter 4, this view is too simplistic. A detailed simulation shows that the velocity (and thus the energy) of the carriers is in fact a distribution; this distribution becomes more spread out as the field increases. High-energy carriers at the high end of the distribution cause impact ionization, generating an electron and a hole; in nFETs, the generated hole is swept away into the bulk, while the electron is injected into the oxide where it can damage the device, leading to long-term reliability problems [33, 34]. This occurrence is known as *hot carrier degradation*, or by some similar name. The rate of degradation has been

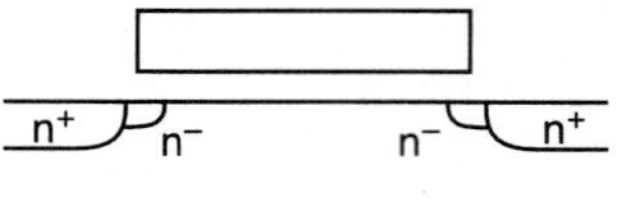

Figure 3.39 An nFET with lightly doped drain (LDD) extensions to the conventional source and drain.

shown to correlate with the substrate current in nFETs [33], and with the gate current in pFETs [35].

The device fabrication technology was modified to mitigate this problem. The field at the drain junction can be decreased by increasing the width of the lateral depletion region between the pinch-off point and the drain. This is done by adding a lightly doped extension to the drain, as shown in Figure 3.39. Normal source and drain diffusions are doped at a concentration of 5×10^{19} cm^{-3} to 1×10^{20} cm^{-3}, while the lightly doped drain (LDD) regions are doped at a concentration of 4×10^{18} cm^{-3} to 8×10^{18} cm^{-3}. This reduces the field, and decreases the generation of hot carriers [36]. A penalty is paid in increased series resistance [37], in process complexity, and, as noted more recently, in an inability to decrease the total device feature size as the effective channel length decreases [38]. Given these difficulties, there is now strong interest in returning to constant field scaling as a method of simplifying the FET structure [24, 39].

Although it is a topic of great interest to the device physicist, in general, a circuit designer does not require a detailed model for hot carrier degradation. The design rules should include constraints on the power supply voltage and operating temperature that guarantee that difficulties with hot carrier effects are not encountered.

Gate-Induced Drain Leakage

Constant voltage scaling has also led to another interesting problem. Consider an nFET drain region, where the gate overlaps the drain diffusion (Figure 3.40a). With the gate grounded and the drain biased, the silicon surface will be depleted, as shown in the band diagram of Figure 3.40b. Since the silicon is doped n$^+$, the depletion

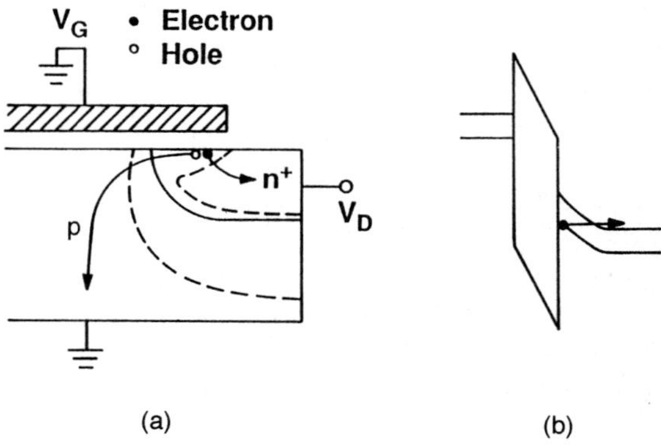

Figure 3.40 The origins of gate-induced drain leakage; (a) gate overlap of the heavily doped drain diffusion; (b) band diagram for $V_{gs} = 0$ and $V_{ds} > 0$. Due to the high field, carriers are generated by band-to-band tunneling. From [40]. © 1987 IEEE. Used by permission.

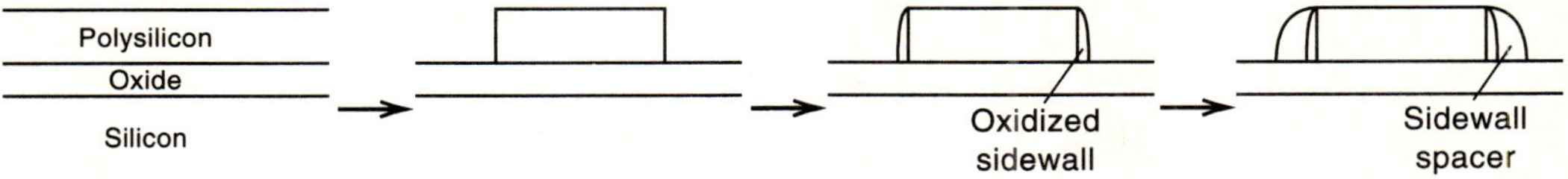

Figure 3.41   Polysilicon gate fabrication process. Sidewall oxidation can introduce point defects into the silicon just below the gate edge.

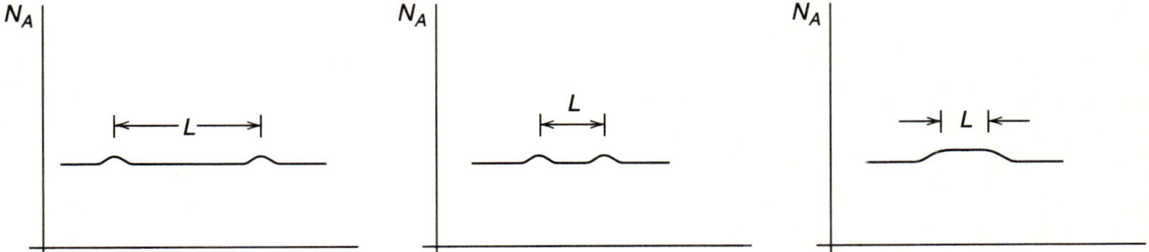

Figure 3.42   Potential profiles in FETs with channel dopant redistribution due to oxidation-enhanced diffusion. In short channel devices, the change becomes significant.

region is very small, and the band bending is confined to a small spatial region; this implies a very high field. Electrons can tunnel from the valence band near the surface into the conduction band, leaving a hole behind. This is a generation mechanism, with the holes swept into the bulk and the electrons into the drain, where they appear as a leakage current [40]. This leakage mechanism is referred to as gate-induced drain leakage (GIDL).

Threshold Voltage Roll-Up

In most FET fabrication processes, after the polysilicon gate is defined by reactive ion etching, its sides are oxidized to "clean them up" for further processing, such as spacer formation (Figure 3.41). Depending on the oxidation process used, mechanical stress at the bottom of the gate can introduce point defects into the silicon just below the gate edge. As processing proceeds, these point defects gather impurities from the substrate, a process referred to as *oxidation-enhanced diffusion* [41]. The effect on the device behavior is depicted in the potential profiles of Figure 3.42. In a long channel device (Figure 3.42a), the two minor "bumps" in the potential profile are not significant when the gate bias forms the inversion layer. As the channel length decreases (Figure 3.42b), the increased potential at the ends of the channel becomes more significant, while in a shorter device (Figure 3.42c), the entire channel now has increased surface doping, leading to a larger channel potential than that found in the longer device. Recalling (3.96) and (3.153),

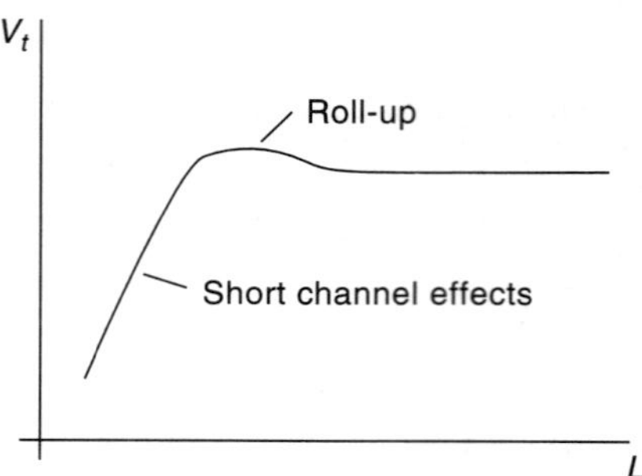

Figure 3.43 Threshold voltage roll-up due to oxidation-enhanced diffusion.

$$V_t = V_{fb} + 2\phi_f + \gamma \cdot (2\phi_f)^{\frac{1}{2}} \tag{3.155}$$

indicates that the increased potential manifests itself as an *increasing* threshold voltage with decreasing channel length (Figure 3.43); this is known as *threshold voltage roll-up* [42].

Substrate Current-Induced Body Effect (SCBE)

As described above, hot carriers generate electron-hole pairs. In an nFET, the holes are swept into the bulk and appear as substrate current. If the generation of holes becomes large, the hole injection current, introducing positive charge into the bulk and charging it, produces an effect similar to that of a small positive substrate bias V_{bs}; hence, the effect is known as the *substrate current-induced body effect* [43]. The result is a decrease in the effective threshold voltage, and therefore an increase in the drain current above the expected value. The total drain current also becomes the sum of the usual drain current and the substrate current, further adding to the observed drain current [44].

Comments on Small Geometry Effects

The above list summarizes the various corrections required to model small geometry effects in FETs. Many of these effects are included in the most popular FET models available in SPICE, and some have been added more recently. A few have yet to be added, and it is not yet clear if their addition will be necessary. As FET dimensions decrease further, it is certain that additional effects will arise that will need to be taken into account.

It is reasonable to ask if the historical and current approach of a "base" long channel model, continually decorated with a growing list of corrections, is still the best method. This is a topic that will require considerable rethinking of the traditional approach to FET modeling. However, as all the present FET models used in SPICE employ this traditional approach, such a discussion will not be undertaken.

BIBLIOGRAPHY

1. G. Neudeck, "The PN Junction Diode," *Modular Series on Solid State Devices* (ed. by G. Neudeck and R. Pierret), volume II, Addison-Wesley, 1983.

2. B. Streetman, *Solid State Electronic Devices* (2nd edition), Prentice-Hall, 1980.

3. A. Vladimirescu and S. Liu, "The Simulation of MOS ICs Using SPICE-2," University of California/Berkeley, Electronics Research Laboratory Document M80/7 (1980).

4. *HSPICE User's Manual*, Meta-Software, Inc., Campbell, California, 1993.

5. Y. Varshni, "Temperature Dependence of the Energy Gap in Semiconductors," *Physica*, vol. 34, pp. 149–154 (1967).

6. H. Barber, "Effective Mass and Intrinsic Concentration in Silicon," *Sol. St. Elec.*, vol. 10, pp. 1039–1051 (1967).

7. W. Shockley, "The Theory of p-n Junctions in Semiconductors and p-n Junction Transistors," *Bell System Technical Journal*, vol. 28, pp. 435–489 (1949).

8. W. Shockley and W. Read, "Statistics of the Recombination of Holes and Electrons," *Phys. Rev.*, vol. 87, pp. 835–849 (1952).

9. R. Hall, "Electron-Hole Recombination in Germanium," *Phys. Rev.*, vol. 87, pp. 387–392 (1952).

10. D. Foty, "The Physics of p-n Junction Leakage," to be published in *Analog Integrated Circuits and Signal Processing*.

11. D. Foty, H. Hanafi, P. Agnello, and H. Ho, "Mixed Schottky/p-n Junction Behavior in Diodes Produced by Outdiffusion from Polycrystalline Cobalt Disilicide," 1992 IEDM Tech. Dig., pp. 841–845.

12. L. Nagel, "SPICE2: A Computer Program to Simulate Semiconductor Circuits," University of California/Berkeley, Memorandum No. UCB/ERL-M520 (1975).

13. G. Massobrio and P. Antognetti, *Semiconductor Device Modeling with SPICE* (2nd ed.), McGraw-Hill, 1993.

14. N. Arora, *MOSFET Models for VLSI Circuit Simulation*, Springer-Verlag, 1993.

15. R. Pierret, "Field Effect Devices," *Modular Series on Solid State Devices* (ed. by G. Neudeck and R. Pierret), volume IV, Addison-Wesley, 1983.

16. A. Grove, *Physics and Technology of Semiconductor Devices*, John Wiley & Sons, 1967.

17. P. Drude, "Zur Elektronentheorie der Metalle. I. Teil," *Ann. Physik*, vol. 1, p. 566 (1900).

18. P. Drude, "Zur Elektronentheorie der Metalle. II. Teil. Galvanomagnetische und Thermomagnetische Effecte," *Ann. Physik*, vol. 3, p. 369 (1900).

19. H. deGraaff and F. Klaassen, *Compact Transistor Modeling for Circuit Design*, Springer-Verlag, 1990.

20. D. Ward and R. Dutton, "A Charge-Oriented Model for MOS Transistor Capacitances," *IEEE J. Sol. St. Circ.*, vol. SC-13, pp. 703–708 (1978).

21. P. Yang, B. Epler, and P. Chatterjee, "An Investigation of the Charge Conservation Problem for MOSFET Circuit Simulation," *IEEE J. Sol. St. Circ.*, SC-18, pp. 128–138 (1983).

22. C. Sah, "Characteristics of the Metal-Oxide-Semiconductor Transistors," *IEEE Trans. Elec. Dev.*, vol. ED-11, pp. 324–345 (1964).

23. H. Ihantola and J. Moll, "Design Theory of a Surface Field-Effect Transistor," *Sol. St. Elec.*, vol. 7, pp. 423–430 (1964).

24. E. Nowak, "Ultimate CMOS ULSI Performance," 1993 IEDM Tech. Dig., pp. 115–118.

25. Y. Tsividis, *The MOS Transistor*, McGraw-Hill, 1987.

26. B. Davari et al., "A Variable Size Shallow Trench Isolation (STI) Technology with Diffused Sidewall Doping for Submicron CMOS," 1988 IEDM Tech. Dig., pp. 92–95.

27. M. Mikoshiba et al., "A New Trench Isolation Technology as a Replacement of LOCOS," 1984 IEDM Tech. Dig., pp. 578–581.

28. M. Lin, "The Classical Versus the Quantum Mechanical Model of Mobility Degradation Due to the Gate Field in MOSFET Inversion Layers," *IEEE Trans. Elec. Dev.*, vol. ED-32, pp. 700–710 (1985).

29. M. Liang, J. Choi, P. Ko, and C. Hu, "Inversion Layer Capacitance and Mobility of Very Thin Gate Oxide MOSFETs," *IEEE Trans. Elec. Dev.*, vol. ED-33, pp. 409–413 (1986).

30. B. Majkusiak and A. Jakubowski, "The Dependence of MOSFET Surface Carrier Mobility on Gate Oxide Thickness," *IEEE Trans. Elec. Dev.*, vol. ED-33, pp. 1717–1721 (1986).

31. Ch. Nguyen-Doc, S. Cristoloveanu, and G. Ghibaudo, "Low Temperature Mobility Behaviour in Submicron MOSFETs and Related Determination of Channel Length and Series Resistance," *Sol. St. Elec.*, vol. 29, pp. 1271–1277 (1986).

32. R. Dennard et al., "Design of Ion Implanted MOSFETs with Very Small Physical Dimensions," *IEEE J. Sol. St. Circ.*, vol SC-9, pp. 256–268 (1974).

33. C. Hu et al., "Hot Electron Induced MOSFET Degradation - Model, Monitor, and Improvement," *IEEE Trans. Elec. Dev.*, vol. ED-32, pp. 375–385 (1985).

34. T. Sakurai, K. Nogami, M. Kakumu, and T. Iizuka, "Hot Carrier Generation in Submicrometer VLSI Environment," *IEEE J. Sol. St. Circ.*, vol. 22, pp. 256–259 (1987).

35. T. Ong, P. Ko, and C. Hu, "Hot Carrier Current Modeling and Device Degradation in Surface Channel pMOSFETs," *IEEE Trans. Elec. Dev.*, vol. ED-37, pp. 1658–1666 (1990).

36. S. Ogura et al., "Design and Characteristics of the Lightly Doped Drain-Source (LDD) Insulated Gate Field Effect Transistor," *IEEE Trans. Elec. Dev.*, vol. ED-27, pp. 1359–1367 (1980).

37. S. Wolf, *Silicon Processing for the VLSI Era* (Volume 2 - Process Integration), Lattice Press, 1990.

38. A. Bryant et al., "A Fundamental Performance Limit of Optimized 3.3V Sub-Quarter-Micrometer Fully Overlapped LDD MOSFETs," *IEEE Trans. Elec. Dev.*, vol. ED-39, pp. 1208–1215 (1992).

39. D. Foty and E. Nowak, "MOSFET Technology for Low-Voltage/Low-Power Applications," *IEEE Micro*, July 1994, pp. 68–77.

40. T. Chan, J. Chen, P. Ko, and C. Hu, "The Impact of Gate-Induced Drain Leakage on MOSFET Scaling," 1987 IEDM Tech. Dig., pp. 718–721.

41. M. Orlowski, C. Mazure, and F. Lau, "Submicron Short Channel Effects Due to Gate Reoxidation Induced Lateral Interstitial Diffusion," 1987 IEDM Tech. Dig., pp. 632–635.

42. C. Mazure and M. Orlowski, "Guidelines for Reverse Short Channel Behavior," *IEEE Elec. Dev. Lett.*, vol. EDL-10, pp. 556–558 (1989).

43. J. Huang et al., "BSIM3 Manual" (version 2.0), The University of California/Berkeley, 1994.

44. F. Hsu et al., "An Analytical Breakdown Model for Short Channel MOSFETs," *IEEE Trans. Elec. Dev.*, vol. ED-29, pp. 1735–1740 (1982).

4

A Comparison of Analytical and Numerical Results

4.1 Introduction

In Chapter 3, basic analytical expressions were derived for the description of two-, three-, and four-terminal MOS structures. These equations form the basis of the MOS models that are implemented in SPICE. As noted during the derivations, some assumptions and approximations were made to allow "clean" analytical expressions to be produced, eliminating any need for inefficient iterative numerical techniques. For completeness, it is worthwhile to compare these expressions with more rigorous one-dimensional numerical simulation results. This is done to verify that the simple analytical equations provide a rather accurate description of MOS structures, while noting that there are some minor differences. These discrepancies are also important information, as they indicate where the analytical expressions might encounter difficulty in the future. The approximations and assumptions used in the derivations will not collapse unceremoniously at some point, but will instead gradually lose accuracy; it is important to know where to look for occurrences of this kind.

Here three different MOS structures will be examined. Numerical simulations of the p-n junction will allow a comparison of the delta depletion region model (and thus the junction capacitance model) with more accurate results. The forward bias current-voltage behavior will also be examined. The inversion and depletion charge layers of a MOS capacitor will be simulated, to compare the charge sheet model and the delta depletion model of the gate-induced depletion region with detailed simulations. Finally, the charged particle behavior of the MOS inversion layer will be described. The basic expression at the root of all the FET current models used in SPICE employs an ohmic expression for the relationship between the lateral field and the carrier velocity; comparison with one-dimensional Monte Carlo simulations provides considerable insight into the more detailed workings of carrier transport.

4.2 The p-n Junction

Although the simple p-n junction can be used for many different applications, as reflected in the enormous depth of the p-n junction model available in SPICE [1], in most MOSFET applications the p-n junction usage is quite simple. In reverse bias, the p-n junction is treated as a simple capacitor, while under small forward biases it is a current-producing element. The SPICE implementation of these phenomena was considered in Chapter 3. Here the simple analytical models will be compared with more accurate numerical results. This will show how the two methods compare, and also how data describing a more complex internal structure can be manipulated so that they fit the parameters of the simpler model.

4.2.1 The Depletion Region

As described in Chapter 3, the analytical model developed to describe the depletion region of a p-n junction assumes that a delta depletion region model is valid; that is, depletion is treated as complete (no free carriers) in the depletion region, while beyond the depletion region edge there is no depletion (the number of free carriers equals the number of ionized dopant impurities). Using this assumption, an abrupt junction doping profile was employed to describe the general depletion width of a p-n junction ((3.30)):

$$W = \left[\frac{2\epsilon_{Si}}{q}(V_{bi} + V_R)\left(\frac{N_A + N_D}{N_A N_D}\right)\right]^{\frac{1}{2}}. \tag{4.1}$$

In MOSFETs, the source and drain junctions are either n^+/p or p^+/n. It was assumed that the junction is abrupt, and that the depletion region does not exist in the heavily doped region; all depletion occurs on the lightly doped side of the junction. The depletion width of an n^+/p junction was thus described by (3.31):

$$W = \left[\frac{2\epsilon_{Si}}{q N_A}(V_{bi} + V_R)\right]^{\frac{1}{2}}. \tag{4.2}$$

By assuming an abrupt junction profile and the formation of a delta depletion region, analytical solutions to the Poisson equation (4.1) and (4.2) can be found. If a more complex doping profile occurs, the Poisson equation cannot be solved analytically; a numerical solution must be found. In modern MOS technology, the abrupt junction assumption is not a good approximation; a shallow junction will inevitably have some degree of grading. A typical junction doping profile is shown in Figure 4.1; this profile will be used here for the comparison of the simple analytical results and the numerical solution. The numerical solution will also allow for examination of the effectiveness of the delta depletion approximation.

With no voltage applied to the diode terminals, no current flows, and only the Poisson equation need be solved to describe the behavior of the structure. The

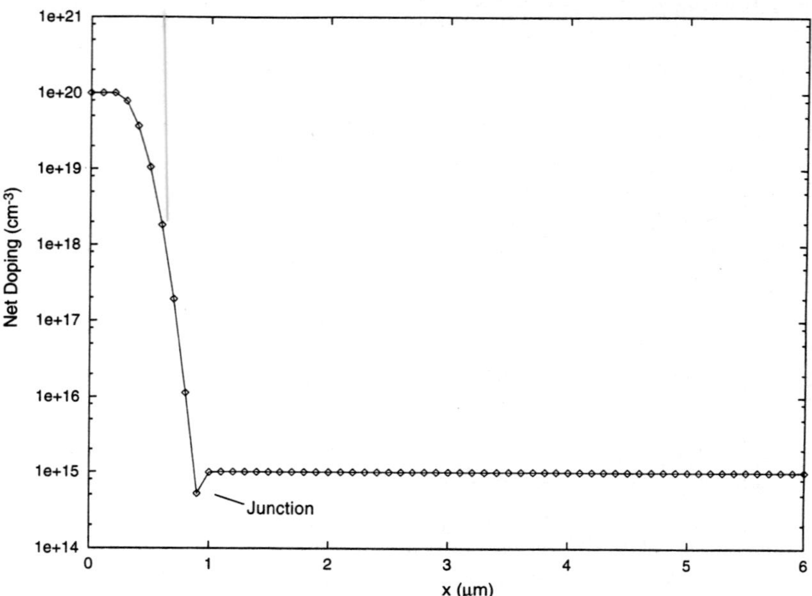

Figure 4.1   A realistic n^+/p junction doping profile, used for the numerical simulations.

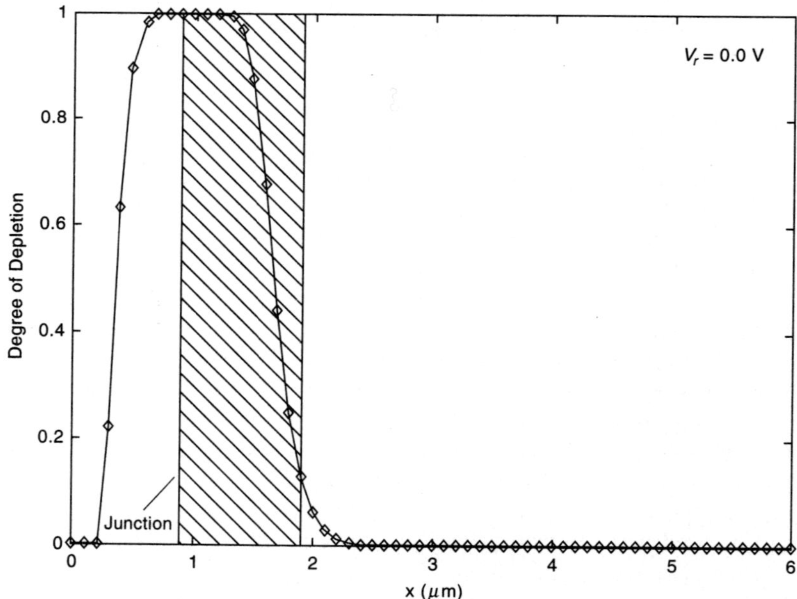

Figure 4.2   A comparison of the analytical delta depletion approximation with the more accurate numerical solution, $V_r = 0.0$ V.

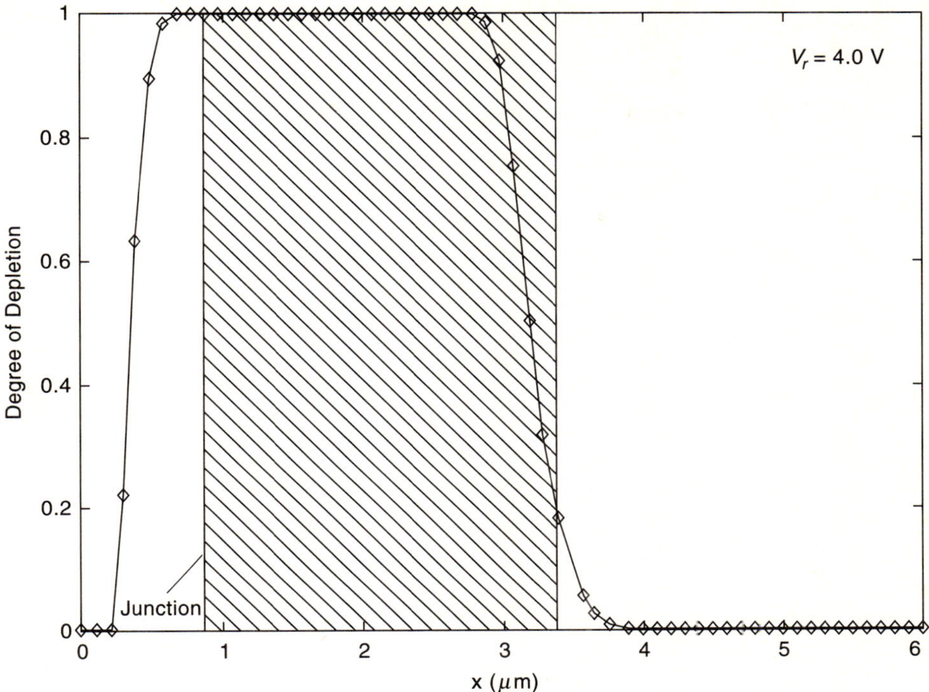

Figure 4.3   A comparison of the analytical delta depletion approximation with the more accurate numerical solution, $V_r = 4.0$ V.

depletion results are shown in Figure 4.2, where the degree of depletion is the total depletion charge (impurity concentration minus the free carrier concentration) divided by the maximum possible depletion charge (the impurity concentration). There are two important considerations in these results. First, since the junction is not abrupt, part of the depletion region is on what is considered to be the n$^+$ side of the junction. Given the doping profile of Figure 4.1, this is not surprising; however, the additional depletion charge will affect the behavior of the structure. Second, the depletion region ends gradually rather than abruptly. Although the agreement is not exact, the simulation results show that the delta depletion approximation does provide a reasonable description of the depletion region on the lightly doped side of the junction.

When a voltage is applied across the diode, the structure is no longer in equilibrium, and some current will flow. To carry out a simulation, the Poisson equation must be solved along with two current continuity equations, one for electrons and one for holes. When this is done for a reverse bias of 4.0 V, the result of Figure 4.3 is produced. Although the reverse bias voltage has expanded the size of the depletion region, the results are qualitatively the same as those described by Figure 4.2. The depletion into the heavily doped side of the junction is virtually identical to the zero bias case; the depletion region penetrates into the n-doped region until it encounters very high doping and can penetrate no further. In the p-doped region, the delta depletion approximation is once again a reasonable approximation.

4.2.2 The Junction Capacitance

While the analytical and numerical descriptions of the p-n junction depletion region are interesting to examine, it must be noted that the depletion region model is developed to provide a description of the junction capacitance. As described in Chapter 3, by using the delta depletion approximation, the junction is treated as a simple two-plate capacitor, with the depletion region forming the dielectric layer between the plates. This approach leads to a simple expression for the capacitance of a reverse biased p-n junction (3.35):

$$C_j = \frac{\epsilon_{Si}}{W}.$$ (4.3)

Taking the tone set in Section 4.2.1, (3.38) was derived to describe the capacitance of an abrupt n^+/p junction:

$$C_j = C_{jo}\left(1 + \frac{V_R}{V_{bi}}\right)^{-m},$$ (4.4)

where C_{jo} is the zero bias capacitance described by (3.37):

$$C_{jo} = \left[\frac{q\epsilon_{Si}N_A}{2V_{bi}}\right]^m.$$ (4.5)

A more rigorous numerical computation of the junction capacitance requires evaluation of the charge distribution in the junction, in which the capacitance is defined by

$$C_j = \frac{\partial Q_j}{\partial V},$$ (4.6)

where Q_j is the charge in the junction. With virtually no exceptions, the diode is much larger in extent than the depletion region, so the applied bias has a negligible effect on the free carrier density. In addition, for all but the highest frequencies, the free carriers will react very rapidly to the applied bias. Thus, the junction capacitance behavior is determined entirely by the change in the total depletion charge as the applied bias changes [2].

In addition, as noted in Section 4.2.1, the junction is not described by a completely abrupt doping profile, but by a more realistic diffused junction, as shown in Figure 4.1. This profile will be used here for the numerical simulation of the junction capacitance. The numerical method used to compute this capacitance is rather straightforward. The coupled Poisson and current continuity equations are solved for a particular bias point

Table 4.1 The capacitance results for a simple abrupt junction (using (4.4) and (4.5) directly) and the corresponding numerical simulation results.

V_r	Strict Analytical Solution	Numerical Solution	Numerical/Analytical
0.0	2.32×10^{-4} F/m^2	9.93×10^{-5} F/m^2	2.34
1.0	1.43×10^{-4} F/m^2	6.73×10^{-5} F/m^2	2.12
2.0	1.12×10^{-4} F/m^2	5.42×10^{-5} F/m^2	2.07
3.0	9.43×10^{-5} F/m^2	4.66×10^{-5} F/m^2	2.02
4.0	8.34×10^{-5} F/m^2	4.15×10^{-5} F/m^2	2.01

as described earlier. By a simple numerical integration along the device length, the total depletion charge is computed. The simulation is then repeated for two additional bias points, one 10 mV larger and one 10 mV smaller than the original bias point, which allows for the numerical evaluation of the change in the charge under a small change in the applied bias, as described by (4.6). This procedure will yield two slightly different values (one from the upper bias, and one from the lower bias) that are averaged to obtain a solution to (4.6).

Repeating this general method for several different reverse bias voltage points allows the junction capacitance to be simulated. The results are described in Table 4.1, where they are compared with the analytical solutions of (4.4). There is a large discrepancy between the numerical and analytical results. By dividing the numerical values by the corresponding analytical solutions, it can be seen that the ratio between the two methods is nearly identical. The departure from that ratio is largest for small reverse bias, and the smallest for large reverse biases, where the depletion region in the lightly doped side of the junction becomes larger. It can thus be concluded that the discrepancy is due to the difference between the abrupt junction assumption (used in the analytical formulation leading to (4.4)) and the more realistic doping profile of Figure 4.1. The ratio becomes nearly constant for large reverse biases, where the much larger depletion region on the lightly doped side of the junction dominates the depletion behavior.

The conclusion of this discussion is that the simple analytical model for the junction capacitance cannot be constructed from knowledge of the substrate and diffusion doping concentrations with the assumption of an abrupt junction between them. It remains to be determined if the analytical form of (4.4) can be put to use in SPICE to model the junction capacitance. The numerical results just determined can be used as data, to be fitted to the form of (4.4). The term C_{jo}, the zero bias junction capacitance, is simply the result for $V_r = 0$, so that $C_{jo} = 2.32 \times 10^{-4}$ F/m^2. As noted in Chapter 3, the built-in potential V_{bi} could be extracted directly from the data; however, this can lead to physically incorrect results, which become a problem when the temperature is varied. Instead, a physically meaningful value of V_{bi} should be determined. Here, since the substrate doping is 10^{15} cm^{-3}, the bulk potential is 0.287 V. Since the potential of the n^+ region is roughly one-half the band gap (about 0.56 V), a reasonable value for the built-in potential is $V_{bi} = 0.847$ V. This leaves the "dopant grading exponent" m to be determined by a simple nonlinear least squares extraction; this yields $m = 0.595$. The numerical data and the model result are plotted in Figure 4.4. The agreement is very

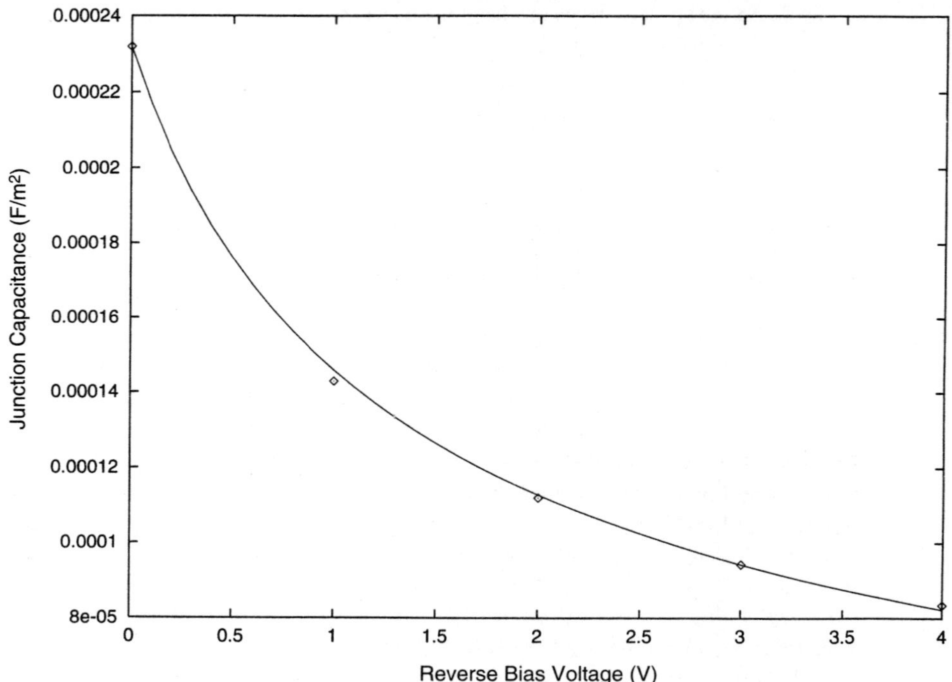

Figure 4.4   The results of fitting the data from the numerical solution to the model equation (4.4).

good; in fact, the model result at each data point is accurate to within 2% of the data. This shows that the form of (4.4) can be used as an empirical expression to describe the junction capacitance, with the parameters determined by extraction. Also note that according to the theory developed in Chapter 3, m should vary between $\frac{1}{3}$ (linearly graded junction) and $\frac{1}{2}$ (abrupt junction), while here it takes on a larger value. This kind of result is common, as the lack of direct knowledge of the junction structure must reduce some part of (4.4) to empirical status.

(As an aside, if both V_{bi} and m are extracted from the data, results of $V_{bi} = 0.660$ and $m = 0.524$ are produced. Although the results are slightly more accurate than those produced by extracting m alone, as discussed in Chapter 3, this value of V_{bi} lacks physical accuracy, and the model will deteriorate when the temperature is varied.)

4.2.3   The Forward Bias Current

In some CMOS circuits, a small forward bias is applied to the source-substrate junction. This injects carriers into the substrate, which can cause latch-up. It is thus important to have an adequate model for the forward bias junction behavior, so that latch-up can be avoided. An analytical model was developed in Chapter 3, resulting in the description of (3.71):

$$J = J_o \left(e^{qV/k_b T} - 1 \right). \tag{4.7}$$

This expression is valid for forward bias voltages less than the built-in potential.

As described in Chapter 3, J_o is a composite term, described within (3.71) as

$$J_o = \left(\frac{q D_p}{L_p} p_n + \frac{q D_n}{L_n} n_p \right). \tag{4.8}$$

An analytical solution of (4.8) is possible but is complicated. The diffusion coefficients D_n and D_p can be related to the carrier mobility by the Einstein relation [3]

$$\frac{D}{\mu} = \frac{k_b T}{q}. \tag{4.9}$$

The diffusion constants are thus related to the applied field, depending on the sophistication of the mobility model chosen. The minority carrier diffusion lengths L_n and L_p are determined through models for the minority carrier lifetime. Finally, the minority carrier concentrations n_p and p_n vary for different bias conditions. All these considerations are handled easily in a numerical simulation, but do not lend themselves well to simple analytical solution. Thus, as described in Chapter 3, the term J_o is treated as an empirical model parameter, to be determined by extraction.

Using the techniques described in Section 4.2.2, the forward bias current can be simulated by solution of the Poisson equation and the two current continuity equations. The results of such a simulation are shown in Figure 4.5. As expected, the ideal log-linear diode slope of 60 mV/decade is produced. In addition, the simulation shows that the log-linear model begins to break down for forward bias voltages less than the built-in potential voltage; specifically, the results depart from the simple model for $V_f > 0.6$ V, while the built-in potential is 0.847 V. Thus, the diode current model should only be trusted for forward bias voltages at least 0.25 V smaller than the built-in potential, as described by the parameters **PB** and **PBSW**. This also reinforces the importance of determining values of **PB** and **PBSW** that are physically meaningful (rather than empirical), as this value communicates an important limit of the model validity to the circuit designer.

The results of Figure 4.5 can also be used to determine the parameter J_o. A value of $J_o = 1.12 \times 10^{-5}$ (A/m^2) is easily extracted, and provides a good fit to the measured data.

4.3   The MOS Capacitor

The MOS capacitor is used principally for routine monitoring of CMOS manufacturing technology. The main development in Chapter 3 focused on the MOS capacitor as the simple basis for the more complicated MOSFET. Along that line, expressions were developed for describing the depletion and inversion charge layers of the MOS capacitor;

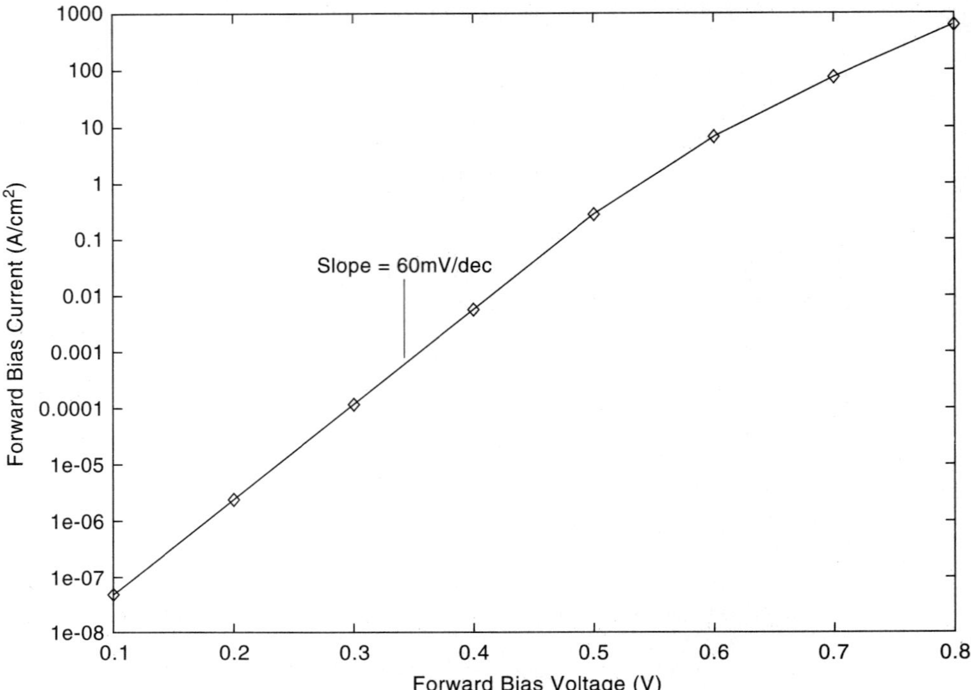

Figure 4.5   The numerical simulation of the forward bias diode characteristic, and the extraction of the model parameter J_o.

these expressions were later used as the basis for developing the simple model for FET current conduction.

In particular, it is assumed that the delta depletion approximation describes the gate-induced depletion layer and that the surface inversion layer is so thin that it can be treated as a two-dimensional charge sheet. More exact results can be obtained through one-dimensional numerical simulation of the MOS capacitor. Although the structure is more complicated than a p-n junction, no current flows; thus, only the Poisson equation must be solved (rather than the Poisson equation and the two current continuity equations). These numerical solutions shed light on the validity of the analytical expressions that are used to describe the MOS capacitor and the MOSFET.

4.3.1   The Depletion Region

The use of the delta depletion region approximation leads to an expression for the depletion region thickness (3.87):

$$x_D = \left(\frac{2\epsilon_{Si}\phi_s}{qN_A}\right)^{\frac{1}{2}}.$$

$$(4.10)$$

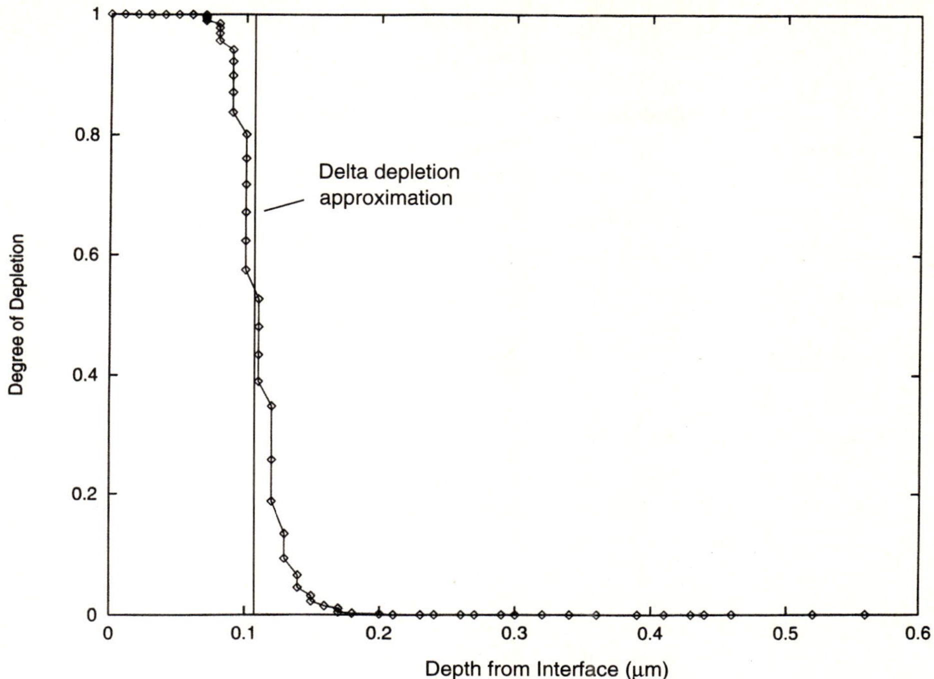

Figure 4.6 A comparison of the numerical simulation of the depletion region in a MOS capacitor with the analytical delta depletion region approximation (4.10).

This expression is considered to be valid for all gate biases under which the MOS capacitor is in strong inversion; in strong inversion, a change in the gate bias affects only the inversion charge, so the depletion charge is undisturbed.

In the numerical solution to be employed here [4], the substrate doping is $N_A = 1.0 \times 10^{17}$ cm^{-3}, the oxide thickness is 10 nm, and the bias on the n$^+$ polysilicon gate is 1.3 V, which places the device far into the strong inversion condition. The results are depicted in Figure 4.6, where the delta depletion approximation is compared with the numerical solution for the degree of depletion, computed from

$$D_d = \left(\frac{N_A - p}{N_A} \right) .$$

(4.11)

The delta depletion approximation compares quite well with the more exact numerical solution. The exact solution shows a gradual end to the depletion region, but the two results are nearly identical. This validates the general use of the delta depletion approximation to describe the MOS gate-induced depletion region.

If the substrate doping is not uniform, the Poisson equation cannot be solved analytically (as it was in Chapter 3), and the solution (4.10) cannot be reached. Direct results can only be obtained by numerical solution of the Poisson equation. Since surface tailor and well implants in modern MOS technology are common, the substrate doping is very rarely found to be uniform. However, the analytical formulations based on (4.10) are still used; this requires that N_A, which appears in the models as the parameter **NSUB**, be treated as a model parameter which is extracted from data to give the closest possible fit to (4.10). In more recent models (such as BSIM3), attempts are made to account for nonuniform substrate doping in a more physically based manner.

4.3.2 The Inversion Layer

In the MOS capacitor computations of Chapter 3, the inversion layer was assumed to be so thin that it could be treated as a two-dimensional charge sheet [5]. This eliminates any need for integration along the axis normal to the silicon surface, greatly simplifying the computations.

Once again, the validity of this approximation can be studied by comparison with the results of numerical simulations. The results here are taken from the simulation described above for the MOS depletion analysis, in which the substrate doping is $N_A = 1.0 \times 10^{17}$ cm^{-3}, the oxide thickness is 10 nm, and the bias on the n$^+$ polysilicon gate is 1.3 V. Figure 4.7a shows that the inversion charge layer is indeed very thin; in fact, there is negligible inversion charge for any point deeper than about 6 nm below the interface. As Figure 4.7b shows, if the inversion layer is defined as the surface region where $n > N_A$, its depth is indeed about 6 nm. Thus, in the general case which applies to MOSFET simulation, the charge sheet approximation for the inversion layer is valid.

For completeness, it should be noted that these solutions to the Poisson equation are classical, which allows the maximum of charge to be found at the interface. If the more rigorous rules of quantum mechanics are applied to the MOS system [6], it is required that the electron wavefunction (and thus the charge) reach zero at the surface; this implies that the charge at the surface is actually zero, and the peak of the inversion charge density is slightly below the interface.

The introduction of quantum mechanics into the study of the MOS system opens up a whole new collection of interesting phenomena, such as the integer [7] and fractional [8] quantum Hall effects. In addition, a set of interesting experiments has been performed [9], in which a positively biased metal plate (the gate) is covered with a layer of liquid helium (the dielectric), and an electron layer is injected above the liquid helium (the inversion layer); the electron layer does not rest at the liquid helium surface (the classical Poisson equation prediction), but is indeed suspended a short distance above the liquid helium (the quantum mechanical prediction).

However fascinating these theoretical predictions and experiments are, this level of detail has yet to be required in describing the MOS system for analytical MOSFET modeling. They are noted here as a caution to the reader that more rigorous physical models of this sort might appear at some time in the future.

4.4 The Conducting MOSFET Inversion Layer

Whether the MOS inversion layer is treated as a two-dimensional charge sheet or in a more rigorous fashion, the behavior of that inversion layer determines the current-carrying

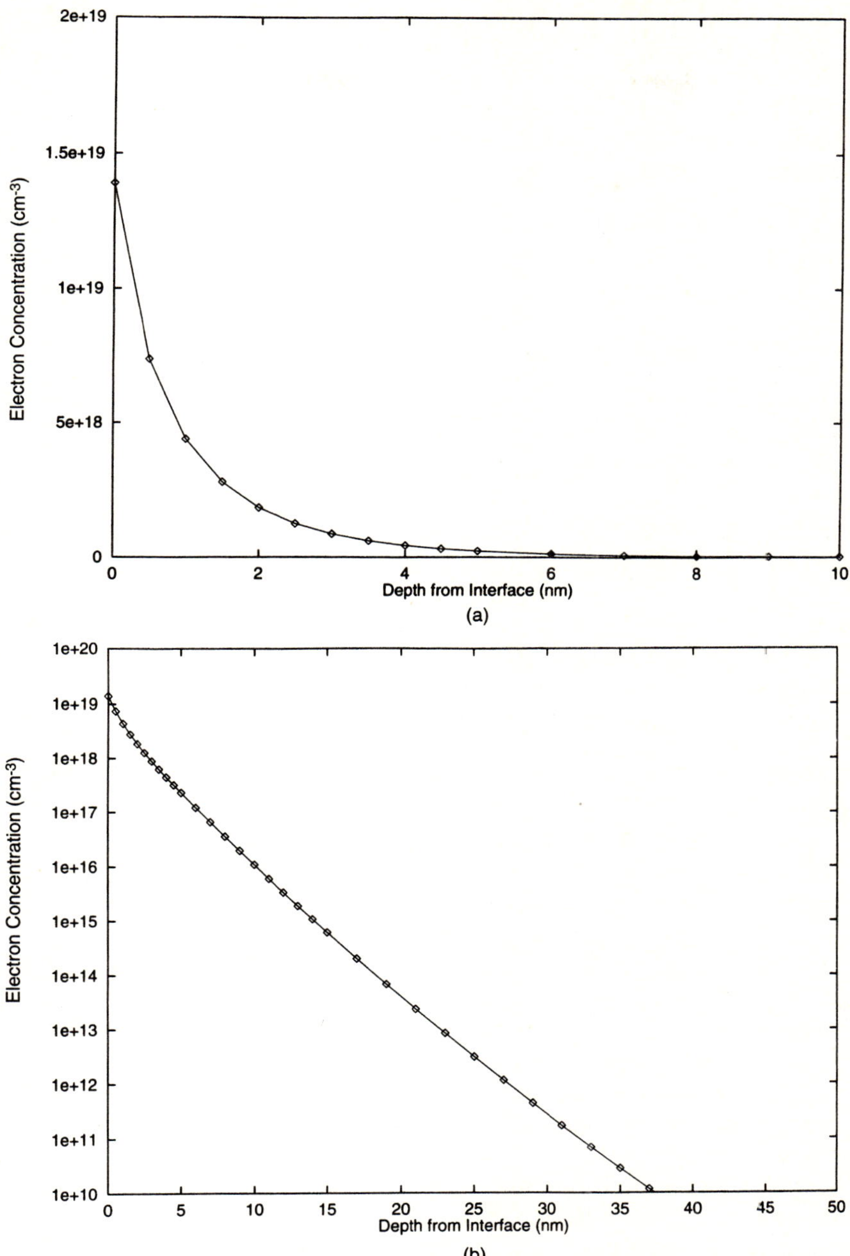

Figure 4.7 Numerical simulation of the MOS inversion layer; (a) linear scale; (b) logarithmic scale.

properties of the FET. From the simplest point of view, each charged particle (electron or hole) in the inversion layer moves under the influence of the lateral field due to the drain-source voltage V_{ds}. If a particle is released into the field caused by this applied voltage, it should be accelerated by that field until it reaches the drain contact.

However, the physical situation is somewhat more complicated. The carriers are not moving in free space, but are instead travelling through a crystalline silicon lattice which also contains the charge centers of ionized dopant impurities. Due to this complexity and the very large number of particles involved, a rigorously complete mathematical and numerical description has yet to be found. Fortunately, with some simplifying approximations, tractable methods of describing conduction in crystalline solids can be reached.

4.4.1   The Ohm's Law Description

The simplest and best known description of the conduction of free charged particles through a crystalline lattice is Ohm's law, which is used as the basis of the FET conduction model developed in Chapter 3. The linear velocity-field relationship of Ohm's law can be derived from basic physical principles, if certain approximations and assumptions are made. These approximations and assumptions are important to note, as they will indicate where this very simple method of describing conduction can break down and lead to inaccurate predictions.

Consider an electron free to move under a constant electric field. Using Newton's second law of motion, the force on the particle is

$$F = -qE = m^*a, \tag{4.12}$$

where m^* is the effective mass of the particle in the material. From (4.12), the acceleration experienced by the particle is

$$a = -\frac{qE}{m^*} = \frac{dv}{dt}. \tag{4.13}$$

If the particle is allowed to start from rest (i.e., assume here that the initial velocity of the particle is zero), the particle velocity is (by integrating (4.13) and setting v_o to zero)

$$v = -\frac{qEt}{m^*}. \tag{4.14}$$

In the absence of any interruption of its acceleration, the particle velocity will increase linearly in time.

Now introduce a drag force (i.e., frictional energy loss to the crystal lattice through which the particle is moving) acting against the trajectory of the particle, as in classical dynamics [10]. Equation (4.13) now becomes

$$\frac{dv}{dt} = -\frac{qE}{m^*} - kv, \tag{4.15}$$

which can be rearranged to

$$\frac{dv}{\frac{qE}{m^*}} + kv = -dt.\tag{4.16}$$

This leads to the integral

$$\int_{v_o}^{v}\frac{dv}{\frac{qE}{m^*}} + kv = -\int_{0}^{t} dt.\tag{4.17}$$

The evaluation of the integrals in (4.17) is rather straightforward [10], leading to

$$v = -\frac{qE}{m^*k} + \frac{kv_o + \frac{qE}{m^*}}{k}e^{-kt}.\tag{4.18}$$

If (4.18) is simplified by letting $v_o = 0$, the expression for the velocity of a particle accelerated from rest in this situation is

$$v = -\frac{qE}{m^*k}\left(1 - e^{-kt}\right).\tag{4.19}$$

For small values of t, a Taylor expansion of the exponential term in (4.19) leads to

$$v = -\frac{qE}{m^*k}\left(1 - (1 - kt)\right);\tag{4.20}$$

(4.19) reduces to the completely free particle expression (4.14). On the other hand, when t becomes large (the conditions under which this assumption is valid will be shown below), (4.19) takes on the limiting behavior

$$\lim_{t\to\infty} v = -\frac{qE}{m^*k}.\tag{4.21}$$

When (4.21) is valid, the carrier velocity is independent of time, and is proportional to the field.

From simple knowledge of semiconductor physics, the mobility can be defined as

$$\mu = \frac{q}{m^*k}.\tag{4.22}$$

With this definition, (4.21) can be rewritten as

$$v = \mu E.\tag{4.23}$$

This is a physical description of Ohm's law, in which the charged particle is giving up its energy to the lattice at the same rate at which it is gaining it from the applied field. As an aside, this description also corresponds to the *terminal velocity*, such as the maximum velocity that a falling body can reach in the earth's atmosphere. The gravitational field of the earth is effectively a constant, and a falling body accelerates to a velocity at which it is losing energy to the fluid of the atmosphere at the same rate that it is gaining energy from the earth's gravitational field. This *terminal velocity* and the *saturation velocity* of charged particles moving through a solid are very different, as will be noted below.

By further analyzing the derivation of Ohm's law, some statements can be made about the range of validity of the approximations that led to such a simple physical description. First, it is appropriate to estimate k, which will allow an assessment of the conditions under which the assumption which led to (4.21) is valid. Note that k has units of s^{-1}, and since it represents frictional damping of the accelerating particle, it can be assumed that k represents the microscopic collision frequency of the particle with the damping force of the lattice. Using the definition in (4.22), and some known quantities ($q \sim 10^{-19}$C, $m^* \sim 10^{-30}$ kg, $\mu \sim 1000$ cm^2/V·s $= 0.1$ m^2/V·s), it is found that

$$k \sim 10^{12} s^{-1}. \tag{4.24}$$

Thus, the assumption used to derive (4.21), which requires that $t \gg \frac{1}{k}$, is valid for $t \geq$ 10 ps. This assumption is certainly valid for the subject to be considered here.

Further physical corroboration may be found by considering the current density:

$$j = nqv = -\frac{nq^2 E}{m^*k}. \tag{4.25}$$

Since

$$j = \sigma E, \tag{4.26}$$

this allows the definition of the electrical conductivity:

$$\sigma \equiv -\frac{nq^2}{m^*k}. \tag{4.27}$$

From the Drude theory [11], the electrical conductivity is

$$\sigma \equiv -\frac{nq^2\tau}{m^*}, \tag{4.28}$$

where τ is the mean free time between collisions of the traveling electron with the lattice. Thus, as postulated above,

$$k \equiv \frac{1}{\tau}, \tag{4.29}$$

where k is indeed the collision frequency for the traveling electron. From (4.24), τ is about 1 ps; this agrees well with Monte Carlo simulations of carrier transport in semiconductor materials [12].

Thus, Ohm's law is a description of the conduction of a particle through a fluid. The particle loses energy to the fluid by friction at the same rate that it gains it from the applied field, so that for a given applied field, the particle maintains a constant velocity. For a large collection of particles, this result is taken to be valid for all the particles; for a given applied field, all the particles move with the same velocity. In this form, the collection of charged particles is itself treated as a fluid, which is "pumped" in a virtually hydraulic fashion through the device.

When the applied field becomes very large, the linear damping description of (4.15) is no longer adequate, but must be modified to include additional nonlinear damping:

$$\frac{dv}{dt} = -\frac{qE}{m^*} - kv - k_2 v^2. \tag{4.30}$$

The solution to (4.30) is quite complicated and will not be discussed here. The results show that for a given field, the particle once again loses energy to the fluid medium at the same rate that it acquires it from the field, and thus moves at a constant velocity. However, if the field is increased, the particle is unable to take up any of this added energy, but merely passes it directly on to the fluid background medium. In this situation, the velocity becomes a constant that is independent of the field strength; this is carrier velocity saturation.

4.4.2 Monte Carlo Simulation Results

Ohm's law is reached by considering the behavior of an individual charged particle, and by assuming that each such particle is affected in exactly the same way. In this manner, a very simple description of charge conduction is developed, in which all the carriers move with the same velocity under a given field strength. The only breakdown of this ohmic behavior occurs at high fields, when the carrier velocity no longer increases with increasing field strength, but becomes constant at the saturation velocity.

A more complicated analysis of individual particle behavior is possible. Monte Carlo simulations [12], mentioned briefly above, consider individual particles which are accelerated by the applied field, and which lose energy to the silicon lattice by various scattering mechanisms. In contrast to the strictly deterministic approach used to derive Ohm's law, various possible scattering events are treated as probabilities; a particle is allowed to accelerate in free flight for some given time, then a random number is generated to choose if the particle is scattered, and by what mechanism. By simulating a large number of particles over a considerable length of time, a remarkably accurate physical description of the system can be reached. This description is gained at the cost of a very large amount of computation time and resources. Although the results are very useful, Monte Carlo simulations are usually considered too slow for two-dimensional device simulations; any use in analytical FET modeling is (at least at present) out of the question.

The rigorous details of Monte Carlo simulation of FETs are beyond the scope of this work; the interested reader is referred to an excellent text on the subject [12] for detailed descriptions. However, some simple Monte Carlo results can serve as a worthwhile analysis of the validity of the Ohm's law method of describing carrier conduction in FETs. For simplicity, the results to be described here are for bulk silicon under an applied field; the nature of the results also applies to the conducting MOSFET inversion layer.

Here a bulk silicon doping of 1.0×10^{15} cm^{-3} is used as the background material, and 100,000 particles are simulated. Each of these particles will have its own velocity; by simulating a large number of particles and examining the distribution of their velocity, insight can be gained into the state of the entire system. The velocity distribution results for a field of 1.0×10^3 V/cm are depicted in Figure 4.8a. In contrast to the Ohm's law approximation, the particles take on quite a distribution of velocities; in fact, due to scattering, many of the particles at any given time are actually moving *opposite* to the direction that the field should be imparting to them. The average velocity of the distribution is clearly positive and, as marked on the graph, is computed from the results to be 1.91×10^6 cm/s. It can also be seen from Figure 4.8a that a large number of particles have velocities (and therefore energies) considerably above this average value.

It is a simple matter to increase the field strength and repeat the simulation. Figures 4.8b and 4.8c depict, respectively, the simulation results for field strengths of 2.0×10^3 V/cm and 8.0×10^3 V/cm. The distribution of velocities is similar to that of Figure 4.8a, with a noticeable shift to the right, indicating an increase in the average particle velocity.

When the field is increased to 2.0×10^4 V/cm, the results (Figure 4.8d) take on some very interesting characteristics. The peak of the distribution has shifted even further to the right, indicating a further increase in the average velocity of the carriers. However, the distribution is somewhat ragged and more spread out. Despite the larger field, there is an increase in the number of particles with very large *negative* velocities; due to the large accelerating field, many scattering events are "more violent," causing the scattered carrier to "bounce back" with a larger velocity. By simply considering momentum and energy conservation, this indicates a much larger transfer of energy to the lattice than occurs at lower field strengths. As discussed above, this is a hallmark of the velocity saturation regime, in which all energy added by the field to the free particles is transferred directly to the lattice.

At the end of this distribution, the number of particles with velocities well above the average velocity has become much larger than that found for lower field strengths. These high-energy carriers represent "hot" electrons, which are accelerated to very high energies. Although they represent a small part of the total number of carriers, they are able to have important effects on device behavior; hot electron-induced MOSFET degradation, discussed briefly in Chapter 3, is the most visible example.

By simulating many different field strengths, the average velocities can be collected and examined. The values of the average carrier velocities are plotted against the field strength in Figure 4.9. For low fields, the ohmic region is clearly visible. For larger field strengths, the carrier velocity departs from the ohmic description, and the carrier velocity begins to saturate.

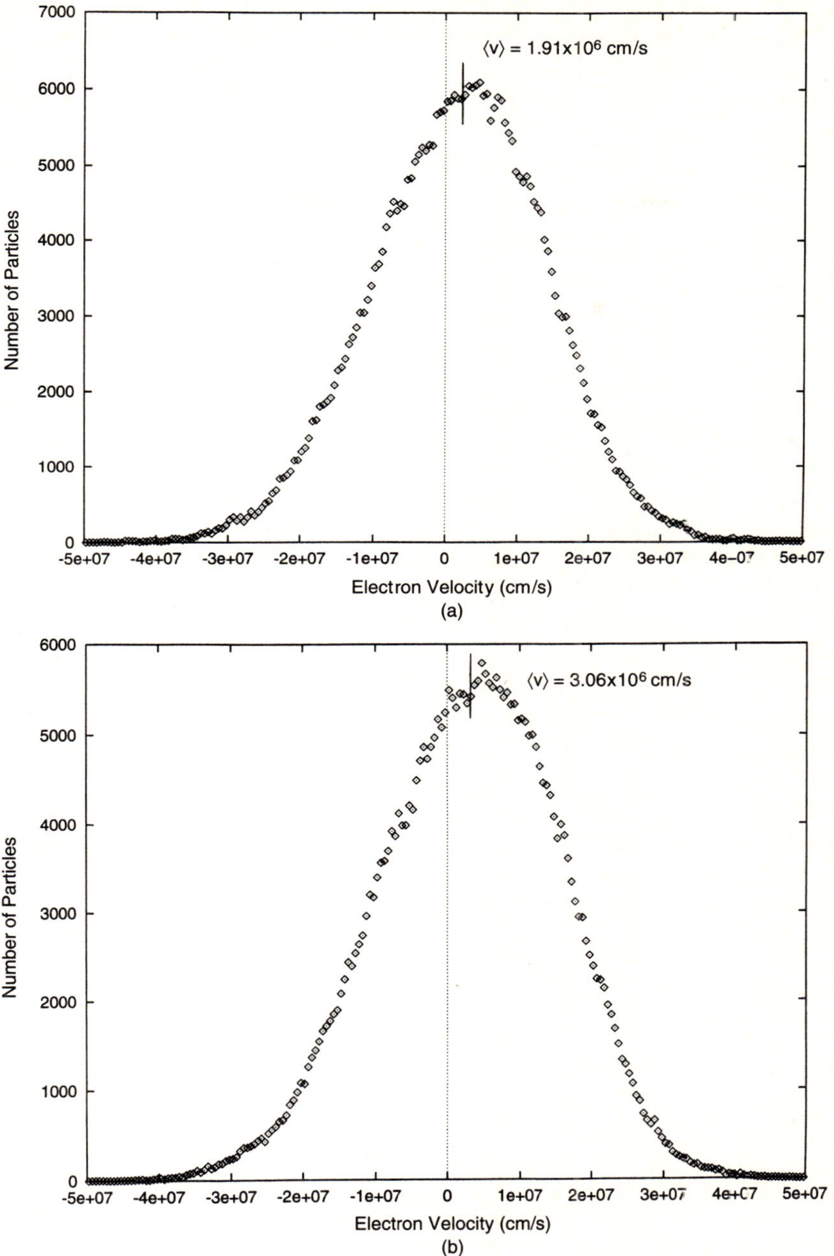

Figure 4.8 Monte Carlo simulation results of the distribution of particle velocities in silicon for several different field strengths; (a) 1.0×10^3 V/cm; (b) 2.0×10^3 V/cm; (c) 8.0×10^3 V/cm; (d) 2.0×10^4 V/cm.

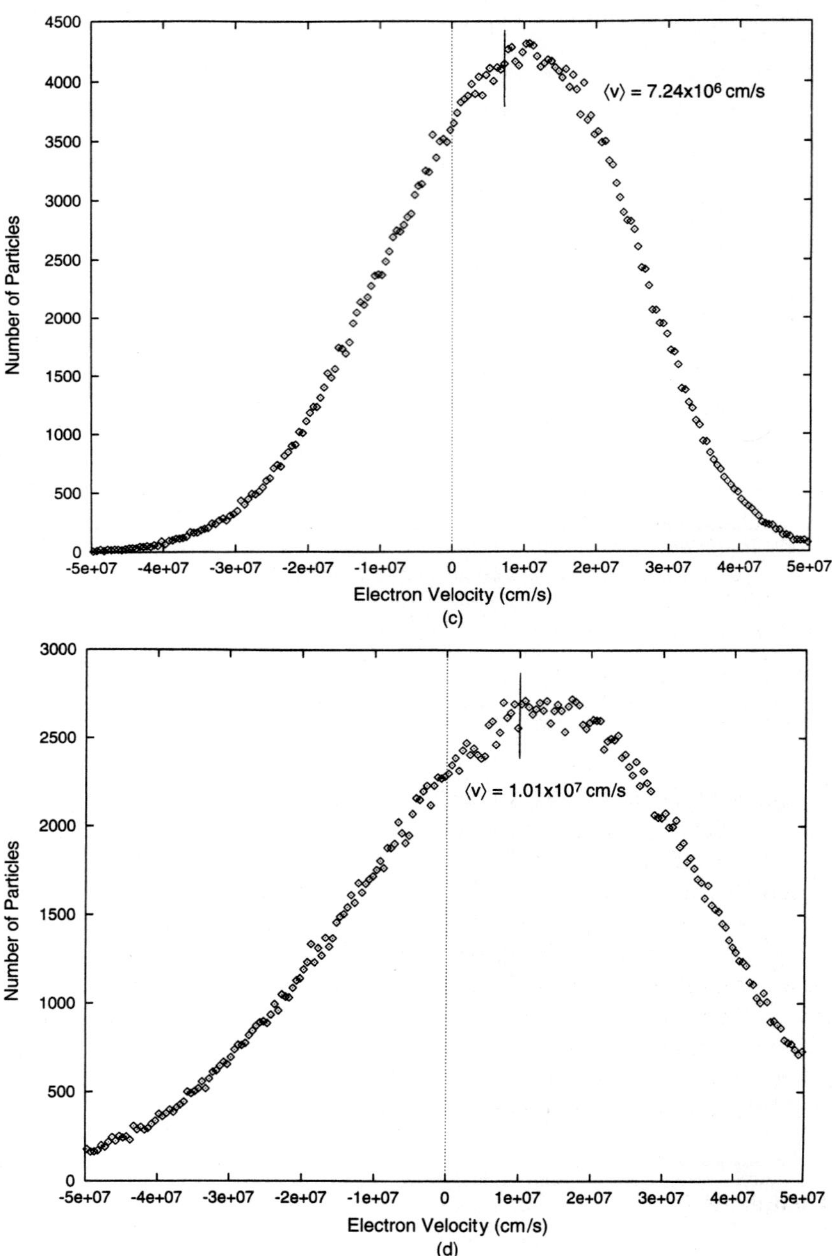

Figure 4.8 *(cont.)*

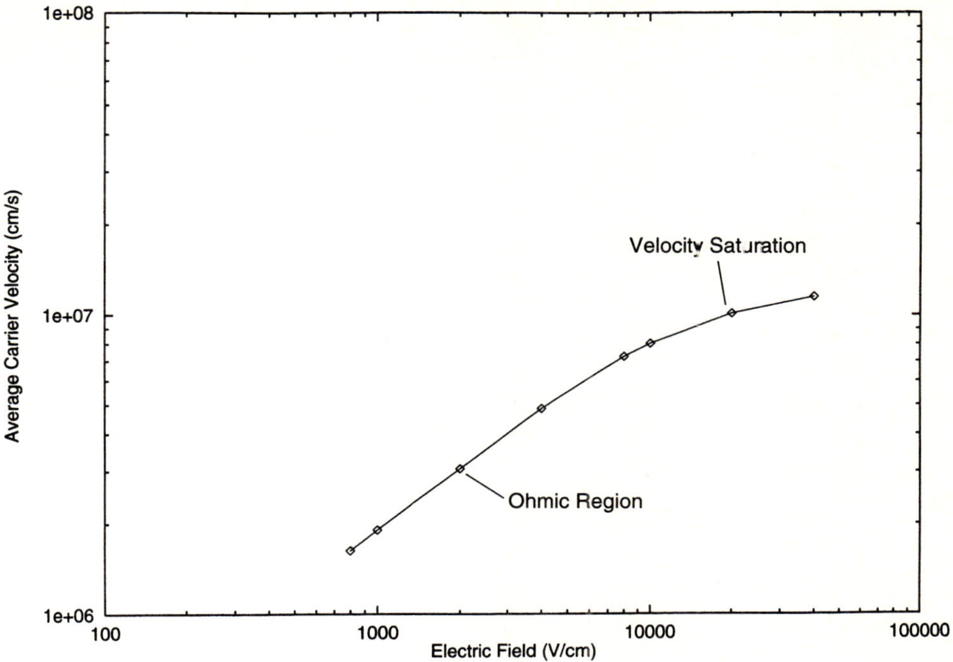

Figure 4.9 The average electron velocity versus the field strength, from the Monte Carlo simulations. Both the ohmic region and velocity saturation are clearly evident.

These simulations indicate that the physical situation in the MOSFET inversion layer is considerably more complex than the ohmic description. That ohmic description, with some modification for carrier velocity saturation, forms the basis for all analytical FET current models. To a reasonable degree of approximation, the behavior of most of the carriers is close enough to the "average" behavior to allow this approach to be valid. The high-energy "tail" of the velocity distribution does introduce some second-order effects, largely related to impact ionization. Some of the more sophisticated FET models (such as BSIM2 and BSIM3) add descriptions of such effects to their drain current and output conductance models. The effect of hot carriers on the device behavior (such as the degradation of the drain current and the threshold voltage) is modeled externally to the SPICE FET models, and need not be considered here.

4.5   Final Comments

As this chapter has demonstrated, the physical behavior of MOS devices is considerably more complex than the simple analytical descriptions of Chapter 3 indicate. Even though these more complex aspects of device physics do not appear in the present SPICE FET models, it is important to note their existence, as they may need to be accounted for in the future. It is also important to note that the approximations and assumptions used in Chapter 3 will not face a sharp limit to their usefulness, but will instead demonstrate a gradual (rather than catastrophic) loss of accuracy in smaller MOS structures. The possible deterioration in the quality of simple physical descriptions must be kept in mind.

While the analytical models may not have the simple physical basis that was intended, they can still be used to describe the behavior of the device; the junction capacitance description of Section 4.2.2 provides an excellent example. The simple description does not serve an attempt to arrive at a model by direct insertion of the known physical parameters; the physical situation is too complex for this approach to work. However, the equation(s) derived from the simple description can be used as a very accurate model for a "real" structure, in which the necessary parameter(s) are determined by curve fitting. Extraction of parameters by curve fitting is a very powerful approach, but this method has many pitfalls of its own. Direct and uneducated curve fitting can produce good results for certain regions of application, but lead to poor results when that model is used elsewhere. For example, the built-in junction potential can be extracted from capacitance-voltage data; however, this can lead to an erroneous value which will give poor model results for different temperatures.

This sets the tone for the extraction of model parameters by curve fitting that is discussed in each FET model chapter. If done carefully and properly, curve fitting can extract very accurate and useful sets of model parameters. However, poorly executed curve fitting will lead to inadequate model parameter sets. The best results are obtained by clearly understanding the model equations, the parameters, and various extraction procedures.

BIBLIOGRAPHY

1. N. Arora, *MOSFET Models for VLSI Circuit Simulation*, Springer-Verlag, 1993.

2. A. Grove, *Physics and Technology of Semiconductor Devices*, John Wiley & Sons, 1967.

3. B. Streetman, *Solid State Electronic Devices* (2nd edition), Prentice-Hall, 1980.

4. S. Selberherr, *Analysis and Simulation of Semiconductor Devices*, Springer-Verlag, 1984.

5. J. Brews, "A Charge Sheet Model of the MOSFET," *Sol. St. Elec.* vol. 21, pp. 345–355 (1978).

6. T. Ando, A. Fowler, and F. Stern, "Electronic Properties of Two Dimensional Systems," *Rev. Mod. Phys.*, vol. 54, pp. 437–672 (1982).

7. K. von Klitzing, K. Dorda, and M. Pepper, "New Method for High Accuracy Determination of the Fine Structure Constant Based on Quantized Hall Resistance," *Phys. Rev. Lett.* vol. 45, pp. 494–497 (1980).

8. D. Tsui, H. Stormer, and A. Gossard, "Two Dimensional Magnetotransport in the Extreme Quantum Limit," *Phys. Rev. Lett.* vol. 48, pp. 1559–1562 (1982).

9. C. Grimes and G. Adams, "Evidence for a Liquid-to-Crystal Phase Transition in a Classical, Two Dimensional Sheet of Electrons," *Phys. Rev. Lett.* vol. 42, pp. 795–798 (1979).

10. J. Marion, *Classical Dynamics of Particles and Systems*, Academic Press, 1970.

11. N. Ashcroft and N. Mermin, *Solid State Physics*, Saunders College Press, 1976.

12. C. Jacoboni and P. Lugli, *The Monte Carlo Method for Semiconductor Device Simulation*, Springer-Verlag, 1989.

5

The Level 1 Model

5.1 Introduction

The Level 1 model is the original MOSFET model included with early versions of the SPICE circuit simulator. In its basic form, it should be familiar to the reader as the simple analytical model for the FET, which is developed in an introductory class on solid-state devices [1, 2].

This model is sometimes referred to as the "Shichman-Hodges" model [3]. Shichman and Hodges did not derive the equations in their 1967 paper, but instead quoted this simple model from a standard reference [4]. However, this paper is historically important, as it marks the first use of an analytical FET model in a computer-aided design simulation of a larger circuit. Since the publication of this paper occurred at the time that early versions of what would become SPICE were being developed at the University of California/Berkeley, its influence was clearly significant.

Thus, the Level 1 model is important in a historical context. In addition, the mathematical approach employed is the basis for the more sophisticated models to be described later; mastering the development of the Level 1 model provides the basic understanding required for the more complex situations to follow. Finally, the simple nature of the Level 1 model allows for a similarly simple extraction of the model parameters; this demonstrates the basic approach to parameter extraction, before the more sophisticated numerical techniques demanded by later models are considered.

A complete listing of the Level 1 model parameters may be found in Table 5.1.

5.2 The Drain Current Model

5.2.1 The Basic Equation

The derivation of the drain current model begins with (3.111):

Table 5.1   The Level 1 model parameters.

Parameter	Units	Description
Process Parameters		
TPG	m	Type of Gate Material
TOX	m	Gate Oxide Thickness
NSUB	cm^{-3}	Substrate Doping Concentration
XJ	m	Source/Drain Junction Depth
Electrical Parameters		
UO	$cm^2/V{\cdot}s$	Zero Bias Low Field Mobility
VTO	V	Threshold Voltage, Long, Wide Device, Zero Substrate Bias
LAMBDA	V^{-1}	Channel Length Modulation/Output Conductance Parameter
CGSO	F/m	Zero Bias Gate-Source Capacitance
CGDO	F/m	Zero Bias Gate-Drain Capacitance
CGBO	F/m	Zero Bias Gate-Bulk Capacitance

$$dV_y = -\frac{I_{ds}\,dy}{\mu W_{eff} Q_{inv}(y)}. \tag{5.1}$$

(Note: The Level 1 development used the drawn channel width W rather than the effective channel width W_{eff}; however, for consistency with the other FET models, W_{eff} is used here.) Also required are (3.119)

$$Q_{inv}(y) = Q_s(y) - Q_{depl}(y), \tag{5.2}$$

and (3.121)

$$Q_{inv}(y) = -C_{ox}\left\{ V_{gs} - V_{fb} - V(y) - 2\phi_f - \gamma \cdot \left[V(y) + 2\phi_f - V_{bs} \right]^{\frac{1}{2}} \right\}. \tag{5.3}$$

Here,

$$\phi_f = \frac{k_b T}{q} \cdot ln\frac{\text{NSUB}}{n_i}, \tag{5.4}$$

where **NSUB** is the substrate doping concentration and n_i is the intrinsic carrier concentration. The flatband voltage V_{fb} was described in Section 3.2; in Level 1 and the other first generation models, V_{fb} is computed as described there. This requires that the parameter **TPG** be specified. If the gate material is polysilicon which is doped the same as the source and drain (e.g., n$^+$ polysilicon in an nFET), **TPG** $= 1$. If the gate material is polysilicon which is doped the opposite to the source and drain (e.g., n$^+$ polysilicon in a pFET), **TPG** $= -1$. It is also possible to include an aluminum gate with **TPG** $= 0$; however, aluminum gates are no longer used in MOS technology, so this setting should not be encountered.

In the derivation of the Level 1 model, it is assumed that the depletion charge does not vary along the channel axis, allowing the $V(y)$ term in $Q_{depl}(y)$ to be discarded. This assumption (the gradual channel approximation) is reasonable for MOSFETs in which, in present terms, the channel length is long, so that the drain bias only affects a very small part of the channel. In this case, (5.3) becomes

$$Q_{inv}(y) = -C_{ox}\left[V_{gs} - V_{fb} - V(y) - 2\phi_f - \gamma \cdot (2\phi_f - V_{bs})^{\frac{1}{2}}\right]. \tag{5.5}$$

To find an expression for the drain current, (5.5) is inserted into (5.1). After some rearrangement,

$$I_{ds}dy = \mu W_{eff}C_{ox}\left[V_{gs} - V_{fb} - V(y) - 2\phi_f - \gamma \cdot (2\phi_f - V_{bs})^{\frac{1}{2}}\right]dV_y. \tag{5.6}$$

Equation (5.6) is then integrated from the source end ($y = 0$, $V_y = 0$) to the drain end ($y = L_{eff}$, $V_y = V_{ds}$):

$$I_{ds}\int_0^{L_{eff}}dy = \mu W_{eff}C_{ox}\left[\int_0^{V_{ds}}(V_{gs} - V_{fb} - 2\phi_f - \gamma \cdot (2\phi_f - V_{bs})^{\frac{1}{2}})dV_y - \int_0^{V_{ds}}V_y dV_y\right]. \tag{5.7}$$

Carrying out the integration and solving for I_{ds} leads to

$$I_{ds} = \frac{\mu W_{eff}C_{ox}}{L_{eff}}\left[(V_{gs} - V_{fb} - 2\phi_f - \gamma \cdot (2\phi_f - V_{bs})^{\frac{1}{2}})V_{ds} - \frac{V_{ds}^2}{2}\right]. \tag{5.8}$$

Equation (5.8) is the basic drain current equation. Using the definition of the threshold voltage (3.89)

$$V_t = V_{fb} + 2\phi_f + \gamma \cdot (2\phi_f - V_{bs})^{\frac{1}{2}}, \tag{5.9}$$

the final basic expression for the linear region drain current is

$$I_{ds,lin} = \frac{\mu W_{eff}C_{ox}}{L_{eff}}\left[(V_{gs} - V_t)V_{ds} - \frac{V_{ds}^2}{2}\right]. \tag{5.10}$$

5.2.2   The Effective Channel Length

The effective channel length, L_{eff}, is the gate material length L minus the total length of the source and drain underdiffusion of the gate. As shown in Figure 5.1, it is assumed that the diffusions are introduced right next to the gate (i.e., sidewall spacers are not considered, as they postdate the Level 1 model), and that the lateral diffusion length is the same as the vertical diffusion length. This vertical diffusion length is the junction depth x_j. Thus, the effective channel length is

$$L_{eff} = L - 2 \cdot x_j. \tag{5.11}$$

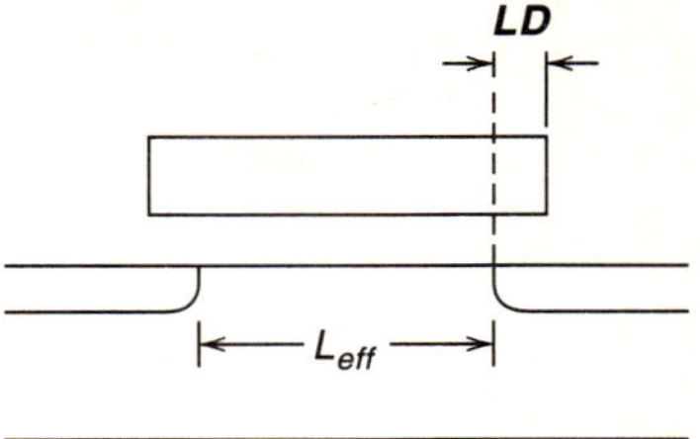

Figure 5.1 Schematic of an FET, showing the underdiffusion of the gate, along with the effective channel length.

Boldfacing parameter names, (5.11) becomes

$$L_{eff} = L - 2 \cdot \mathbf{XJ}. \tag{5.12}$$

This is how the effective channel length was computed in the original implementation of Level 1. However, the HSPICE implementation of Level 1 uses a slightly different approach, in which

$$L_{eff} = L - 2 \cdot (0.75 \cdot \mathbf{XJ}). \tag{5.13}$$

It is also possible to employ a more sophisticated approach employing the parameters **LD** and **XL**, as described in Appendix B.

5.2.3 Channel Pinch-Off and the Saturation Voltage

As described in Chapter 3, when the drain bias V_{ds} in an FET is increased, the gate to drain bias V_{gd} decreases. Due to this decreasing voltage, the inversion charge $Q_{inv}(y)$ at the drain decreases. When V_{ds} becomes sufficiently large, $Q_{inv}(y)$, to some degree of approximation, becomes zero, and the conducting inversion layer no longer reaches the drain. The carriers now cross a depleted space charge region, and the current no longer increases with V_{ds}. The value of the drain voltage at which this occurs is defined as the saturation voltage V_{dsat}; at V_{dsat}, the channel pinches off. As V_{ds} increases further above V_{dsat}, the pinch-off point moves toward the source.

Using (5.5), the inversion charge at the drain end of the channel can be written as

$$Q_{inv}(L_{eff}) = -C_{ox}(V_{gs} - V_{fb} - V_{ds} - 2\phi_f - \gamma \cdot [2\phi_f - V_{bs}]^{\frac{1}{2}}). \tag{5.14}$$

Using the definition of the threshold voltage (5.8), (5.11) can be rewritten as

$$Q_{inv}(L_{eff}) = -C_{ox}(V_{gs} - V_t - V_{ds}). \tag{5.15}$$

As just described, when $V_{ds} = V_{dsat}$, the inversion charge at the drain, to some degree of approximation, vanishes. Thus, (5.15) becomes

$$Q_{inv}(L_{eff}) = 0 = -C_{ox}(V_{gs} - V_t - V_{dsat}). \tag{5.16}$$

Solving (5.16) for V_{dsat} produces

$$V_{dsat} = (V_{gs} - V_t).$$

(5.17)

Therefore, it can be concluded that (5.10) is valid for $V_{ds} < V_{dsat}$. For $V_{ds} \geq V_{dsat}$, (5.10) becomes

$$I_{ds} = \frac{\mu W_{eff} C_{ox}}{L_{eff}} \left[(V_{gs} - V_t)V_{dsat} - \frac{V_{dsat}^2}{2} \right].$$

(5.18)

Substituting the definition of V_{dsat} from (5.17) into (5.18) produces the saturation region current expression

$$I_{ds,sat} = \frac{\mu W_{eff} C_{ox}}{2L_{eff}} (V_{gs} - V_t)^2.$$

(5.19)

This is the rather well-known "square law" description of the saturated drain current. In this form, I_{dsat} is constant for $V_{ds} > V_{dsat}$.

5.2.4   Channel Length Modulation

As just noted, the format of (5.19) is such that the drain current does not increase any further once it has reached I_{dsat}, the drain current at $V_{ds} = V_{dsat}$. However, as V_{ds} increases above V_{dsat}, the point at which $Q_{inv}(y) = 0$ moves toward the source, as shown in Figure 5.2. At the pinch-off point, $V(y) = V_{dsat}$. As the pinch-off point moves toward the source, the distance from the pinch-off point to the source decreases; since V_{dsat} remains constant, the lateral channel field increases. This causes the current to increase roughly in proportion to V_{ds}, as shown in Figure 5.3. A simple expression separating the saturation current from the current above saturation to the point I_{ds} is

$$I_{ds} = I_{dsat} + (I_{ds} - I_{dsat}).$$

(5.20)

The slope of the boldfaced line segment is

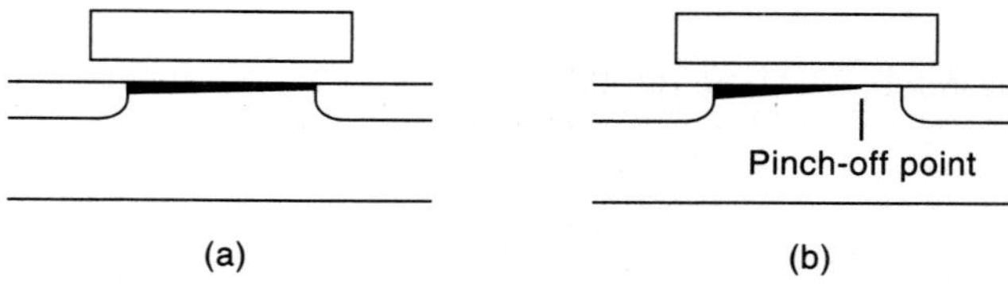

Figure 5.2   Pinch-off and current saturation in an FET; (a) $V_{ds} = V_{dsat}$, and the channel pinches off at the drain junction; (b) $V_{ds} > V_{dsat}$; the pinch-off point has moved toward the source.

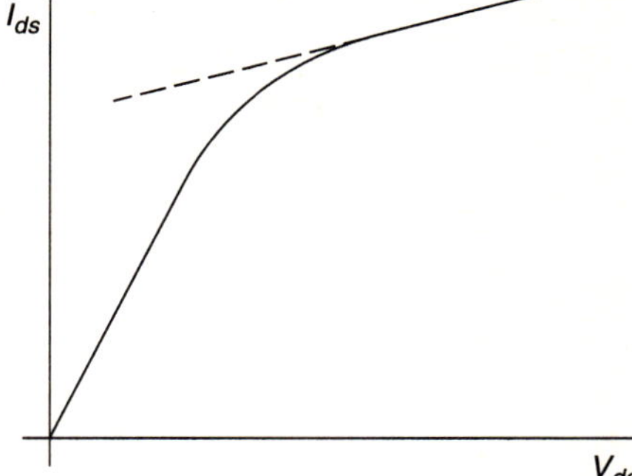

Figure 5.3   FET drain current characteristics, shcwing the increase in the drain current roughly in proportion to V_{ds} when $V_{ds} > V_{dsat}$.

$$\alpha = \frac{I_{ds} - I_{dsat}}{V_{ds}}. \tag{5.21}$$

Substituting (5.21) into the right side of (5.20) produces

$$I_{ds} = I_{dsat} - \alpha V_{ds}. \tag{5.22}$$

To allow a simpler expression and produce a dimensionless constant, α can be redefined as

$$\alpha = \lambda I_{dsat}. \tag{5.23}$$

This allows (5.22) to be written as

$$I_{ds} = I_{dsat}(1 + \lambda V_{ds}). \tag{5.24}$$

Thus, the final form of the drain current equation for $V_{ds} > V_{dsat}$ (by modifying (5.19)) is

$$I_{ds,sat} = \frac{\mu W_{eff} C_{ox}}{2L_{eff}}(V_{gs} - V_t)^2(1 + \lambda V_{ds}). \tag{5.25}$$

Although it is a negligible modification, this expression is also added to the linear region expression (5.14), to guarantee continuity of the drain current at $V_{ds} = V_{dsat}$:

$$I_{ds,lin} = \frac{\mu W_{eff} C_{ox}}{L_{eff}}\left[(V_{gs} - V_t)V_{ds} - \frac{V_{ds}^2}{2}\right](1 + \lambda V_{ds}). \tag{5.26}$$

Boldfacing parameter names, (5.25) and (5.26) are rewritten as the final Level 1 drain current expressions:

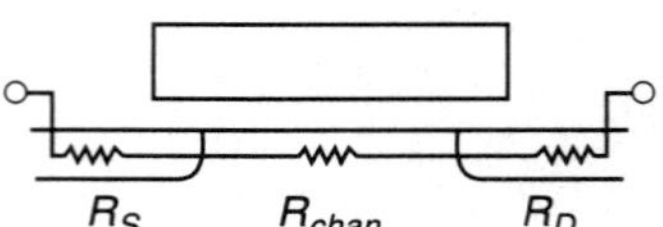

Figure 5.4 Resistances in the FET drain-source current path, showing how the source and drain resistance is added to the channel resistance; this increases the total resistance in the current path.

$$I_{ds,lin} = \frac{\mu W_{eff} C_{ox}}{L_{eff}} \left[(V_{gs} - V_t)V_{ds} - \frac{V_{ds}^2}{2} \right] (1 + \textbf{LAMBDA} \cdot V_{ds}), \qquad (5.27)$$

and

$$I_{ds,sat} = \frac{\mu W_{eff} C_{ox}}{2L_{eff}} (V_{gs} - V_t)^2 (1 + \textbf{LAMBDA} \cdot V_{ds}). \qquad (5.28)$$

5.2.5 The Source/Drain Series Resistance

The integration described in (5.7) assumes that the drain to source voltage V_{ds} is dropped entirely across the channel region (from the drain-channel junction to the source-channel junction). However, as shown in Figure 5.4, this is not strictly the case, as resistances introduced by the source and drain regions drop part of the applied terminal voltage. This is accounted for by the introduction of two additional parameters, **RS** and **RD**. These parameters are available in most implementations of the Level 1 model; however, their introduction in SPICE occurred with the development of later FET models (such as Levels 2 and 3), rather than with the original Level 1 derivation. If the use of **RS** and **RD** is required, it is likely that a more sophisticated FET model is needed. Thus, the general availability of **RS** and **RD** will be merely noted here, and a full discussion deferred to the later consideration of more sophisticated models.

5.2.6 Interface States

In addition to depletion and inversion charges, interface states (at the oxide-silicon boundary) can contribute to the total MOS charge density. This can be accounted for in the Level 1 model with the parameter **NSS**, the areal surface state density. While a parameter of this sort used to be required to describe the MOS system, in modern MOS technology surface state densities have been reduced to less than 10^{10} cm^{-2} and rendered negligible. Therefore, it is not necessary to consider their effects.

5.3 The Charge Model

As described in Chapter 3, the evaluation of the gate capacitance is a two-step process. First, the node charges are computed (from the FET model), after which the gate capacitance is calculated. The gate capacitance computation depends on its specific implementation in SPICE at the circuit simulation level. Here the device charge equations will be derived from the current model; in the chapters which follow, this will be done with each particular FET model's current equations. It is important to note the separation of the *charge* computation (which is model specific) and the *capacitance* computation (which is a circuit simulation implementation problem). This situation is considered in detail in Chapter 13.

Since the gate charge is imaged in the silicon substrate, charge neutrality dictates that

$$Q_{GATE} + Q_{INV} + Q_{DEPL} = 0, \tag{5.29}$$

where the subscripts in capital letters (e.g., Q_{INV}) are meant to indicate that the term describes the *total* charge, rather than the charge per unit area [5]. Solving (5.29) for the gate charge leads to

$$Q_{GATE} = -(Q_{INV} + Q_{DEPL}). \tag{5.30}$$

The charge can be taken to be constant across the channel, but not along the channel; thus, (5.30) must be recast with an integral along the channel:

$$Q_{GATE} = -W_{eff} \int_0^{L_{eff}} Q_{inv}(y)dy - W_{eff} \int_0^{L_{eff}} Q_{depl}(y)dy, \tag{5.31}$$

where the terms for charge per unit area have been introduced inside the integrals. As in the surface charge derivation presented above (see Section 5.2.1), it is assumed that the depletion charge does not vary along the channel. Under this assumption, as derived in (3.84), the depletion charge per unit area is

$$Q_{depl} = C_{ox} \cdot \gamma \cdot (2\phi_f - V_{bs})^{\frac{1}{2}}, \tag{5.32}$$

leading to

$$Q_{DEPL} = W_{eff} \cdot L_{eff} \cdot Q_{depl} = W_{eff} \cdot L_{eff} \cdot C_{ox} \cdot \gamma \cdot (2\phi_f - V_{bs})^{\frac{1}{2}}. \tag{5.33}$$

An expression for the inversion charge along the channel was also derived earlier, in (3.121):

$$Q_{inv}(y) = -C_{ox}(V_{gs} - V_t - V(y)). \tag{5.34}$$

There will also be a need for the drain current expression employed earlier in (5.1):

$$dy = -\frac{\mu W_{eff} Q_{inv}(y)}{I_{ds}} dV. \tag{5.35}$$

Substituting (5.33)–(5.35) into (5.31) and using (5.35) to change the integration variable leads to

$$Q_{GATE} = \left[\frac{\mu W_{eff}^2 C_{ox}^2}{I_{ds}} \int_0^{V_{ds}} (V_{gs} - V_t - V)^2 dV \right] - Q_{DEPL}. \tag{5.36}$$

The Linear Region

To evaluate (5.36), the expression for the linear region drain current (5.10) is included:

$$I_{ds} = \frac{\mu W_{eff} C_{ox}}{L_{eff}} \left(V_{gs} - V_t - \frac{V_{ds}^2}{2} \right) V_{ds}. \tag{5.37}$$

Since in this model any direct drain-source interactions are being ignored, V_{ds} is replaced (see Figure 3.17) in (5.37) with

$$V_{gd} = (V_{gs} - V_{ds}), \tag{5.38}$$

which transforms (5.37) to

$$I_{ds} = \frac{\mu W_{eff} C_{ox}}{2 L_{eff}} \left[(V_{gs} - V_t)^2 - (V_{gd} - V_t)^2 \right] \tag{5.39}$$

Substituting (5.39) into (5.36) and integrating leads to the final expression for the linear region gate charge:

$$Q_{GATE} = \frac{2}{3} W_{eff} L_{eff} C_{ox} \left[\frac{(V_{gd} - V_t)^3 - (V_{gs} - V_t)^3}{(V_{gd} - V_t)^2 - (V_{gs} - V_t)^2} \right] - Q_{DEPL}, \tag{5.40}$$

where Q_{DEPL} is given by (5.33).

The Saturation Region

Equation (5.40) was derived for the linear region of MOSFET operation. In the saturation region, the drain voltage V_{ds} in (5.37) is replaced by the saturation voltage V_{dsat}. From (5.17), the saturation voltage is

$$V_{dsat} = (V_{gs} - V_t), \tag{5.41}$$

so that the gate to drain voltage becomes

$$V_{gd} = V_{gs} - V_{dsat} = V_t; \tag{5.42}$$

in saturation, the gate to drain voltage does not exceed the threshold voltage. Substituting (5.42) into (5.40) produces an expression for the gate charge in the saturation region:

$$Q_{GATE} = \left[\frac{2}{3} W_{eff} L_{eff} C_{ox} (V_{gs} - V_t) \right] - Q_{DEPL}; \tag{5.43}$$

once again, Q_{DEPL} is given by (5.33).

The Subthreshold Region

Although the Level 1 current model ignored the subthreshold region, it is possible to expand the original development to include the charge conditions for $V_{gs} < V_t$. In this regime, $Q_{inv} \ll Q_{depl}$, so in contrast to the case of strong inversion, the charge characteristics will be determined by the depletion charge. Neglecting the inversion charge, the gate charge is

$$Q_{GATE} = -Q_{DEPL}.$$ (5.44)

In the weak inversion regime ($V_{fb} < V_{gs} < V_t$), the depletion charge was described earlier, by (3.130):

$$Q_{DEPL} = W_{eff} \cdot L_{eff} \cdot C_{ox} \cdot \gamma \cdot (\phi_s - V_{bs})^{\frac{1}{2}},$$ (5.45)

where the strong inversion condition $\phi_s = 2\phi_f$ has been replaced with the more general term ϕ_s.

The Gate Capacitance Computation

As implemented in SPICE, the Level 1 charge expressions are used in the Meyer model [6] to compute the active gate capacitance. However, it should be noted that the active gate capacitance models are essentially independent of each particular FET current model. Therefore, the Meyer model and the various other gate capacitance models are discussed in detail in Chapter 13.

5.3.1   The Zero Bias Gate Capacitance

At the time that the Level 1 model was developed, it was reasonable to assume that a comparitively large channel area dominated the gate capacitance, and that the small zero bias capacitance between the gate and the other nodes could be neglected. As device geometries have decreased, these capacitances are no longer negligible. This difficulty is solved empirically in SPICE by adding three parameters (**CGSO**, **CGDO**, and **CGBO**), which represent the zero bias gate-source, gate-drain, and gate-bulk capacitances. These parameters and their usage are described in greater detail in Chapter 13.

Although these parameters are available in Level 1, their situation is similar to that of the series resistance parameters described in Section 5.2.5. These zero bias gate capacitance parameters can be used in Level 1, but a need for their use indicates a small FET geometry is under consideration; it is likely that a more sophisticated FET model is needed to adequately describe the device characteristics.

5.4   Parameter Extraction

The final form of the Level 1 drain current model, as described by (5.27) and (5.28), contains six basic parameters: **TOX**, **NSUB**, **XJ**, **VTO**, **UO**, and **LAMBDA**; other

Table 5.2 Final model parameter set (including both provided process parameters and extracted electrical parameters).

Parameter	Value
TPG	1
TOX	1.0E-07
NSUB	1.0E+15
XJ	1.0E-06
UO	659
VTO	0.618
LAMBDA	0.0134

commonly quoted parameters (such as **GAMMA** and **KP**) are actually secondary parameters, computed from one or more of the basic parameters. The parameter **TOX**, the gate oxide thickness, is usually extracted from a simple C-V measurement [7, 8], and is considered to be a known, provided process parameter. In later models, the substrate doping **NSUB** and the junction depth **XJ** have more complex implications, due to both more modern process technology and nuances of the model structure. However, for the technology of the time of the Level 1 model, both of these parameters can also be treated as known, provided process parameters.

In the example to be considered here, **TOX** $= 100$ nm, **NSUB** $= 1.0 \times 10^{15}$ cm^{-3}, and **XJ** $= 1.0$ μm. This leaves three electrical parameters to be extracted: **VTO**, **UO**, and **LAMBDA**. A complete list of the extracted parameters may be found in Table 5.2.

5.4.1 The Linear Region

The first two parameters, **VTO**, the zero bias threshold voltage, and **UO**, the channel mobility, can be extracted from the linear region of the characteristics of a long channel device; in this example, $W/L = 100$ μm/50 μm. With a very low drain bias ($V_{ds} = 0.1$ V or less) and $V_{bs} = 0$, (5.27) becomes

$$I_{ds} = \frac{\mu W_{eff} C_{ox}}{L_{eff}} \left[(V_{gs} - V_t) V_{ds} \right]. \tag{5.46}$$

For purposes of parameter extraction, V_t can be expanded to the form described by (5.9):

$$V_t = V_{to} + \gamma \cdot (2\phi_f - V_{bs})^{\frac{1}{2}} - \gamma \cdot (2\phi_f)^{\frac{1}{2}}, \tag{5.47}$$

where V_{to} is the threshold voltage with no substrate bias applied:

$$V_{to} = V_{fb} + 2\phi_f + \gamma \cdot (2\phi_f)^{\frac{1}{2}}. \tag{5.48}$$

This arrangement is convenient, for when $V_{bs} = 0$, $V_t = V_{to}$.

Since the case of $V_{bs} = 0$ is being considered here, the drain current expression (5.46) becomes

$$I_{ds} = \frac{\mu W_{eff} C_{ox}}{L_{eff}} \left[(V_{gs} - V_{to}) V_{ds} \right]. \tag{5.49}$$

Adopting the practice of boldfacing parameter names, (5.49) is rewritten as

$$I_{ds} = \frac{\mathbf{UO} \cdot W_{eff} \cdot C_{ox}}{L_{eff}} \left[(V_{gs} - \mathbf{VTO}) V_{ds} \right]. \tag{5.50}$$

This expression can be rearranged to

$$I_{ds} = \frac{\mathbf{UO} \cdot W_{eff} \cdot C_{ox}}{L_{eff}} V_{gs} \cdot V_{ds} - \frac{\mathbf{UO} \cdot W_{eff} \cdot C_{ox}}{L_{eff}} \mathbf{VTO} \cdot V_{ds}. \tag{5.51}$$

By using I_{ds} as y and V_{gs} as x, (5.51) is in slope-intercept form. The slope expression is used to extract **UO**, and then the value of **UO** is used in the y-intercept expression to extract **VTO**. The extraction for the example employed here is shown in Figure 5.5, where **UO** and **VTO** are determined (**UO** $= 659$ cm^2/V·s and **VTO** $= 0.618$ V).

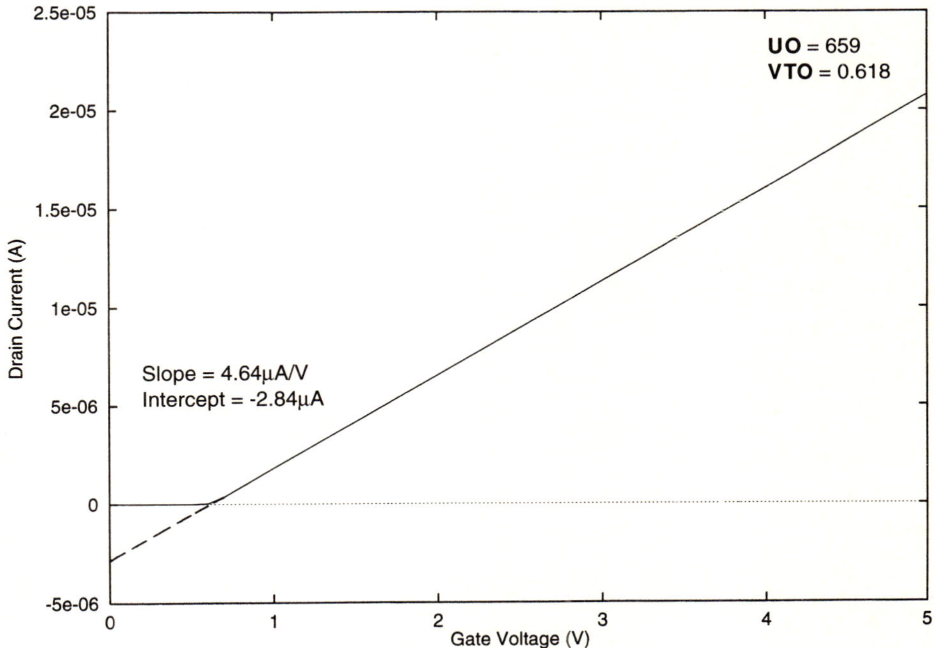

Figure 5.5   Linear FET characteristic, $W/L = 100$ µm/50 µm, $V_{bs} = 0$, showing the extraction of the parameters **UO** and **VTO** using (5.51).

5.4.2 The Saturation Region

Once **VTO** and **UO** have been determined from the linear region characteristics, a single parameter, **LAMBDA**, remains to describe the saturation region characteristics. For a short device with $V_{ds} > V_{dsat}$ and $V_{bs} = 0$, (5.28) becomes

$$I_{ds} = \frac{\mathbf{UO} \cdot W_{eff} \cdot C_{ox}}{2L_{eff}}(V_{gs} - \mathbf{VTO})^2(1 + \mathbf{LAMBDA} \cdot V_{ds}), \qquad (5.52)$$

where L_{eff} is computed from (5.12). A useful mathematical interpretation of this equation can be found by manipulating (5.52) into the form

$$\frac{I_{ds}}{\mathbf{LAMBDA}}\left[\frac{2 \cdot L_{eff}}{\mathbf{UO} \cdot W_{eff} \cdot C_{ox}(V_{gs} - \mathbf{VTO})^2}\right] - \frac{1}{\mathbf{LAMBDA}}. \qquad (5.53)$$

With V_{ds} as x and I_{ds} as y, (5.53) is in slope-intercept form, with an x-intercept of $-(1/\mathbf{LAMBDA})$. The implications are depicted graphically in Figure 5.6, where the mathematical form of (5.53) indicates that an extrapolation of all the saturation region characteristics should converge to a single x-intercept (where $I_{ds} = 0$), yielding a value for **LAMBDA**. (This is identical to the use of the Early voltage [9] for the

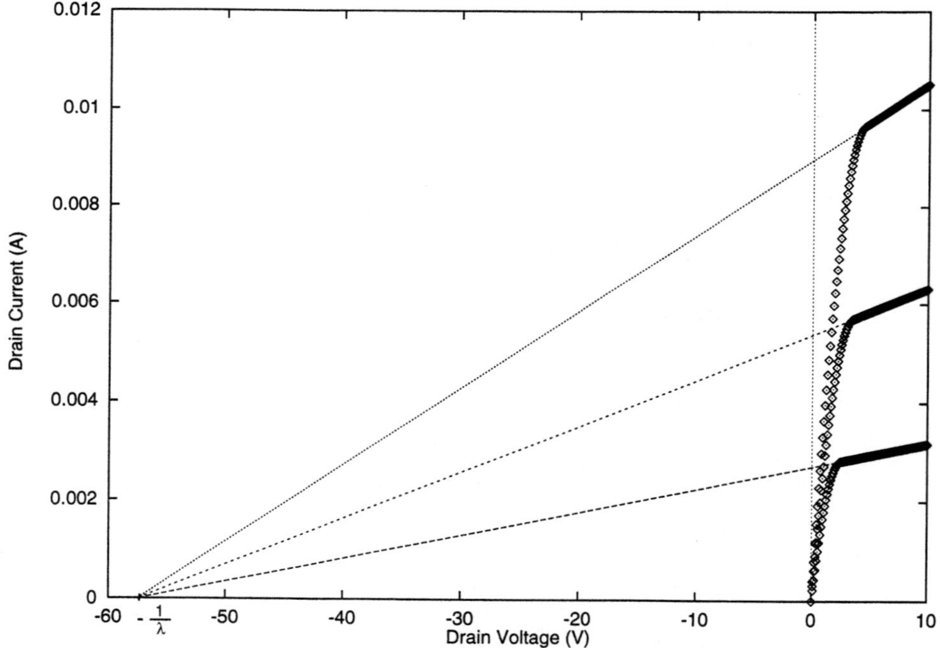

Figure 5.6 Saturated FET characteristics, showing how (5.52) implies that all the saturated current curves have the same x-intercept at $-(1/\mathbf{LAMBDA})$.

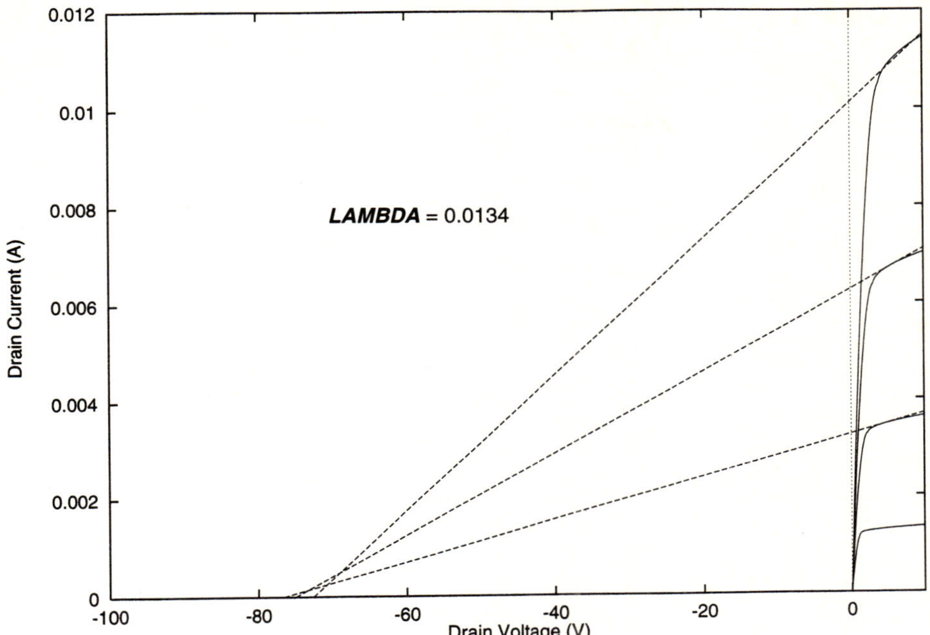

Figure 5.7   Extraction of **LAMBDA** from saturated FET characteristics, $W/L = 100$ μm/4.0 μm, $V_{bs} = 0$, showing that several values of the x-intercept must be averaged to determine a single parameter value.

description of similar behavior in bipolar transistors.) This is the extraction method which is used to determine **LAMBDA**. As shown in Figure 5.7, when employed on real data from an $L = 4.0$ μm device, the situation is not quite as clear-cut as the model structure suggests. However, a reasonable result is produced, and a value of **LAMBDA** $= 0.0134$ is extracted by averaging three different values of the x-intercept.

5.4.3   The Gate Capacitance

The gate capacitance of an FET is quite small; for an FET with $W = L = 100$ μm and $t_{ox} = 100$ nm, the gate capacitance is approximately

$$C_{gs} \sim \frac{2}{3} W \cdot L \cdot C_{ox} \approx 25 fF. \tag{5.54}$$

This represents the *maximum* gate capacitance; in the transition region around the threshold voltage, its value will be lower.

Thus, it is very difficult to measure the active gate capacitance for parameter extraction. Efforts are being made to improve this situation (see, for example, [10];

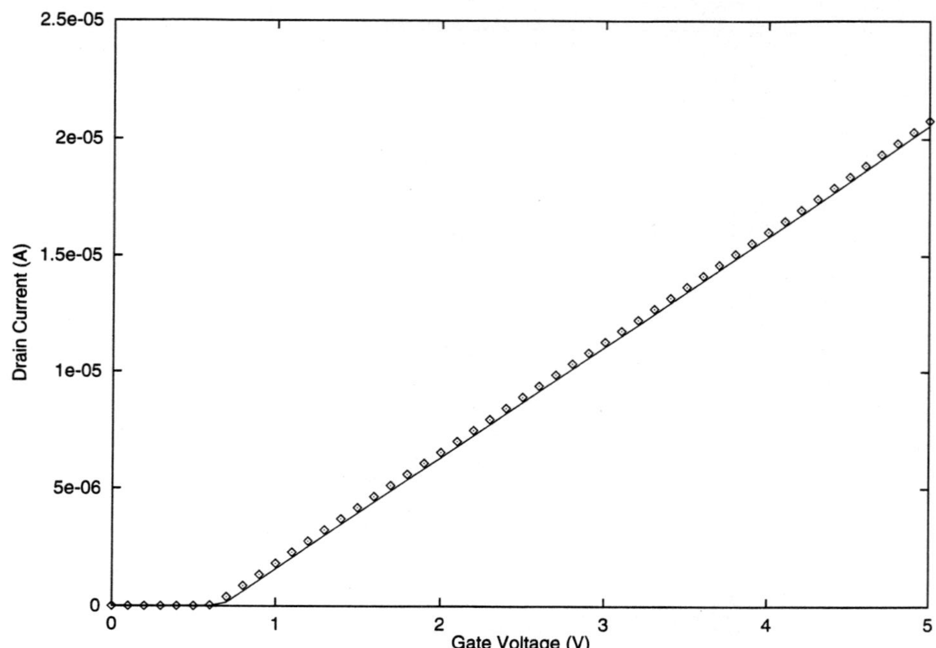

Figure 5.8   Comparison of the final model and data for the linear region of an FET with $W/L = 100$ μm/50 μm; the agreement is good, but breaks down for higher gate biases, where the mobility is affected by the gate field.

however, most models in the literature instead concentrate on improving the expressions for the charge values that are used to compute the gate capacitance. This situation is considered in more detail in Chapter 13.

5.4.4   Model Playback

Once the electrical parameters have been determined, it is worthwhile to compare the final model with the measured data. Figure 5.8 compares the model and data of the $I_{ds} - V_{gs}$ characteristic for the $W/L = 100$ μm/50 μm device. The agreement is quite good. Figure 5.9 compares the $I_{ds} - V_{ds}$ characteristics (model and data) of the $W/L = 100$ μm/4.0 μm device. A rather mediocre result is produced; the shortcomings of the very simple model for channel length modulation are clearly evident.

5.5   Final Comments

It would not be difficult to take advantage of the simplicity of the Level 1 model to launch into a long and ultimately pedantic discussion. However, this model is far too

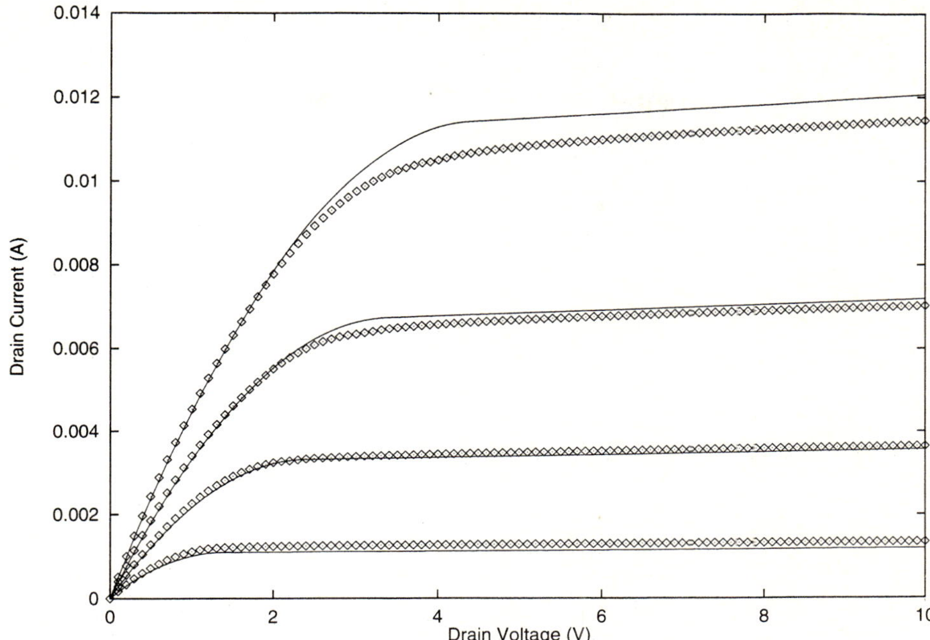

Figure 5.9 Comparison of the final model and data for the saturation region of an FET with $W/L = 100$ μm/4.0 μm; the agreement is reasonably good; however, the simplicity of the λ model requires averaging over several of the saturated regions of the curves, leading to some loss of accuracy.

simple to now be of practical use with modern MOSFET technology. From an industrial design standpoint, the Level 1 model is obsolete.

However, the model does have some instructive utility. First, the derivation is simple, and demonstrates the basic approach to developing workable analytic models for the MOS transistor. All the more sophisticated models start with the same approach, and add corrections for small geometry effects. Second, the uncomplicated mathematical formulation allows parameter extraction to be carried out in a very simple fashion, using linear regression fitting to slope-intercept equations. The more complicated models require nonlinear fitting of several parameters at a time to more complex equations.

BIBLIOGRAPHY

1. R. Pierret, "Field Effect Devices," *Modular Series on Solid State Devices* (ed. by G. Neudeck and R. Pierret), volume IV, Addison-Wesley, 1983.
2. B. Streetman, *Solid State Electronic Devices* (2nd edition), Prentice-Hall, 1980.

3. H. Shichman and D. Hodges, "Modeling and Simulation of Insulated-Gate Field-Effect Transistor Switching Circuits," *IEEE J. Sol. St. Circ.* vol. SC-3, pp. 285–289 (1968).

4. *Field-Effect Transistors* (ed. by J. Wallmark and H. Johnson), Prentice-Hall, 1966.

5. N. Arora, *MOSFET Models for VLSI Circuit Simulation*, Springer-Verlag, 1993.

6. J. Meyer, "MOS Models and Circuit Simulation," *RCA Review* vol. 32, pp. 42–63 (1971).

7. G. Neudeck, "The PN Junction Diode," *Modular Series on Solid State Devices* (ed. by G. Neudeck and R. Pierret), volume II, Addison-Wesley, 1983.

8. Y. Tsividis, *The MOS Transistor*, McGraw-Hill, 1987.

9. J. Early, "Effects of Space Charge Layer Widening in Junction Transistors," *Proc. IRE* vol. 40, pp. 1401–1406 (1952).

10. N. Arora, D. Bell, and L. Bair, "An Accurate Method of Determining MOSFET Gate Overlap Capacitance," *Sol. St. Elec.* vol. 35, pp. 1817–1822 (1992).

6

The Level 2 Model

6.1   Introduction

The Level 1 model described in Chapter 5 is of a very simple form; the physical description is essentially that of a MOS capacitor with a thick gate oxide. A single parameter, **LAMBDA**, was added to account for only one type of small geometry effect, the increase in the drain current due to channel length modulation.

As gate oxides became thinner and device geometries decreased, it was apparent that the Level 1 model was inadequate to describe FET behavior. A number of small geometry effects are unaccounted for in the Level 1 model, and the Level 2 model was developed to address these shortcomings. The basic approach is to begin with the Level 1 model, and add equations and parameters to include the small geometry effects as corrections to the base model. Unlike the Level 1 model, it is assumed that the depletion charge varies along the length of the channel; this results in a complex but more accurate expression for the drain current. In short channel devices the overlap of the gate-induced depletion region with the source and drain depletions regions becomes significant, while in narrow devices, the gate-induced depletion region spreads beyond the channel edges. High normal (gate) and lateral (drain) fields reduce the channel mobility. In short channel devices, in addition to the effect of channel length modulation, carrier velocity saturation reduces the saturation voltage from its classical value of $V_{dsat} = (V_{gs} - V_t)$. The Level 2 model attempts to account for all these phenomena.

Unfortunately, the additions in the Level 2 model were allowed to become very complex mathematically. This makes the model rather inefficient, and convergence problems are often encountered. This has limited the widespread use of the model, as attention soon shifted to the semi-empirical but computationally more efficient Level 3 model. The Level 2 model is most frequently encountered in legacy situations, where an FET model parameter set was developed and has remained in use. Many of the shortcomings of the Level 2 model will be demonstrated at the end of this chapter, when parameter extraction and model development are discussed.

Table 6.1   The Level 2 model parameters.

Parameter	Units	Description
Process Parameters		
TPG	m	Type of Gate Material
TOX	m	Gate Oxide Thickness
LD	m	Channel Length Reduction From Drawn Value
WD	m	Channel Width Reduction From Drawn Value
Electrical Parameters		
VTO	V	Threshold Voltage, Long, Wide Device, Zero Substrate Bias
UO	$cm^2/V \cdot s$	Zero Bias Low Field Mobility
UCRIT	V/cm	Critical Vertical Field For Mobility Reduction
UEXP		Exponent in the Mobility Model
RS	ohm	Source Series Resistance
RD	ohm	Drain Series Resistance
DELTA		Narrow Channel Effect on the Threshold Voltage
NSUB	cm^{-3}	Substrate Sensitivity Parameter (Effective Substrate Doping)
XJ	m	Short Channel Correction to the Substrate Sensitivity
VMAX		Maximum Carrier Velocity
NEFF		Fractional Depletion Charge Reduction Due to Channel Length Modulation
NFS	cm^{-2}	Subthreshold Region Fitting Parameter
CGSO	F/m	Zero Bias Gate-Source Capacitance
CGDO	F/m	Zero Bias Gate-Drain Capacitance
CGBO	F/m	Zero Bias Gate-Bulk Capacitance
XQC		Charge Partitioning Parameter
Additional Mobility Parameter (HSPICE Implementation)		
UTRA	V^{-1}	Drain Bias Effect on the Vertical Field For Mobility Reduction

A list of the Level 2 model parameters may be found in Table 6.1.

6.2   The Electrostatic Model of the MOSFET Structure

As described in Chapter 5, the Level 1 model employs the simple MOS capacitor model for the depletion charge, neglecting any effects due to the source and drain depletion regions and short or narrow channels. However, as MOSFET technology advanced, it became obvious that these effects would have to be included.

The Level 2 model introduces small geometry effects into the depletion charge model. Being due to different causes, short and narrow channel effects can be examined separately.

6.2.1   The Short Channel Model

The basic depletion charge model for the FET (along the channel axis) is shown in Figure 6.1. The source and drain junctions each introduce their own depletion regions into the bulk, and there is also the gate-induced depletion region. In a long device, the gate-induced depletion region is not greatly affected by the presence of the source and drain depletion regions. However, as the channel is shortened, the total depletion charge

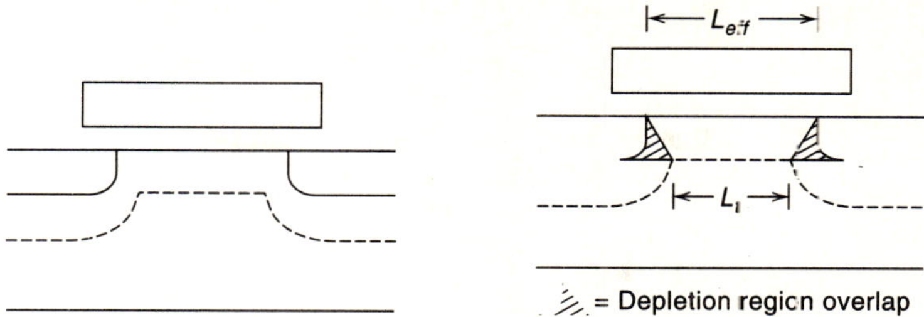

Figure 6.1 The general (large FET) bulk charge model.

Figure 6.2 The short channel bulk charge model used in Level 2.

in the device decreases due to overlap of the gate-induced depletion region with the source and drain depletion regions. For a given gate bias, this decrease in the depletion charge causes an increase in the inversion charge; as the channel length decreases, this has the effect of decreasing the threshold voltage.

To account for this behavior, the Level 2 model employs the depletion charge model described by Yau [1], as depicted in Figure 6.2. Generically, the shaded triangles represent the portions of the gate-induced depletion region that is shared with the source and drain diffusions, leaving a trapezoid-shaped depletion region due entirely to the gate; the value of the depletion charge in this region will determine the device behavior.

Several assumptions were made in Yau's model. First, it is assumed that under the gate, the source and drain junctions can be described as one-quarter cylinders with radii equal to the junction depth x_j. It is also assumed that the gate-induced depletion depth is equal to the junction depth. As described in Chapter 3, the junction depletion depth is

$$W_j = \left(\frac{2\epsilon_{Si}(V_{bi} - V_{bs} + V_r)}{q N_{sub}} \right)^{\frac{1}{2}}, \tag{6.1}$$

where V_r is the reverse bias applied to the junction. The gate-induced depletion depth in strong inversion is

$$W_j = \left(\frac{2\epsilon_{Si}(2\phi_f - V_{bs})}{q N_{sub}} \right)^{\frac{1}{2}}. \tag{6.2}$$

Thus, this approximation is only strictly valid for $2\phi_f = V_{bi}$, which is a reasonable assumption for a strong inversion bias, for which this model was developed. Finally, it is assumed that charge is shared equally by the source and the drain.

In the absence of the source and drain, the gate-induced depletion charge, as described by (3.88), is

$$Q_{depl} = q N_{sub} W_d. \tag{6.3}$$

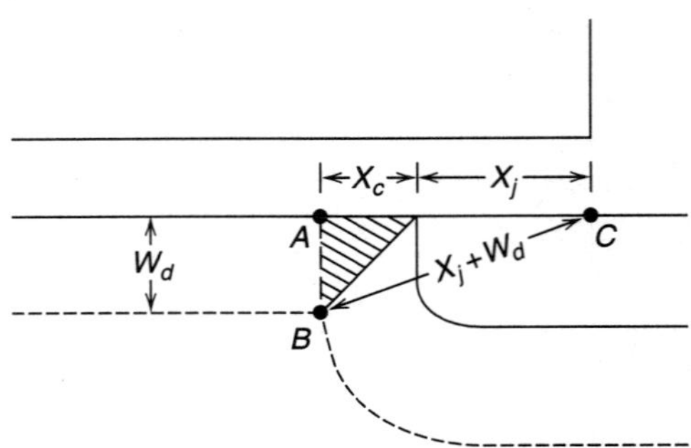

Figure 6.3 The computation of the trapezoid length L_t.

In contrast, the depletion charge in the trapezoid is

$$Q'_{depl} = q N_{sub} W_d \left(\frac{L_{eff} + L_t}{2L_{eff}} \right). \tag{6.4}$$

The reduction of the depletion charge by the source and drain can then be computed as a fractional ratio:

$$f_s \equiv \frac{Q'_{depl}}{Q_{depl}} = \left(\frac{L_{eff} + L_t}{2L_{eff}} \right). \tag{6.5}$$

The approach here will be to compute the depletion charge using (6.3), and then to multiply the result by (6.5).

This requires a computation that includes L_t. The method of determining L_t is described in detail in Figure 6.3. Since

$$L_t = L_{eff} - 2x_c, \tag{6.6}$$

a computation of x_c is required to find the area of the shaded triangle. The larger triangle ABC contains x_c and terms of W_d and x_j. The Pythagorean theorem leads to

$$W_d^2 + (x_c + x_j)^2 = (x_j + W_d)^2. \tag{6.7}$$

Solving for x_c produces

$$x_c = x_j \left[\left(1 + \frac{2W_d}{x_j} \right)^{\frac{1}{2}} - 1 \right]. \tag{6.8}$$

Substituting (6.8) into (6.6), and (6.6) into (6.5), yields

$$f_s = 1 - \frac{x_c}{L_{eff}} = 1 - \frac{x_j}{L_{eff}} \left[\left(1 + \frac{2W_d}{x_j} \right)^{\frac{1}{2}} - 1 \right]. \tag{6.9}$$

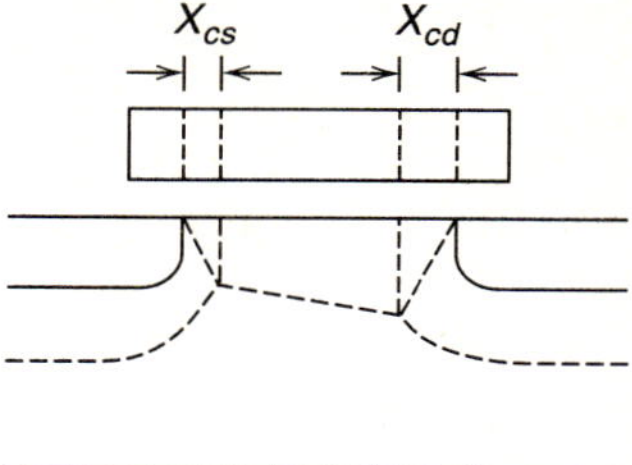

Figure 6.4 Extension of the Level 2 bulk charge model to include the effect of a nonzero drain bias.

The gate-induced depletion charge is easily computed from (6.3) and (6.9):

$$Q'_{depl} = f_s \cdot Q_{depl}. \tag{6.10}$$

The Level 2 implementation then extends Yau's model by adding the effect of a nonzero drain bias, as shown in Figure 6.4. In contrast to the case just described, three different depletion depths must be computed. With the source as the reference, the source depletion depth is

$$W_{js} = \left(\frac{2\epsilon_{Si}(V_{bi} - V_{bs})}{q N_{sub}} \right)^{\frac{1}{2}}, \tag{6.11}$$

and the drain depletion depth is

$$W_{jd} = \left(\frac{2\epsilon_{Si}(V_{bi} - V_{bs} + V_{ds})}{q N_{sub}} \right)^{\frac{1}{2}}, \tag{6.12}$$

At the source, the situation as described above is unchanged, and

$$W_{ds} = \left(\frac{2\epsilon_{Si}(2\phi_f - V_{bs})}{q N_{sub}} \right)^{\frac{1}{2}}. \tag{6.13}$$

However, at the drain, $W_{dd} = W_{jd}$. As also described above, the approximation is made here that $2\phi_f = V_{bi}$, so that

$$W_{dd} = W_{jd} = \left(\frac{2\epsilon_{Si}(2\phi_f - V_{bs} + V_{ds})}{q N_{sub}} \right)^{\frac{1}{2}}. \tag{6.14}$$

Without charge sharing, the gate-induced depletion charge would be

$$Q_{depl} = q N_{sub} \left(\frac{W_{ds} + W_{dd}}{2} \right). \tag{6.15}$$

However, when charge sharing is included, the gate-induced depletion charge is

$$Q_{depl} = q N_{sub} \left(\frac{W_{ds} + W_{dd}}{2} \right) \left(\frac{L_{eff} + L_t}{2L_{eff}} \right). \tag{6.16}$$

Dividing (6.16) by (6.15) produces an expression identical to (6.5):

$$\frac{Q'_{depl}}{Q_{depl}} = f_s = \left(\frac{L_{eff} + L_t}{2L_{eff}} \right). \tag{6.17}$$

However, instead of (6.6), now

$$L_t = L_{eff} - x_{cs} - x_{cd}. \tag{6.18}$$

The terms x_{cs} and x_{cd} are computed as x_c was earlier:

$$x_{cs} = x_j \left[\left(1 + \frac{2W_{ds}}{x_j} \right)^{\frac{1}{2}} - 1 \right], \tag{6.19}$$

$$x_{cd} = x_j \left[\left(1 + \frac{2W_{dd}}{x_j} \right)^{\frac{1}{2}} - 1 \right]. \tag{6.20}$$

Substituting (6.19) and (6.20) into (6.18), and (6.18) into (6.17) produces

$$f_s = 1 - \frac{x_c}{2L_{eff}}(x_{cs} + x_{cd})$$

$$= 1 - \frac{x_j}{2L_{eff}} \left\{ \left[\left(1 + \frac{2W_{ds}}{x_j} \right)^{\frac{1}{2}} - 1 \right] + \left[\left(1 + \frac{2W_{dd}}{x_j} \right)^{\frac{1}{2}} - 1 \right] \right\}. \tag{6.21}$$

Note that for $V_{ds} = 0$, $W_{dd} = W_{ds}$, and (6.21) reduces to (6.9). This is the model that is implemented in Level 2. Boldfacing parameter names, (6.21) becomes

$$f_s = 1 - \frac{x_c}{2L_{eff}}(x_{cs} + x_{cd})$$

$$= 1 - \frac{\mathbf{XJ}}{2L_{eff}} \left\{ \left[\left(1 + \frac{2W_{ds}}{\mathbf{XJ}} \right)^{\frac{1}{2}} - 1 \right] + \left[\left(1 + \frac{2W_{dd}}{\mathbf{XJ}} \right)^{\frac{1}{2}} - 1 \right] \right\}. \tag{6.22}$$

It is also assumed in the original model [1] that the junction depth is equal to the lateral diffusion under the gate, as in Level 1. However, the SPICE implementation of this model introduces a separate parameter, **LD**, to describe gate underdiffusion. This has the unintended effect of changing the meaning of the parameter **XJ**. When a

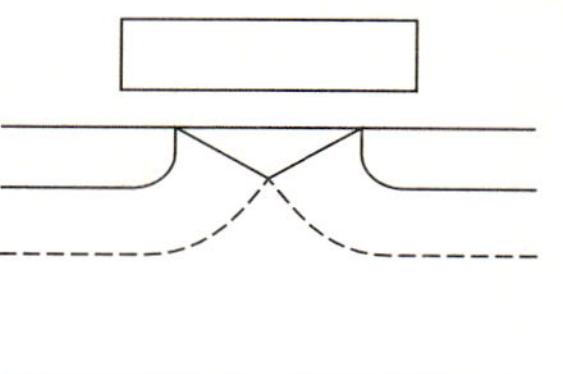

Figure 6.5 Source-drain charge sharing, in which the trapezoidal region becomes a triangle. This model is **not** implemented in Level 2.

substrate-source bias is applied, the depletion region around the source diffusion expands. In a long channel device, this expansion is not large compared to the total FET depletion region size. In a short channel device, the change to the total FET depletion region size is significant. Thus, **XJ** in reality serves as a short channel correction to the substrate sensitivity. This situation is considered further when parameter extraction is examined.

There is, however, a serious flaw inherent in this approach. As pointed out by Yau [1], if the source and drain depletion regions become large, they will merge, and the trapezoid of Figure 6.2 becomes a triangle, as shown in Figure 6.5. The simple model developed here breaks down, requiring modifications, as described in [1].

However, none of these modifications are implemented in the Level 2 model. Yau [1] noted that this merging of the source and drain depletion regions could occur under a very high substrate bias. However, this can also occur in short channel FETs. At the time that the Level 2 model was developed, channel lengths were not small enough for this to occur, and only the simpler charge sharing described above was considered. However, neglect of this phenomenon is not valid in more modern short channel FETs; this severely limits the usefulness of the Level 2 model.

6.2.2 Narrow Channel Effects

In addition to the decrease of the total depletion charge due to overlap of the gate-induced depletion region with the source and drain depletion regions, the depletion charge is also increased by the spreading of the gate-induced depletion region outside the channel edges and under the isolation, as shown in Figure 6.6. In the Level 2 model, the situation is simplified to that shown in Figure 6.7, where a gate electrode over a flat uniformly doped silicon surface is assumed. The total depletion volume in the main channel is

$$D_{ch} = W_{eff} \cdot L \cdot x_n, \tag{6.23}$$

while the total depletion volume in the two extra regions is

$$D_{ed} = A \cdot L, \tag{6.24}$$

where A is the cross-sectional area of the extra regions. The fraction of depletion volume in the edge regions relative to the total depletion volume is

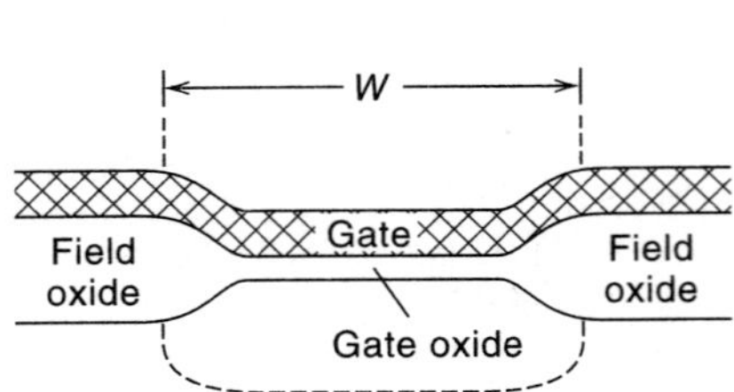

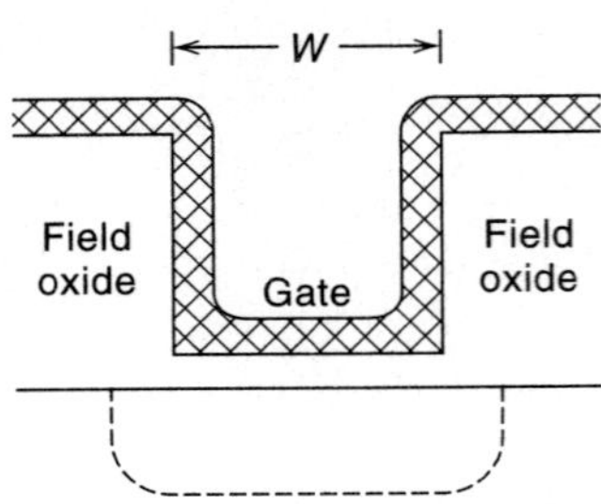

Figure 6.6 Spreading of the depletion region outside the width of the channel and under the LOCOS isolation.

Figure 6.7 The simplified structure used in Level 2 to model the spreading of the depletion region outside the defined channel width.

$$P = \frac{D_{ed}}{D_{ed} + D_{ch}} = \frac{A \cdot L}{A \cdot L + W_{eff} \cdot L \cdot x_n} = \frac{A}{A + W_{eff} \cdot x_n}. \tag{6.25}$$

As shown in Figure 6.7, the Level 2 model approximates each edge region as a quarter-circle with radius x_n [2, 3]. The total area of the edge regions is then

$$A = 2 \times \frac{\pi x_n^2}{4} = \frac{\pi x_n^2}{2}. \tag{6.26}$$

Equation (6.25) then becomes

$$P = \frac{\frac{\pi x_n^2}{2}}{\frac{\pi x_n^2}{2} + W_{eff} \cdot x_n} = \frac{1}{1 + \frac{2W_{eff}}{\pi x_n}}. \tag{6.27}$$

The gate depletion depth x_n is usually less than 1 μm, so for the technology of the time that the Level 2 model was developed, it was reasonable to assume that $W \gg x_n$. This reduces (6.27) to

$$P = \frac{\pi x_n}{2W_{eff}}. \tag{6.28}$$

The additional depletion charge due to the edge regions is then computed from

$$Q_{edge} = P \cdot Q_{depl} = \left(\frac{\pi x_n}{2W_{eff}}\right)(qN_{sub}x_n) = \frac{qN_{sub}\pi x_n^2}{2W_{eff}}. \tag{6.29}$$

From (3.87),

$$x_n = \left(\frac{2\epsilon_s \phi_s}{qN_{sub}}\right)^{\frac{1}{2}}. \tag{6.30}$$

In strong inversion, $\phi_s = (2\phi_f - V_{bs})$, so (6.19) becomes

$$Q_{edge} = \frac{qN_{sub}\pi}{2W_{eff}}\left(\frac{2\epsilon_s(2\phi_f - V_{bs})}{qN_{sub}}\right) = \frac{\pi\epsilon_s(2\phi_f - V_{bs})}{W_{eff}}. \tag{6.31}$$

The Level 2 implementation of (6.31) adds an erroneous 4 to the denominator, and, since the structure described by Figure 6.7 is an approximation of the physical situation shown in Figure 6.6, the model parameter δ is also added (which covers up the erroneous 4 by sweeping the error into the extraction of δ). Equation (6.31) is then redefined to

$$\frac{Q_{edge}}{C_{ox}} = \delta\frac{\pi\epsilon_s(2\phi_f - V_{bs})}{4C_{ox}W_{eff}} = f_n \cdot (2\phi_f - V_{bs}). \tag{6.32}$$

Using the convention of bold-facing parameter names, f_n is

$$f_n = \textbf{DELTA} \cdot \frac{\pi\epsilon_s}{4C_{ox}W_{eff}}. \tag{6.33}$$

6.2.3   The Total Depletion Charge

The computation of the total depletion charge is a three step process. First, the total depletion charge of the three independent regions (source junction, drain junction, and MOS capacitor) is computed. Next, the fractional reduction term f_s, which accounts for the reduction of the total depletion charge by the overlap of the gate-induced depletion region with the source and drain depletion regions, is determined and multiplied by the simple total depletion charge. Finally, the additional depletion charge due to depletion region spreading outside the channel edges is computed and added to the short channel depletion charge.

Combining the effects of short and narrow channels, the Level 2 depletion charge model for a small FET is

$$\frac{Q_{depl,small}}{C_{ox}} = f_s \cdot \frac{Q_{depl,large}}{C_{ox}} + \frac{Q_{depl,edge}}{C_{ox}}$$

$$= \left(1 - \frac{\textbf{XJ}}{2L_{eff}}\left\{\left[\left(1 + \frac{2W_{ds}}{\textbf{XJ}}\right)^{\frac{1}{2}} - 1\right] + \left[\left(1 + \frac{2W_{dd}}{\textbf{XJ}}\right)^{\frac{1}{2}} - 1\right]\right\}\right) \cdot \gamma \cdot (2\phi_f - V_{bs})^{\frac{1}{2}}$$

$$+ \textbf{DELTA} \cdot \frac{\pi\epsilon_s}{4C_{ox}W_{eff}} \cdot (2\phi_f - V_{bs})$$

$$= f_s \cdot \gamma \cdot (2\phi_f - V_{bs})^{\frac{1}{2}} + f_n \cdot (2\phi_f - V_{bs}), \tag{6.34}$$

where f_s is given by (6.21) and f_n by (6.32).

6.3 The Threshold Voltage Model

In the three-terminal MOS capacitor case described in (3.113), the threshold voltage is defined as

$$V_t = V_{fb} + 2\phi_f + \frac{Q_{depl}}{C_{ox}} = V_{fb} + 2\phi_f + \gamma \cdot (2\phi_f - V_{bs})^{\frac{1}{2}}, \qquad (6.35)$$

where any effects due to non-uniform bulk doping have been neglected. The flatband voltage V_{fb} was described in Section 3.2; in Level 2 and the other first generation models, V_{fb} is computed as described there. This requires that the parameter **TPG** be specified. If the gate material is polysilicon with the same doping as the source and drain (e.g., n^+ polysilicon in an nFET), **TPG** $= 1$. If the gate material is polysilicon with doping opposite to that of the source and drain (e.g., n^+ polysilicon in a pFET), **TPG** $= -1$. It is also possible to include an aluminum gate with **TPG** $= 0$; however, aluminum gates are no longer used in MOS technology, so this setting should not be encountered.

Since this model was developed for a three-terminal MOS structure, it takes no account of either narrow channel effects or the drain terminal. In shorter channel FETs, the sizes of the source and drain depletion regions become important with respect to the channel length, and the drain bias renders the gradual channel approximation invalid. The simple threshold voltage model must be modified to account for these effects.

For the simple three-terminal MOS capacitor, the total depletion charge is described by

$$\frac{Q_{depl}}{C_{ox}} = \gamma(2\phi_f - V_{bs})^{\frac{1}{2}}. \qquad (6.36)$$

The Level 2 threshold model modifies (6.36) to account for charge sharing between the gate-induced depletion region and the source and drain depletion regions, and for additional depletion charge outside the channel edges, as described in Section 6.2. With these additions, (6.36) becomes

$$\frac{Q_{depl}}{C_{ox}} = f_s \cdot \gamma(2\phi_f - V_{bs})^{\frac{1}{2}} + f_n \cdot (2\phi_f - V_{bs}). \qquad (6.37)$$

This is the depletion charge model that is implemented in Level 2, and the expression for the threshold voltage becomes

$$V_t = V_{fb} + 2\phi_f + f_s \cdot \gamma(2\phi_f - V_{bs})^{\frac{1}{2}} + f_n \cdot (2\phi_f - V_{bs}). \qquad (6.38)$$

For long, wide FETs, $f_s \to 1$, $f_n \to 0$, and (6.35) is recovered. This also allows for the definition of the model parameter **VTO**, the long, wide threshold voltage with zero substrate bias:

$$\mathbf{VTO} = V_{fb} + 2\phi_f + \gamma \cdot (2\phi_f)^{\frac{1}{2}}. \qquad (6.39)$$

This parameter is easily determined, as will be shown when parameter extraction is considered. However, **VTO** never appears explicitly in any of the Level 2 model equations; the threshold voltage is computed using (6.38). In (6.38), ϕ_f is determined from the substrate doping parameter **NSUB** and V_{fb} is computed as described above. This is in contrast to the situation in Level 3, where **VTO** appears in the mobility model.

In this model, all drain bias effects on the threshold voltage are ascribed to charge sharing. However, threshold voltage reduction due to drain-induced barrier lowering (DIBL) (described in Chapter 3) is neglected. This is not a good approximation in modern short channel FETs.

6.4   The Mobility Model

In the Level 1 model, the channel mobility is treated as a constant. In long FETs with thick gate oxides, this is a good approximation. However, it was observed by Leistiko et al. [4] that when the normal field at the silicon surface exceeds 6×10^4 V/cm, the mobility begins to decrease. This led Frohman-Bentchkowsky [5] to suggest a mobility model in which the surface mobility μ_s equals μ_o only when the vertical field is less than the critical value of 6×10^4 V/cm. When the surface field exceeds this value, the mobility is instead computed from

$$\mu_s = \mu_o \left(\frac{E_{crit}}{E_s} \right)^C ,$$

(6.40)

where C is an empirical constant; Frohman-Bentchkowsky found good results for sample data when $C = 0.15$. The model described by (6.40) forms the basis of the mobility computation used in the Level 2 model.

To compute the surface field E_s, the continuity relation at the oxide-silicon interface

$$\epsilon_{ox} E_{ox} = \epsilon_{Si} E_s$$

(6.41)

is used, with

$$E_{ox} = \frac{V_{ox}}{t_{ox}}.$$

(6.42)

The oxide voltage is computed directly from the gate voltage and the threshold voltage, while the contribution of the drain voltage is averaged along the channel by introducing the parameter U_d:

$$V_{ox} = (V_{gs} - V_t - U_d \cdot V_{ds}).$$

(6.43)

After using (6.42) and (6.43) to solve (6.41) for E_s, E_s is then inserted into (6.40) to yield the final expression for the mobility:

$$\mu_s = \mu_o \left(\frac{\epsilon_{Si} t_{ox}}{\epsilon_{ox}} \frac{E_{crit}}{(V_{gs} - V_t - U_d \cdot V_{ds})} \right)^C$$

$$= \mu_o \left(\frac{\epsilon_{Si}}{C_{ox}} \frac{E_{crit}}{(V_{gs} - V_t - U_d \cdot V_{ds})} \right)^C, \tag{6.44}$$

which, as described above, is only valid for $E_s > E_{crit}$. Three parameters (E_{crit}, U_d, and C) are extracted to account for the effect of a large vertical field. Rewriting (6.44) with parameter names boldfaced leads to

$$\mu_s = \mathbf{UO} \left(\frac{\epsilon_{Si}}{C_{ox}} \frac{\mathbf{UCRIT}}{(V_{gs} - V_t - \mathbf{UTRA} \cdot V_{ds})} \right)^{\mathbf{UEXP}}. \tag{6.45}$$

The parameter **UTRA** can cause negative resistance for drain biases slightly larger than the saturation voltage [6]. Since this can cause convergence problems during circuit simulation, **UTRA** was discarded and never coded into the Berkeley SPICE implementation of the Level 2 model. Thus, **UTRA** is set to zero, and the implemented version of (6.45) is

$$\mu_s = \mathbf{UO} \left(\frac{\epsilon_{Si}}{C_{ox}} \frac{\mathbf{UCRIT}}{(V_{gs} - V_t)} \right)^{\mathbf{UEXP}}. \tag{6.46}$$

The HSPICE implementation of Level 2 [6] does make **UTRA** available, so (6.45) is used there.

This model does not include the effect of high lateral fields on the mobility. This was a reasonable approximation when the Level 2 model was introduced; however, in more modern FETs, this approximation is poor. This is another factor contributing to the limited utility of the Level 2 model.

6.5   The Drain Current Equations

The Level 2 derivation of the main drain current equation uses the same mathematical approach as that employed for the Level 1 model. However, the more complex approach for computing the depletion charge in small geometry devices, described in Section 6.2, is used.

6.5.1   The Drain Current Above Threshold

With the expressions for small geometry effects added, the simpler form of the depletion charge (3.130) becomes

$$Q_{depl}(y) = f_s \cdot [2\epsilon_s q N_{sub} \phi_s(y)]^{\frac{1}{2}} + f_n \cdot \phi_s(y). \tag{6.47}$$

As described in (3.131),

$$\phi_s = V(y) + 2\phi_f - V_{bs}, \tag{6.48}$$

so (6.47) becomes

$$Q_{depl}(y) = f_s \cdot \left[2\epsilon_s q N_{sub}(V(y) + 2\phi_f - V_{bs})\right]^{\frac{1}{2}} + f_n \cdot (V(y) + 2\phi_f - V_{bs}). \tag{6.49}$$

The total surface charge (3.134) is

$$Q_s(y) = -C_{ox}(V_{gb} - V_{fb} - V(y) - 2\phi_f + V_{bs}), \tag{6.50}$$

so the expression for the inversion charge ((3.135) and (3.137)) becomes

$$Q_{inv}(y) = Q_s(y) - Q_{depl}(y) = -C_{ox}\left[V_{gs} - V_{fb} + V(y) - 2\phi_f\right]$$

$$- f_s \cdot \left[2\epsilon_s q N_{sub}(V(y) + 2\phi_f - V_{bs})\right]^{\frac{1}{2}} - f_n \cdot (V(y) + 2\phi_f - V_{bs}). \tag{6.51}$$

Equation (6.51) is then inserted into the drain current expression (3.127):

$$dV_y = -\frac{I_{ds}dy}{\mu W_{eff} Q_{inv}(y)}. \tag{6.52}$$

Integrating (6.52) as in Chapter 5 (see (5.6) and (5.7)) leads to

$$I_{ds} = \frac{\mu W_{eff} C_{ox}}{L_{eff}} \left\{ (V_{gs} - (V_{fb} + 2\phi_f))V_{ds} - \frac{V_{ds}^2}{2} \right.$$

$$- \frac{2}{3} f_s \cdot \gamma \cdot \left[(V_{ds} + 2\phi_f - V_{bs})^{\frac{3}{2}} - (2\phi_f - V_{bs})^{\frac{3}{2}}\right]$$

$$\left. - f_n \cdot \left[\frac{V_{ds}^2}{2} + (2\phi_f - V_{bs})V_{ds}\right] \right\}. \tag{6.53}$$

This is the basic drain current equation employed by the Level 2 model.

(The reader can verify that this form of the drain current equation is in fact identical to the equation which appears in the University of California Berkeley documentation [7]:

$$I_{ds} = \frac{\mu W_{eff} C_{ox}}{L_{eff}} \left\{ \left[V_{gs} - (V_{fb} + 2\phi_f + f_n \cdot (2\phi_f - V_{bs})) - (1 + f_n) \cdot \frac{V_{ds}^2}{2}\right] \right.$$

$$\left. - \frac{2}{3} f_s \cdot \gamma \cdot \left[(V_{ds} + 2\phi_f - V_{bs})^{\frac{3}{2}} - (2\phi_f - V_{bs})^{\frac{3}{2}}\right]. \tag{6.54}$$

When written in the form of (6.53), short and narrow channel effects (introduced as corrections through, respectively, f_s and f_n) are more clearly separated from the main current equation.)

6.5.2 Channel Pinch-Off (Saturation Voltage)

The Level 2 computation of the pinch-off voltage V_{dsat} employs the approach used in Level 1, with modification to include the extended depletion charge model. At $V_{ds} = V_{dsat}$, the inversion charge at the drain end of the channel $Q_{inv}(L_{eff})$, to some degree of approximation, becomes zero. Examining (6.52), it can be seen that this definition is somewhat unsatisfactory; if $Q_{inv}(y) \to 0$, the lateral field E_y and the carrier velocity become infinite. The Level 2 model provides a correction for this situation which is described below.

"First Principles" Computation

Recall from (3.79) that in an MOS structure, the total surface charge is computed from

$$Q_{surf} = Q_{inv} + Q_{depl}, \tag{6.55}$$

which can be solved for Q_{inv}:

$$Q_{inv} = Q_{surf} - Q_{depl}. \tag{6.56}$$

For a four-terminal MOSFET structure, as described by (3.134),

$$Q_{surf} = -C_{ox}\left[V_{gs} - V_{fb} - 2\phi_f - V_y(y)\right], \tag{6.57}$$

where $V_y(y)$ is the lateral voltage at any point along the channel due to the drain-source bias V_{ds}. The depletion charge was computed earlier in (6.34). This expression will require modifications to account for the lateral field due to V_{ds}, becoming

$$Q_{depl} = C_{ox}\left[f_s \cdot \gamma \cdot (2\phi_f - V_{bs} + V_x(x))^{\frac{1}{2}} + f_n \cdot (2\phi_f - V_{bs} + V_x(x))\right]. \tag{6.58}$$

Here the goal is to compute $Q_{inv}(y)$ for $y = L_{eff}$, the inversion charge at the drain end of the channel, when $V_{ds} = V_{dsat}$. Thus, $V_y(y)$ in (6.57) and (6.58) is replaced with V_{dsat}, and (6.56) is solved:

$$Q_{inv}(L_{eff}) = 0 = -C_{ox}\left[V_{gs} - V_{fb} - 2\phi_f - V_{dsat}\right]$$

$$- C_{ox}\left[f_s \cdot \gamma \cdot (2\phi_f - V_{bs} + V_{dsat})^{\frac{1}{2}} + f_n \cdot (2\phi_f - V_{bs} + V_{dsat})\right]. \tag{6.59}$$

Equation (6.59) is quadratic in $(V_{dsat})^{\frac{1}{2}}$, and can thus be solved [7]. A lengthy and tedious exercise in algebraic manipulation leads to

$$V_{dsat} = \frac{V_{gs} - (V_{fb} + 2\phi_f + f_n \cdot (2\phi_f - V_{bs}))}{(1 + f_n)} + \frac{1}{2}\left[\frac{f_s \cdot \gamma}{1 + f_n}\right]^2 \left\{1 \pm \left[1 + 4\left(\frac{1 + f_n}{f_s \cdot \gamma}\right)^2\right.\right.$$

$$\left.\left. \times \left(\frac{V_{gs} - (V_{fb} + 2\phi_f + f_n(2\phi_f - V_{bs}))}{1 + f_n} + 2\phi_f - V_{bs}\right)\right]^{\frac{1}{2}}\right\}. \tag{6.60}$$

Since the quadratic form introduces a double solution ($\pm$), it must be decided which choice is physically meaningful. If short and narrow channel effects are neglected ($f_s \to 1$, $f_n \to 0$), (6.60) becomes

$$V_{dsat} = V_{gs} - (V_{fb} + 2\phi_f) + \frac{1}{2}\gamma^2$$

$$\times \left\{1 \pm \left[1 + 4\left(\frac{1}{\gamma^2}\right)\left(V_{gs} - (V_{fb} + 2\phi_f) + (2\phi_f - V_{bs})\right)\right]^{\frac{1}{2}}\right\}. \tag{6.61}$$

By further manipulation, (6.61) can be converted to

$$V_{dsat} = (V_{gs} - V_{tl}) + \gamma \cdot (2\phi_f - V_{bs})^{\frac{1}{2}} + \frac{1}{2}\gamma^2 \left\{1 \pm \left[1 + \frac{4}{\gamma^2}(V_{gs} - V_{fb} - V_{bs})\right]^{\frac{1}{2}}\right\}, \tag{6.62}$$

where V_{tl} is the threshold voltage expression (6.38) with $f_s \to 1$ and $f_n \to 0$:

$$V_{tl} = V_{fb} + 2\phi_f. \tag{6.63}$$

It would be hoped that (6.62) would be similar to the simpler Level 1 expression (5.17):

$$V_{dsat} = (V_{gs} - V_{tl}). \tag{6.64}$$

However, note by examining (6.62) that the "+" solution would cause all the terms to be additive, increasing the value of V_{dsat}. Added effects should decrease rather than increase the saturation voltage. It can thus be concluded that only the "−" solution is physically valid, so with slight modification to (6.60) the final expression for V_{dsat} is

$$V_{dsat} = \frac{V_{gs} - (V_{fb} + 2\phi_f + f_n \cdot (2\phi_f - V_{bs}))}{(1 + f_n)} + \frac{1}{2}\left[\frac{f_s \cdot \gamma}{1 + f_n}\right]^2 \left\{1 - \left[1 + 4\left(\frac{1 + f_n}{f_s \cdot \gamma}\right)^2\right.\right.$$

$$\left.\left. \times \left(\frac{V_{gs} - (V_{fb} + 2\phi_f + f_n \cdot (2\phi_f - V_{bs}))}{1 + f_n} + 2\phi_f - V_{bs}\right)\right]^{\frac{1}{2}}\right\}. \tag{6.65}$$

Computation with the Critical Field Specified

Equation (6.65) was derived assuming that the critical field for the onset of velocity saturation $E_{crit,l}$ was not specified. If $E_{crit,l}$ is specified, but **VMAX** is not specified, then the saturation voltage (written here as V'_{dsat}) is instead computed using a slightly different method [7]. As briefly described above, it was noted that if pinch-off is defined by allowing the inversion charge $Q_{inv}(y)$ to become zero at the drain end of the channel, both the lateral field (as described by (6.52)) and the carrier velocity become infinite. This, of course, makes no physical sense, so a different approach is required in which $Q_{inv}(y)$ becomes very small at the pinch-off point but does not vanish.

A simple expression for the saturation current can be written by multiplying the inversion charge density at the drain by the carrier velocity (which is taken as being saturated):

$$I_{dsat} = v_{SAT} \cdot W_{eff} \cdot Q_{inv}(L_{eff}), \tag{6.66}$$

where $Q_{inv}(L_{eff})$ is the areal (two-dimensional) inversion charge density at the pinch-off point, as computed in (6.59); multiplying by the effective channel width W_{eff} gives the proper current result. The saturation velocity can be described by a model for the carrier velocity proposed by Caughey and Thomas [8]:

$$v = \frac{\mu E_y}{1 + \dfrac{E_y}{E_{crit,l}}}, \tag{6.67}$$

where E_y is the lateral field in the channel and $E_{crit,l}$ is the critical field for the onset of velocity saturation (the Level 2 parameter **ECRIT**). The reader can easily verify that for low lateral fields ($E_y \ll E_{crit,l}$), (6.67) reduces to the simple ohmic expression $v = \mu E_y$, while for high lateral fields ($E_y \gg E_{crit,l}$), (6.67) becomes a constant velocity expression $v = \mu E_{crit,l}$. In this model, the saturation voltage is directly connected with the saturation velocity; this allows the following definitions:

$$E_y = \frac{V'_{dsat}}{L_{eff}}, \tag{6.68}$$

$$E_{crit,l} = \mathbf{ECRIT} = \frac{V_{crit}}{L_{eff}}. \tag{6.69}$$

The saturation voltage is written here as V'_{dsat}, to keep it separate from the V_{dsat} computed above.

The saturation current can also be expressed by replacing V_{ds} with V'_{dsat} in (6.53):

$$I_{dsat} = \frac{\mu W_{eff} C_{ox}}{L_{eff}} \left\{ \left[(V_{gs} - (V_{fb} + 2\phi_f)) \right] V'_{dsat} - \frac{V'_{dsat}}{2} \right.$$

$$-\frac{2}{3} f_s \cdot \gamma \cdot \left[(V_{ds} + 2\phi_f - V_{bs})^{\frac{3}{2}} - (2\phi_f - V_{bs})^{\frac{3}{2}} \right]$$

$$\left. - f_n \left[\frac{V'_{dsat}}{2} + (2\phi_f - V_{bs}) V'_{dsat} \right] \right\}. \tag{6.70}$$

Substituting (6.67) into (6.66) and setting the resulting expression equal to (6.70) (eliminating I_{dsat}) leads to

$$\frac{V'_{dsat}}{1 + \frac{V'_{dsat}}{V_{crit}}} \left\{ -\left[V_{gs} - V_{fb} - 2\phi_f - V'_{dsat} \right] \right.$$

$$\left. - \left[f_s \cdot \gamma \cdot (2\phi_f - V_{bs} + V'_{dsat})^{\frac{1}{2}} + f_n \cdot (2\phi_f - V_{bs} + V'_{dsat}) \right] \right\}$$

$$= \left\{ \left[V_{gs} - (V_{fb} + 2\phi_f) \right] V'_{dsat} - \frac{V'_{dsat}}{2} \right.$$

$$-\frac{2}{3} f_s \cdot \gamma \cdot \left[(V'_{dsat} + 2\phi_f - V_{bs})^{\frac{3}{2}} - (2\phi_f - V_{bs})^{\frac{3}{2}} \right]$$

$$\left. - f_n \cdot \left[\frac{V'_{dsat}}{2} + (2\phi_f - V_{bs}) V'_{dsat} \right] \right\}. \tag{6.71}$$

Equation (6.71), like (6.59), is in quadratic form, and can be solved for V'_{dsat}. Another lengthy and tedious algebraic exercise produces

$$V'_{dsat} = V_{dsat} + V_{crit} - (V_{dsat}^2 + V_{crit}^2)^{\frac{1}{2}}, \tag{6.72}$$

where V_{dsat} is the expression of (6.65) and

$$V_{crit} = \mathbf{ECRIT} \cdot L_{eff}. \tag{6.73}$$

This expression for the saturation voltage is more complex, but does provide a more physically realistic description of its relationship to the saturation velocity.

Computation with the Saturation Velocity Specified

Although the use of a "critical field" to describe velocity saturation (and thus the saturation voltage) is fairly fundamental, in practice it is somewhat difficult to specify a value for $E_{crit,l}$ (i.e., a value for the parameter **ECRIT**). An alternative approach is to treat v_{SAT} as a parameter to be extracted.

The method employed here [7] begins by solving (6.66) for v_{SAT}:

$$v_{SAT} = \frac{I_{dsat}}{W_{eff} \cdot Q_{inv}(L_{eff})}.$$

(6.74)

Here v_{SAT} is replaced by the parameter **VMAX**:

$$\mathbf{VMAX} = \frac{I_{dsat}}{W_{eff} \cdot Q_{inv}(L_{eff})}.$$

(6.75)

Next (6.59) and (6.70) are inserted for $Q_{inv}(L_{eff})$ and I_{dsat} respectively, yielding

$$
\begin{aligned}
\mathbf{VMAX} = \Bigg[\mu \bigg\{ & \left[V_{gs} - (V_{fb} + 2\phi_f) \right] V'_{dsat} - \frac{V'_{dsat}}{2} \\
& - \frac{2}{3} f_s \cdot \gamma \cdot \left[(V'_{dsat} + 2\phi_f - V_{bs})^{\frac{3}{2}} - (2\phi_f - V_{bs})^{\frac{3}{2}} \right] \\
& - f_n \cdot \left(\frac{V'_{dsat}}{2} + (2\phi_f - V_{bs})V'_{dsat} \right) \bigg\} \Bigg] \\
\div \Bigg[\bigg\{ & - \left[V_{gs} - V_{fb} - 2\phi_f - V'_{dsat} \right] \\
& - \left[f_s \cdot \gamma \cdot (2\phi_f - V_{bs} + V'_{dsat})^{\frac{1}{2}} + f_n \cdot (2\phi_f - V_{bs} + V'_{dsat}) \right] \bigg\} \Bigg].
\end{aligned}
$$

(6.76)

By defining

$$x \equiv \left(V'_{dsat} + 2\phi_f - V_{bs} \right)^{\frac{1}{2}},$$

(6.77)

(6.76) can, after a great deal of unpleasant algebra, be rewritten as a quartic equation of the form

$$x^4 + ax^3 + bx^2 + cx + d = 0.$$

(6.78)

The coefficients are defined by

$$a = \frac{4}{3} f_s \cdot \gamma,$$

(6.79)

$$b = -2(G_1 + G_3),$$

(6.80)

$$c = -2f_s \cdot \gamma \cdot G_3, \qquad (6.81)$$

$$d = -G_2{}^2 - \frac{4}{3} f_s \cdot \gamma \cdot G_2{}^{\frac{3}{2}} + 2G_1 G_2 + 2G_1 G_3, \qquad (6.82)$$

with secondary coefficients

$$G_1 = \frac{V_{gs} - \left(V_{fb} + 2\phi_f + f_n \cdot (2\phi_f - V_{bs})\right)}{1 + f_n} + 2\phi_f - V_{bs}, \qquad (6.83)$$

$$G_2 = 2\phi_f - V_{bs}, \qquad (6.84)$$

$$G_3 = \frac{\text{VMAX} \cdot L_{eff}}{\mu}. \qquad (6.85)$$

The general quartic equation has four solutions on the complex plane, which occur in pairs. A suggested form for the solutions is [9]

$$x = -\frac{a}{4} + \frac{1}{2}\left[\frac{a^2}{4} - b - \left(A + B - \frac{p}{3}\right)\right]^{\frac{1}{2}}$$

$$\pm \frac{1}{2}\left\{\frac{a^2}{2} - b - \left(A + B - \frac{p}{3}\right) - \frac{4ab - 8c - a^3}{4\left[\frac{a^2}{4} - b + \left(A + B - \frac{p}{3}\right)\right]^{\frac{1}{2}}}\right\}^{\frac{1}{2}}, \qquad (6.86)$$

and

$$x = -\frac{a}{4} - \frac{1}{2}\left[\frac{a^2}{4} - b - \left(A + B - \frac{p}{3}\right)\right]^{\frac{1}{2}}$$

$$\pm \frac{1}{2}\left\{\frac{a^2}{2} - b - \left(A + B - \frac{p}{3}\right) + \frac{4ab - 8c - a^3}{4\left[\frac{a^2}{4} - b + \left(A + B - \frac{p}{3}\right)\right]^{\frac{1}{2}}}\right\}^{\frac{1}{2}}. \qquad (6.87)$$

The subsidiary coefficients are given by

$$A = \left[-\frac{\beta}{2} + \left(\frac{\beta^2}{4} + \frac{\alpha^3}{27}\right)^{\frac{1}{2}}\right]^{\frac{1}{3}}, \qquad (6.88)$$

$$B = -\left[\frac{\beta}{2} + \left(\frac{\beta^2}{4} + \frac{\alpha^3}{27}\right)^{\frac{1}{2}}\right]^{\frac{1}{3}}, \qquad (6.89)$$

$$\alpha = \frac{1}{3}(3q - p^2), \qquad (6.90)$$

$$\beta = \frac{1}{27}(2p^3 - 9pq + 27r), \tag{6.91}$$

$$p = -b, \tag{6.92}$$

$$q = (ac - 4d), \tag{6.93}$$

$$r = -a^2 d + 4bd - c^2. \tag{6.94}$$

Since there are four roots of (6.78) on the complex plane, there can be four, two, or no real solutions for V'_{dsat}. Since only one (real) value for V'_{dsat} is sought, the following approach is implemented [7]: If there are four or two real solutions, the smallest (positive) value for V'_{dsat} is chosen, as carrier velocity saturation serves to reduce the saturation voltage. If there are no real roots, the much simpler computation of V_{dsat} discussed above (where it is assumed that $Q_{inv}(L_{eff}) \to 0$ at the pinch-off point, leading to (6.65)) is used.

Comments on the Computation of the Saturation Voltage

As will be described below when parameter extraction is considered, the most common approach employed in dealing with the saturation characteristics is to extract the parameter **VMAX**. This indicates that the third method of computing the saturation voltage is generally employed. This method is so complicated that it is not fully described in the University of California/Berkeley documentation [7], and is not even dealt with in the documentation accompanying HSPICE [6]. The approach is mathematically inefficient; the many computations and postcomputation choices (possibly followed by additional computations) give the impression that an election is held every time a value of the saturation voltage is computed. This is a major weakness of the Level 2 model in circuit simulation. Furthermore, higher-order polynomial equations (such as the quartic equation) are notoriously ill-behaved, and frequently cause nonconvergence in simulations using Level 2. Finally, despite the attempt to use an intricate and fundamental model, the results obtained are not particularly good, especially in modern short channel devices; this will be shown below when parameter extraction and model construction are considered.

6.5.3   Channel Length Modulation

The use of the saturation voltage in (6.53) has the effect of pinning the drain current I_{ds} at its value (I_{dsat}) for $V_{ds} = V_{dsat}$. For larger values of V_{ds}, the model allows I_{ds} to increase very slightly, due to the presence of the charge sharing parameter f_s.

However, the drain current in short channel devices tends to increase more substantially as V_{ds} increases above V_{dsat}. The situation is described in Figure 6.8. As shown in Figure 6.8a, when $V_{ds} = V_{dsat}$, the conducting inversion layer just reaches the drain. As V_{ds} increases above V_{dsat}, the pinch-off point moves toward the source, leaving a space-charge region between the pinch-off point and the drain. The lateral channel voltage (V_{dsat}) is dropped between the source and the pinch-off point, regardless of the latter's location. When $V_{ds} = V_{dsat}$ (Figure 6.8a),

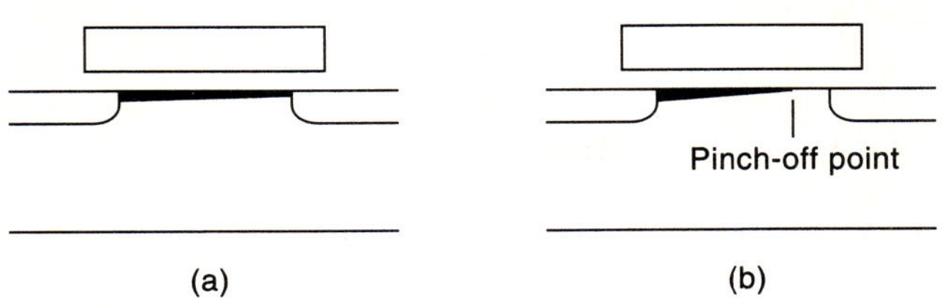

Figure 6.8 Channel pinch-off; (a) $V_{ds} = V_{dsat}$; the inversion layer just reaches the drain; (b) $V_{ds} > V_{dsat}$; the pinch-off point has moved toward the source.

$$E_y = \frac{V_{dsat}}{L_{eff}}, \tag{6.95}$$

while for $V_{ds} > V_{dsat}$ (Figure 6.8b),

$$E_y = \frac{V_{dsat}}{L_{eff} - L'}. \tag{6.96}$$

Thus, as V_{ds} increases, L' increases; by (6.96), this increases the lateral channel field E_y, which in turn increases I_{ds}. Thus, (6.70) should be written as

$$
\begin{aligned}
I_{dsat} = \frac{\mu W_{eff} C_{ox}}{L_{eff} - L'} &\left\{ \left[(V_{gs} - (V_{fb} + 2\phi_f)) \right] V'_{dsat} - \frac{V'_{dsat}}{2} \right. \\
&- \frac{2}{3} f_s \cdot \gamma \cdot \left[(V_{ds} + 2\phi_f - V_{bs})^{\frac{3}{2}} - (2\phi_f - V_{bs})^{\frac{3}{2}} \right] \\
&\left. - f_n \cdot \left[\frac{V'_{dsat}}{2} + (2\phi_f - V_{bs}) V'_{dsat} \right] \right\}.
\end{aligned}
\tag{6.97}
$$

The Modified "Lambda" Method

The Level 2 model provides several options to account for this phenomenon [7]. The simplest option is a modified version of the "λ-method" used in Level 1 (Section 5.2.3). For illustrative purposes, the solution for L' will be derived using the simpler Level 1 equations; the result in Level 2 is essentially the same. Recall the Level 1 equation for the drain current when $V_{ds} > V_{dsat}$ (5.25):

$$I_{ds} = \frac{\mu W_{eff} C_{ox}}{2L_{eff}} (V_{gs} - V_t)^2 (1 + \lambda \cdot V_{ds}). \tag{6.98}$$

The objective is to recast (6.98) so that the increase in I_{ds} when $V_{ds} > V_{dsat}$ is accounted for entirely by channel length modulation:

$$I_{ds} = \frac{\mu W_{eff} C_{ox}}{2(L_{eff} - L')} (V_{gs} - V_t)^2. \tag{6.99}$$

Setting (6.98) equal to (6.99) and solving for L' leads to

$$L' = \frac{\lambda V_{ds} L_{eff}}{1 - \lambda V_{ds}}.$$

(6.100)

Since $\lambda V_{ds} \ll 1$, (6.100) is simplified to

$$L' = \lambda V_{ds} L_{eff} = \mathbf{LAMBDA} \cdot V_{ds} \cdot L_{eff}.$$

(6.101)

In this simplest approach, the Level 2 method of accounting for channel length modulation is identical to that employed in Level 1. However, note that for $V_{ds} = V_{dsat}$, L' is **not** equal to zero, which is not the result expected from the earlier discussion. This indicates that the "λ-method" may be workable, but lacks physical accuracy.

If **LAMBDA** is not specified, the Level 2 model attempts to calculate it by reworking (6.101) into the form

$$\mathbf{LAMBDA} = \frac{L'}{V_{ds} L_{eff}}.$$

(6.102)

This requires some solution for L'. The simplest approach, due to Reddi and Sah [10], considers only the lateral field across the space charge region, which is treated as being fully depleted of free carriers. In this case, the distance L' is

$$L' = \frac{V_{ds} - V_{dsat}}{E_{yp}},$$

(6.103)

where E_{yp} is the field across the space charge region; since the space charge region is treated as being full depleted, E_{yp} is a constant. Equation (6.103) can be used to solve the Poisson equation in one dimension, yielding

$$L' = x_D \cdot (V_{ds} - V_{dsat})^{\frac{1}{2}},$$

(6.104)

where

$$x_D = \left(\frac{2\epsilon_{Si}}{q \cdot N_{sub}} \right)^{\frac{1}{2}} = \left(\frac{2\epsilon_{Si}}{q \cdot \mathbf{NSUB}} \right)^{\frac{1}{2}}.$$

(6.105)

Note that here, when $V_{ds} = V_{dsat}$, $L' = 0$ as expected.

This simple model has its shortcomings. As pointed out by Frohman-Bentchkowsky and Grove [11], it neglects the fringing fields between the gate electrode and, respectively, the channel and the drain (Figure 6.9). This causes the simple model (6.104) to overpredict the output conductance g_{ds}, where

$$g_{ds} = \frac{\partial I_{ds}}{\partial V_{ds}}.$$

(6.106)

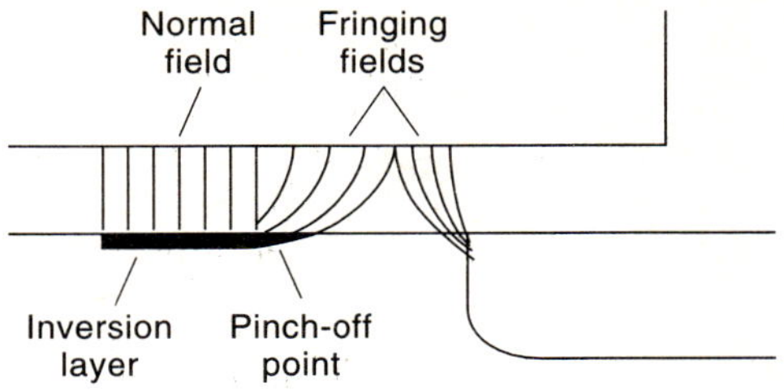

Figure 6.9 Fringing fields between the gate and, respectively, the channel and the drain.

Frohman-Bentchkowsky and Grove [11] developed a more complex model to address this problem.

In the interest of simplicity, the Level 2 model basically incorporates the simpler Reddi-Sah model. However, use of (6.104) and (6.102) for $V_{ds} \geq V_{dsat}$ will cause a discontinuity in both the drain current equation and in the first derivative (6.106) at $V_{ds} = V_{dsat}$ (at the linear to saturation transition point). The drain current discontinuity is particularly strange; while (6.104), as noted above, correctly produces $L' = 0$ for $V_{ds} = V_{dsat}$, the use of **LAMBDA** in (6.101) for $V_{ds} \leq V_{dsat}$ (linear region) produces a nonzero value for L'. The first derivative discontinuity requires some tedious mathematics, but can also be demonstrated. The Level 2 model solves these problems by using a very much ad hoc expression for L' that is loosely based on the Reddi-Sah model but appears to have no formal physical basis; it serves only to eliminate the discontinuities described above:

$$L' = x_D \left\{ \frac{V_{ds} - V_{dsat}}{4} + \left[1 + \left(\frac{V_{ds} - V_{dsat}}{4} \right)^2 \right]^{\frac{1}{2}} \right\}^{\frac{1}{2}}. \qquad (6.107)$$

Note, however, that the discontinuity problems are solved by once again employing a physically unrealistic solution in which $L' \neq 0$ at $V_{ds} = V_{dsat}$; more specifically, when $V_{ds} = V_{dsat}$, (6.107) yields $L' = x_D$.

This entire discussion points out the shortcomings of the "λ-method" for describing channel length modulation and the increase of I_{ds} for $V_{ds} > V_{dsat}$. Proper mathematical forms are only attainable by using physically unrealistic models for channel length modulation; as will be noted in Chapter 7, these deficiencies caused the "λ-method" to be discarded when the Level 3 model was developed. Clearly, a more physically reasonable description is required.

An Improved Semi-Empirical Method

A new and improved method for describing channel length modulation is also available in Level 2. This model employs a derived expression for L' proposed by Baum and Beneking [12], which is taken from a one-dimensional solution of the Poisson equation with the assumption that the space charge density in the pinched-off region is $q N_{sub}$:

$$L' = \left[\left(\frac{E_{sat}}{2a} \right)^2 + \frac{(V_{ds} - V_{dsat})}{a} \right]^{\frac{1}{2}} - \frac{E_{sat}}{2a}, \tag{6.108}$$

where E_{sat} is the field at the pinch-off point, and

$$a = \frac{q \cdot N_{sub}}{2\epsilon_s} = \frac{q \cdot \mathbf{NSUB}}{2\epsilon_s}. \tag{6.109}$$

Note that from a first principles point of view, the correct expression for E_{sat} is

$$E_{sat} = \frac{V_{dsat}}{L_{eff} - L'}. \tag{6.110}$$

However, (6.108) and (6.110) depend on each other; a solution would be found by alternately solving (6.108) and (6.110) until correct results for E_{sat} and L' are obtained. Any iterative approach is computationally inefficient in a circuit simulation environment.

In contrast to the complicated "brute force" approach used to compute the saturation voltage, the Level 2 model uses a simplified approach here. The saturation field is set as

$$E_{sat} = \frac{V_{dsat}}{L_{eff}}, \tag{6.111}$$

in which case E_{sat} is less than it should be when $V_{ds} > V_{dsat}$. Also, E_{sat} no longer depends on V_{ds} (through the V_{ds}-induced change in L'). In and of itself, this approach produces poor results for the output conductance [7].

This problem is corrected by making a semi-empirical alteration to (6.108), introducing a parameter to account for the behavior. (The introduction of any parameter in this situation is somewhat ad hoc; as will be described in Chapter 7, a parameter is introduced in a different location in (6.108) for use in the Level 3 model, with essentially the same result.) The Level 2 model attempts to maintain a physical connection by ascribing the poor output conductance result to a change in the total depletion charge from qN_{sub} to some fraction thereof, in this case $q \cdot N_{eff} \cdot N_{sub}$, where N_{eff} represents the ratio between the actual and expected depletion charge. With this modification, a in (6.109) becomes

$$a = \frac{q \cdot N_{eff} \cdot N_{sub}}{2\epsilon_s}. \tag{6.112}$$

Rewriting (6.112) with parameter names boldfaced leads to

$$a = \frac{q \cdot \mathbf{NEFF} \cdot \mathbf{NSUB}}{2\epsilon_s}. \tag{6.113}$$

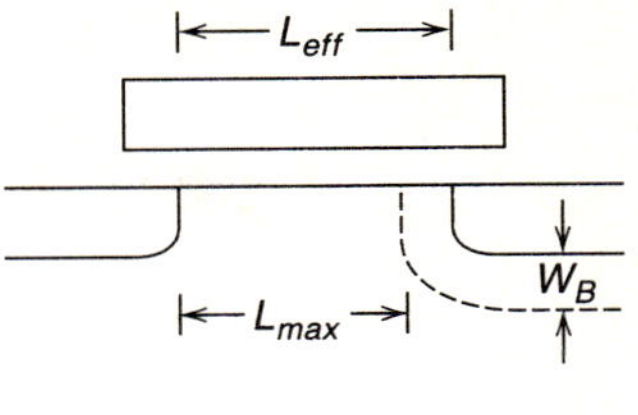

Figure 6.10 Description of the method used by the University of California/Berkeley implementation of SPICE which prevents L' from becoming larger than L_{eff} during the computation of channel length modulation.

This expression is used in (6.108) instead of the simpler expression (6.109). The implementation suggests a physical basis; however, the introduction of this parameter is entirely arbitrary, as is the approach used in the Level 3 model.

A major weakness of (6.108) is that it is possible for L' to become larger than L_{eff} (i.e., the inversion channel length becomes negative), which is physically impossible. In reality, the extension of the pinch-off region to the source side of the device places the FET into a punchthrough mode which is not accounted for by the Level 2 model. Instead, the Level 2 model includes a mathematical method of preventing L' from exceeding L_{eff}.

The Berkeley version of SPICE uses a scheme based on that depicted in Figure 6.10 [7]. The term W_B is the zero bias depletion region surrounding the drain diffusion (the source depletion region is neglected), allowing the definition of L_{max}, the maximum inversion channel length:

$$L_{max} = L_{eff} - W_B, \tag{6.114}$$

with

$$W_B = \sqrt{\frac{2\epsilon_{Si} \cdot \mathbf{PB}}{q \cdot \mathbf{NSUB}}}. \tag{6.115}$$

In this scheme, if the inversion channel length $L'_{eff} = L_{eff} - L'$ becomes less than W_B, L'_{eff} is instead computed from

$$L'_{eff} = \frac{W_B}{1 + \frac{L' - L_{max}}{W_B}}. \tag{6.116}$$

The largest value that L' can take on is L_{eff}; when this is inserted into (6.116) along with (6.114), the minimum value possible for L'_{eff} is found to be

$$L'_{eff} = \frac{W_B}{2}. \tag{6.117}$$

A somewhat different scheme is implemented in HSPICE [6].

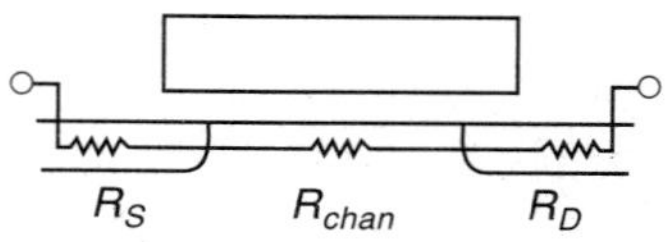

Figure 6.11 Lumped element approach to describing the series resistance.

6.5.4 The Source/Drain Series Resistance

The basic Level 2 drain current equation (6.97) was derived assuming that the conducting channel represents the major resistance element in the drain current path. However, as discussed in Chapter 3, in shorter channel FETs the resistance of the source and drain regions becomes significant when compared to the channel resistance; this phenomenon must be accounted for as another short channel correction.

Although the physical basis of the extrinsic series resistance is quite complex, the Level 2 model uses the simple lumped element approach depicted in Figure 6.11, in which two extra resistances due to the source and drain are added, yielding

$$R_{total} = R_{chan} + R_s + R_d. \tag{6.118}$$

In this model, part of the total drain to source bias (V_{ds}) is dropped across R_s and R_d; this has the effect of decreasing the voltage across the conducting channel, which reduces the drain current from its expected value. Since the extrinsic resistance is usually inversely proportional to the device width, this simple, lumped element approach has the drawback of not being useful for different device widths. Some simulators (such as HSPICE) have modified these models to allow for more general geometric applicability.

In a long channel device, the channel resistance is much larger than the extrinsic resistance, and (6.118) reduces to

$$R_{total} = R_{chan}. \tag{6.119}$$

The unmodified current model (6.97) should describe the device characteristics. In a short channel device at very low drain bias, (6.118) accounts for changes in the drain current due to short channel effects; any departure of the measured current from this predicted value is deemed to be due to series resistance. This explanation will become clearer below when parameter extraction is considered.

6.5.5 The Subthreshold Current Model

In the Level 1 model, the drain current is treated as being zero for any gate bias less than the threshold voltage; any subthreshold current is neglected. The Level 2 model introduces a separate current model for subthreshold conduction. While a worthwhile addition, the physical basis of the model is poor, which limits its usefulness.

The subthreshold current equation is based on the model of Swanson and Meindl [13], in which the current is described by

$$I_{ds} = I_{on}e^{q(V_{gs} - V_{on})/nk_bT}, \tag{6.120}$$

where I_{on} is the drain current at $V_{gs} = V_{on}$, as will be described below. The term V_{on} is a modified threshold voltage; as V_{gs} decreases below V_{on}, the current decreases exponentially. The introduction of V_{on} is an attempt to bridge the gap between the fully "on" region of device operation, where drift dominates conduction, and the subthreshold region, where diffusion dominates conduction. As described in Chapter 3, there is a gate bias region where neither mechanism dominates, which complicates analytical modeling; the term V_{on} is used in this region to maintain current continuity at the boundary between subthreshold and superthreshold conduction.

The use of V_{on} has the effect of ascribing all conduction to diffusion. This V_{on} is

$$V_{on} = V_t + n\frac{k_b T}{q}. \tag{6.121}$$

The term n is similar to the diode ideality factor, but is defined as

$$n = 1 + \frac{q \cdot \mathbf{NFS}}{C_{ox}} + \frac{C_{depl}}{C_{ox}}, \tag{6.122}$$

where **NFS** is the fast interface state density; any variation in the subthreshold behavior is ascribed to this parameter. As described in the discussion of parameter extraction, good results for the subthreshold region are found only by allowing **NFS** to take on very large and physically unrealistic values. Thus, **NFS** can be regarded as a purely empirical fitting constant. Another problem is that **NFS** has no drain or substrate bias dependence, and no temperature dependence, further limiting its usefulness.

The final term in (6.122) can be defined as

$$\frac{C_{depl}}{C_{ox}} = \frac{1}{C_{ox}}\frac{\partial Q_{depl}}{\partial V_{bs}}. \tag{6.123}$$

Recalling (6.34),

$$\frac{Q_{depl}}{C_{ox}} = f_s \cdot \gamma \cdot (2\phi_f - V_{bs})^{\frac{1}{2}} + f_n \cdot (2\phi_f - V_{bs}) \tag{6.124}$$

leads to

$$\frac{C_{depl}}{C_{ox}} = \frac{1}{C_{ox}}\left[\frac{f_s \cdot \gamma \cdot (2\phi_f - V_{bs})^{\frac{1}{2}}}{2(2\phi_f - V_{bs})} - f_n\right]. \tag{6.125}$$

The Level 2 implementation simplifies (6.125) by employing a common denominator; it contains an error of dropping a "2" during the simplification. Thus, (6.125) should be

$$\frac{C_{depl}}{C_{ox}} = \frac{1}{C_{ox}}\left[\frac{f_s \cdot \gamma \cdot (2\phi_f - V_{bs})^{\frac{1}{2}} - 2f_n \cdot (2\phi_f - V_{bs})}{2(2\phi_f - V_{bs})}\right], \tag{6.126}$$

but is instead actually implemented as

$$\frac{C_{depl}}{C_{ox}} = \frac{1}{C_{ox}} \left[\frac{f_s \cdot \gamma \cdot (2\phi_f - V_{bs})^{\frac{1}{2}} - f_n \cdot (2\phi_f - V_{bs})}{2(2\phi_f - V_{bs})} \right]. \tag{6.127}$$

Substituting (6.127) into (6.122) leads to the final expression for the ideality factor:

$$n = 1 + \frac{q \cdot \mathbf{NFS}}{C_{ox}} + \frac{1}{C_{ox}} \left[\frac{f_s \cdot \gamma \cdot (2\phi_f - V_{bs})^{\frac{1}{2}} - f_n \cdot (2\phi_f - V_{bs})}{2(2\phi_f - V_{bs})} \right]. \tag{6.128}$$

With V_{on} defined by (6.121), I_{on}, the drain current at $V_{gs} = V_{on}$, can now be considered. The condition is stipulated that the drain current must be continuous at $V_{gs} = V_{on}$. At $V_{gs} = V_{on}$, from (6.120), $I_{ds} = I_{on}$. To provide continuity, each superthreshold equation is forced to equal I_{on} when $V_{gs} = V_{on}$; this allows I_{on} to be defined. In the linear region, substituting $V_{gs} = V_{on}$ into (6.53) yields

$$
\begin{aligned}
I_{on} = \frac{\mu W_{eff} C_{ox}}{L_{eff}} \Bigg\{ & (V_{on} - (V_{fb} + 2\phi_f))V_{ds} - \frac{V_{ds}^2}{2} \\
& - \frac{2}{3} f_s \cdot \gamma \cdot \left[(V_{ds} + 2\phi_f - V_{bs})^{\frac{3}{2}} - (2\phi_f - V_{bs})^{\frac{3}{2}} \right] \\
& - f_n \cdot \left[\frac{V_{ds}^2}{2} + (2\phi_f - V_{bs})V_{ds} \right] \Bigg\}.
\end{aligned}
\tag{6.129}
$$

In the saturation region, substituting $V_{gs} = V_{on}$ into (6.97) leads to

$$
\begin{aligned}
I_{on} = \frac{\mu W_{eff} C_{ox}}{L_{eff} - L'} \Bigg\{ & \left[(V_{on} - (V_{fb} + 2\phi_f)) \right] V'_{dsat} - \frac{V'_{dsat}}{2} \\
& - \frac{2}{3} f_s \cdot \gamma \cdot \left[(V_{ds} + 2\phi_f - V_{bs})^{\frac{3}{2}} - (2\phi_f - V_{bs})^{\frac{3}{2}} \right] \\
& - f_n \left[\frac{V'_{dsat}}{2} + (2\phi_f - V_{bs})V'_{dsat} \right] \Bigg\}.
\end{aligned}
\tag{6.130}
$$

Note that when either (6.129) or (6.130) is substituted into (6.120), I_{on} is a constant for all gate biases when $V_{gs} < V_{on}$. Also note that the current is guaranteed to be continuous at $V_{gs} = V_{on}$, but that the first derivative is not continous at that point; this can cause convergence problems during circuit simulation [14]. Finally, note that the use of V_{on} defines which drain current is used in the model. For $V_{gs} < V_{on}$, the subthreshold equation (6.120) is used, while for $V_{gs} > V_{on}$, either the linear region expression (6.53) or the saturation region equation (6.97) is employed.

6.6 The Charge Model

As described in Chapter 5, the models implemented in SPICE to describe the active gate capacitance are essentially independent of each particular FET current model. Each FET model can yield a charge model which is used in some manner (at the circuit simulation level) to compute the gate capacitance. The intention in this text is to present the node charge model which is developed from each particular current model with that current model, while postponing a general discussion of gate capacitance computations to Chapter 13.

Among the FET models, the Level 2 model is unique in that its current model is **not** used to derive a node charge model. It was decided that the complexity of the Level 2 model structure would lead to a computationally inefficient charge model; the simpler Level 1 charge model, derived in Section 5.3, is used instead. For completeness, the final charge equations are repeated here. In all cases, the charge neutrality condition

$$Q_{GATE} + Q_{INV} + Q_{DEPL} = 0 \qquad (6.131)$$

must be satisfied; the capitalized subscripts indicate *total* charge rather than charge per unit area. By computing two of the three expressions (Q_{GATE} and Q_{DEPL}) in (6.131), the third (Q_{INV}) is then found.

The Linear Region

In the linear region,

$$Q_{DEPL} = W_{eff} \cdot L_{eff} \cdot C_{ox} \cdot \gamma \cdot (2\phi_f - V_{bs})^{\frac{1}{2}}, \qquad (6.132)$$

and

$$Q_{GATE} = \frac{2}{3} W_{eff} L_{eff} C_{ox} \left[\frac{(V_{gd} - V_t)^3 - (V_{gs} - V_t)^3}{(V_{gd} - V_t)^2 - (V_{gs} - V_t)^2} \right] - Q_{DEPL}. \qquad (6.133)$$

The Saturation Region

In the saturation region, the depletion charge is given by (6.132), while the gate charge is

$$Q_{GATE} = \left[\frac{2}{3} W_{eff} L_{eff} C_{ox} (V_{gs} - V_t) \right] - Q_{DEPL}. \qquad (6.134)$$

The Subthreshold Region

In the subthreshold region, the inversion charge Q_{INV} is taken to be negligible and is set to zero. The depletion charge is found from

$$Q_{DEPL} = W_{eff} \cdot L_{eff} \cdot C_{ox} \cdot \gamma \cdot (\phi_s - V_{bs})^{\frac{1}{2}}, \qquad (6.135)$$

so the gate charge is simply

$$Q_{GATE} = -Q_{DEPL}. \tag{6.136}$$

6.6.1 Charge Partitioning

As described in Chapter 13, the Level 1 FET charge model was used in the Meyer model [15] to compute the active gate capacitance. This model assumes that the node capacitance elements are all reciprocal (e.g., $C_{gd} = C_{dg}$), and computes the gate capacitance elements by considering only the gate charge. While this method is computationally efficient and adequate for most digital circuits, it became known that during transients in some circuits, the charge neutrality requirement (6.131) was not maintained. This is referred to as the inability to conserve charge.

Although the node charge model is that of Level 1, the Level 2 implementation in SPICE made use of the newer Ward-Dutton model [16], also described in Chapter 13. This model treats the capacitive elements as nonreciprocal (e.g., $C_{gd} \neq C_{dg}$), and uses the charge (rather than the voltage) for the description of the situation at the device terminals. To accomplish this, a penalty is paid in computational efficiency.

In the Meyer model, the inversion charge Q_{INV} was taken to be a single element. In the Ward-Dutton model, it is necessary to split up Q_{INV}, and assign it in portions to the source and the drain nodes. This is done by introducing the parameter **XQC**, defined by

$$Q_D = \mathbf{XQC} \cdot Q_{INV}, \tag{6.137}$$

and

$$Q_S = (1 - \mathbf{XQC}) \cdot Q_{INV}. \tag{6.138}$$

The default value of **XQC** is 0.5, representing equal charge partitioning. Often, **XQC** is specified as 0.4, as Ward and Dutton [16] found the best agreement with numerical simulations with this value of **XQC**.

Selecting a Gate Capacitance Model

When the Level 2 model was implemented in Berkeley SPICE, the Meyer model was discarded and replaced with the Ward-Dutton model [7]. HSPICE [6] allows the user to choose either the Meyer model or the Ward-Dutton model. Since the model parameter set is unaffected by the selection (other than by the need for **XQC** in the Ward-Dutton model), this choice is usually made by the circuit designer, who can choose the Meyer model when computational efficiency is the dominant consideration, or the Ward-Dutton model when more detailed accuracy is required.

6.6.2 The Zero Bias Gate Capacitance

In a large FET, the zero bias capacitance between the gate and the other nodes could be neglected. In smaller FETs, this capacitance is more significant and must be included. As in the Level 1 model, the Level 2 model contains the zero bias capacitance parameters **CGSO**, **CGDO**, and **CGBO**; these parameters represent, respectively, the zero bias gate-source, gate-drain, and gate-bulk capacitances. More specifically, **CGSO** and **CGDO** represent the capacitance per unit length (in F/m) across the width of the device at the gate edge, where it overlaps each diffusion; the total capacitance is found by multiplying **CGSO** and **CGDO** by the effective channel width W_{eff}. Meanwhile, **CGBO** is the capacitance per unit length (in F/m) along the channel direction; the total capacitance is found by multiplying **CGBO** by the effective channel length L_{eff}. These parameters are examined more closely in Chapter 13.

(There are deficiencies in both the Meyer model and the Ward-Dutton model in submicron devices. To compensate, **CGSO** and **CGDO** are often treated as variable parameters, allowing FET models to match some particular set of representative circuits, such as a group of ring oscillators. This situation is considered in more detail in Chapters 13 and 15.)

6.7   Parameter Extraction and Model Development

Given the mathematical structure of the Level 2 model, that model must now be exercised in parameter extraction, to demonstrate its capabilities and shortcomings. Due to the similarity of the Level 2 model to the more widely used Level 3 model, the reader is encouraged to compare the extraction results here with those of Level 3, which may be found in Section 7.7.

6.7.1   Basic Parameter Extraction

In contrast to the very simple Level 1 model, the Level 2 model is relatively complex. As a result, parameters cannot be extracted using simple linear least squares (slope-intercept) line fitting. Instead, a more involved method, iterative nonlinear least squares fitting, is employed. This approach is numerically intensive, but is appropriate for the situation.

The basic structure of the Level 2 model indicates how parameter extraction will proceed. A long, wide FET is taken as providing the base model, and small geometry effects are added to this model as corrections. In addition, short and narrow channel effects are separated from each other in the model; thus, the extraction procedure will also treat them separately.

The basic sequence of parameter extraction is as follows [17]:

- The oxide thickness, as well as the differences between the drawn and effective channel dimensions, are provided as process input.
- A long, wide FET is used to extract the base parameters.
- Parameters are extracted to describe the behavior of short and narrow devices at low drain bias.
- Parameters which describe high drain bias operation are extracted.
- A subthreshold parameter is extracted.

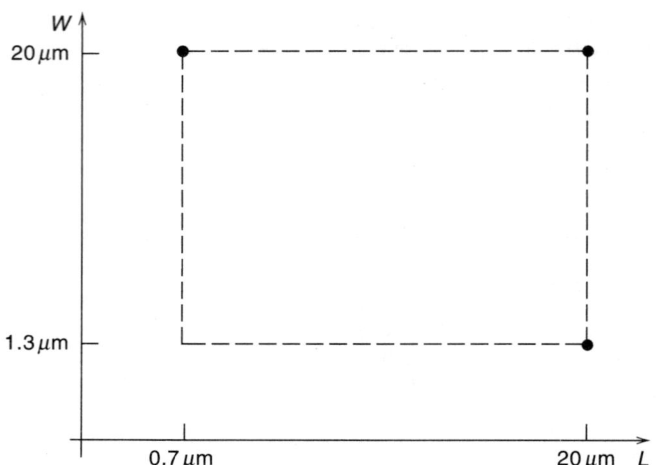

Figure 6.12   The geometric scheme for extracting the Level 2 model parameters. A long, wide device is used to extract the basic parameters, while a short, wide device and a long, narrow device are used to extract the small geometry corrections.

Three FETs are required for model construction, as shown in Figure 6.12: (1) a long, wide FET for extraction of the base parameters; (2) a wide, short device for determination of the short channel parameters; (3) a long, narrow device for the extraction of the narrow channel parameters. In this example, the device geometries employed are 20 μm/20 μm, 20 μm/0.7 μm, and 1.3 μm/20 μm. The gate oxide thickness is 13.0 nm. The use of submicron FETs will clearly demonstrate the capabilities and shortcomings of the Level 2 model.

First, the base parameters **VTO**, **UO**, **UCRIT**, and **UEXP** are extracted using the long, wide device at low drain bias. By using a large device with a low drain bias, all small geometry effects are negated. The results of the extraction are depicted in Figure 6.13. The results are quite good over most of the gate bias range. However, note that the model begins to fail for the highest gate biases.

Next the source and drain series resistance parameters **RS** and **RD** are determined. In most FETs, the source and drain contacts are equidistant from the gate edge, and with a self-aligned source/drain process, it is reasonable to assume that **RS** = **RD**. In the original implementation of FET models in Berkeley SPICE, **RS** and **RD** were treated as lumped resistances; in truth, the series resistance scales with channel width in the form

$$\mathbf{RS} = \mathbf{RD} = \mathbf{R}/W_{eff}, \tag{6.139}$$

where **R** is independent of the channel width. Some simulators, such as HSPICE [6] allow **R** to be specified as a model parameter, so that **RS** and **RD** are computed for each device width. If the simulator does not allow this, different values of **RS** and **RD** must be computed for each channel width which is used.

The results of this extraction, using a wide, short device with a low drain bias, are depicted in Figure 6.14. The resulting values of **RS** and **RD** are lumped resistance values, which are valid only when $W = 20$ μm. To find values for other channel lengths, (6.139) must be employed.

Next the narrow channel parameter **DELTA** is extracted from a long, narrow device with a low drain bias; the results are presented in Figure 6.15. Note that **DELTA** is the only parameter in the Level 2 model which accounts for narrow channel behavior.

Now two steps are used to describe the effect of the substrate bias on the device behavior. When a substrate bias is applied, the bulk charge is changed, and thus the threshold voltage is also changed. As a result, the current-voltage characteristics are altered. A long, wide device with a low drain bias and several substrate biases is used to determine the parameter **NSUB**. The larger the value of **NSUB**, the larger will be the change in the current-voltage characteristics under an applied substrate bias.

The results of the extraction are shown in Figure 6.16. The results are relatively good, especially considering that the Level 2 structure assumes the substrate doping, as represented by **NSUB**, to be uniform; this is rarely the case in modern FETs, which have shallow channel impurity concentration adjustments to control the threshold voltage. As Figure 6.16 indicates, the effects of assuming a uniform substrate doping are usually minor, and a good result is produced. It is also worthwhile to contrast the Level 2 extraction with the identical step in the Level 3 model, as depicted in Figure 7.13. The Level 2 mobility model includes the substrate bias (through the use of the full threshold voltage expression), while the Level 3 model neglects the substrate bias (and employs the parameter **VTO** rather than the full threshold voltage expression). As a result, the Level 2 result is actually superior to the Level 3 result.

An additional extraction step is required for the description of the effect of the substrate bias on short channel devices. This is done by determining the parameter **XJ** using a wide, short device with a low drain bias. In older models (such as Level 1), the junction depth was used to describe the lateral encroachment of the source and drain depletion regions into the channel area, where they overlapped with the gate-induced depletion region. However, in the Level 2 formulation, this lateral encroachment is instead described by the model parameter **LD**. Thus, **XJ** does **not** represent the physical junction depth, but is instead an empirical description of the effect of a short channel length on the substrate sensitivity. The results of the extraction are depicted in Figure 6.17. The extracted value of **XJ** of 77.7 nm clearly shows that this parameter does not contain a physical value for the junction depth.

Next the parameters which describe high drain bias effects are determined. In the Level 2 model, two parameters are employed. The parameter **VMAX** represents the carrier saturation velocity, which has the effect of decreasing the saturation voltage, while the parameter **NEFF** describes the reduction of the depletion charge due to channel length modulation. A wide, short device with a high drain bias is used, producing the results shown in Figure 6.18. These results indicate a serious shortcoming of the Level 2 model, which becomes extreme in submicron FETs. There is a severe first derivative discontinuity at the linear-saturation transition point, which is clearly visible as a kink in the current-voltage characteristics. This is due to the computed reduction of the saturation

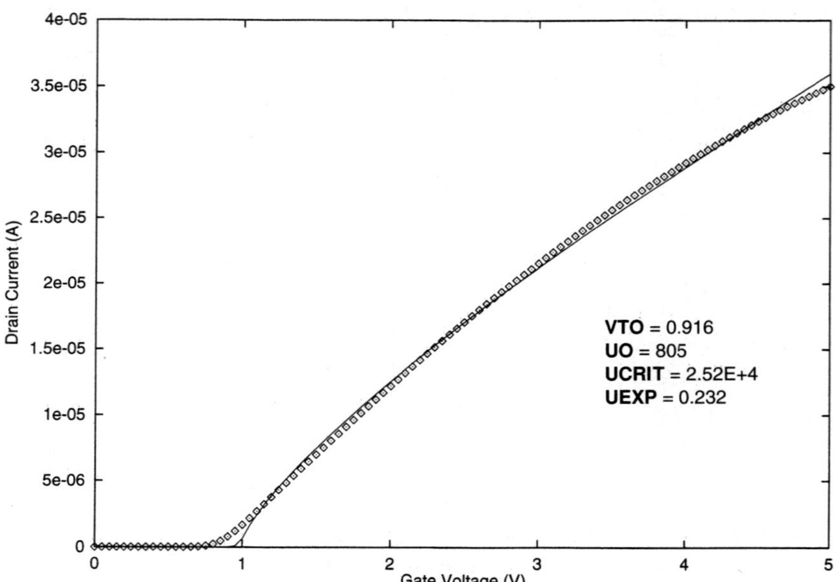

Figure 6.13 The extraction of the base device parameters **VTO, UO, UCRIT,** and **UEXP,** $W/L = 20$ μm/20 μm.

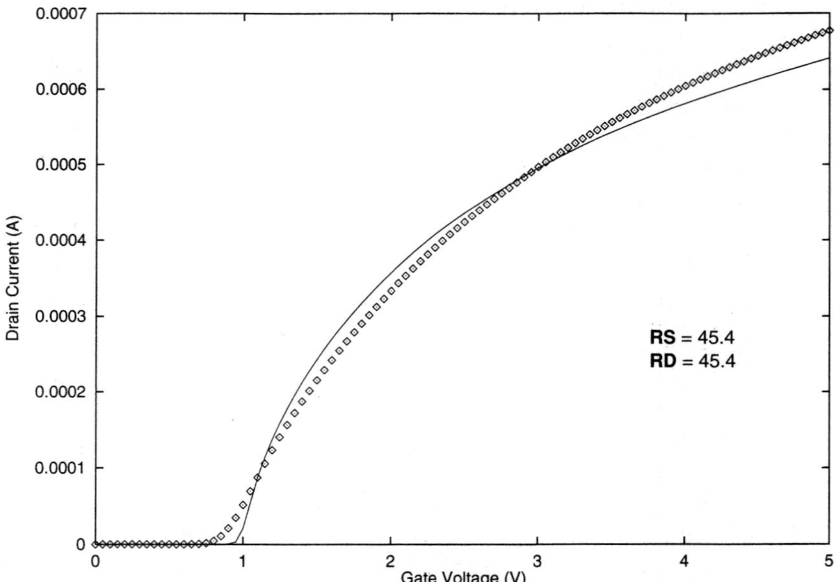

Figure 6.14 The extraction of the parasitic series resistance parameters **RS** and **RD,** $W/L = 20$ μm/0.7 μm.

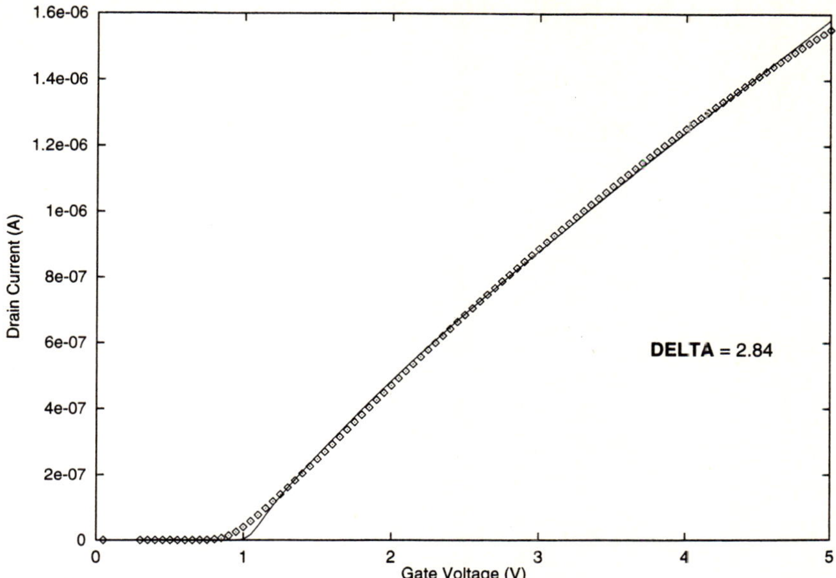

Figure 6.15 The extraction of the parameter **DELTA**, which accounts for the spreading of the gate-induced depletion region outside of the channel edges, $W/L = 1.3$ μm/20 μm.

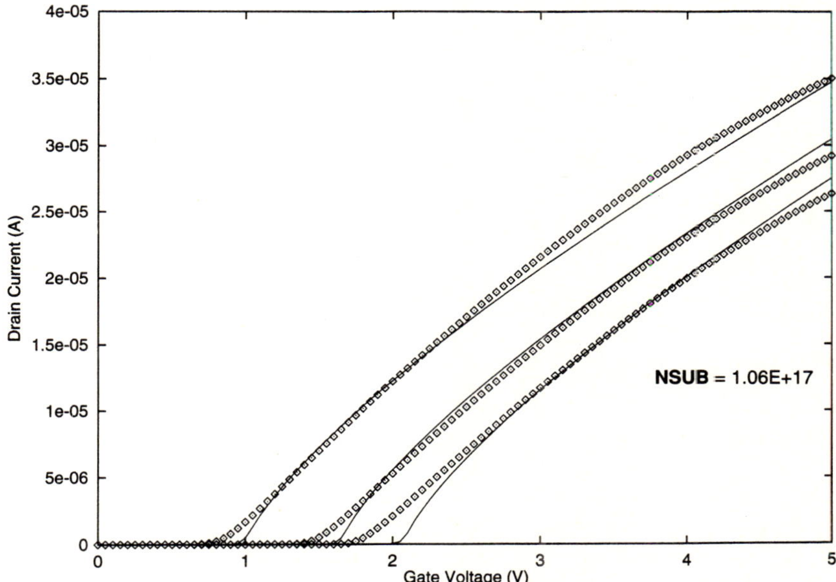

Figure 6.16 The extraction of the substrate sensitivity/doping parameter **NSUB**, $W/L = 20$ μm/20 μm.

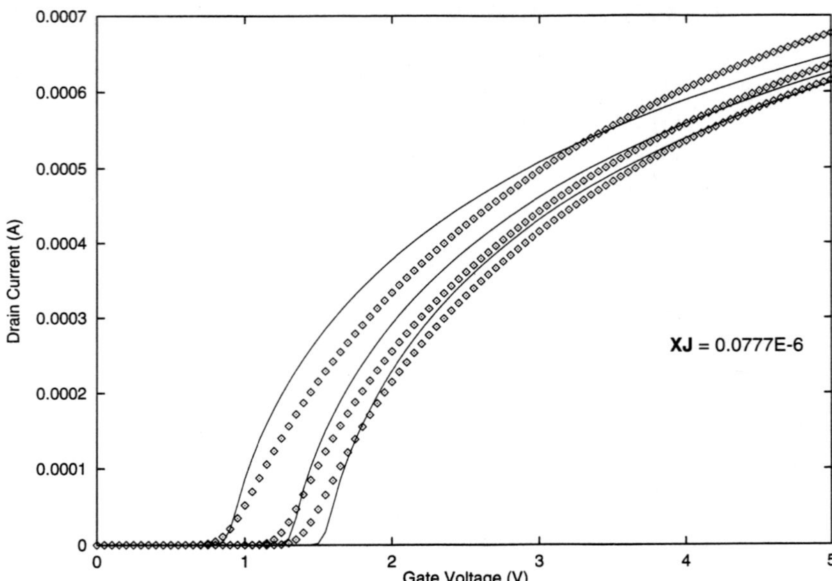

Figure 6.17 The extraction of the parameter **XJ**, which is **not** the junction depth, but the short channel correction to the substrate sensitivity, $W/L = 20$ μm/0.7 μm.

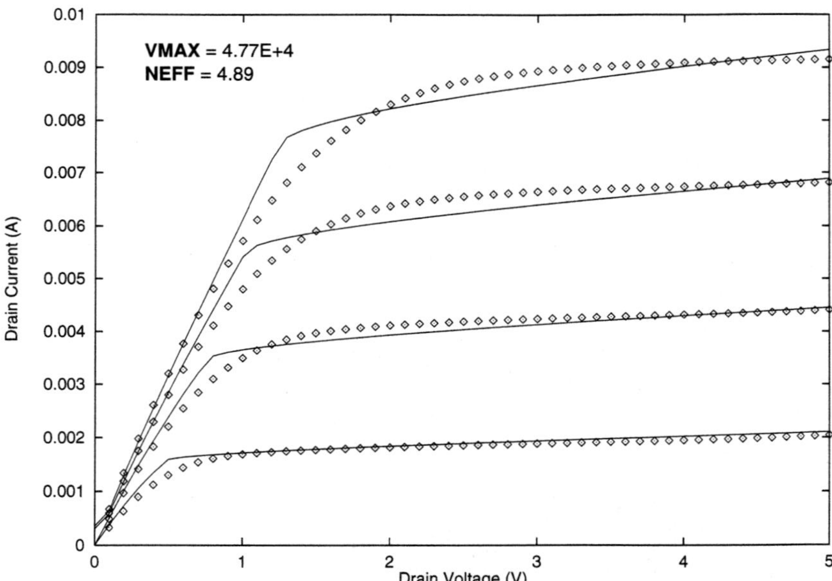

Figure 6.18 The extraction of the high drain bias parameters **VMAX** and **NEFF**, $W/L = 20$ μm/0.7 μm. Note the severe first derivative discontinuity in the model at $V_{ds} = V_{dsat}$.

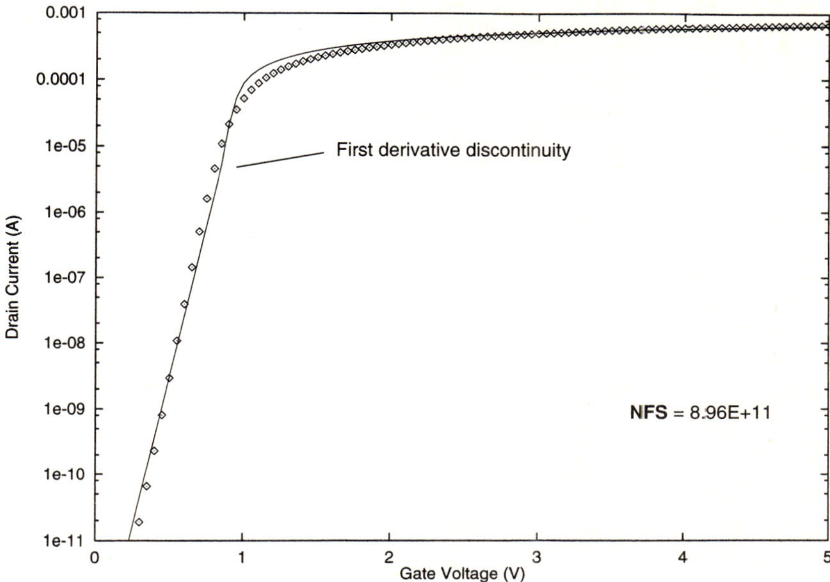

Figure 6.19   The extraction of the subthreshold region fitting parameter **NFS**, $W/L = 20$ μm/0.7 μm.

voltage by carrier velocity saturation. The difference between the classical (no velocity saturation) value of the saturation voltage and the adjusted (velocity saturation) value becomes large; this forces a sharp change from the linear region model to the saturation region model, which causes the severe kink.

It should be noted that this poor result is obtained despite the very involved description of the saturation voltage which the Level 2 model employs (see Section 6.5.2). This should be compared to the Level 3 result, depicted in Figure 7.15. In Level 3, a better result is obtained with a much simpler and more efficient model for the saturation voltage. It should also be noted that the extracted value of **VMAX** $= 4.77 \times 10^4$ m/s is roughly half of the generally accepted value of the carrier saturation velocity. Thus, **VMAX** must be regarded as a fitting parameter.

Finally, the lone parameter for the description of the subthreshold region can be determined. As the subthreshold equation (6.120) indicates, with only one parameter available (**NFS**), the model will be quite limited; only one combination of geometry, temperature, and biases can be chosen. The results of the extraction are presented in Figure 6.19. A reasonable fitting outcome is obtained, although the first derivative discontinuity at $V_{gs} = V_{on}$ can be clearly seen. The extracted value of **NFS** $= 8.96 \times 10^{11}$ cm^{-2} is much larger than the physical value of less than 10^{10} cm^{-2} which is found in modern MOS devices; **NFS** has strictly empirical meaning. Since the Level 2 subthreshold model is very limited, it is common to merely ignore the subthreshold part of the model, as this region is not very important to most digital circuit designs; the same statement applies to the Level 3 model as well. As is also the case with Level 3, if a good subthreshold model is required, this indicates that the Level 2 model will in general be inadequate; an FET model with improved overall detail should be employed.

The complete set of extracted model parameters is tabulated in Table 6.2.

Table 6.2 Final model parameter set (including both provided process parameters and extracted electrical parameters).

Parameter	Value
TOX	13.0E-09
LD	0.120E-6
WD	0.200E-6
VTO	0.916
UO	805
UCRIT	2.52E+4
UEXP	0.232
RS	45.4
RD	45.4
DELTA	2.84
NSUB	1.06E+17
XJ	0.0777E-06
VMAX	4.77E+04
NEFF	4.89
NFS	8.96E+11

6.7.2 Model "Playback"

Once the model has been constructed, it is useful to compare the simulated results with device data. This will be useful for further demonstrating the strengths and weaknesses of the Level 2 model, while indicating how the parameter extraction process affects the final results.

In Figure 6.20, the final model is compared with data for the linear and saturation regions of the 20/20 device. Rather poor results are obtained; this may be a particular surprise for the linear region, as the first extraction step involved fitting to these data. This highlights an important aspect of the Level 2 model structure. The long, wide device is used in the first step to extract a set of base parameters for the model; this long, wide device is treated as being infinitely long and wide. However, this is only an approximation; when the short channel parameters are included in the model, the result for the long, wide device is corrupted. The long, wide device is used to build the model, but in the end is not properly described by that model. Some iterative scheme could be undertaken to synthesize a model result that would provide a better *overall* result; however, this would be done at the price of decreasing the quality of the short channel model, where the highest accuracy is desired. Since the first-generation models should

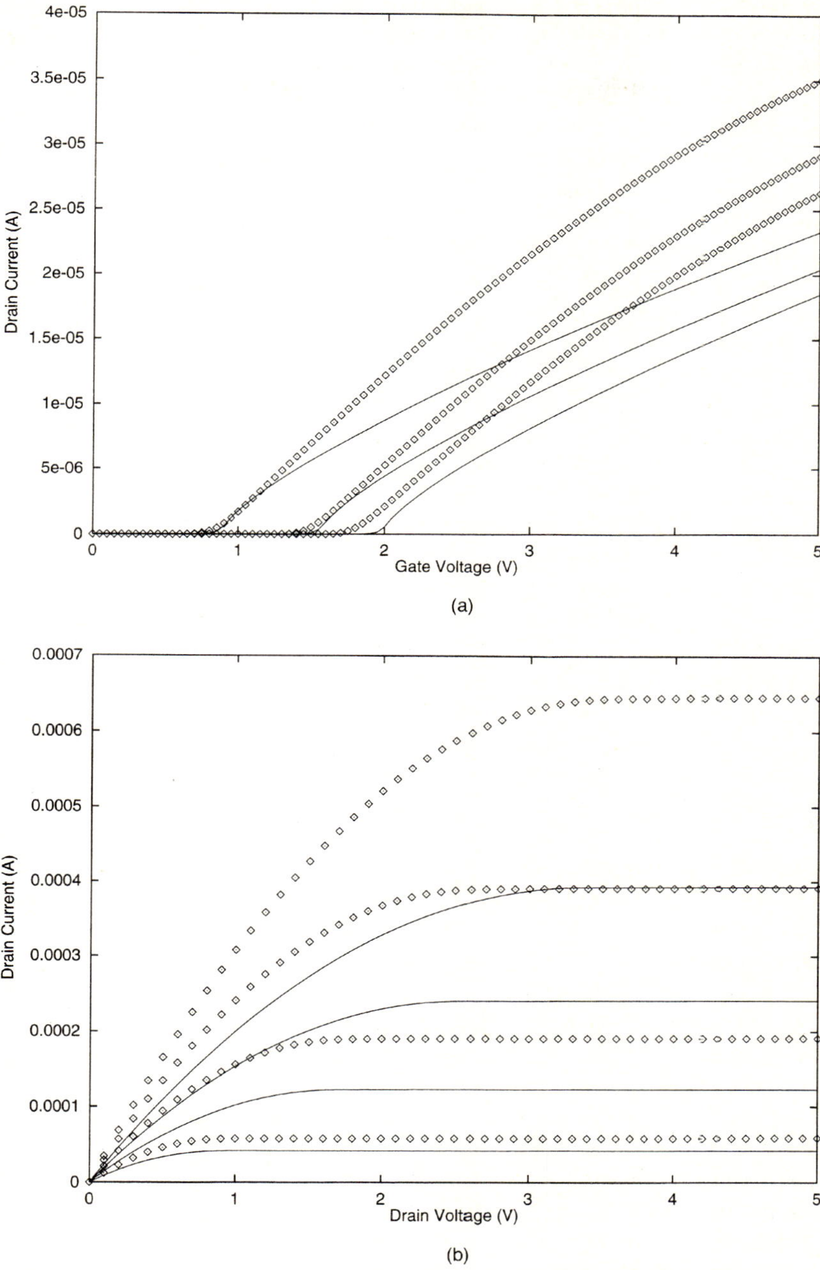

Figure 6.20 Comparison of measured data and the final model for the 20/20 device; (a) linear characteristic; (b) saturation characteristic.

only be used for digital circuit design, the final accuracy for the long, wide device should not be a consideration. If accuracy *is* required for large geometries, the circuit design likely has other requirements which the Level 2 model will be unable to meet.

Final model results for the 1.3/20 device are presented in Figure 6.21. When compared to the extraction step of Figure 6.15, it can be seen that the zero substrate bias linear region characteristic has been slightly corrupted by the short channel extraction, while the nonzero substrate bias results are somewhat mediocre. The saturation characteristics are acceptable for most situations in which the Level 2 model might be used. In the case of the 20/0.7 device (Figure 6.22), since the final extraction steps were carried out on this device, the results are unchanged from Figure 6.17 and 6.18. Once again, it is important to note the very severe first derivative discontinuity at the linear-saturation transition.

Thus far, model-data comparisons have only been made for device geometries which were used for parameter extraction. It is also important to examine other device geometries, to provide insight into the general model behavior.

Figures 6.23–6.25 contain model and data results for device geometries of, respectively, 20/0.9, 2.0/20, and 1.3/0.7. Overall, the model provides a less accurate description at these device dimensions. By targeting specific device geometries (in this case, $L = 0.7$ μm and $W = 1.3$ μm), parameter extraction is able to overcome the various model shortcomings and produce reasonable results. However, when the geometry moves away from these two anchor points, the results begin to deteriorate. For the 20/0.9 and 2.0/20 devices, one anchor point is maintained, and the results are not seriously compromised. This is an important consideration in digital designs which employ a single drawn channel length (e.g., for this situation, $L_{drawn} = 0.7$ μm), since small changes in the channel length will occur due to process variations; the model should be able to cope with this requirement. When a short, narrow device is examined, the model results are quite poor.

A solution to this problem is to build submodels which are only valid for certain device geometries; this use of model binning is considered in detail in Chapter 7.

6.8   Final Comments

As noted throughout this chapter, the Level 2 model was developed as the first attempt to describe the behavior of small geometry MOSFETs. The approach taken set the tone for succeeding developments, in which the original large geometry FET model was supplemented with corrections for small geometry effects. However, the Level 2 model was constructed in a mathematically complex form; the model is slow and inefficient, and frequently encounters convergence problems. For these reasons, the Level 2 model has largely fallen into disuse, and has been replaced by the similar but simpler and more robust Level 3 model.

The most important shortcomings of the Level 2 model will be briefly recapitulated here. The short channel depletion charge model only takes account of the overlap of the source and drain depletion regions with the gate-induced depletion region; charge sharing

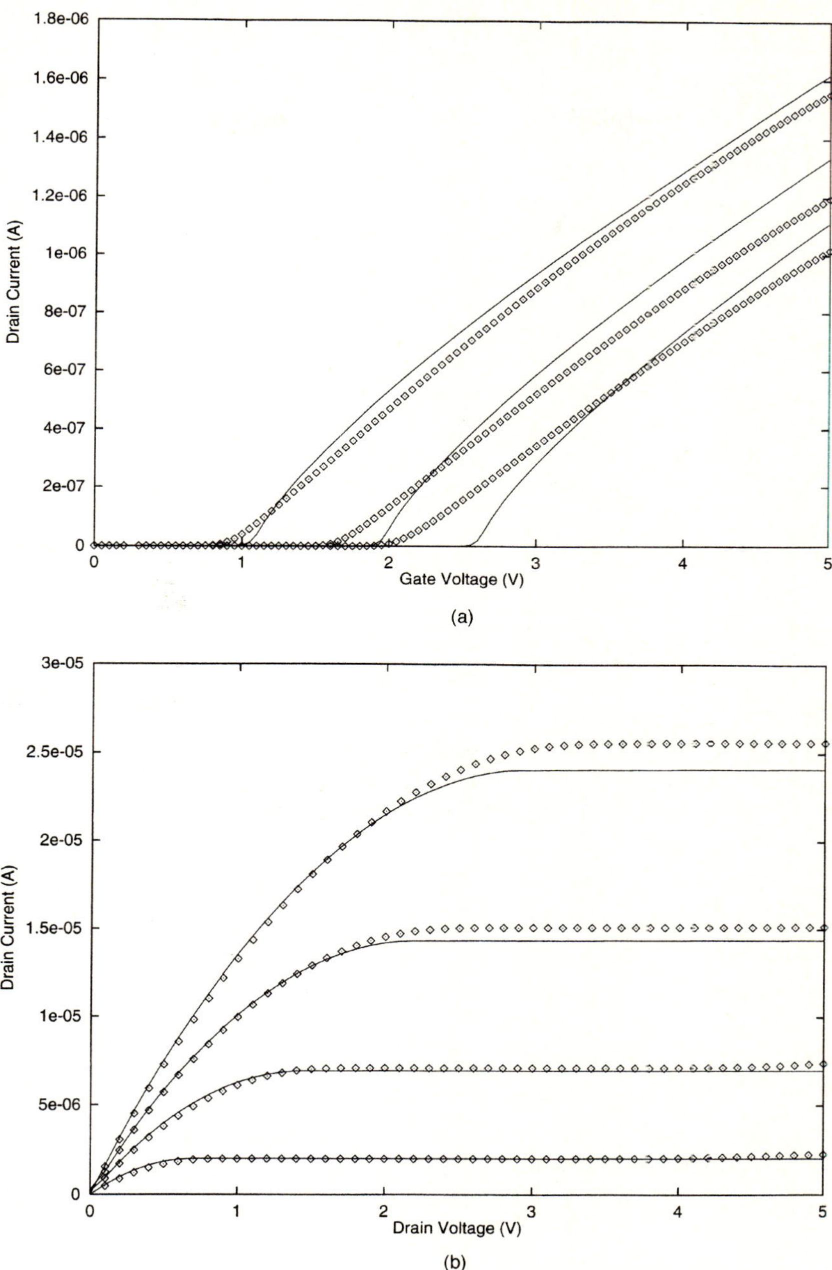

Figure 6.21   Comparison of measured data and the final model for the 1.3/20 device; (a) linear characteristic; (b) saturation characteristic.

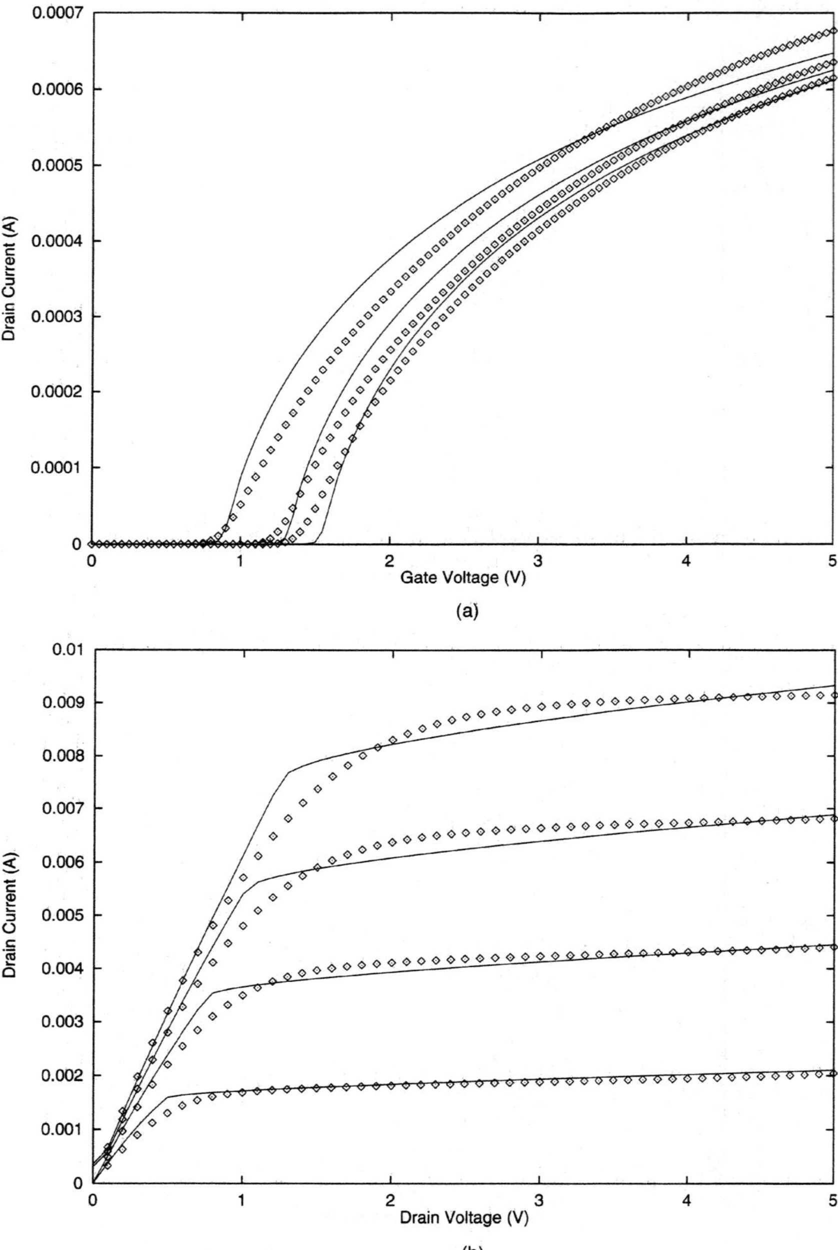

Figure 6.22 Comparison of measured data and the final model for the 20/0.7 device; (a) linear characteristic; (b) saturation characteristic.

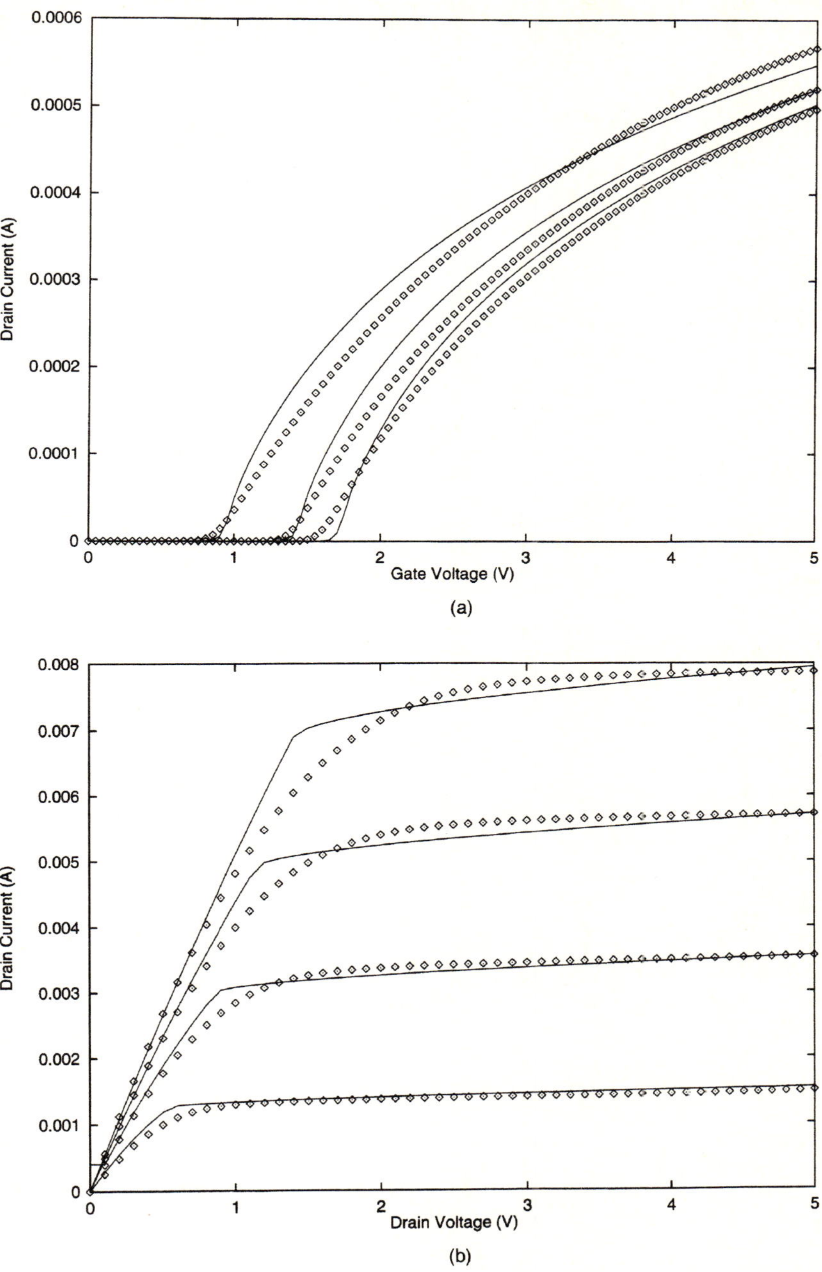

Figure 6.23 Comparison of measured data and the final model for the 20/0.9 device; (a) linear characteristic; (b) saturation characteristic.

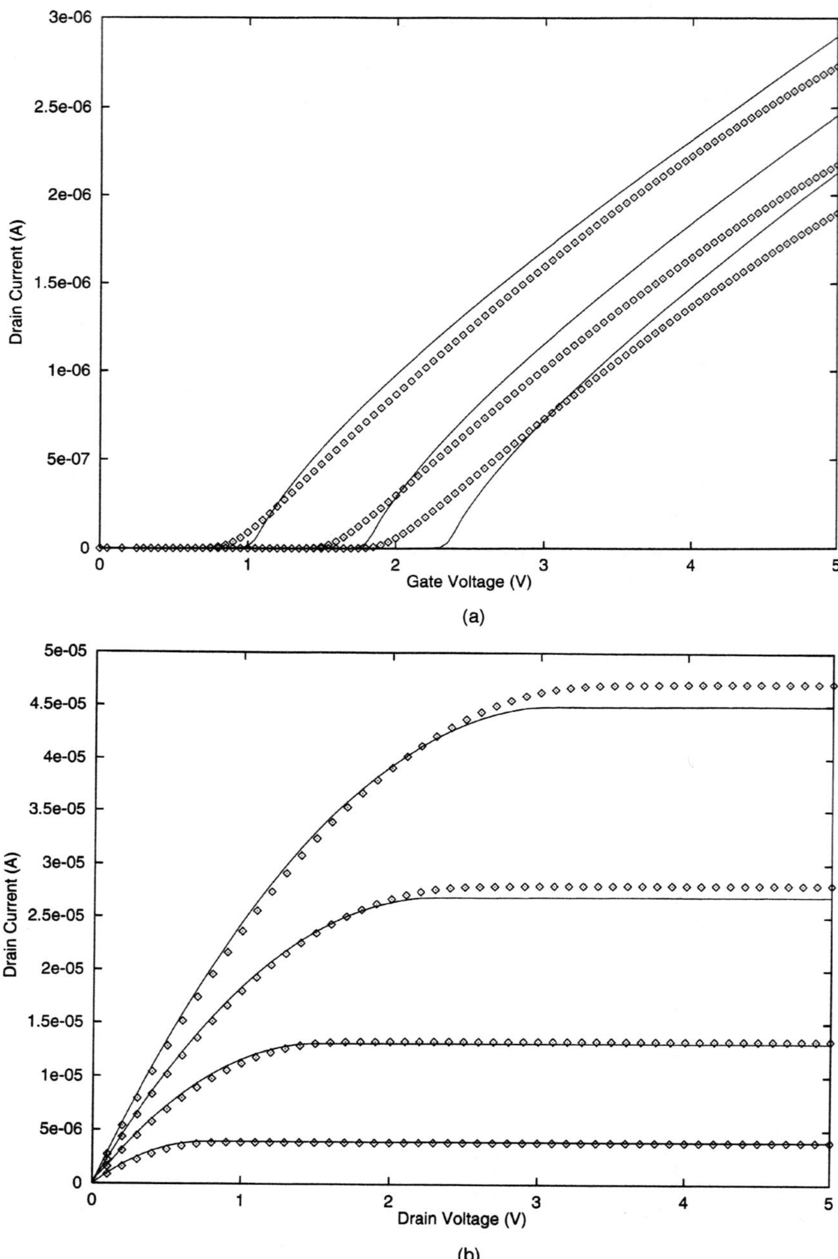

Figure 6.24 Comparison of measured data and the final model for the 2.0/20 device; (a) linear characteristic; (b) saturation characteristic.

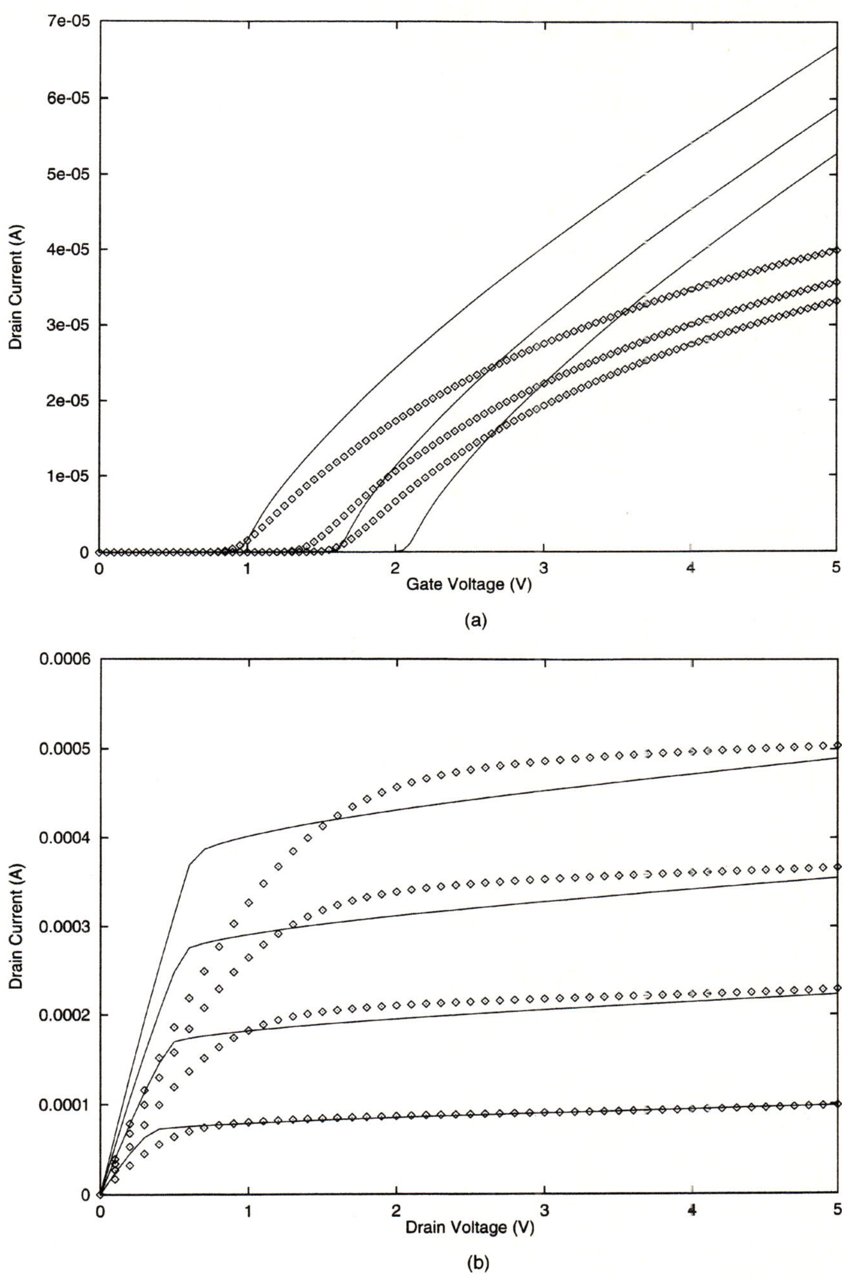

Figure 6.25 Comparison of measured data and the final model for the 1.3/0.7 device; (a) linear characteristic; (b) saturation characteristic.

between the source and drain is not considered. The threshold voltage model includes only this simple depletion charge model, and does not include any terms to account for drain-induced barrier lowering (DIBL). The channel mobility model, while workable, uses a rather complicated critical field method; during parameter extraction, one of the model parameters (**UCRIT**) is usually deemed redundant and set at a fixed value instead of being extracted. The parameter **UTRA**, which describes the effect of the drain bias on the vertical field reduction of the mobility, can cause negative resistance for drain biases slightly larger than the saturation voltage [6]. Since this can cause convergence problems, **UTRA** was never even coded in the Berkeley SPICE implementation of Level 2, and is thus de facto forced to be zero. Note that **UTRA** is available in the HSPICE implementation of Level 2 [6], but the risk of using this parameter far outweighs any possible improvements in the model accuracy.

Perhaps the most bizarre aspect of the Level 2 model is the saturation voltage model, with its many possible choices. The most commonly employed approach is to extract **VMAX** as a parameter; in this case, a quartic expression for the saturation voltage V_{dsat} is used. Four possible values of V_{dsat} occur, requiring one to be selected. If no acceptable value is found, a "back-up" computation must be carried out. Higher-order polynomial equations are notoriously ill-behaved; this is the root of many of the convergence problems that occur when the Level 2 model is used in circuit simulation. Despite all the effort and complexity, the saturation voltage model works rather poorly, producing a very noticeable kink (and badly discontinuous first derivative) in this operating region.

The "λ-method" of describing the device characteristics when $V_{ds} > V_{dsat}$ is available in Level 2 as a leftover from the Level 1 model. However, this model is rather weak, as it assumes that all the saturated current characteristics will converge at some (negative) drain bias x-intercept (see Section 5.2.4). With the more detailed models developed here, it becomes clear that the "λ-method" can only be made to work by forcing the use of a physically incorrect description of pinch-off and channel length modulation. Ironically, though, the availability of the "λ-method" has provided one of the few incentives for the use of the Level 2 model. To a crude approximation, the parameter **LAMBDA** describes the saturated drain conductance g_{ds}. This is a key piece of information in CMOS analog circuit design, and the "λ-method" is often encountered in texts on the subject [18]. The reader is cautioned, though, that this model is very weak; analog designers are better served by use of HSPICE Level 28 or BSIM2.

A subthreshold model has been included with Level 2, but the model is physically unrealistic. It contains one parameter (**NFS**) which has no geometry, temperature, or drain bias dependence. Thus, the model can be made to work at one choice of all three of these variables, and for no other conditions. In addition, although current continuity is guaranteed at the subthreshold-superthreshold transition point, the first derivative is not continuous; this can be a source of convergence problems during circuit simulation.

Finally, as noted during model "playback," the small geometry corrections can be made to work well at one particular value each of the channel length and width; for intermediate geometries, where parameter extraction was not performed, the model gives poor results. The addition of the correction parameters also corrupts the results for the base long-wide device. To build a model which is valid over a range of geometries, the model must be split up into a set of sub-models, a practice referred to as binning. This method will be described in Chapter 7, as it is frequently encountered in the more popular Level 3 model. If needed, the same technique can be applied to Level 2 model development.

BIBLIOGRAPHY

1. L. Yau, "A Simple Theory to Predict the Threshold Voltage of Short Channel IGFETs," *Sol. St. Elec.* vol. 17, pp. 1059–1063 (1974).

2. K. Jeppson, "Influence of the Channel Width on the Threshold Voltage Modulation in MOSFETs," *Elec. Lett.* vol. 11, pp. 297–299 (1975).

3. J. Copeland, "Diode Edge Effect on Doping Profile Measurements," *IEEE Trans. Elec. Dev.* vol. ED-17, pp. 404–407 (1970).

4. O. Leistiko, A. Grove, and C. Sah, "Electron and Hole Mobilities in Inversion Layers on Thermally Oxidized Silicon Surfaces," *IEEE Trans. Elec. Dev.* vol. ED-12, pp. 248–254 (1965).

5. D. Frohman-Bentchkowsky, "On the Effect of Mobility Variation on MOS Device Characteristics," *Proc. IEEE* vol. 56, pp. 217–218 (1968).

6. *HSPICE User's Manual*, Meta-Software, Inc., Campbell, California, 1993.

7. A. Vladimirescu and S. Liu, "The Simulation of MOS ICs Using SPICE-2," University of California/Berkeley, Electronics Research Laboratory Document M80/7 (1980).

8. D. Caughey and R. Thomas, "Carrier Mobilities in Silicon Empirically Related to Doping and Field," *Proc. IEEE* vol. 55, pp. 2192–2193 (1967).

9. *CRC Standard Mathematical Tables*, 26th Edition (ed. by W. Beyer), CRC Press, 1981.

10. V. Reddi and C. Sah, "Source to Drain Resistance Beyond Pinch-off in MOS Transistors," *IEEE Trans. Elec. Dev.* vol. ED-12, pp. 139–141 (1965).

11. D. Frohman-Bentchkowsky and A. Grove, "Conductance of MOS Transistors in Saturation," *IEEE Trans. Elec. Dev.* vol. ED-16, pp. 108–113 (1969).

12. G. Baum and H. Beneking, "Drift Velocity Saturation in MOS Transistors," *IEEE Trans. Elec. Dev.* vol. ED-17, pp. 481–482 (1970).

13. R. Swanson and J. Meindl, "Ion-Implanted Complementary MOS Transistors in Low-Voltage Circuits," *IEEE J. Sol. St. Circ.* vol. SC-7, pp. 146–153 (1972).

14. P. Antognetti, D. Cavigia, and E. Profumo, "CAD Model for Threshold and Subthreshold Conduction in MOSFETs," *IEEE J. Sol. St. Circ.* vol. SC-17, pp. 454–458 (1982).

15. J. Meyer, "MOS Models and Circuit Simulation," *RCA Review* vol. 32, pp. 42–63 (1971).

16. D. Ward and R. Dutton, "A Charge-Oriented Model for MOS Transistor Capacitances," *IEEE J. Sol. St. Circ.* vol. SC-13, pp. 703–708 (1978).

17. *Aurora User's Manual*, Technology Modeling Associates, Inc., Palo Alto, California, 1994.

18. P. Allen and D. Holberg, *CMOS Analog Circuit Design*, Holt, Rinehart and Winston, 1987.

7

The Level 3 Model

7.1 Introduction

As described in Chapters 5 and 6, the Level 1 model was derived for long, wide FETs with thick gate oxides, and could not account for short channel behavior. The Level 2 model attempted to address these shortcomings, but the approach is mathematically complex, suffers from a severe first derivative discontinuity at the linear to saturation transition point, and all too frequently encounters convergence problems during circuit simulation. In addition, the possible overlap of the source and drain depletion regions in very short channel devices, lateral field effects on the mobility, and drain-induced barrier lowering (DIBL) are neglected; short channel effects are only partially considered.

The Level 3 model was developed to address these shortcomings. The basic approach is similar to Level 2, in that all geometry dependence is coded in the model equations, and many of the model equations and parameters are the same (or very nearly the same). However, in contrast to Level 2, the Level 3 model adopts a semi-empirical approach, in which the model for mobility reduction due to the normal field, using a critical field (represented in Level 2 through the parameters **UEXP**, **UCRIT**, and **UTRA**), is discarded and replaced by a much simpler model employing one parameter. The threshold voltage model is augmented with a parameter (**ETA**) describing drain-induced barrier lowering. In addition, the channel length modulation model employed in Levels 1 and 2, using the parameter **LAMBDA**, is replaced by a slightly more complex model involving the parameter **VMAX** and another semi-empirical parameter. Finally, a correction is introduced to account for the reduction of the mobility by high lateral fields.

Due to its simplicity and semi-empirical nature, the Level 3 model has been very successful. Parameter extraction is quite straightforward, and the model itself has proven to be more efficient than the Level 2 model; it thus provides better accuracy than the Level 2 model, while running significantly faster and rarely encountering convergence problems. In addition, despite the fact that the Level 3 model was originally developed

Table 7.1   The Level 3 model parameter set.

Parameter	Units	Description
Process Parameters		
TPG	m	Type of Gate Material
TOX	m	Gate Oxide Thickness
LD	m	Channel Length Reduction From Drawn Value
WD	m	Channel Width Reduction From Drawn Value
Electrical Parameters		
UO	$cm^2/V \cdot s$	Zero Bias Low Field Mobility
VTO	V	Threshold Voltage, Long, Wide Device, Zero Substrate Bias
THETA	V^{-1}	Gate Field Induced Mobility Reduction Parameter
RS	ohm	Source Series Resistance
RD	ohm	Drain Series Resistance
DELTA		Narrow Channel Effect on the Threshold Voltage
NSUB	cm^{-3}	Substrate Sensitivity Parameter (Effective Substrate Doping)
XJ	m	Short Channel Correction to the Substrate Sensitivity
VMAX		Maximum Carrier Velocity
ETA		DIBL Coefficient
KAPPA	V^{-1}	Channel Length Modulation Effect on the Drain Current
NFS	cm^{-2}	Subthreshold Region Fitting Parameter
CGSO	F/m	Zero Bias Gate-Source Capacitance
CGDO	F/m	Zero Bias Gate-Drain Capacitance
CGBO	F/m	Zero Bias Gate-Bulk Capacitance
XQC		Charge Partitioning Parameter
Temperature Parameters (HSPICE Implementation)		
BEX		Temperature Effect on the Low Field Mobility

for channel lengths greater than 1 μm, it can, with appropriate care, be extended for use at much shorter dimensions.

A complete listing of the Level 3 model parameters may be found in Table 7.1.

7.2   The Electrostatic Model of the MOSFET Structure

In the Level 1 model discussed in Chapter 5, it was assumed that the FET was very large, and the simple MOS capacitor expression for the depletion charge could be used. In the Level 2 model (Chapter 6), expressions had to be derived for changes in the depletion charge due to short and narrow channel effects. The Level 3 model includes similar modifications, but extends them to account for the behavior of very short channel FETs.

Although the physical model for the depletion charge used in Level 3 is different from the one employed in Level 2, the implementation of the results is identical in the two models. Short channel effects are described by f_s, the fractional reduction of the depletion charge by the overlap of the gate-induced depletion region with the source and drain depletion regions. Narrow channel effects are described by f_n, the fractional increase in the depletion charge caused by spreading of the depletion region beyond the defined channel edges.

7.2.1 Short Channel Effects

The Level 2 short channel model employs a trapezoidal approximation for the depletion charge. This approach neglects the possible overlap of the source and drain depletion regions. In the Level 3 development, it was deemed desirable to improve the model to account for this behavior.

The electrostatic structure of the short channel MOSFET implemented in Level 3 is based on the model of Dang [1]. Dang's model describes the depletion region sizes in the four-terminal MOSFET structure, and also accounts for the merging of the source and drain depletion regions.

The electrostatic model is based on the MOSFET structure shown in Figure 7.1. Dang's treatment divides the depletion region into three separate subregions. First, there is a simple planar p-n junction between the source/drain diffusions and the substrate. Second, there is a depletion region under the channel region due to the gate bias, which is also assumed to be planar. Third, there is a lateral depletion region into the channel region from the source and drain diffusions. This third space charge region is approximated by a cylindrical surface; Poisson's equation is solved in cylindrical coordinates for a one-sided junction. Assuming that the potential drop across the planar junction is identical to that across the curved part of the junction, and using a numerical approximation for a one-sided cylindrical junction, the depletion region width in the curved region can be related to the width of the planar junction by

$$W_C = 0.0631353 + 0.8013292 W_P - 0.01110777 \frac{W_P{}^2}{x_j}, \qquad (7.1)$$

where x_j is the metallurgical junction depth and W_P is the depletion width across a planar junction:

$$W_P = \left(\frac{2\epsilon_s V_x}{q N_{sub}} \right)^{\frac{1}{2}}. \qquad (7.2)$$

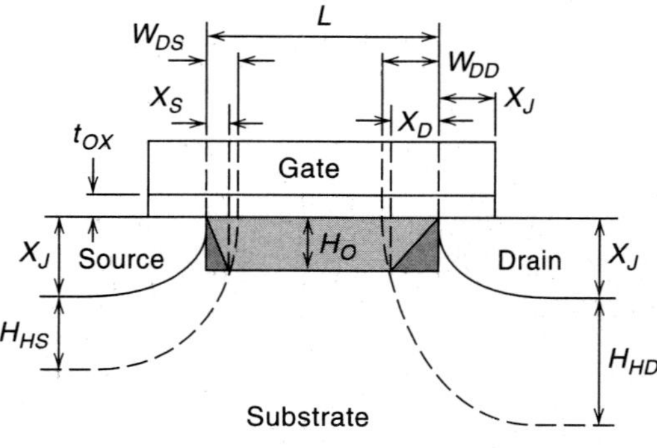

Figure 7.1   The depletion region of a long channel MOSFET, showing the imaginary division into junction, gate-induced, and lateral depletion regions. From [1]. © 1977 IEEE. Used by permission.

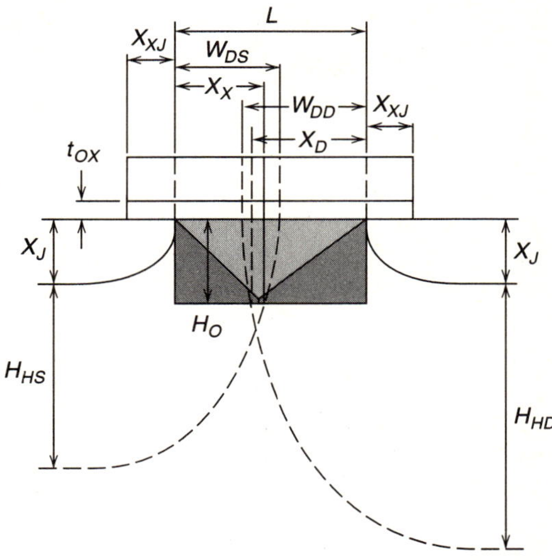

Figure 7.2   The depletion region of a short channel MOSFET, showing the overlap of the lateral source and drain depletion regions. From [1]. © 1977 IEEE. Used by permission.

Adopting the convention here of boldfacing model parameter names, (7.1) becomes

$$W_C = 0.0631353 + 0.8013292 W_P - 0.01110777 \frac{W_P{}^2}{\mathbf{XJ}}. \tag{7.3}$$

The quadratic approximation (7.3) has been shown to result in a maximum error of about 5% when compared to the full analytical solution. However, the resulting decrease in computation time is significant. The term V_x accounts for the applied bias across the diffusion. Assuming that the source diffusion is grounded (or the reference node), for the source,

$$V_x = V_{bi}, \tag{7.4}$$

while for the drain,

$$V_x = (V_{bi} - V_{ds}), \tag{7.5}$$

where V_{bi} is the built-in junction potential.

The short channel model is described in Figure 7.2. Here the short channel length allows the lateral source and drain depletion regions to overlap and share depletion charge. This sharing is accounted for by the expression

$$K = 1 - \frac{X_S - X_D}{2L}. \tag{7.6}$$

In the long channel limit, $K \rightarrow 1$. In addition, it is arbitrarily assumed that the lateral underdiffusion of the gate by the source and drain regions is equal to $0.7 \cdot \mathbf{XJ}$. With this approximation, the full expression for K is

$$K = 1 - \frac{1}{2L} \left\{ \left[(W_{DS} + 0.7 \cdot \mathbf{XJ}) \left[1 - \frac{H_o^2}{(H_{HS} + \mathbf{XJ})^2} \right]^{\frac{1}{2}} - 0.7 \cdot \mathbf{XJ} \right] \right.$$

$$\left. + \left[(W_{DD} + 0.7 \cdot \mathbf{XJ}) \left[1 - \frac{H_o^2}{(H_{HD} + \mathbf{XJ})^2} \right]^{\frac{1}{2}} - 0.7 \cdot \mathbf{XJ} \right] \right\}. \tag{7.7}$$

The term H_o is the gate depletion depth at the onset of strong inversion:

$$H_o = \left[\frac{2\epsilon_{Si}}{q N_{sub}} (V_o - V_{bs}) \right]^{\frac{1}{2}}, \tag{7.8}$$

where N_{sub} is the substrate doping concentration and $V_o = 2\phi_f$, which is the condition for the onset of strong inversion. Again assuming that the source is the reference node, the expression for the depletion region size under the source diffusion is

$$H_{HS} = \left[\frac{2\epsilon_{Si}}{q N_{sub}} (V_{bi} - V_{sub}) \right]^{\frac{1}{2}}, \tag{7.9}$$

while for the drain diffusion,

$$H_{HD} = \left[\frac{2\epsilon_{Si}}{q N_{sub}} (V_{bi} + V_{ds} - V_{sub}) \right]^{\frac{1}{2}}. \tag{7.10}$$

The terms W_{DS} and W_{DD} are W_C for the source and drain respectively. The first term, W_{DS}, is computed using (7.3) with W_P computed from (7.4) (source as reference), while W_{DD} is computed using (7.5).

This is the form of the model for the MOSFET depletion regions developed by Dang [1]. The actual implementation in the Level 3 model equations makes several further changes and simplifications. First, as noted earlier, in Dang's model, the a priori assumption is made that the extent of underdiffusion of the gate is equal to 0.7 times the junction depth $\mathbf{XJ}$. The Level 3 model replaces this assumption with a separate parameter $\mathbf{LD}$, which decouples the diffusion spreading under the gate from the diffused junction depth. This makes the model more flexible and applicable to a wider variety of processes. In addition, as will be discussed in Section 7.7, the parameter $\mathbf{LD}$ actually accounts *in total* for the reduction to the *metallurgical* channel length from the *drawn* channel length due to processing. Thus, in the absence of independent measurements, $\mathbf{LD}$ accounts for both the source/drain underdiffusion of the gate *and* the offset between the *drawn* gate length and the *printed* gate length due to processing during gate definition (e.g., overetching of the polysilicon under the defined photoresist dimension).

(As an aside, this has the unintended effect of changing the meaning of the parameter **XJ**. When a substrate-source bias is applied, the depletion region around the source diffusion expands. In a long channel device, this expansion is not large compared to the total FET depletion region size. In a short channel device, the change to the total FET depletion region size is significant. Thus, **XJ** in reality serves as a short channel correction to the substrate sensitivity. This situation is considered further when parameter extraction is examined.)

Second, the simplification is made of neglecting the effect of the drain bias on the drain depletion region. This reduces both W_{DS} and W_{DD} to the simple expression for W_C, in which W_P is computed using the condition described in (7.4). This allows the setting of $H_{HS} = H_{HD} = W_P$. (It is interesting to note that this simplification was **not** made in the development of the Level 2 model; it was thus likely made in Level 3 to improve the computational efficiency. This approximation is questionable in modern short channel FETs, in which the drain depletion region becomes relatively large under an applied drain bias.) The additional simplification is made of setting the gate depletion depth H_o equal to the diffused junction depletion depth W_P. Using the approximation discussed above of neglecting the effect of the drain bias on the drain junction depletion region, by setting $H_o = W_P$, it is found that this approximation is strictly valid when

$$2\phi_f = V_{bi}. \tag{7.11}$$

For substrates that are doped to a reasonable level, this approximation is relatively good when the device is biased into strong inversion; this is the region of operation for which the Level 3 model is designed.

With these changes, Dang's term for charge sharing (K) reduces to the Level 3 expression f_s:

$$f_s = 1 - \frac{1}{L_{eff}} \left[(W_C + \mathbf{LD}) \left[1 - \left(\frac{W_P}{\mathbf{XJ} + W_P} \right)^2 \right]^{\frac{1}{2}} - \mathbf{LD} \right], \tag{7.12}$$

where L_{eff} is the metallurgical channel length described as $L_{eff} = L_{drawn} - 2 \cdot \mathbf{LD}$. This expression describes the fractional reduction of the total depletion volume due to the overlap of the source and drain depletion regions:

$$Q_{depl,short} = f_s \cdot Q_{depl,long}. \tag{7.13}$$

7.2.2   Narrow Channel Effects

In addition to the decrease of the total bulk charge due to the merging of the source and drain depletion regions, the bulk charge is also increased by the spreading of the depletion region outside the channel edges and under the LOCOS bird's beak, as shown in Figure 7.3. In the Level 3 model, the situation is simplified to that shown in Figure

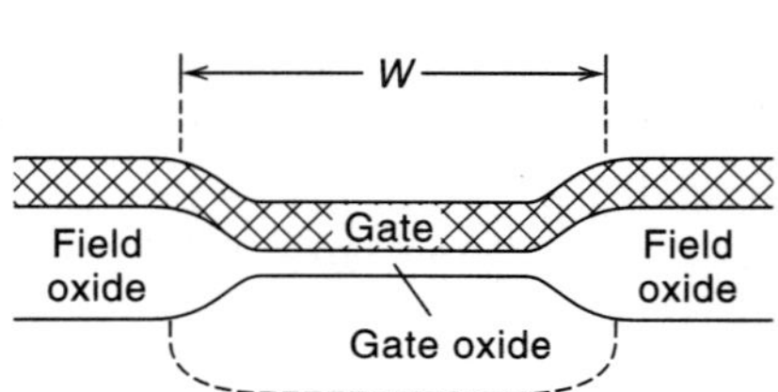

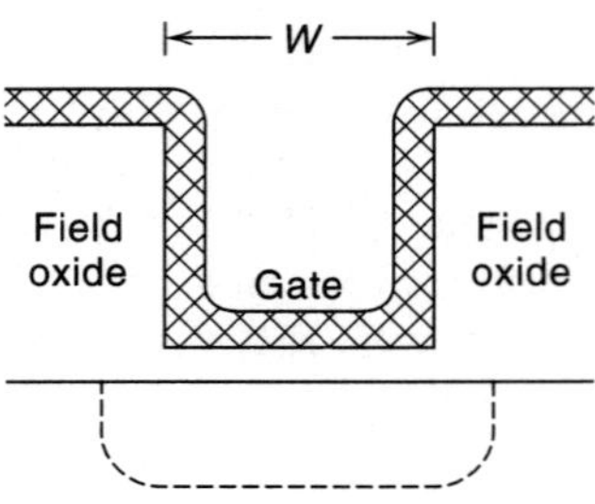

Figure 7.3 The spreading of the gate-induced depletion region outside the channel edges and under the isolation.

Figure 7.4 The simplified scheme employed by Level 3 for describing the spreading of the gate-induced depletion region outside the channel edges.

7.4, where a gate electrode over a flat uniformly doped silicon surface is assumed. The total depletion volume in the main channel is

$$D_{ch} = W_{eff} \cdot L \cdot x_n,$$ (7.14)

while the total depletion volume in the two extra regions is

$$D_{ed} = A \cdot L,$$ (7.15)

where A is the cross-sectional area of the extra regions. The fraction of depletion volume in the edge regions relative to the total depletion volume is

$$P = \frac{D_{ed}}{D_{ed} + D_{ch}} = \frac{A \cdot L}{A \cdot L + W_{eff} \cdot L \cdot x_n} = \frac{A}{A + W_{eff} \cdot x_n}.$$ (7.16)

As shown in Figure 7.4, the Level 3 model approximates each edge region as a quarter-circle with radius x_n [2, 3]. The total area of the edge regions is then

$$A = 2 \times \frac{\pi x_n^2}{4} = \frac{\pi x_n^2}{2}.$$ (7.17)

Equation (7.16) then becomes

$$P = \frac{\frac{\pi x_n^2}{2}}{\frac{\pi x_n^2}{2} + W_{eff} \cdot x_n} = \frac{1}{1 + \frac{2W_{eff}}{\pi x_n}}.$$ (7.18)

The gate depletion depth x_n is usually less than 1 µm, so for the technology of the time that the Level 3 model was developed, it was reasonable to assume that $W_{eff} \gg x_n$. This reduces (7.18) to

$$P = \frac{\pi x_n}{2W_{eff}}. \tag{7.19}$$

The additional bulk charge contributed by the edge regions is then computed from

$$Q_{edge} = P \cdot Q_{bulk} = \left(\frac{\pi x_n}{2W_{eff}}\right)(qN_Ax_n) = \frac{qN_A\pi x_n^2}{2W_{eff}}, \tag{7.20}$$

where the gate depletion depth x_n is taken from (3.87):

$$x_n = \left(\frac{2\epsilon_{Si}\phi_s}{qN_A}\right)^{\frac{1}{2}}. \tag{7.21}$$

In strong inversion, $\phi_s = (2\phi_f - V_{bs})$, so (7.20) becomes

$$Q_{edge} = \frac{qN_A\pi}{2W_{eff}}\left[\frac{2\epsilon_{Si}(2\phi_f - V_{bs})}{qN_A}\right] = \frac{\pi\epsilon_{Si}(2\phi_f - V_{bs})}{W_{eff}}. \tag{7.22}$$

The Level 3 implementation of (7.22) adds an erroneous 2 to the denominator (as opposed to the erroneous 4 of the Level 2 model), and, since the structure described by Figure 7.4 is an approximation of the physical situation shown in Figure 7.3, the model parameter δ is also added (which has the effect of covering up the erroneous 2 by sweeping the error into the extraction of δ). Equation (7.22) is then redefined to

$$\frac{Q_{edge}}{C_{ox}} = \delta\frac{\pi\epsilon_{Si}(2\phi_f - V_{bs})}{2C_{ox}W_{eff}} = f_n \cdot (2\phi_f - V_{bs}). \tag{7.23}$$

Using the convention of bold-facing parameter names, f_n is

$$f_n = \mathbf{DELTA} \cdot \frac{\pi\epsilon_{Si}}{2C_{ox}W_{eff}}. \tag{7.24}$$

7.2.3   The Total Depletion Charge

The computation of the total depletion charge is a three-step process. First, the depletion charge associated with each of the three independent device terminals (source, drain, and gate) is computed and summed to give the total depletion charge of a large FET. Next, the short channel fractional reduction term f_s, which accounts for the overlap of these three regions, is computed and introduced as a modification to the large geometry depletion charge. Finally, the additional charge due to the lateral spreading of the gate-induced depletion region outside of the channel edges is computed and added to the short channel depletion result.

By thus combining the effects of short and narrow channels, the Level 3 depletion charge model for a small FET is

$$\frac{Q_{bulk,small}}{C_{ox}} = \frac{f_s \cdot Q_{bulk,large}}{C_{ox}} + \frac{Q_{edge}}{C_{ox}}$$

$$= \left\{ 1 - \frac{1}{L_{eff}} \left[(W_C + \mathbf{LD}) \left[1 - \left(\frac{W_P}{\mathbf{XJ} + W_P} \right)^2 \right]^{\frac{1}{2}} - \mathbf{LD} \right] \right\} \cdot \gamma \cdot (2\phi_f - V_{bs})^{\frac{1}{2}}$$

$$+ \mathbf{DELTA} \cdot \frac{\pi \epsilon_{Si}}{2 C_{ox} W_{eff}} \cdot (2\phi_f - V_{bs})$$

$$= f_s \cdot \gamma \cdot (2\phi_f - V_{bs})^{\frac{1}{2}} + f_n \cdot (2\phi_f - V_{bs}), \tag{7.25}$$

where f_s is given by (7.12) and f_n by (7.24).

7.3   The Threshold Voltage Model

For the three-terminal MOS capacitor case described in Section 3.4, the threshold voltage is defined as

$$V_t = V_{fb} + 2\phi_f + \frac{Q_{depl}}{C_{ox}} = V_{fb} + 2\phi_f + \gamma \cdot (2\phi_f - V_{bs})^{\frac{1}{2}}, \tag{7.26}$$

where any effects due to nonuniform substrate doping have been neglected. The flatband voltage V_{fb} was described in Section 3.2; in Level 3 and the other first generation models, V_{fb} is computed as described there. This requires that the parameter **TPG** be specified. If the gate material is polysilicon with the same doping as the source and drain (e.g., n^+ polysilicon in an nFET), $\mathbf{TPG} = 1$. If the gate material is polysilicon with doping opposite to that of the source and drain (e.g., n^+ polysilicon in a pFET), $\mathbf{TPG} = -1$. It is also possible to include an aluminum gate with $\mathbf{TPG} = 0$; however, aluminum gates are no longer used in MOS technology, so this setting should not be encountered.

Since this model was developed for a three-terminal MOS structure, it takes no account of either narrow channel effects or the drain terminal. In shorter channel FETs, the sizes of the source and drain depletion regions become important with respect to the channel length, and the drain bias renders the gradual channel approximation invalid. The simple threshold voltage model must be modified to account for these effects.

For the simple three-terminal MOS capacitor, the total depletion charge is described by

$$\frac{Q_{depl}}{C_{ox}} = \gamma \cdot (2\phi_f - V_{bs})^{\frac{1}{2}}. \tag{7.27}$$

The Level 3 threshold voltage model modifies (7.27) by including the small geometry depletion charge model described in Section 7.2. To account for these effects, (7.27) is replaced with (7.25):

$$\frac{Q_{depl}}{C_{ox}} = f_s \cdot \gamma \cdot (2\phi_f - V_{bs})^{\frac{1}{2}} + f_n \cdot (2\phi_f - V_{bs}). \tag{7.28}$$

This is the depletion charge model that is implemented in Level 3, and the expression for the threshold voltage becomes

$$V_t = V_{fb} + 2\phi_f + f_s \cdot \gamma \cdot (2\phi_f - V_{bs})^{\frac{1}{2}} + f_n \cdot (2\phi_f - V_{bs}). \tag{7.29}$$

For long, wide FETs, $f_s \rightarrow 1$, $f_n \rightarrow 0$, and (7.26) is recovered.

As described in Section 3.5, in short channel FETs with an applied drain bias, the depletion width at the drain becomes significant compared to the effective channel length, resulting in a reduction of the channel potential barrier (drain-induced barrier lowering, or DIBL) [4]. In contrast to Level 2, the Level 3 development accounts for this behavior. The Level 3 model for DIBL makes use of the empirical observation [4] that the threshold voltage decreases linearly with increasing drain voltage:

$$\Delta V_{t,DIBL} = \sigma \cdot V_{ds}. \tag{7.30}$$

Following Masuda et al. [5], (7.30) is rewritten as

$$\Delta V_{t,DIBL} = \frac{\eta}{C_{ox} L_{eff}{}^n} V_{ds}; \tag{7.31}$$

η is often referred to as the *coefficient of static feedback*. In the work of Masuda et al. [5], n was found to range from 2.6 to 3.2 (for MOSFETs with 10^{15} cm$^{-3} \leq N_A \leq 10^{16}$ cm^{-3} and 0.15 μm $\leq x_j \leq 0.41$ μm). In the Level 3 implementation, n is chosen to be 3. In addition, the coefficient η is multiplied by an arbitrary constant to make the extracted numerical value of η more palatable. Equation (7.31) then reaches its final form of

$$\Delta V_{t,DIBL} = \eta \frac{8.15 \times 10^{-22}}{C_{ox} L_{eff}{}^3} V_{ds}. \tag{7.32}$$

Combining (7.32) with (7.29) leads to the final Level 3 expression for the threshold voltage:

$$V_t = V_{fb} + 2\phi_f + f_s \cdot \gamma \cdot (2\phi_f - V_{bs})^{\frac{1}{2}} + f_n \cdot (2\phi_f - V_{bs}) + \eta \frac{8.15 \times 10^{-22}}{C_{ox} L_{eff}{}^3} V_{ds}. \tag{7.33}$$

The only model parameter explicitly contained in (7.33) is the DIBL coefficient η; **XJ** and **LD** enter implicitly through f_s, while **NSUB** enters implicitly through ϕ_f and γ. Using the convention of boldfacing parameter names, (7.33) becomes

$$V_t = V_{fb} + 2\phi_f + f_s \cdot \gamma \cdot (2\phi_f - V_{bs})^{\frac{1}{2}} + f_n \cdot (2\phi_f - V_{bs}) + \textbf{ETA} \frac{8.15 \times 10^{-22}}{C_{ox} L_{eff}{}^3} V_{ds}. \tag{7.34}$$

The long, wide channel threshold voltage with zero back bias is a model parameter that is defined as

$$\mathbf{VTO} = V_{fb} + 2\phi_f + \gamma \cdot (2\phi_f)^{\frac{1}{2}}. \tag{7.35}$$

Notably, **VTO** is **not** used in computing the threshold voltage as described by (7.34), but only appears in the expression for the channel mobility. It should also be noted that the use of (7.35) deletes any substrate bias dependence from the mobility model.

7.4   The Mobility Model

As described in Chapter 5, Level 1 implemented the simplest possible model for the channel mobility, in which μ is a constant. The mobility model in Level 2, discussed in Chapter 6, attempts to cope with the effect of the vertical field by introducing a model employing a *critical vertical field*, above which the mobility decreases. The resulting model is rather unwieldy, with three parameters (only two of which are used, as **UTRA** was never implemented). In addition, in modern thin oxide devices, the vertical field is usually larger than the critical field for all values of $V_{gs} > V_t$. Mobility reduction by the lateral field was not even considered in Level 2.

The Level 3 mobility model describes the effect of both the vertical and the lateral field. As implemented, the mobility is separated into three components, which are treated as being physically independent of each other. These three components are:

1. The low field mobility μ_o, which is identical to the mobility used in Level 1;
2. The mobility μ_v, which represents the modification to μ_o due to the vertical field;
3. The mobility μ_{eff}, which represents the modification to μ_v due to the lateral (channel) field.

The effects of vertical and lateral fields are taken to be independent of each other. In this manner, the final value for the mobility is constructed sequentially. First, a value for μ_o is found. Next, μ_v is computed from μ_o to account for the effect of the vertical field. Finally, μ_{eff} is computed from μ_v to account for the effect of the lateral field.

7.4.1   The Effect of the Vertical Field

The semi-empirical model for the effect of the vertical field on the mobility traces itself back to the classic work of Schrieffer [6]. Schrieffer showed that for a classically treated surface charge layer,

$$\mu_v \propto E^{-1}, \tag{7.36}$$

where μ_v is the mobility and E is the normal (vertical) electric field. The implications of (7.36) can be illustrated graphically, as shown in Figure 7.5. The linear proportionality

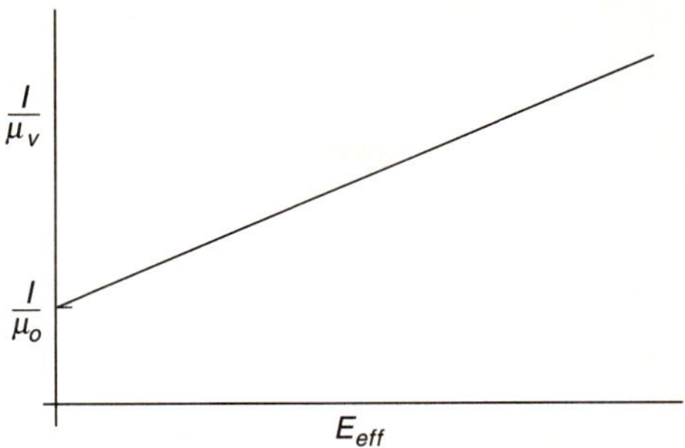

Figure 7.5 A plot constructed from (7.36), showing the slope and intercept interpretation of Schrieffer's model.

between $1/\mu_v$ and E also implies a nonzero y-intercept, since the mobility does not become infinite as E goes to zero. Instead, the mobility goes to its low field value μ_o, which is the value used in the much simpler Level 1 model. This identification of μ_o is reasonable from a device physics standpoint. Before the gate bias has reached the threshold voltage (the onset of strong inversion), the inversion charge density increases with increasing gate bias, and the mobility increases. At and above the threshold voltage, the inversion layer behaves as a very dense space charge layer, and its intrinsic mobility no longer increases. Thus, at $(V_{gs} - V_t) = 0$, the mobility reaches an approximate maximum, where the inversion layer has a high charge density and the mobility has yet to be reduced by the vertical field for $(V_{gs} - V_t) > 0$. This is the situation described by Figure 7.5.

From Figure 7.5, the mobility-field relationship can be written as

$$\frac{1}{\mu_v} = \alpha \cdot E + \frac{1}{\mu_o}, \tag{7.37}$$

where α is a proportionality constant. From Gauss' law, it can readily be shown that in strong inversion, when the depletion charge remains constant with gate bias,

$$E \propto (V_{gs} - V_t). \tag{7.38}$$

This allows (7.37) to be rewritten as

$$\frac{1}{\mu_v} = \alpha' \cdot (V_{gs} - V_t) + \frac{1}{\mu_o}, \tag{7.39}$$

where α' represents the modified α from the proportionality in (7.38). By simple algebraic manipulation, and allowing the definition

$$\theta = \mu_o \cdot \alpha', \tag{7.40}$$

(7.39) can be rewritten as

$$\frac{1}{\mu_v} = \frac{1 + \theta \cdot (V_{gs} - V_t)}{\mu_o}, \tag{7.41}$$

or in the more familiar form of

$$\mu_v = \frac{\mu_o}{1 + \theta \cdot (V_{gs} - V_t)}. \tag{7.42}$$

(The same result could be obtained by noting that the gate field reduces the mobility, taking a Taylor expansion in powers of $(V_{gs} - V_t)$, and retaining only the linear term.) The terms μ_o, θ, and V_t are model parameters, which are determined by extraction. Using the standard notation employed here of boldfacing model parameter names, (7.42) is rewritten as

$$\mu_v = \frac{\mathbf{UO}}{1 + \mathbf{THETA} \cdot (V_{gs} - \mathbf{VTO})}. \tag{7.43}$$

As noted earlier, this approach to including the effect of the vertical field contains no substrate bias dependence. The implications of this omission will be examined later, when models are constructed from specific data.

7.4.2 The Effect of the Lateral Field

As noted in Chapter 4, the mobility only has true physical meaning when carriers are giving up energy to the lattice at the same rate that they are gaining it from the electric field, in which case the Ohm's law approximation of

$$v = \mu E, \tag{7.44}$$

where v is the carrier velocity, is valid. As also discussed in Chapter 4, for higher field strengths, the lattice acts as a nonlinear damping force on the carriers, and despite an increasing field, the carriers are unable to travel faster than the saturation velocity (v_{SAT}). This indicates that the simple model for low values of the lateral field (7.43) should be modified. Level 3 improves on Level 2 by adding a model for these lateral field effects into both the mobility model and the saturation voltage model; in Level 2, only the saturation voltage computation is modified by high lateral field effects.

The implementation of the effect of the lateral field begins with the Caughey-Thomas model [7] for carrier velocity in an FET channel:

$$v = \frac{\mu E}{\left(1 + \frac{E}{E_C}\right)}, \tag{7.45}$$

where E_C is a *critical field* that will be defined shortly. In the limit $E \ll E_C$, (7.45) reduces to

$$v = \mu E, \tag{7.46}$$

which is the ohmic approximation for the carrier velocity (7.44). In the limit $E \gg E_C$, (7.45) becomes

$$v = \mu E_C. \tag{7.47}$$

Equation (7.47) defines a constant; since it represents the upper limit of (7.45), it is defined as the saturation velocity v_{SAT}, which is the highest velocity that the carriers can reach. This allows the critical field E_C to be defined from (7.47):

$$E_C = \frac{v_{SAT}}{\mu}. \tag{7.48}$$

Now divide both sides of (7.47) by E, and, for reasons that will be explained below, define $\mu_{eff} = \frac{v}{E}$. This yields

$$\mu_{eff} = \frac{\mu}{\left(1 + \frac{E}{E_C}\right)}. \tag{7.49}$$

In the limit $E \ll E_C$, (7.49) reduces to $\mu_{eff} = \mu$; the low field mobility is recovered. However, in the limit $E \gg E_C$, (7.49) becomes

$$\mu_{eff} = \frac{\mu E_C}{E}. \tag{7.50}$$

This is an interesting result, as for $E \gg E_C$, μ_{eff} decreases monotonically with increasing E. The reason for this result becomes clear by returning to the definition of μ_{eff} and writing

$$v = \mu_{eff} E. \tag{7.51}$$

For $E \ll E_C$, (7.51) reduces to the ohmic expression $v = \mu E$. However, for $E \gg E_C$,

$$v = \left(\frac{\mu E_C}{E}\right) E = \mu E_C = v_{SAT}. \tag{7.52}$$

This approach allows the saturation velocity to be imbedded in the mobility model, which is a very useful and practical choice for device modeling.

In the conducting FET channel, the lateral electric field cannot exceed some value E_{sat}. At $E = E_{sat}$, the drain bias has reached the value where the channel pinches off, and a higher *channel* field cannot be achieved. (It should be noted that E_{sat} and v_{SAT} are not specifically related, other than by an unfortunate choice of terminology; the actual connection between the two terms is discussed during the derivation of the drain current equations. To separate the two different phenomena here, the lower case subscript *sat* will be used for terms associated with FET saturation due to channel pinch-off, while the upper case subscript *SAT* will be used for terms associated with the carrier velocity saturation.) Defining

$$E' = min(E, E_{sat}), \tag{7.53}$$

(7.49) can be rewritten as

$$\mu_{eff} = \frac{\mu}{\left(1 + \frac{E'}{E_C}\right)}. \tag{7.54}$$

Assuming that the effective channel length L_{eff} is much larger than the length of the pinched-off (space charge) region, (7.54) can be changed to

$$\mu_{eff} = \frac{\mu}{\left(1 + \frac{E \cdot L_{eff}}{E_C \cdot L_{eff}}\right)} = \frac{\mu}{\left(1 + \frac{V'_{ds}}{V_C}\right)}, \tag{7.55}$$

where, following (7.53)

$$V'_{ds} = min(V_{ds}, V_{dsat}), \tag{7.56}$$

and V_C is a *critical voltage* to be described below. V_{dsat} is the drain saturation voltage, the value of V_{ds} at which the channel pinches off; the form of V_{dsat} used in Level 3 is derived along with the drain current equations. Using the definition (7.48) for E_C leads to

$$V_C = E_C \cdot L_{eff} = \frac{v_{SAT} \cdot L_{eff}}{\mu}. \tag{7.57}$$

As in (7.55), it is assumed that L_{eff} is much larger than the length of the pinched-off region, so that the distance from the source to the pinch-off point can be taken to be L_{eff}. Placing (7.57) in (7.55) produces

$$\mu_{eff} = \left(\frac{\mu}{1 + \frac{\mu \cdot V'_{ds}}{v_{SAT} \cdot L_{eff}}}\right). \tag{7.58}$$

In the limit $V'_{ds} \ll V_C$, (7.58) becomes $\mu_{eff} = \mu$, and the simple ohmic behavior is recovered. In the limit $V'_{ds} \gg V_C$, (7.58) becomes

$$\mu_{eff} = \frac{v_{SAT} \cdot L_{eff}}{V'_{ds}}. \tag{7.59}$$

Using (7.56), the largest value that V'_{ds} can reach is V_{dsat}. In this case,

$$\mu_{eff} = \frac{v_{SAT} \cdot L_{eff}}{V_{dsat}} = \frac{v_{SAT}}{E_{sat}}. \tag{7.60}$$

These expressions allow the difference between v_{SAT} and (V_{dsat}, E_{sat}) to be explained. Recall from Chapter 5 (the Level 1 model) that, to some degree of approximation

$$V_{dsat} \doteq (V_{gs} - V_t). \tag{7.61}$$

For low gate biases, V_{ds} reaches its maximum value of V_{dsat} such that $V_{dsat} \ll V_C$. In this case, the device "saturates" (the channel pinches off and the device changes from the linear to the saturation region of operation), and carrier velocity saturation need not be considered; the simple mobility model $\mu_{eff} = \mu$ is valid. For large values of V_{gs}, $V_{dsat} \gg V_C$, and the mobility is reduced as the carriers reach the saturation velocity before the channel pinches off. In this case, device saturation corresponds to carrier velocity saturation, and V_{dsat} is reduced from the classical value of (7.61) [3]. This situation is described more carefully in the following section, when the Level 3 expression for V_{dsat} is derived.

The mobility model (7.58) is the one that is used in Level 3. As described earlier, μ_{eff} represents the modification of μ_v by the lateral field; thus, μ in (7.58) should be replaced by μ_v, with μ_v having already been computed using (7.42). Also, the saturation velocity v_{SAT} is represented in Level 3 by the parameter **VMAX**, the maximum carrier velocity. Thus, (7.58) is rewritten as

$$\mu_{eff} = \frac{\mu_v}{1 + \frac{\mu_v \cdot V'_{ds}}{\textbf{VMAX} \cdot L_{eff}}}. \tag{7.62}$$

Although **VMAX** is supposed to represent the saturation velocity, extraction generally produces a quantity 1.2 to 2 times the accepted value.

If **VMAX** is not specified, it defaults to zero (and, of course, a user could set **VMAX** to zero). In this case, the Level 3 implementation of μ_{eff} treats the situation as one where (**VMAX** $\cdot$ L_{eff}) is very large; thus, μ_{eff} becomes μ_v. This is the situation in a very long channel device, where the carrier velocity in the inversion layer never reaches the saturation velocity; this causes **VMAX** to be of no consequence.

7.4.3 Temperature Dependence

In Berkeley SPICE, the temperature dependence of the mobility is described by [9]

$$\mu(T) = \mu(T_{nom}) \cdot \left(\frac{T}{T_{nom}} \right)^{-1.5}, \tag{7.63}$$

where T and T_{nom} are in **degrees Kelvin**. While this expression is correct from a first principles point of view [10], in practice it does not work well, and the user has no method of modifying (7.63). Thus, even if a good model is constructed at one particular temperature (e.g., 25°C), it will lose accuracy at different temperatures (e.g., 125°C). The HSPICE implementation of Level 3 [11] modifies (7.63) to

$$\mu(T) = \mu(T_{nom}) \cdot \left(\frac{T}{T_{nom}}\right)^{\mathbf{BEX}},\tag{7.64}$$

where **BEX** is a parameter. As will be discussed in Section 7.7, **BEX** is easily extracted from data.

7.5   The Drain Current Equations

The Level 3 drain current model uses the approach developed in the Level 2 model. Thus, for the current in strong inversion, the same equation that appears in the Level 2 model ((6.53)) occurs in the Level 3 formulation. However, as described in Chapter 6, this equation contains depletion charge terms in the power of $\frac{3}{2}$ which are inefficient during circuit simulation. The Level 3 formulation uses this Level 2 equation as a starting point, and introduces a Taylor series expansion for the depletion charge terms. This serves to eliminate the $\frac{3}{2}$-power terms, which greatly improves the model efficiency. In addition, an entirely new expression for the saturation voltage V_{dsat} is developed. Rather than leading to an "election" system based on a quartic equation, a simple quadratic expression is produced. This approach is more robust and efficient, and also gives a much better result at the linear-saturation transition point, where a severe first derivative discontinuity occurs in Level 2. The remaining drain current model components (channel length modulation, series resistance, and the subthreshold current) are virtually identical to those developed for Level 2.

7.5.1   The Drain Current Above Threshold

To begin, the expression for the drain current in the linear region ($V_{gs} > V_t$, small V_{ds}) can be derived. When small geometry effects are included, the expression for the depletion charge (3.130) becomes

$$Q_{depl}(y) = f_s \cdot [2\epsilon_{Si}qN_A\phi_s(y)]^{\frac{1}{2}} + f_n \cdot \phi_s(y).\tag{7.65}$$

As described in Chapter 3, in strong inversion

$$\phi_s(y) = V(y) + 2\phi_f - V_{bs},\tag{7.66}$$

so (7.65) becomes

$$Q_{depl}(y) = f_s \cdot \left[2\epsilon_{Si}qN_A(V(y) + 2\phi_f - V_{bs})\right]^{\frac{1}{2}} + f_n \cdot (V(y) + 2\phi_f - V_{bs}).\tag{7.67}$$

The total surface charge (3.130) is

$$Q_s(y) = -C_{ox}(V_{gb} - V_{fb} - V(y) - 2\phi_f + V_{bs}), \qquad (7.68)$$

so that the expression for the inversion charge ((3.135) and (3.137)) becomes

$$Q_{inv}(y) = Q_s(y) - Q_{depl}(y) = -C_{ox}\left[V_{gs} - V_{fb} + V(y) - 2\phi_f\right]$$

$$- f_s \cdot \left[2\epsilon_{Si}qN_A(V(y)+2\phi_f-V_{bs})\right]^{\frac{1}{2}} - \frac{\pi\epsilon_{Si}}{W_{eff}}(V(y)+2\phi_f-V_{bs}). \quad (7.69)$$

Equation (7.69) is then inserted into the drain current expression (3.127):

$$dV_y = \frac{-I_{ds}dy}{\mu W_{eff}Q_{inv}(y)}. \qquad (7.70)$$

Integrating (7.70) as in Chapter 6 leads to

$$I_{ds} = \frac{\mu W_{eff}C_{ox}}{L_{eff}}\left\{(V_{gs} - (V_{fb} + 2\phi_f))V_{ds} - \frac{V_{ds}^2}{2}\right.$$

$$- \frac{2}{3}f_s \cdot \gamma \cdot \left[(V_{ds} + 2\phi_f - V_{bs})^{\frac{3}{2}} - (2\phi_f - V_{bs})^{\frac{3}{2}}\right]$$

$$\left. - f_n \cdot \left[\frac{V_{ds}^2}{2} + (2\phi_f - V_{bs})V_{ds}\right]\right\}. \qquad (7.71)$$

This expression is identical to the Level 2 drain current equation, with the exception of the different expressions contained in f_s and f_n (see Section 7.2). However, use of this equation in Level 2 demonstrated that it was mathematically inefficient, largely due to the $\frac{3}{2}$-power terms. To make better use of computing resources, it was decided to simplify (7.71).

The method chosen for the Level 3 current model is based on a Taylor series expansion approach originally described by Merckel, Borel, and Cupcea [12]. For the discussion which follows, it will be useful to rewrite (7.71) as

$$I_{ds} = \frac{\mu_{eff}C_{ox}W_{eff}}{L_{eff}}\left\{(V_{gs} - (V_{fb} + 2\phi_f))V_{ds} - \frac{V_{ds}^2}{2}\right.$$

$$- f_n \cdot \frac{V_{ds}^2}{2} - f_n \cdot (V_{bs} - 2\phi_f)V_{ds}$$

$$\left. + \frac{2}{3}f_s \cdot \gamma \cdot (2\phi_f - V_{bs})^{\frac{3}{2}} - \frac{2}{3}f_s \cdot \gamma \cdot (V_{ds} + 2\phi_f - V_{bs})^{\frac{3}{2}}\right\}. \qquad (7.72)$$

The approach taken by Merckel et al. [12] was to assume that $(2\phi_f - V_{bs}) \gg V_{ds}$ and perform a Taylor series expansion of the last term in (7.72); the resulting expression is then allowed to be valid for all V_{ds} (the reasonableness of this approximation will be discussed below). The Taylor expansion is

$$f(x_o + h) = f(x_o) + h\frac{\partial f}{\partial x}\bigg|_{x_o} + \frac{h^2}{2}\frac{\partial^2 f}{\partial x^2}\bigg|_{x_o} + \cdots, \tag{7.73}$$

where $x_o = (2\phi_f - V_{bs})$ and $h = V_{ds}$. For mathematical conciseness, let $x = (V_{ds} + 2\phi_f - V_{bs})$, so that

$$f(x) = \frac{2}{3}f_s \cdot \gamma \cdot x^{\frac{3}{2}}. \tag{7.74}$$

The derivatives are

$$\frac{\partial f}{\partial x} = f_s \cdot \gamma \cdot x^{\frac{1}{2}}, \tag{7.75}$$

and

$$\frac{\partial^2 f}{\partial x^2} = \frac{1}{2}f_s \cdot \gamma \cdot x^{-\frac{1}{2}}. \tag{7.76}$$

The specific terms in the Taylor expansion (7.73) then become

$$f(x_o) = \frac{2}{3}f_s \cdot \gamma \cdot (2\phi_f - V_{bs})^{\frac{3}{2}}, \tag{7.77}$$

$$\frac{\partial f}{\partial x}\bigg|_{x_o} = f_s \cdot \gamma \cdot (2\phi_f - V_{bs})^{\frac{1}{2}}, \tag{7.78}$$

and

$$\frac{\partial^2 f}{\partial x^2}\bigg|_{x_o} = \frac{1}{2}f_s \cdot \gamma \cdot (2\phi_f - V_{bs})^{-\frac{1}{2}}. \tag{7.79}$$

If only the first order terms are retained, (7.73) yields

$$f(x) = f(x_o) + h\frac{\partial f}{\partial x}\bigg|_{x_o} = \frac{2}{3}f_s \cdot \gamma \cdot (2\phi_f - V_{bs})^{\frac{3}{2}} + V_{ds} \cdot f_s \cdot \gamma \cdot (V_{bs} + 2\phi_f)^{\frac{1}{2}}. \tag{7.80}$$

Replacing the final term in (7.72) with this expression produces

$$I_{ds} = \frac{\mu_{eff} C_{ox} W_{eff}}{L_{eff}} \left\{ \left[V_{gs} - (V_{fb} + 2\phi_f) - f_s \cdot \gamma \cdot (2\phi_f - V_{bs})^{\frac{1}{2}} - f_n \cdot (2\phi_f - V_{bs}) \right] \right.$$

$$\left. V_{ds} - (1 + f_n)\frac{V_{ds}^2}{2} \right\}, \tag{7.81}$$

which is identical to the Level 1 drain current equation, with the minor addition of the f_s and f_n terms. As discussed in Chapter 6, (7.81) is inadequate for shorter channel FETs.

For improved accuracy, it is necessary to include the second order expansion term:

$$f(x) = f(x_o) + h \frac{\partial f}{\partial x}\bigg|_{x_o} + \frac{h^2}{2}\frac{\partial^2 f}{\partial x^2}\bigg|_{x_o}$$

$$= \frac{2}{3} f_s \cdot \gamma \cdot (2\phi_f - V_{bs})^{\frac{3}{2}} + V_{ds} \cdot f_s \cdot \gamma \cdot (2\phi_f - V_{bs})^{\frac{1}{2}}$$

$$+ \frac{V_{ds}^2}{2} \cdot \frac{f_s \cdot \gamma}{2}(2\phi_f - V_{bs})^{-\frac{1}{2}}. \tag{7.82}$$

Now (7.72) becomes

$$I_{ds} = \frac{\mu_{eff} C_{ox} W_{eff}}{L_{eff}} \left\{ \left[V_{gs} - (V_{fb} + 2\phi_f) - f_s \cdot \gamma \cdot (2\phi_f - V_{bs})^{\frac{1}{2}} - f_n \cdot (2\phi_f - V_{bs}) \right] \right.$$

$$\left. V_{ds} - \frac{V_{ds}^2}{2}\left[1 + \frac{f_s \cdot \gamma}{2(2\phi_f - V_{bs})^{\frac{1}{2}}} + f_n \right] \right\}. \tag{7.83}$$

Equation (7.83) is the basic drain current expression that is implemented in Level 3. However, two minor modifications are in order:

1. The original Berkeley derivation contains an error, in which the 2 in the rightmost denominator was replaced with a 4. (It should be noted that for consistency this error has never been corrected, and the same expression is used in HSPICE and PSPICE.)
2. Equation (7.83) has recovered the entire Level 3 threshold voltage expression, with the exception of the DIBL term. This is easily included, so that the entire Level 3 threshold voltage expression (7.34) is placed in (7.83).

With these changes, (7.73) is transformed to the final expression for the drain current in the linear region:

$$I_{ds} = \frac{\mu_{eff} C_{ox} W_{eff}}{L_{eff}} \left\{ (V_{gs} - V_t)V_{ds} - \frac{V_{ds}^2}{2}\left[1 + \frac{f_s \cdot \gamma}{4(2\phi_f - V_{ts})^{\frac{1}{2}}} + f_n \right] \right\}. \tag{7.84}$$

In contrast to the Level 2 linear region expression (6.53), the Level 3 equation is more compact, and contains none of the $\frac{3}{2}$-power terms which appear in the Level 2 drain current model. Thus, the Level 3 expression is more efficient. Equation (7.84) does not account for channel pinch-off.

7.5.2 Channel Pinch-Off (Saturation Voltage)

As described earlier, the Taylor expansion used to derive (7.83) employs the assumption that

$$(2\phi_f - V_{bs}) \gg V_{ds}, \tag{7.85}$$

but the resulting expression is allowed to be valid for all V_{ds}. In (7.83), V_{ds} can increase until it reaches V_{dsat}, at which point the channel pinches off at the drain end of the channel and the voltage at the pinch-off point does not increase above V_{dsat}. For a quick examination of this problem, consider the Level 1 expression $V_{dsat} = (V_{gs} - V_t)$; in a typical FET, this value can reach several volts. The left-hand side of the inequality (7.85) is at its minimum for $V_{bs} = 0$, in which case $2\phi_f$ is about 1 V. Thus, the approximation underlying (7.85) seems to be of poor validity. However, experimental results indicate that these shortcomings are covered up during parameter extraction, allowing good fits to experimental data.

At the onset of pinch-off, $V_{ds} = V_{dsat}$. As an example, first examine the result from the first-order Taylor expansion, which yielded (7.81). In this case, (7.81) (with the changes to the definitions described earlier) becomes

$$I_{dsat} = \frac{\mu_{eff} C_{ox} W_{eff}}{L_{eff}} \left[(V_{gs} - V_t)V_{dsat} - \frac{V_{dsat}^2}{2} \right]. \tag{7.86}$$

The saturation current I_{dsat} can also be generically defined as the charge passing into the drain per unit time, which can be written as

$$I_{dsat} = Q_{dw} \cdot v, \tag{7.87}$$

where Q_{dw} is the areal charge density at the drain times the device width,

$$Q_{dw} = W C_{ox}(V_{gs} - V_t - V_{ds}), \tag{7.88}$$

and v is the carrier velocity. *Pinch-off* can be defined as the drain bias at which $Q_{inv}(y)$ becomes zero. If this definition is used in (7.70), the lateral field E_y and the carrier velocity become infinite. Instead, the carrier velocity is not allowed to exceed v_{SAT}, which establishes the de facto relationship between E_{sat} and v_{SAT}; this is used in (7.87).

To find an expression for V_{dsat}, (7.86) is set equal to (7.87), yielding

$$\frac{\mu_{eff} C_{ox} W_{eff}}{L_{eff}} \left[(V_{gs} - V_t) V_{dsat} - \frac{V_{dsat}^2}{2} \right] = W_{eff} C_{ox} (V_{gs} - V_t - V_{ds}). \qquad (7.89)$$

Examination reveals that (7.89) is quadratic in V_{dsat}, and thus can be solved for V_{dsat}. This leads to the double solution

$$V_{dsat} = (V_{gs} - V_t) + \frac{v_{SAT} \cdot L_{eff}}{\mu_{eff}} \pm \sqrt{(V_{gs} - V_t)^2 + \left(\frac{v_{SAT} \cdot L_{eff}}{\mu_{eff}} \right)^2}. \qquad (7.90)$$

Only one of the two solutions will be physically valid; it is necessary to determine which one that is. For a very long channel device, recovery of the simple Level 1 expression for V_{dsat} would be expected; under this condition,

$$\left(\frac{v_{SAT} \cdot L_{eff}}{\mu_{eff}} \right) \gg (V_{gs} - V_t), \qquad (7.91)$$

and (7.90) becomes

$$V_{dsat} = (V_{gs} - V_t) + \frac{v_{SAT} \cdot L_{eff}}{\mu_{eff}} \pm \frac{v_{SAT} \cdot L_{eff}}{\mu_{eff}}. \qquad (7.92)$$

For the "+" solution, V_{dsat} becomes very large, while for the "−" solution, the Level 1 expression $V_{dsat} = (V_{gs} - V_t)$ is recovered. It can be concluded that only the "−" solution is physically valid, and thus (7.90) can be selectively rewritten as

$$V_{dsat} = (V_{gs} - V_t) + \frac{v_{SAT} \cdot L_{eff}}{\mu_{eff}} - \sqrt{(V_{gs} - V_t)^2 + \left(\frac{v_{SAT} \cdot L_{eff}}{\mu_{eff}} \right)^2}. \qquad (7.93)$$

This same procedure can now be applied to the modified second-order drain current expression (7.84). For I_{dsat}, (7.84) becomes

$$I_{ds} = \frac{\mu_{eff} C_{ox} W}{L_{eff}} \left\{ (V_{gs} - V_t) V_{dsat} - \frac{V_{dsat}^2}{2} \left[1 + \frac{f_s \cdot \gamma}{4(2\phi_f - V_{bs})^{\frac{1}{2}}} + f_n \right] \right\}. \qquad (7.94)$$

Due to charge sharing (the overlap of the source and drain depletion regions), the expression for Q_{dw} becomes

$$Q_{dw} = W_{eff} C_{ox} \left[(V_{gs} - V_t) - \left(1 + \frac{f_s \cdot \gamma}{4(2\phi_f - V_{bs})^{\frac{1}{2}}} + f_n \right) V_{dsat} \right], \qquad (7.95)$$

where, once again, the erroneous "4" appears. Multiplying (7.95) by v_{SAT} to yield I_{dsat} (as in (7.87)) and setting that expression equal to (7.94) produces

$$\frac{\mu_{eff} C_{ox} W_{eff}}{L_{eff}} \left\{ (V_{gs} - V_t)V_{dsat} - \frac{V_{dsat}^2}{2} \left[1 + \frac{f_s \cdot \gamma}{4(2\phi_f - V_{bs})^{\frac{1}{2}}} \right] + f_n \right\}$$

$$= W_{eff} C_{ox} \left[(V_{gs} - V_t) - \left(1 + \frac{f_s \cdot \gamma}{4(2\phi_f - V_{bs})^{\frac{1}{2}}} + f_n \right) V_{dsat} \right] v_{SAT}. \qquad (7.96)$$

As before, this expression is quadratic in V_{dsat}, and can be solved. Discarding the physically meaningless "+" solution leads to

$$V_{dsat} = \frac{(V_{gs} - V_t)}{\left(1 + \frac{f_s \cdot \gamma}{4(2\phi_f - V_{bs})^{\frac{1}{2}}} + f_n \right)} + \frac{v_{SAT} \cdot L_{eff}}{\mu_{eff}}$$

$$- \sqrt{\left[\frac{(V_{gs} - V_t)}{\left(1 + \frac{f_s \cdot \gamma}{4(2\phi_f - V_{bs})^{\frac{1}{2}}} + f_n \right)} \right]^2 + \left[\frac{v_{SAT} \cdot L_{eff}}{\mu_{eff}} \right]^2}. \qquad (7.97)$$

The effect of velocity saturation, briefly described earlier, now becomes clear. The inclusion of the terms accounting for velocity saturation in (7.97) causes V_{dsat} to decrease from its classical value, described by the very first term in (7.97). If the classical expression is used for V_{dsat}, the reader can verify that the first derivative of the current equation (7.94) is continuous at $V_{ds} = V_{dsat}$. However, when (7.97) is used for V_{dsat}, the first derivative becomes discontinuous at this point. The implications of this situation will be demonstrated during the discussion of model development in Section 7.7.

Replacing v_{SAT} with the parameter **VMAX** produces the expression used by Level 3 for V_{dsat}:

$$V_{dsat} = \frac{(V_{gs} - V_t)}{\left(1 + \frac{f_s \cdot \gamma}{4(2\phi_f - V_{bs})^{\frac{1}{2}}} + f_n \right)} + \frac{\mathbf{VMAX} \cdot L_{eff}}{\mu_{eff}}$$

$$- \sqrt{\left[\frac{(V_{gs} - V_t)}{\left(1 + \frac{f_s \cdot \gamma}{4(2\phi_f - V_{bs})^{\frac{1}{2}}} + f_n \right)} \right]^2 + \left[\frac{\mathbf{VMAX} \cdot L_{eff}}{\mu_{eff}} \right]^2}. \qquad (7.98)$$

By defining

$$\alpha = \frac{(V_{gs} - V_t)}{\left(1 + \frac{f_s \cdot \gamma}{4(2\phi_f - V_{bs})^{\frac{1}{2}}} + f_n \right)}, \qquad (7.99)$$

and

$$\beta = \frac{\mathbf{VMAX} \cdot L_{eff}}{\mu_{eff}}, \tag{7.100}$$

(7.98) can be written in the more concise form of

$$V_{dsat} = \alpha + \beta + \left(\alpha^2 + \beta^2\right)^{\frac{1}{2}}. \tag{7.101}$$

In the Level 3 implementation, if **VMAX** is not specified, it defaults to zero (or, of course, it can be set to zero). If this setting is used literally in (7.98), a nonphysical result of $V_{dsat} = 0$ is produced. Under this special condition, another expression for V_{dsat} is used, which can be found by taking the omission of **VMAX** to mean that (**VMAX** · L_{eff})/μ_{eff} is so large that the specific value of **VMAX** is inconsequential. In this case, (7.98) becomes

$$V_{dsat} = \frac{(V_{gs} - V_t)}{\left(1 + \frac{f_s \cdot \gamma}{4(2\phi_f - V_{bs})^{\frac{1}{2}}} + f_n\right)}; \tag{7.102}$$

this expression is identical to the simple Level 1 model for the saturation voltage, with the inclusion of the modifications which account for small geometry effects on the depletion regions.

With the inclusion of pinch-off, the drain current expression (7.84) can be rewritten in a general form for both the linear and saturation regions:

$$I_{ds} = \frac{\mu_{eff} C_{ox} W_{eff}}{L_{eff}} \left\{ (V_{gs} - V_t)V'_{ds} - \frac{V'^2_{ds}}{2} \left[1 + \frac{f_s \cdot \gamma}{4(2\phi_f - V_{bs})^{\frac{1}{2}}} + f_n \right] \right\}, \tag{7.103}$$

where

$$V'_{ds} = min(V_{ds}, V_{dsat}). \tag{7.104}$$

Note that, in and of itself, (7.103) pins I_{ds} at I_{dsat} for $V_{ds} > V_{dsat}$. In Level 3, the observed increase in I_{ds} for increasing V_{ds} when $V_{ds} > V_{dsat}$ is accounted for entirely through the expression for channel length modulation.

7.5.3 Channel Length Modulation

As just described, aside from the minor effect of f_s, (7.103) prevents I_{ds} from increasing as V_{ds} increases above V_{dsat}. However, in short channel FETs, the drain current actually increases slightly with V_{ds} as V_{ds} increases above V_{dsat}. The physical explanation of this behavior is shown in Figure 7.6. In Figure 7.6a, $V_{ds} = V_{dsat}$, and the conducting channel reaches to the drain. In Figure 7.6b, $V_{ds} > V_{dsat}$, and the pinch-off point has moved toward the source. Since the voltage at the pinch-off point is V_{dsat} regardless of where the pinch-off point is located, in Figure 7.6a, the lateral channel field is

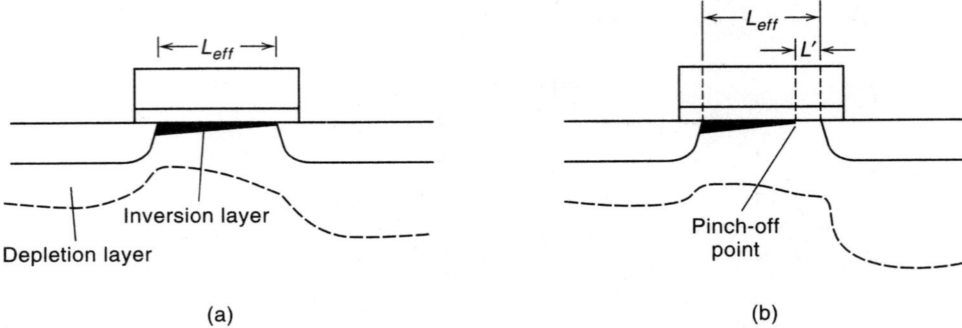

Figure 7.6 Pinch-off and channel length modulation; (a) $V_{ds} = V_{dsat}$, conducting channel reaches drain; (b) $V_{ds} > V_{dsat}$, pinch-off point moves toward the source.

$$E_{chan} = \frac{V_{dsat}}{L_{eff}}, \tag{7.105}$$

while in Figure 7.6b,

$$E_{chan} = \frac{V_{dsat}}{L_{eff} - L'}. \tag{7.106}$$

As V_{ds} increases, L' increases, which increases E_{chan} and thus I_{ds}.

To account for this behavior, an expression for L' is required. The Level 3 expression for L' is based on the model of Baum and Beneking [13]. This model is taken from a one-dimensional solution of the Poisson equation with the assumption that the space charge density in the pinched-off region is qN_{sub}:

$$L' = \left[\left(\frac{E_{sat}}{2a} \right)^2 + \frac{(V_{ds} - V_{dsat})}{a} \right]^{\frac{1}{2}} - \frac{E_{sat}}{2a}, \tag{7.107}$$

where E_{sat} is the field at the pinch-off point, and

$$a = \frac{q \cdot N_{sub}}{2\epsilon_{Si}} = \frac{q \cdot \mathbf{NSUB}}{2\epsilon_{Si}}. \tag{7.108}$$

From a first principles point of view, the correct expression for E_{sat} is

$$E_{sat} = \frac{V_{dsat}}{L_{eff} - L'}. \tag{7.109}$$

However, (7.109) and (7.107) depend on each other; a solution would be found by alternately solving (7.107) and (7.109) until correct results for E_{sat} and L' are obtained.

Any iterative approach is computationally inefficient in a circuit simulation environment. Instead, the Level 3 development sets

$$E_{sat} = \frac{V_{dsat}}{L_{eff}},$$ (7.110)

in which case E_{sat} is less than it should be when $V_{ds} > V_{dsat}$. Also, E_{sat} no longer depends on V_{ds} (through the V_{ds}-induced change in L'). This shortcoming is accounted for by adding this dependence into the second term in the brackets of (7.107) (which already has V_{ds} dependence) via the introduction of the parameter κ:

$$L' = \left[\left(\frac{V_{dsat}}{2a \cdot L_{eff}}\right)^2 + \frac{\kappa \cdot (V_{ds} - V_{dsat})}{a}\right]^{\frac{1}{2}} - \frac{V_{dsat}}{2a \cdot L_{eff}}.$$ (7.111)

Rewriting (7.111) with κ identified as a Level 3 parameter produces

$$L' = \left[\left(\frac{V_{dsat}}{2a \cdot L_{eff}}\right)^2 + \frac{\mathbf{KAPPA} \cdot (V_{ds} - V_{dsat})}{a}\right]^{\frac{1}{2}} - \frac{V_{dsat}}{2a \cdot L_{eff}}.$$ (7.112)

If **VMAX** is not specified (and defaults to **VMAX** $= 0$), V_{dsat} is, in effect, very large, and (7.112) becomes

$$L' = \left[\frac{\mathbf{KAPPA} \cdot (V_{ds} - V_{dsat})}{a}\right]^{\frac{1}{2}}.$$ (7.113)

Equations (7.112) and (7.113) are the model for channel length modulation that is implemented in Level 3. With channel length modulation included, the current expression (7.103) becomes

$$I_{ds} = \frac{\mu_{eff} C_{ox} W_{eff}}{L_{eff} - L'} \left\{ (V_{gs} - V_t) V'_{ds} - \frac{{V'_{ds}}^2}{2} \left[1 + \frac{f_s \cdot \gamma}{4(2\phi_f - V_{bs})^{\frac{1}{2}}} + f_n\right] \right\},$$ (7.114)

A major weakness of (7.112) is that it is possible for L' to become larger than L_{eff} (i.e., the inversion channel length becomes negative), which is physically impossible. In reality, the extension of the pinch-off region to the source side of the device places the FET into a punchthrough mode which is not accounted for by the Level 3 model. Instead, the Level 3 model includes a mathematical method of preventing L' from exceeding L_{eff}.

The Berkeley version of SPICE uses the method depicted in Figure 7.7 [9]. The term W_B is the zero bias depletion region surrounding the drain diffusion (the source depletion region is neglected), allowing the definition of L_{max}, the maximum inversion channel length:

$$L_{max} = L_{eff} - W_B, \tag{7.115}$$

with

$$W_B = \sqrt{\frac{2\epsilon_{Si} \cdot \mathbf{PB}}{q \cdot \mathbf{NSUB}}}. \tag{7.116}$$

In this scheme, if the inversion channel length $L'_{eff} = L_{eff} - L'$ becomes less than W_B, L'_{eff} is instead computed from

$$L'_{eff} = \frac{W_B}{1 + \frac{L' - L_{max}}{W_B}}. \tag{7.117}$$

The largest value that L' can take on is L_{eff}; when this is inserted into (7.117) along with (7.115), the minimum value possible for L'_{eff} is found to be

$$L'_{eff} = \frac{W_B}{2}. \tag{7.118}$$

A somewhat different scheme is implemented in HSPICE [11].

7.5.4 Source/Drain Series Resistance

The Level 3 drain current equation (7.103) was derived assuming that the conducting channel represents the major resistance element in the drain current path. However, as discussed in Chapter 3, in shorter channel FETs, the resistance of the source and drain regions becomes significant when compared to the channel resistance. The model must be modified to account for this phenomenon.

Although the physical basis of the extrinsic series resistance is quite complex, the Level 3 model uses the simple lumped element approach depicted in Figure 7.8, in which

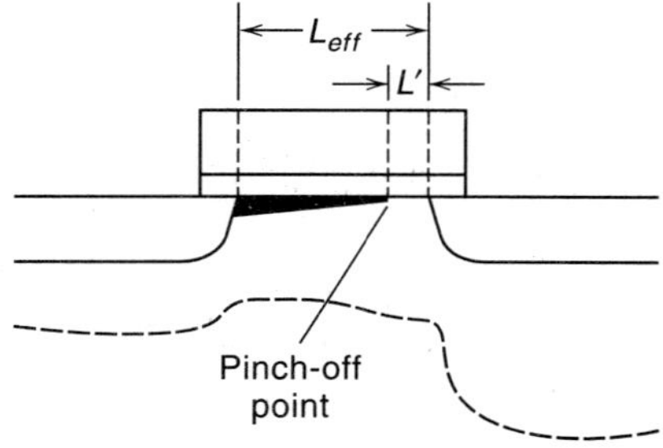

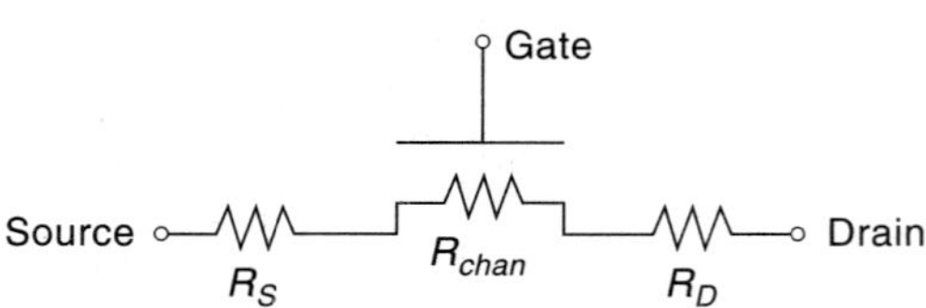

Figure 7.7 Method used by Berkeley SPICE to keep the computed pinch-off point from reaching the source.

Figure 7.8 The introduction of parasitic series resistance into Level 3.

two extra resistances due to the source and drain are added, yielding

$$R_{total} = R_{chan} + R_s + R_d. \tag{7.119}$$

In this model, part of the total drain to source bias (V_{ds}) is dropped across R_s and R_d; this has the effect of decreasing the voltage across the conducting channel, which reduces the drain current from its expected value. Since the extrinsic resistance is usually inversely proportional to the device width, this simple, lumped element approach has the drawback of not being useful for different device widths. Some simulators (such as HSPICE) have modified these models to allow more general geometric applicability.

In a long channel device, the channel resistance is much larger than the extrinsic resistance, and (7.119) reduces to

$$R_{total} = R_{chan}. \tag{7.120}$$

The unmodified current model (7.103) should describe the device characteristics. In a short channel device at very low drain bias, (7.103) accounts for changes in the drain current due to short channel effects (the merging of the source and drain depletion regions, and DIBL). Any departure of the measured current from this predicted value is deemed to be due to series resistance. This is examined more fully below in the discussion of parameter extraction.

7.5.5   The Subthreshold Current Model

The drain current model implemented in Level 3 contains a separate model for subthreshold conduction. While a worthwhile addition, the physical basis of the model is poor, which limits its usefulness.

The subthreshold current equation is based on the model of Swanson and Meindl [14], in which the current is described by

$$I_{ds} = I_{on} e^{q(V_{gs} - V_{on})/nk_b T}, \tag{7.121}$$

where I_{on} is the drain current at $V_{gs} = V_{on}$, as will be described below. The term V_{on} is a modified threshold voltage; as V_{gs} decreases below V_{on}, the current decreases exponentially. The introduction of V_{on} is an attempt to bridge the gap between the fully "on" region of device operation, where drift dominates conduction, and the subthreshold region, where diffusion dominates conduction. As described in Chapter 3, there is a gate bias region where neither mechanism dominates, which complicates analytical modeling. The use of V_{on} has the effect of ascribing all conduction to diffusion. This V_{on} is

$$V_{on} = V_t + n\frac{k_b T}{q}. \tag{7.122}$$

The term n is similar to the diode ideality factor, but is defined as

$$n = 1 + \frac{q \cdot \mathbf{NFS}}{C_{ox}} + \frac{C_{depl}}{C_{ox}}, \tag{7.123}$$

where **NFS** is the fast interface state density; any variation in the subthreshold behavior is attributed to this parameter. As will be described below in the discussion of parameter extraction, good results for the subthreshold region are found only by allowing **NFS** to take on a very large and physically unrealistic value. Thus, **NFS** can be regarded as a purely empirical fitting constant. Another problem is that **NFS** has no drain or substrate bias dependence, and no temperature dependence, further limiting its usefulness.

The final term in (7.123) can be defined as

$$\frac{C_{depl}}{C_{ox}} = \frac{1}{C_{ox}} \frac{\partial Q_{depl}}{\partial V_{bs}}. \tag{7.124}$$

Recalling (7.25) (which neglects DIBL),

$$\frac{Q_{depl}}{C_{ox}} = f_s \cdot \gamma \cdot (2\phi_f - V_{bs})^{\frac{1}{2}} + f_n \cdot (2\phi_f - V_{bs}) \tag{7.125}$$

leads to

$$\frac{C_{depl}}{C_{ox}} = \frac{1}{C_{ox}} \left[\frac{f_s \cdot \gamma \cdot (2\phi_f - V_{bs})^{\frac{1}{2}}}{2(2\phi_f - V_{bs})} - f_n \right]. \tag{7.126}$$

The Level 3 implementation simplifies (7.126) by employing a common denominator; it contains an error of dropping a "2" during the simplification. Thus, (7.126) should be

$$\frac{C_{depl}}{C_{ox}} = \frac{1}{C_{ox}} \left[\frac{f_s \cdot \gamma \cdot (2\phi_f - V_{bs})^{\frac{1}{2}} - 2f_n \cdot (2\phi_f - V_{bs})}{2(2\phi_f - V_{bs})} \right], \tag{7.127}$$

but is instead actually implemented as

$$\frac{C_{depl}}{C_{ox}} = \frac{1}{C_{ox}} \left[\frac{f_s \cdot \gamma \cdot (2\phi_f - V_{bs})^{\frac{1}{2}} - f_n \cdot (2\phi_f - V_{bs})}{2(2\phi_f - V_{bs})} \right]. \tag{7.128}$$

Substituting (7.128) into (7.123) leads to the final expression for the ideality factor:

$$n = 1 + \frac{q \cdot \mathbf{NFS}}{C_{ox}} + \frac{1}{C_{ox}} \cdot \left[\frac{f_s \cdot \gamma \cdot (2\phi_f - V_{bs})^{\frac{1}{2}} - f_n \cdot (2\phi_f - V_{bs})}{2(2\phi_f - V_{bs})} \right]. \tag{7.129}$$

The term I_{on} can now be considered. From (7.121), it is readily seen that for $V_{gs} = V_{on}$, $I_{ds} = I_{on}$. To guarantee current continuity at V_{on}, I_{on} is defined from the superthreshold current expression (7.114), where V_{gs} is replaced with V_{on} as given by (7.122). This produces

$$I_{on} = \frac{\mu_{eff} C_{ox} W_{eff}}{L_{eff} - L'} \left\{ (V_{on} - V_t) V'_{ds} - \frac{V'^2_{ds}}{2} \left[1 + \frac{f_s \cdot \gamma}{4(2\phi_f - V_{bs})^{\frac{1}{2}}} + f_n \right] \right\}. \qquad (7.130)$$

As defined by (7.130), I_{on} is a constant for all gate biases, and is used in (7.121) to compute the subthreshold current. Although this approach guarantees current continuity at $V_{gs} = V_{on}$, it does not guarantee continuity of the first derivative, as will be seen when parameter extraction is carried out. This discontinuity can cause convergence problems during circuit simulation. Note that the value of V_{on} defines which equation is used to compute the drain current. When $V_{gs} < V_{on}$, the subthreshold equation (7.121) is used; when $V_{gs} > V_{on}$, the superthreshold equation (7.114) is employed.

7.6   The Charge Model

In the Level 2 development, it was decided that the drain current equations were too complicated to be used in the computation of the node charges; the Level 1 node charge equations were used instead. In contrast, the simpler Level 3 current expressions lead to a tractable set of node charge equations. These node charges are then used to compute the gate capacitance.

When Level 3 was introduced into Berkeley SPICE, the Meyer model for the gate capacitance [15] was discarded and replaced with the Ward-Dutton model [16] (see Chapter 13). Thus, there is no choice of gate capacitance model in Berkeley SPICE, or in PSPICE. HSPICE [11] allows the user to choose between the Meyer model and the Ward-Dutton model. This choice depends on the particular circuit application. The Ward-Dutton model has an improved physical basis, and is generally the better choice. Its major drawback is that its iterative computational structure causes the model to be slow. In a large circuit where simulation efficiency is a major consideration, the Meyer model is a reasonable option.

The Level 3 charge computation uses the same basic approach described in Chapters 5 and 6. Charge neutrality requires that the total charge in the device must sum to zero:

$$Q_{INV} + Q_{DEPL} + Q_{GATE} = 0. \qquad (7.131)$$

Here, as in the earlier descriptions, the capitalized subscripts indicate the use of the total charge (not the charge per unit length or per unit area).

For the charges inside the device (Q_{INV} and Q_{DEPL}), the effective channel dimensions (W_{eff} and L_{eff}) should be used. For the gate charge Q_{GATE}, the gate dimensions (W_{gate} and L_{gate}) should be used. However, this will complicate the final equations considerably. In the University of California/Berkeley version of Level 3, nothing is done to account for reduction of the channel width by either processing or bird's beak encroachment; only the drawn value W is available, and this is used in the model. While a model is available to account for channel length reduction from the drawn value, it lumps together both the effect of processing on the drawn gate dimension and the underdiffusion of the gate by the source and drain. Thus, for capacitance modeling, only L_{eff} is available. Since this is the original formulation, it is the one that will be used in this derivation. HSPICE includes parameters to separate the drawn gate dimensions, the actual gate dimensions, and the effective channel dimensions. This allows a more accurate gate capacitance calculation, particularly for smaller devices in which the differences among the three values become significant. The method of accounting for this behavior in HSPICE and in MOS Model 9 is described in more detail in Appendix B.

Assuming uniformity across the width of the device, the total gate charge can be written as

$$Q_{GATE} = W \int_0^{L_{eff}} Q_{gate}(y)dy, \tag{7.132}$$

where Q_{gate}, with the lower case subscript, now indicates the gate charge per unit length along the length of the channel. Using (7.70), a description of the drain current can be included as

$$I_{ds} = \mu W Q_{inv}(V_y)\frac{dV_y}{dy}. \tag{7.133}$$

This expression can be used to replace dy in (7.132), leading to

$$Q_{GATE} = \frac{\mu W^2}{I_{ds}} \int_0^{V_{ds}} Q_{gate}(V_y)Q_{inv}(V_y)dV_y \tag{7.134}$$

The gate charge along the channel should be the charge "mirror" of the surface charge along the channel, which was defined in (7.68). Therefore,

$$Q_{gate}(V_y) = -Q_s(V_y) = C_{ox}(V_{gs} - V_t - V_y). \tag{7.135}$$

However, the implementation of the threshold voltage in (7.135) is somewhat strange, as the small geometry modifications to the depletion charge (see (7.25)) are completely neglected; both f_s and f_n are set to zero. While setting $f_n = 0$ is the limiting case when edge effects are neglected, setting $f_s = 0$ seems to be a very poor approximation, as it neglects the effect on the threshold voltage of both the substrate bias and the depletion charge. In contrast, the drain-induced barrier lowering (DIBL) term in the threshold voltage expression is retained. Examining the Level 3 threshold voltage equation (7.34), if the more reasonable approximation of letting $f_s \rightarrow 1$ and $f_n \rightarrow 0$ is made, the threshold voltage expression would be

$$V_t = V_{fb} + 2\phi_f + \gamma(2\phi_f - V_{bs})^{\frac{1}{2}} - \sigma \cdot V_{ds}. \tag{7.136}$$

When V_{bs} is set to zero in (7.136), the very simple expression **VTO**, defined in (7.35) for use in the mobility model, is recovered.

However, for some reason, $f_s \to 0$ is used here, so the threshold voltage expression is

$$V_t = V_{fb} + 2\phi_f - \sigma \cdot V_{ds}.$$ (7.137)

Thus, the final expression for the gate charge along the channel is

$$Q_{gate}(V_y) = C_{ox}\left[V_{gs} - (V_{fb} + 2\phi_f - \sigma \cdot V_{ds}) - V_y\right].$$ (7.138)

A description of the inversion charge along the channel $Q_{inv}(V_y)$ is now required. An expression can be found by working backwards from the linear region drain current expressions. Using (7.133), the drain current can be written as

$$I_{ds} = \frac{\mu W}{L_{eff}} \int_0^{V_{ds}} Q_{inv}(V_y) dV_y.$$ (7.139)

The Level 3 equation for the linear region current was derived in (7.84):

$$I_{ds} = \frac{\mu W C_{ox}}{L_{eff}}\left[(V_{gs} - V_t)V_{ds} - \frac{V_{ds}^2}{2}(1 + f_b)\right],$$ (7.140)

where, for convenience in what follows, the term f_b has been used:

$$f_b = \left[\frac{\gamma \cdot f_s}{4(2\phi_f - V_{bs})^{\frac{1}{2}}} + f_n\right].$$ (7.141)

By taking the derivative of (7.140) with respect to V_{ds} and manipulating the result in (7.139), it can be shown that the inversion charge is described by

$$Q_{inv}(V_y) = -C_{ox}\left[V_{gs} - V_t - (1 + f_b)V_y\right],$$ (7.142)

where V_t is the full threshold voltage expression (7.34). Using (7.138), (7.140), and (7.142), the total gate charge can now be found by carrying out the integral (7.134), leading to

$$Q_{GATE} = W \cdot L_{eff} \cdot C_{ox}\left[V_{gs} - (V_{fb} + 2\phi_f - \sigma \cdot V_{ds}) - \frac{V_{ds}}{2} + \frac{1 + f_b}{12 \cdot f_i} \cdot V_{ds}^2\right],$$ (7.143)

where the term f_i is

$$f_i = V_{gs} - V_t - \frac{1 + f_b}{2} \cdot V_{ds}.$$ (7.144)

The total depletion charge Q_{DEPL} can be computed in a similar fashion. Following (7.132), the total depletion charge is

$$Q_{DEPL} = W \int_0^{L_{eff}} Q_{depl}(y)\,dy. \tag{7.145}$$

Using (7.133) to replace dy in (7.145), Q_{DEPL} can be rewritten as

$$Q_{DEPL} = \frac{\mu W^2}{I_{ds}} \int_0^{V_{ds}} Q_{depl}(V_y) \cdot Q_{inv}(V_y) \cdot dV_y. \tag{7.146}$$

Using (7.25), the depletion charge along the channel is

$$Q_{depl}(V_y) = f_s \cdot \gamma \cdot (2\phi_f - V_{bs})^{\frac{1}{2}} + f_n \cdot (2\phi_f - V_{bs}). \tag{7.147}$$

Using (7.140), (7.142), and (7.147), the integral (7.146) can be evaluated, leading to the final expression for the total depletion charge:

$$Q_{DEPL} = -W \cdot L_{eff} \cdot C_{ox} \left[f_s \cdot \gamma \cdot (2\phi_f - V_{bs})^{\frac{1}{2}} \right.$$

$$\left. + f_n \cdot (2\phi_f - V_{bs}) + \frac{f_b}{2} \cdot V_{ds} - \frac{f_b \cdot (1 + f_b)}{12 \cdot f_i} \cdot V_{ds}^2 \right]. \tag{7.148}$$

Using the charge conservation expression (7.131), the total inversion (or channel) charge is

$$Q_{INV} = -(Q_{GATE} + Q_{DEPL}), \tag{7.149}$$

where Q_{GATE} and Q_{DEPL} are inserted from (7.143) and (7.148), respectively. Equation (7.149) is used in either the Meyer model or the Ward-Dutton model to compute the gate capacitance. In the linear region, the charge is divided equally between the source and the drain:

$$Q_S = Q_D = \frac{1}{2} Q_{INV}. \tag{7.150}$$

In the saturation region, V_{ds} is replaced in the model equations with the saturation voltage V_{dsat}, and the charge is divided between the source and the drain using the parameter **XQC**:

$$Q_D = \mathbf{XQC} \cdot Q_{INV}, \tag{7.151}$$

$$Q_S = (1 - \mathbf{XQC}) \cdot Q_{INV}. \tag{7.152}$$

If not specified, **XQC** defaults to 0.5, and Q_{INV} is divided equally between the source and the drain. Ward and Dutton [16] noted that their model results came closest to numerical simulations when **XQC** $= 0.4$; this value appears frequently in Level 3 model parameter sets.

As noted earlier, in the University of California/Berkeley implementation of the Level 3 model [9], only the Ward-Dutton gate capacitance model [16] is available. In HSPICE [11], several variations of both the Meyer model and the Ward-Dutton model can be used with Level 3.

7.6.1   The Zero Bias Gate Capacitance

At the time that both the Meyer and Ward-Dutton models were introduced, FET sizes were large (by present standards). This caused the active gate capacitance to be much larger than the zero bias capacitance between the gate electrode and, respectively, the source and drain diffusions; the zero bias capacitance could be neglected. In smaller devices, the zero bias capacitance is relatively large and must be taken into account.

As described in Chapters 5 and 6, this modification is carried out by including the parameters **CGSO** and **CGDO**, the zero bias capacitance (in F/m) between the gate and, respectively, the source diffusion and the drain diffusion. There is also a parameter to describe the capacitance between the gate electrode and the substrate, **CGBO**, but this capacitance tends to be very small compared to the gate to diffusion capacitances. The total gate capacitance is computed from

$$C_{total} = C_{active} + \textbf{CGSO} \cdot W_{eff} + \textbf{CGDO} \cdot W_{eff} + \textbf{CGBO} \cdot L_{eff}. \tag{7.153}$$

If the gate to substrate capacitance is to be ignored, it is best to set **CGBO** $= 0$ in the model parameter listing, as some simulators will attempt to compute a value of **CGBO** if it is not specified.

7.7   Parameter Extraction and Model Development

As noted in Chapter 2, the mathematical structure of an FET model represents an upper limit of the capabilities of that model. To reach that limit, proper parameter extraction and model construction are necessary.

Since Level 3 is similar to Level 2, parameter extraction will follow the same general form. The reader is encouraged to compare the Level 3 results with the corresponding parts of the Level 2 extraction described in Chapter 6, as this clearly demonstrates the superior results that are possible with Level 3. In contrast to Level 2, Level 3 is currently a widely used model in digital circuit design. Although Level 3 was not intended to serve as a model for submicron FETs, it will be shown here that with proper care, this is feasible. Part of this success is due to model binning, the division of the length/width geometry space into submodels that describe narrowly defined regions of that space. Careful model building and binning allow Level 3 to be effectively used for smaller FET channel lengths than was originally intended; however, binning introduces problems of its own that will be demonstrated here.

7.7.1 Basic Parameter Extraction

As in the Level 1 and Level 2 models, the focus of parameter extraction for Level 3 is on matching the gathered data to the model equations, within the mathematical constraints of that model. Due to the semi-empirical nature of Level 3, many of the parameters have, at least in name, a physical basis.

The basic structure of the Level 3 model is the same as that of the Level 2 model; short and narrow channel effects are separated from each other. Therefore, the approach to parameter extraction will follow the same general set of steps [17]:

- The oxide thickness, along with the differences between the drawn and effective channel dimensions, are provided as process input.
- A base set of parameters is extracted from a long, wide FET.
- Parameters describing the behavior of short and narrow devices at low drain bias are extracted.
- Parameters describing the effect of the substrate bias at low drain bias are extracted.
- Parameters related to high drain bias operation are extracted.
- The subthreshold parameter is extracted.
- If available, the temperature dependent parameter is extracted.

As in Level 2, the long, wide device is treated as providing the base model, with corrections appended to it to account for short and narrow channel behavior.

Once again, three FETs are required for model construction, as shown in Figure 7.9: (1) a long, wide FET for the base parameters; (2) a wide, short device for the short channel corrections; (3) a long, narrow device for the narrow channel corrections. Here, measured data from FETs with channel widths/lengths of 20 μm/20 μm, 20 μm/0.7 μm, and 1.3 μm/20 μm will be employed. For this die, the gate oxide thickness is 13.0 nm. The use of small geometry devices with a very thin gate oxide will not only allow for

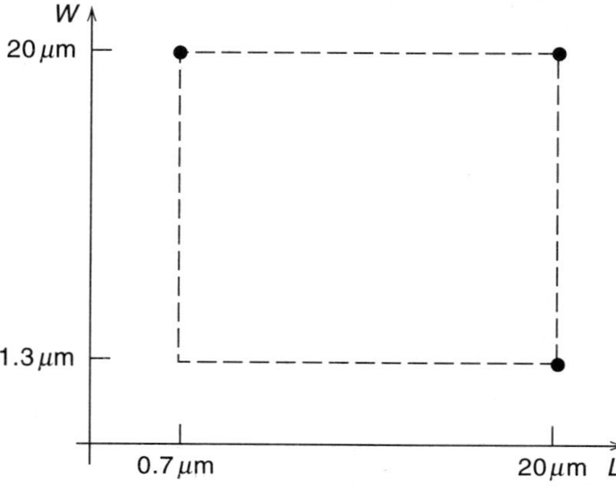

Figure 7.9 The geometric scheme for extracting the Level 3 model parameters. A long, wide device is used to extract the basic parameters, while a short, wide device and a long, narrow device are used to extract the small geometry corrections.

development of an appropriate Level 3 model; it will also demonstrate the submicron capabilities of Level 3, as well as point out some of the deficiencies encountered when building such models for contemporary FET technology.

The first step is to extract the base device parameters **UO**, **VTO**, and **THETA** from a long, wide device under low drain bias. The long, wide device removes the effect of small geometry, while the use of a low drain bias negates short channel effects (channel pinch-off, carrier velocity saturation, etc.). The extraction is done using a nonlinear least squares algorithm, with the result shown in Figure 7.10. Only data well above the threshold voltage are used for fitting; inclusion of subthreshold data corrupts the extraction. The theta model for mobility degradation (7.42) provides good results over most of the gate bias range. Note, however, how it begins to fail at high gate bias, as the mobility decreases more rapidly with gate bias than is predicted by the theta model. This indicates a shortcoming of the Level 3 mobility model, which becomes more important as the gate oxide thickness decreases and the magnitude of the normal field increases.

Next values for the parasitic source and drain series resistances **RS** and **RD** are extracted. Most FET layouts have the source and drain contacts equidistant from the gate edge, so that it can generally be assumed that **RS** = **RD**. In practice, the series resistance varies with channel width, so that a simple lumped resistance is inadequate when various channel widths are used. A superior approach is to compute **RS** and **RD** as secondary parameters using

$$\mathbf{RS} = \mathbf{RD} = \mathbf{R}/W_{eff}, \tag{7.154}$$

where **R** is the series resistance in ohm-μm. In the Berkeley SPICE version of Level 3, only the simple lumped parameters **RS** and **RD** are available; if different channel widths are employed, the user must make multiple copies of the FET models, compute appropriate values of **RS** and **RD** for each W_{eff}, and ensure that the appropriate model is called by each FET in the circuit netlist. This is cumbersome, and complicates the situation for the circuit designer, who would rather have a single FET model to call from the netlist. The HSPICE simulator provides a variety of schemes which address this shortcoming, as well as additional features to account for contact placement, LDD implants, and stacked devices [11].

Figure 7.11 shows an extraction of the series resistance using a wide, short device in the linear region. Note once again that some further work must be done to make the resistance values more generally applicable to a variety of channel widths.

The following step is to extract the parameter **DELTA**, which describes the effect of a narrow channel on the threshold voltage. As described earlier in this chapter, the depletion region spreads outside the channel edges, adding to the total bulk charge; this causes the threshold voltage to increase as the channel narrows. As shown in Figure 7.12, **DELTA** is extracted from a long, narrow device operating in the linear region.

Two steps follow which account for the effect of the substrate bias on the device characteristics. As described in Chapter 3, the application of a substrate bias changes the depletion charge and thus the threshold voltage. In a long, wide device (20/20), the shift of the I_d-V_g curves with applied substrate bias is determined by the substrate doping

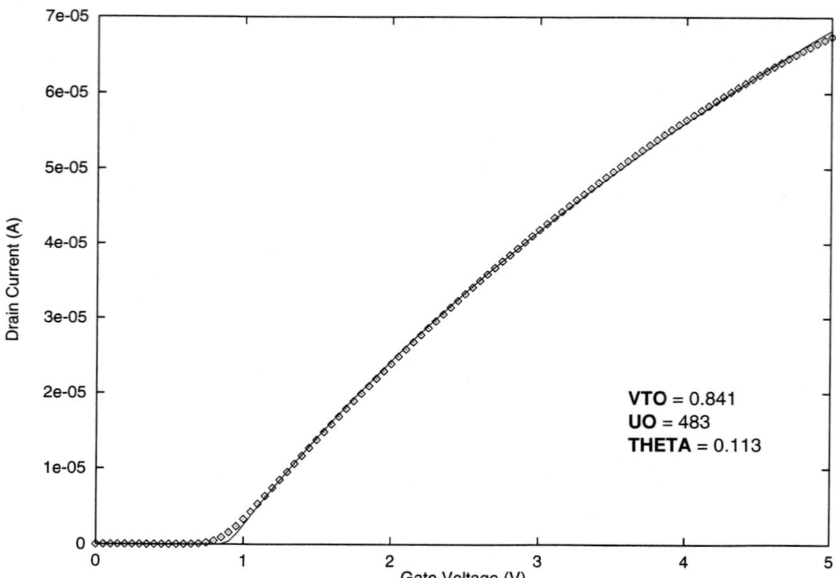

Figure 7.10 The extraction of the base device parameters **VTO**, **UO**, and **THETA**, $W/L = 20 \ \mu m/20 \ \mu m$.

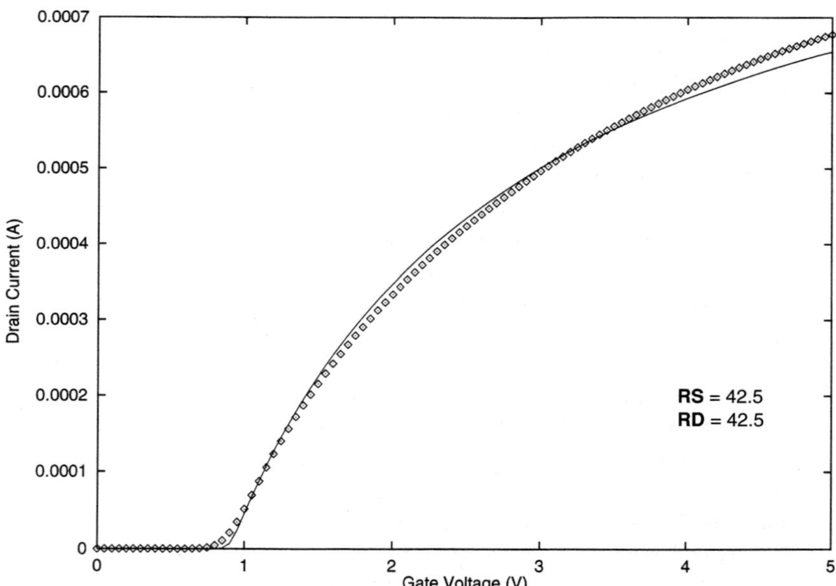

Figure 7.11 The extraction of the parasitic series resistance parameters **RS** and **RD**, $W/L = 20 \ \mu m/0.7 \ \mu m$.

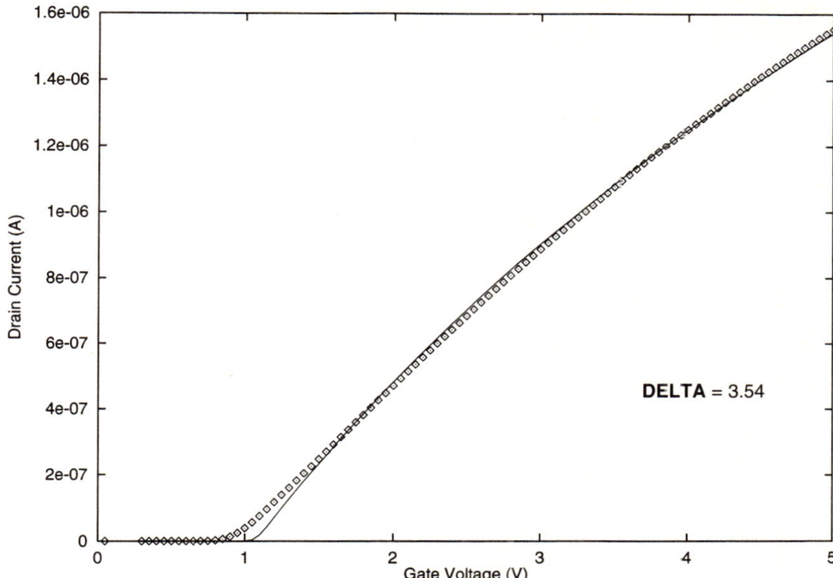

Figure 7.12   The extraction of the parameter **DELTA**, which accounts for the spreading of the gate-induced depletion region outside of the channel edges, $W/L = 1.3\ \mu m/20\ \mu m$.

NSUB; the larger the value of **NSUB**, the greater the shift in the I_d-V_g curves from the zero substrate bias value. Thus, **NSUB** is extracted from linear region data with several applied substrate biases, as shown in Figure 7.13. Note that the fit is somewhat mediocre for nonzero substrate biases; this is due to the absence of the substrate bias from the mobility model, as detailed in Section 7.4. In addition, the model assumes a single uniform substrate doping **NSUB**, which is rarely the case in contemporary FET designs. In practice, as the gate bias is increased, the gate-induced depletion region expands into a region where the doping varies with vertical location; this possibility is not accounted for in the Level 3 model. Fortunately, the effect of nonuniform channel doping is usually minor, and good results can be obtained for FETs with implanted channels, such as those used here.

The effect of the substrate bias on short channel devices requires a second extraction step. As noted in Section 7.2, with the use of the channel shortening parameter **LD**, the parameter describing the junction depth (**XJ**) has no effect laterally into the channel region; **XJ** only has meaning in the vertical direction. When a substrate bias V_{bs} is applied, the depletion region around the source diffusion increases in extent, which changes the total depletion charge and thus the threshold voltage. This effect is negligible in long channel devices, where the effect of the substrate bias on the gate-induced depletion charge is much more important. In a short channel device, the change in the depletion charge around the source diffusion becomes a significant fraction of the additional depletion charge caused by the substrate bias. Thus, in practice, **XJ** does not represent the junction depth, but is instead a short channel correction to the substrate

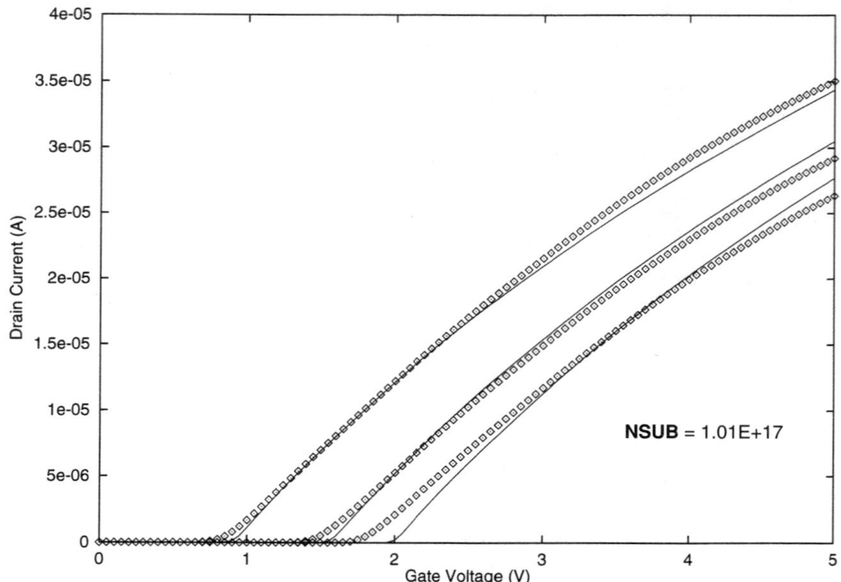

Figure 7.13 The extraction of the substrate sensitivity/doping parameter **NSUB**, $W/L = 20$ μm/20 μm. Note how the absence of substrate bias dependence in the mobility model causes an inaccurate fit at nonzero values of the substrate bias.

sensitivity. It is extracted from a wide, short device in the linear region with several substrate biases (Figure 7.14). Note that the attempt to fit the data to several substrate biases decreases the accuracy of the zero substrate bias fit. In this case, the result of **XJ** = 0.154 μm is actually a reasonable value for a junction depth. However, this result is **not** physically meaningful; much smaller values are frequently extracted for **XJ**.

Next, the parameters describing the behavior of the device at high drain bias (the saturation region) are determined. The three parameters (**VMAX**, **ETA**, and **KAPPA**) each serve a different purpose. The parameter **VMAX** represents the saturation velocity (the maximum velocity which the carriers can attain), **ETA** describes the effect of the drain bias on the threshold voltage (due to drain-induced barrier lowering, or DIBL), and **KAPPA** describes the increase in the saturated current with increasing drain bias due to channel length modulation. The results, obtained from a wide, short device (20/0.7) are shown in Figure 7.15. The results show a serious shortcoming of the Level 3 model. Note the severe first derivative discontinuity at the linear/saturation transition point (the visible kink in the modeled characteristic). When $V_{ds} = V_{dsat}$, it is assumed that the carrier velocity is limited to **VMAX**; this scheme works well for low gate biases, in which V_{dsat} is reached for a low drain bias. However, at larger gate biases, V_{dsat}, as computed from the classical expression (7.102), becomes large; the carrier velocity saturates before V_{ds} reaches V_{dsat}. The saturation voltage, as described by (7.97), is reduced from the classical value, and the first derivative of the current equation becomes discontinuous at $V_{ds} = V_{dsat}$; this leads to the kink in the model results of Figure 7.15. This effect

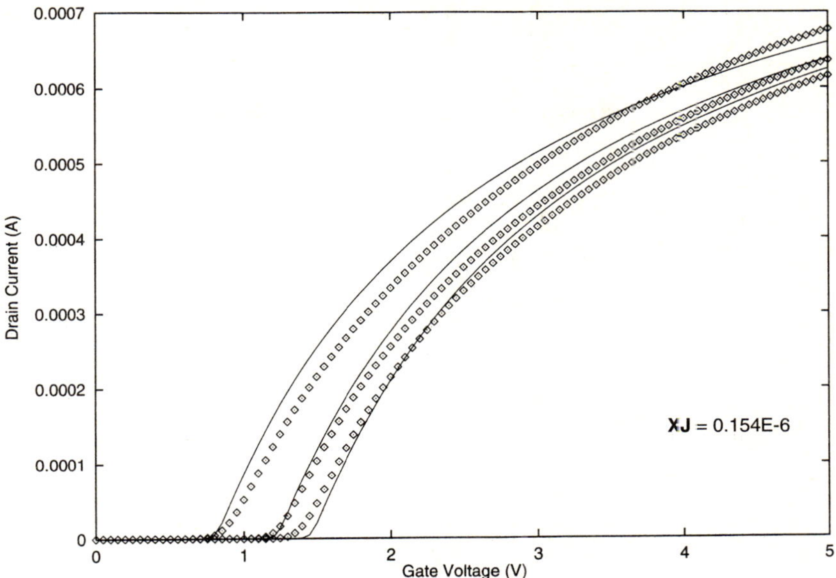

Figure 7.14 The extraction of the parameter **XJ**, which is **not** the junction depth, but the short channel correction to the substrate sensitivity, $W/L = 20$ μm/0.7 μm.

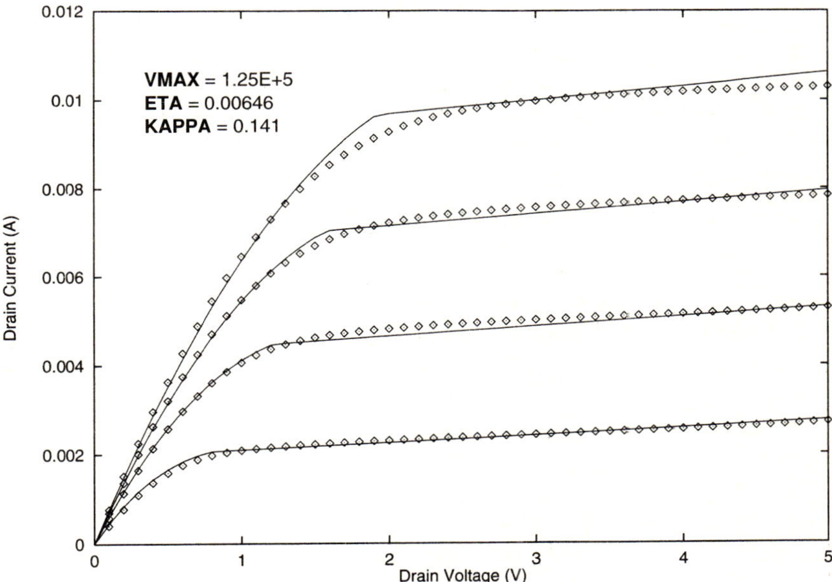

Figure 7.15 The extraction of the high drain bias parameters **VMAX**, **ETA**, and **KAPPA**, $W/L = 20$ μm/0.7 μm. Note the first derivative discontinuity in the model at $V_{ds} = V_{dsat}$.

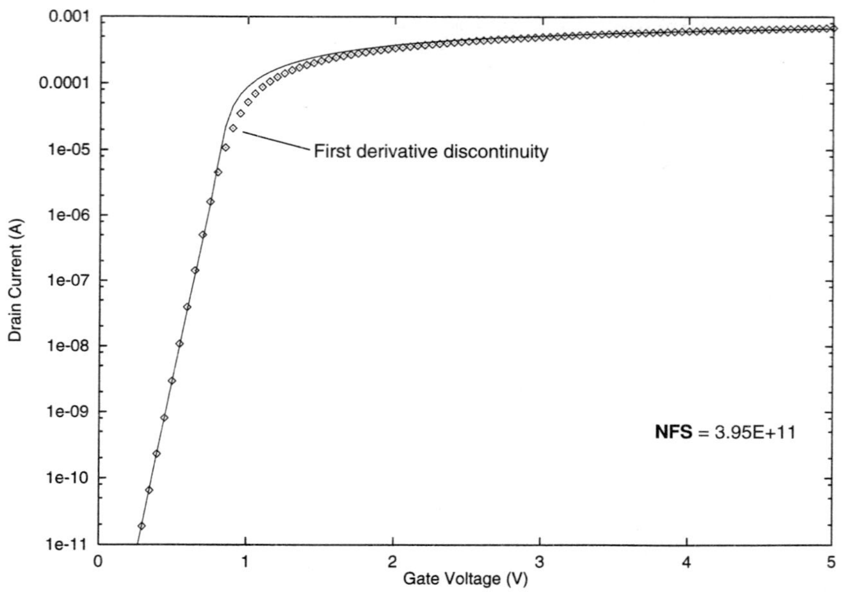

Figure 7.16   The extraction of the subthreshold region fitting parameter **NFS**, $W/L =$ 20 μm/0.7 μm.

becomes increasingly severe as L_{eff} decreases, as the difference between the classically computed V_{dsat} and the modified V_{dsat} (which takes into account **VMAX**) becomes larger. Also note that the extracted value of **VMAX** $= 1.25 \times 10^5$ m/s is actually about 25% larger than generally accepted physical values of the carrier saturation velocity (roughly 1.0×10^5 m/s). **VMAX** is in truth a fitting parameter. Note also that the modeled current in the saturation region is a straight line and does not account for the finer detail which occurs. The model for the saturation region current, as described in Section 7.5.3, employs the parameter **KAPPA** in a manner nearly identical to that involving **LAMBDA** in Level 1 and Level 2. This simple straight-line model for the saturation region current is insufficient when high lateral fields are present.

Finally, a model for the subthreshold current can be constructed. As described by (7.121), only one parameter, **NFS**, can be used in the extraction. Furthermore, **NFS** can be extracted for only one combination of geometry, temperature, and drain bias. Such an extraction is shown in Figure 7.16. A reasonable fit is obtained, although the first derivative discontinuity at $V_{gs} = V_{on}$ is clearly evident. The value of **NFS** $= 3.95 \times 10^{11}$ cm^{-2} is very large and unrealistic in contemporary MOS devices, where negligible values of 10^{10} cm^{-2} or less usually occur. In addition, as just noted, the applicability of this model is severely restricted. Because of these limitations, it is common to ignore the subthreshold region in Level 3 model construction. Generally, if a good subthreshold model is required, there are additional model requirements which Level 3 cannot meet.

The complete set of extracted model parameters is listed in Table 7.2.

Table 7.2   The extracted Level 3 parameters.

Parameter	Value
TOX	13.0E-09
LD	0.120E-6
WD	0.200E-6
VTO	0.841
UO	483
THETA	0.113
RS	42.5
RD	42.5
DELTA	3.54
NSUB	1.01E+17
XJ	0.154E-06
VMAX	1.25E+05
ETA	6.46E-03
KAPPA	0.141
NFS	3.95E+11
BEX	-1.63

Temperature Dependence

As described in Section 7.4, the Berkeley implementation of Level 3 includes no parameters to model the temperature dependence of the device characteristics. The HSPICE implementation of Level 3 adds the parameter **BEX** (see (7.64)) to account for this temperature dependence. In Berkeley SPICE, the model itself contains the equivalent of **BEX** $= -1.5$.

This parameter (**BEX**) is easily determined. After the main model has been completed, **BEX** is extracted from a long, wide device in the linear region at a different temperature (Figure 7.17). Here a temperature of 125°C was used, and a value of **BEX** $= -1.63$ was extracted. Although it is quite simple, this approach works remarkably well over a large range of temperatures, geometries, and bias conditions.

7.7.2   Model Playback

The parameter extraction scheme described above is fairly simple. With the model completed, it is appropriate to exercise it over a range of geometry and operating conditions to determine its accuracy and integrity.

Figure 7.18 compares the model with, respectively, the linear and saturation characteristics of the 20/20 device. The results are not particularly good, which may seem surprising for the linear case (since an extraction step was carried out on this data). However, recall that the linear region was used to extract the parameters **UO**, **VTO**, and **THETA**. These parameters take no account of the drain bias and/or short channel effects; when the short channel parameters are included, they reduce the accuracy of the fit for the long device. In this form, the approach to developing a Level 3 model is to treat the long, wide device as an infinitely long, infinitely wide FET from which the base parameters are extracted. Since no device can be infinitely long and wide, some reduction to the fitting accuracy due to the introduction of the short channel parameters

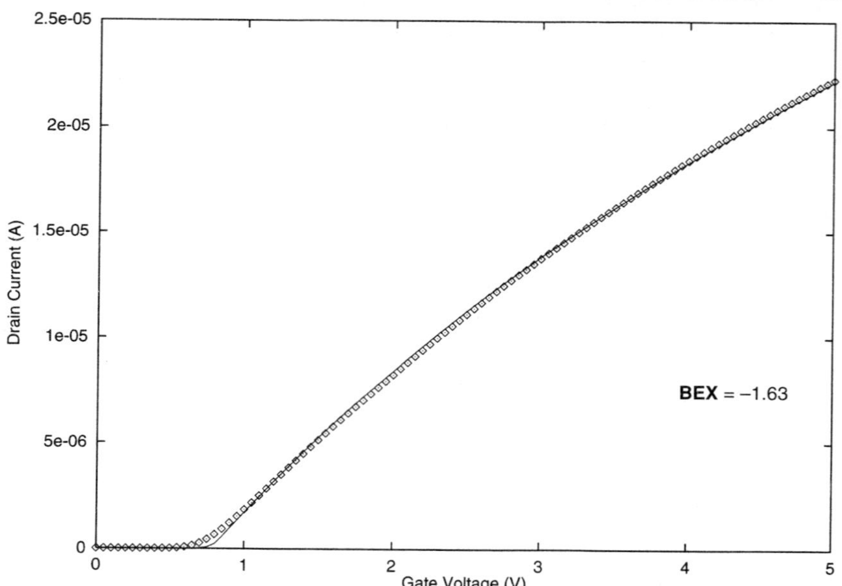

Figure 7.17 The extraction of the HSPICE parameter **BEX** at 125°C, $W/L = 20$ μm/20 μm.

is inevitable. It is possible to come up with some sort of iterative scheme to improve the accuracy of the result, but this kind of process is time consuming, and will inevitably degrade the results for the more important short channel region. Since Level 3 is only of use in digital design, a high degree of accuracy is not required for long devices. The only use long devices have in digital design is as load resistors, and the level of accuracy shown here is adequate. If a more accurate model of long devices is required, it is likely that other shortcomings of Level 3 make it an inappropriate model choice.

Figure 7.19 depicts results for the 1.3/20 device. Since the linear region of this narrow device was modeled after the series resistance was extracted (but before the saturation parameters were determined), the zero substrate bias result is still quite good; the nonzero substrate bias results are poor, due to the absence of the substrate bias from the mobility model. The saturation characteristics show a poor result; the discussion above of the 20/20 device applies here as well. Figure 7.20 shows results for the 20/0.7 device. The linear result is somewhat poor, having been corrupted by fitting to nonzero substrate biases to extract **XJ** (Figure 7.14). Given the constraints of the model, the saturation fit is quite reasonable.

However, if device geometries other than those employed for parameter extraction are used, the results lose accuracy. Figures 7.21–7.24 depict model results for geometries of 20/0.9, 2.0/20, 1.3/0.7, and 2.0/2.0. The model provides a less accurate representation of the measured data. As described earlier, Level 3 treats the short and narrow channel parameters as corrections to a long, wide device model. The model neglects some

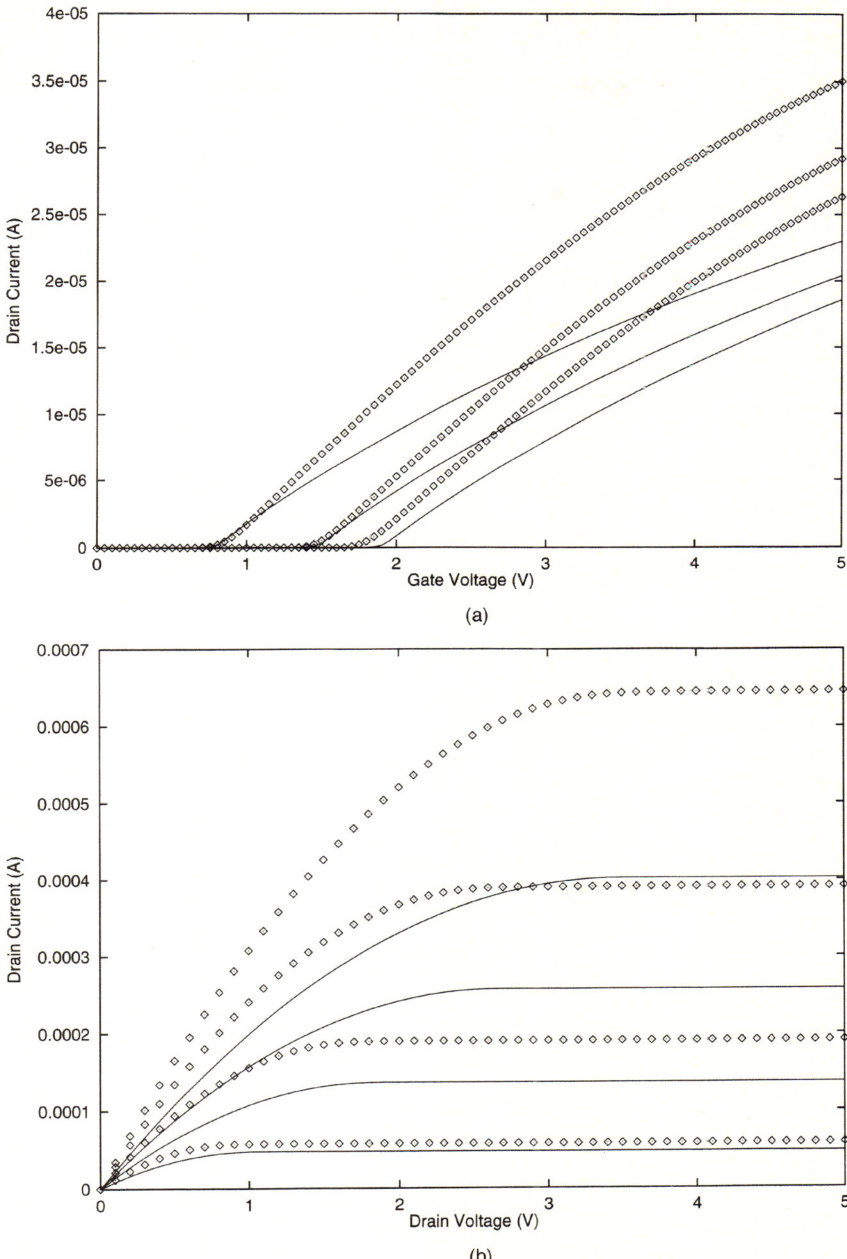

Figure 7.18 Comparison of measured data and the final model for the 20/20 device; (a) linear characteristic; (b) saturation characteristic.

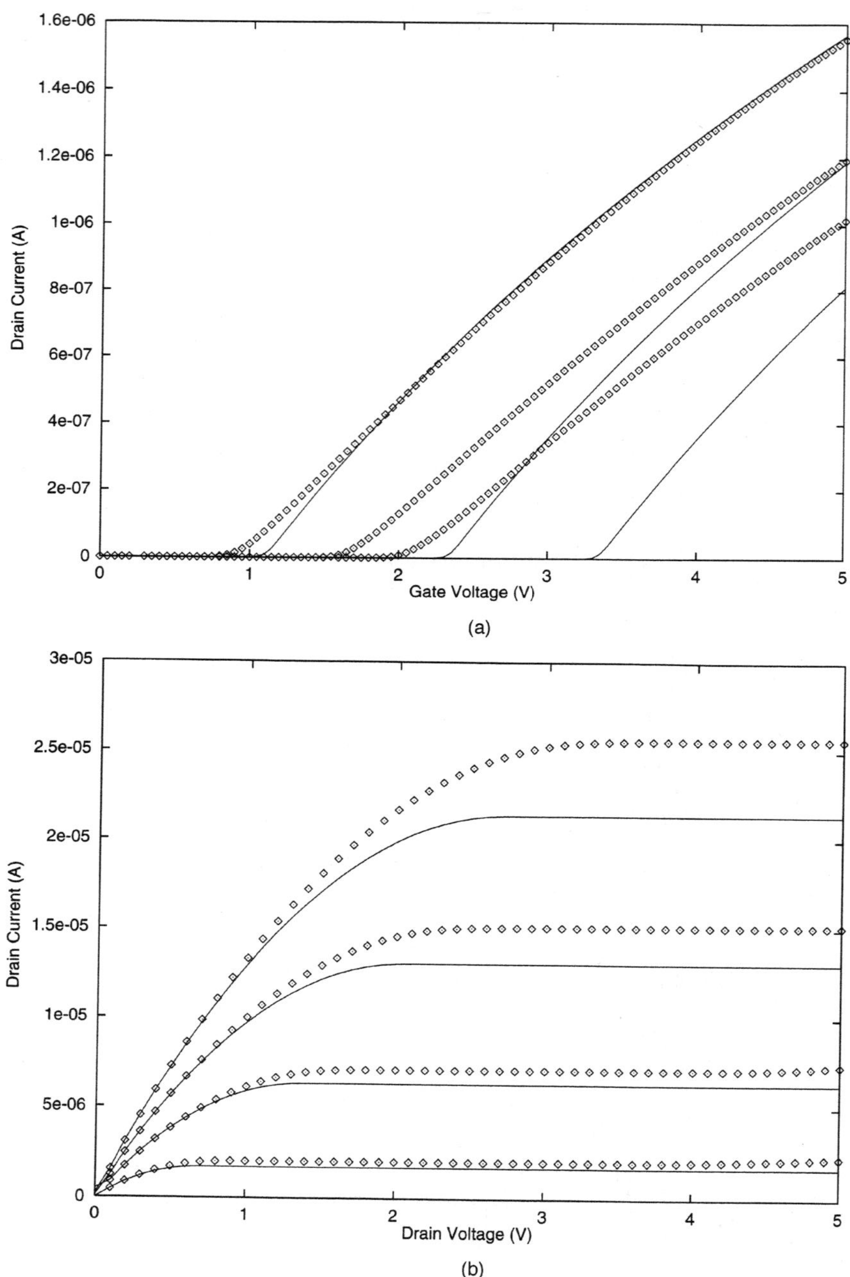

Figure 7.19 Comparison of measured data and the final model for the 1.3/20 device; (a) linear characteristic; (b) saturation characteristic.

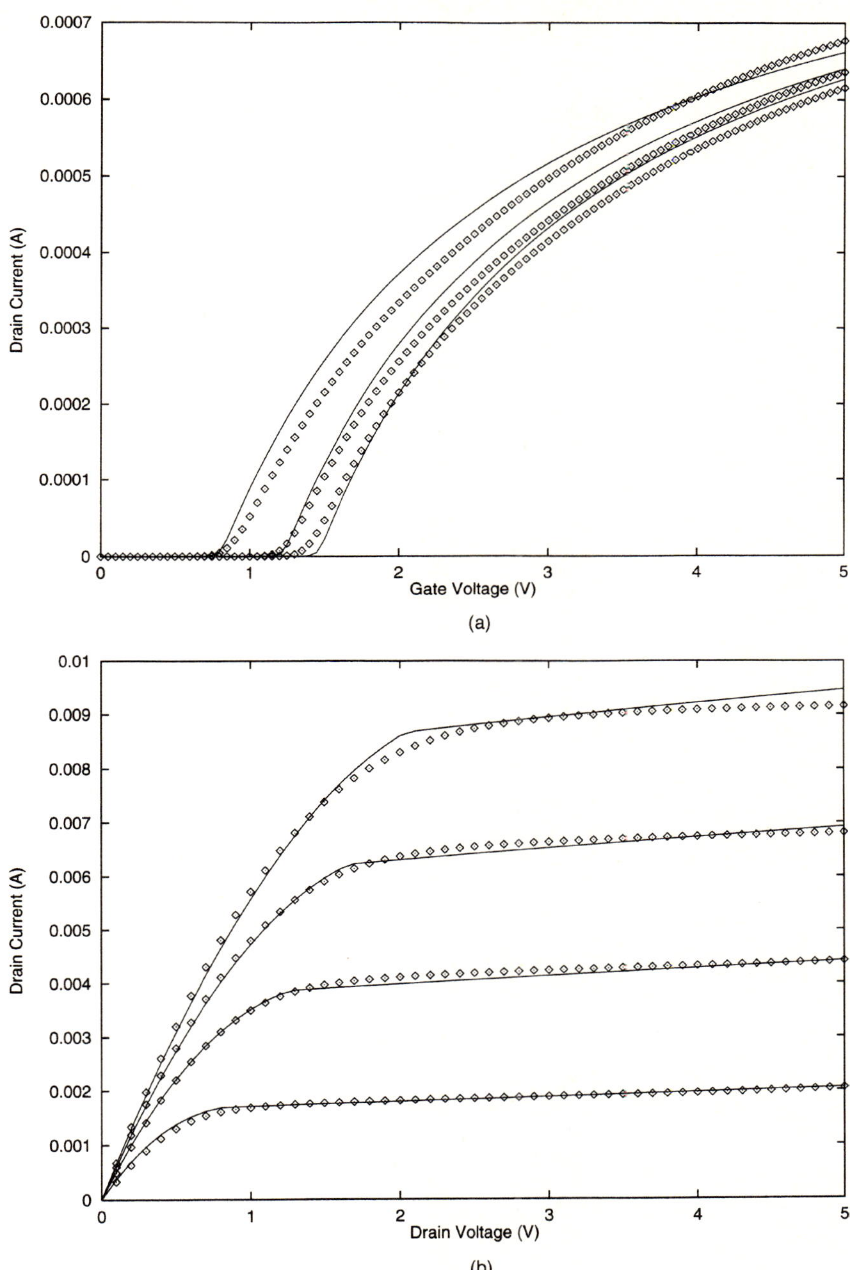

Figure 7.20   Comparison of measured data and the final model for the 20/0.7 device; (a) linear characteristic; (b) saturation characteristic.

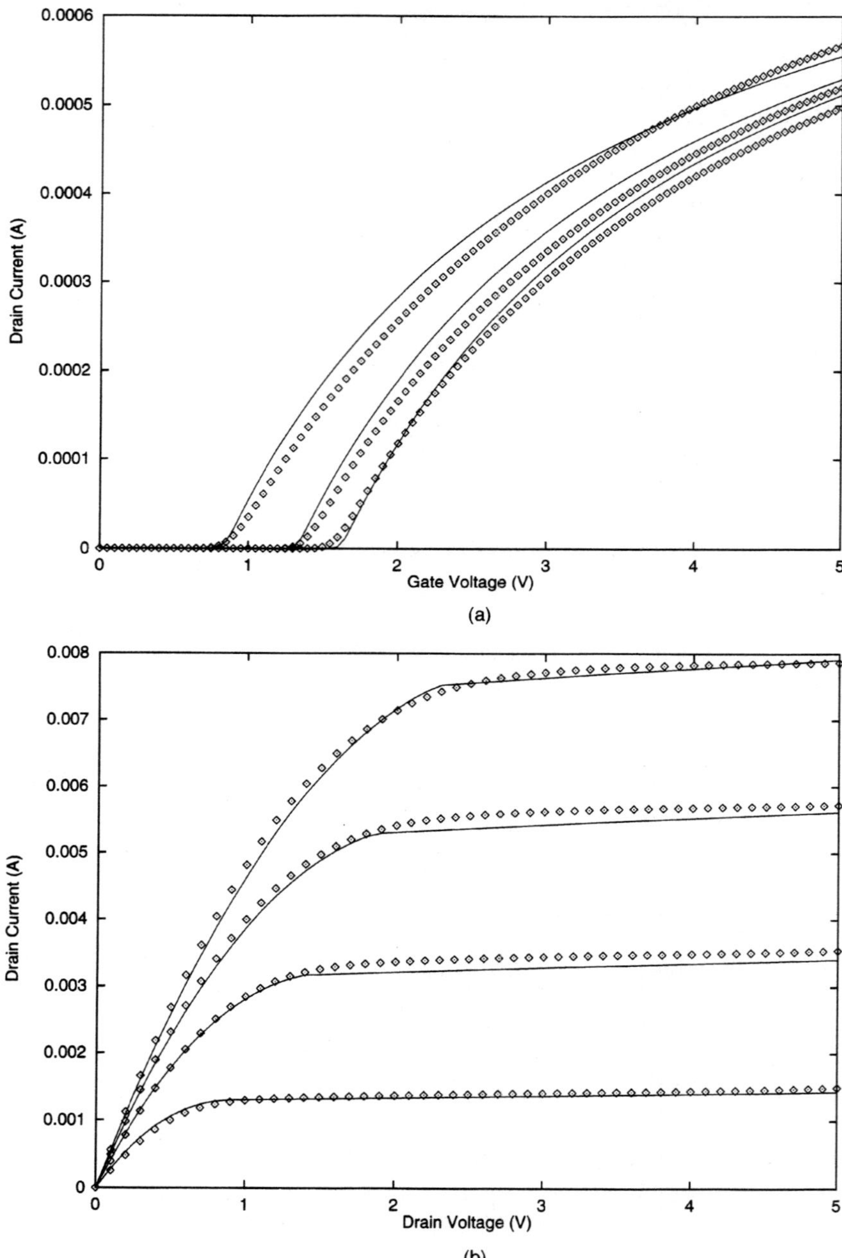

Figure 7.21 Comparison of measured data and the final model for the 20/0.9 device; (a) linear characteristic; (b) saturation characteristic.

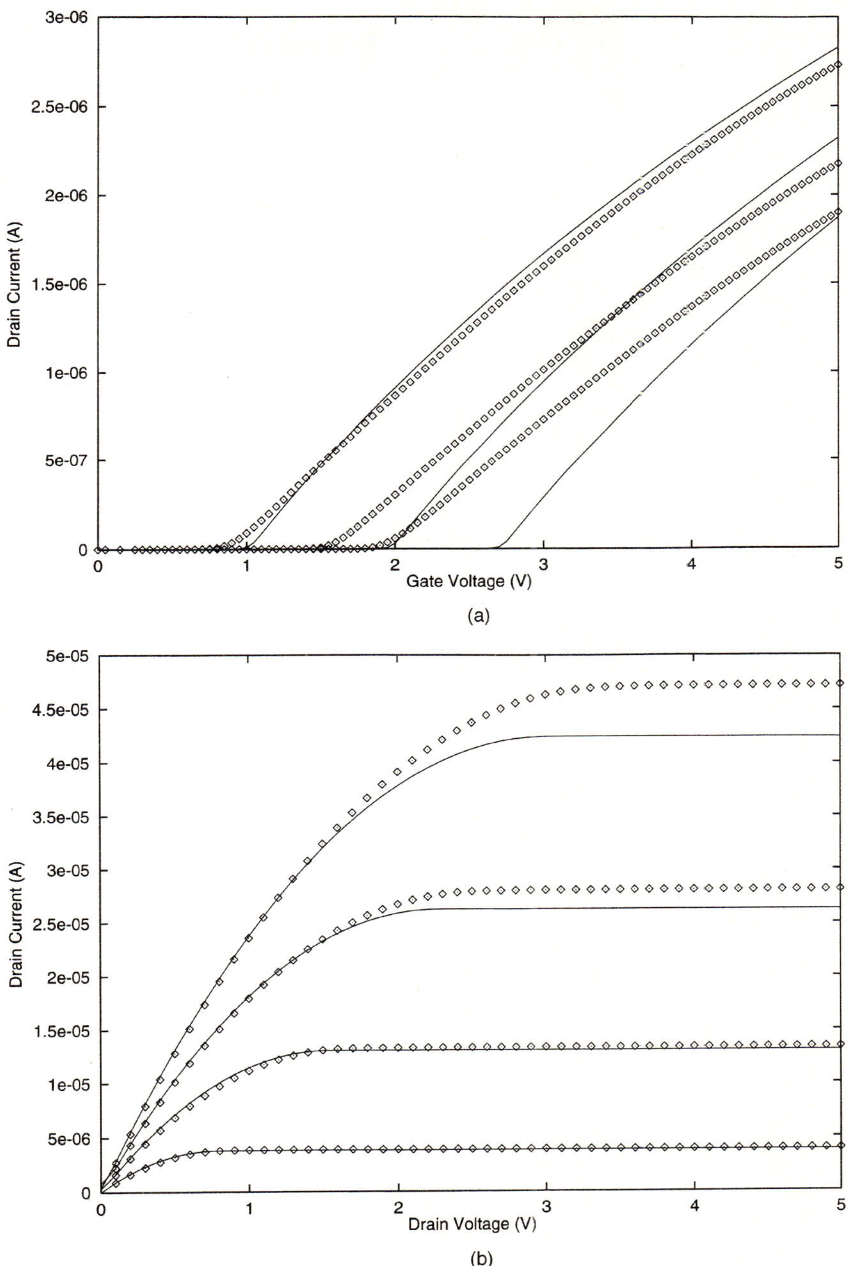

Figure 7.22 Comparison of measured data and the final model for the 2.0/20 device; (a) linear characteristic; (b) saturation characteristic.

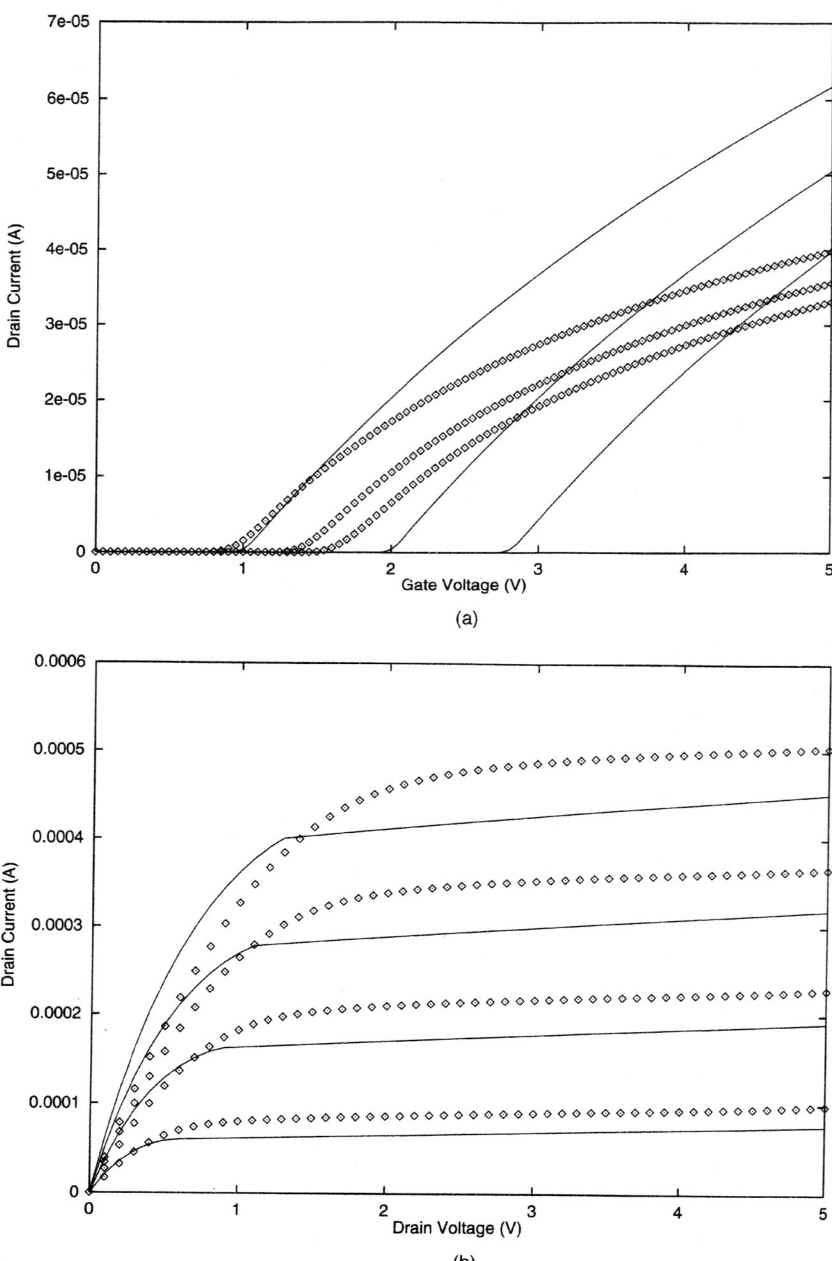

Figure 7.23 Comparison of measured data and the final model for the 1.3/0.7 device; (a) linear characteristic; (b) saturation characteristic.

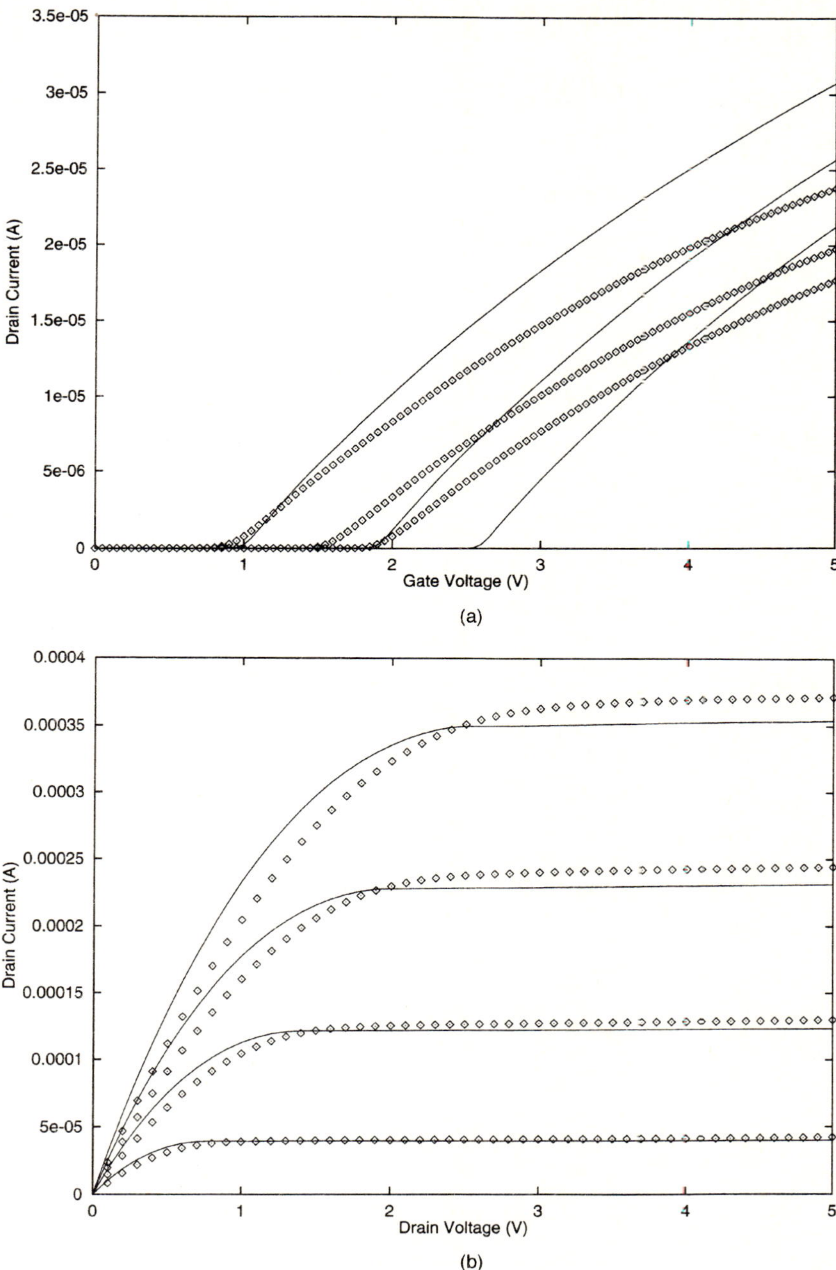

Figure 7.24   Comparison of measured data and the final model for the 2.0/2.0 device; (a) linear characteristic; (b) saturation characteristic.

important physical effects, and several mathematical simplifications were made to ease the computational burden. The net result is that these omissions are overcome with parameter extraction, but these parameters are only valid at the specific geometry at which they were extracted. When the geometry is changed from the specific target values (here, $L = 0.7$ μm and $W = 1.3$ μm), the simulation results deteriorate.

7.7.3 Model Binning

In a digital design, a single drawn channel length, equal to the minimum poly line width allowed, is generally the only size employed. Given the above discussion, it might appear that a model which is accurate for a single poly dimension might be adequate. However, it is not uncommon for some designs to include a few FETs which are up to 0.2 μm longer than the minimum dimension. In addition, systematic process variations, discussed in Chapter 14, cause the channel length to vary, even if this is not intended in the design.

As described above, the Level 3 model treats short and narrow channel effects as corrections to a long, wide device. To account for other geometries (besides L_{min} and W_{min}), it is possible to build separate models at larger values of L and W. This is illustrated in Figure 7.25, where it is shown that separate models will be constructed at $L = 0.7$ μm and $L = 0.9$ μm, and $W = 1.3$ μm and $W = 2.0$ μm. This leads to separate models in four subregions of the geometry space. These submodel regions are referred to as *bins*, and this practice is thus known as *binning*. Models are constructed at a particular point and allowed to be valid across an entire bin. For example, the model developed above was constructed at $L = 0.7$ μm and $W = 1.3$ μm; this now becomes submodel 4, which is considered to be valid for 0.7 μm $\leq L <$ 0.9 μm and 1.3 μm $\leq W <$ 2.0 μm. (Note that the lower limit of the bin is defined by "$\leq$", while the upper limit is defined by "$<$".) Since the Level 3 model completely decouples length and width effects, the short channel corrections will be identical in bins 1 and 3, and 2 and 4, while the narrow channel corrections will be identical in bins 1 and 2, and 3 and 4.

The model extraction procedure is identical to the one described earlier. Results are shown in Figures 7.26–7.28, and a complete listing of the binned model may be

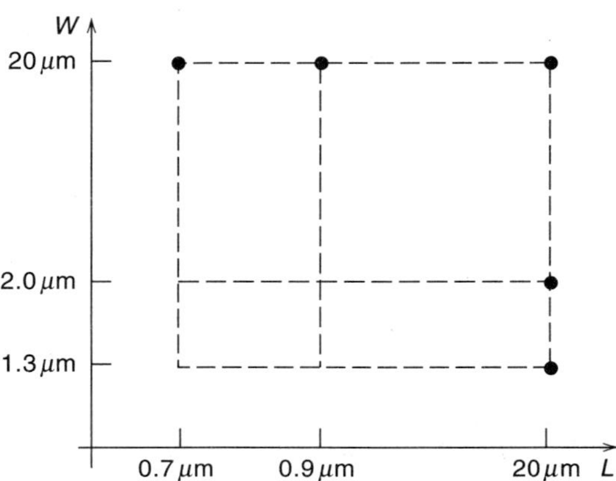

Figure 7.25 The modified geometric scheme for construction of a binned Level 3 model. Five devices are used here, rather than three.

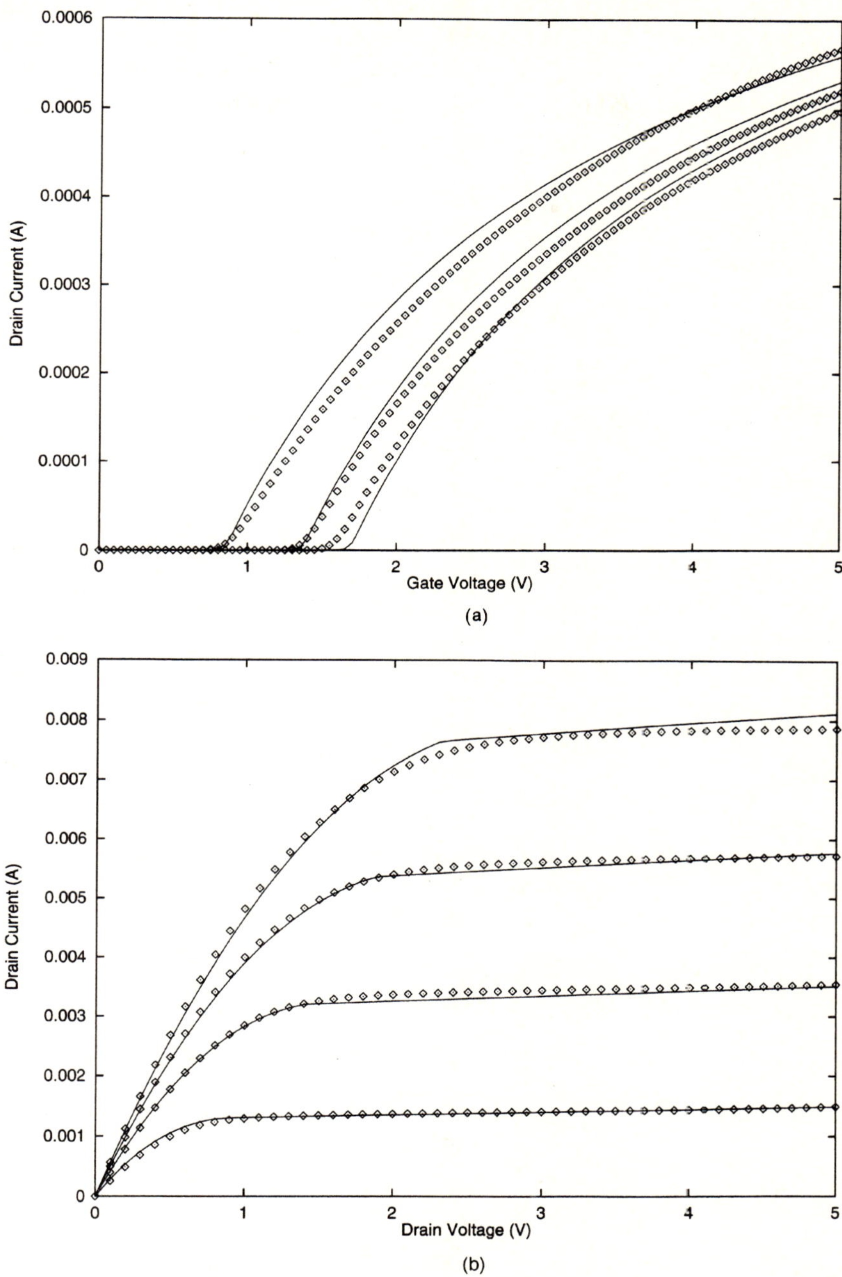

Figure 7.26 Comparison of measured data and the binned model for the 20/0.9 device; (a) linear characteristic; (b) saturation characteristic.

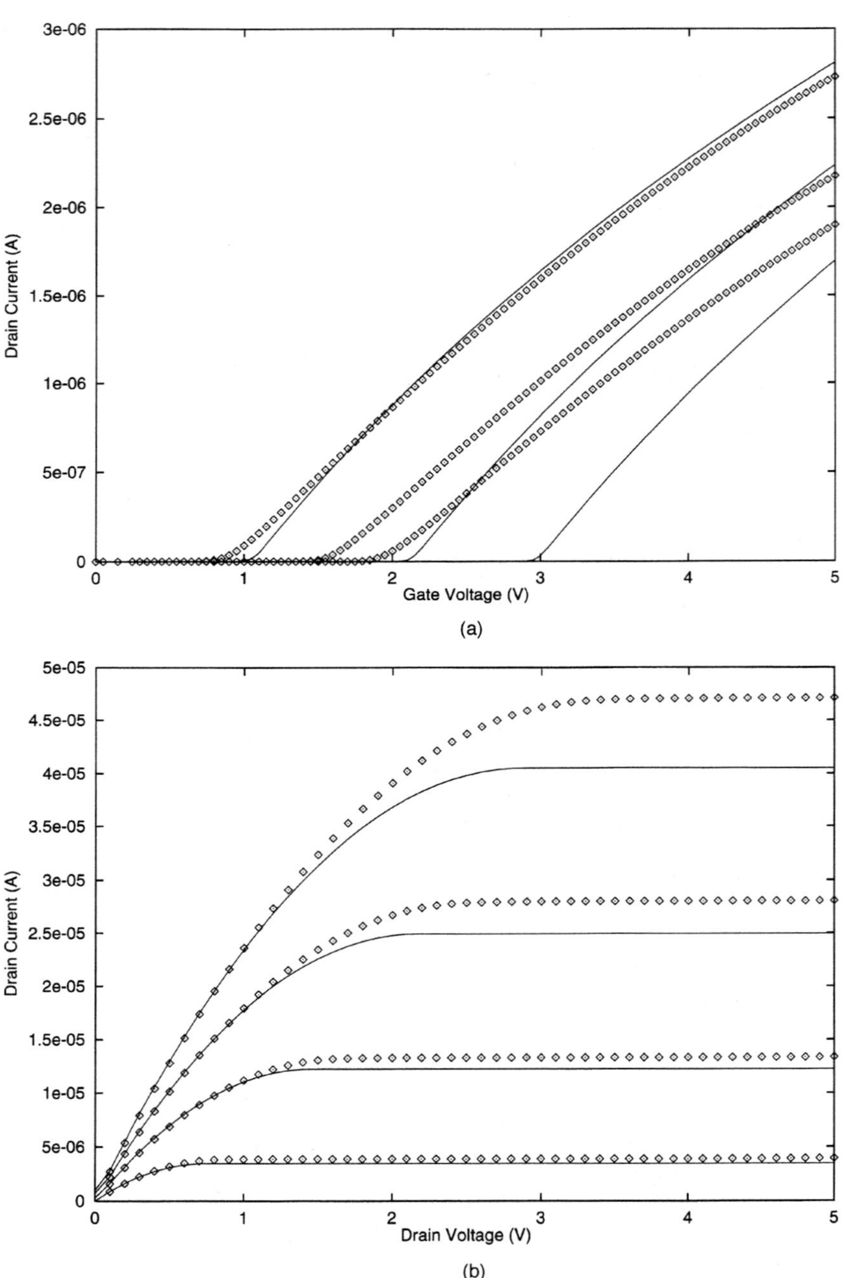

Figure 7.27 Comparison of measured data and the binned model for the 2.0/20 device; (a) linear characteristic; (b) saturation characteristic.

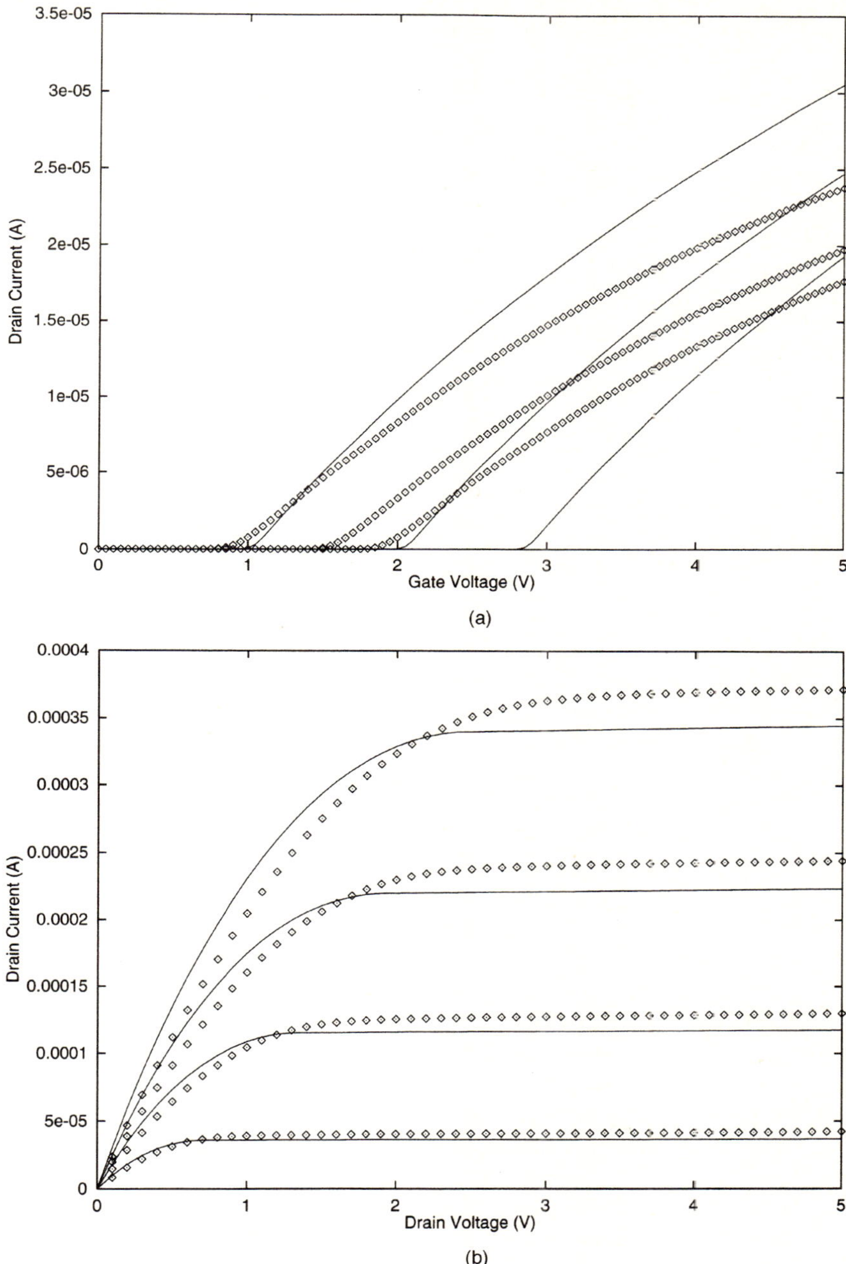

Figure 7.28 Comparison of measured data and the binned model for the 2.0/2.0 device; (a) linear characteristic; (b) saturation characteristic.

Table 7.3   The extracted Level 3 parameters for the binned model. Note that the model of Table 7.2 now corresponds to bin 4.

Parameter	Bin 1	Bin 2	Bin 3	Bin 4
LMIN	0.9 µm	0.7 µm	0.9 µm	0.7 µm
LMAX	1000 µm	0.9 µm	1000 µm	0.9 µm
WMIN	2.0 µm	2.0 µm	1.3 µm	1.3 µm
WMAX	1000 µm	1000 µm	2.0 µm	2.0 µm
TOX	13.0E-09	13.0E-09	13.0E-09	13.0E-09
LD	0.120E-6	0.120E-6	0.120E-6	0.120E-6
WD	0.200E-6	0.200E-6	0.200E-6	0.200E-6
VTO	0.841	0.841	0.841	0.841
UO	483	483	483	483
THETA	0.113	0.113	0.113	0.113
RS	41.9	42.5	41.9	42.5
RD	41.9	42.5	41.9	42.5
DELTA	4.70	4.70	3.54	3.54
NSUB	1.01E+17	1.01E+17	1.01E+17	1.01E+17
XJ	0.124E-6	0.154E-6	0.124E-6	0.154E-6
VMAX	1.30E+5	1.25E+5	1.30E+5	1.25E+5
ETA	12.7E-3	6.46E-3	12.7E-3	6.46E-3
KAPPA	0.164	0.141	0.164	0.141
NFS	4.69E+11	3.95E+11	4.69E+11	3.95E+11
BEX	−1.63	−1.63	−1.63	−1.63

found in Table 7.3. The 20/0.9 characteristics are slightly improved, as a model specific to $L = 0.9$ µm has been constructed. The results for the 2.0/20 device also show a slight improvement over the earlier results. Finally, the results for a 2.0/2.0 device are not particularly good. The difference between the model constructed at $L = 0.9$ µm and the behavior of a device with $L = 2.0$ µm is too great to allow a good result.

The choice of the number and sizes of bins depends on the needs of the model user. A basic digital circuit usually employs a small selection of short channel lengths; any need to model different channel lengths is usually due to the requirement to model process variations. The binned model developed here provides good results from $L = 0.7$ µm up to at least $L = 0.9$ µm; this is sufficient for most cases.

It is possible, if desired, to extend the applicability of the model by constructing additional bins. This can be done by trial and error, where more bins are added until some desired accuracy over the geometric range of interest is achieved. The logically absurd limit of this approach is "blind overkill," in which a very large number of bins of a very small size are constructed to force the model to be accurate. Either approach requires that a very large number of device geometries be available for model construction.

Finally, there is one major difficulty inherent in any binned Level 3 model. Since each submodel is constructed at one particular geometry (the minimum length and width of the bin), it is built in isolation from every other submodel. This leads to the problem depicted in Figure 7.29, where the drain current of a $W = 20$ µm device is plotted against channel length. At the bin boundary ($L = 0.9$ µm), the current is discontinuous with channel length. This can be a problem if a designed channel length is near a bin boundary,

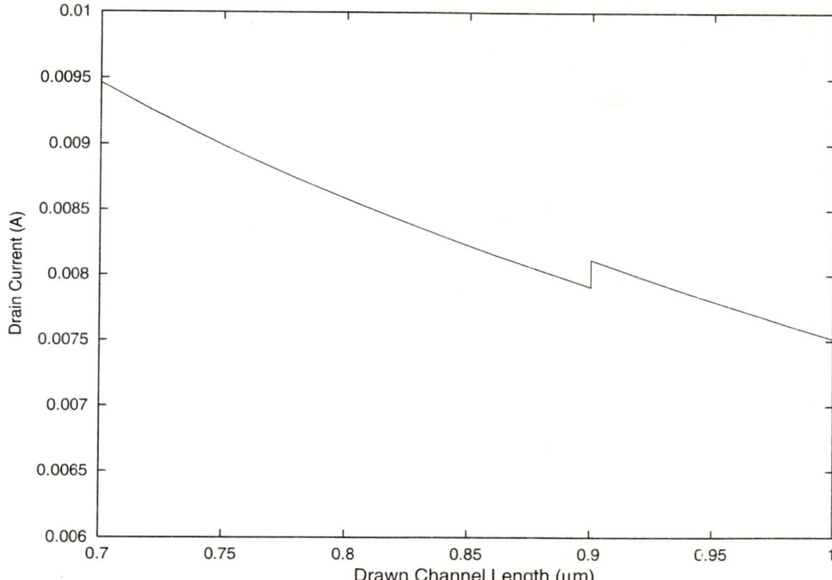

Figure 7.29   The modeled drain current versus channel length, $V_{gs} = V_{ds} = 5.0$ V, showing the discontinuity at the $L = 0.9$ μm bin boundary.

as different process variations cause the channel length to move across the region. (For example, if the typical channel length is 0.92 μm and process variations allow the channel length to reach 0.88 μm, the bin boundary will be crossed.) During statistical simulation runs, in which the channel length is allowed to vary over its possible range, there will be a region where the drain current is very different for a very small change in the channel length. In addition, there is increasing interest in optimizing the channel length in certain designs. Attempting to optimize the channel length near a bin boundary (here, $L = 0.9$ μm) can be a severe problem, as the current is poorly represented for any channel length slightly less than the bin boundary value.

As Figure 7.29 shows, at the bin boundary, the modeled drain current actually *decreases* as the channel length decreases. This artifact, introduced by the use of binning, complicates the statistical analysis of device and circuit behavior, or any attempt to closely optimize the channel length.

7.7.4   Binning for Temperature

As described above, the temperature dependence in the Berkeley version of the Level 3 model, described by (7.63), does not work well in practice. Separate models must be constructed for all temperatures of interest (e.g., 25°C for the basic design, and −55°C and 125°C to allow the functionality of a design to be verified at the possible extremes of temperature). The scheme employed by HSPICE, described by (7.64), works very well over a large temperature range (at least $-55°C$ to $125°C$), so there is no need for separate models for different temperatures.

7.8   Final Comments

As noted at the beginning of this chapter, the simplicity, operational reliability, and extendability of the Level 3 model have made it very popular for digital circuit design. In static logic operation, the device terminals are biased at the power supply or at ground; model accuracy at these two conditions is the most important consideration for accurate results. Bias levels between these two extremes occur only in transient conditions, when capacitive loading prevents an instantaneous transition between the power supply voltage and ground.

However, analog circuit applications are growing in importance, and the Level 3 model is unable to adequately meet this modeling requirement for two reasons. First, the form of the current model, with an abrupt change from the linear to saturation regions, leads to a first derivative discontinuity in the current at $V_{ds} = V_{dsat}$ that becomes increasingly severe for shorter channel lengths and higher fields; this discontinuity was discussed earlier and clearly noted in Figure 7.15. When the I_d-V_d curve is differentiated $\left(\frac{\partial I_d}{\partial V_d} \right)$ to yield the differential output conductance, a severe discrepancy between the model and the data is produced, with a very strange model result, as shown in Figure 7.30. The model provides a poor fit to the data. More importantly, the first derivative discontinuity at $V_{ds} = V_{dsat}$ is strikingly evident. In analog circuit design, an accurate model of the output conductance is required, as carefully chosen FET biases allow a long device to be used as a controllable differential output resistance element with a large dynamic range; Level 3 is clearly inadequate for the task. In addition, Level 3 is not designed to be accurate at longer channel lengths, as the largest available device is used as merely a starting point for developing the short and narrow channel corrections. In digital circuit design, only the shortest available channel lengths are of interest, and accuracy for longer devices is not required. However, analog circuits make use of long channel devices, where Level 3 is not accurate. If necessary, Level 3 can be made accurate for long channel devices at the sacrifice of accuracy at other geometries; however, the problems with the output conductance remain.

Furthermore, Level 3 does not include a physically realistic subthreshold model. Many analog circuits require a detailed model of the device turn-off into the subthreshold region; once again, Level 3 is not adequate. This is the most acute symptom of a major difference between the modeling requirements for digital and analog circuit design. As noted above, in static digital logic circuits, nodes are biased at either the power supply or at ground; other biases are present only under transient conditions, when capacitive loading prevents an instantaneous transition between logic states (V_{dd} and ground). For other biases, the accuracy requirements on the current model are not as stringent. In analog circuits, static voltage biases other than the power supply voltage and ground are used; accuracy is required over the entire applied bias range. A similar situation is arising with dynamic digital logic circuits, in which biases between the power supply voltage

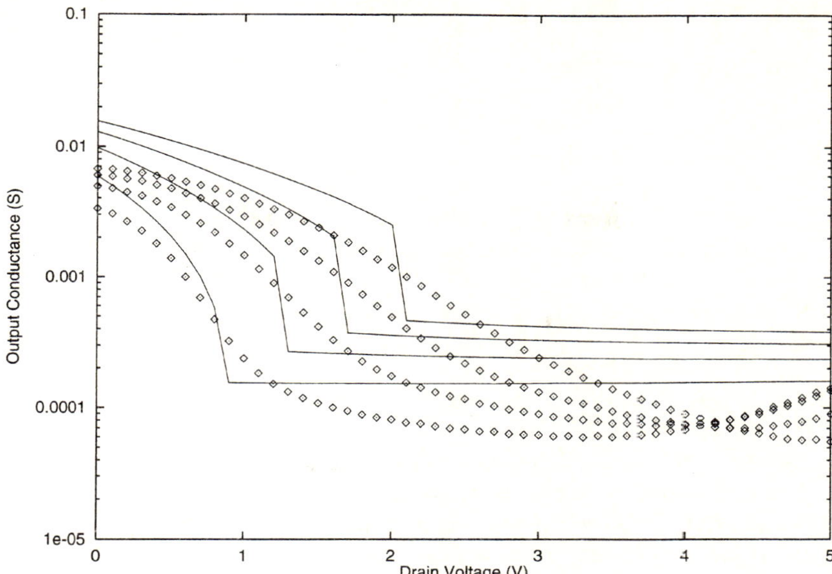

Figure 7.30 A comparison of the modeled and measured output conductance of the
20/0.7 device. Due to the discontinuity in the first derivative introduced by the model
for V_{dsat}, the Level 3 model gives a very poor result, rendering it unsuitable for analog
applications.

and ground are used to drive current into the output node of a logic gate. The absence of
a "genuine" subthreshold model and the problems of obtaining an accurate drain current
result near the linear to saturation transition point make Level 3 unsuitable as a model in
these situations.

Even in static digital logic circuits, the Level 3 model is likely to become more
vulnerable in the future as power supply voltages decrease. A major problem facing
low-voltage/low-power technology is the inability to decrease the threshold voltage as the
power supply voltage is decreased [18, 19]. As the magnitude of $(V_{dd} - V_t)$ decreases,
the extent of the regions where the Level 3 model has difficulty becomes relatively larger.
Since any transient will spend a larger fraction of time in these vulnerable regions, the
need for accuracy away from V_{dd} and ground becomes stronger.

The poor intrinsic geometry dependence of the Level 3 model, and the failure to
properly model the output conductance and the subthreshold region, provided the impetus
for the development of the second generation of FET models. These models are discussed
in the next section.

BIBLIOGRAPHY

1. L. Dang, "A Simple Current Model for Short-Channel IGFET and Its Application to Circuit Simulation," *IEEE J. Sol. St. Circ.* vol. SC-14, pp. 358–367 (1979).

2. K. Jeppson, "Influence of the Channel Width on the Threshold Voltage Modulation in MOSFETs," *Elec. Lett.* vol. 11, pp. 297–299 (1975).

3. J. Copeland, "Diode Edge Effect on Doping Profile Measurements," *IEEE Trans. Elec. Dev.* vol. ED-17, pp. 404–407 (1970).

4. R. Troutman, "VLSI Limitations from Drain-Induced Barrier Lowering," *IEEE J. Sol. St. Circ.* vol. SC-14, pp. 383–390 (1979).

5. H. Masuda, M. Nakai, and M. Kubo, "Characteristics and Limitation of Scaled-Down MOSFETs Due to Two-Dimensional Field Effect," *IEEE Trans. Elec. Dev.* vol. ED-26, pp. 980–986 (1979).

6. J. Schrieffer, "Effective Carrier Mobility in Surface Space Charge Layers," *Phys. Rev.* vol. 97, pp. 641–646 (1955).

7. D. Caughey and R. Thomas, "Carrier Mobilities in Silicon Empirically Related to Doping and Field," *Proc. IEEE* vol. 55, pp. 2192–2193 (1967).

8. C. Sodini, P. Ko, and J. Moll, "The Effect of High Fields on MOS Device and Circuit Performance," *IEEE Trans. Elec. Dev.* vol. ED-31, pp. 1386–1393 (1984).

9. A. Vladimirescu and S. Liu, "The Simulation of MOS ICs Using SPICE-2," University of California/Berkeley, Electronics Research Laboratory Document M80/7 (1980).

10. K. Hess, *Advanced Theory of Semiconductor Devices*, Prentice-Hall, 1988.

11. *HSPICE User's Manual*, Meta-Software, Inc., Campbell, California, 1993.

12. G. Merckel, J. Borel, and N. Cupcea, "An Accurate Large-Signal MOS Transistor Model for Use in Computer-Aided Design," *IEEE Trans. Elec. Dev.* vol. ED-19, pp. 681–690 (1972).

13. G. Baum and H. Beneking, "Drift Velocity Saturation in MOS Transistors," *IEEE Trans. Elec. Dev.* vol. ED-17, pp. 481–482 (1970).

14. R. Swanson and J. Meindl, "Ion-Implanted Complementary MOS Transistors in Low-Voltage Circuits," *IEEE J. Sol. St. Circ.* vol. SC-7, pp. 146–153 (1972).

15. J. Meyer, "MOS Models and Circuit Simulation," *RCA Review* vol. 32, pp. 42–63 (1971).

16. D. Ward and R. Dutton, "A Charge-Oriented Model for MOS Transistor Capacitances," *IEEE J. Sol. St. Circ.* vol. SC-13, pp. 703–708 (1978).

17. *Aurora User's Manual*, Technology Modeling Associates, Inc., Palo Alto, California, 1994.

18. E. Nowak, "Ultimate CMOS ULSI Performance," 1993 IEDM Tech. Dig., pp. 115–118.

19. D. Foty and E. Nowak, "MOSFET Technology for Low-Voltage/Low-Power Applications," *IEEE Micro*, July 1994, pp. 68–77.

8

BSIM

8.1 Introduction

As described in Chapter 2, the Level 1, 2, and 3 models comprise the first generation of FET models implemented in SPICE. The basic model formulation is very much concentrated on the analytical expressions, and parameter extraction is fairly simple.

The BSIM (**B**erkeley **S**hort-Channel **IGFET M**odel) model [1], [2] (sometimes referred to as Level 4), however, represents a major departure from the first generation approach, and is more properly considered as the founding member of the group of second-generation models. These models put less effort into developing physically exacting analytical models formed without regard for complexity, but instead concentrate on including mathematical conditioning for faster and more robust circuit simulation.

More importantly, instead of trying to formulate very complex equations to describe subtle submicron FET phenomena, empirical parameters are introduced to handle the various nuances. This has the effect of shifting the "action" in the second-generation models from the model formulation to parameter extraction. The use of numerous parameters simplifies the model equation set and allows for more accurate fitting of the model to measured data, but this is done at the price of greatly weakening the link between the model parameter set and the fabrication process. This is perhaps the most significant philosophical change introduced by the second-generation models, and is a matter of considerable debate at this time [3].

8.2 BSIM Background

BSIM is an outgrowth of CSIM (Compact Short-Channel IGFET Model), introduced by AT&T [4], [5]. The tone of CSIM was to concentrate on the use of the FET model in a circuit simulator, and thus to do as much as possible to mathematically condition the model equations for robust and efficient use; this is done by extensively reworking the model equations, at a cost of eliminating what had been considered to be analytically exact descriptions of device behavior. A first step in this direction was taken in Level 3, where the exact but complex depletion charge expression was replaced with a simplifying Taylor expansion [6]. The BSIM formulation continues this trend, but extends it to a greater degree.

In addition, BSIM contains further improvements over the first generation models. The expression describing the reduction of the channel mobility by the vertical field includes substrate bias dependence. The threshold voltage expression is expanded to include another term which, although basically ad hoc, is able to provide better results for nonuniformly doped substrates. An improved subthreshold current model is introduced, with mathematical conditioning to guarantee that both the drain current expressions and their first derivatives are continuous at the subthreshold-superthreshold transition point; this eliminates a potential source of convergence problems. Completely different approaches to describing velocity saturation, reduction of the channel mobility by the lateral field, and the saturation voltage are introduced, greatly simplifying the drain current model; in particular, the description of the saturated current is considerably simpler than those found in the first-generation models. A more robust and accurate charge model [7] is also introduced; this new model conserves charge, but is more mathematically efficient in circuit simulation than the Ward-Dutton model [8]. Finally, the entire approach to the use of the model parameters is a radically different one from that employed by the first-generation models. Since this is an important basis for all the second-generation models, it will be discussed more fully below.

Temperature Dependence

As detailed in Chapter 2, SPICE FET models include a nominal temperature (T_{nom}), the temperature at which the parameters were extracted and the FET model was constructed. For operating temperatures other than T_{nom}, the parameters must be modified to properly describe the device behavior.

As originally formulated, BSIM contains no method of accounting for variations in the operating temperature. A particular nominal temperature is selected; the model is constructed and used at that temperature. The model formulation does contain some built-in temperature dependence; however, as noted in Chapters 6 and 7, this is inadequate. A method of adding temperature dependence to the BSIM structure was developed later [9], but it was never introduced into the general SPICE code. The HSPICE implementation of BSIM (HSPICE Level 13) includes a workable method of accounting for temperature dependence [10].

8.3   The Philosophy of the Second-Generation Models

The second generation of SPICE FET models includes BSIM and its two descendants, HSPICE Level 28 [10] and BSIM2 [11]. In them, an entirely new approach to the model parameters is introduced.

In the first-generation models, a long, wide FET is defined as providing a base model as a starting point; small geometry effects are treated as corrections to the base model. All the geometry dependence (channel length and width) is coded into the model equations; it is assumed that the model parameter set is valid over the entire geometry range. As described in Chapters 6 and 7, this last assumption is generally not valid; the model description can be made to be valid at one particular value of a short channel length and at one particular value of a narrow channel width. Away from these two points, the model results deteriorate. As described in Chapter 7, this situation is dealt with by extensive model binning, the introduction of a set of submodels that is valid only in particular narrowly defined geometric ranges.

The second-generation models take a completely different approach. A core set of equations is derived to describe the device behavior; these equations contain the channel length and width. This intrinsic part of the model formulation is identical in basic structure to that of the first-generation models. However, in sharp contrast to the first-generation models, an additional construct, the extrinsic model structure, is created on top of the intrinsic structure. This structure contains additional geometry dependence, embodied in the expression

$$X = X_o + \frac{LX}{L_{eff}} + \frac{WX}{W_{eff}}. \tag{8.1}$$

In the first-generation models, parameters (such as X) were taken to be isolated and independent. In the second-generation models, X is a composite parameter, synthesized from three parameters (X_o, LX, and WX) which actually appear in the model parameter list. Thus, in the second-generation models, parameters do not appear singly (X), but occur in triplets (X_o, LX, and WX).

Since each model parameter to be computed and inserted into the model equations (e.g., X) requires three parameters (e.g., X_o, LX, and WX) in the FET model parameter set, the number of parameters supplied in a particular model parameter set becomes quite large. For example, the original BSIM model [2] claims to require 17 parameters. However, this is only the set of the X parameters. Since all but one of these 17 parameters is actually composed of triplets of X_o, LX, and WX, the final model parameter set actually contains 49 separate parameters.

An unfortunate choice in the original BSIM documentation is the use of the same name for both the *summed* parameter X and the first corresponding *triplet* parameter X_o. Here the convention will be adopted of italicizing the summed parameter name X, and boldfacing the triplet parameter names X_o, LX, and WX, since these actually appear in the model parameter list. For example, the summed parameter *X3MS* is computed from

$$X3MS = \mathbf{X3MS} + \frac{\mathbf{LX3MS}}{L_{eff}} + \frac{\mathbf{WX3MS}}{W_{eff}}. \tag{8.2}$$

The extrinsic model structure is discussed in Section 8.8.

Table 8.1   The BSIM model parameter set.

Parameter	Units	Description
Process Parameters		
TOX	m	Gate Oxide Thickness
LD	m	Source/Drain Underdiffusion of Gate
WD	m	Isolation Reduction of Channel Width
VDD	V	Maximum Applied Supply Voltage
Electrical Parameters		
MUZ	cm^2/V·s	Zero Bias Low Field Mobility
X2MZ	cm^2/V^2·s	Substrate Bias Effect on the Low Field Mobility
MUS	cm^2/V·s	High Drain Bias Mobility
X2MS	cm^2/V^2·s	Substrate Bias Effect on the High Drain Bias Mobility
X3MS	cm^2/V^2·s	Slope of Mobility at $V_{ds} = $ **VDD**
U0	V^{-1}	Gate Field Induced Mobility Reduction Parameter
X2U0	V^{-2}	Substrate Bias Effect on the Gate Field Mobility Reduction
VFB	V	Flatband Voltage
PHI	V	Surface Potential $(= 2\phi_f)$
K1	V$^{\frac{1}{2}}$	Body Effect on Threshold Voltage (First-Order Term)
K2		Body Effect on Threshold Voltage (Second-Order Term)
ETA		DIBL Coefficient When $V_{ds} = $ **VDD** and $V_{bs} = 0$
X2E	V$^{-\frac{1}{2}}$	Substrate Bias Effect on DIBL Coefficient
X3E	V^{-1}	Drain Bias Effect on DIBL Coefficient
U1	V^{-1}	High Drain Field Mobility Reduction Parameter
X2U1	μm/V^{-2}	Substrate Bias Effect on the High Drain Field Mobility Reduction
X3U1	V^{-2}	Drain Bias Effect on the High Drain Field Mobility Reduction
N0		Low Field Subthreshold Ideality Factor
NB		Substrate Bias Effect on the Subthreshold Ideality Factor
ND		Drain Bias Effect on the Subthreshold Ideality Factor
CGSO	F/m	Zero Bias Gate-Source Capacitance
CGDO	F/m	Zero Bias Gate-Drain Capacitance
CGBO	F/m	Zero Bias Gate-Bulk Capacitance

Table 8.2   Temperature-dependent parameters introduced into the HSPICE implementation of BSIM (HSPICE Level 13).

Parameter	Units	Description
BEX		Temperature Effect on the Low Drain Bias Mobility
TCV	V/K	Threshold Voltage Variation with Temperature
FEX		Temperature Effect on the High Drain Bias Behavior

Parameter Listings

A list of the BSIM model parameters may be found in Table 8.1. In addition, a list of the temperature dependent parameters added to the model by the HSPICE implementation of BSIM (HSPICE Level 13) is presented in Table 8.2. Finally, it should be noted that several parameters in HSPICE Level 13 have different names for the first of the three

**Table
8.3** Differences in parameter names between standard BSIM
and the HSPICE implementation of BSIM (HSPICE Level 13).

BSIM Parameter Name	HSPICE Level 13 Parameter Name
VDD	VDDM
U0	U00
VFB	VFB0
PHI	PHI0
ETA	ETA0
NB	NB0
ND	ND0

triplet parameters; these parameters are listed in Table 8.3. In all cases, the corresponding
L and W parameter names are unchanged.

8.4   The Depletion Charge Model

The BSIM model for the FET depletion charge is similar in intent to those of Level 2
(Chapter 6) and Level 3 (Chapter 7). The goal is to account for the reduction of the
total depletion charge by the overlap of the gate-induced depletion region with the source
and drain depletion regions. The approach is actually a simplified version of the earlier
methods, leading to a similar result—the computation of a fractional reduction of the
total depletion charge. However, that expression is later absorbed into a model parameter
($K1$), which helps to negate any shortcomings that this approach might have.

The BSIM computation of the depletion charge begins by considering the individual
terminal contributions separately, as shown in Figure 8.1. With only the gate considered
(Figure 8.1a), there is the usual gate-induced depletion region of a MOS capacitor. With
only the source and drain considered (Figure 8.1b), there are junction depletion regions
surrounding the source and drain diffusions. When all three terminals are considered
(Figure 8.1c), there is a region where the gate-induced depletion region overlaps the
source and drain depletion regions. This will reduce the total depletion charge from what
would be expected following a simple summation of the three independent depletion
regions.

As in the earlier (Level 2 and Level 3) models, BSIM computes a fractional
reduction of Q_{depl}, the depletion charge of a simple two-terminal MOS capacitor. As in
the Level 2 and Level 3 models, the overlap region will be modeled as a subtraction from
the gate-induced depletion charge only, allowing the computation of the source and drain
depletion regions to remain unaffected. Defining that fraction of the depletion charge as
t leads to

$$Q_{depl,BSIM} = Q_{depl,2D} - t \cdot Q_{depl,2D} = (1 - t) \cdot Q_{depl,2D}. \qquad (8.3)$$

Several comments can be made about this approach. Examining (8.3), the use of t has
the same effect as that of lowering the substrate doping N_A. Also, by inference from

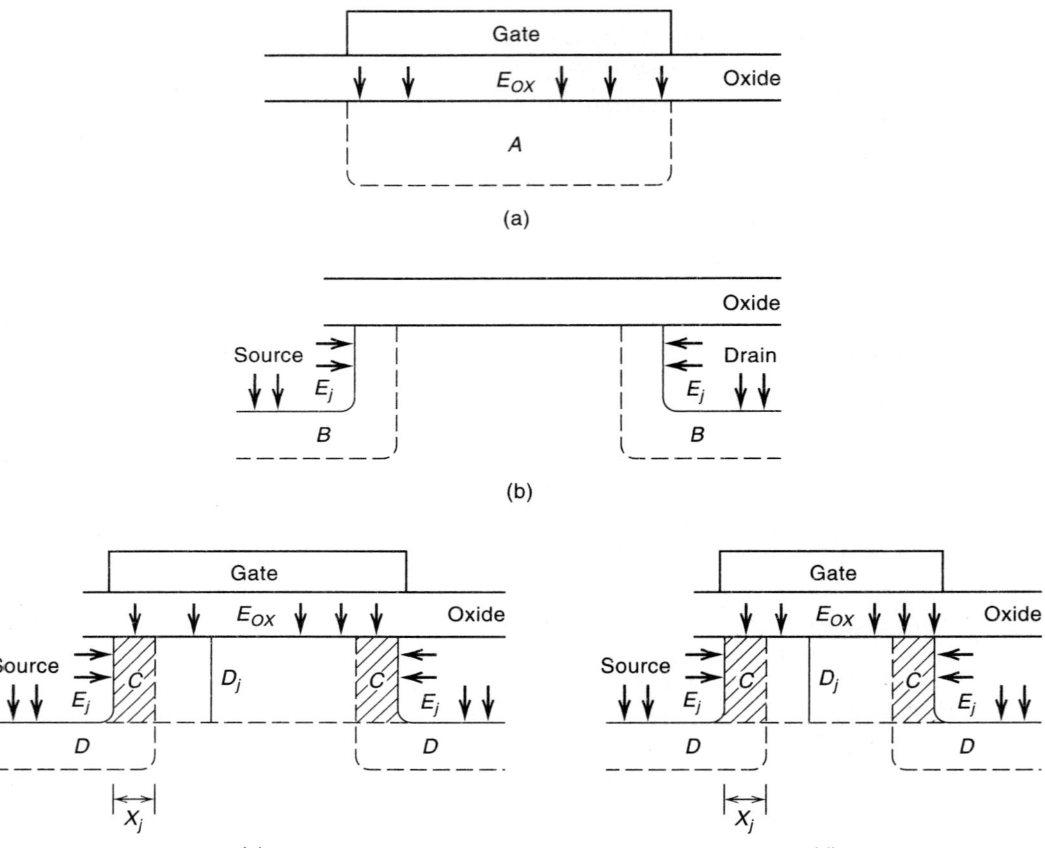

Figure 8.1   MOSFET depletion regions; (a) gate-induced depletion region, as in a MOS capacitor; (b) source and drain p-n junction depletion regions; (c) combined gate-induced, source, and drain depletion regions, showing the overlapping portions; (d) depletion regions in a shorter channel device, showing how the overlap regions retain their size while their portion of the total depletion region increases. From [1]. © 1985. The Regents of the University of California. Used by permission.

Figures 8.1c and 8.1d, t will be a function of the effective channel length L_{eff}. Although the overlap region is not affected by the channel length, the nonoverlap part of the gate-induced depletion region decreases with decreasing channel length; this increases the fractional value t. Thus, t is the ratio of the charge in the overlap regions to the total two-dimensional gate-induced depletion charge. From Figure 8.1c, the total two-dimensional depletion charge is

$$Q_{depl} = q \cdot N_A \cdot L_{eff} \cdot x_d,$$

(8.4)

while the depletion charge in the overlap region is

$$Q_{depl,overlap} = 2 \cdot q \cdot N_A \cdot W_d \cdot x_d; \tag{8.5}$$

the "2" is included to account for the two (source and drain) overlap regions. The fraction t is computed from this ratio:

$$t = \frac{Q_{depl,overlap}}{Q_{depl}} = \frac{2 \cdot W_d}{L_{eff}}. \tag{8.6}$$

Recall from Chapter 3 that the depletion depth of a one-sided p-n junction (e.g., n^+/p) is described by (3.31)

$$W = \left[\frac{2\epsilon_{Si}}{qN_A} (V_{bi} + V_R) \right]^{\frac{1}{2}} \propto N_A^{-\frac{1}{2}}. \tag{8.7}$$

Substituting (8.7) into (8.6) leads to

$$t \propto \frac{1}{(L_{eff} \cdot N_A^{\frac{1}{2}})}. \tag{8.8}$$

From (8.8) and (8.3), the fractional reduction of the depletion charge can be written as

$$f(L_{eff}, N_A) = (1 - t) = 1 - \frac{K_f}{(L_{eff} N_A^{\frac{1}{2}})}, \tag{8.9}$$

where K_f is a constant. From the first principles viewpoint used to develop (8.6), K_f should be 2, reflecting the existence of the two overlap regions. However, in BSIM, K_f is left as an arbitrary constant. In addition, K_f is never defined, since, as will be shown below when the drain current model is developed, the entire expression (8.9) is eventually absorbed into the model parameter $K1$. For completeness, the BSIM depletion charge can be expressed as

$$Q_{depl,BSIM} = f(L_{eff}, N_A) \cdot Q_{depl,2D}. \tag{8.10}$$

This absorption of the depletion charge model into $K1$ demonstrates a major philosophical change in the second-generation models. As (8.9) indicates, $f(L_{eff}, N_A)$ can, at least in intent, be computed from the physically meaningful process parameters L_{eff} and N_A; this is the approach taken in the Level 2 and Level 3 models. However, by absorbing $f(L_{eff}, N_A)$ into a model parameter, the burden of obtaining good model fitting to data is shifted away from (apparently) physically meaningful quantities and equations and toward parameter extraction.

8.5   The Threshold Voltage Model

Recall from Chapter 3 that the threshold voltage of a three-terminal MOS structure is

$$V_t = V_{fb} + 2\phi_f + \frac{Q_{depl}}{C_{ox}} = V_{fb} + 2\phi_f + \gamma \cdot (2\phi_f - V_{bs})^{\frac{1}{2}}. \qquad (8.11)$$

This basic expression neglects nonuniform doping, small geometry effects, and the effect of the drain bias. In the Level 3 model, (8.11) was modified to include the effect of charge sharing between the source and drain through the coefficient f_s, and the effect of depletion charge spreading outside the defined channel width with the coefficient f_n:

$$V_t = V_{fb} + 2\phi_f + f_s \cdot \gamma \cdot (2\phi_f - V_{bs})^{\frac{1}{2}} + f_n \cdot (2\phi_f - V_{bs}). \qquad (8.12)$$

A term for drain-induced barrier lowering (DIBL) was also added, using the empirical observation [12] that the threshold voltage decreases linearly with an increasing drain bias:

$$\Delta V_{t,DIBL} = \sigma \cdot V_{ds}. \qquad (8.13)$$

Following [12], (8.13) is cast into a more detailed form of

$$\Delta V_{t,DIBL} = \frac{\eta}{C_{ox} L_{eff}^{\;n}} V_{ds}. \qquad (8.14)$$

For implementation in Level 3, n was chosen to be 3, and an empirical constant was included to change η to a more reasonable value:

$$\Delta V_{t,DIBL} = \eta \frac{8.15 \times 10^{-22}}{C_{ox} L_{eff}^{\;3}} V_{ds}. \qquad (8.15)$$

Using (8.13), (8.12) is modified to the final Level 3 threshold voltage expression

$$V_t = V_{fb} + 2\phi_f + f_s \cdot \gamma \cdot (2\phi_f - V_{bs})^{\frac{1}{2}} + f_n \cdot (2\phi_f - V_{bs}) - \sigma \cdot V_{ds}. \qquad (8.16)$$

The BSIM threshold voltage expression is very similar to the Level 3 expression (8.16). An important change is that the Level 3 term $2\phi_f$, which is computed from the substrate doping parameter **NSUB**, is instead replaced with a model parameter ϕ_s; ϕ_s accounts for the effective substrate doping, as there is no explicit substrate doping parameter in BSIM. In the approximation of uniform substrate doping, $\phi_s = 2\phi_f$. The BSIM threshold voltage equation is

$$V_t = V_{fb} + \phi_s + K_1 \cdot (\phi_s - V_{bs})^{\frac{1}{2}} - K_2 \cdot (\phi_s - V_{bs}) - \eta \cdot V_{ds}. \qquad (8.17)$$

In the first-generation models, the flatband voltage V_{fb} was computed from the description of the gate material and from the substrate doping, and remained a constant; however, in BSIM, V_{fb} is treated as a model parameter. The parameters K_1 and η are obviously directly related to the Level 3 expressions $(f_s \cdot \gamma)$ and σ, respectively. The K_1 term provides, in part, an expression for the depletion charge and its reduction by short channel effects. In the Level 3 formulation, f_s is a composite term developed from the charge sharing model of Dang [13]. In BSIM, K_1 is simply an extracted parameter, with length and width sensitivities. The parameter η is the same as the Level 3 DIBL parameter σ. However, in Level 3, σ is a composite parameter, which also contains a dependence on the effective channel length $(\sigma \propto L_{eff}^{-3})$. In BSIM, η is also a composite parameter, but this is instead due to the inclusion of substrate and drain bias dependence in the form

$$\eta = \eta_z + \eta_b \cdot V_{bs} + \eta_d \cdot (V_{ds} - V_{dd}). \tag{8.18}$$

The K_2 term in (8.17) is very similar to the f_n term in the Level 3 expression (8.16); it is unclear if the BSIM expression is in some way a direct descendant of the Level 3 expression. In the Level 3 formulation, the f_n term is clearly designed to account for narrow channel effects (see Section 7.2.2). In BSIM, the addition of the K_2 term is stated to serve two purposes: to account for nonuniform substrate doping, and charge sharing between the source and the drain. To account for nonuniform substrate doping, the depletion charge expression should be based on a form similar to [14]

$$K_1 \cdot (\phi_{s1} - V_{bs})^{\frac{1}{2}} + K_2 \cdot (\phi_{s2} - V_{bs})^{\frac{1}{2}}, \tag{8.19}$$

which describes the spreading of the gate-induced depletion region first into the surface doping region (the K_1 term), then later, as the gate bias increases, into the doping region beneath the surface (the K_2 term). To describe source-drain charge sharing, the Level 3 model uses the composite expression $(f_s \cdot \gamma)$, which corresponds to K_1 in BSIM. From this discussion, it can be inferred that the K_2 term does indeed provide greater accuracy in describing these device phenomena, but this occurs mostly by the expedient of adding an additional term to the threshold voltage expression. Since the BSIM length and width sensitivities are also included, additional accuracy is gained.

These equations can now be rewritten with the parameter names identified. Equation (8.18) becomes

$$\eta = ETA + X2E \cdot V_{bs} + X3E \cdot (V_{ds} - \mathbf{VDD}). \tag{8.20}$$

Equation (8.17) becomes

$$V_t = VFB + PHI + K1 \cdot (PHI - V_{bs})^{\frac{1}{2}} - K2 \cdot (PHI - V_{bs}) - \eta \cdot V_{ds}, \tag{8.21}$$

where η was defined in (8.20). It is interesting to note once again that in BSIM (and the other second-generation models), the flatband voltage is an extracted parameter, whereas in the first-generation models it is a composite parameter computed from other parameters.

Later, when the drain current model is developed, it will be shown that both K_1 and η can be derived as composite parameters composed of other apparently more physically based parameters. However, K_1 and the composite η are treated as basic model parameters and are extracted from data.

Temperature Dependence

As noted at the beginning of this chapter, the original formulation of BSIM contains no temperature dependence. The HSPICE implementation of BSIM (HSPICE Level 13) contains modifications to include temperature dependence, in a manner that is available for all the HSPICE implementations of the second-generation models.

In HSPICE Level 13, the temperature dependence of the threshold voltage comes in two parts: the independent temperature parameter **TCV**, and the built-in temperature dependence of the surface potential parameter *PHI*. Using the simple definition of the Fermi potential (3.50), *PHI* implies a substrate doping concentration via

$$PHI(T_{nom}) = 2 \cdot \frac{k_b T_{nom}}{q} \cdot ln\left(\frac{N_A}{n_i}\right);\qquad(8.22)$$

Equation (8.22) can be solved for N_A; using this value of N_A, a new surface potential is easily computed from

$$PHI(T) = 2 \cdot \frac{k_b T}{q} \cdot ln\left(\frac{N_A}{n_i}\right).\qquad(8.23)$$

Equation (8.23) is inserted into the threshold voltage expression (8.21).

The use of the temperature parameter **TCV** is also quite simple. The temperature dependence of the threshold voltage is defined from

$$V_t(T) = V_t(T_{nom}) - \textbf{TCV} \cdot (T - T_{nom}),\qquad(8.24)$$

where $V_t(T_{nom})$ is given by (8.21). Equation (8.24) assumes that the threshold voltage varies linearly with temperature, which is a reasonable assumption. The model parameter **TCV** is easily extracted from measured data.

8.6   The Mobility Model

As in the first generation models, the mobility in BSIM is defined by a basic low field mobility μ_o, which is altered by the effect of vertical and lateral fields. However, in contrast to the earlier models, where μ_o is a single parameter with an implied physical basis, in BSIM μ_o is instead a composite parameter constructed from five other model parameters.

The composite μ_o is created by a quadratic interpolation through three points. The first point is a low drain field contribution:

$$\mu_o\big|_{V_{ds}=0} = \mu_z + \mu_{zb} \cdot V_{bs}. \tag{8.25}$$

Note that this introduces substrate bias dependence into the mobility expression, and that for $V_{bs} = 0$, the low field mobility is described by μ_z. Thus, it is really μ_z that represents the basic low field mobility, defined as μ_o in the first-generation models. The second interpolation expression is for a high drain field:

$$\mu_o\big|_{V_{ds}=V_{dd}} = \mu_s + \mu_{sb} \cdot V_{bs}. \tag{8.26}$$

Note here that (8.26) (and other expressions to be discussed below) force the maximum supply voltage V_{dd} to be introduced as a parameter. The final contribution is the sensitivity of μ_o to the drain bias at $V_{ds} = V_{dd}$; this is represented by a single parameter, to be discussed below.

The convention of identifying parameter names will now be adopted. The maximum power supply voltage is represented by the parameter **VDD**. The need for this parameter should concern the reader, as different model parameter results will be obtained for the same process technology if different values of **VDD** are used. Equation (8.25) is rewritten as

$$\mu_o\big|_{V_{ds}=0} = \mathbf{MUZ} + \mathit{X2MZ} \cdot V_{bs}. \tag{8.27}$$

As described above, **MUZ** represents the low field channel mobility, and thus corresponds to the parameter **UO** of Levels 2 and 3. Due to this relationship, in BSIM **MUZ** does **not** have length and width sensitivity, and is treated as an independent parameter (rather than taking the form of (8.1)). Equation (8.26) is recast as

$$\mu_o\big|_{V_{ds}=\mathbf{VDD}} = \mathit{MUS} + \mathit{X2MS} \cdot V_{bs}. \tag{8.28}$$

The final contribution, the sensitivity of μ_o to the drain bias at $V_{ds} = \mathbf{VDD}$, is described by the parameter $\mathit{X3MS}$.

The parameter **VDD** is treated as a provided process parameter. For example, for a 3.3V technology, the JEDEC standard is $V_{dd} = 3.3 \pm 0.3$ V, so **VDD** $= 3.6$ V, and data for parameter extraction should be available at that voltage.

8.6.1 Mobility Reduction by the Vertical Field

The BSIM model for the channel mobility reduction by the vertical field is essentially the same as that constructed for Level 3 (see Section 7.4.1), based on the description of surface charge of Schrieffer [15]. In BSIM, this is written as

$$\mu_v = \frac{\mu_o}{\left[1 + U_o(V_{gs} - V_t)\right]}, \tag{8.29}$$

where U_o replaces the Level 3 term θ. However, U_o is a composite term, which includes substrate bias dependence:

$$U_o = U_{oz} + U_{ob} \cdot V_{bs}. \tag{8.30}$$

Rewriting (8.30) with parameter names identified leads to

$$U_o = U0 + X2U0 \cdot V_{bs}. \tag{8.31}$$

Note that $U0$ is the direct equivalent of the Level 3 parameter **THETA**.

The mobility model also contains terms to describe the reduction of the mobility by the lateral field. However, these terms are intertwined with the drain current and saturation voltage models. They will be discussed when these items are considered.

Temperature Dependence

As originally constructed, BSIM contains no parameters to account for temperature dependence; there is only built-in temperature dependence in certain model parameters, such as *PHI*. The HSPICE implementation of BSIM (HSPICE Level 13) includes temperature dependence for the channel mobility in a manner essentially identical to that introduced into the Level 3 model (see Section 7.4). This takes the form

$$\mu(T) = \mu(T_{nom}) \cdot \left(\frac{T}{T_{nom}}\right)^{\textbf{BEX}}, \tag{8.32}$$

where T and T_{nom} are in **degrees Kelvin**. In BSIM, μ is a composite parameter; from (8.32) it is clear that certain model parameters will make use of **BEX**. In the original implementation, three model parameters, **MUZ**, *MUS*, and *X3MS* are subject to the use of **BEX**, in the form

$$\textbf{MUZ}(T) = \textbf{MUZ}(T_{nom}) \cdot \left(\frac{T}{T_{nom}}\right)^{\textbf{BEX}}. \tag{8.33}$$

In a later update, the option is provided of also making the parameters *X2MZ* and *X2MS* subject to (8.33). As will be described when parameter extraction is discussed, **BEX** is easily extracted from measured data.

8.7   The Drain Current Model

As might be expected, the BSIM drain current model was developed in essentially the same manner as that used in the first-generation models. However, a number of modifications are made. A rather different method of including the effect of the drain bias is used, and a completely new expression for the saturation voltage is developed.

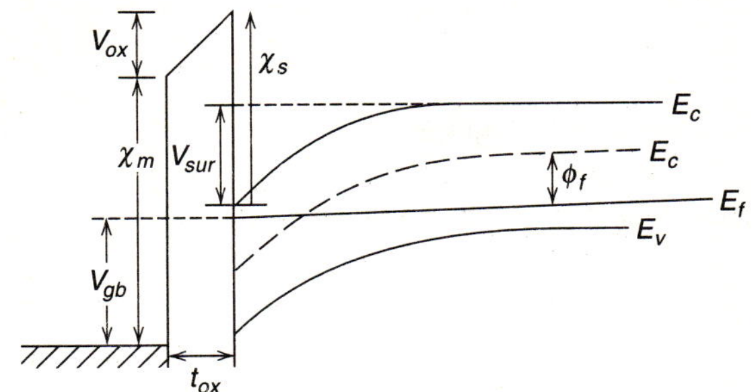

Figure 8.2 Band diagram of a MOS structure. From [1]. © 1985. The Regents of the University of California. Used by permission.

As in the Level 3 model, the $\frac{3}{2}$-power terms which result from the integration of the depletion charge expression are replaced; here, rather than a Taylor series expansion, a polynomial expression is used. Finally, a much more realistic subthreshold current model is introduced.

8.7.1   The Linear Region

The BSIM drain current formulation begins by considering the band diagram for the MOS structure, depicted in Figure 8.2. The summations of the voltages on opposite sides of the oxide must equal each other; that is,

$$V_{ox} + X_m = V_{gb} + \phi_f + \frac{E_g}{2} - V_{sur} + X_s. \tag{8.34}$$

(Note that the surface potential, described in the "textbook" fashion of ϕ_s in Chapters 3–7, is instead written here as V_{sur}. This is to avoid confusion with the BSIM model parameter ϕ_s, or *PHI*, described earlier. The choice of terminology is unfortunate, but the original sources will be followed here for consistency, and to allow the reader to more easily examine the original references.) The workfunction difference between the gate material and the silicon substrate is defined in the standard manner [16]:

$$\phi_{ms} = X_m - \left(X_s + \phi_f + \frac{E_g}{2} \right). \tag{8.35}$$

Using (8.35), (8.34) can be solved for the oxide voltage:

$$V_{ox} = V_{gb} - V_{sur} - \phi_{ms}. \tag{8.36}$$

Assuming that the oxide is essentially a perfect dielectric (with no mobile charge to respond to the gate field), the oxide voltage and field are related by

$$V_{ox} = \epsilon_{ox} E_{ox}. \tag{8.37}$$

To develop a drain current model, an expression must be found which relates the oxide field E_{ox} and the charge in the silicon substrate. To simplify this task, two assumptions are made. First, it is assumed that the gate oxide thickness is much less than the effective channel length. This assumption is reasonable for any generation of MOSFET technology; for example, common present values are $t_{ox} \approx 10$ nm and $L_{eff} \approx 0.25$ μm. This allows the oxide field to be taken as normal to the silicon surface ($E_{ox} = E_{ox}(x)$, $E_{ox}(y) = 0$). Second, it is assumed that on the silicon side of the oxide-silicon interface, the lateral field E_y is much less than the vertical field E_x. This assumption is quite reasonable in the region where the inversion charge density Q_{inv} is large, as the field lines from the gate have many charges upon which to terminate. However, when $V_{ds} > V_{dsat}$, a space charge region, created by channel length modulation, exists near the drain. Here, Q_{inv} is small, and the influence of the drain field becomes more important. Note, however, that the drain current expression is found by integrating along the entire length of the channel. In all but the most extreme circumstances, the channel length modulation region is much smaller than the channel length, so the effect of this region on the integral will be negligible.

Using these assumptions, from Gauss' law the fields at the interface must satisfy the continuity relation

$$\epsilon_{Si} E_{Si} = \epsilon_{ox} E_{ox} + Q_{ss}, \qquad (8.38)$$

where Q_{ss} is the fixed interface charge density. In modern MOS technology, Q_{ss} is negligible and can in general be ignored; it will be "carried along" here for completeness. The vertical field at the interface can be written as

$$\epsilon_{Si} E_{Si} = Q_{inv} + Q_{depl,BSIM} + \epsilon_{Si} E_1, \qquad (8.39)$$

where $Q_{depl,BSIM}$ is the modified depletion charge as described by (8.10). For a MOS capacitor, Q_{inv} and Q_{depl} are the standard contributors to the field expression. The BSIM formulation adds another term, E_1, which represents a contribution to the vertical field at the interface due to the drain bias.

To make use of (8.39), an expression for E_1 must be developed. Consider an FET with biases applied, except that $V_{ds} = 0$. The term $\epsilon_{Si} E_1$ in (8.39) vanishes, leaving the simplified charge-field expression; the device depletion regions are described by a solution of the Poisson equation. Now allow a drain bias to be applied; V_{ds} is no longer zero. A new potential is added to the device; however, this potential occurs in a region that is already depleted. No new charge is added to the Poisson equation, so this new potential can be described by a solution of the Laplace equation, and the result simply added by superposition to the already determined solution of the Poisson equation; this justifies the simple addition of the E_1 term to the standard expression, as was done in (8.39). The Laplace equation can then be solved for V_1, the potential associated with the field E_1.

To solve the Laplace equation, it is assumed that the junction depth x_j is much less than the effective channel length L_{eff}. In recent years, the effective channel length

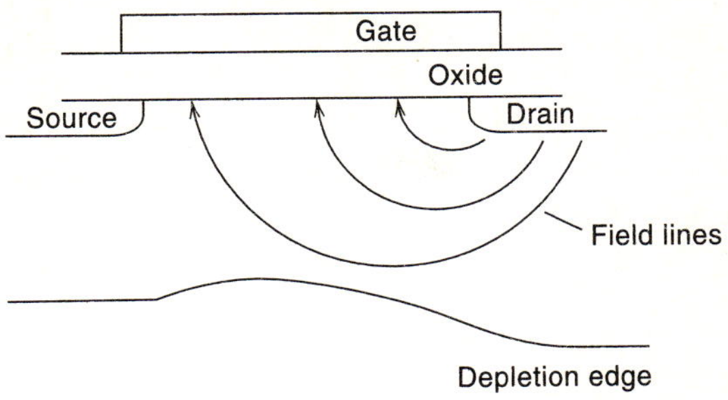

Figure 8.3   Field description of the influence of the drain bias on the surface field. From [1]. © 1985. The Regents of the University of California. Used by permission.

has decreased more rapidly than has the junction depth; for example, common present values are $x_j \approx 0.10$ µm and $L_{eff} \approx 0.25$ µm, so it should be noted that this approximation is questionable. However, it does allow the field lines associated with V_1 to be taken as fitting a cylindrical profile, as shown in Figure 8.3. The boundary conditions require that $V_1 = V_{ds}$ at the drain. A boundary condition must also be imposed at the interface, where the field lines terminate. Since those field lines terminate at the interface, a boundary condition of $V_1 = 0$ can be imposed there. The two-dimensional Laplace equation in cylindrical coordinates can then be solved, resulting in

$$V_1 = V_{ds} \cdot \frac{\theta}{\pi}, \tag{8.40}$$

where θ is the coordinate noted in Figure 8.3.

To find an expression for E_1, (8.40) must be differentiated. The coordinate θ must be chosen so as to lie along the channel, while r must be chosen to reach the source, so that the entire channel is considered. In effect, the coordinates are transformed, so that the derivative becomes

$$E_1 = \left.\frac{dV_1}{dy}\right|_{y=source} . \tag{8.41}$$

Carrying out the coordinate transformation and solving (8.41) leads to

$$E_1 = -\frac{V_{ds}}{\pi \cdot L_{eff}}. \tag{8.42}$$

This added surface field will induce additional inversion charge. which increases I_{ds} from its expected value. Also, since E_1 is proportional to V_{ds}, this added field will account for the behavior of the output conductance g_{ds} in the saturation region.

To develop a drain current equation, an expression for the inversion charge Q_{inv} must be found. The process is begun by solving (8.39) for Q_{inv}:

$$Q_{inv} = \epsilon_{Si} E_{Si} - Q_{depl,BSIM} - \epsilon_{Si} E_1. \tag{8.43}$$

Using the continuity relation (8.38) and replacing $\epsilon_{Si} E_{Si}$ leads to

$$Q_{inv} = \epsilon_{ox} E_{ox} + Q_{ss} - Q_{depl,BSIM} - \epsilon_{Si} E_1. \tag{8.44}$$

Using (8.36) and (8.37), E_{ox} is replaced, producing

$$Q_{inv} = C_{ox}(V_{gb} - V_{sur} - \phi_{ms}) + Q_{ss} - Q_{depl,BSIM} - \epsilon_{Si} E_1. \tag{8.45}$$

With some further rearrangement, and substituting (8.10) for $Q_{depl,BSIM}$ and (3.88) for Q_{depl}, (8.45) becomes

$$Q_{inv} = C_{ox}\left(V_{gb} - V_{sur} - \phi_{ms} + \frac{Q_{ss}}{C_{ox}}\right) - f(L_{eff}, N_A) \cdot (2\epsilon_{Si} q N_A V_{sur})^{\frac{1}{2}} - \epsilon_{Si} E_1, \tag{8.46}$$

or

$$Q_{inv} = C_{ox}\left[V_{gb} - V_{sur} - \phi_{ms} + \frac{Q_{ss}}{C_{ox}} - f(L_{eff}, N_A) \cdot \frac{(2\epsilon_{Si} q N_A V_{sur})^{\frac{1}{2}}}{C_{ox}}\right] - \epsilon_{Si} E_1. \tag{8.47}$$

Some definitions can now be introduced. From simple MOS capacitor theory [16], the flatband voltage is defined as

$$V_{fb} = \phi_{ms} - \frac{Q_{ss}}{C_{ox}}. \tag{8.48}$$

As noted earlier, in modern MOS technology, the fixed interface charge Q_{ss} is negligible, and the flatband voltage is usually equal to the gate-substrate workfunction difference. In addition, the composite term K_1 can be defined as

$$K_1 = f(L_{eff}, N_A) \cdot \frac{(2\epsilon_{Si} q N_A)^{\frac{1}{2}}}{C_{ox}}. \tag{8.49}$$

Substituting (8.48) and (8.49) into (8.47) leads to

$$Q_{inv} = C_{ox}\left[V_{gb} - V_{sur} - V_{fb} - K_1 \cdot V_{sur}^{\frac{1}{2}}\right] - \epsilon_{Si} E_1. \tag{8.50}$$

As (8.48) and (8.49) indicate, the terms V_{fb} and K_1 can be constructed from knowledge of certain physically based quantities. However, in most cases, these terms are treated as model parameters which are determined directly from parameter extraction. This explains the comment in Section 8.4 that although $f(L_{eff}, N_A)$ could be defined physically, it is instead absorbed, along with the constant K_f, into the extracted parameter *K1*. Both *VFB* and *K1* have already been introduced in the threshold voltage and mobility expressions. Using the convention of identifying model parameter names, (8.50) becomes

$$Q_{inv} = C_{ox}\left[V_{gb} - V_{sur} - VFB - K1 \cdot V_{sur}^{\frac{1}{2}}\right] - \epsilon_{Si}E_1. \tag{8.51}$$

Next, an expression for the surface potential V_{sur} must be found. BSIM takes the more physically correct approach of including the electron quasi-Fermi level, which must be different from the Fermi level in the channel for conduction to occur. Modifying the simple strong inversion expression $\phi_s = 2\phi_f - V_{bs}$, the surface potential can be described by

$$V_{sur} = \phi_n + (PHI - V_{bs}), \tag{8.52}$$

where ϕ_n is the electron quasi-Fermi level. Using (8.52) and redefining the gate bias from

$$V_{gs} = V_{gb} + V_{bs}, \tag{8.53}$$

(8.51) becomes the final working expression for Q_{inv}:

$$Q_{inv} = C_{ox}\left[(V_{gs} - VFB - \phi_n - PHI) - K1 \cdot (\phi_n + PHI - V_{bs})^{\frac{1}{2}}\right]. \tag{8.54}$$

Equation (8.54) will be used to derive an expression for the drain current in the linear region. To begin the derivation, it will be assumed that the current flows in the y-direction (along the channel) only. This allows the current density to be written as the standard one-dimensional equation [17]

$$J_y = q\mu_v n\left(\frac{d\phi_n}{dy}\right), \tag{8.55}$$

where μ_v is the mobility including the vertical field reduction as described by (8.29). To convert (8.55) into an expression for the drain current, this equation must be integrated into the inversion layer (x-direction) and across the channel (z-direction). To integrate in the x-direction, it will be assumed that the inversion layer terminates abruptly at some depth x_i, and that this layer is so thin that it can be treated as a two-dimensional charge layer (this is the charge sheet approximation [18]). That is,

$$q \int_0^{x_i} n \cdot dx = Q_{inv}. \tag{8.56}$$

In the z-direction, it is assumed that the device is completely uniform across the channel; mathematically, this assumption results in

$$\int_0^{W} dz = W. \tag{8.57}$$

Substituting (8.56) and (8.57) into (8.55) produces

$$I_{ds} = \mu_v W_{eff} Q_{inv} \left(\frac{d\phi_n}{dy}\right). \tag{8.58}$$

Equation (8.58) can be rearranged to the integral expression

$$I_{ds} \int_0^{L_{eff}} dy = \mu_v W_{eff} \int_0^{V_{ds}} Q_{inv} \cdot d\phi_n. \tag{8.59}$$

(As noted in the original BSIM documentation [1], the upper integral limit L_{eff} should possibly be modified to $L_{eff} - L'$ when channel length modulation occurs, where L' is the length of the space charge region created by channel length modulation. However, this computation is for the linear region, so channel length modulation is not a concern here. Also, BSIM, like the earlier models, derives its equations without considering channel length modulation; a correction is added later to account for this behavior.)

Inserting (8.54) into (8.59) and also making the ad hoc but reasonable addition of the DIBL term (as in (8.17)), (8.59) becomes

$$I_{ds} = \frac{\mu_v C_{ox} W_{eff}}{L_{eff}} \int_0^{V_{ds}} \left[(V_{gs} - VFB - \phi_n - PHI)\right.$$
$$\left. - K1 \cdot (\phi_n + PHI - V_{bs})^{\frac{1}{2}} + \eta \cdot V_{ds}\right]d\phi_n. \tag{8.60}$$

This expression can be integrated term by term as

$$I_{ds} = \frac{\mu_o C_{ox} W_{eff}}{L_{eff}} \left[\int_0^{V_{ds}} (V_{gs} - VFB - PHI) \cdot d\phi_n - \int_0^{V_{ds}} \phi_n \cdot d\phi_n\right.$$
$$\left. - K1 \cdot \int_0^{V_{ds}} (\phi_n + PHI - V_{bs})^{\frac{1}{2}} \cdot d\phi_n + (\eta \cdot V_{ds}) \int_0^{V_{ds}} d\phi_n\right]. \tag{8.61}$$

The integral with the $\frac{1}{2}$-power term can be solved by a simple change of variables; the remaining integrals are straightforward. Carrying out the integrations leads to

$$I_{ds} = \frac{\mu_o C_{ox} W_{eff}}{L_{eff}} \left\{(V_{gs} - VFB - PHI) \cdot V_{ds} + (\eta - \frac{1}{2}) \cdot V_{ds}^2\right.$$
$$\left. - \frac{2}{3} \cdot K1 \cdot \left[(V_{ds} + PHI - V_{bs})^{\frac{3}{2}} - (PHI - V_{bs})^{\frac{3}{2}}\right]\right\}. \tag{8.62}$$

Note that if DIBL is neglected ($\eta \to 0$) and depletion region overlap is similarly ignored ($f(L_{eff}, N_A) \to 1$, so that $K1 \to \gamma$, where γ is the standard bulk charge expression (3.94)), (8.62) reduces to the simpler $\frac{3}{2}$-power linear current expression used in the Level 2 model (see Section 6.5).

As described in Chapter 6, an expression with $\frac{3}{2}$-power terms is computationally inefficient, and can also lead to convergence problems during circuit simulation. In the Level 3 formulation (see Section 7.5), a Taylor series expansion was used to simplify the current expression and eliminate the $\frac{3}{2}$-power terms. BSIM also eliminates the $\frac{3}{2}$-power terms, but in a somewhat different manner. For purposes of convenience, define a mathematical function from the $\frac{3}{2}$-power terms of (8.62) as

$$F(V_{ds}, PHI - V_{bs}) \equiv \frac{2}{3} \cdot K1 \cdot \left[(V_{ds} + PHI - V_{bs})^{\frac{3}{2}} - (PHI - V_{bs})^{\frac{3}{2}} \right]. \qquad (8.63)$$

For further simplicity, define

$$V_A \equiv PHI - V_{bs}, \qquad (8.64)$$

which transforms (8.62) to

$$F(V_{ds}, V_A) \equiv \frac{2}{3} \cdot K1 \cdot \left[(V_{ds} + V_A)^{\frac{3}{2}} - (V_A)^{\frac{3}{2}} \right]. \qquad (8.65)$$

An expansion for (8.65) is

$$F(V_{ds}, V_A) = K1 \cdot \left[V_A^{\frac{1}{2}} \cdot V_{ds} + \frac{0.25 \cdot V_{ds}^2}{V_A^{\frac{1}{2}}} + \dots \right]. \qquad (8.66)$$

However, this expression is a very poor approximation when $V_A \gg V_{ds}$, which is certainly the case for the low values of V_{ds} in the linear region. Instead, the expansion (8.66) is changed to

$$F(V_{ds}, V_A) = K1 \cdot \left[V_A^{\frac{1}{2}} \cdot V_{ds} + \frac{0.25 \cdot g \cdot V_{ds}^2}{V_A^{\frac{1}{2}}} \right]. \qquad (8.67)$$

The objective is to determine an expression for g which gives good results over a wide range of possible operating conditions. This is done by using (8.67) to plot $F(V_{ds}, V_A)$ versus V_{ds} for many different values of V_A, as shown in Figure 8.4. It was found empirically that a workable expression for g is

$$\frac{1}{1 - g} = P_1 + P_2 \cdot V_A. \qquad (8.68)$$

The terms P_1 and P_2 are then determined by numerical fitting; as shown in Figure 8.4, the extracted results are $P_1 = 1.744$ and $P_2 = 0.8364$. Including these values, (8.68) can be solved for g:

$$g = 1 - \frac{1}{1.744 + 0.8364 \cdot V_A}. \qquad (8.69)$$

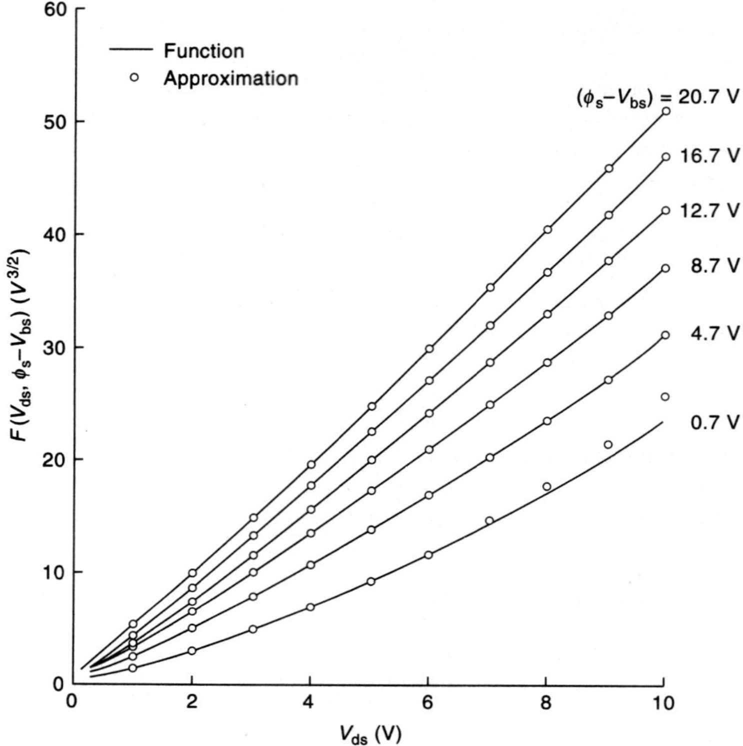

Figure 8.4 Plot of the function $F(V_{ds}, V_A)$ versus V_{ds} for several values of V_A, and the results of numerical fitting to that data. From [1]. © 1985. The Regents of the University of California. Used by permission.

Using (8.64), (8.67), and (8.69), the drain current expression (8.62) becomes

$$
I_{ds} = \frac{\mu_v C_{ox} W_{eff}}{L_{eff}} \left\{ (V_{gs} - VFB - PHI) \cdot V_{ds} + (\eta - \frac{1}{2}) \cdot V_{ds}^2 \right.
$$

$$
\left. - K1 \cdot \left[(PHI - V_{bs})^{\frac{1}{2}} + \frac{g \cdot V_{ds}^2}{4(PHI - V_{bs})^{\frac{1}{2}}} \right] \right\}.
\tag{8.70}
$$

This expression can be rearranged to

$$
I_{ds} = \frac{\mu_v C_{ox} W_{eff}}{L_{eff}} \left\{ (V_{gs} - VFB - PHI) \cdot V_{ds} + (\eta \cdot V_{ds}) \cdot V_{ds} \right.
$$

$$
\left. - K1 \cdot (PHI - V_{bs})^{\frac{1}{2}} \cdot V_{ds} - \frac{1}{2} \cdot V_{ds}^2 - K1 \cdot \frac{g \cdot V_{ds}^2}{4(PHI - V_{bs})^{\frac{1}{2}}} \right\},
\tag{8.71}
$$

then further manipulated to the form

$$I_{ds} = \frac{\mu_v C_{ox} W_{eff}}{L_{eff}} \left\{ \left[V_{gs} - (VFB + PHI + K1 \cdot (PHI - V_{bs})^{\frac{1}{2}} - \eta \cdot V_{ds}) \right] \cdot V_{ds} \right.$$

$$\left. - \frac{V_{ds}^2}{2} \left[1 + \frac{g \cdot K1}{2(PHI - V_{bs})^{\frac{1}{2}}} \right] \right\}. \tag{8.72}$$

Some simplifications can now be made. The term a can be introduced, defined as

$$a = 1 + \frac{g \cdot K1}{2(PHI - V_{bs})^{\frac{1}{2}}}. \tag{8.73}$$

In addition, (8.72) contains a sequence of terms that is very nearly the BSIM threshold voltage expression (8.21), except for the $K2$ term. Once again, an ad hoc but reasonable addition is made, this time involving the inclusion of the $K2$ term, so that (8.72) becomes

$$I_{ds} = \frac{\mu_v C_{ox} W_{eff}}{L_{eff}} \left\{ \left[V_{gs} - (VFB + PHI + K1 \cdot (PHI - V_{bs})^{\frac{1}{2}} \right. \right.$$

$$\left. \left. + K2 \cdot (PHI - V_{bs}) - \eta \cdot V_{ds}) \right] \cdot V_{ds} - \frac{V_{ds}^2}{2} \left[1 + \frac{g \cdot K1}{2(PHI - V_{bs})^{\frac{1}{2}}} \right] \right\}. \tag{8.74}$$

Including the BSIM threshold voltage expression (8.21) simplifies (8.74) to the final expression for the drain current in the linear region:

$$I_{ds} = \frac{\mu_v C_{ox} W_{eff}}{L_{eff}} \left[(V_{gs} - V_t) \cdot V_{ds} - \frac{a}{2} \cdot V_{ds}^2 \right]. \tag{8.75}$$

This expression is very simple, and is virtually identical to the Level 1 linear region expression. However, the simplicity of (8.75) is reached by hiding a great deal of complexity; in particular, the terms μ_v, V_t, and a are quite sophisticated. However, since these three terms represent numerical values that are computed from various model parameters during circuit simulation, the simplicity of (8.75) is clearly beneficial to the circuit design user of any BSIM FET model.

8.7.2 The Saturation Region

In Level 1, a very simple model for the saturation voltage was developed, leading to

$$V_{dsat} = V_{gs} - V_t. \tag{8.76}$$

This expression also guarantees the continuity of the drain current and its first derivative at $V_{ds} = V_{dsat}$. However, in shorter channel length devices, carrier velocity saturation serves to decrease the saturation voltage from its classical value as given by (8.76). The first attempt to deal with this phenomenon was introduced into Level 2, leading to a very complex expression based on a quartic polynomial equation. Despite the complexity, the results are very poor, as there is a severe discontinuity in the first derivative of the current at $V_{ds} = V_{dsat}$. Not only does this produce a poor fit to data; the derivative discontinuity frequently prevents circuit simulations from converging. A more empirical but improved (and simplified) expression was introduced into the Level 3 model. Although the fitting results are considerably better than those found using Level 2, and convergence problems rarely occur, there is still an unpleasant discontinuity in the first derivative of the drain current at $V_{ds} = V_{dsat}$.

The route to the BSIM expression for the drain current in saturation begins with a reevaluation of the channel charge Q_{inv}. The available BSIM documents [1], [2] are quite terse on this subject; for more detailed information and an involved explanation of the basics of what is to follow, the reader is instead referred to a reference text [17].

Neglecting the drain bias contribution to the vertical surface field from E_1, the inversion charge can be expressed as

$$Q_{inv} = C_{ox}(V_{gs} - VFB - PHI - \phi_n) - Q_{depl,BSIM}. \tag{8.77}$$

The BSIM depletion charge is included in a form slightly modified from (8.10):

$$Q_{depl,BSIM} = C_{ox} \cdot f(L_{eff}, N_A) \cdot \gamma \cdot (\phi_n + PHI - V_{bs})^{\frac{1}{2}}, \tag{8.78}$$

which, using (8.49), becomes

$$Q_{depl,BSIM} = C_{ox} \cdot K1 \cdot (\phi_n + PHI - V_{bs})^{\frac{1}{2}}. \tag{8.79}$$

Using (8.79), (8.77) becomes

$$Q_{inv} = C_{ox} \left[(V_{gs} - VFB - PHI - \phi_n) - K1 \cdot (\phi_n + PHI - V_{bs})^{\frac{1}{2}} \right]. \tag{8.80}$$

To make use of (8.80), some approximation for the depletion charge expression (8.79) must be developed.

Using a Taylor expansion of the square root term in (8.80) and retaining only the first two terms leads to

$$(\phi_n + PHI - V_{bs})^{\frac{1}{2}} = (PHI - V_{bs})^{\frac{1}{2}} + \frac{\phi_n}{2(PHI - V_{bs})^{\frac{1}{2}}}. \tag{8.81}$$

Note that this expansion is virtually identical to that described earlier in (8.66). As in that situation, (8.81) is a poor approximation when $(PHI - V_{bs})^{\frac{1}{2}} \gg \phi_n$, which is certainly the case for low values of V_{ds}. Thus, (8.81) is modified in the same manner that was used to reach (8.67):

$$(\phi_n + PHI - V_{bs})^{\frac{1}{2}} = (PHI - V_{bs})^{\frac{1}{2}} + \frac{g \cdot \phi_n}{2(PHI - V_{bs})^{\frac{1}{2}}}. \tag{8.82}$$

Substituting (8.82) into (8.80) leads to

$$Q_{inv} = C_{ox} \left\{ (V_{gs} - VFB - PHI - \phi_n) - K1 \cdot \left[(PHI - V_{bs})^{\frac{1}{2}} + \frac{g \cdot \phi_n}{2(PHI - V_{bs})^{\frac{1}{2}}} \right] \right\}, \tag{8.83}$$

which can be rearranged to

$$Q_{inv} = C_{ox} \left\{ \left[V_{gs} - VFB - PHI - K1 \cdot (PHI - V_{bs})^{\frac{1}{2}} \right] - \phi_n \cdot \left[1 + \frac{g \cdot K1}{2(PHI - V_{bs})^{\frac{1}{2}}} \right] \right\}. \tag{8.84}$$

This expression can be greatly simplified. First, note that the BSIM threshold voltage expression (8.21)

$$V_t = VFB + PHI + K1 \cdot (PHI - V_{bs})^{\frac{1}{2}} - K2 \cdot (PHI - V_{bs}) - \eta \cdot V_{ds} \tag{8.85}$$

is nearly available in the term

$$V_t = VFB + PHI + K1 \cdot (PHI - V_{bs})^{\frac{1}{2}}. \tag{8.86}$$

As has occurred previously in the development of the BSIM formulation, an ad hoc but reasonable approach is taken, and the $K2$ and η terms are introduced into (8.84), allowing the partial threshold voltage (8.86) to be replaced with the full BSIM threshold voltage expression (8.85). Second, the term a, introduced in (8.73),

$$a = 1 + \frac{g \cdot K1}{2(PHI - V_{bs})^{\frac{1}{2}}}, \tag{8.87}$$

can be substituted into (8.84). This transforms (8.84) into the very simple form

$$Q_{inv} = C_{ox}(V_{gs} - V_t - a \cdot \phi_n). \tag{8.88}$$

Next, a way must be found to include velocity saturation in the drain current formulation. Recall from (8.58) that the drain current is described by

$$I_{ds} = \mu_v W_{eff} Q_{inv} \left(\frac{d\phi_n}{dy} \right). \tag{8.89}$$

Noting that the lateral field is defined by

$$E_y = \frac{d\phi_n}{dy}, \tag{8.90}$$

and using the simple Ohm's law definition of the carrier velocity

$$v = \mu_v \cdot E_y, \tag{8.91}$$

(8.89) can be rewritten as

$$I_{ds} = W_{eff} Q_{inv} v. \tag{8.92}$$

To proceed, an expression for the carrier velocity must be employed. The BSIM formulation uses an expression suggested by Sze [19] and others [20]:

$$v = \frac{\mu_v \cdot E_y}{\left[1 + \frac{E_y}{E_c}\right]}, \tag{8.93}$$

where E_c is a critical field. In the limit of $E_y \to \infty$, (8.93) becomes

$$v = \mu_v \cdot E_c \equiv v_{SAT}, \tag{8.94}$$

which defines the saturation velocity v_{SAT}. Using (8.94) to define E_c, (8.93) can be reformulated as

$$v = \frac{\mu_v \cdot E_y}{\left[1 + \frac{\mu_v \cdot E_y}{v_{SAT}}\right]}. \tag{8.95}$$

Substituting this expression for the carrier velocity into the drain current equation (8.92) leads to

$$I_{ds} = \mu_v W_{eff} Q_{inv} \left[\frac{E_y}{1 + \frac{\mu_v \cdot E_y}{v_{SAT}}}\right]. \tag{8.96}$$

This drain current expression can be further developed by inserting the expression for the inversion charge Q_{inv} (8.88), producing

$$I_{ds} = \mu_v W_{eff} C_{ox} (V_{gs} - V_t - a \cdot \phi_n) \left[\frac{E_y}{1 + \frac{\mu_v \cdot E_y}{v_{SAT}}}\right]. \tag{8.97}$$

The lateral field E_y can be replaced with its definition (8.90), which transforms (8.97) to

$$I_{ds} = \mu_v W_{eff} C_{ox} (V_{gs} - V_t - a \cdot \phi_n) \left[\frac{\frac{d\phi_n}{dy}}{1 + \frac{\mu_v}{v_{SAT}} \frac{d\phi_n}{dy}}\right], \tag{8.98}$$

which can be algebraically rearranged to

$$I_{ds} = \frac{d\phi_n}{dy}\left[\mu_v W_{eff} C_{ox}(V_{gs} - V_t - a \cdot \phi_n) - I_{ds} \cdot \frac{\mu_v}{v_{SAT}}\right]. \tag{8.99}$$

This equation can now be integrated; writing out the integration term by term leads to

$$I_{ds} \cdot \int_0^{L_{eff}} dy = \mu_v W_{eff} C_{ox} \int_0^{V_{ds}} (V_{gs} - V_t) \cdot d\phi_n$$

$$-\mu_v W_{eff} C_{ox} \cdot a \int_0^{V_{ds}} \phi_n d\phi_n - I_{ds} \frac{\mu_v}{v_{SAT}} \int_0^{V_{ds}} d\phi_n. \tag{8.100}$$

After integration, (8.100) becomes

$$I_{ds} \cdot L_{eff} = \mu_v W_{eff} C_{ox}(V_{gs} - V_t) \cdot V_{ds} - \mu_v W_{eff} C_{ox} \cdot a \cdot \frac{V_{ds}^2}{2} - I_{ds}\frac{\mu_v}{v_{SAT}}V_{ds}. \tag{8.101}$$

To simplify (8.101), the definition

$$R_{sat} \equiv \frac{1}{W_{eff} \cdot C_{ox} \cdot v_{SAT}} \tag{8.102}$$

can be introduced into the last term of (8.101). This transforms (8.101) to

$$I_{ds} = \frac{\mu_v C_{ox} W_{eff}}{L_{eff}}\left[(V_{gs} - V_t - I_{ds}R_{sat}) \cdot V_{ds} - \frac{a}{2} \cdot V_{ds}^2\right]. \tag{8.103}$$

Note that (8.103) contains the drain current I_{ds} on both sides of the equation. By regrouping terms, (8.103) becomes

$$I_{ds}\left[1 + \left(\frac{\mu_v C_{ox} W_{eff}}{L_{eff}}\right) \cdot R_{sat} \cdot V_{ds}\right] = \frac{\mu_v C_{ox} W_{eff}}{L_{eff}}\left[(V_{gs} - V_t) \cdot V_{ds} - \frac{a}{2} \cdot V_{ds}^2\right]. \tag{8.104}$$

This allows the definition of the composite parameter U_1 as

$$U_1 = \mu_v C_{ox} W_{eff} \cdot R_{sat}. \tag{8.105}$$

Using (8.105), (8.104) is re-written as a description of the linear region drain current:

$$I_{ds} = \left(\frac{\mu_v C_{ox} W_{eff}}{L_{eff}}\right)\left(\frac{1}{1 + \frac{U_1}{L_{eff}} \cdot V_{ds}}\right)\left[(V_{gs} - V_t) \cdot V_{ds} - \frac{a}{2} \cdot V_{ds}^2\right], \tag{8.106}$$

where the second term in (8.106) describes the degradation of the channel mobility by the lateral field.

In a fashion similar to what has occurred several times in the development of the BSIM formulation, U_1 could be computed from a collection of physically based values. However, U_1 is instead treated as a composite model parameter, defined by

$$U_1 = U1 + X2U1 \cdot V_{bs} + X3U1 \cdot (V_{ds} - \mathbf{VDD}). \tag{8.107}$$

The saturation voltage behavior can now be considered. If the simplistic definition of the saturation voltage as the drain voltage at which $Q_{inv} \rightarrow 0$ is used in the channel charge expression (8.88), the classical expression

$$V_{dsat} = V_{gs} - V_t \tag{8.108}$$

is recovered. However, this approach is physically unrealistic, since, as discussed in Chapters 6 and 7, it implies that the lateral field E_y becomes infinite at the pinch-off point. A more realistic method, also discussed in Chapters 6 and 7, defines the saturation of the drain current as being attained when the carriers reach the saturation velocity. Using (8.92),

$$I_{dsat} = W_{eff} Q_{inv} v_{SAT}, \tag{8.109}$$

which can be solved for Q_{inv}:

$$Q_{inv} = \frac{I_{dsat}}{W_{eff} \cdot v_{SAT}}. \tag{8.110}$$

Substituting the channel charge Q_{inv} from (8.88) into (8.110) leads to

$$\frac{I_{dsat}}{W_{eff} \cdot v_{SAT}} = C_{ox}(V_{gs} - V_t - a \cdot \phi_n). \tag{8.111}$$

At the saturation voltage, $\phi_n \rightarrow V_{dsat}$, so (8.111) can be solved for V_{dsat}:

$$I_{dsat} = C_{ox} W_{eff} v_{SAT} (V_{gs} - V_t - a \cdot V_{dsat}). \tag{8.112}$$

Using the definition of R_{sat} (8.102), (8.112) can be solved for V_{dsat}:

$$V_{dsat} = \frac{1}{a} \left[V_{gs} - V_t - I_{dsat} R_{sat} \right]. \tag{8.113}$$

Next, recall the expression for the linear region current (8.103), derived earlier in this section. By making the usual definition of the saturation current I_{dsat} as the drain current when the drain bias $V_{ds} = V_{dsat}$, and replacing V_{ds} with V_{dsat}, (8.103) becomes

$$I_{dsat} = \frac{\mu_v C_{ox} W_{eff}}{L_{eff}} \left[(V_{gs} - V_t - I_{dsat} R_{sat}) \cdot V_{dsat} - \frac{a}{2} \cdot V_{dsat}^2 \right]. \tag{8.114}$$

Substituting the definition of V_{dsat} (8.113) into (8.114) produces, after some algebraic manipulation,

$$I_{dsat} = \frac{\mu_v C_{ox} W_{eff}}{L_{eff}} \left(\frac{1}{2a} \right) \left(V_{gs} - V_t - I_{dsat} R_{sat} \right)^2. \tag{8.115}$$

The liberty can also be taken of defining an expression for I_{dsat} which has the appearance of the standard expression:

$$I_{dsat} = \frac{\mu_v C_{ox} W_{eff}}{L_{eff}} \left(\frac{1}{2 \cdot a \cdot K} \right) (V_{gs} - V_t)^2. \tag{8.116}$$

(Note that (8.116) will ultimately be used as the model equation describing the saturation region drain current.) This equation can be substituted into both places in (8.115) where I_{dsat} appears. After a very lengthy and tedious algebraic exercise, an expression which is quadratic in K is produced. Recalling the definition of the parameter U_1 from (8.105), the quadratic equation is

$$K^2 - K \cdot \left[1 + \left(\frac{U_1}{L_{eff}} \right) \cdot \frac{V_{gs} - V_t}{a} \right] + \left[\left(\frac{U_1}{L_{eff}} \right)^2 \cdot \frac{(V_{gs} - V_t)^2}{(2a)^2} \right] = 0. \tag{8.117}$$

By defining a *critical voltage* as

$$V_c \equiv \left(\frac{U_1}{L_{eff}} \right) \cdot \frac{V_{gs} - V_t}{a}, \tag{8.118}$$

(8.117) is transformed to

$$K^2 - K \cdot (1 + V_c) + \left(\frac{V_c^2}{4} \right) = 0. \tag{8.119}$$

This quadratic expression can be solved for K, yielding

$$K = \frac{1 + V_c \pm (1 + 2V_c)^{\frac{1}{2}}}{2}. \tag{8.120}$$

Only one choice of the "$\pm$" will be valid; this choice must be determined. This is done very simply by inserting K as defined by (8.120) into the defined expression for the saturation current (8.116), producing

$$I_{dsat} = \frac{\mu_v C_{ox} W_{eff}}{L_{eff}} \left(\frac{1}{2a}\right) \left(\frac{2}{1 + V_c \pm (1 + 2V_c)^{\frac{1}{2}}}\right) (V_{gs} - V_t)^2. \qquad (8.121)$$

In the long channel limit ($L_{eff} \rightarrow \infty$), $V_c \rightarrow 0$. If V_c is allowed to go to zero in (8.121), the "$-$" choice of "$\pm$" yields an infinite current, while the "$+$" choice leads to a finite current. It can thus be concluded that the "$+$" solution is the physically valid one, allowing the final definition of K as

$$K = \frac{1 + V_c + (1 + 2V_c)^{\frac{1}{2}}}{2}. \qquad (8.122)$$

It is worthwhile to examine the limiting behavior of V_c. In the long channel limit (large L_{eff}), $V_c \ll 1$. Using a simple expansion of $(1 + 2V_c)^{\frac{1}{2}}$ for small V_c, (8.122) becomes

$$K = 1 + V_c = 1 + \left(\frac{U_1}{L_{eff}}\right) \cdot \frac{(V_{gs} - V_t)}{a}. \qquad (8.123)$$

Substituting this definition of K into the saturation current expression (8.116) leads to

$$I_{dsat} = \frac{\mu_v C_{ox} W_{eff}}{2L_{eff}} \left(\frac{1}{1 + \frac{U_1}{L_{eff}} \cdot (V_{gs} - V_t)}\right) (V_{gs} - V_t)^2, \qquad (8.124)$$

which maintains the basic structure of the classical expression for I_{dsat}, in which

$$I_{dsat} \propto (V_{gs} - V_t)^2. \qquad (8.125)$$

In the short channel limit (small L_{eff}), $V_c \gg 1$. The square root term in (8.122) becomes insignificant, and the equation for K thus becomes

$$K = \frac{V_c}{2} = \left(\frac{U_1}{L_{eff}}\right) \cdot \frac{(V_{gs} - V_t)}{a}. \qquad (8.126)$$

When this definition of K is substituted into (8.116), the result is

$$I_{dsat} = \mu_v C_{ox} W_{eff} \cdot \left(\frac{1}{U_1}\right) (V_{gs} - V_t). \qquad (8.127)$$

Replacing U_1 with its definition (8.105) and then replacing R_{sat} with its definition (8.102) transforms (8.127) to

$$I_{dsat} = v_{SAT} C_{ox} W_{eff} (V_{gs} - V_t). \qquad (8.128)$$

This is an important result for short channel devices, in which velocity saturation dominates the drain current behavior. First, note that the channel length L_{eff} does not appear in (8.128); the drain current remains constant with decreasing channel length, rather than increasing as described by the classical model. Second, in (8.128), the drain current I_{dsat} does not follow (8.125), but instead has the form

$$I_{dsat} \propto (V_{gs} - V_t). \tag{8.129}$$

This indicates that the drain current will not increase as rapidly with increasing $(V_{gs} - V_t)$; in short channel devices, the drain current increases *linearly* with $(V_{gs} - V_t)$, rather than *quadratically*, as predicted by the classical model.

Finally, the BSIM description of the saturation voltage can be considered. An expression for the saturation voltage V_{dsat} is found by inserting the saturation current equation (8.116) into the definition of V_{dsat} (8.113), producing

$$V_{dsat} = \frac{1}{a}\left[V_{gs} - V_t - \frac{\mu_v C_{ox} W_{eff}}{L_{eff}} \cdot \frac{R_{sat}}{2 \cdot a \cdot K}(V_{gs} - V_t)^2 \right]$$

$$= \frac{V_{gs} - V_t}{a} - \frac{\mu_v C_{ox} W_{eff}}{L_{eff}} \cdot \frac{R_{sat}}{2K}\left(\frac{V_{gs} - V_t}{a}\right)^2. \tag{8.130}$$

Using (8.105), (8.130) becomes

$$V_{dsat} = \frac{V_{gs} - V_t}{a} - \frac{1}{2K} \cdot \frac{U_1}{L_{eff}}\left(\frac{V_{gs} - V_t}{a}\right)^2$$

$$= \frac{V_{gs} - V_t}{a} - \frac{1}{2K} \cdot \frac{U_1}{L_{eff}}\left(\frac{V_{gs} - V_t}{a}\right)\left(\frac{V_{gs} - V_t}{a}\right). \tag{8.131}$$

Consolidating some terms using the definition of the critical voltage V_c from (8.118) transforms (8.131) to

$$V_{dsat} = \frac{V_{gs} - V_t}{a} - \frac{V_c}{2K} \cdot \left(\frac{V_{gs} - V_t}{a}\right)$$

$$= \frac{V_{gs} - V_t}{a} \cdot \left(1 - \frac{V_c}{2K}\right) = \frac{V_{gs} - V_t}{a} \cdot \left(\frac{2K - V_c}{2K}\right). \tag{8.132}$$

To solve (8.132), an expression for V_c in terms of K must be found. This is done by using the definition of K in (8.120), which can be algebraically manipulated into the quadratic equation

$$V_c^2 - 4KV_c + 4K(K + 1) = 0. \tag{8.133}$$

The quadratic solution for V_c is

$$V_c = 2K \pm 2K^{\frac{1}{2}}. \tag{8.134}$$

Inserting this result into (8.132) leads to

$$V_{dsat} = \frac{V_{gs} - V_t}{a} \cdot \left(\frac{\mp K^{\frac{1}{2}}}{K} \right). \tag{8.135}$$

Of the two choices, the "$-$" solution of (8.135) will produce a negative value for the saturation voltage, which is not physically valid. The "$+$" solution is the one sought, producing the final BSIM expression for the saturation voltage:

$$V_{dsat} = \frac{V_{gs} - V_t}{a \cdot K^{\frac{1}{2}}}. \tag{8.136}$$

Note that this expression is considerably simpler than the expressions developed for use in Level 2 and Level 3. The physical effects of importance (chiefly the influence of carrier velocity saturation) are accounted for with an equation that is simple, efficient, and less likely to cause convergence problems during circuit simulation.

Temperature Dependence

As noted above several times, the original BSIM formulation contains no parameters to allow the model to have temperature dependence. The HSPICE implementation of BSIM (HSPICE Level 13) adds temperature dependence to the model structure.

In HSPICE Level 13, the temperature dependence of the high drain bias behavior is described by the parameter **FEX**. Three parameters, *U1*, *X2U1*, and *X3U1*, make use of **FEX**, in the form

$$U1(T) = U1(T_{nom}) \cdot \left(\frac{T}{T_{nom}} \right)^{\textbf{FEX}}, \tag{8.137}$$

where T and T_{nom} are in **degrees Kelvin**. **FEX** is easily determined during parameter extraction.

8.7.3 The Subthreshold Region

The Level 2 and Level 3 drain current models include a model for subthreshold conduction which is very simple. As implemented, this model contains only one parameter, **NFS**, which is supposed to represent the interface state density. In reality, the model fits data for unreasonably large values of **NFS** ($\sim 10^{-12}$ cm^{-2}, two or more orders of magnitude larger than measured values). In addition, **NFS** has no geometry, bias, or temperature dependence; an accurate subthreshold model can only be constructed for one choice of these three variables. Finally, the use of the term V_{on}, designed to guarantee the continuity of the drain current across the weak to strong inversion boundary, introduces a discontinuity in the first derivative $\left(\frac{\partial I_{ds}}{\partial V_{gs}}\right)$ [21]; this can cause convergence problems during circuit simulation.

In BSIM, the subthreshold model is substantially upgraded. The derivation begins with the standard equation for the subthreshold current [17]

$$I_{exp} = I_o \cdot e^{\frac{q}{k_b T}[(V_{gs} - V_t)/n]} \left(1 - e^{-\frac{q}{k_b T} \cdot V_{ds}}\right),\qquad(8.138)$$

where I_o is a prefactor and n is an ideality factor given by

$$n \doteq 1 + \frac{C_{depl}}{C_{ox}}.\qquad(8.139)$$

In the ideal case, $C_{depl} \ll C_{ox}$, and $n \to 1$. In BSIM, the prefactor is determined empirically from fitting to measured data [1], [22] as

$$I_o = \frac{\mu_v C_{ox} W_{eff}}{L_{eff}} \cdot \left(\frac{k_b T}{q}\right)^2 e^{1.8}.\qquad(8.140)$$

Substituting (8.140) into (8.138) produces the final expression for the subthreshold current:

$$I_{exp} = \frac{\mu_v C_{ox} W_{eff}}{L_{eff}} \cdot \left(\frac{k_b T}{q}\right)^2 e^{1.8} \cdot e^{\frac{q}{k_b T}[(V_{gs} - V_t)/n]} \left(1 - e^{-\frac{q}{k_b T} \cdot V_{ds}}\right).\qquad(8.141)$$

The ideality factor n could, to some degree of approximation, be computed from (8.139). However, BSIM instead treats n as a composite parameter, defined by

$$n \equiv N0 + NB \cdot V_{bs} + ND \cdot V_{ds}.\qquad(8.142)$$

The description of the subthreshold current (8.141) works well in the deep subthreshold region, where, by definition, only the subthreshold current is present. As shown in Figure 8.5, the subthreshold current is due to diffusion, while the strong inversion current is due to drift. In the region where both mechanisms are present, something must be done to link the two separate models together. The scheme used in the Level 2 and Level 3 models is able to guarantee continuity of the current in this region, but contains a discontinuity in the first derivative [21].

The BSIM formulation adopts a method of dealing with this problem that was suggested by Antognetti *et al.* [21]. As shown in Figure 8.5, for gate biases larger than the threshold voltage, the diffusion current becomes nearly constant. From examination of simulations, this point can be defined as the transition voltage, chosen at a gate bias of $V_{gs} = V_{tr}$:

$$V_{tr} = V_t + 3 \left(\frac{k_b T}{q}\right).\qquad(8.143)$$

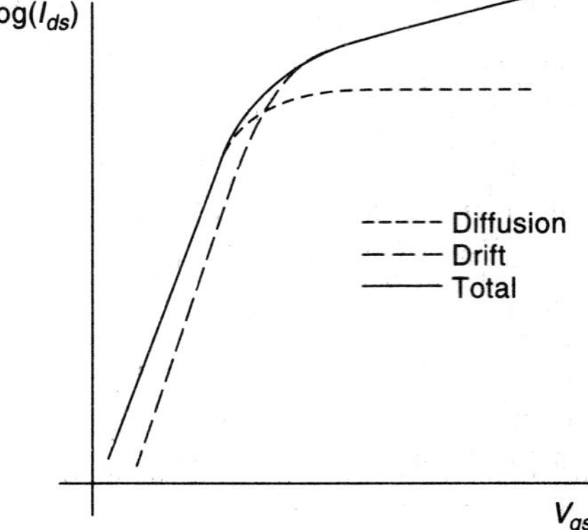

Figure 8.5 Generalized description of the drain current. In the subthreshold region conduction is dominated by diffusion, while in the strong inversion region it is dominated by drift. Near the subthreshold-superthreshold transition, both mechanisms are important.

Using the classical definition of the saturation region current

$$I_{ds} = \frac{\mu_v C_{ox} W_{eff}}{2 L_{eff}} (V_{gs} - V_t)^2,$$ (8.144)

substitution of (8.143) into (8.144) produces an expression for the upper limit of the subthreshold current:

$$I_{limit} = \frac{\mu_v C_{ox} W_{eff}}{2 L_{eff}} \left(3 \cdot \frac{k_b T}{q} \right)^2.$$ (8.145)

Next, some method must be introduced to guarantee that the current will be described by I_{exp} (8.141) in the deep subthreshold region and by I_{limit} (8.145) at the transition voltage V_{tr}. This is done empirically by defining the weak inversion current for values of $V_{gs} \leq V_{tr}$ as

$$I_{weak} = \frac{I_{exp} \cdot I_{limit}}{I_{exp} + I_{limit}}.$$ (8.146)

In the deep subthreshold region, $I_{exp} \ll I_{limit}$, and (8.146) reduces to

$$I_{weak} = I_{exp}.$$ (8.147)

At the transition voltage V_{tr}, $I_{exp} \gg I_{limit}$, and (8.146) reduces to

$$I_{weak} = I_{limit}.$$ (8.148)

Examination of (8.147) and (8.148) indicates that (8.146) is able to satisfy the two required conditions at the extremes of its range.

8.7.4   Connecting the Weak and Strong Inversion Models

The BSIM method of linking the weak inversion current model to the strong inversion current model can now be considered. The strong inversion current can be defined as a current component, either in the linear or the saturation region, which is computed when $V_{gs} > V_t$. For $V_{gs} \leq V_t$,

$$I_{strong} = 0. \tag{8.149}$$

The weak inversion current I_{weak}, as described by (8.146), has a finite value for all gate biases. For large values of V_{gs}, it is assumed that $I_{strong} \gg I_{weak}$. (Note that this approximation is a poor one for gate biases only slightly larger than the threshold voltage; this is a major problem when BSIM is used for applied gate biases in this region, as occurs in many analog circuit designs. A similar problem occurs in low-voltage digital circuits.) This allows the total drain current to be written as the simple sum of the strong and weak inversion components:

$$I_{ds,total} = I_{strong} + I_{weak}. \tag{8.150}$$

8.8   The Charge Model

As described in Chapter 3, in the general case, the computation of the node charges for each FET is separate from the computation of the gate capacitance. The FET model equations produce expressions for the node charges, which are then used to compute the gate capacitance. However, the gate capacitance computation is actually independent of the FET model; it is instead a matter of implementation in the SPICE circuit simulator. For that reason, the charge models are discussed with each particular FET model, while the computation of the gate capacitance is discussed in Chapter 13.

In BSIM, the charge model is virtually identical to that suggested by Yang *et al.* [7]. The charge is computed for four different regions of operation (accumulation, subthreshold, linear, and saturation), and continuity is guaranteed at each boundary so that the gate capacitance will also be continuous across these boundaries. This makes it possible to guarantee that charge is conserved during circuit simulation; however, the final guarantee of charge conservation is determined by the general gate capacitance calculations, as described in Chapter 13.

8.8.1   The Accumulation Region

In accumulation, the inversion density Q_{inv} is zero; the gate charge is mirrored in the substrate. The charge expressions are easily written down from simple MOS capacitor considerations [19]:

$$\begin{aligned} Q_{GATE} &= W_{eff} \cdot L_{eff} \cdot C_{ox}(V_{gb} - VFB) \\ &= W_{eff} \cdot L_{eff} \cdot C_{ox}(V_{gs} - VFB - V_{bs}), \end{aligned} \tag{8.151}$$

$$Q_{BULK} = -Q_{GATE}, \tag{8.152}$$

$$Q_{INV} = 0. \tag{8.153}$$

(The capitalized subscripts indicate total charge, rather than charge per unit area.) Here the accumulation charge is written as Q_{BULK} for lexical accuracy; in the other regions of operation, this term will be described by the depletion charge Q_{DEPL}.

8.8.2   The Subthreshold Region

In the subthreshold region, it is assumed that $Q_{INV} \ll Q_{DEPL}$; this allows Q_{INV} to be set to zero, as in the accumulation region. Once again, the bulk/depletion charge is taken as mirroring the gate charge. Yang *et al.* [7] employed the simple device charge model developed by Ihantola and Moll [23] in their derivation, which includes the simple body effect expression γ. The BSIM description is much more complicated, including two body effect parameters ($K1$ and $K2$), as well as the composite DIBL parameter η. However, for simplicity, the BSIM formulation uses the description of Yang *et al.* [7], replacing the term γ with its close equivalent, the parameter $K1$. The subthreshold charge equations are thus

$$Q_{GATE} = W_{eff} \cdot L_{eff} \cdot C_{ox} \cdot \frac{K1}{2}\left\{-1 + \left[1 + \frac{4(V_{gs} - VFB - V_{bs})}{K1^2}\right]^{\frac{1}{2}}\right\}, \tag{8.154}$$

$$Q_{DEPL} = -Q_{GATE} \tag{8.155}$$

$$Q_{INV} = 0. \tag{8.156}$$

8.8.3   The Strong Inversion Region

In the strong inversion region, the inversion charge is no longer negligible, and all three components of the charge (Q_{GATE}, Q_{DEPL}, and Q_{INV}) must be computed. With inversion charge present, the total charge in the device is now dependent on the drain bias; the components must be computed by integration of simpler charge per unit area expressions. The full BSIM expression for the inversion charge density per unit area was developed earlier in (8.54), but is not used here; instead, the simpler expressions employed by Yang *et al.* [7] are used directly. In all cases, charge is treated as uniform across the width of the channel.

The Linear Region

In the linear region, the gate charge per unit area is

$$Q_{gate}(y) = C_{ox}(V_{gs} - VFB - PHI - \phi_n), \tag{8.157}$$

and the inversion charge per unit area is

$$Q_{inv}(y) = -C_{ox}(V_{gs} - V_t - \alpha_x \cdot \phi_n). \tag{8.158}$$

This inversion charge density expression is very similar to the modified form of the inversion charge density in the linear region described by (8.88). However, the term α_x, as described in [7], is actually a composite expression describing short channel effects. (The use of γ as a short channel parameter in [7] is an unfortunate and confusing choice of a symbol.) In BSIM, α_x is instead defined as

$$\alpha_x = a \cdot \left[1 + \frac{U_1}{L_{eff}}(V_{gs} - V_t)\right].$$ (8.159)

Since charge neutrality must be preserved,

$$Q_{gate}(y) + Q_{inv}(y) + Q_{depl}(y) = 0.$$ (8.160)

This allows an expression for the depletion charge to be found from

$$Q_{depl}(y) = -\left[Q_{gate}(y) + Q_{depl}(y)\right] = -C_{ox}\left[V_t - VFB - PHI - (1 - x_x) \cdot \phi_n\right].$$ (8.161)

The total charge for each component is found by integrating (8.157), (8.158), and (8.161) across the width and along the length of the channel. As mentioned above, the charge is treated as completely uniform across the width of the channel; the integral in that direction produces the effective channel width W_{eff}. The integral for the total gate charge is thus

$$Q_{GATE} = W_{eff} \int_0^{L_{eff}} Q_{gate}(y) \cdot dy.$$ (8.162)

Substituting (8.157) for $Q_{gate}(y)$, changing variables from dy to $d\phi_n$, and carrying out the integration leads to

$$Q_{GATE} = W_{eff} \cdot L_{eff} \cdot C_{ox} \cdot \left[V_{gs} - VFB - PHI - \frac{V_{ds}}{2} + \frac{1}{12} \cdot \frac{\alpha_x \cdot V_{ds}^2}{V_{gs} - V_t - \frac{\alpha_x}{2} \cdot V_{ds}}\right].$$ (8.163)

By a similar approach, the total depletion charge is

$$Q_{DEPL} = W_{eff} \cdot L_{eff} \cdot C_{ox} \cdot \left[-V_t + VFB + PHI + \frac{1 - \alpha_x}{2} \cdot V_{ds}\right.$$

$$\left. - \frac{1}{12} \cdot \frac{(1 - \alpha_x)\alpha_x V_{ds}^2}{V_{gs} - V_t - \frac{\alpha_x}{2} \cdot V_{ds}}\right],$$ (8.164)

while the total inversion charge is

$$Q_{INV} = -W_{eff} \cdot L_{eff} \cdot C_{ox} \left[V_{gs} - V_t - \frac{\alpha_x}{2} \cdot V_{ds} + \frac{1}{12} \cdot \frac{(x_x \cdot V_{ds})^2}{V_{gs} - V_t - \frac{\alpha_x}{2} \cdot V_{ds}}\right].$$ (8.165)

With significant inversion charge present, some way of dividing that charge between the source and the drain terminals must be introduced, as discussed below.

The Saturation Region

The charge terms in the saturation region are computed in the same fashion as those of the linear region, leading to

$$Q_{GATE} = W_{eff} \cdot L_{eff} \cdot C_{ox} \cdot \left[V_{gs} - VFB - PHI - \frac{V_{gs} - V_t}{3\alpha_x} \right], \tag{8.166}$$

$$Q_{DEPL} = W_{eff} \cdot L_{eff} \cdot C_{ox} \cdot \left[VFB + PHI - V_t + \frac{(1 - \alpha_x)(V_{gs} - V_t)}{3\alpha_x} \right], \tag{8.167}$$

and

$$Q_{INV} = -\frac{2}{3} \cdot W_{eff} \cdot L_{eff} \cdot C_{ox} \cdot (V_{gs} - V_t). \tag{8.168}$$

As in the case of the linear region, the presence of significant inversion charge forces the introduction of some scheme for partitioning the inversion charge between the source and the drain.

Charge Partitioning

The problem of partitioning the inversion charge between the source and the drain terminals was first considered in detail by Ward and Dutton [8]. Although the simplest approach is to divide the inversion charge density equally between the source and the drain, Ward and Dutton found that the best agreement between their analytical model and numerical simulations occurred when 60% of the inversion charge was assigned to the source while the remaining 40% was assigned to the drain. This partitioning scheme is generally used in the first-generation models, and is implemented by setting the partitioning parameter **XQC** to 0.4 (instead of the default value of 0.5). As an alternative, the physically based argument can be made [7] that in saturation, the drain is electrically disconnected from the inversion layer by the pinch-off region, so that all the inversion charge should be assigned to the source.

Instead of allowing the user to choose an arbitrary value for the partitioning parameter **XQC**, BSIM provides three choices of partitioning between the drain and the source—40%/60%, 0%/100%, and 50%/50%. The choice among these three is made by setting the flag parameter **XPART** to 0 for the 40%/60% method, 1 for the 0%/100% method, and 0.5 for the 50%/50% method. The partitioning is computed as described by Ward and Dutton [8] from

$$Q_S = W_{eff} \int_0^{L_{eff}} \left(1 - \frac{y}{L_{eff}} \right) \cdot Q_{inv}(y) \cdot dy, \tag{8.169}$$

and

$$Q_D = W_{eff} \int_0^{L_{eff}} \frac{y}{L_{eff}} \cdot Q_{inv}(y) \cdot dy, \tag{8.170}$$

where y is the point along the channel where the partition point is deemed to reside. This approach also guarantees that

$$Q_S + Q_D = Q_{INV}. \tag{8.171}$$

The 40%/60% Charge Partitioning Method

This choice of a partitioning scheme was suggested by the numerical results of Ward and Dutton [8]. Here the integrals (8.169) and (8.170) are evaluated with the partition point set at 60% of the distance from the source to the drain. In the linear region, the results are

$$Q_S = -W_{eff} \cdot L_{eff} \cdot C_{ox} \left\{ \frac{V_{gs} - V_t}{2} + \frac{1}{12} \cdot \frac{(\alpha_x \cdot V_{ds})^2}{V_{gs} - V_t - \frac{\alpha_x}{2} \cdot V_{ds}} - \frac{\alpha_x \cdot V_{ds}}{(V_{gs} - V_t - \frac{\alpha_x}{2} \cdot V_{ds})^2} \right.$$

$$\left. \times \left[\frac{(V_{gs} - V_t)^2}{6} - \frac{\alpha_x \cdot V_{ds} \cdot (V_{gs} - V_t)}{8} + \frac{(\alpha_x \cdot V_{ds})^2}{40} \right] \right\}, \tag{8.172}$$

and

$$Q_D = -W_{eff} \cdot L_{eff} \cdot C_{ox} \left\{ \frac{V_{gs} - V_t}{2} - \frac{1}{2} \cdot \alpha_x \cdot V_{ds} + \frac{\alpha_x \cdot V_{ds}}{(V_{gs} - V_t - \frac{\alpha_x}{2} \cdot V_{ds})^2} \right.$$

$$\left. \times \left[\frac{(V_{gs} - V_t)^2}{6} - \frac{\alpha_x \cdot V_{ds} \cdot (V_{gs} - V_t)}{8} + \frac{(\alpha_x \cdot V_{ds})^2}{40} \right] \right\}. \tag{8.173}$$

Note that for $V_{ds} = 0$,

$$Q_S = Q_D = -\frac{1}{2} \cdot W_{eff} \cdot L_{eff} \cdot C_{ox} \cdot (V_{gs} - V_t), \tag{8.174}$$

and the inversion charge in the linear region is divided equally between the source and the drain; this would be expected with no drain bias applied.

In the saturation region,

$$Q_S = -\frac{2}{5} \cdot W_{eff} \cdot L_{eff} C_{ox} \cdot (V_{gs} - V_t), \tag{8.175}$$

and

$$Q_S = -\frac{4}{15} \cdot W_{eff} \cdot L_{eff} C_{ox} \cdot (V_{gs} - V_t). \tag{8.176}$$

Note from (8.175) and (8.176) that the total inversion charge (8.168) is recovered from the simple summation condition (8.171); also note that these expressions for the source and drain charges in saturation are indeed divided 40%/60% between the drain and the source.

The 0%/100% Charge Partitioning Method

This choice of a partitioning method was suggested by Yang *et al.* [7]. The development of this method includes several assumptions and requirements. First, the charge expressions will be forced to be continuous across the linear region-saturation region boundary. Second, when the drain bias is zero, the inversion charge will be equally divided between the source and drain (i.e., $Q_S = Q_D$). Third, in saturation, the drain will be taken to be electrically disconnected from the inversion layer by the pinch-off region, and all the inversion charge will be assigned to the source (i.e., $Q_D = 0$, $Q_S = Q_{INV}$). Finally, the expressions will be required to use only the terms in the expressions for Q_{INV} ((8.165) and (8.168)); no new terms will be added.

With these constraints, the results in the linear region are

$$Q_S = -W_{eff} \cdot L_{eff} \cdot C_{ox} \left[\frac{V_{gs} - V_t}{2} + \frac{1}{4} \cdot \alpha_x \cdot V_{ds} - \frac{1}{24} \cdot \frac{(\alpha_x \cdot V_{ds})^2}{V_{gs} - V_t - \frac{\alpha_x}{2} \cdot V_{ds}} \right], \quad (8.177)$$

and

$$Q_D = -W_{eff} \cdot L_{eff} \cdot C_{ox} \left[\frac{V_{gs} - V_t}{2} + \frac{3}{4} \cdot \alpha_x \cdot V_{ds} + \frac{1}{8} \cdot \frac{(\alpha_x \cdot V_{ds})^2}{V_{gs} - V_t - \frac{\alpha_x}{2} \cdot V_{ds}} \right]. \quad (8.178)$$

As in the 40%/60% partitioning scheme, when $V_{ds} = 0$,

$$Q_S = Q_D = -\frac{1}{2} \cdot W_{eff} \cdot L_{eff} \cdot C_{ox} \cdot (V_{gs} - V_t), \quad (8.179)$$

and the inversion charge is divided equally between the source and the drain.

In the saturation region, using (8.168),

$$Q_S = Q_{INV} = -\frac{2}{3} \cdot W_{eff} \cdot L_{eff} \cdot C_{ox} \cdot (V_{gs} - V_t), \quad (8.180)$$

and

$$Q_D = 0. \quad (8.181)$$

All the inversion charge is now assigned to the source.

The 50%/50% Charge Partitioning Method

This choice of a partitioning method is the simplest possible, and is considered the default approach in the first-generation models. In both the linear region and the saturation region, the inversion charge is divided equally between the source and the drain.

In the linear region,

$$Q_S = Q_D = -\frac{1}{2} W_{eff} \cdot L_{eff} \cdot C_{ox} \left[V_{gs} - V_t - \frac{\alpha_x}{2} \cdot V_{ds} + \frac{1}{12} \cdot \frac{(\alpha_x \cdot V_{ds})^2}{V_{gs} - V_t - \frac{\alpha_x}{2} \cdot V_{ds}} \right]. \qquad (8.182)$$

As in the other two partitioning methods, when $V_{ds} = 0$, the expected result of

$$Q_S = Q_D = -\frac{1}{2} \cdot W_{eff} \cdot L_{eff} \cdot C_{ox} \cdot (V_{gs} - V_t) \qquad (8.183)$$

is found. In the saturation region, using (8.168),

$$Q_S = Q_D = \frac{Q_{INV}}{2} = -\frac{1}{3} W_{eff} \cdot L_{eff} \cdot C_{ox} \cdot (V_{gs} - V_t). \qquad (8.184)$$

8.9 The Extrinsic Model Structure

As noted at the beginning of this chapter, the second-generation models are unique in possessing a twofold mathematical structure. The first part is the intrinsic structure; that is, the basic mathematical equations that constitute the core of the model. This part of any second-generation model closely resembles the entire structure of a first-generation model. The second-generation models also contain a second part, an extrinsic structure that imposes an independent and external geometry dependent model on top of the intrinsic formulation. This added structure is represented by the composite parameter expression (8.1).

The set of device geometries used to construct a BSIM model parameter set bears an initial resemblance to the requirements of the first-generation models; however, this set turns out to be somewhat different; this is shown in Figure 8.6. As in the first-generation models (e.g., Level 3), a long, wide device is used as a base reference point. Short and narrow channel effects are once again treated as being linearly independent of each other; wide, short devices are used to describe the short channel behavior, while long, narrow devices are used to describe narrow channel effects. In theory, this could be done with three devices, as in Level 2 and Level 3—a long, wide device as the base FET, a short, wide device for short channel effects, and a long, narrow device for narrow channel effects. However, for reasons that will be explained below, BSIM includes several intermediate geometries, as shown in Figure 8.6. Here, six devices are employed, with three short channel devices and two narrow channel devices.

For this discussion, let Z be some BSIM parameter. Using (8.1), Z is a composite parameter, described by

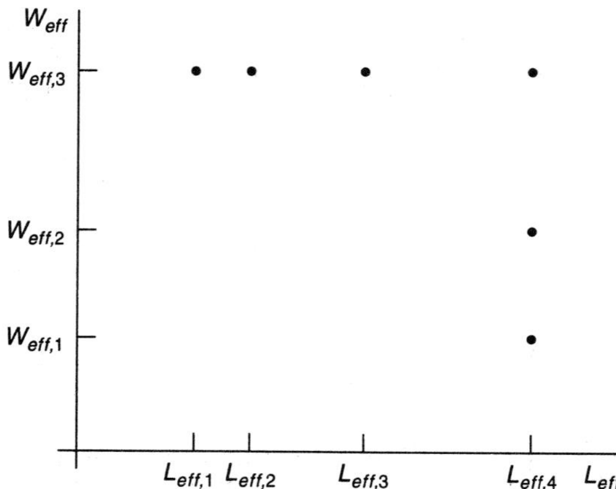

Figure 8.6   An example of a typical set of device geometries used to extract the data for construction of a BSIM FET model.

$$Z = \mathbf{Z} + \frac{\mathbf{LZ}}{L_{eff}} + \frac{\mathbf{WZ}}{W_{eff}}. \tag{8.185}$$

For $L_{eff} \to \infty$ and $W_{eff} \to \infty$, Z reduces to

$$Z = \mathbf{Z}. \tag{8.186}$$

For large devices, the computed model parameters approach long channel values; the extrinsic geometry dependence no longer affects the device model parameters.

In the first phase of model construction, a complete set of independent intrinsic model parameters is extracted for each of the six devices depicted in Figure 8.6. For illustrative purposes, let the particular value of Z extracted at each point be subscripted to identify the particular length and width; for example, Z_{43} indicates the extracted value for $L_{eff} = L_4$ and $W_{eff} = W_3$.

To properly describe the extrinsic model structure of (8.185), consider the set of devices with $W_{eff} = W_3$ and various channel lengths. Choosing the very wide devices allows the limit of $W_{eff} \to \infty$ to be taken; the $\mathbf{WZ}$ term in (8.185) goes to zero and can be discarded. This simplifies (8.185) to

$$Z = \mathbf{Z} + \frac{\mathbf{LZ}}{L_{eff}}. \tag{8.187}$$

Examination of (8.187) indicates that it is in linear slope-intercept form. By taking $\frac{1}{L_{eff}}$ as x and Z as y, a straight line is produced, with $\mathbf{LZ}$ as the slope and $\mathbf{Z}$ as the intercept. This is depicted graphically in Figure 8.7, where the extracted values of Z (Z_{43}, Z_{33}, ...) are plotted against $\frac{1}{L_{eff}}$. The role of the y-intercept point becomes clear. Since $\frac{1}{L_{eff}} \to 0$, then $L_{eff} \to \infty$, and $\mathbf{Z}$ is the long channel value of Z. It would be possible to use (8.187) to extract $\mathbf{Z}$ and $\mathbf{LZ}$ with only two devices ($L_{eff} = L_4$ and $L_{eff} = L_1$), as in the Level

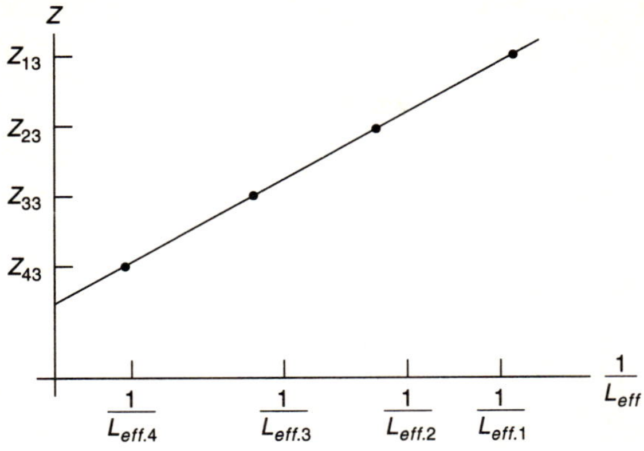

Figure 8.7 The extraction of **Z** and **LZ** from the values of Z extracted for individual devices.

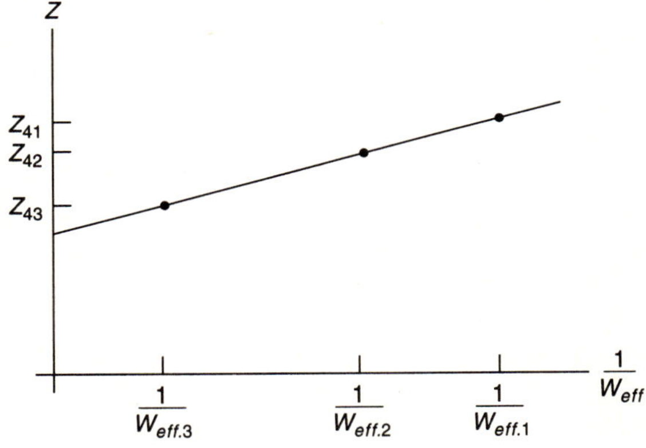

Figure 8.8 The extraction of **Z** and **WZ** from the values of Z extracted for individual devices.

3 approach. However, the additional data points are used to improve the model accuracy for intermediate geometries.

In a similar fashion, the set of devices with $L_{eff} = L_4$ and different widths can be considered. Taking L_4 to be very large (and thus $L_{eff} \rightarrow \infty$) allows the **LZ** term in (8.185) to be discarded. This simplifies (8.185) to

$$Z = \mathbf{Z} + \frac{\mathbf{WZ}}{W_{eff}}. \tag{8.188}$$

Using the reasoning just described leads to the plot of Figure 8.8. A straight line is produced, with the slope yielding the model parameter **WZ**. At the intercept, $W_{eff} \rightarrow \infty$, and Z reduces to the wide channel value **Z**.

Two comments should be made here regarding this scheme. If a first principles approach is adopted, (8.187) and (8.188) indicate that *none* of the extracted values of Z

(e.g., Z_{43}, Z_{33}, ...) actually appear in the final device model parameter set. Instead, they are used as data to determine $\mathbf{Z}$, $\mathbf{LZ}$, and $\mathbf{WZ}$. An inherent danger in this approach is that (8.187) and (8.188) can yield different values of $\mathbf{Z}$. Since the final model parameter set can only accommodate one value of $\mathbf{Z}$, some method of choosing one of the two values, or averaging between them, would have to be adopted. In practice, this problem is avoided. Instead of dealing with two values for $\mathbf{Z}$, (8.187) and (8.188) are used only to extract, respectively, $\mathbf{LZ}$ and $\mathbf{WZ}$. The long, wide base device (here, $L_{eff} = L_4$ and $W_{eff} = W_3$) is taken to represent $L_{eff} \rightarrow \infty$ and $W_{eff} \rightarrow \infty$. The extracted value of Z at this geometry (Z_{43}) is used for the model parameter $\mathbf{Z}$.

In addition, examination of (8.187) and (8.188) (and Figures 8.7 and 8.8) shows that it is implicitly assumed that the independently extracted values of Z (Z_{43}, Z_{33}, ...) will follow the linear relationships of (8.187) and (8.188). This assumption actually turns out to be reasonable in FETs with effective channel lengths larger than about 1 µm and effective channel widths larger than about 2 µm [14]. However, for smaller geometries, this method breaks down. It becomes particularly poor in submicron channel length devices, as will be shown later.

Some methods of coping with the latter problem are feasible. It is possible to implement a binning scheme that is similar to that described in Section 7.7 for the Level 3 model. This approach is more commonly encountered with BSIM2, so a discussion of the method is postponed until Chapter 10. A possible alternative is the use of a modified extrinsic structure available in the HSPICE implementation of BSIM (HSPICE Level 13). This approach is most commonly encountered in HSPICE Level 28, so that discussion may be found in Chapter 9.

8.10   Parameter Extraction and Model Development

As noted as the start of this chapter, BSIM (and the other second-generation models) employ extensive mathematical conditioning and an empirical model structure. As a result, BSIM provides more of a shell than was the case with its predecessors. This shifts more of the weight of providing good model results to the parameter extraction process.

Due to this empirical character and the larger number of model parameters, there are many possible approaches to carrying out parameter extraction. The method discussed here provides one possible route; other approaches are commonly used. However, the presentation here will be sufficient to demonstrate the behavior of BSIM in model building, and what strengths and weaknesses BSIM presents to the model user. Of particular note is the twofold intrinsic-extrinsic model structure; the model features associated with each of these two facets must be clearly separated.

8.10.1   Basic Parameter Extraction

In BSIM, as in the first-generation models, a long, wide device is used to provide some base parameters; corrections are then included to describe short and narrow channel effects. In contrast to the first-generation models, BSIM contains a larger number of parameters; in addition, model building must cope with the extrinsic model structure.

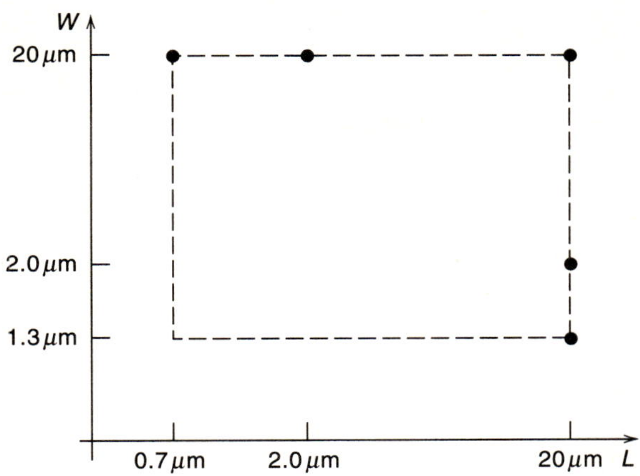

Figure 8.9   The specific geometry scheme used to extract the parameters for the BSIM model developed here.

The basic approach is quite similar to that of the first-generation models, although there are more specific steps (which are also more involved). The structure of the approach used here is:

- The oxide thickness and the effective-versus-drawn channel geometry parameters are included as process input.
- In the first phase, a long, wide device is used to determine a few base parameters; these parameters are used as the starting point for each of the individual extractions of the second phase. Note that in the model formulation, only the low field channel mobility parameter **MUZ** has no length and width parameters and is treated as globally meaningful. It is up to the model builder to decide if any other parameters will be determined here and used in a global fashion, rather than extracted for each individual device.
- In the second phase, a BSIM parameter set is extracted *independently* for each device. In essence, this phase involves the fitting of each individual device to the intrinsic model structure.
- In the third phase, the compiled parameters from the second phase are used to determine the geometry dependent parameters (the L and W parameters).

In this example, five devices will be used for model construction; this is detailed in Figure 8.9. At least three lengths and three widths are required to provide sufficient data for a linear regression analysis when the extrinsic model structure is developed.

The First Phase

In the first phase, some base parameters are determined; these parameters are then used at the start of each individual device extraction set in the second phase. As noted above, the low field channel mobility parameter **MUZ** is a global parameter which is determined

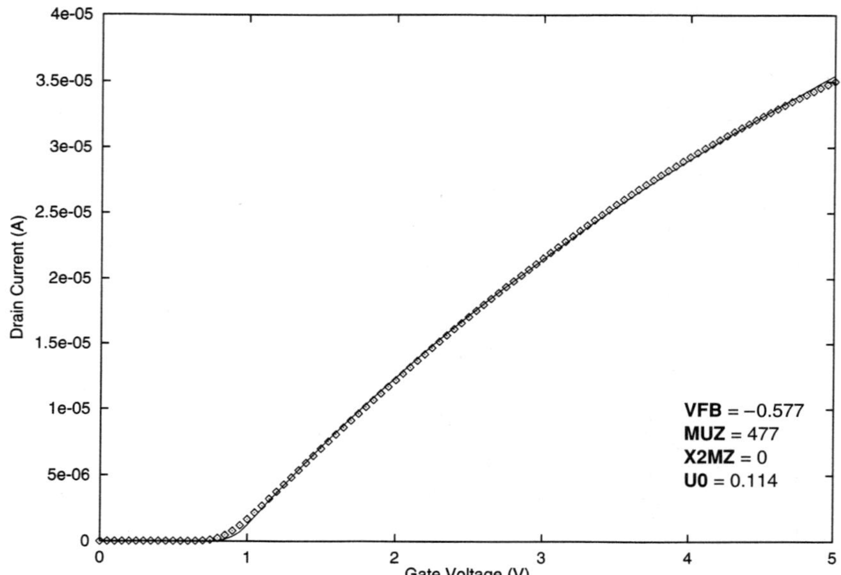

Figure 8.10   Phase I: The extraction of the base device parameters **VFB**, **MUZ**, **X2MZ**, and **U0**, $W/L = 20$ μm/20 μm.

during this phase and used in the final model. Here, the parameter *PHI* will also be treated as global; a value will be determined in this phase and used in the final model.

The first step involves fitting the model to a linear region characteristic with no applied substrate bias, as shown in Figure 8.10; a good result is obtained. In this step, the parameter **MUZ** is determined and then left untouched throughout the rest of the model construction process. In the second step, a fit is done with linear region data with substrate biases applied (Figure 8.11); while the results are not particularly good, note that the purpose here is to obtain a good starting point for the individual device extractions of the second phase. A third step is used to determine *PHI* (Figure 8.12). Experience shows that *PHI* varies very little from device to device; therefore, *PHI* is determined here and used unchanged for the complete model.

The Second Phase

The steps of the second phase of model construction are repeated for each individual device; in each case, the results of the first phase are used as the starting point. This phase will permit an examination of the behavior of the intrinsic model equations without interference from the complications introduced by the extrinsic model structure.

The first step involves fitting the model to data for linear region operation with substrate biases applied, as shown in Figure 8.13. For the short (20/0.7) device (Figure

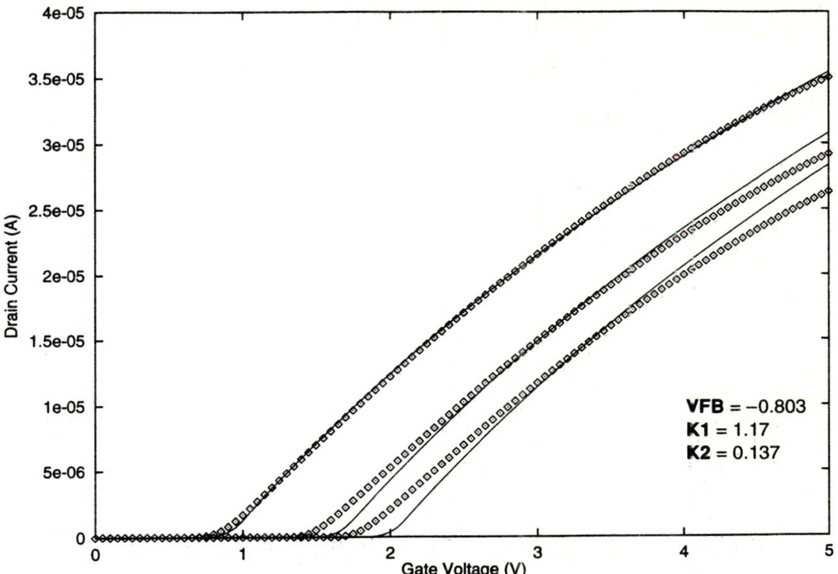

Figure 8.11 Phase I: The extraction of the base device parameters **VFB**, **K1**, and **K2**, $W/L = 20$ μm/20 μm.

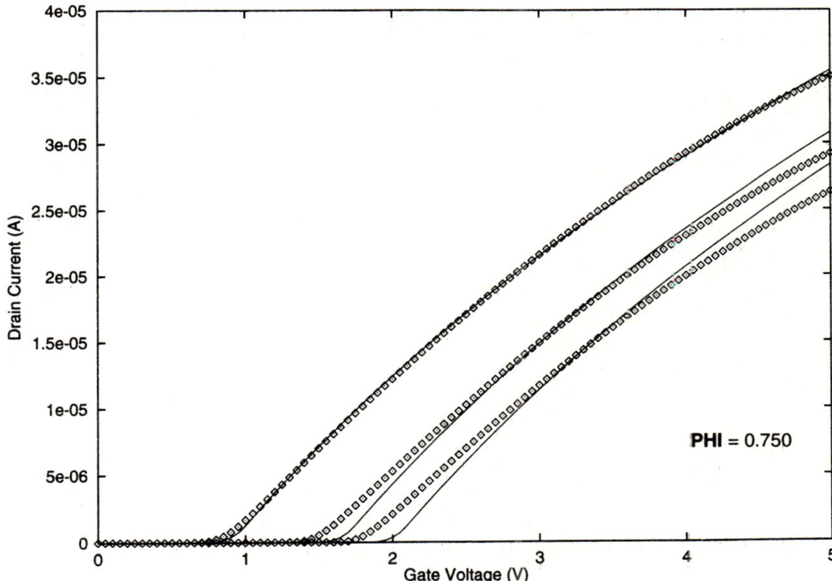

Figure 8.12 Phase I: The extraction of the base device parameter **PHI**, $W/L = 20$ μm/20 μm.

8.13a), the results are not particularly good. In contrast, the results for a longer (20/2.0) device (Figure 8.13b) and a narrow (1.3/20) device are good. It has been noted that BSIM encounters difficulty when it is used for submicron channel length devices; these results clearly illustrate this problem.

In the second step, parameters are extracted for the description of saturation region operation with no substrate bias applied; results are depicted in Figure 8.14. By inspection, it can be seen that the results are much improved over the first-generation results (e.g., compare these results with those for Level 3 in Chapter 7), showing a smooth linear-saturation transition and a more accurate and detailed description of the saturation region (versus the straight-line approach of the first-generation models). However, note that for the 20/0.7 device (Figure 8.14a), the model has trouble fitting the data near the linear-saturation transition. In contrast, the result for the 20/2.0 device (Figure 8.14b) is quite good. This once again confirms the earlier observation that BSIM encounters difficulties for submicron channel lengths. The result for a 1.3/20 device is interesting, as the model has difficulty with the saturation region; the model ascribes a much larger slope to the saturation region than is present in the data. This is a shortcoming of BSIM in narrow devices.

In the third step, the model is fit to data for operation in the saturation region with an applied substrate bias (Figure 8.15); this is an improvement over what is available in the first-generation models. The results are reasonable, but not particularly good. However, it is important to note some details here. In the case of the 20/2.0 device (Figure 8.15b), the overall behavior is qualitatively correct; the drain current decreases as the magnitude of the substrate bias increases. If detailed modeling of the behavior of the drain current versus the substrate bias at high drain biases is not required, this model should be adequate. In contrast, for the 20/0.7 device (Figure 8.15a), for a high gate bias, the quantitative behavior is incorrect, showing an *increase* in the drain current for an increasing magnitude of the substrate bias. This is another problem which is frequently encountered when BSIM is used with submicron channel length devices.

The fourth and final step of the second phase is a subthreshold extraction, as shown in Figure 8.16. When compared with the first-generation models, much better results are obtained. BSIM also includes the substrate bias in the model, and the subthreshold parameters have geometry dependence; this allows the subthreshold current model to have more general applicability. The careful observer of Figure 8.16 will note that there is a minor kink at the weak-strong inversion transition point. As described earlier, this is due to the simple addition of the weak inversion current to the strong inversion current under the assumption that the latter is much larger than the former in the strong inversion region. This assumption does not work well under low drain biases near the transition point, as will be shown in detail later.

The complete set of parameters extracted in the second phase is compiled in Table 8.4.

The Third Phase

With the second phase of parameter extraction completed, the compiled sets of parameters from the individual devices are used to construct the extrinsic model structure, as described

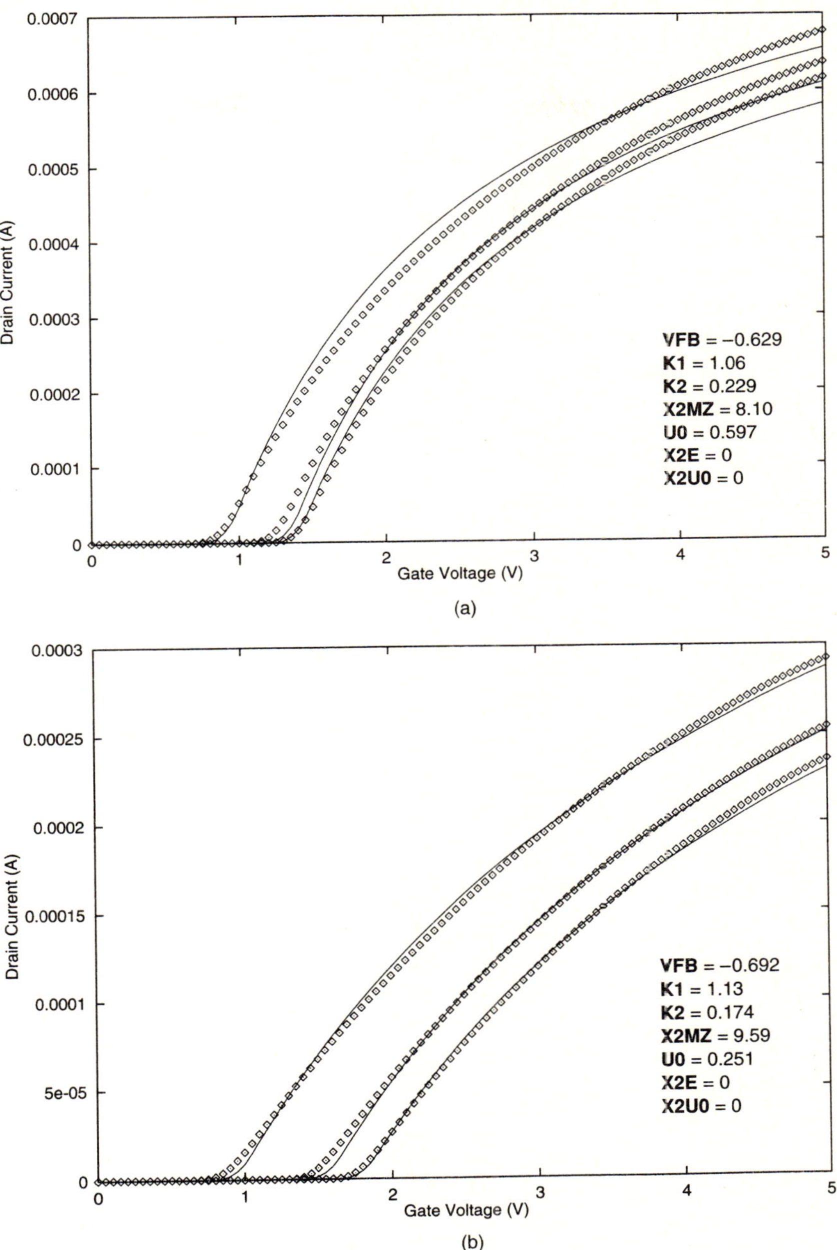

Figure 8.13 Phase II: The extraction of the device parameters **VFB**, **K1**, **K2**, **X2MZ**, **U0**, **X2E**, and **X2U0**; (a) $W/L = 20$ μm/0.7 μm; (b) $W/L = 20$ μm/2.0 μm; (c) $W/L = 1.3$ μm/20 μm.

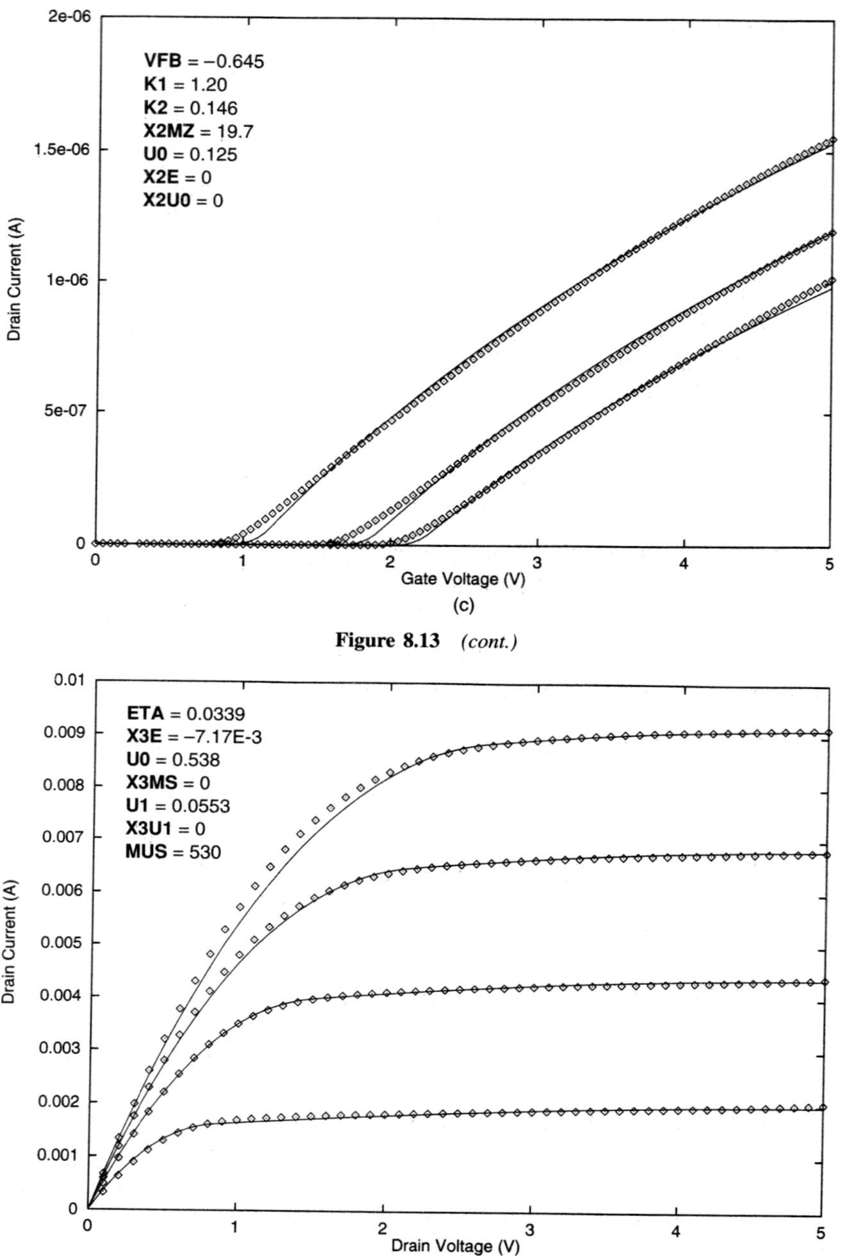

Figure 8.13   *(cont.)*

Figure 8.14   Phase II: The extraction of the device parameters **ETA**, **X3E**, **U0**, **X3MS**, **U1**, **X3U1**, and **MUS**; (a) $W/L = 20$ μm/0.7 μm; (b) $W/L = 20$ μm/2.0 μm; (c) $W/L = 1.3$ μm/20 μm.

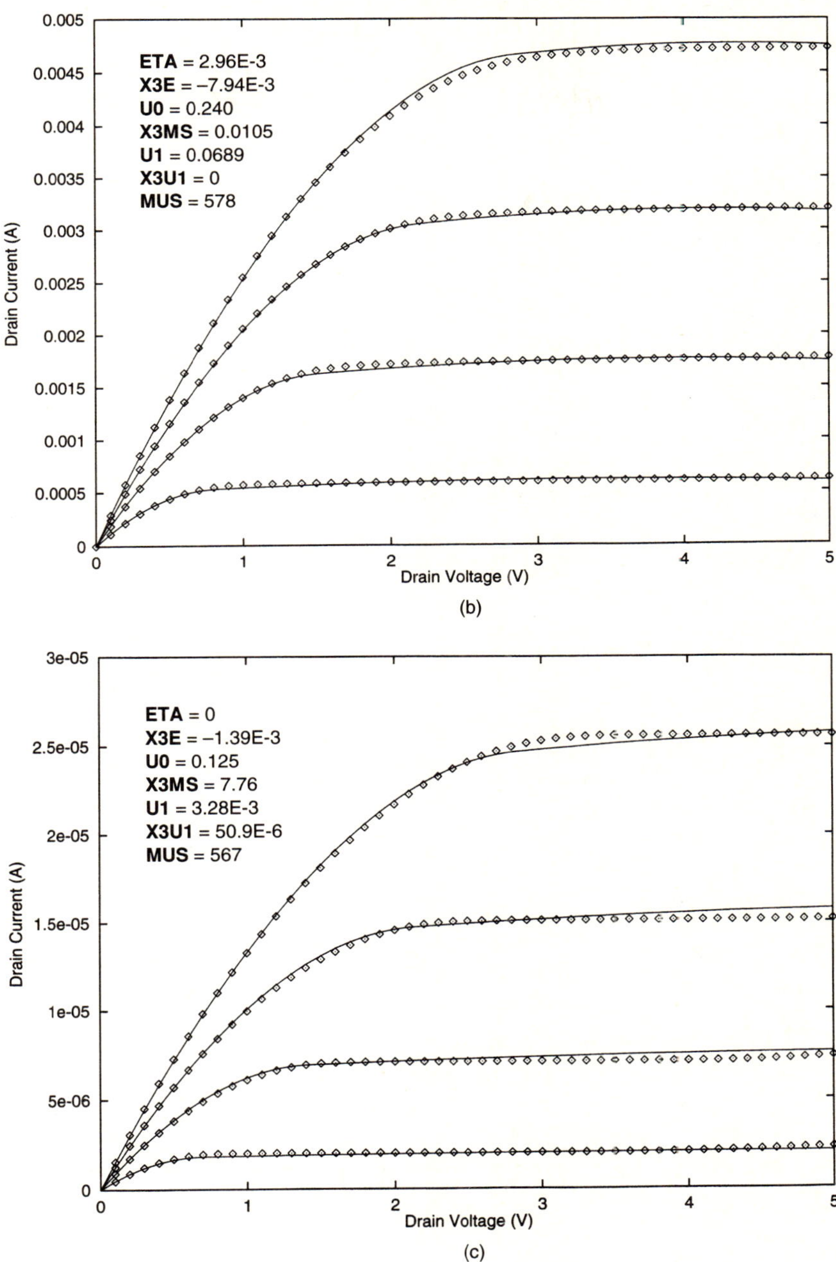

(b)

(c)

Figure 8.14   *(cont.)*

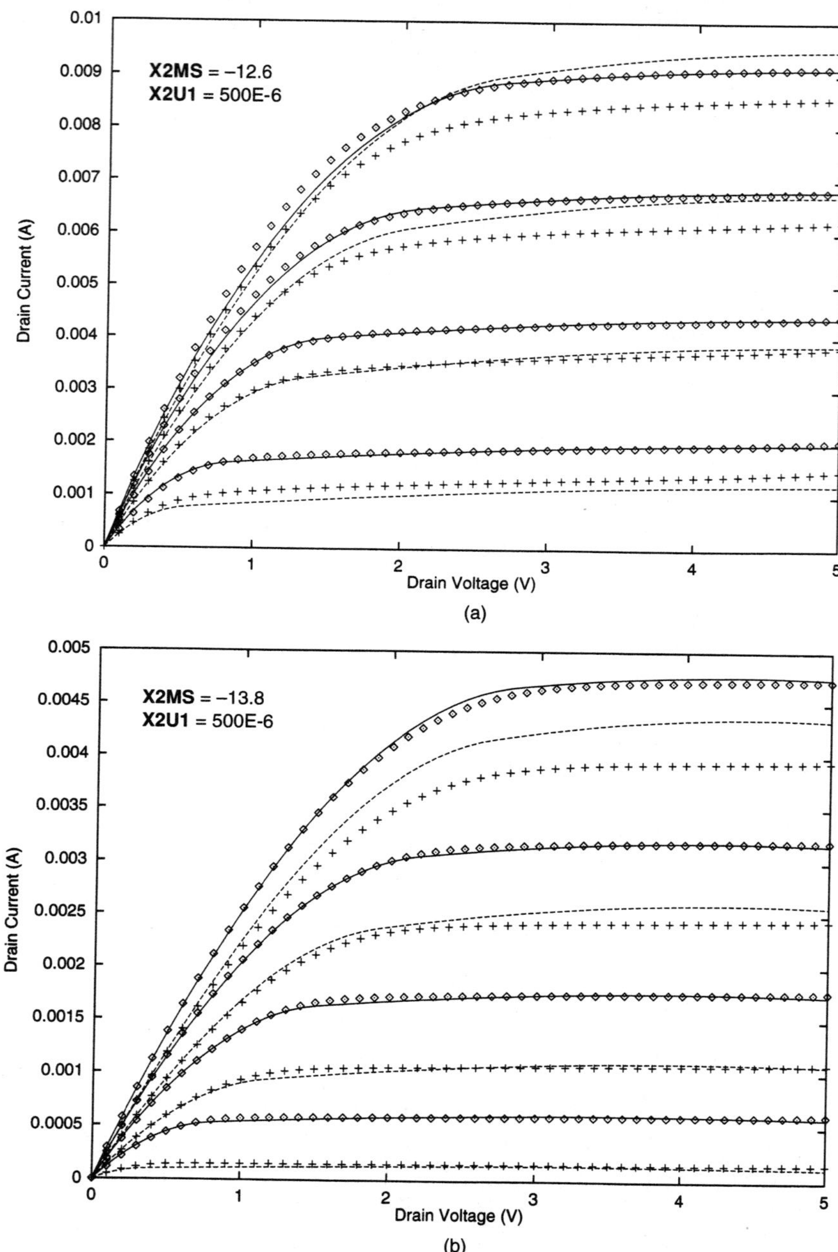

Figure 8.15 Phase II: The extraction of the device parameters **X2MS** and **X2U1**; (a) $W/L = 20$ μm/0.7 μm; (b) $W/L = 20$ μm/2.0 μm.

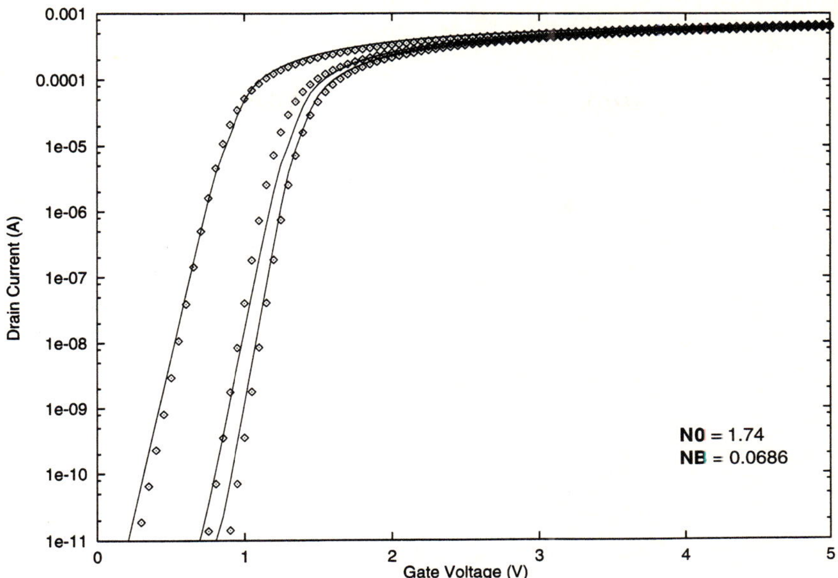

Figure 8.16   Phase II: The extraction of the device parameters **N0** and **NB**, $W/L =$ 20 µm/0.7 µm.

in Section 8.9. This involves fitting the L and W parameters so as to most closely match the extrinsic model equation

$$Z = \mathbf{Z} + \frac{\mathbf{LX}}{L_{eff}} + \frac{\mathbf{WX}}{W_{eff}}, \qquad (8.189)$$

where $\mathbf{Z}$ is the large geometry limit of Z. For a very wide set of devices, (8.189) simplifies to

$$Z = \mathbf{Z} + \frac{\mathbf{LX}}{L_{eff}}. \qquad (8.190)$$

As just noted, **Z** is taken from the individual parameter set for the longest device (here, $L = 20$ µm); a linear regression is then carried out on the data from the second phase to determine **LX**. It is assumed, of course, that the data from the second phase will closely follow the model of (8.189). However, this is not the case, at least in submicron devices. Figure 8.17 shows the linear regression fit to the length-specific model equation (8.190) for the model parameter set **VFB**. It is easily seen that the data points do not describe a straight line. The resulting model (the line in the figure) is the best linear regression fit. With three data points (from three different channel lengths), the best results are obtained

Table 8.4 The individual parameters extracted independently for each device during the second phase of parameter extraction.

Parameter	20/20	20/2.0	20/0.7	2.0/20	1.3/20
TOX	13.0E-9	13.0E-9	13.0E-9	13.0E-9	13.0E-9
LD	0.240E-6	0.240E-6	0.240E-6	0.240E-6	0.240E-6
WD	0.400E-6	0.400E-6	0.400E-6	0.400E-6	0.400E-6
VDD	5.0	5.0	5.0	5.0	5.0
MUZ	476	476	476	476	476
PHI	0.750	0.750	0.750	0.750	0.750
VFB	−0.747	−0.692	−0.629	−0.657	−0.645
K1	1.12	1.13	1.06	1.17	1.20
K2	0.156	0.174	0.229	0.157	0.146
X2MZ	9.69	9.59	8.10	17.2	19.7
X2E	0	0	0	0	0
X2U0	1.09E-3	0	0	0	0
ETA	0	2.96E-3	0.0339	0	0
X3E	−46.8E-6	−7.94E-3	−7.17E-3	−6.75E-6	−1.39E-3
U0	0.117	0.240	0.538	0.135	0.125
X3MS	1.56	0.0105	0	4.65	7.76
U1	0.0511	0.0689	0.0553	6.85E-3	3.28E-3
X3U1	−18.1E-6	0	0	−13.6E-6	−50.9E-6
MUS	586	578	530	589	567
X2MS	−26.7	−13.8	−12.6	12.9	14.3
X2U1	500E-6	500E-6	500E-6	0.0366	0.0105
N0	1.57	1.53	1.74	1.60	1.53
NB	18.1E-3	100E-6	0.0686	100E-6	100E-6
ND	0	0	0	0	0

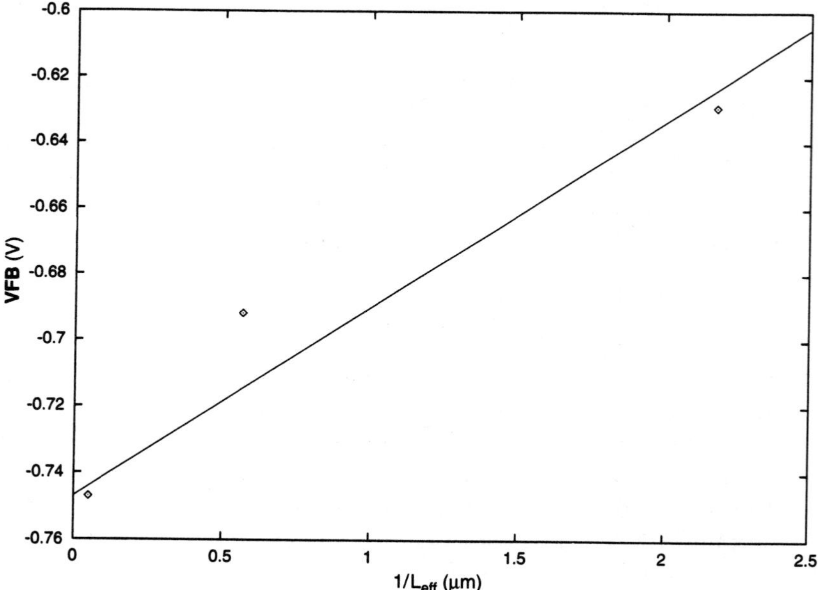

Figure 8.17 The individual device results for the model parameter **VFB** versus $\frac{1}{L_{eff}}$, and the linear regression fit to the model equation (8.190) for the determination of **LVFB**. Note how the data points do not follow a straight line.

Table 8.5   The final BSIM model parameter set.

Base Parameter	Length Parameter	Width Parameter
TOX = 13.0E-9	—	—
LD = 0.240E-6	—	—
WD = 0.400E-6	—	—
VDD = 5.0	—	—
MUZ = 476	—	—
PHI = 0.750	LPHI = 0	WPHI = 0
VFB = −0.747	LVFB = 5.70E-2	WVFB = 1.04E-1
K1 = 1.12	LK1 = −2.47E-2	WK1 = 7.38E-2
K2 = 0.156	LK2 = 3.34E-2	WK2 = −6.44E-3
X2MZ = 9.69	LX2MZ = −6.96E-1	WX2MZ = 9.72
X2E = 0	LX2E = 0	WX2E = 0
X2U0 = 1.09E-3	LX2U0 = −5.92E-4	WX2U0 = −1.16E-3
ETA = 0	LETA = 1.49E-2	WETA = 0
X3E = −46.8E-6	LX3E = −3.95E-3	WX3E = −9.01E-4
U0 = 0.117	LU0 = 1.95E-1	WU0 = 1.24E-2
X3MS = 1.56	LX3MS = −8.46E-1	WX3MS = 5.42
U1 = 0.0511	LU1 = 3.81E-3	WU1 = −4.95E-2
X3U1 = −18.1E-6	LX3U1 = 9.83E-6	WX3U1 = −2.07E-5
MUS = 586	LMUS = −25.0	WMUS = −11.8
X2MS = −26.7	LX2MS = 7.52	WX2MS = 43.2
X2U1 = 500E-6	LX2U1 = 0	WX2U1 = 2.07E-2
N0 = 1.57	LN0 = 6.87E-2	WN0 = −1.58E-2
NB = 18.1E-3	LNB = 1.97E-2	WNB = −1.92E-2
ND = 0	LND = 0	WND = 0

at the extremes of the data (the shortest and longest devices), while the results are quite poor in the middle region. Of course, more devices could be used; while this would improve the results for the middle region, the accuracy for the shorter devices would suffer.

The data of Table 8.4 are used with (8.189) to extract the geometry dependent parameters of the extrinsic model structure. The final model parameter set is tabulated in Table 8.5.

Temperature Dependence

As noted at the beginning of the chapter, the original BSIM structure made no provision for any variation of the temperature from that used for data collection. Methods to include temperature dependence were later published, but these were never included in BSIM. The HSPICE implementation of BSIM (HSPICE Level 13) includes a temperature dependence scheme that is essentially identical to that used for HSPICE Level 28 (see Chapter 9).

8.10.2   Model Playback

As is the case with all models, it is important to compare the predictions of the final model with data for BSIM. This is particularly important for the second-generation models, as this comparison will allow the observer to separate shortcomings originating with the intrinsic model structure (discussed during the second-phase extractions) from those that are due to the extrinsic model structure.

Final results for the 20/0.7 device may be found in Figure 8.18. These final results suffer from the same submicron shortcomings that were described earlier. In addition, note that in comparison with Figure 8.13a and Figure 8.14a, the extrinsic geometry structure has caused a further loss of accuracy, as predicted by Figure 8.17. This is an important consideration in the second-generation models, which can suffer a deterioration of results from two different sources. Figure 8.19 depicts final model results for the 20/2.0 device. When compared with the results of Figure 8.13b and 8.13c, it can be seen that the extrinsic model structure has seriously damaged the accuracy of the results for this middle region device geometry. The 20/0.9 device was not used for parameter extraction; final model results are presented in Figure 8.20. Like the 20/2.0 device, the results are not particularly good. This further serves to indicate that the geometry dependent extrinsic model structure, described by (8.189), is not particularly valid for submicron channel lengths.

Figure 8.21 depicts final model results for the 1.3/20 device, which should be compared with Figure 8.13c and Figure 8.14c. Once again, it is clear that the imposition of the extrinsic geometry structure has led to a serious impairment of the final model accuracy. Similar problems are encountered for the 2.0/20 device (Figure 8.22), although in this case the results are actually better than those of the 1.3/20 device.

A narrow, short (1.3/0.7) device is examined in Figure 8.23. In this case, very poor results are obtained. This indicates that the method of treating short and narrow effects independently does not work well; this was also the case for the first-generation models. There is nothing in the BSIM extrinsic model structure to account for the effect of both short *and* narrow channels at the same time.

Finally, model playback can be used to highlight two other model problems. Figure 8.24 is a detail of a drain bias characteristic for a low gate bias in a short (0.7 μm) device. At a high drain bias, a negative output conductance occurs. This is frequently a problem with BSIM, and the model user should be aware that this situation might be encountered. Figure 8.25 is a detail of the weak-strong inversion transition region under a low drain bias in a short device. A distinct "jump" in the characteristics is noted in the neighborhood of $V_{gs} = 0.9$ V. As described earlier, in BSIM, the total current is found from the simple summation

$$I_{ds,total} = I_{strong} + I_{weak}. \qquad (8.191)$$

It is assumed that for $V_{gs} \leq V_t$, $I_{strong} = 0$, while for $V_{gs} \geq V_t$, $I_{strong} \gg I_{weak}$. However, just above the transition under a low drain bias, this last assumption is poor; this causes the jump which is observed in the model result. This kink can be a source of convergence problems during circuit simulation.

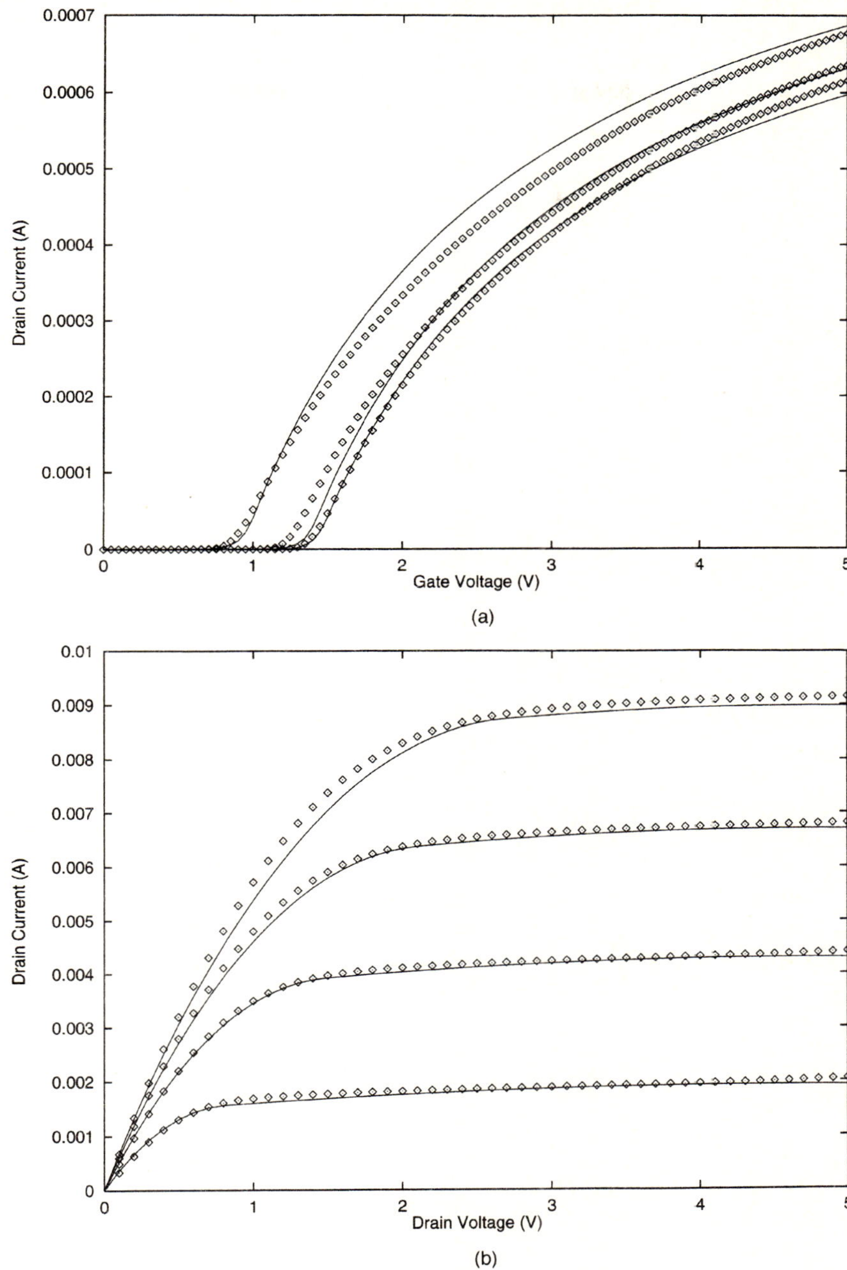

Figure 8.18 Comparison of measured data and the final model for the 20/0.7 device; (a) linear characteristic; (b) saturation characteristic.

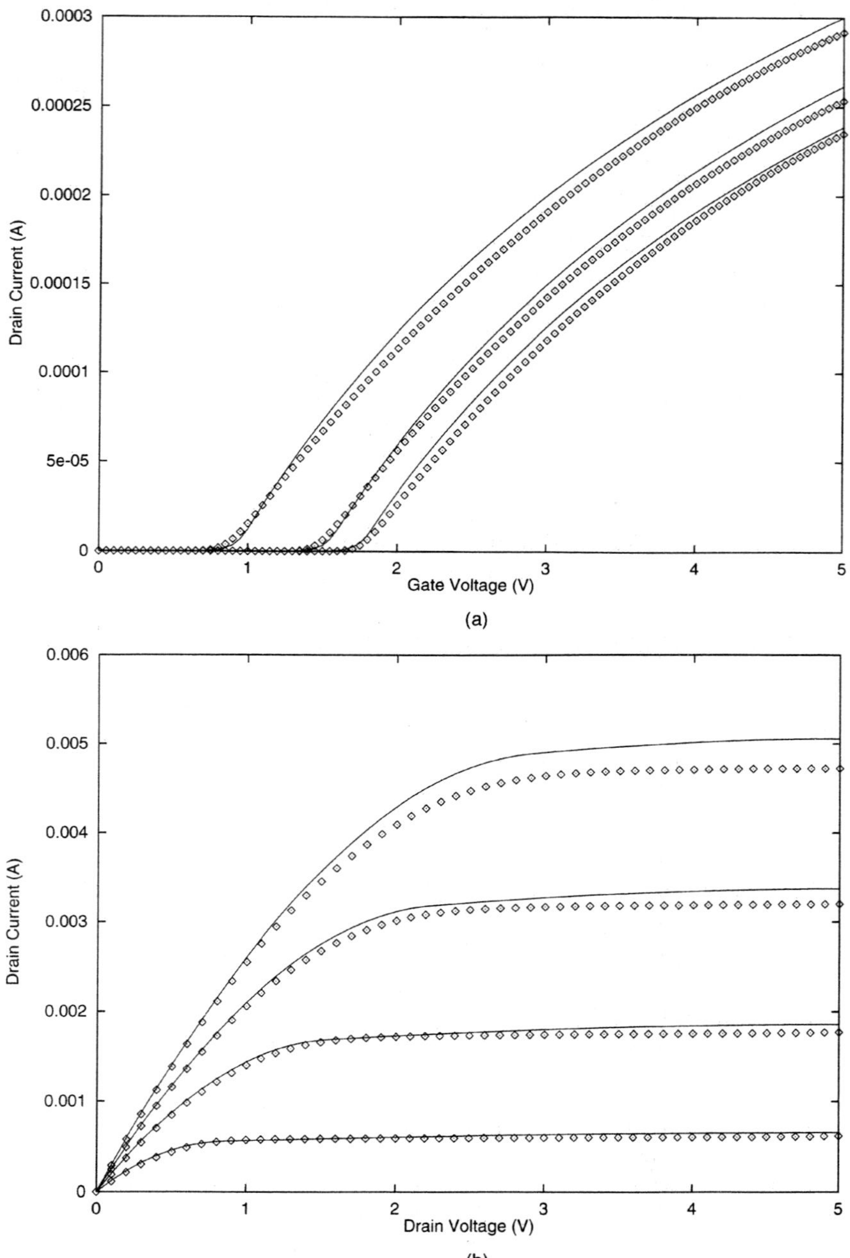

Figure 8.19 Comparison of measured data and the final model for the 20/2.0 device; (a) linear characteristic; (b) saturation characteristic.

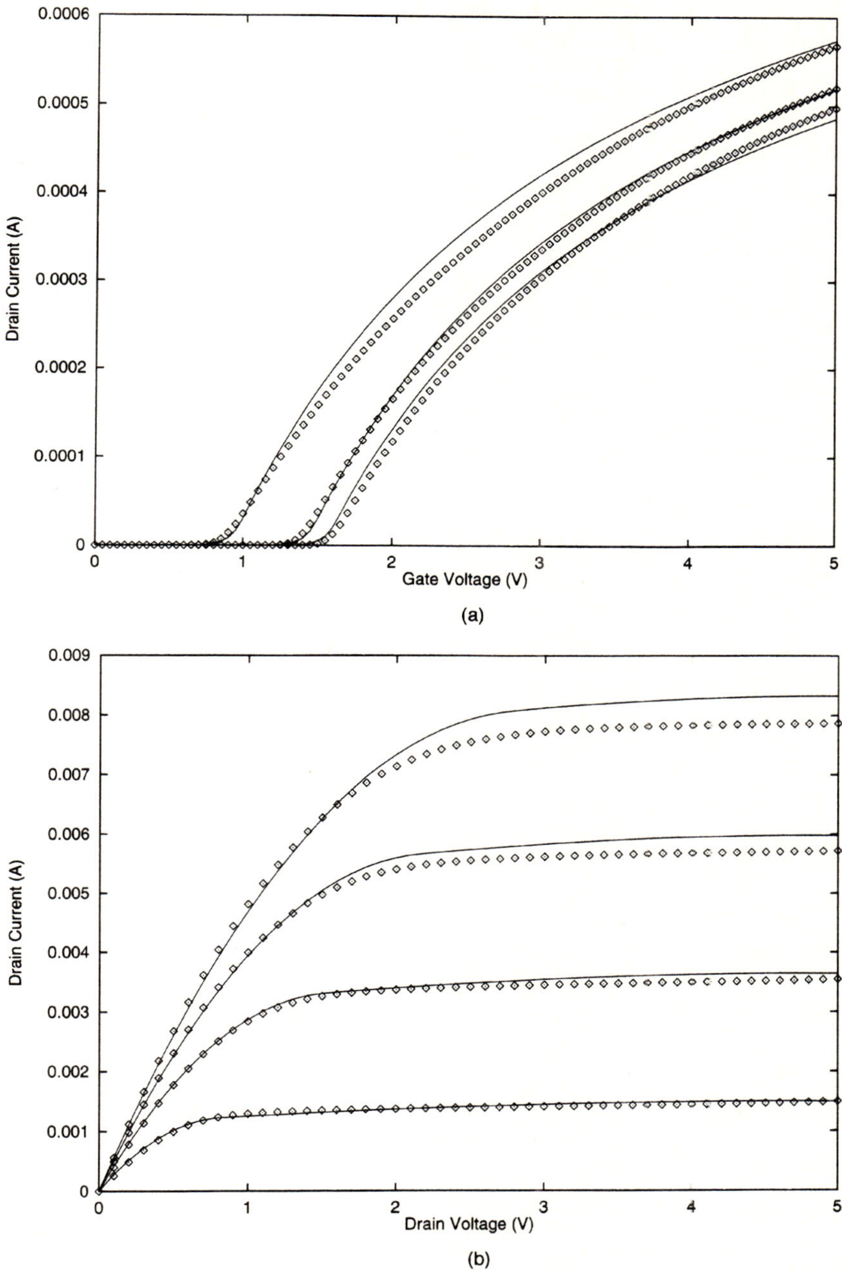

Figure 8.20 Comparison of measured data and the final model for the 20/0.9 device; (a) linear characteristic; (b) saturation characteristic.

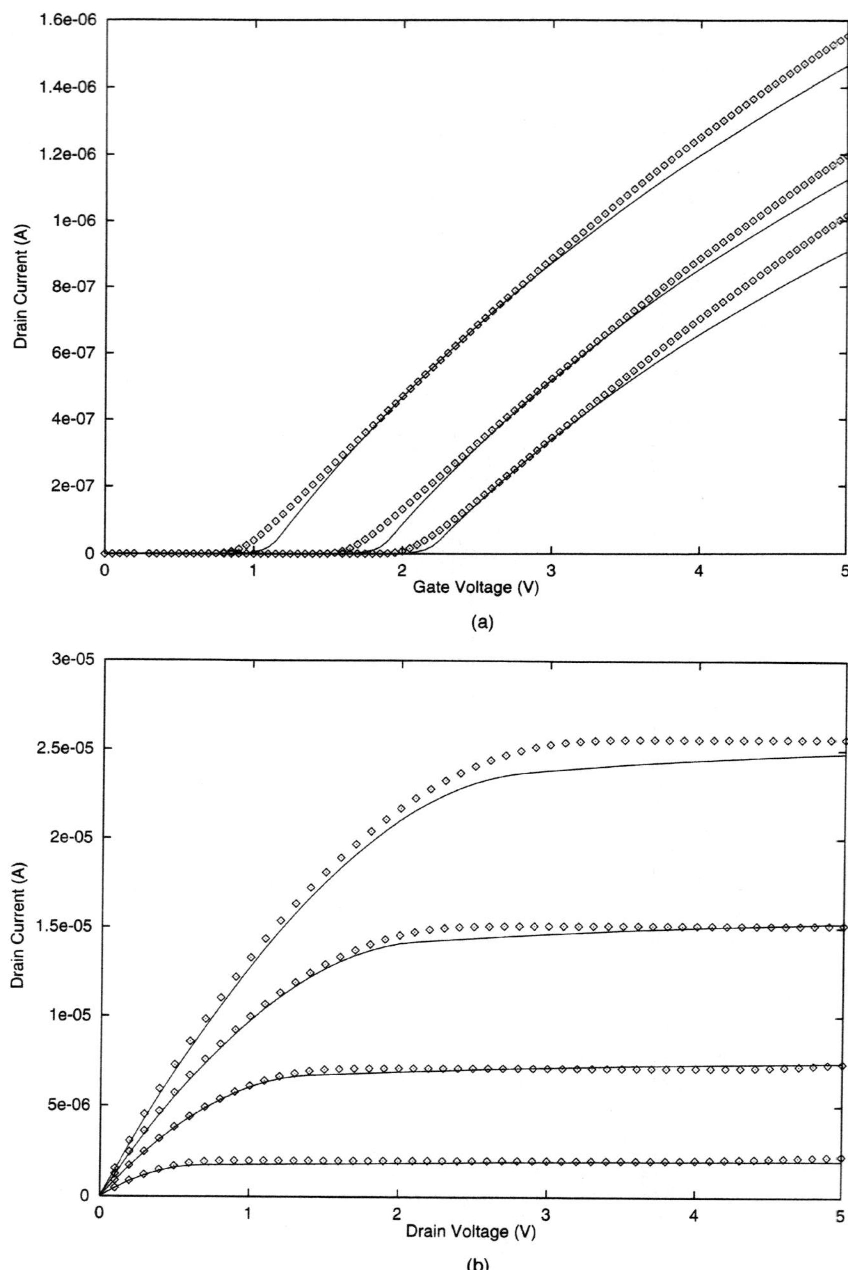

Figure 8.21 Comparison of measured data and the final model for the 1.3/20 device; (a) linear characteristic; (b) saturation characteristic.

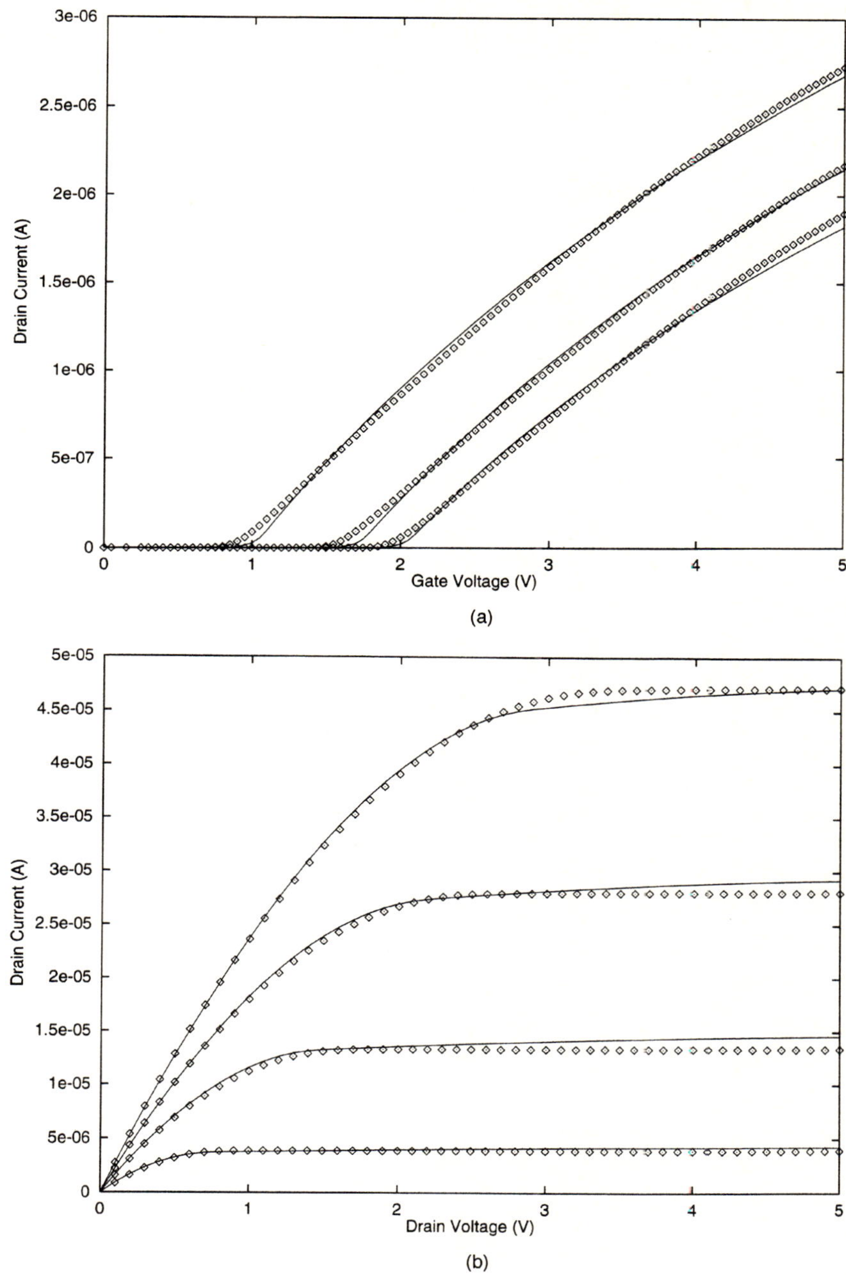

Figure 8.22 Comparison of measured data and the final model for the 2.0/20 device; (a) linear characteristic; (b) saturation characteristic.

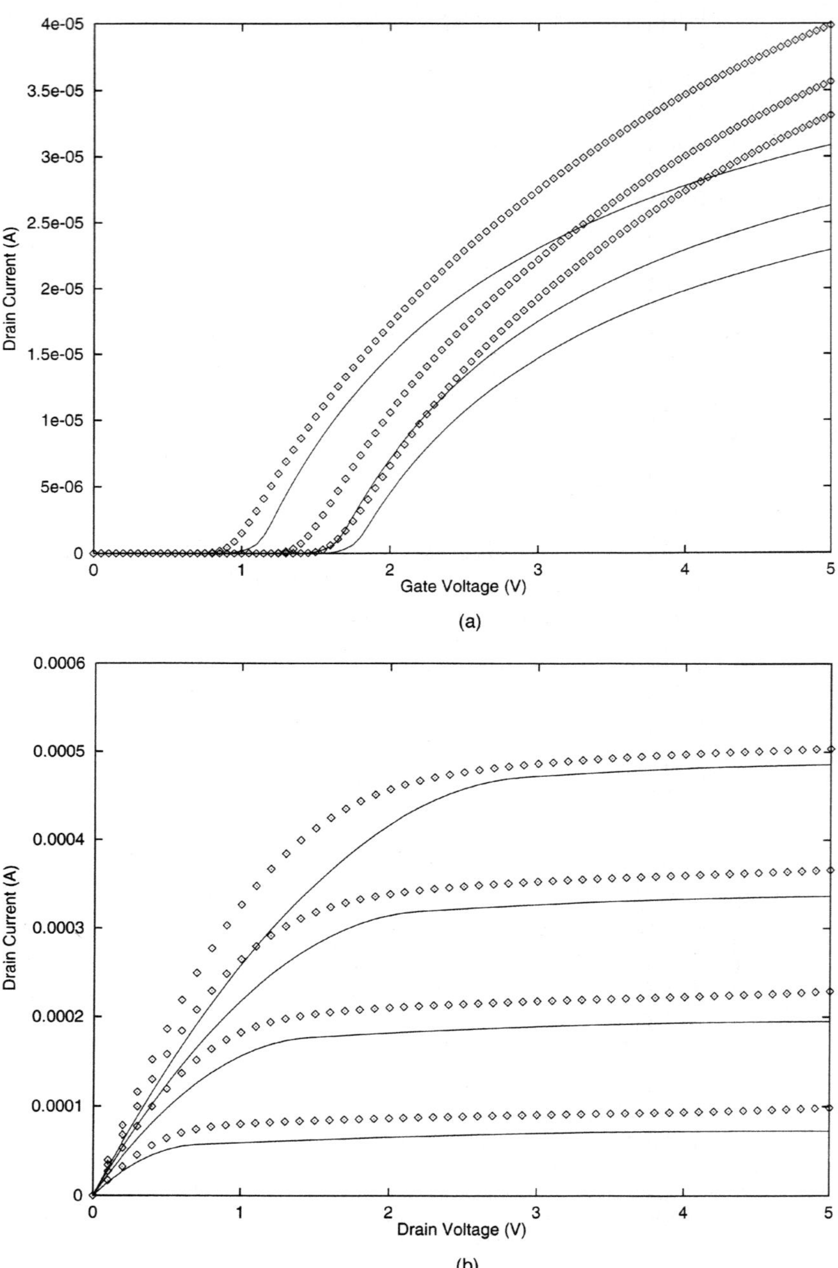

Figure 8.23   Comparison of measured data and the final model for the 1.3/0.7 device; (a) linear characteristic; (b) saturation characteristic.

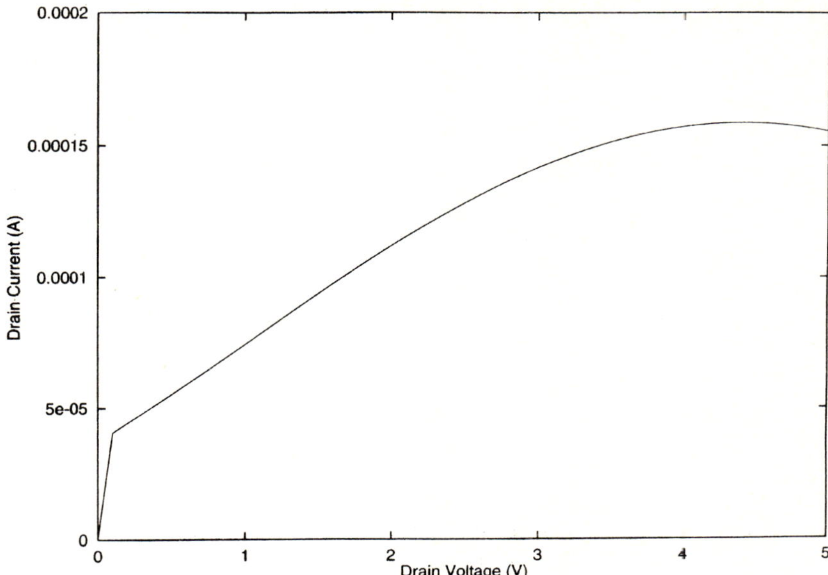

Figure 8.24 Final model result for the 20/0.7 device with a gate bias of $V_{gs} = 1.0$ V. Note the negative output conductance for high drain biases.

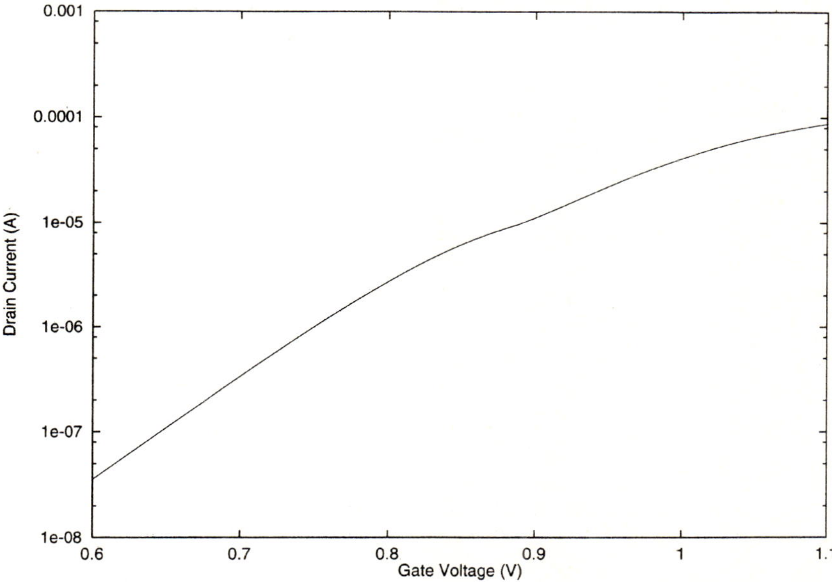

Figure 8.25 Detail of the final model result for the 20/0.7 device in the weak-strong inversion transition region, $V_{ds} = 0.1$ V. Note the jump in the characteristic near $V_{gs} = 0.9$ V due to the simple addition of the weak and strong inversion currents.

8.11   Final Comments

As described throughout this chapter, BSIM represented a major change in the philosophy behind the analytical modeling of FETs. Rather than attempt to construct a physically based description of a particular process technology, BSIM strongly shifted the modeling emphasis to the circuit design user. This approach is manifested through extensive mathematical conditioning to improve efficiency during circuit simulation; this is done at the cost of weakening the physical connection to the underlying fabrication process. As a result, BSIM has a rather empirical mathematical structure, with an increased number of model parameters (versus the first-generation models) which have considerable empirical character. This shift of emphasis reflects the true industrial requirements of analytical FET modeling; the main goal is accurate circuit simulation, not process characterization. By shifting the modeling emphasis and introducing considerable empirical character to the model structure, BSIM started a trend which continues to this day.

Furthermore, BSIM introduced the twofold intrinsic-extrinsic model structure which is the distinguishing feature of the second-generation models; this structure is also partially implemented in the third-generation MOS Model 9 (see Chapter 12). This further increases the empirical character of the model. In this form, the FET model becomes a malleable shell; this greatly increases the burden on parameter extraction to provide good results. Of particular note, with this type of model structure, poor parameter extraction will have more serious consequences.

As was also noted, BSIM does have some shortcomings that have become more apparent in recent years. Most of BSIM's mathematical conditioning involved the use of polynomial functions as replacements for more awkward expressions. However, these polynomial functions can behave quite badly, particularly if the model is stretched for use outside the geometric range for which it was designed. For example, as shown in Figure 8.24, under certain conditions in submicron devices, a negative output conductance is predicted by the model. As detailed in Section 8.10, BSIM has difficulty fitting the current-voltage characteristics of submicron devices.

The extrinsic model structure assumes that individual device parameters will follow a geometry dependence of $\frac{1}{L_{eff}}$ and $\frac{1}{W_{eff}}$; as shown in Figure 8.17, this is not the case in submicron channel length devices. As a result, the use of this extrinsic structure causes a loss of accuracy over geometry in the final model. Like the first-generation models, binned models can be constructed; however, as described in Chapter 7, binning can introduce its own problems. (A method of binning both BSIM and BSIM2 is discussed in Chapter 10.)

As shown in Figure 8.25, under low drain biases, there is a jump in the current-voltage characteristic at the weak-strong inversion transition point; this can cause convergence problems during circuit simulation. The negative conductance depicted in Figure 8.24 can cause problems during circuit simulation, and also makes BSIM unsuitable for analog circuit design.

For the description of individual device characteristics, BSIM represents a major improvement over the capabilities of the first-generation models. However, BSIM is more complex mathematically, and is slower and less efficient during circuit simulation.

At present, BSIM has largely been superceded by its second-generation successors, HSPICE Level 28 and BSIM2. BSIM is very important for historical reasons, in that it introduced the second-generation structure; in addition, BSIM's mathematical formulation is for the most part still present in HSPICE Level 28 and BSIM2. Ironically, BSIM is now little used because of its role as a transitional model. If a very efficient digital circuit simulation is required (for either rapid analysis or the simulation of a very large digital circuit), Level 3 can be used. If a detailed description of submicron devices is required, or an analog circuit must be simulated, then HSPICE Level 28, BSIM2, or perhaps one of the newer third-generation models is used.

BIBLIOGRAPHY

1. B. Sheu, "MOS Transistor Modeling and Characterization for Circuit Simulation," University of California/Berkeley, Electronics Research Laboratory Memorandum No. UCB/ERL M85/85 (1985).

2. B. Sheu, D. Scharfetter, P. Ko, and M. Jeng, "BSIM: Berkeley Short Channel IGFET Model for MOS Transistors," *IEEE J. Sol. St. Circ.* vol. SC-22, pp. 558–566 (1987).

3. Y. Tsividis and K. Suyama, "MOSFET Modeling for Analog Circuit CAD: Problems and Prospects," *IEEE J. Sol. St. Circ.* vol. SC-29, pp. 210–216 (1994).

4. S. Liu and L. Nagel, "Small Signal MOSFET Models for Analog Circuit Design," *IEEE J. Sol. St. Circ.* vol. SC-17, pp. 983–998 (1982).

5. B. Sheu, D. Scharfetter, and H. Poon, "Compact Short Channel IGFET Model (CSIM)," University of California/Berkeley, Electronics Research Laboratory Memorandum No. UCB/ERL M84/20 (1984).

6. G. Merckel, J. Borel, and N. Cupcea, "An Accurate Large-Signal MOS Transistor Model for Use in Computer-Aided Design," *IEEE Trans. Elec. Dev.* vol. ED-19, pp. 681–690 (1972).

7. P. Yang, B. Epler, and P. Chatterjee, "An Investigation of the Charge Conservation Problem for MOSFET Circuit Simulation," *IEEE J. Sol. St. Circ.* vol. SC-18, pp. 128–138 (1983).

8. D. Ward and R. Dutton, "A Charge-Oriented Model for MOS Transistor Capacitances," *IEEE J. Sol. St. Circ.* vol. SC-13, pp. 703–708 (1978).

9. C. Wan and B. Sheu, "Temperature Dependence Modeling for MOS VLSI Circuit Simulation," *IEEE Trans. Comp. Aid. Des.* vol. CAD-8, pp. 1065–1072 (1989).

10. *HSPICE User's Manual*, Meta-Software, Inc., Campbell, California, 1993.

11. M. Jeng, "Design and Modeling of Deep Submicrometer MOSFETs," University of California/Berkeley, Electronics Research Laboratory Memorandum No. UCB/ERL M90/90 (1990).

12. H. Masuda, M. Nakai, and M. Kubo, "Characteristics and Limitation of Scaled-Down MOSFETs Due to Two-Dimensional Field Effect," *IEEE Trans. Elec. Dev.* vol. ED-26, pp. 980–986 (1979).

13. L. Dang, "A Simple Current Model for Short-Channel IGFET and Its Application to Circuit Simulation," *IEEE J. Sol. St. Circ.* vol. SC-14, pp. 358–367 (1979).

14. H. deGraaff and F. Klaassen, *Compact Transistor Modeling for Circuit Design*, Springer-Verlag, 1990.

15. J. Schrieffer, "Effective Carrier Mobility in Surface Space Charge Layers," *Phys. Rev.* vol. 97, pp. 641–646 (1955).

16. B. Streetman, *Solid State Electronic Devices* (2nd ed.), Prentice-Hall, 1980.

17. N. Arora, *MOSFET Models for VLSI Circuit Simulation*, Springer-Verlag, 1993.

18. J. Brews, "A Charge Sheet Model of the MOSFET," *Sol. St. Elec.* vol. 21, pp. 345–355 (1978).

19. S. Sze, *Physics of Semiconductor Devices* (2nd ed.), John Wiley & Sons, 1981.

20. D. Caughey and R. Thomas, "Carrier Mobilities in Silicon Empirically Related to Doping and Field," *Proc. IEEE* vol. 55, pp. 2192–2193 (1967).

21. P. Antognetti, D. Cavigia, and E. Profumo, "CAD Model for Threshold and Subthreshold Conduction in MOSFETs," *IEEE J. Sol. St. Circ.* vol. SC-17, pp. 454–458 (1982).

22. A. Fung, "A Subthreshold Conduction Model for BSIM," University of California/Berkeley, Electronics Research Laboratory Memorandum No. UCB/ERL M85/22 (1985).

23. H. Ihantola and J. Moll, "Design Theory of a Surface Field-Effect Transistor," *Sol. St. Elec.* vol. 7, pp. 423–430 (1964).

9

HSPICE Level 28

9.1 Introduction

As noted throughout Chapter 8, BSIM represents a radical departure from the first-generation models. Whereas the earlier models had focused upon developing physically based formulations, BSIM instead concentrates on mathematical conditioning of the equations, to improve the final accuracy of the FET model and the circuit simulation behavior. The model accounts for various effects not through complex, derived expressions for various phenomena, but by a somewhat ad hoc addition of a parameter (or parameters) to describe a particular effect. As a result, improved accuracy is obtained at the cost of losing connections to the specific process technology and detailed device physics. In contrast to the first-generation models, the emphasis is on circuit simulation usage, rather than on a very detailed analytical description of the FET.

Although it represents a major improvement over the first-generation models, BSIM has a number of shortcomings that have become more apparent as FET dimensions have continued to decrease. This has led to two developments in second-generation models. The University of California/Berkeley responded with BSIM2 [1], an improved version of BSIM that will be described in Chapter 10. Another response has been HSPICE Level 28 from Meta-Software [2]. Due to the widespread industrial use of Meta-Software's HSPICE simulator, HSPICE has served as a vehicle for the Level 28 model; this has helped Level 28 to become a very important general use FET model. HSPICE Level 28 is based on BSIM, but improves upon many of the intrinsic shortcomings. In addition, BSIM predates the explosion of interest in analog CMOS circuits; the modifications included in HSPICE Level 28 make it suitable for analog design. Although widely used, the details of the model formulation remain largely proprietary, with only a few hints of what modifications were made to the original BSIM equations. However, much can be learned from the small amount of information available, and with knowledge of BSIM and some careful effort, it is possible to extract parameters and construct HSPICE Level

28 FET models. By constructing models, comparisons can be made with BSIM results; the mathematical improvements introduced into HSPICE Level 28 become apparent.

9.2   Background

Although a general improvement over the first-generation models, BSIM does have shortcomings, as noted at the end of Chapter 8. For channel lengths below $L_{eff} \simeq$ 0.8 μm and/or gate oxide thicknesses less than $t_{ox} \simeq 15$ nm, BSIM loses accuracy [1]. BSIM simplifies the mathematical formulation by replacing various expressions (such as the $\frac{3}{2}$-power terms in the original bulk charge model) with polynomial expressions. Unfortunately, polynomial expressions can behave quite badly, leading to convergence problems during circuit simulation. For example [2], the BSIM threshold voltage expression (see (8.21)) is quadratic in the drain bias V_{ds} and the substrate bias expression ($PHI - V_{bs}$); these quadratics can behave badly. In addition, as demonstrated in Section 8.10, negative output conductance results can be produced, particularly for low gate biases. This is, of course, physically incorrect, and such a model cannot be used for analog circuit designs in which g_{ds} is important. Even if g_{ds} is not important to the model user, this problem can cause the iterative matrix solution to oscillate between two states and not converge to a solution. A negative g_{ds} can be removed [2] by ensuring that *X3MS* ≥ 0 and *MUZ* + (**VDDM** · *X33M*) < *MUS*, and setting the parameter *X3U1* to zero (by setting **X3U1**, **LX3U1**, and **WX3U1** to zero). These conditions should be imposed during parameter extraction. If they are imposed after the fact on the completed FET model, the exercise is rather arbitrary and runs the risk of corrupting the accuracy of the FET model. The consequences of such a change are hard to predict; model accuracy in the linear region may be reduced [2], and there may be additional unpleasant consequences. In HSPICE Level 28, the low field mobility model is reformulated to eliminate the quadratic dependence on V_{ds} and obviate this problem. This is done by removing the parameters *MUS* and *X2MS* and their associated equation (see (8.28)), and instead employing an expression based on the low drain bias mobility description and the mobility slope at $V_{ds} = V_{dd}$. A new parameter, *X33M*, is added to describe the effect of the gate bias on that slope.

Another improvement in BSIM (versus the first-generation models) is the continuity of the derivatives of all the current equations for the whole range of gate and drain biases. Of particular note, in BSIM, the transconductance g_m ($\frac{I_{ds}}{V_{gs}}$) is continuous at the threshold voltage $V_{gs} = V_t$. However, $\frac{g_m}{I_{ds}}$ is nonmonotonic as V_{gs} crosses V_t, and the second derivative of I_{ds} with respect to V_{gs} is not continuous. This is physically incorrect, and can be another source of convergence problems. HSPICE Level 28 removes this problem with a completely different model for weak inversion conduction and the weak inversion-strong inversion transition. Two new parameters (*WFAC* and *WFACU*) are added to the original BSIM model as part of this improvement. In addition, BSIM computes the total current by merely adding together the strong inversion and weak inversion contributions. While this approach is reasonable for large gate biases, problems are encountered for gate biases only slightly larger than the threshold voltage, when both contributions are similar in magnitude.

Finally, the linear-saturation transition region can be modeled much more accurately in HSPICE Level 28. In BSIM, the saturation voltage V_{dsat} is defined, and the two current expressions (linear region and saturation region) are connected at that point. In HSPICE Level 28, a transition region is defined around V_{dsat}, and a smoothing function is used to describe the current in this transition region. This brings another two new parameters (*B1* and *B2*) into the model.

An important philosophical difference between BSIM and HSPICE Level 28 should also be noted. In BSIM (and BSIM2 as well), the model is derived with the intent of providing one set of model parameters for the entire geometry range of lengths and widths. Binning, already described for Level 3 (see Section 7.7), is regarded as an unpleasant occurrence, for which the model is not designed. Binning must be accommodated in a sloppy fashion, with possible current discontinuities at the bin boundaries. In contrast, HSPICE Level 28 modifies the BSIM extrinsic model structure, and is thus designed with an admission that model binning into subregions will be required. Additional expressions are added to cope with the side effects of binning, improving the usefulness of the model. Of particular note, in constrast to the general method of constructing parameters in the second-generation models (as described by (8.1)), HSPICE Level 28 employs an expression of the form

$$
X = X_o + LX \cdot \left[\frac{1}{L_{eff}} - \frac{1}{L_{ref,eff}} \right] + WX \cdot \left[\frac{1}{W_{eff}} - \frac{1}{W_{ref,eff}} \right]
$$

$$
+ PX \cdot \left[\frac{1}{L_{eff}} - \frac{1}{L_{ref,eff}} \right] \cdot \left[\frac{1}{W_{eff}} - \frac{1}{W_{ref,eff}} \right]. \tag{9.1}
$$

When compared to (8.1), two innovations appear in (9.1). First, rather than simply using L_{eff} and W_{eff} as in BSIM, the new terms $L_{ref,eff}$ and $W_{ref,eff}$ are introduced. As will be detailed in Section 9.8, this modification makes HSPICE Level 28 more amenable to binning. (As described in Chapters 8 and 10, this modification is also available in the HSPICE implementations of BSIM (HSPICE Level 13) and BSIM2 (HSPICE Level 39) for the same reason.) Second, in addition to the usual LX and WX terms, HSPICE Level 28 is unique in its addition of a product term PX. Thus, parameters come in quadruplets, rather than the triplets of BSIM. For example, the parameter *X3MS* is comprised of the parameters **X3MS**, **LX3MS**, **WX3MS**, and **PX3MS**. (The product term is also available in HSPICE Level 39, but **not** in HSPICE Level 13.) This improves the model continuity across the bin boundaries, as will also be described in Section 9.8.

(As noted in Chapter 8, the original BSIM documentation and implementation used the same name for the summed parameter X and the first of the triplet parameters X_o. As in Chapter 8, the convention will be adopted here of italicizing the summed parameter name X, and boldfacing the first parameter name X_o. The length and width parameters (LX and WX), along with all the other general model parameters, will also be boldfaced. The boldfaced parameters actually appear in the FET model's list of parameters, while the italicized parameters are computed in the form of (9.1). For example, the summed parameter *X3MS* is computed from

$$X3MS = \mathbf{X3MS} + \mathbf{LX3MS} \cdot \left[\frac{1}{L_{eff}} - \frac{1}{L_{ref,eff}} \right] + \mathbf{WX3MS} \cdot \left[\frac{1}{W_{eff}} - \frac{1}{W_{ref,eff}} \right]$$

$$+ \mathbf{PX3MS} \cdot \left[\frac{1}{L_{eff}} - \frac{1}{L_{ref,eff}} \right] \cdot \left[\frac{1}{W_{eff}} - \frac{1}{W_{ref,eff}} \right]. \tag{9.2}$$

The italicized summed parameter $X3MS$ thus serves the same function as the parameters beginning with z (e.g., $zx3ms$) in the HSPICE documentation [2].)

The parameters used in the HSPICE Level 28 model are summarized in Table 9.1. Some of the model parameters do not follow the simple naming convention described in (9.2); these parameters are listed in Table 9.2.

Temperature Dependence

The HSPICE Level 28 documentation [2] contains little specific information on the temperature dependence of the model parameters. However, the temperature-fitting parameters **BEX**, **TCV**, and **FEX** are employed. Their applicability can be deduced from the discussions of HSPICE Level 13 (BSIM) and HSPICE Level 39 (BSIM2). In addition to these parameters, the usual built-in temperature dependence is found in certain parameters.

9.3   The Mobility Model

As in the other models, HSPICE Level 28 defines a base low field channel mobility, which is then reduced by the vertical and lateral fields. This computation of the low field mobility is very similar to the BSIM computation; however, as noted above, the quadratic function of V_{ds} used to determine the mobility at $V_{ds} = V_{dd}$ often behaves poorly, and can lead to negative values of the output conductance g_{ds}. This particular expression (of the three used to define the low field mobility μ_o) is replaced in HSPICE Level 28.

The first point ($V_{ds} = 0$) is described in the same manner as that used in BSIM (see (8.27)):

$$\mu_o\big|_{V_{ds}=0} = MUZ + X2M \cdot V_{bs}, \tag{9.3}$$

where $X2M$ is identical to the BSIM parameter $X2MZ$. Note, however, that in contrast to **MUZ** in BSIM, MUZ in HSPICE Level 28 has geometric sensitivity; the model includes the parameters **MUZ**, **LMUZ**, **WMUZ**, and **PMUZ**. In theory, MUZ represents the most basic low field channel mobility, and should not have geometric sensitivity. However, by allowing this addition, the final model accuracy is improved.

The third contribution, the sensitivity of μ_o to the drain bias at $V_{ds} = V_{dd}$, is also the same as that used in BSIM, and is represented by the parameter $X3MS$. However, HSPICE Level 28 adds another parameter, $X33M$, which describes the reduction of $X3MS$ by the gate field. The parameter $X33M$ will appear in the next paragraph.

Table 9.1   The HSPICE Level 28 model parameter set.

Parameter	Units	Description
Process Parameters		
TOX	m	Gate Oxide Thickness
XL	m	Process Bias on the Polysilicon Channel Length
XW	m	Process Bias on the Channel Width
LD	m	Source/Drain Underdiffusion of Gate
WD	m	Isolation Reduction of Channel Width
VDDM	V	Maximum Applied Supply Voltage
Electrical Parameters		
MUZ	cm^2/V·s	Zero Bias Low Field Mobility
X2M	cm^2/V^2·s	Substrate Bias Effect on the Low Field Mobility
X3MS	cm^2/V^2·s	Slope of Mobility at $V_{ds} =$ **VDDM**
X33M	cm^2/V^2·s	Gate Field Reduction of *X3MS*
U0	V^{-1}	Gate Field Induced Mobility Reduction Parameter
X2U0	V^{-2}	Substrate Bias Effect on the Gate Field Mobility Reduction
VFB	V	Flatband Voltage
PHI	V	Surface Potential ($= 2\phi_f$)
K1	V$^{\frac{1}{2}}$	Body Effect on Threshold Voltage (First-Order Term)
K2		Body Effect on Threshold Voltage (Second-Order Term)
ETA		DIBL Coefficient When $V_{ds} =$ **VDDM** and $V_{bs} = 0$
X2E	V$^{-\frac{1}{2}}$	Substrate Bias Effect on DIBL Coefficient
X3E	V^{-1}	Drain Bias Effect on DIBL Coefficient
GAMMN	V$^{-\frac{1}{2}}$	Threshold Voltage Linear Conditioning Parameter 1
ETAMN		Threshold Voltage Linear Conditioning Parameter 2
U1	V^{-1}	High Drain Field Mobility Reduction Parameter
X2U1	μm/V^{-2}	Substrate Bias Effect on the High Drain Field Mobility Reduction
X3U1	V^{-2}	Drain Bias Effect on the High Drain Field Mobility Reduction
B1		Lower V_{dsat} Transition Region Parameter
B2		Upper V_{dsat} Transition Region Parameter
WFAC		Weak Inversion Region Conditioning Parameter 1
WFACU		Weak Inversion Region Conditioning Parameter 2
N0		Low Field Subthreshold Ideality Factor
NB		Substrate Bias Effect on the Subthreshold Ideality Factor
ND		Drain Bias Effect on the Subthreshold Ideality Factor
CGSO	F/m	Zero Bias Gate-Source Capacitance
CGDO	F/m	Zero Bias Gate-Drain Capacitance
CGBO	F/m	Zero Bias Gate-Bulk Capacitance
Temperature Parameters		
BEX		Temperature Effect on the Low Drain Bias Mobility
TCV	V/K	Threshold Voltage Variation with Temperature
FEX		Temperature Effect on the High Drain Bias Behavior

Table 9.2 Composite parameters which have modified names for the geometry-dependent parameters.

Composite Parameter	Corresponding Model Parameters			
U0	U00	LU0	WU0	PU0
VFB	VFB0	LVFB	WVFB	PVFB
PHI	PHI0	LPHI	WPHI	PPHI
ETA	ETA0	LETA	WETA	PETA
GAMMN	GAMMN	LGAMN	WGAMN	PGAMN
NB	NB0	LNB	WNB	PNB
ND	ND0	LND	WND	PND

The second of the three points used for the quadratic extrapolation of μ_o is determined in a manner quite different from that of BSIM (see (8.28)). A modification to the low drain bias expression (9.3) is employed:

$$\mu_o\big|_{V_{ds}=\mathbf{VDDM}} = (MUZ + X2M \cdot V_{bs})$$

$$\cdot \left\{ 1 + \frac{X3MS}{MUZ + X33M \cdot (V_{gs} - V_t)} \left[(2 - \sqrt{2}) \cdot \mathbf{VDDM} \right] \right\}, \quad (9.4)$$

where $X33M$, mentioned above, makes its appearance. As detailed in Chapter 8, **VDDM** represents the maximum power supply voltage V_{dd} for which the model is supposed to be used. As such, **VDDM** can be treated as a provided process parameter. For example, if the JEDEC standard of $V_{dd} = 3.3$ V $\pm$ 0.3 V is employed, **VDDM** $= 3.6$ should be used, and data (for parameter extraction) should be available up to and including this voltage.

9.3.1   Mobility Reduction Due to the Vertical Field

The HSPICE Level 28 expression for the reduction of the mobility by the vertical field is nearly identical to the BSIM formulation, which has its roots in the Level 3 implementation. The expression used in HSPICE Level 28 is

$$\mu_{eff} = \frac{m_{eff}}{1 + U_o \cdot (V_{gs} - V_t)}. \qquad (9.5)$$

As in BSIM, U_o is a composite parameter which includes substrate bias dependence, and is given by

$$U_o = U0 + X2U0 \cdot V_{bs}. \qquad (9.6)$$

The term m_{eff} is actually a more generalized version of the high field mobility expression (9.4):

$$m_{eff} = (MUZ + X2M \cdot V_{bs})$$

$$\times \left\{ 1 + \frac{X3MS}{MUZ + X33M \cdot (V_{gs} - V_t)} \left[\mathbf{VDDM} + V_{ds} - \left(\mathbf{VDDM}^2 + V_{ds}^2 \right)^{\frac{1}{2}} \right] \right\}. \qquad (9.7)$$

Note that in contrast to (8.28), (9.7) explicitly contains V_{ds}, and thus partially accounts for mobility reduction by the lateral field. As in BSIM, it is likely that there are additional expressions to describe the reduction of the mobility by the lateral field, intertwined with the drain current model.

Temperature Dependence

As detailed in Chapter 2, SPICE FET models include a nominal temperature (T_{nom}), the temperature at which the parameters were extracted and the FET model was constructed. For operating temperatures other than T_{nom}, the parameters must be modified to properly describe the device behavior.

As originally constructed, BSIM (and BSIM2) contain no parameters to account for temperature dependence; there is only built-in temperature dependence in certain model parameters, such as *PHI*. HSPICE Level 28 includes temperature dependence for the channel mobility in a manner essentially identical to that introduced into Level 3 (see Section 7.4). This takes the form

$$\mu_{eff}(T) = \mu_{eff}(T_{nom}) \cdot \left(\frac{T}{T_{nom}}\right)^{\mathbf{BEX}}, \tag{9.8}$$

where T and T_{nom} are in **degrees Kelvin**. Since μ_{eff} is a composite parameter, from (9.8) it is clear that certain model parameters will make use of **BEX**. Four model parameters, *MUZ*, *X2M*, *X3MS*, and *X33M*, are subject to the use of **BEX**, in the form

$$MUZ(T) = MUZ(T_{nom}) \cdot \left(\frac{T}{T_{nom}}\right)^{\mathbf{BEX}}. \tag{9.9}$$

As will be described when parameter extraction is discussed, **BEX** is easily determined from measured data.

9.4   The Threshold Voltage Model

The HSPICE Level 28 threshold voltage model is nearly identical to the BSIM threshold voltage model, and is given by (see (8.21))

$$V_t = VFB + PHI + K1 \cdot (PHI - V_{bs})^{\frac{1}{2}} - K2 \cdot (PHI - V_{bs}) - \eta \cdot V_{ds}. \tag{9.10}$$

As in BSIM (see (8.20)), the drain-induced barrier lowering (DIBL) expression η is a composite parameter, described by

$$\eta = ETA + X2E \cdot V_{bs} + X3E \cdot (V_{ds} - \mathbf{VDDM}). \tag{9.11}$$

With the BSIM threshold voltage model as a base, HSPICE Level 28 includes a few additions to improve the mathematical conditioning. Close examination of (9.10) and (9.11) indicates that the threshold voltage expression (9.10) is quadratic in both $(PHI - V_{bs})^{\frac{1}{2}}$ and V_{ds}. One aspect of the possibly poor behavior of these polynomial expressions in BSIM is that these quadratics can go in the wrong direction. To prevent this from happening, HSPICE Level 28 is supplemented with two additional parameters to join the quadratic expression to linear equations at

$$\frac{dV_t}{d\left[(PHI - V_{bs})^{\frac{1}{2}}\right]} = GAMMN \tag{9.12}$$

and

$$\frac{dV_t}{dV_{ds}} = -ETAMN \tag{9.13}$$

Specifically, *GAMMN* and *ETAMN* set lower limits, respectively, on *K1* and *ETA* to achieve the desired purpose. The addition of *GAMMN* and *ETAMN* is probably not strictly necessary for model construction, but their availability will improve the mathematical behavior.

Temperature Dependence

As in the case of the mobility parameters, there is no documentation for the temperature dependence of the threshold voltage. However, since the HSPICE Level 28 threshold voltage model is virtually identical to the BSIM threshold voltage model, the documentation for HSPICE Level 13 likely provides the required information. The temperature dependence comes in two parts: the built-in temperature dependence of the surface potential parameter *PHI*, and the independent temperature parameter **TCV**.

Using the simple definition of the Fermi potential (3.50), *PHI* implies a substrate doping concentration via

$$PHI(T_{nom}) = 2 \cdot \frac{k_b T_{nom}}{q} \cdot ln\left(\frac{N_A}{n_i}\right); \tag{9.14}$$

Equation (9.14) can be solved for N_A; using this value of N_A, a new surface potential is easily computed from

$$PHI(T) = 2 \cdot \frac{k_b T}{q} \cdot ln\left(\frac{N_A}{n_i}\right). \tag{9.15}$$

Equation (9.15) is inserted into the threshold voltage expression (9.10).

The use of the temperature parameter **TCV** is also quite simple. The temperature dependence of the threshold voltage is defined from

$$V_t(T) = V_t(T_{nom}) - \textbf{TCV} \cdot (T - T_{nom}), \tag{9.16}$$

where $V_t(T_{nom})$ is given by (9.10). Equation (9.16) assumes that the threshold voltage varies linearly with temperature, which is a reasonable assumption. The model parameter **TCV** is easily extracted from measured data.

9.5   The Saturation Voltage

Before discussing the HSPICE Level 28 drain current model, it is necessary to consider the model for the saturation voltage. In contrast to BSIM (and the earlier models), HSPICE Level 28 supplements the standard approach to the saturation voltage with secondary voltages defined above and below V_{dsat}. These secondary voltages define a transition region around V_{dsat} in which a third current expression (in addition to the usual linear region and saturation region expressions) is employed. As will be seen when parameter extraction and model development are described later, this allows very good results to be obtained for the linear to saturation transition region.

HSPICE Level 28 employs the same method of simplifying the $\frac{3}{2}$-power bulk charge terms that is used in BSIM (see Section 8.7). The bulk charge coefficient a is used,

$$a = 1 + \frac{g \cdot K1}{2(PHI - V_{bs})^{\frac{1}{2}}}, \qquad (9.17)$$

where g is a polynomial fit for a more complicated expression and is given by

$$g = 1 - \frac{1}{1.744 + 0.8364(PHI - V_{bs})}. \qquad (9.18)$$

The HSPICE Level 28 saturation voltage expression is

$$V_{dsat} = \frac{2(V_{gs} - V_t)}{(a + R)}, \qquad (9.19)$$

where the term R is given by

$$R = \left[a^2 + 2 \cdot U1 \cdot a \cdot (V_{gs} - V_t) + 4 \cdot X3U1 \cdot (V_{gs} - V_t)^2\right]^{\frac{1}{2}}. \qquad (9.20)$$

No specific details of the origin of (9.19) are given, other than that when $V_{ds} = V_{dsat}$, the partial derivative with respect to V_{ds} of

$$f(V_{ds}, (V_{gs} - V_t), V_{bs}) = \left[\frac{(V_{gs} - V_t) - a \cdot V_{ds}}{2}\right] \cdot \frac{V_{ds}}{[1 + (U1 + X3U1 \cdot V_{ds}) \cdot V_{ds}]} \qquad (9.21)$$

is equal to zero. This expression reappears in the drain current model, and suggests that the saturation voltage is defined in the simplest manner from

$$\frac{\partial I_{ds,lin}}{\partial V_{ds}} = 0. \tag{9.22}$$

As described above, in addition to the saturation voltage, two secondary voltages are defined near V_{dsat} for use in a smoothing function to improve the description of the linear-saturation transition. The voltages are V_1 below V_{dsat} and V_2 above V_{dsat}, which are given by

$$V_1 = V_{dsat} - B1 \cdot \frac{V_{dsat}}{1 + V_{dsat}} \tag{9.23}$$

and

$$V_2 = V_{dsat} + B2 \cdot (V_{gs} - V_t). \tag{9.24}$$

9.6   The Drain Current Model

The HSPICE Level 28 drain current model is quite different from its BSIM forebearer. However, like much of the HSPICE Level 28 model, the details of the drain current model are for the most part proprietary, with only a few pieces documented. The most important details can be inferred from the results obtained during parameter extraction.

9.6.1   The Strong Inversion Current

In all the earlier models, the strong inversion current is divided into the linear region and the saturation region, and the two separate models are connected at $V_{ds} = V_{dsat}$. In HSPICE Level 28, the strong inversion current is separated into three regions; as described above, a transition region around V_{dsat}, defined by V_1 from (9.23) and V_2 from (9.24), is introduced to improve the model accuracy.

For $V_{ds} < V_1$, the drain current is given by

$$I_{ds} = \frac{\mu_{eff} \cdot W_{eff} \cdot C_{ox}}{L_{eff}} \cdot \left[\frac{(V_{gs} - V_t) - a \cdot V_{ds}}{2} \right] \cdot \left[\frac{V_{ds}}{1 + (U1 + X3U1 \cdot V_{ds}) \cdot V_{ds}} \right]. \tag{9.25}$$

As described above, V_{dsat} is defined as the drain bias at which the derivative of I_{ds} in (9.25) with respect to V_{ds} becomes zero. Thus, (9.25) is used to define V_{dsat}, but not to determine the drain current at V_{dsat}, since I_{ds} for $V_{ds} = V_{dsat}$ is computed from the expression used in the transition region. In that transition region ($V_1 < V_{ds} < V_2$), no details of the current model are available, except that the derivative $\frac{\partial I_{ds}}{\partial V_{ds}}$ varies approximately linearly. This implies that some sort of virtually parabolic (quadratic) expression is used. For $V_{ds} > V_2$, no model equation is available either; a comment is made that I_{ds} is a function of $\left(\frac{\mu_{eff} \cdot W_{eff} \cdot C_{ox}}{L_{eff}} \right)$ and $(V_{gs} - V_t)$ only. This implies the use of a very simple model for the saturation current, similar to the expression developed in BSIM (8.116). However, the additional comment is made that if $B1$ and $B2$ are positive (see (9.23) and (9.24)), this increases the drain current in saturation.

Temperature Dependence

The temperature dependence of the high drain bias behavior is described by the parameter **FEX**. As with **BEX** and **TCV**, the use of **FEX** in HSPICE Level 28 is barely documented; however, documentation is available for HSPICE Level 13 (BSIM) and HSPICE Level 39 (BSIM2). Three parameters, *U1*, *X2U1*, and *X3U1*, make use of **FEX**, in the form

$$U1(T) = U1(T_{nom}) \cdot \left(\frac{T}{T_{nom}} \right)^{\textbf{FEX}} , \qquad (9.26)$$

where T and T_{nom} are in **degrees Kelvin**. **FEX** is easily determined during parameter extraction.

9.6.2 The Weak Inversion Current

As noted in Chapter 8, BSIM contains the first genuine expression for the subthreshold current to be included in a SPICE FET model; the formulation available in Levels 2 and 3 is very flimsy and appears to have been included virtually as an afterthought. Although it is a major improvement, the BSIM subthreshold current model is susceptible to difficulties. The most serious problem, caused by the simple addition of the strong and weak inversion currents under the assumption that $I_{strong} \gg I_{weak}$, is a kink in the current in the neighborhood of the threshold voltage ($V_{gs} = V_t$); the derivative of I_{ds} with respect to V_{gs} is not continuous. This is a major hazard during the simulation of certain analog circuits which are designed to operate near the threshold voltage. In addition, this kink is a mathematical problem that can prevent a simulation from converging. HSPICE Level 28 adds to the original BSIM subthreshold model to correct these problems by modifying the model equations, defining several subregions, and adding two parameters (*WFAC* and *WFACU*).

In a fashion similar to that found in BSIM, HSPICE Level 28 defines a weak inversion current that is added to the strong inversion current. The expression used in HSPICE Level 28 is

$$I_{ds,total} = I_{strong} + I_{weak} \cdot \left[1 - exp \left(-\frac{q V_{ds}}{k_b T} \right) \right], \qquad (9.27)$$

which is similar to the BSIM expression (8.150); the term in brackets in (9.27) is easily derived from basic device physics [3]. BSIM employs a somewhat complex multiplicative method of keeping the weak inversion expression from interfering with the strong inversion current at higher gate biases, using a limiting current. An exponential factor ($e^{1.8}$) is also included as an empirical method of matching the model to data.

In contrast, HSPICE Level 28 approaches this problem entirely through the computation of I_{weak} in (9.27). A strictly mathematical approach to dealing with the problems discussed above is adopted. First, I_{weak} is computed only when *NO* is less than 200. In real devices, this is virtually always the case; as will be seen when parameter extraction is considered, values of *NO* are usually in the range of 1.0– 1.7. As in BSIM, the composite ideality parameter n is used, and is defined by

$$n = N0 + NB \cdot V_{bs} + ND \cdot V_{ds}. \tag{9.28}$$

To fulfill the mathematical goals detailed above, I_{weak} is computed differently under four separately defined conditions (subregions). When $\left[\left(\frac{q}{k_bT}\right)\left(\frac{V_{gs}-V_t}{n}\right)\right] < -WFAC + V_{A0}$, the device is in the deep subthreshold regime, and I_{weak} is computed from a basic subthreshold equation (see Section 3.5.1):

$$I_{weak} = C \cdot exp\left[\left(\frac{q}{k_bT}\right)\left(\frac{V_{gs}-V_t}{n}\right)\right], \tag{9.29}$$

where C is an unspecified constant. In this region, I_{strong} is zero, and only I_{weak} contributes to the current. For a larger V_{gs}, operation moves closer to the threshold voltage. In the region where V_{gs} is just less than V_t, $-WFAC + V_{A0} < \left[\left(\frac{q}{k_bT}\right)\left(\frac{V_{gs}-V_t}{n}\right)\right] < 0$, and I_{weak} is computed from

$$I_{weak} = C \cdot exp\left[\left(\frac{q}{k_bT}\right)\left(\frac{V_{gs}-V_t}{n}\right) - C \cdot W_f\right], \tag{9.30}$$

where W_f is the integral with respect to $\left[\left(\frac{q}{k_bT}\right)\left(\frac{V_{gs}-V_t}{n}\right)\right]$ of

$$dW_f = \left[\left(\frac{q}{k_bT}\right)\left(\frac{V_{gs}-V_t}{n}\right) + WFAC - V_{A0}\right]^2$$

$$\div \left[\left\{1 + \left[\left(\frac{q}{k_bT}\right)\left(\frac{V_{gs}-V_t}{n}\right)\right] + WFAC - V_{A0}\right\}\right.$$

$$\left. \cdot \left\{1 + WFACU \cdot \left[\left(\frac{q}{k_bT}\right)\left(\frac{V_{gs}-V_t}{n}\right)\right] + WFAC - V_{A0}\right\}\right]. \tag{9.31}$$

In this region, once again, $I_{strong} = 0$, and all the current is contributed by I_{weak}. Clearly this is a mathematical construction designed to guarantee the continuity of the current and its first derivative at the boundaries with the previous region and the next one.

When the gate bias is slightly larger than the threshold voltage, (9.30) requires modification. For a gate bias V_{gs} slightly larger than the threshold voltage V_t, the strong inversion current is no longer zero, but is now described by

$$I_{strong} = C \cdot \left[\left(\frac{q}{k_bT}\right)\left(\frac{V_{gs}-V_t}{n}\right)\right]^2. \tag{9.32}$$

To eliminate the aforementioned kink in I_{ds} at $V_{gs} = V_t$, I_{weak} requires inclusion of a term of the form $\left\{-\left[\left(\frac{q}{k_bT}\right)\left(\frac{V_{gs}-V_t}{n}\right)\right]^2\right\}$. This is done by arranging for I_{weak} to go to zero when $\left[\left(\frac{q}{k_b T}\right)\left(\frac{V_{gs}-V_t}{n}\right)\right] > V_{A0}$, where V_{A0} is a voltage slightly larger than V_t. No details are available on how V_{A0} is determined, other than that it is defined from continuity constraints at the boundary between the subthreshold and the superthreshold regions. In this form, V_{A0} appears to be similar to the term V_{on} used in the simpler subthreshold current model employed in Level 2 and Level 3 and the offset voltage V_{off} used in BSIM. Thus, when $V_{gs} > V_t$ and $V_{gs} < V_{A0}$, $0 < \left[\left(\frac{q}{k_b T}\right)\left(\frac{V_{gs}-V_t}{n}\right)\right] < V_{A0}$; the expression for I_{weak} is a modified form of (9.30):

$$I_{weak} = C \cdot exp\left[\left(\frac{q}{k_b T}\right)\left(\frac{V_{gs}-V_t}{n}\right) - C \cdot W_f\right] - C \cdot \left[\left(\frac{q}{k_b T}\right)\left(\frac{V_{gs}-V_t}{n}\right)\right]^2. \quad (9.33)$$

In this region, both I_{weak} and I_{strong} are significant, and great care must be taken to ensure that the current and its first derivative remain continuous at the region boundaries. The additional term in (9.33) is used to exactly cancel the contribution of I_{strong}, described by (9.32), in this region; this is required to eliminate the kink in the drain current at $V_{gs} = V_t$, by guaranteeing that the current and its first and second derivatives remain continuous across this region. Finally, when $V_{gs} > V_{A0}$, $\left[\left(\frac{q}{k_b T}\right)\left(\frac{V_{gs}-V_t}{n}\right)\right] > V_{A0}$. In this case, I_{weak} is allowed to vanish:

$$I_{weak} = 0. \quad (9.34)$$

In this regime, the drain current is determined entirely from the expression for I_{strong}.

Final Comments on the HSPICE Level 28 Formulation

The publicly available information on the HSPICE Level 28 drain current model is sketchy and incomplete. Enough is known, though, to allow the observer to note what problems of BSIM the HSPICE Level 28 model was designed to correct. These improvements will be more clearly described when parameter extraction and model construction are discussed.

9.7   The Charge Model

No specific charge model was developed for use with HSPICE Level 28. The most common choice (and the model default) is the BSIM charge model, described in Chapter 8. It is also possible to use the BSIM2 charge model, which is described in Chapter 10.

9.8   The Extrinsic Model Structure

Up to now, the intrinsic mathematical structure of the HSPICE Level 28 model has been considered. However, as noted in Chapter 8, the unique structure of the second-generation

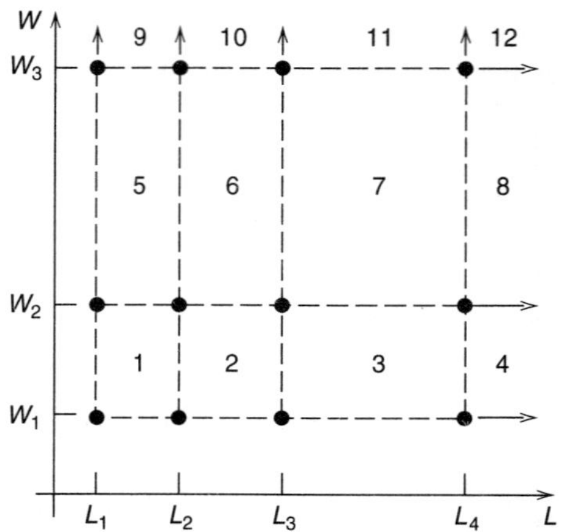

Figure 9.1 The length/width geometry space, showing the twelve devices used for parameter extraction, and the twelve bins thus implied.

models requires that an entirely separate discussion of the extrinsic model structure, revolving around the usage of the geometry parameters (the LX and WX terms in (9.1)), be undertaken. Also, as noted in Section 9.2, there are several modifications and additions which make Level 28 more amenable to binning than is the case with BSIM. To make full use of these additions, there is a requirement for more FET data than are generally used in constructing BSIM-based models. If such data are available, better model results over the entire geometry range can be obtained.

In particular, HSPICE Level 28 has been designed to include binning as part of the formal model structure. This is done through the introduction of the terms $L_{ref,eff}$ and $W_{ref,eff}$, as will be discussed below. In BSIM, parameter extraction is done with several devices with a fixed (wide) width and several channel lengths, and with several devices with a (long) fixed channel length and several channel widths. HSPICE Level 28 employs a two-dimensional grid of devices, as shown in Figure 9.1. (BSIM would only employ the top row and the right-hand column of this grid.) This, of course, requires that such a test grid be available, stressing once again the importance of adequate test pattern design if proper model building is to be done.

The grid of Figure 9.1 uses a cross-coupled combination of four channel lengths and three channel widths for a total of twelve devices. As shown, this permits the geometry space to be divided into twelve subregions (bins). As with Level 3 (see Section 7.7), a separate submodel will be constructed in each bin. However, in contrast to the situation with Level 3, the model structure allows the submodels to be constructed in communication with each other, so that the simulated FET characteristics can be made to be continuous at the various bin boundaries.

For the illustrative discussion to follow, let Z be some particular Level 28 model parameter. Using (9.1), Z is a composite parameter, computed from

$$Z = \mathbf{Z} + \mathbf{LZ} \cdot \left[\frac{1}{L_{eff}} - \frac{1}{L_{ref,eff}} \right] + \mathbf{WZ} \cdot \left[\frac{1}{W_{eff}} - \frac{1}{W_{ref,eff}} \right]$$

$$+ \mathbf{PZ} \cdot \left[\frac{1}{L_{eff}} - \frac{1}{L_{ref,eff}} \right] \cdot \left[\frac{1}{W_{eff}} - \frac{1}{W_{ref,eff}} \right]. \qquad (9.35)$$

In the binned model, $\mathbf{Z}$, $\mathbf{LZ}$, $\mathbf{WZ}$, and $\mathbf{PZ}$ will take on different values in each particular bin. For purposes here, they can be subscripted as $\mathbf{Z}_i$, $\mathbf{LZ}_i$, $\mathbf{WZ}_i$, and $\mathbf{PZ}_i$, where i is the number of the bin under discussion. For reasons that will become clear below, the terms $L_{ref,eff}$ and $W_{ref,eff}$ are chosen as the lower left corner of the bin. For example, for bin **6**, $L_{ref,eff} = L_2$ and $W_{ref,eff} = W_2$. The goal of what follows is to guarantee that Z, as computed by (9.35), is continuous across all the bin boundaries depicted in Figure 9.1.

To begin, a complete set of independent model parameters is extracted at each grid point. For descriptive purposes, each independently extracted Z will be subscripted to identify its particular length and width; for example, Z_{32} would indicate the value of Z extracted at $L_{eff} = L_3$ and $W_{eff} = W_2$.

Now consider the situation along the bottom of bin **1**, where $W_{eff} = W_1$, and L_{eff} varies from L_1 to L_2. At $L_{eff} = L_1$, (9.35) reduces to

$$Z_{11} = \mathbf{Z}_1. \qquad (9.36)$$

Equation (9.36) represents a notably simple feature of this model structure as embodied in (9.35)—at the lower left corner of each bin, all the terms in brackets in (9.35) go to zero, leaving only the first term from the right-hand side of the equation. Similarly, $Z_{12} = \mathbf{Z}_2$, $Z_{13} = \mathbf{Z}_3$, etc. Thus, in this scheme, the various extracted values (Z_{11}, Z_{12}, ...) will appear explicitly in the model. By contrast, in the standard BSIM structure, the extracted values (Z_{44}, Z_{43}, ...) are used as data for a linear regression computation, in which $\mathbf{Z}$, $\mathbf{LZ}$, and $\mathbf{WZ}$ are merely the linear equation slope and intercept values; the specific extracted values (Z_{44}, Z_{43}, ...) do not appear in the final list of model parameters.

Now consider the situation at $L_{eff} = L_2$. Solving (9.35) for that point in bin **2** produces

$$Z_{12} = \mathbf{Z}_2. \qquad (9.37)$$

However, since this point is on a boundary, (9.35) can also be solved for that point in bin **1**. Since in bin **1**, $W_{ref,eff} = W_1$ and only $W_{eff} = W_1$ is being considered here,

$$\left[\frac{1}{W_{eff}} - \frac{1}{W_{ref,eff}} \right] \rightarrow 0. \qquad (9.38)$$

This simplifies (9.35) for this point in bin **1** to

$$Z_{12} = \mathbf{Z}_1 + \mathbf{LZ}_1 \left[\frac{1}{L_2} - \frac{1}{L_1} \right]. \qquad (9.39)$$

Setting (9.39) equal to (9.37) allows the parameter $\mathbf{LZ}_1$ to be determined:

$$\mathbf{LZ}_1 = \frac{\mathbf{Z}_2 - \mathbf{Z}_1}{\left[\frac{1}{L_2} - \frac{1}{L_1}\right]}. \tag{9.40}$$

Equation (9.40) can easily be generalized for the extraction of $\mathbf{LZ}$ for each bin. Note that taken together, (9.37) and (9.39) show that the same (extracted) value of Z_{12} is computed from each side of the boundary (from bin **1** and from bin **2**). This guarantees that for $W_{eff} = W_1$, the general Z and all the other electrical parameters will be continuous across each bin boundary. If all the parameters are continuous across the boundary, this also guarantees that the drain current (and all the other simulated device characteristics) will be continuous across the boundary. This is an improvement over the Level 3 binning situation, described in Section 7.7.

By a similar derivation, it is easily shown that

$$\mathbf{WZ}_1 = \frac{\mathbf{Z}_5 - \mathbf{Z}_1}{\left[\frac{1}{W_2} - \frac{1}{W_1}\right]}. \tag{9.41}$$

This expression can also be generalized to determine $\mathbf{WZ}$ for each bin.

The scheme described so far is available for HSPICE Level 28, as well as HSPICE Level 13 (BSIM) and HSPICE Level 39 (BSIM2). It guarantees continuity of the model parameter set and the simulated device characteristics when exiting a particular bin at either its *upper left* or *lower right* corner (see Figure 9.2). Such continuity cannot be guaranteed at either the *lower left* or *upper right* corner, or across any of the bin faces.

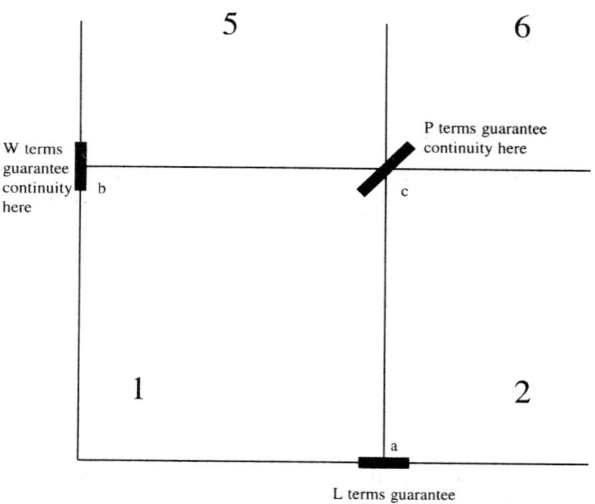

Figure 9.2 A description of how the L, W, and P parameters affect model continuity across bin boundaries.

While guaranteeing parameter and characteristic continuity across the bin faces is rather difficult, HSPICE Level 28 (and Level 39, but **not** Level 13) also includes a more recent modification to guarantee continuity at the lower left and upper right bin corners; this also guarantees continuity at the bin faces. Consider the upper right corner of bin **1**, where $L_{eff} = L_2$ and $W_{eff} = W_2$. Here (9.35) becomes

$$Z_{22} = \mathbf{Z}_1 + \mathbf{LZ}_1 \cdot \left[\frac{1}{L_2} - \frac{1}{L_1} \right] + \mathbf{WZ}_1 \cdot \left[\frac{1}{W_2} - \frac{1}{W_1} \right]$$

$$+ \mathbf{PZ}_1 \cdot \left[\frac{1}{L_2} - \frac{1}{L_1} \right] \cdot \left[\frac{1}{W_2} - \frac{1}{W_1} \right], \tag{9.42}$$

where $\mathbf{LZ}_1$ and $\mathbf{WZ}_1$ were computed as described above and are thus already known. At this point, from bin **6**,

$$Z_{22} = \mathbf{Z}_6. \tag{9.43}$$

Setting (9.42) equal to (9.43) allows an expression for $\mathbf{PZ}_1$ to be found:

$$\mathbf{PZ}_1 = \frac{\mathbf{Z}_6 - \mathbf{Z}_1 - \mathbf{LZ}_1 \cdot \left[\frac{1}{L_2} - \frac{1}{L_1} \right] - \mathbf{WZ}_1 \cdot \left[\frac{1}{W_2} - \frac{1}{W_1} \right]}{\left[\frac{1}{L_2} - \frac{1}{L_1} \right] \cdot \left[\frac{1}{W_2} - \frac{1}{W_1} \right]}. \tag{9.44}$$

This expression is also easily generalized for the computation of $\mathbf{PZ}$ for each bin.

The situation described up to now applies very nicely to all the "closed" bins shown in Figure 9.1. However, note that there are several bins (**4**, **8**, **9**, **10**, and **11**) that are open on one side, and one (**12**) that is open on two sides. Each of these bins is charged with describing a region of either a length or a width (or, in one case, both) that is greater than the largest device size available. It is feasible to decide that the largest available device geometry (here, $L_{eff} = L_4$ and $W_{eff} = W_3$) exceeds any device size that will be used, and to thus merely decline to even build models for these bins. Usually, L_4 and W_3 will have values similar to 10 or 20 μm, so models are usually constructed for these open bins. Fortunately, the geometry dependence at these large geometries is well described by the W_{eff} and L_{eff} terms embedded in the model equations, so geometrically fine parameter sets are not required for model construction.

Consider bin **8** in Figure 9.1. Along the left side of the bin, $L_{eff} = L_4$ and W_{eff} varies from W_2 to W_3. Using a generalized form of (9.41), the parameter $\mathbf{WZ}_8$ can be extracted for this bin. However, along the length axis, the situation is more complicated, as there is no right boundary for the bin. For $W_{eff} = W_2$, all that can be written down is a general expression for Z anywhere along that line:

$$Z = \mathbf{Z}_8 + \mathbf{LZ}_8 \cdot \left[\frac{1}{L_{eff}} - \frac{1}{L_4} \right]. \tag{9.45}$$

Allowing $L_{eff} \to \infty$ simplifies (9.45) to

$$Z = \mathbf{Z}_8 - \frac{\mathbf{LZ}_8}{L_4}. \tag{9.46}$$

This expression has two unknowns, Z (which is unknown away from the specific data points) and $\mathbf{LZ}_8$. Without another reference point, a second equation cannot be found. Instead, L_4 is treated as also being very large, allowing the $\mathbf{LZ}_8$ term in (9.45) and (9.46) to vanish. Thus, in bin **8** the $\mathbf{LZ}_8$ term is regarded as meaningless; it is either set to zero or simply neglected (in which case it takes on the default value of zero). With, essentially

$$\left[\frac{1}{L_{eff}} - \frac{1}{L_4} \right] \to 0, \tag{9.47}$$

the product term also vanishes, and $\mathbf{PZ}_8$ is also set to zero or neglected. In total, Z in bin **8** is described by the simplified equation

$$Z = \mathbf{Z}_8 + \mathbf{WZ}_8 \cdot \left[\frac{1}{W_{eff}} - \frac{1}{W_2} \right]. \tag{9.48}$$

This expression is generalized to describe the situation in bin **4**. The full models for bins **4** and **8** contain W parameters but no L or P parameters.

Now consider bin **11**. Here, along the lower edge of the bin, $W_{eff} = W_3$. Using a generalized form of (9.40), $\mathbf{LZ}_{11}$ can be found. For $L_{eff} = L_3$,

$$Z = \mathbf{Z}_{11} + \mathbf{WZ}_{11} \cdot \left[\frac{1}{W_{eff}} - \frac{1}{W_3} \right]; \tag{9.49}$$

Allowing $W_{eff} \to \infty$, the problem of two unknowns is once again encountered. As in the case of (9.45) and (9.46), W_3 is assumed to be large, and the right-hand term in (9.49) vanishes. Here, in effect,

$$\left[\frac{1}{W_{eff}} - \frac{1}{W_3} \right] \to 0, \tag{9.50}$$

so the product term also vanishes; the parameter $\mathbf{PZ}_{11}$ is regarded as meaningless, and is either set to zero or neglected. Therefore, Z in bin **11** is described by

$$Z = \mathbf{Z}_{11} + \mathbf{WZ}_{11} \cdot \left[\frac{1}{L_{eff}} - \frac{1}{L_3} \right]. \tag{9.51}$$

This model can be generalized to describe what occurs in bins **9** and **10**. The full models for bins **9**, **10** and **11** contain L parameters but no W or P parameters. The W parameters (e.g., $\mathbf{WZ}_{11}$) are set to zero or neglected.

In bin **12**, the general expression for Z is

$$Z = \mathbf{Z}_{12} + \mathbf{LZ}_{12} \cdot \left[\frac{1}{L_{eff}} - \frac{1}{L_4} \right] + \mathbf{WZ}_{12} \cdot \left[\frac{1}{W_{eff}} - \frac{1}{W_3} \right]$$

$$+ \mathbf{PZ}_{12} \cdot \left[\frac{1}{L_{eff}} - \frac{1}{L_4} \right] \cdot \left[\frac{1}{W_{eff}} - \frac{1}{W_3} \right]. \tag{9.52}$$

Allowing $L_{eff} \rightarrow \infty$ and $W_{eff} \rightarrow \infty$, (9.52) contains five unknowns. However, by assuming that L_4 and W_3 are large, (9.52) simplifies greatly to

$$Z = \mathbf{Z}_{12}. \tag{9.53}$$

Here, $\mathbf{LZ}_{12}$, $\mathbf{WZ}_{12}$, and $\mathbf{PZ}_{12}$ are all regarded as meaningless, and are set to zero or neglected. Bin **12** will contain only the independent parameter values extracted at $L_{eff} = L_4$ and $W_{eff} = W_3$ (e.g., Z_{43}), but no L, W, or P parameters.

9.9 Parameter Extraction and Model Development

HSPICE Level 28 has the basic second-generation model structure, with intrinsic (equation) and extrinsic (geometric) components. Thus, the model itself forms a shell, and the quality of any final model depends very heavily on the process of parameter extraction. As noted in Chapter 8, there are many possible parameter extraction routes; the approach described here represents one possible route.

Given the semi-proprietary nature of the HSPICE Level 28 model formulation, parameter extraction is particularly important for the evaluation of the model's strengths and weaknesses. As the description proceeds, it will be worthwhile for the reader to compare the results with those obtained for BSIM (Section 8.10) and BSIM2 (Section 10.9).

As was the case with BSIM, care must be taken to separate the behavior of the intrinsic model structure and the extrinsic model structure. In contrast to BSIM (and most other models), the HSPICE Level 28 extrinsic structure was constructed to accommodate model binning; this added feature must be examined with some care.

9.9.1 Basic Parameter Extraction

Like BSIM and BSIM2, parameter extraction for HSPICE Level 28 proceeds in three phases. However, in contrast to the other second-generation models, the long, wide device does not serve as full a role as a reference. A long, wide device *is* used to obtain a few starting values for the individual device extraction steps, but each device is treated quite independently.

The basic approach to parameter extraction is as follows:

- The oxide thickness and the differences between the drawn and effective channel dimensions are provided as process input.

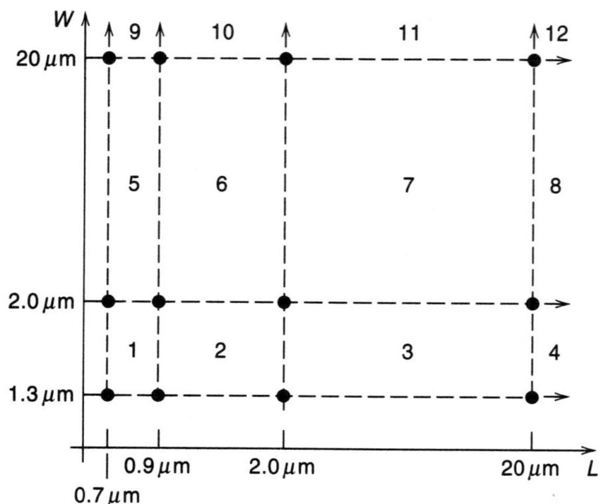

Figure 9.3 The specific geometry scheme used to extract the parameters for the HSPICE Level 28 model developed here.

- In the first phase, a long, wide device is used to define a few base parameters, which are used as the starting point for each of the individual device extractions which are carried on in the second phase. Unlike BSIM, in HSPICE Level 28, *all* parameters have geometric sensitivity; the model builder can choose to select a few parameters as global and determine them permanently during this phase.

- In the second phase, an HSPICE Level 28 parameter set is extracted independently for each device; this represents the fitting of data to the intrinsic model structure. Note once again that HSPICE Level 28 is unique in that it employs a matrix of device geometries (Figure 9.1), allowing for both model binning and a more detailed description of short, narrow devices.

- In the third phase, the compiled sets of parameters from the second phase are used to determine the extrinsic model structure. This involves using the extrinsic model structure to create a binned model; note also that HSPICE Level 28 has not only length and width parameters, but also the product term described in Section 9.8.

The matrix of device geometries to be employed here is detailed in Figure 9.3. The individual devices are represented by nodes, and the numbered bins describe subregions of the geometry space. This requires that such a device matrix be available among the test structures which are used for data collection. It is also interesting to note that the structure of the matrix in Figure 9.3 bears a strong resemblance to a two-dimensional finite difference analysis of a real space (x-y) situation.

The First Phase

Parameter extraction begins by using a long, wide (20/20) device to extract a set of base parameters for use as the starting point for the individual device extractions of the

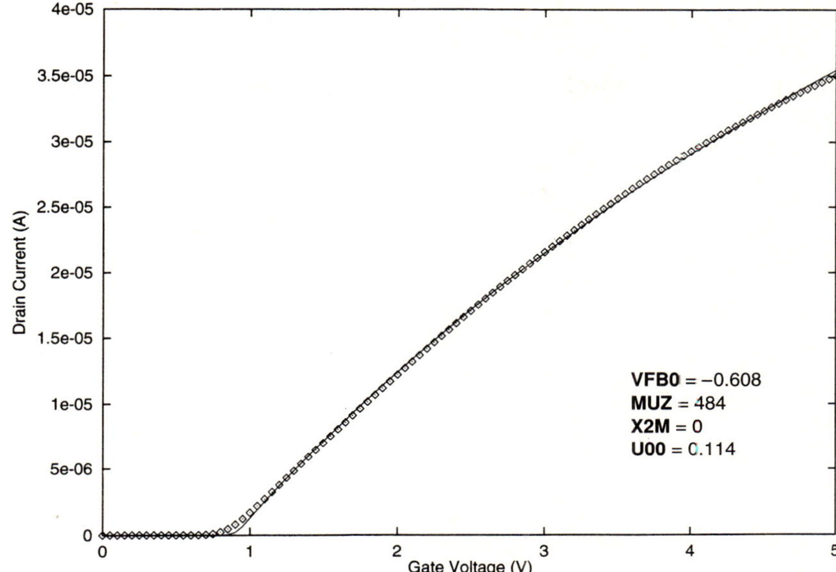

Figure 9.4   Phase I: The extraction of the base device parameters **VFB0**, **MUZ**, **X2M**, and **U00**, $W/L = 20\,\mu\text{m}/20\,\mu\text{m}$.

following phase. The first step involves fitting the linear characteristic with no applied substrate bias; as shown in Figure 9.4, the results are quite good. The second step extends the process by fitting the linear characteristic with applied substrate biases. As shown in Figure 9.5, the results are somewhat poor; however, as was the case with BSIM, this extraction represents an initial guess for use in the following phase, so the details of this step are not significant. The third step, portrayed in Figure 9.6, is an extraction of the parameter **PHI0** using the linear characteristic with substrate biases applied. Experience has shown that **PHI0** tends not to vary greatly for each device; thus, the value determined here is used for *PHI0*, and is taken to be a global value, fixed for use throughout the remainder of model construction.

The Second Phase

In the second phase of parameter extraction, a set of individual device parameters is determined for *every* device (here, for 12 devices), using the base parameters from the first phase as the starting point for each device. As noted above, this represents a fitting of each individual device to the intrinsic model structure.

The first step is a fitting of the linear region characteristic, as depicted in Figure 9.7. The results are quite good for a short (20/0.7) device (Figure 9.7a) and for a narrow (1.3/20) device (Figure 9.7b). The result is also very good for a short, narrow (1.3/0.7)

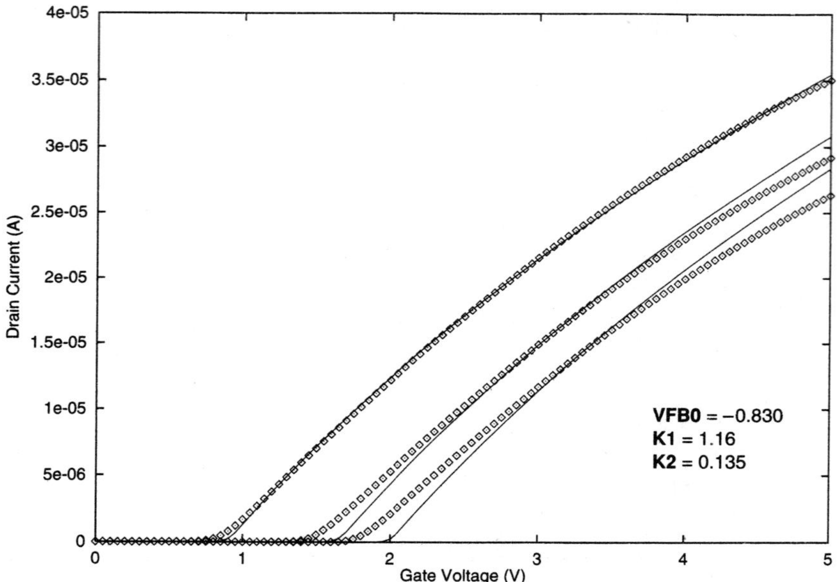

Figure 9.5 Phase I: The extraction of the base device parameters **VFB0**, **K1**, and **K2**, $W/L = 20$ μm/20 μm.

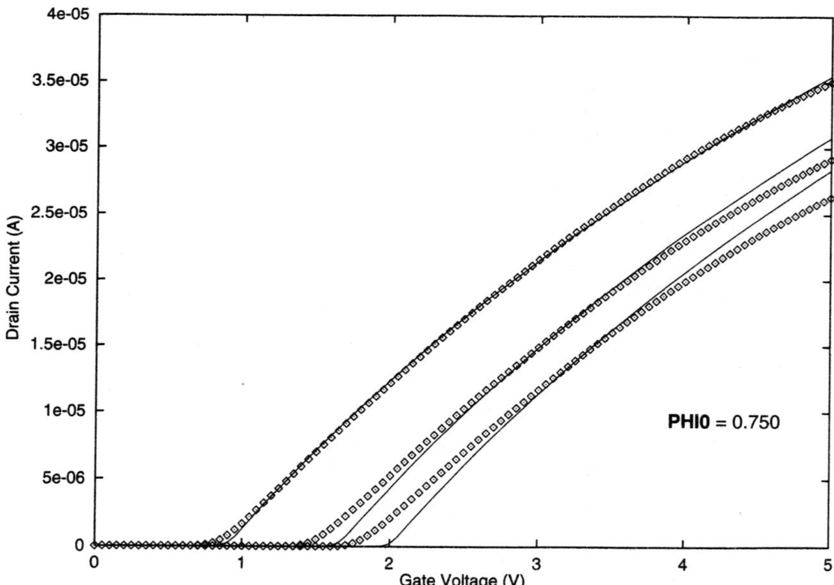

Figure 9.6 Phase I: The extraction of the base device parameter **PHI**, $W/L = 20$ μm/20 μm.

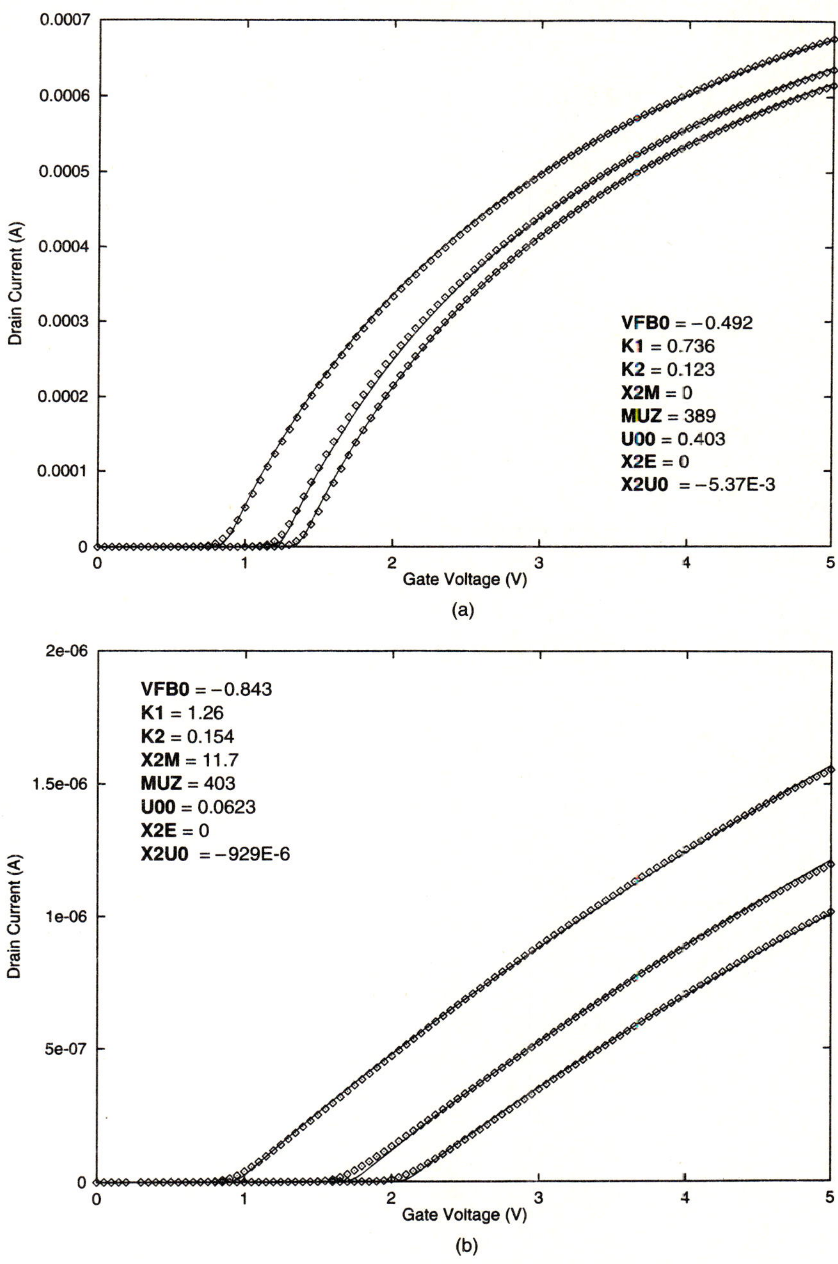

Figure 9.7   Phase II: The extraction of the device parameters **VFB0**, **K1**, **K2**, **X2M**, **MUZ**, **U00**, **X2E**, and **X2U0**; (a) $W/L = 20$ μm/0.7 μm; (b) $W/L = 20$ μm/2.0 μm; (c) $W/L = 1.3$ μm/20 μm.

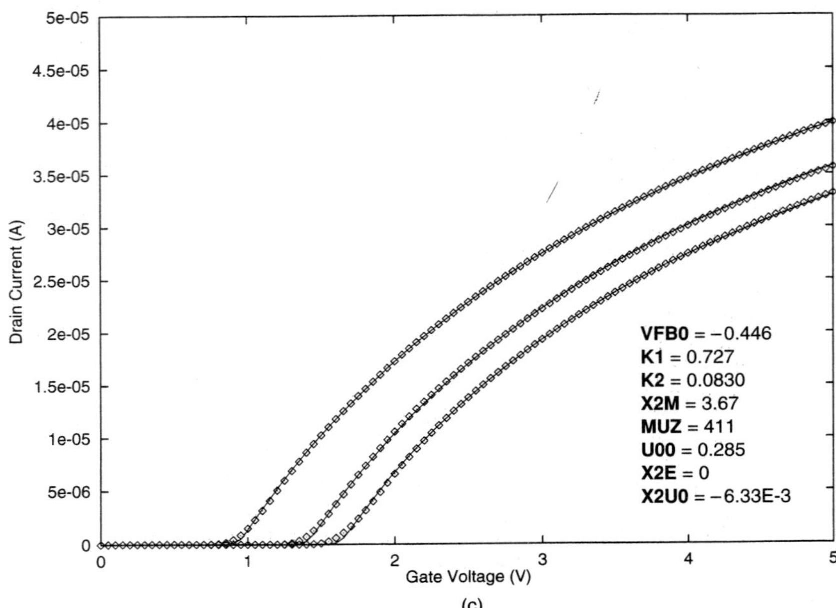

Figure 9.7   *(cont.)*

device. One empirically noted feature of HSPICE Level 28 is that it permits very good results for narrow devices.

In the second step, data for the saturation region characteristics are employed; as shown in Figure 9.8, very good results are obtained. A detailed examination of the short devices (20/0.7 and 1.3/0.7, in Figures 9.8a and 9.8c, respectively) shows that there is some difficulty near the linear-saturation transition, but this is an improvement over the BSIM results (see Figure 8.14). The narrow (1.3/20) device result (Figure 9.8b) is very good.

In the third step, the saturation characteristic is used with substrate biases applied (Figure 9.9). As with BSIM, the results are moderately good, but this degree of accuracy is usually adequate for most applications. Of particular note, the results do not degrade when moving from a longer device (20/2.0) to a submicron device (20/0.7). This is in contrast to the BSIM result of Figure 8.15; here, in a submicron (20/0.7) device, the drain current *decreases* with an increasing magnitude of the substrate bias, which is the qualitatively correct result.

In the fourth step, the subthreshold characteristics are employed, with a result as presented in Figure 9.10. A very good result is obtained, providing an accurate description of the current for several decades below the weak-strong inversion transition. Like BSIM, the HSPICE Level 28 subthreshold region model accounts for substrate bias dependence, and also has geometric dependence through the extrinsic model structure.

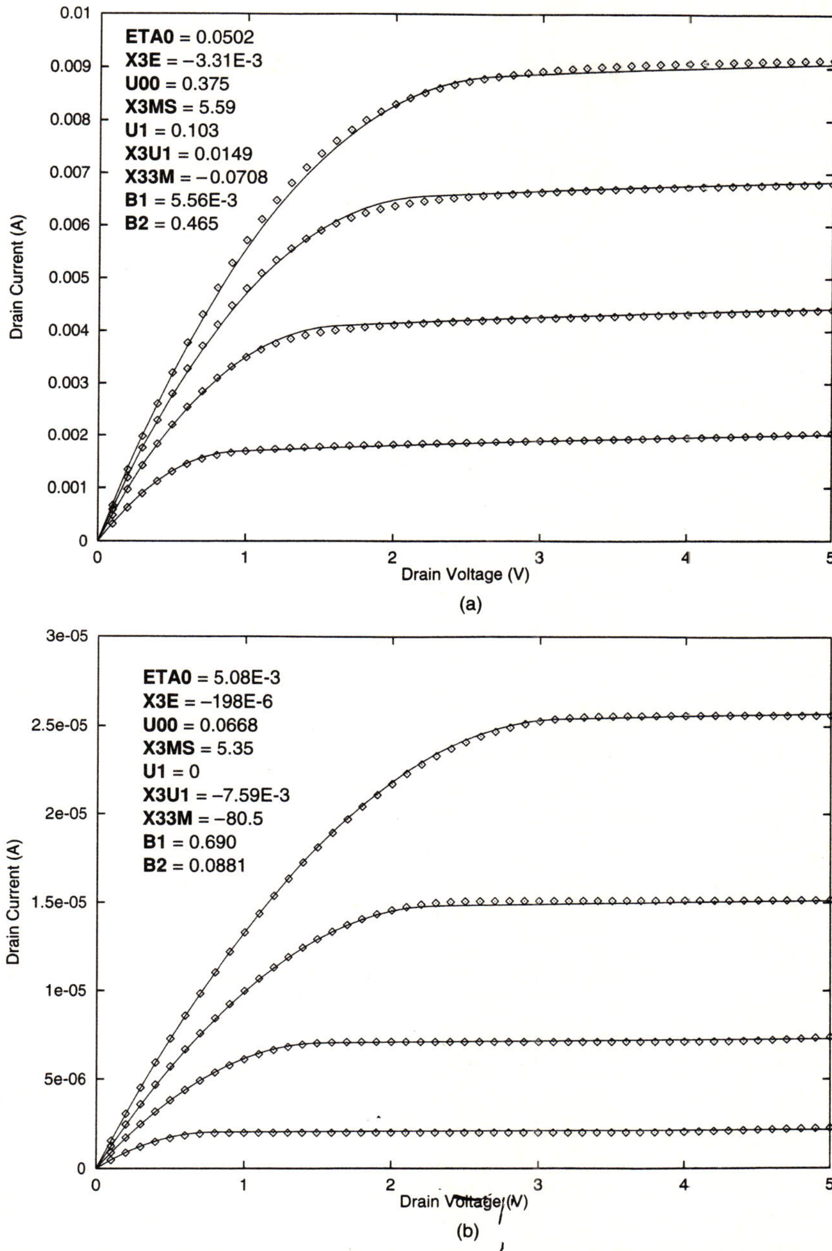

Figure 9.8 Phase II: The extraction of the device parameters **ETA0, X3E, U00, X3MS, U1, X3U1, X33M, B1** and **B2**; (a) $W/L = 20\,\mu\text{m}/0.7\,\mu\text{m}$; (b) $W/L = 20\,\mu\text{m}/2.0\,\mu\text{m}$; (c) $W/L = 1.3\,\mu\text{m}/20\,\mu\text{m}$.

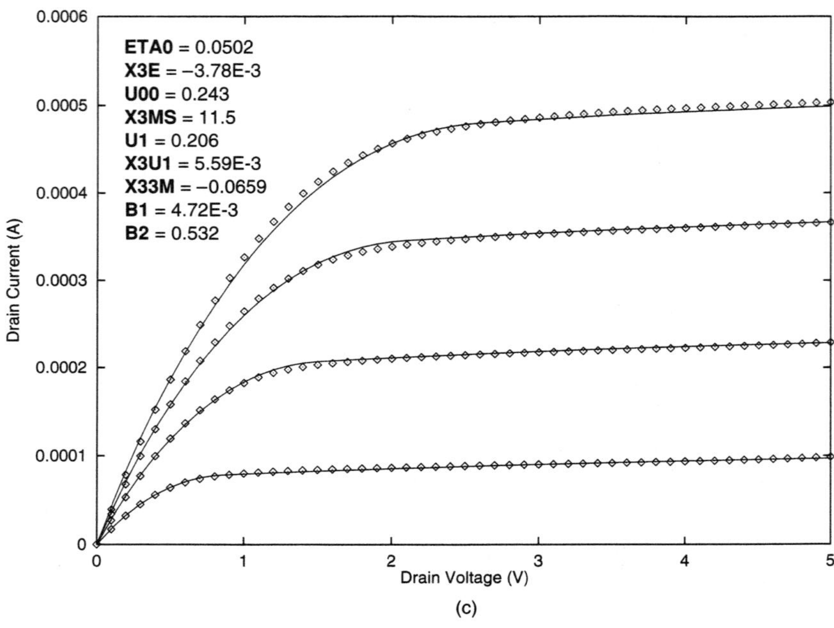

Figure 9.8   *(cont.)*

Thus, the subthreshold model is more generally applicable (and more useful) than the first-generation subthreshold model.

The last possible extraction step involves the fitting of output conductance data, and is potentially very tricky. The description of equation derivatives involves very fine detail, and represents a major challenge, both in the use of presently available models and in the development of new models. With no changes, the output conductance result for the 20/0.7 device is as shown in Figure 9.11a. A result of this type is actually reasonable for most analog circuit designs. An extraction step can be carried out in an attempt to improve the output conductance result; as shown in Figure 9.11b, a slight improvement for high drain biases is noted. However, this attempt to improve the output conductance result badly corrupts the current result, as shown in Figure 9.12.

The description of the output conductance must be approached with great caution, and must be based on an understanding of the specific accuracy requirements of the circuit design user. When attempting to improve the model's output conductance result, it is easy to very badly corrupt the drain current result. If the output conductance result must be improved, the extraction must be done using sophisticated weighting methods that fit the drain current and the output conductance at the same time. It must be stressed that great care should be employed, as any such procedure is heavily dependent on the particular fabrication technology which is under study. Clear communication with the

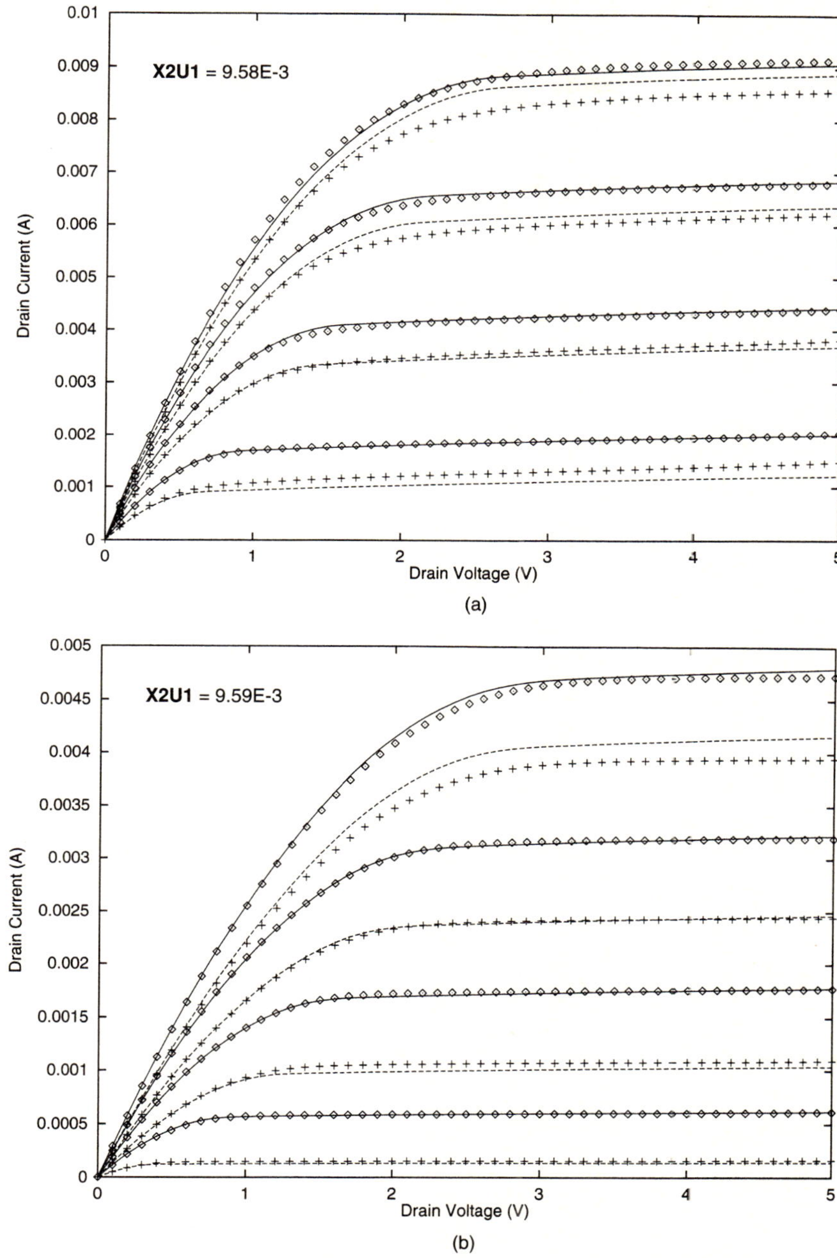

Figure 9.9   Phase II: The extraction of the device parameter **X2U1**; (a) $W/L = 20\,\mu\text{m}/0.7\,\mu\text{m}$; (b) $W/L = 20\,\mu\text{m}/2.0\,\mu\text{m}$.

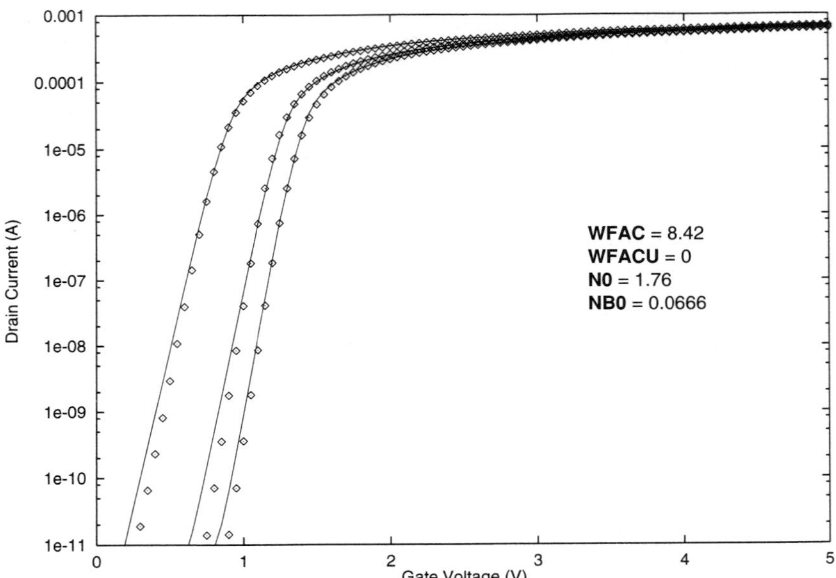

Figure 9.10 Phase II: The extraction of the device parameters **N0**, **NB**, **WFAC**, and **WFACU**, $W/L = 20$ μm/0.7 μm.

circuit design user to understand accuracy requirements is also an important aid, as small improvements in model accuracy are obtained at a very large cost in time and effort.

Temperature Dependence

HSPICE Level 28 includes parameters for the description of temperature dependence. While the approach is quite simple, good results are obtained.

Two extraction steps are used to determine the relevant parameters. In the first step, the long, wide (20/20) device's linear characteristic is used at 125°C; this extraction is similar to that used in the HSPICE implementation of Level 3. As shown in Figure 9.13, a good result is obtained. In the second step, the saturation characteristic of a short (20/0.7) device is used. The result, depicted in Figure 9.14, is reasonable for the description of the high-temperature worst-case behavior of a circuit.

Tabulated Parameters

The results of the second phase of parameter extraction are compiled in Appendix D. Due to the structure of HSPICE Level 28, the individual device parameters appear as the nongeometry parameter in the appropriate bin of the final model.

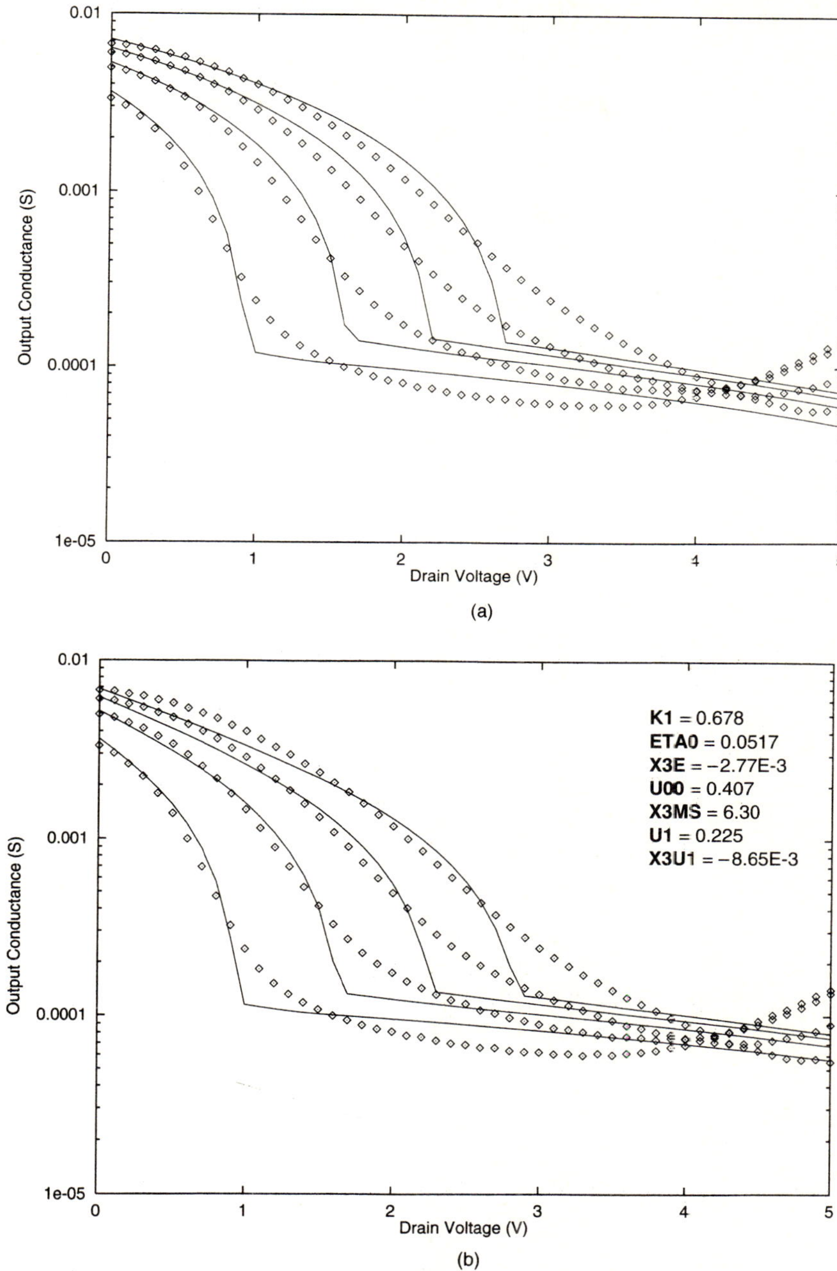

Figure 9.11 Phase II: The extraction of **K1**, **ETA0**, **X3E**, **U00**, **X3MS**, **U1**, and **X3U1** for output conductance fitting, $W/L = 20$ μm/0.7 μm; (a) before the extraction; (b) after the extraction.

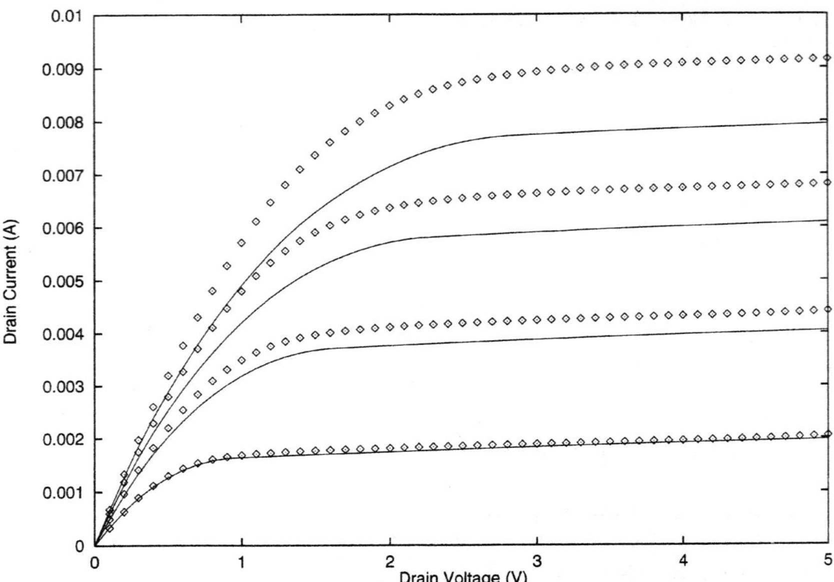

Figure 9.12   The drain current-drain voltage characteristics of a W/L = 20 μm/0.7 μm device after the output conductance extraction depicted in Figure 9.11(b).

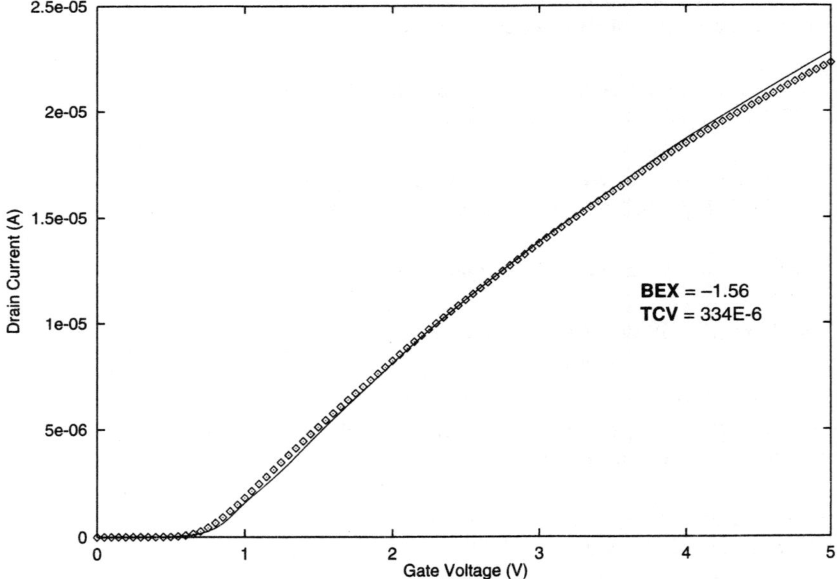

Figure 9.13   Phase II: The extraction of the device parameters **BEX** and **TCV**, $W/L = 20$ μm/20 μm, $T = 125°$C.

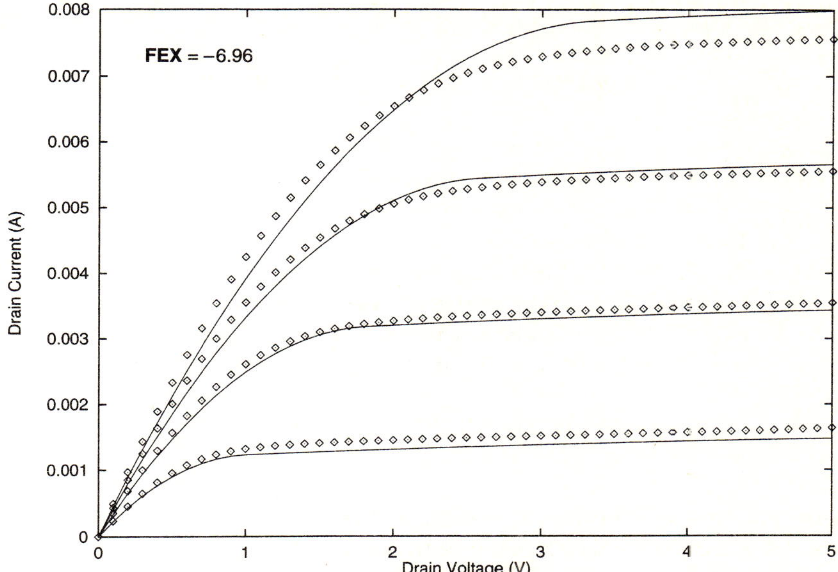

Figure 9.14 Phase II: The extraction of the device parameter **FEX**, $W/L = 20\,\mu\text{m}/0.7\,\mu\text{m}$, $T = 125°\text{C}$.

The Third Phase

In the third phase of model construction, the parameters compiled for each device in the second phase are used to build the extrinsic model structure. This involves the determination of the geometry parameters for each bin, using the extrinsic model equation

$$Z = \mathbf{Z} + \mathbf{LZ} \cdot \left[\frac{1}{L_{eff}} - \frac{1}{L_{ref,eff}} \right] + \mathbf{WZ} \cdot \left[\frac{1}{W_{eff}} - \frac{1}{W_{ref,eff}} \right]$$

$$+ \mathbf{PZ} \cdot \left[\frac{1}{L_{eff}} - \frac{1}{L_{ref,eff}} \right] \cdot \left[\frac{1}{W_{eff}} - \frac{1}{W_{ref,eff}} \right] \tag{9.54}$$

for the determination of **LZ**, **WZ**, and **PZ**, where Z represents an arbitrary model parameter. Due to the large size of this final model, the results are collected in Appendix D.

9.9.2 Model "Playback"

With the final model completed, it is important to compare the predictions of that final model with data. This will permit a brief examination of the robustness of both the intrinsic and extrinsic structure of HSPICE Level 28.

Figure 9.15 depicts the final model results for the 20/0.7 device. The linear characteristic (Figure 9.15a) shows some degradation due to later fitting steps (when compared with Figure 9.7a). The saturation characteristic (Figure 9.15b) is similar to the original result (Figure 9.8a), showing the minor difficulty near the linear-saturation transition. The 20/2.0 device (Figure 9.16) has better results, although the linear characteristic (Figure 9.16a) has deteriorated slightly due to later extraction steps. Of particular note, this result is much better than that obtained for the same device in the final BSIM model. Since HSPICE Level 28 is built to include model binning, the intermediate geometry ranges are well described by the final model.

Figure 9.17 is the final result for a 20/0.9 device. The linear characteristic (Figure 9.17a) shows less deterioration due to later extraction steps than was the case for the 20/0.7 device. The comments made above about intermediate geometries apply here as well; with binning included in the model structure, the result is much improved over the final BSIM result. Figure 9.18 shows that a very good result is obtained for the 1.3/20 device; as noted earlier, an empirical observation is that HSPICE Level 28 can be made to produce very good results for narrow devices. The same comments apply to the 2.0/20 device (Figure 9.19).

Due to the use of the matrix of devices, good results can be obtained for short, narrow devices. The final results for the 1.3/0.7 device are depicted in Figure 9.20. Since this is a short device, the linear characteristic (Figure 9.20a) shows some degradation due to later extraction steps, as was the case for the 20/0.7 device. The saturation characteristic (Figure 9.20b) shows good results. These results are in marked contrast to the situation in the first-generation models and in BSIM, where short and narrow geometry effects were described separately but not combined. Due to the extrinsic structure with a matrix of devices, HSPICE Level 28 is able to produce good results for short, narrow devices.

Some of the shortcomings of BSIM should now be briefly revisited. Figure 9.21 shows the model result for the drain current behavior of a short (20/0.7) device with a gate bias just above the threshold voltage; in contrast to the BSIM result (Figure 8.24), there is no negative output conductance for high drain biases. Figure 9.22 is a detail of the weak-strong transition region of a short (20/0.7) device; unlike the final result in BSIM (Figure 8.25), the transition region near 0.9 V is smooth.

Finally, as noted earlier, HSPICE Level 28 is designed to accommodate binning and permit continuity of the model across bin boundaries. This is demonstrated in Figure 9.23, where the model prediction for the drain current in a short (20/0.7) device is plotted against the channel length. The current is continuous across the bin boundary at $L_{drawn} = 0.9$ µm, although a minor kink is discernable at that boundary. When compared with the result for a binned Level 3 model (Figure 7.29), this is obviously a better representation of the situation. This ability to ensure model continuity becomes important when employing statistical simulations of the effect of process variations on the effective channel length in the neighborhood of this boundary, and when using the model to optimize the channel length in certain circuit designs.

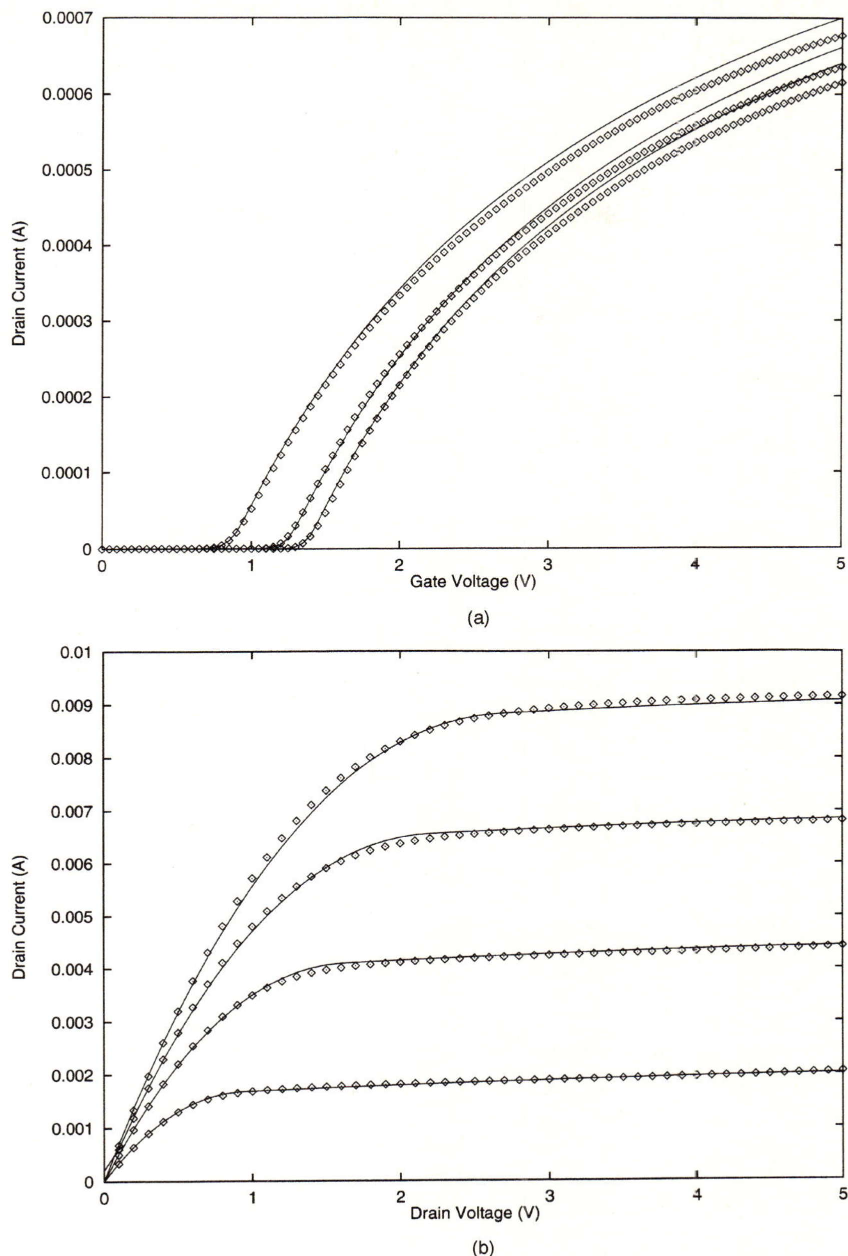

Figure 9.15 Comparison of measured data and the final model for the 20/0.7 device; (a) linear characteristic; (b) saturation characteristic.

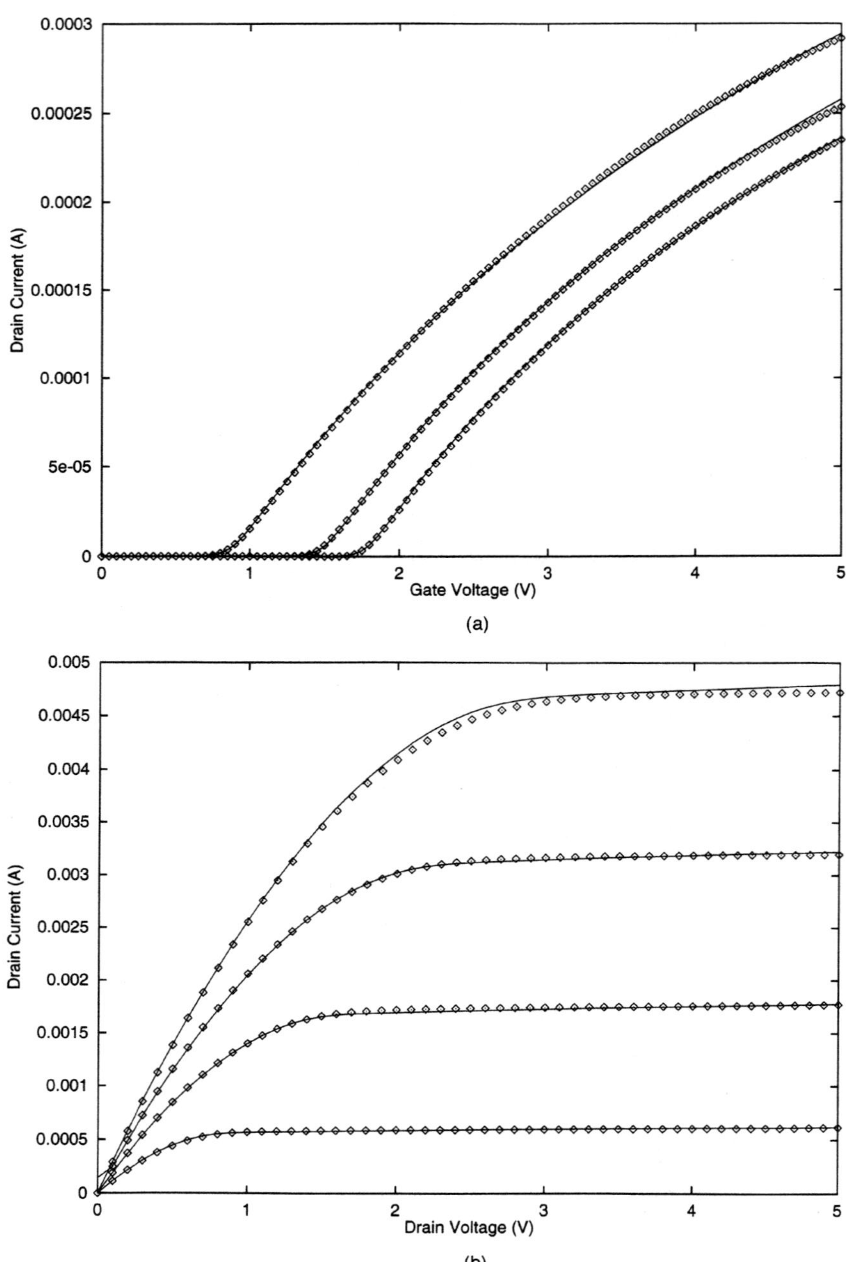

Figure 9.16 Comparison of measured data and the final model for the 20/2.0 device; (a) linear characteristic; (b) saturation characteristic.

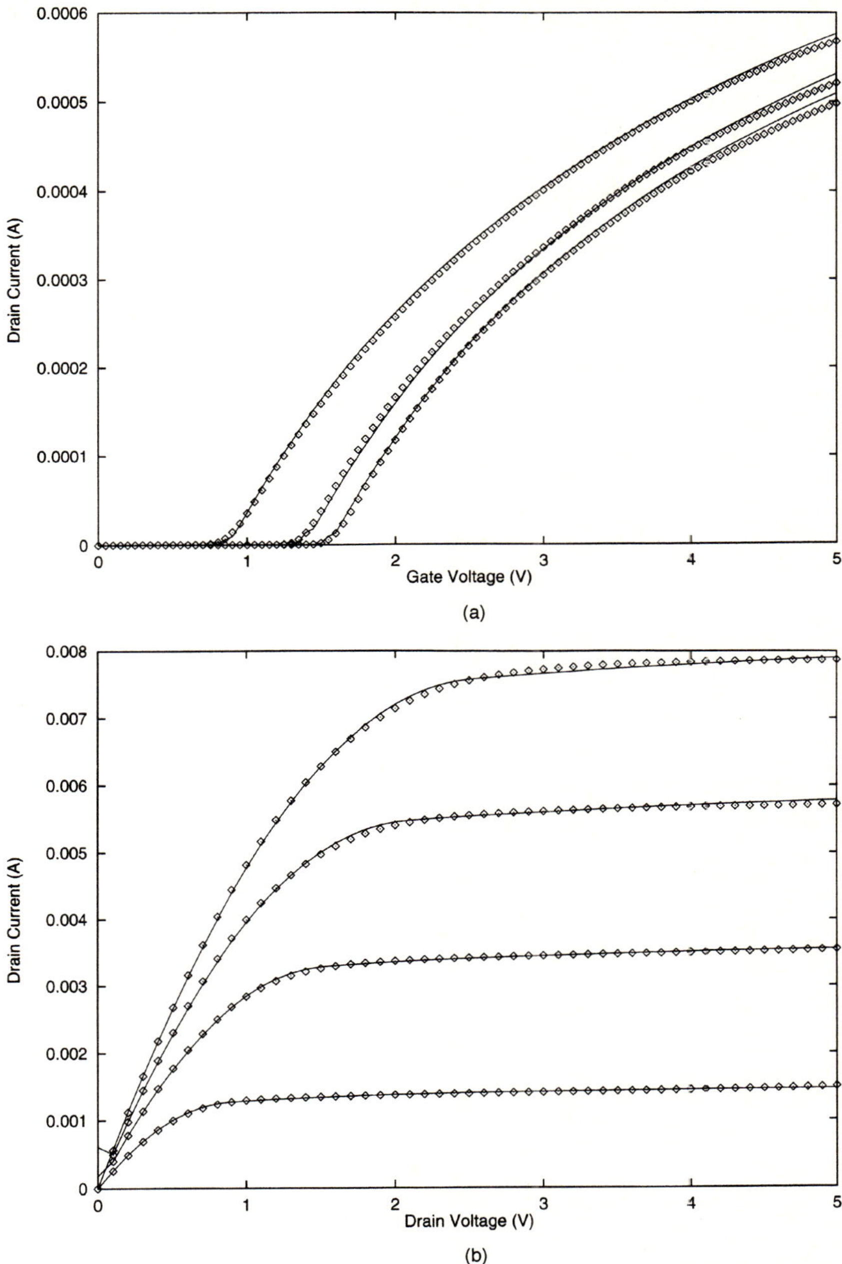

Figure 9.17   Comparison of measured data and the final model for the 20/0.9 device; (a) linear characteristic; (b) saturation characteristic.

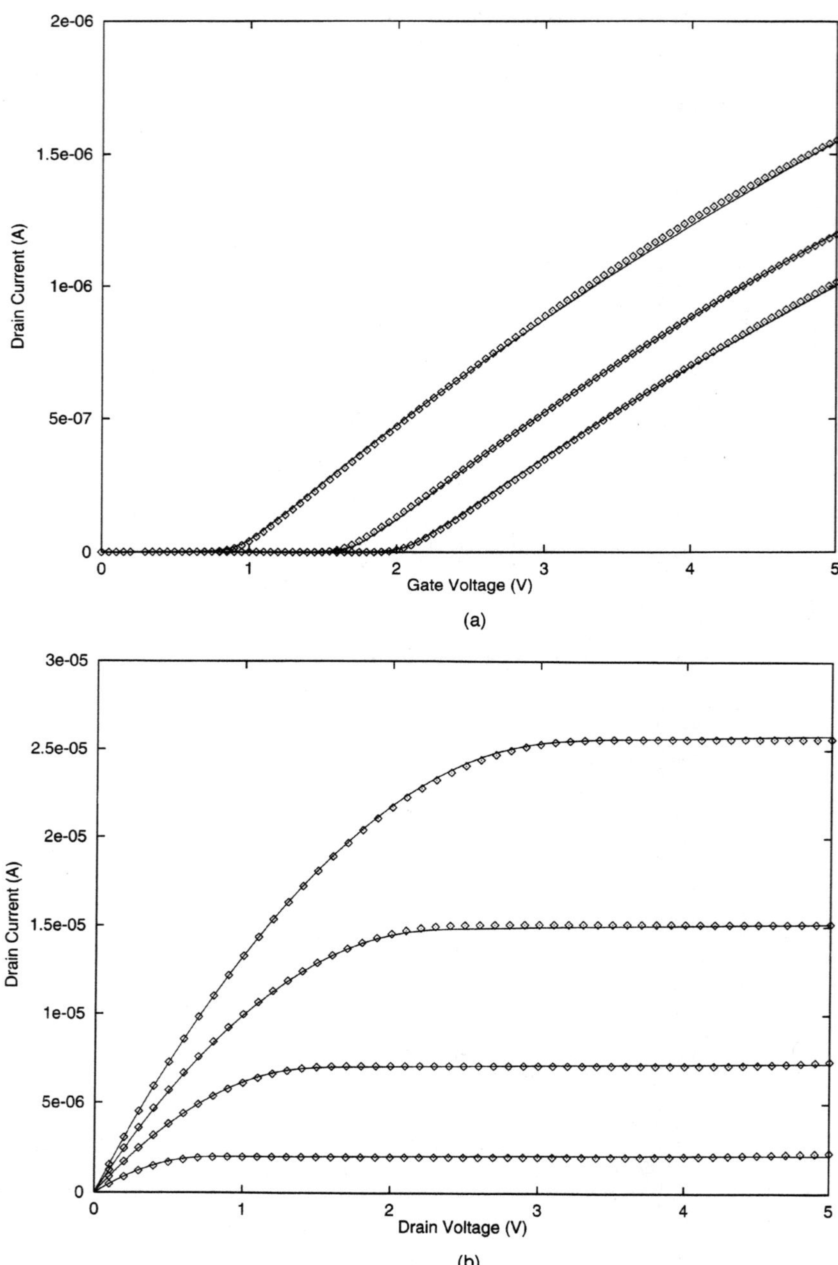

Figure 9.18 Comparison of measured data and the final model for the 1.3/20 device;
(a) linear characteristic; (b) saturation characteristic.

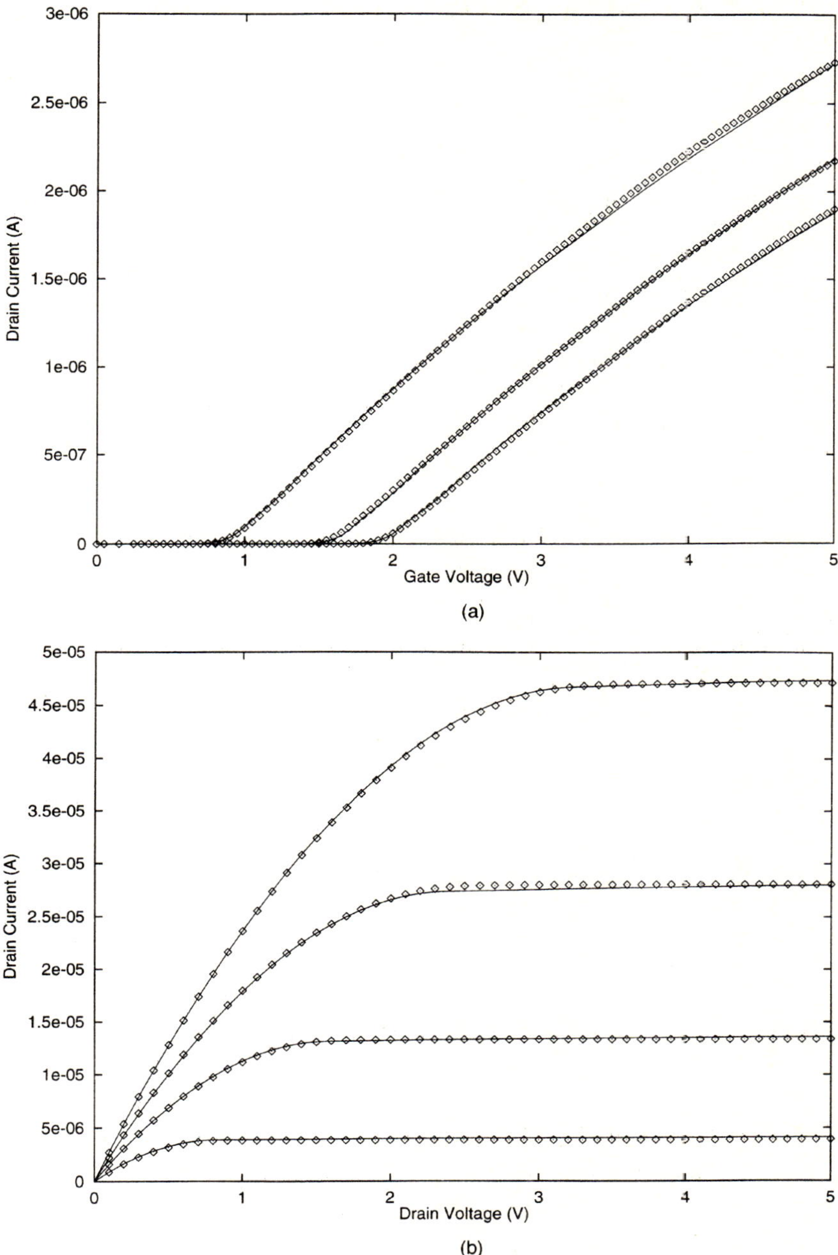

Figure 9.19 Comparison of measured data and the final model for the 2.0/20 device; (a) linear characteristic; (b) saturation characteristic.

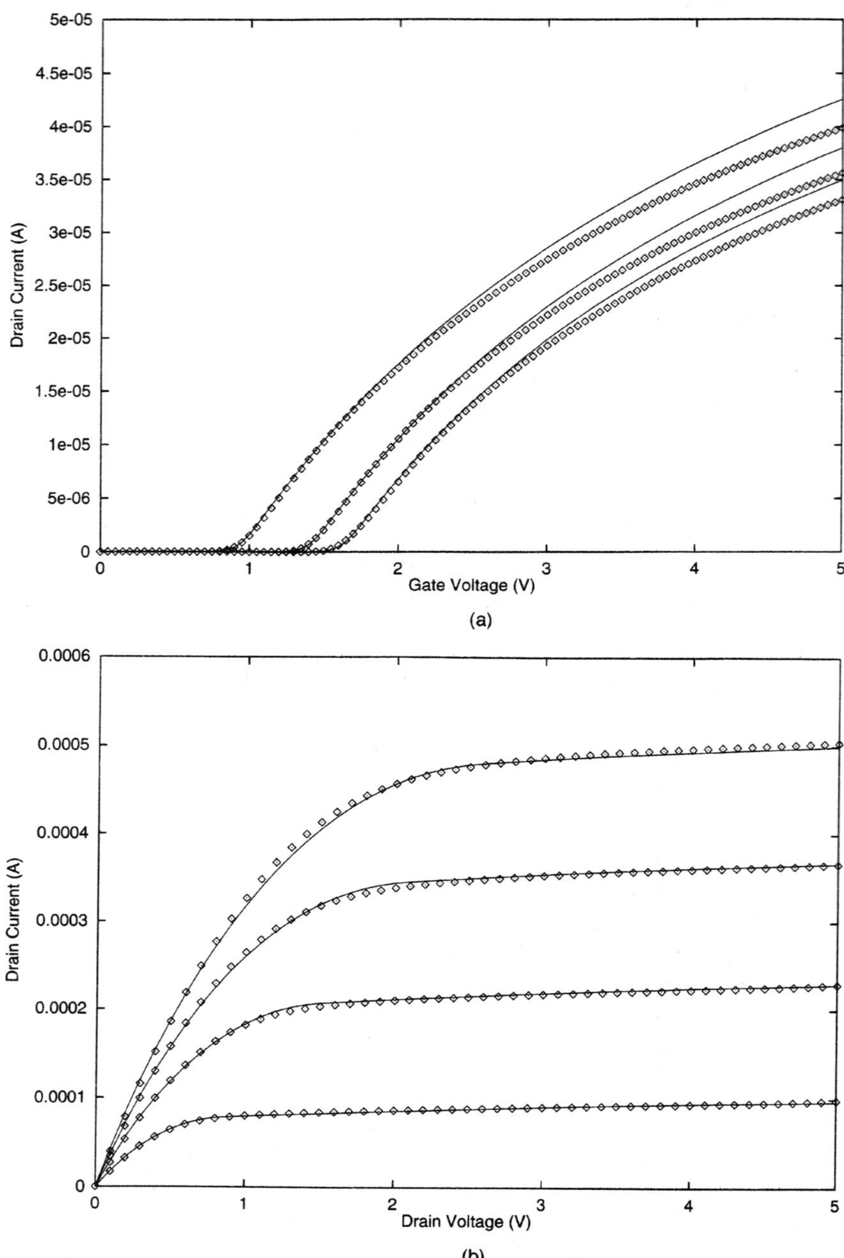

Figure 9.20 Comparison of measured data and the final model for the 1.3/0.7 device; (a) linear characteristic; (b) saturation characteristic.

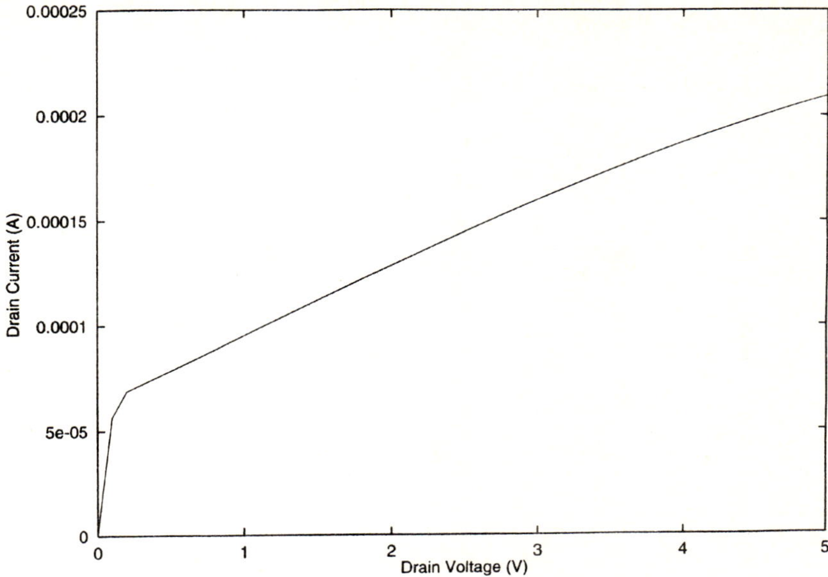

Figure 9.21 Final model result for the 20/0.7 device with a gate bias of $V_{gs} = 1.0$ V. Note that in contrast to the BSIM result (Figure 8.24), there is no negative output conductance for high drain biases.

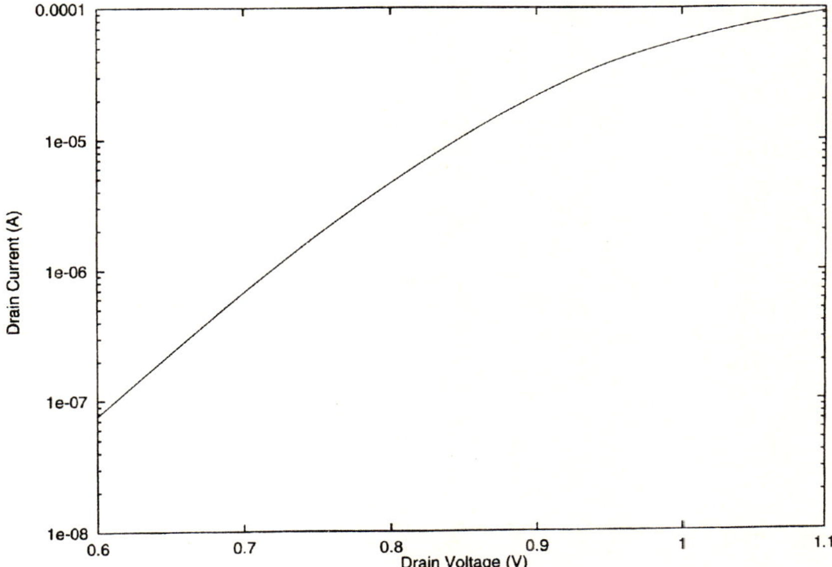

Figure 9.22 Detail of the final model result for the 20/0.7 device in the weak-strong inversion transition region, $V_{ds} = 0.1$ V. Note that the transition region is smooth, while in the BSIM result (Figure 8.25), there is a jump in the characteristic near $V_{gs} = 0.9$ V.

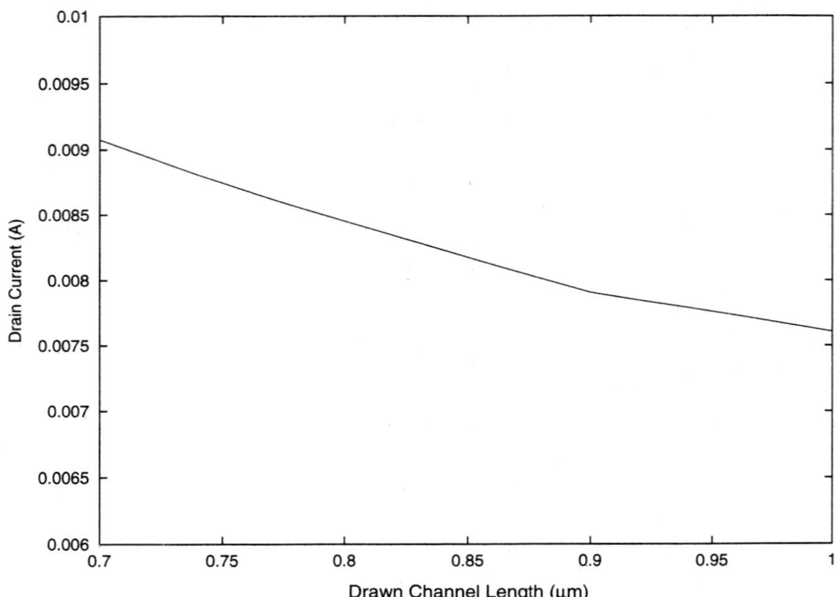

Figure 9.23 The modeled drain current versus channel length, $V_{gs} = V_{ds} = 5.0$ V, showing a continuous result across the bin boundary at $L = 0.9$ μm. This is in contrast to the Level 3 result (Figure 7.29), which shows a discontinuity at this bin boundary.

9.10 Final Comments

As described at the end of Chapter 8, BSIM contains some mathematical weaknesses, and has difficulty in properly describing submicron FETs. HSPICE Level 28 makes progress on these difficulties by taking an empirical approach of mathematically conditioning the BSIM equations. This is essentially a "brute force" method which adds further to the largely empirical character of the second-generation models. A fringe benefit is that the mathematical conditioning results in a model formulation that is suitable for analog circuit design. As will be described in Chapter 10, similar results can be obtained with BSIM2, which also employs mathematical conditioning to reach a model that is suitable for analog circuit design.

Like BSIM, HSPICE Level 28 is a shell, and heavy reliance is placed on parameter extraction to attain the proper model quality. However, in its basic formulation, HSPICE Level 28 does improve on BSIM in several ways. As noted during parameter extraction, better results are obtained in the second phase of parameter extraction, when the individual devices are fitted to the intrinsic model structure. There is no negative output conductance for low gate and high drain biases, and the weak-strong inversion transition region is smooth; these two items are problems in BSIM. There is still some difficulty in obtaining an accurate fit in the linear-saturation transition region of submicron channel length devices; however, the situation is improved over BSIM. The substrate bias dependence of the drain current under high gate and drain biases is quantitatively correct for all device geometries; this is a problem for BSIM in submicron devices. HSPICE Level 28 can also provide useful results for the output conductance; however, *great care* must be taken when fitting the output conductance, as the drain current results are easily corrupted by this effort. Finally, HSPICE Level 28 adds temperature dependence to the basic BSIM structure.

However, the unique innovation in HSPICE Level 28 is that a structure is built around the basic model formulation to accommodate binning in a planned fashion, rather than as an unpleasant after-the-fact necessity. (This approach is also implemented in HSPICE Level 13 (BSIM) and HSPICE Level 39 (BSIM2)). The use of $L_{ref,eff}$ and $W_{ref,eff}$ permits the simple division of the length and width geometry space into subregions (bins), and also provides drain current continuity as, respectively, the channel length and width are varied along the lines of the device length/width matrix. The further introduction of the product term improves the overall accuracy, and also guarantees current continuity across all bins. A guarantee of current continuity as the device geometry changes is particularly important if the channel length and/or width are to be optimized for a particular design (an increasingly common practice), and when statistical simulations are carried out near a bin boundary.

The extrinsic structure, with its matrix of devices, allows HSPICE Level 28 to provide a good description of short, narrow devices. In addition, in contrast to BSIM, this extrinsic structure allows for much better results for intermediate geometry values. However, to properly make use of this extrinsic structure, a large number of devices are required. These devices must be available in the test structures which are used for model building. In addition, the use of a large number of devices increases the workload in data collection and parameter extraction.

Since HSPICE Level 28 is a second-generation model, it suffers from an important second-generation shortcoming. These models are very empirical and very complex, and as a result contain a very large number of equations. As a result, HSPICE Level 28, like BSIM and BSIM2, can be slow and inefficient during circuit simulation.

The largest practical difficulty with HSPICE Level 28 is that it locks the user to the HSPICE circuit simulator. This is more than a problem with business choices or the aesthetics of a particular circuit simulator. Many computer-aided design (CAD) packages will not interface with HSPICE, preventing use of HSPICE Level 28 with them. In addition, several analog circuit design CAD packages make further use of the FET model formulation, and therefore must contain the model equations; this is not currently possible with HSPICE Level 28. BSIM2 provides a second-generation alternative in this situation.

BIBLIOGRAPHY

1. M. Jeng, "Design and Modeling of Deep Submicrometer MOSFETs," University of California/Berkeley, Electronics Research Laboratory Memorandum No. UCB/ERL M90/90 (1990).

2. *HSPICE User's Manual*, Meta-Software, Inc., Campbell, California, 1993.

3. H. deGraaff and F. Klaassen, *Compact Transistor Modeling for Circuit Design*, Springer-Verlag, 1990.

10
BSIM2

10.1 Introduction

As described in Chapters 8 and 9, although BSIM represents a major improvement over the first-generation models, it has some deficiencies that have become more significant as device dimensions have continued to decrease. In particular, as shown in Section 8.10, the ability to fit measured data erodes for effective channel lengths below about 1 μm and/or oxide thicknesses below about 15 nm. In addition, the BSIM formulation attempts to simplify the model equations for improved circuit simulation performance by replacing various physical expressions with polynomial equations. The polynomial equations can behave poorly, often preventing a circuit simulation from converging and leading to a negative output conductance for low gate biases. Finally, the weak inversion current expression is simply added to the strong inversion current equation; this causes a kink in the drain current I_{ds} at the threshold voltage V_t, which can lead to convergence problems. Furthermore, BSIM was developed before serious interest in analog CMOS circuit design appeared. The difficulties described above are major problems which inhibit the use of BSIM for analog designs.

These shortcomings and omissions of BSIM prompted two responses. Meta-Software used BSIM as a base to develop its very successful HSPICE Level 28 model, as detailed in Chapter 9. The University of California/Berkeley responded with BSIM2 [1].

BSIM2 is closely based on BSIM; like HSPICE Level 28, it can be regarded as BSIM with extensions. Much of the intrinsic BSIM structure is retained, while the extrinsic structure is identical, using the composite expression

$$X = X_o + \frac{LX}{L_{eff}} + \frac{WX}{W_{eff}} \tag{10.1}$$

to define the model parameters. For the first time in a SPICE FET model, extensive use is made of two-dimensional analysis to arrive at simpler analytical expressions for various types of device behavior. The BSIM bulk charge model is used, and the BSIM threshold voltage model is employed with only minor modifications. Although the BSIM base is taken as the starting point, extensive modifications are made to both the mobility model and the drain current equations. An entirely new model for the subthreshold current is included, and a transition region around the weak inversion-strong inversion boundary, similar to that developed in HSPICE Level 28, is introduced. Finally, several new equations and parameters are added to describe the output conductance/resistance. This addition to the intrinsic model structure makes BSIM2 suitable for analog circuit design.

Like BSIM, BSIM2 contains no additions to account for variations in the operating temperature. Some parameters contain built-in temperature dependence, but this is inadequate to properly describe the device behavior. Several years after BSIM was introduced, a method of adding temperature dependence to the model formulation was published [2]; however, this scheme was never added to the general SPICE code. It is likely that this approach would also work with BSIM2. The HSPICE implementation of BSIM2 (HSPICE Level 39) includes parameters to account for the temperature dependence of the device behavior [3], in a manner similar to that already described for HSPICE Level 28.

The parameters used in BSIM2 are listed in Table 10.1. The temperature parameters included in HSPICE Level 39 are listed in Table 10.2. It should also be noted that several parameters in HSPICE Level 39 have different names for the first of the three triplet parameters; these parameters are listed in Table 10.3. In all cases, the corresponding L and W parameter names are unchanged. In addition, HSPICE Level 39 makes available the product terms described in Chapter 9 (see (9.1)).

10.2   The Depletion Charge Model

The BSIM2 depletion charge model is identical to that developed for BSIM. As in all the earlier models, a fraction is computed to describe the reduction of the bulk charge from that of a simple two-dimensional MOS capacitor. In this case,

$$Q_{depl,BSIM} = f_d \cdot Q_{depl,2D} = \left(1 - \frac{K_f}{L_{eff} N_A^{\frac{1}{2}}}\right) \cdot (2\epsilon_{Si} q N_A \phi_s)^{\frac{1}{2}}, \qquad (10.2)$$

where K_f is a constant which need not be defined, as it is later absorbed into the model parameter $K1$. As in BSIM, there is no consideration of the depletion charge along the channel edges; this effect is accounted for during parameter extraction and model development.

10.3   The Threshold Voltage

The BSIM2 threshold voltage expression is virtually identical to the BSIM description. It employs the same basic threshold voltage expression,

Table 10.1   The BSIM2 model parameter set.

Parameter	Units	Description
Process Parameters		
TOX	m	Gate Oxide Thickness
LD	m	Source/Drain Underdiffusion of Gate
WD	m	Isolation Reduction of Channel Width
VDD	V	Maximum Applied Supply Voltage
Electrical Parameters		
MU0	$\text{cm}^2/\text{V·s}$	Zero Bias Low Field Mobility
MU0B	$\text{cm}^2/\text{V}^2\text{·s}$	Substrate Bias Effect on the Low Field Mobility
MUS0	$\text{cm}^2/\text{V·s}$	High Drain Bias Mobility
MUSB	$\text{cm}^2/\text{V}^2\text{·s}$	Substrate Bias Effect on the High Drain Bias Mobility
UA0	V^{-1}	Gate Field Induced Mobility Reduction Parameter (First Order)
UAB	V^{-2}	Substrate Bias Effect on the Gate Field Mobility Reduction (First Order)
UB0	V^{-2}	Gate Field Induced Mobility Reduction Parameter (Second Order)
UBB	V^{-3}	Substrate Bias Effect on the Gate Field Mobility Reduction (Second Order)
VFB	V	Flatband Voltage
PHI	V	Surface Potential $(= 2\phi_f)$
K1	$\text{V}^{\frac{1}{2}}$	Body Effect on Threshold Voltage (First-Order Term)
K2		Body Effect on Threshold Voltage (Second-Order Term)
ETA		DIBL Coefficient When $V_{ds} = $ **VDDM** and $V_{bs} = 0$
ETAB	V^{-1}	Substrate Bias Effect on DIBL Coefficient
U10	V^{-1}	High Drain Field Mobility Reduction Parameter
U1B	V^{-2}	Substrate Bias Effect on the High Drain Field Mobility Reduction
U1D	V^{-2}	Drain Bias Effect on the High Drain Field Mobility Reduction
VOF	V	Threshold Voltage Offset
VOFB	V^{-2}	Substrate Bias Effect on the Threshold Voltage Offset
VOFD	V^{-2}	Drain Bias Effect on the Threshold Voltage Offset
VGLOW	V	Lower Weak to Strong Inversion Transition Voltage
VGHIGH	V	Upper Weak to Strong Inversion Transition Voltage
N0		Low Field Subthreshold Ideality Factor
NB	$\text{V}^{\frac{1}{2}}$	Substrate Bias Effect on the Subthreshold Ideality Factor
ND	V^{-1}	Drain Bias Effect on the Subthreshold Ideality Factor
MU20		Output Conductance Parameter 2
MU2B	V^{-1}	Substrate Bias Effect on Output Conductance Parameter 2
MU2G	V^{-1}	Gate Bias Effect on Output Conductance Parameter 2
MU30	$\text{cm}^2/\text{V}^2\text{·s}$	Output Conductance Parameter 3
MU3B	$\text{cm}^2/\text{V}^3\text{·s}$	Substrate Bias Effect on Output Conductance Parameter 3
MU3G	$\text{cm}^2/\text{V}^3\text{·s}$	Gate Bias Effect on Output Conductance Parameter 3
MU40	$\text{cm}^2/\text{V}^3\text{·s}$	Output Conductance Parameter 4
MU4B	$\text{cm}^2/\text{V}^4\text{·s}$	Substrate Bias Effect on Output Conductance Parameter 4
MU4G	$\text{cm}^2/\text{V}^4\text{·s}$	Gate Bias Effect on Output Conductance Parameter 4
AI0		Impact Ionization Coefficient Parameter
AIB	V^{-1}	Substrate Bias Effect on Impact Ionization Coefficient Parameter
BI0		Impact Ionization Exponent Parameter
BIB	V^{-1}	Substrate Bias Effect on Impact Ionization Exponent Parameter
CGSO	F/m	Zero Bias Gate-Source Capacitance
CGDO	F/m	Zero Bias Gate-Drain Capacitance
CGBO	F/m	Zero Bias Gate-Bulk Capacitance

Table 10.2 Temperature-dependent parameters introduced into the HSPICE implementation of BSIM2 (HSPICE Level 39).

Parameter	Units	Description
BEX		Temperature Effect on the Low Drain Bias Mobility
TCV	V/K	Threshold Voltage Variation with Temperature
FEX		Temperature Effect on the High Drain Bias Behavior

Table 10.3 Differences in parameter names between standard BSIM2 and the HSPICE implementation of BSIM2 (HSPICE Level 39).

BSIM2 Parameter Name	HSPICE Level 39 Parameter Name
MU0	*MU0*
ETA	**ETA0**
VOF	**VOF0**

$$V_t = VFB + PHI + K1 \cdot (PHI - V_{bs})^{\frac{1}{2}} - K2 \cdot (PHI - V_{bs}) - \eta \cdot V_{ds}. \qquad (10.3)$$

As in BSIM, the addition of another term, quadratic in $(PHI - V_{bs})^{\frac{1}{2}}$, improves the model accuracy. However, a change is made to the drain-induced barrier lowering (DIBL) term η. In BSIM, η is written as

$$\eta = ETA + X2E \cdot V_{bs} + X3E \cdot (V_{ds} - \mathbf{VDD}). \qquad (10.4)$$

Use of (10.4) in (10.3) causes the threshold voltage expression to be quadratic in V_{ds}. This quadratic can go in the wrong direction, leading to problems during circuit simulation. In particular, if $X3E$ has the wrong sign, the output conductance g_{ds} can be negative at low currents. In BSIM2, the V_{ds} term in (10.4) is simply dropped, and $X2E$ is renamed. The BSIM2 expression for η is

$$\eta = ETA + ETAB \cdot V_{bs}. \qquad (10.5)$$

Temperature Dependence

As noted at the beginning of this chapter, the original formulation of BSIM2 contains no temperature dependence. The HSPICE implementation of BSIM2 (HSPICE Level 39) contains modifications to include temperature dependence, in a manner that is available for all the HSPICE implementations of the second-generation models.

The temperature dependence of the threshold voltage comes in two parts: the built-in temperature dependence of the surface potential parameter *PHI*, and the independent temperature parameter **TCV**. Using the simple definition of the Fermi potential (3.50), *PHI* implies a substrate doping concentration via

$$PHI(T_{nom}) = 2 \cdot \frac{k_b T_{nom}}{q} \cdot ln\left(\frac{N_A}{n_i}\right);$$
(10.6)

Equation (10.6) can be solved for N_A; using this value of N_A, a new surface potential is easily computed from

$$PHI(T) = 2 \cdot \frac{k_b T}{q} \cdot ln\left(\frac{N_A}{n_i}\right).$$
(10.7)

Equation (10.7) is inserted into the threshold voltage expression (10.3).

The use of the temperature parameter **TCV** is quite simple, although its implementation is more convoluted than the method used in HSPICE Level 13 and HSPICE Level 28. For a given temperature T, the threshold voltage is given by

$$V_t(T) = V_{bi}(T) + K1 \cdot (\phi_s - V_{bs})^{\frac{1}{2}} - K2 \cdot (\phi_s - V_{bs}) - ETA \cdot V_{ds},$$
(10.8)

where V_{bi} is the built-in voltage, to be defined below. Next define V_{to} as the threshold voltage for zero substrate bias. Setting $V_{bs} = 0$ in (10.3) leads to

$$V_{to}(T) = VFB + PHI + K1 \cdot (\phi_s)^{\frac{1}{2}} - K2 \cdot \phi_s = V_{bi}(T) + K1 \cdot (\phi_s)^{\frac{1}{2}} - K2 \cdot \phi_s.$$
(10.9)

This allows the built-in voltage V_{bi} in (10.8) to be written as

$$V_{bi}(T) = V_{to}(T) - K1 \cdot (\phi_s)^{\frac{1}{2}} + K2 \cdot \phi_s.$$
(10.10)

The temperature parameter **TCV** is introduced into the expression for the zero substrate bias threshold voltage; this is in contrast to the approach in HSPICE Level 13 and HSPICE Level 28, where **TCV** was introduced into the full threshold voltage expression. As in the previous cases, **TCV** describes the variation of the threshold voltage with temperature:

$$V_{to}(T) = V_{to}(T_{nom}) - \mathbf{TCV} \cdot (T - T_{nom}).$$
(10.11)

As in the earlier cases, the simple but reasonable assumption inherent in (10.11) is that the threshold voltage varies linearly with temperature. Equation (10.11) is inserted into (10.10), which is then placed into (10.8) to compute the threshold voltage $V_t(T)$. The parameter **TCV** is easily extracted from data.

10.4   The Mobility Model

The BSIM2 mobility model is also very similar to the BSIM description. However, an important addition is made which improves the model accuracy for high vertical fields.

As in BSIM, the low gate bias mobility μ_o is computed by extrapolation. In contrast to the three points used by BSIM, BSIM2 uses only two points. For a low drain bias, an expression identical to that used in BSIM (with the parameters renamed in BSIM2) is used:

$$\mu_o\Big|_{V_{ds}=0} = \mathbf{MU0} + MU0B \cdot V_{bs}. \tag{10.12}$$

In BSIM2, **MU0** is treated as the basic low field mobility for a long, wide FET, and has no length and width sensitivity parameters. In the HSPICE implementation of BSIM2 (HSPICE Level 39), the same term is treated as $MU0$, with the associated length, width, and product sensitivity terms.

The second point is also identical to the corresponding BSIM expression, with the parameter names changed slightly in BSIM2:

$$\mu_o\Big|_{V_{ds}=\mathbf{VDD}} = MUS0 + MUSB \cdot V_{bs}. \tag{10.13}$$

The parameter **VDD** represents the maximum voltage for the technology under consideration. For example, if a JEDEC standard of 3.3 V $\pm$ 0.3 V is being employed, **VDD** = 3.6, and data for model building should be available up to and including this voltage. BSIM2 also introduces parameters for the maximum gate and substrate voltages (**VGG** and **VBB**), but these do not appear in the model equations, and the rationale for their introduction is not clear.

In BSIM, a third extrapolation point was used; this is the slope of the mobility computed from (10.13), as described by the parameter $X3MS$. This third point and the associated parameter are not used in BSIM2.

10.4.1   Mobility Reduction by the Vertical Field

BSIM2 introduces a new term into the description of mobility reduction by the vertical field which improves the model accuracy. In BSIM, this phenomenon was described by a model virtually identical to that introduced in Level 3:

$$\mu = \frac{\mu_o}{1 + U_o \cdot (V_{gs} - V_t)}, \tag{10.14}$$

where U_o is a composite parameter, as described by (8.31). In BSIM2, another term is added to the denominator, which has the effect of adding one more term to a power series expansion:

$$\mu = \frac{\mu_o}{1 + U_a \cdot (V_{gs} - V_t) + U_b \cdot (V_{gs} - V_t)^2}. \tag{10.15}$$

The terms U_a and U_b are composite parameters, described by

$$U_a = UA0 + UAB \cdot V_{bs}, \tag{10.16}$$

and

$$U_b = UB0 + UBB \cdot V_{bs}.$$ (10.17)

Equation (10.15) describes the behavior of the drain current versus gate voltage for a low drain bias. However, all four parameters in (10.16) and (10.17) have length and width dependence, and the parameters are extracted for long and short channel devices. This absorbs the series resistance into the mobility model, which actually simplifies model construction.

10.4.2   Mobility Reduction by the Lateral Field

The BSIM2 mobility model includes terms which describe the reduction of the mobility by the lateral field. However, as in BSIM, these terms are intertwined with the drain current model, and thus will be considered during that discussion.

Temperature Dependence

Like BSIM, BSIM2 has no parameters to describe the behavior of the device characteristics for operating temperatures other than the nominal temperature T_{nom}, the temperature at which the parameters were extracted and the model constructed. The HSPICE implementation of BSIM2 (HSPICE Level 39) accounts for the temperature dependence of the mobility in the same manner as that used in the other second-generation models:

$$\mu(T) = \mu(T_{nom}) \cdot \left(\frac{T}{T_{nom}} \right)^{\textbf{BEX}},$$ (10.18)

where T and T_{nom} are in **degrees Kelvin**. In BSIM2, the mobility model is modified to allow a description of the output resistance/conductance; a full discussion of the temperature dependence of the mobility is thus postponed until that situation is considered in Section 10.6.

10.5   The Drain Current Model

Although loosely based on BSIM, the BSIM2 drain current model is substantially revised. The changes introduced through the mobility model were already discussed above. In addition, new expressions for velocity saturation and the subthreshold current are introduced, and modifications are made to the strong inversion current equations. These items will all be considered in turn.

10.5.1   Velocity Saturation

In the earlier models (Level 2, Level 3, and BSIM), a simple expression for the relationship between the carrier velocity and the mobility was employed [4, 5]:

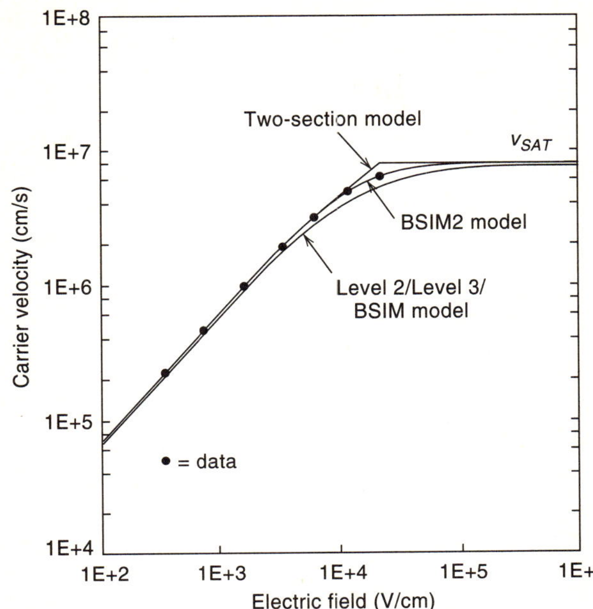

Figure 10.1   The carrier velocity versus the lateral channel field, computed using three different methods. The compromise approach used in BSIM2 provides improved agreement with the available data. From [1]. © 1990. The Regents of the University of California. Used by permission.

$$v = \frac{\mu}{1 + \frac{E_y}{E_c}}. \tag{10.19}$$

This allowed a definition of the critical field E_c as

$$E_c = \frac{v_{SAT}}{\mu}, \tag{10.20}$$

as discussed in detail in Chapter 7. This description of velocity saturation leads to relatively simple drain current equations, but has two shortcomings. First, as shown in Figure 10.1, this approach underestimates the carrier velocity in the region just below E_c. Second, note by examining (10.19) that when the mobility μ is reduced by the gate field, the carrier velocity is decreased. This implies that as the gate bias increases, the saturation velocity v_{SAT} decreases. Measurements show that the saturation velocity is not a function of the gate bias, but is constant for a given temperature.

An alternative description [6] uses a two-section approach to avoid these problems. The critical field is defined as

$$E_c = \frac{2v_{SAT}}{\mu}. \tag{10.21}$$

Equation (10.19) is employed for $E_y \leq E_c$, while for $E_y > E_c$ the velocity is simply set equal to v_{SAT}. As shown in Figure 10.1, this method is more accurate than the previously used model, but the ad hoc change at E_c causes the derivative of the drain current $\frac{\partial I_{ds}}{\partial V_{ds}}$ to be discontinuous at E_c; this can lead to computational difficulties during circuit simulation.

BSIM2 adopts a mathematically empirical compromise that avoids the difficulties of both alternatives discussed above. The relationship between the critical field and the saturation velocity is

$$E_c = E_{co}\left[1 + E_{cd} \cdot \frac{(V_{ds} - V_{dsat})^2}{V_{dsat}^2}\right] \qquad V_{ds} \le V_{dsat}, \qquad (10.22)$$

$$E_c = E_{co} \qquad V_{ds} > V_{dsat}. \qquad (10.23)$$

This produces a better fit to available data, as shown in Figure 10.1. The term E_{co} is a composite parameter, defined as

$$E_{co} = U10 + U1B \cdot V_{bs}. \qquad (10.24)$$

The term E_{cd} is a model parameter, $U1D$, so (10.22) should be rewritten as

$$E_c = E_{co}\left[1 + U1D \cdot \frac{(V_{ds} - V_{dsat})^2}{V_{dsat}^2}\right] \qquad V_{ds} \le V_{dsat}. \qquad (10.25)$$

The quadratic term $(V_{ds} - V_{dsat})^2$ in (10.25) guarantees that the derivative $\frac{\partial I_{ds}}{\partial V_{ds}}$ will be continuous at $V_{ds} = V_{dsat}$.

10.5.2 The Linear Region Drain Current

In BSIM, the expression for the drain current in the linear region is

$$I_{ds} = \frac{\mu_o C_{ox} W_{eff}}{L_{eff}} \cdot \frac{\left[(V_{gs} - V_t) \cdot V_{ds} - \frac{a}{2} \cdot V_{ds}^2\right]}{\left[1 + U_o \cdot (V_{gs} - V_t)\right]\left[1 + \frac{U_1}{L_{eff}} \cdot V_{ds}\right]}. \qquad (10.26)$$

Note that the U_o and U_1 terms in the denominator of (10.26) are multiplied together. It has been shown [1] that this approach loses accuracy for effective channel lengths below about 1 μm; this is one of the major weaknesses of BSIM. In BSIM2, this multiplicative approach is replaced with a more accurate [1] additive expression; the modified mobility reduction equation (10.15) is also included. Thus, the BSIM2 linear region current expression, as modified from (10.26), is

$$I_{ds} = \frac{\mu_o C_{ox} W_{eff}}{L_{eff}} \cdot \frac{\left[(V_{gs} - V_t) \cdot V_{ds} - \frac{a}{2} \cdot V_{ds}^2\right]}{\left[1 + U_a \cdot (V_{gs} - V_t) + U_b \cdot (V_{gs} - V_t)^2\right] + \frac{U_1}{L_{eff}} \cdot V_{ds}}, \qquad (10.27)$$

where U_1 is a composite parameter, described by (10.22) and (10.23) as

$$U_1 = (U10 + U1B \cdot V_{bs}) \cdot \left[1 + U1D \cdot \frac{(V_{ds} - V_{dsat})^2}{V_{dsat}^2}\right] \qquad (10.28)$$

for the linear region ($V_{ds} \leq V_{dsat}$). The term a was derived in Chapter 8 (see (8.73)):

$$a = 1 + \frac{g \cdot K1}{2(PHI - V_{bs})^{\frac{1}{2}}}, \qquad (10.29)$$

with

$$g = 1 - \frac{1}{1.744 + 0.8364 \cdot (PHI - V_{bs})}. \qquad (10.30)$$

These terms were introduced in BSIM as a mathematical method of simplifying the $\frac{3}{2}$-power terms that appear in the full derivation of the FET bulk charge (see Chapters 6 and 8).

10.5.3   The Saturation Region Drain Current

In BSIM, a separate expression for the drain current in saturation was derived:

$$I_{ds} = \frac{\mu_o C_{ox} W_{eff}}{L_{eff}} \cdot \frac{1}{2 \cdot a \cdot K} \cdot \frac{(V_{gs} - V_t)^2}{[1 + U_o \cdot (V_{gs} - V_t)]}, \qquad (10.31)$$

where a is defined by (10.29), while K is given by

$$K = \frac{1}{2}\left[1 + V_c + (1 + 2V_c)^{\frac{1}{2}}\right]. \qquad (10.32)$$

The expression for the critical voltage V_c is

$$V_c = \frac{U_1}{L_{eff}} \cdot \frac{(V_{gs} - V_t)}{a}. \qquad (10.33)$$

The use of the linear region expression (10.26) is separated from the use of the saturation region expression by the saturation voltage description:

$$V_{dsat} = \frac{(V_{gs} - V_t)}{a \cdot K^{\frac{1}{2}}}. \qquad (10.34)$$

In BSIM2, a separate saturation current expression is not used. Instead, the linear region expression (10.27) is employed, with V_{ds} replaced by V_{dsat}:

$$I_{ds} = \frac{\mu_o C_{ox} W_{eff}}{L_{eff}} \cdot \frac{\left[(V_{gs} - V_t) \cdot V_{dsat} - \frac{a}{2} \cdot V_{dsat}^2\right]}{[1 + U_a \cdot (V_{gs} - V_t) + U_b \cdot (V_{gs} - V_t)^2] + \frac{U_-}{L_{ef}} \cdot V_{dsat}}, \qquad (10.35)$$

where V_{dsat} is given by (10.34) and K by (10.32). Here, the composite parameter U_1 is described by (10.23) and (10.24):

$$U_1 = U10 + U1B \cdot V_{bs}.$$ (10.36)

(Note, in comparing (10.28) and (10.36), that U_1 takes very different forms in the linear region and the saturation region.) However, since a different drain current equation (versus BSIM) is used, a different expression for V_c is developed. By working through a derivation similar to that carried out in Chapter 8 to produce (10.33), it can be shown that

$$V_c = \frac{U_1}{L_{eff} \cdot \left[1 + U_a \cdot (V_{gs} - V_t) + U_b \cdot (V_{gs} - V_t)^2\right]} \cdot \frac{(V_{gs} - V_t)}{a},$$ (10.37)

where U_1 is given by (10.36).

Temperature Dependence

As noted above several times, the original formulations of both BSIM and BSIM2 contain no parameters to allow the model to have temperature dependence. The HSPICE implementation of BSIM2 (HSPICE Level 39) adds temperature dependence to the model structure in a manner similar to that already described for HSPICE Level 13 (Chapter 8) and HSPICE Level 28 (Chapter 9).

The temperature dependence of the high drain bias behavior is described by the parameter **FEX**. Two parameters, $U10$ and $U1B$, make use of **FEX**, in the form

$$U1(T) = U1(T_{nom}) \cdot \left(\frac{T}{T_{nom}}\right)^{\textbf{FEX}},$$ (10.38)

where T and T_{nom} are in **degrees Kelvin**. **FEX** is easily determined during parameter extraction.

10.5.4 The Subthreshold Current

In a departure from the earlier FET models, BSIM2 implements a well-respected charge sheet model for the subthreshold current developed by Brews [7]. The original expression describes the current in the deep subthreshold region, where conduction can be attributed entirely to diffusion:

$$I_{ds} = \frac{3}{2} \frac{\mu_o C_{ox} W_{eff}}{L_{eff}} \cdot \left(\frac{k_b T}{q}\right)^2 \frac{C_{depl}}{C_{ox}} \left(\frac{q}{k_b T}\right) e^{(\phi_s - 2\phi_p)} \cdot \left[1 - e^{\frac{q}{k_b T} \cdot V_{ds}}\right].$$ (10.39)

Here ϕ_p is the bulk Fermi potential, given by the simple expression

$$\phi_p = \frac{k_b T}{q} ln \frac{N_A}{n_i},$$ (10.40)

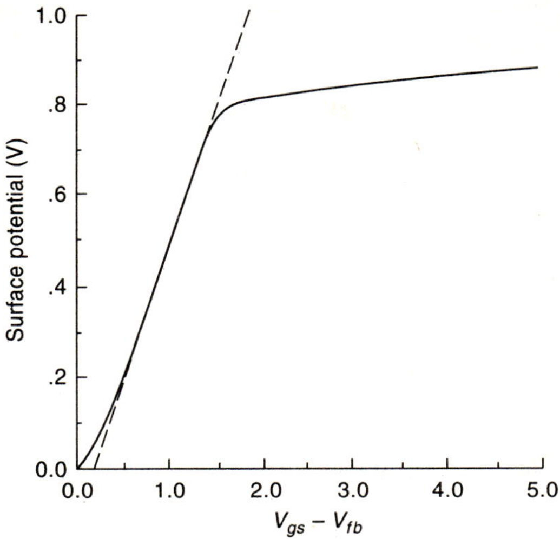

Figure 10.2 The computed surface potential versus ($V_{gs} - V_{fb}$). The dashed line shows a linear expression that is valid in the subthreshold region. From [1]. © 1990. The Regents of the University of California. Used by permission.

while C_{depl} is the gate-induced depletion capacitance, given by

$$C_{depl} = -\frac{dQ_{depl}}{d\phi_s} = \left(\frac{q\epsilon_{Si}N_A}{2\phi_s}\right)^{\frac{1}{2}}.$$

(10.41)

Both (10.40) and (10.41) assume that the substrate doping is uniform.

Examination of (10.39) indicates that the current is expressed in terms of the surface potential ϕ_s rather than the gate voltage V_{gs}. In order for (10.39) to be implemented in a circuit simulation environment, an expression linking V_{gs} and ϕ_s must be included. Sze [4] derived such an expression, but it is too complicated for use in circuit simulation.

To find a simplified expression, BSIM2 uses a two-dimensional analysis of the surface potential versus the gate bias, as shown in Figure 10.2. In the subthreshold region, where $\phi_s < 2\phi_p$, the surface potential increases virtually linearly with increasing gate bias. This allows a slope-intercept expression for the surface potential as a function of the gate bias to be found:

$$\phi_s = \left[\left(\frac{q}{k_bT}\right)\left(\frac{V_{gs} - V_{fb}}{n}\right)\right] - V_o.$$

(10.42)

The term n is an ideality factor which relates the depletion capacitance and the oxide capacitance; when $\phi_s \sim 2\phi_p$, n can be shown to a reasonable approximation [4] to be

$$n = 1 + \frac{C_{depl}}{C_{ox}}.$$

(10.43)

By substituting (10.43) into (10.42) and carrying out some further manipulations, V_o can be computed to be

$$V_o = 2\phi_p \cdot \frac{C_{depl}}{C_{depl} + C_{ox}}.$$

(10.44)

Note that (10.44) requires a specific value of the substrate doping concentration to uniquely determine ϕ_p.

When (10.42) is used to replace ϕ_s, (10.39) becomes

$$I_{ds} = \frac{3}{2}\frac{\mu_o C_{ox} W_{eff}}{L_{eff}} \cdot \left(\frac{k_b T}{q}\right)^2 \frac{C_{depl}}{C_{ox}}\left(\frac{q}{k_b T}\right)\frac{e^{V_{gs}-V_t-2\phi_p-n\cdot V_o}}{n} \cdot \left[1 - e^{\frac{q}{k_b T}\cdot V_{ds}}\right].$$

(10.45)

By further algebraic manipulation, (10.45) can be simplified to

$$I_{ds} = \frac{\mu_o C_{ox} W_{eff}}{L_{eff}} \cdot \left(\frac{k_b T}{q}\right)\frac{e^{V_{gs}-V_t-V_{off}}}{n} \cdot \left[1 - e^{\frac{q}{k_b T}\cdot V_{ds}}\right].$$

(10.46)

Here many pieces of (10.45) have been subsumed into the offset voltage V_{off}, eliminating the ambiguous and hard to define terms C_{depl} and ϕ_p. The offset voltage V_{off} is a model parameter, and accounts for the offset between the end of the subthreshold region (diffusion conduction) and the beginning of the strong inversion region (drift conduction). It is thus similar to the computed term V_{on} used in the subthreshold current model of Levels 2 and 3, and V_{off}, used in BSIM. In contrast to BSIM, in BSIM2 V_{off} is a composite parameter, defined by

$$V_{off} = VOF + VOFB \cdot V_{bs} + VOFD \cdot V_{ds}.$$

(10.47)

The ideality factor n is also a composite parameter, with the bias dependence defined slightly differently than usual:

$$n = N0 + \frac{NB}{(PHI - V_{bs})^{\frac{1}{2}}} + ND \cdot V_{ds}.$$

(10.48)

Thus far, the BSIM2 subthreshold region model is quite similar to those found in the earlier FET models. However, BSIM2 extends the development with a more detailed description of the transition region between the subthreshold and superthreshold regions. Recalling (3.114), the inversion charge and the gate bias can be simply described by

$$Q_{inv} = C_{ox} \cdot (V_{gs} - V_t).$$

(10.49)

When compared with a numerical simulation (Figure 10.3), (10.49) agrees quite well for gate biases well above the threshold voltage. However, near the threshold voltage, (10.49) becomes less accurate. For gate biases slightly below the threshold voltage, (10.49)

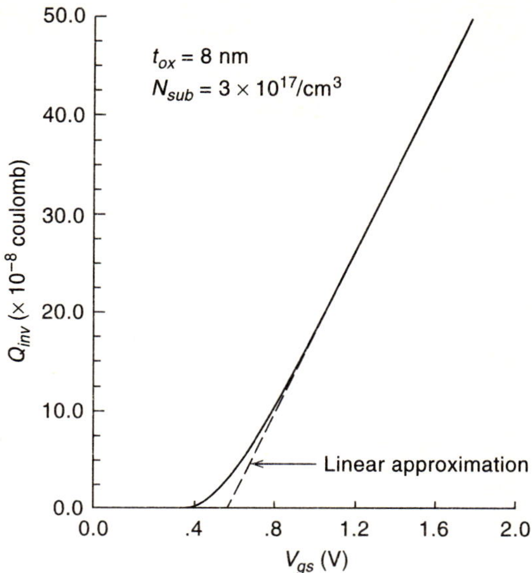

Figure 10.3 The inversion charge density versus the gate bias; the solid line is the result computed from two-dimensional analysis, while the dashed line shows the simple approximation. The error in the linear approximation near the threshold voltage is apparent. From [1]. © 1990. The Regents of the University of California. Used by permission.

Figure 10.4 The equivalent circuit for a MOS capacitor with an inversion layer present. From [1]. © 1990. The Regents of the University of California. Used by permission.

predicts $Q_{inv} = 0$, while the numerical simulation shows that some inversion charge does in fact exist.

BSIM2 includes a model for this transition region in the following manner. Treating the problem as a one-dimensional charge situation between the gate and the substrate, the equivalent circuit of Figure 10.4 can be used. Here, the oxide capacitance C_{ox} exists between the gate and the silicon surface, while the inversion layer capacitance C_{inv} is found between the inversion layer and the silicon bulk. As noted in Figure 10.4, this system is constrained to have the surface potential ϕ_s at the node between C_{ox} and C_{inv}. The total capacitance across this circuit is simply written down from the expression for two capacitors in series:

$$C_{ox}^{\star} = \frac{C_{ox} \cdot C_{inv}}{C_{ox} + C_{inv}}. \tag{10.50}$$

The inversion layer capacitance can be defined as [6]

$$C_{inv} = \frac{d Q_{inv}}{d \phi_s} = C_{depl} \cdot e^{\left(\frac{q}{k_b T}\right) \cdot (\phi_s - 2\phi_p)}. \tag{10.51}$$

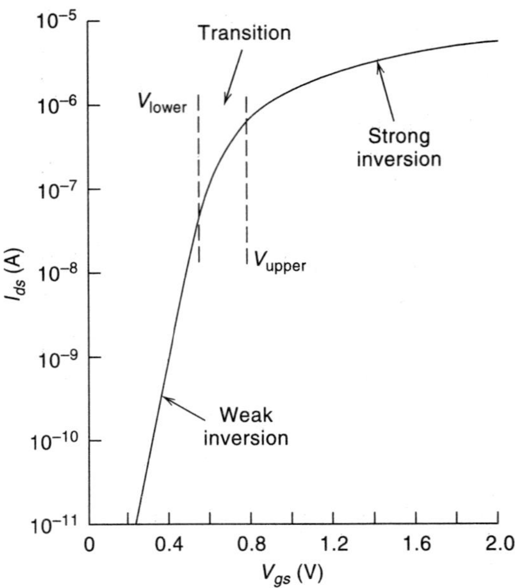

Figure 10.5   The drain current versus the gate bias, showing the weak inversion region, the strong inversion region, and the transition region between them. The terms V_{lower} and V_{upper} define the extent of the transition region. From [1]. © 1990. The Regents of the University of California. Used by permission.

By substituting the expression (10.42) into (10.51) and carrying out the algebraic manipulations used to derive (10.46), it is found that

$$C_{inv} = C_{depl} \cdot e^{\left(\frac{q}{k_b T}\right) \cdot \left(\frac{V_{gs} - V_t - V_{off}}{n}\right)}. \tag{10.52}$$

In strong inversion, $C_{inv} \gg C_{ox}$, and $C_{ox}^{\star} \cong C_{ox}$; the linear approximation (10.49) is valid. In the deep subthreshold region, $C_{inv} \ll C_{ox}$, and $C_{ox}^{\star} \cong C_{inv}$. In the transition region, C_{inv} and C_{ox} are of a comparable magnitude, and are used to define the extent of that region (Figure 10.5). The lower limit of the transition region can be defined by $C_{ox} = (100 \cdot C_{inv})$, leading to [1]

$$V_{gl} = V_t - V_{off} + n \cdot \frac{k_b T}{q} \cdot \ln\left(\frac{C_{ox}}{100 \cdot C_{depl}}\right). \tag{10.53}$$

In a similar fashion, the upper limit of the transition region can be defined by $C_{inv} = (100 \cdot C_{ox})$, producing [1]

$$V_{gu} = V_t + n \cdot \frac{k_b T}{q} \cdot \ln\left(\frac{100 \cdot C_{ox}}{C_{depl}}\right). \tag{10.54}$$

The transition region is generally 0.2 V–0.3 V wide. However, V_{gl} and V_{gu} are rarely determined from (10.53) and (10.54). Instead, V_{gl} and V_{gu} are model parameters (*VGLOW* and *VGHIGH*), which are determined during parameter extraction.

To compute the drain current in the transition region, BSIM2 uses a cubic spline function to define an auxiliary gate voltage V'_{gs} from the actual applied gate bias V_{gs}:

$$V'_{gs} = \alpha_0 + \alpha_1 \cdot V_{gs} + \alpha_2 \cdot V_{gs}^2 + \alpha_3 \cdot V_{gs}^3. \tag{10.55}$$

The strong inversion expression for I_{ds} is then used, but with V'_{gs} taken from (10.55) rather than the actual value of V_{gs}. This allows a single current equation to be used unambiguously in the transition region. The coefficients in (10.55) must be computed in a manner which ensures that the drain current and its derivative $\frac{\partial I_{ds}}{\partial V_{ds}}$ are continuous at both *VGLOW* and *VGHIGH*. At $V_{gs} = VGLOW$, the boundary conditions are [1]

$$V'_{gs} = (2 \cdot a \cdot K)^{\frac{1}{2}} \left(\frac{k_b T}{q} \right) e^{\left(\frac{q}{k_b T} \right)\left(\frac{VGLOW - V_{off}}{2 \cdot n} \right)} \left[1 - e^{-\left(\frac{q}{k_b T} \right) \cdot V_{ds}} \right]^{\frac{1}{2}} \equiv R_c, \tag{10.56}$$

and

$$\frac{dV'_{gs}}{dV_{gs}} = \left(\frac{q}{k_b T} \right) \left(\frac{VGLOW - V_t - V_{off}}{2 \cdot n} \right) \equiv R_d. \tag{10.57}$$

At $V_{gs} = VGHIGH$, the boundary conditions are

$$V'_{gs} = V_{gu}, \tag{10.58}$$

and

$$\frac{dV'_{gs}}{dV_{gs}} = 1. \tag{10.59}$$

The coefficients are determined by solving the determinants of several 3×3 matrices [1]:

$$\alpha_0 = VGHIGH - \alpha 1 \cdot VGHIGH - \alpha_2 \cdot VGHIGH^2 - \alpha_3 \cdot VGHIGH^3, \tag{10.60}$$

$$\alpha_1 = \frac{1}{\Delta} \begin{vmatrix} 1 & 2VGHIGH & 3VGHIGH^2 \\ R_d & 2VGLOW & 3VGLOW^2 \\ R_c & VGLOW^2 + VGHIGH^2 & 2VGHIGH^3 + VGLOW^3 \end{vmatrix}, \tag{10.61}$$

$$\alpha_2 = \frac{1}{\Delta} \begin{vmatrix} 1 & 1 & 3VGHIGH^2 \\ 1 & R_d & 3VGLOW^2 \\ VGLOW & R_c & 2VGHIGH^3 + VGLOW^3 \end{vmatrix}, \tag{10.62}$$

$$\alpha_3 = \frac{1}{\Delta} \begin{vmatrix} 1 & 2VGHIGH & 1 \\ 1 & 2VGLOW & R_d \\ VGLOW & VGLOW^2 + VGHIGH^2 & R_c \end{vmatrix}, \tag{10.63}$$

$$\Delta = \begin{vmatrix} 1 & 2VGHIGH & 3VGHIGH^2 \\ 1 & 2VGLOW & 3VGLOW^2 \\ VGLOW & VGLOW^2 + VGHIGH^2 & 2VGHIGH^3 + VGLOW^3 \end{vmatrix}. \quad (10.64)$$

Examining (10.55)–(10.64), it is clear that this description of the subthreshold to superthreshold transition region is very complicated, and requires intensive matrix computations for each instance in which V_{gs} happens to fall into the transition region between *VGLOW* and *VGHIGH*. Although this approach is able to produce good results, the method is cumbersome and inefficient.

10.6   The Output Resistance Model

Since BSIM was developed before the explosion of interest in analog CMOS circuits, no modifications were included for the modeling of the output conductance/resistance, defined by

$$g_{ds} = \frac{\partial I_{ds}}{\partial V_{ds}}, \quad (10.65)$$

and

$$R_{out} = \frac{1}{g_{ds}} = \frac{\partial V_{ds}}{\partial I_{ds}}. \quad (10.66)$$

This provided a major impetus for the development of HSPICE Level 28 and BSIM2. As described in Chapter 9, HSPICE Level 28 is able to describe the output conductance by properly conditioning the model equations. No parameters are added to the drain current model specifically to deal with the output conductance; instead, those already present are modified to describe g_{ds}. In contrast, BSIM2 adds several new parameters to the model solely to describe the output conductance. In both HSPICE Level 28 (see Chapter 9) and BSIM2, this added requirement puts a great deal of strain on the parameter extraction procedure, as it can be very difficult to properly fit the output conductance without corrupting the original drain current result.

To model the output conductance/resistance, BSIM2 considers three influences: drain-induced barrier lowering, channel length modulation, and the change in the effective substrate potential induced by hot carrier generation. DIBL is modeled by the composite parameter η, which was described in Section 10.3. Additional equations and model parameters are added to the model to describe channel length modulation and the hot carrier induced change in the substrate potential. Their influence on the output resistance is shown in Figure 10.6.

Channel length modulation has been described several times in earlier chapters; it is reprised in Figure 10.7, so as to define the terms to be used below. When the drain bias V_{ds} is slightly greater than the saturation voltage V_{dsat}, channel length modulation

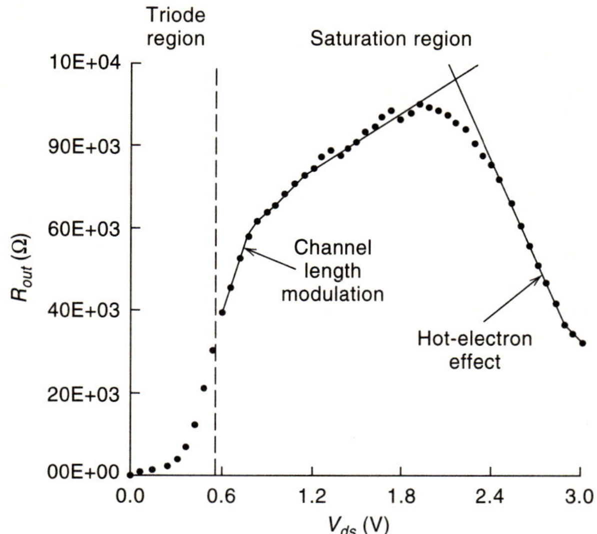

Figure 10.6 The output resistance of a short channel MOSFET, showing the major influences on the various regions of the characteristic. From [1]. © 1990. The Regents of the University of California. Used by permission.

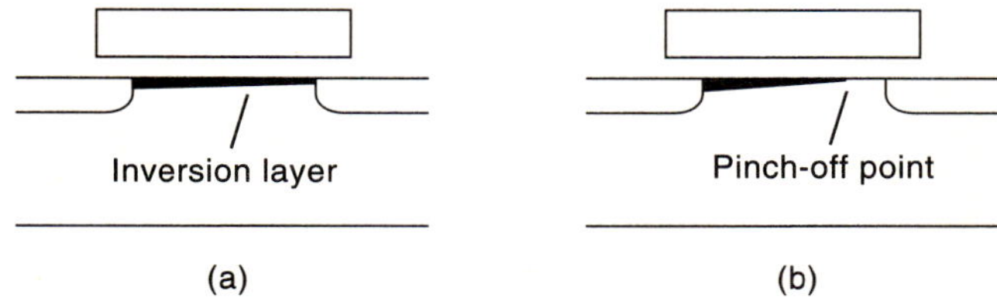

Figure 10.7 Channel length modulation; (a) $V_{ds} = V_{dsat}$; the inversion layer just reaches the drain; (b) $V_{ds} > V_{dsat}$; the inversion layer no longer reaches the drain, and the pinch-off region is present.

dominates the output resistance behavior, as L' increases. BSIM2 adds this effect, but works its way to a somewhat empirical expression with several parameters. To begin, the prefactor which appears in the current equations can be defined as

$$\beta_0 \equiv \frac{\mu_o\Big|_{V_{ds}=0} C_{ox} W_{eff}}{L_{eff}}, \tag{10.67}$$

where the mobility is found from (10.12). When channel length modulation occurs, L_{eff} in (10.67) is replaced by $L_{eff} - L'$. After some algebraic manipulation, a new term, β_0', can be defined by

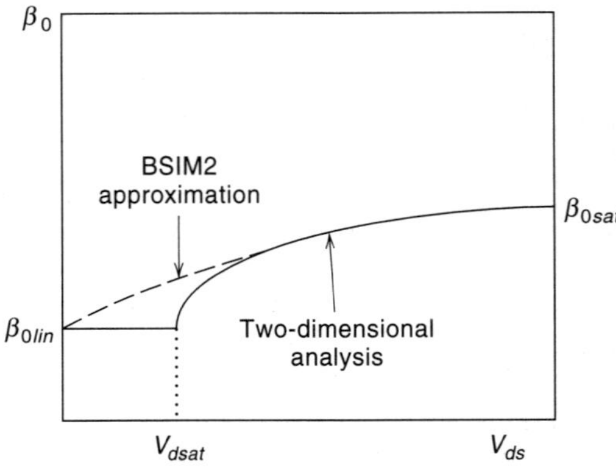

Figure 10.8 A description of the conductance variable β_0 versus the drain bias. The solid line was computed from two-dimensional analysis, while the dashed line is the approximation used in BSIM2. From [1]. © 1990. The Regents of the University of California. Used by permission.

$$\beta_0' = \frac{\mu_o\big|_{V_{ds}=0} C_{ox} W_{eff}}{L_{eff} - L'} = \beta_0 \left(\frac{1}{1 - \frac{L'}{L_{eff}}} \right). \tag{10.68}$$

Allowing for the general case of $L' \ll L_{eff}$, (10.68) can be rewritten as

$$\beta_0' = \beta_0 \left(1 + \frac{L'}{L_{eff}} \right). \tag{10.69}$$

Equation (10.69) is fairly simple, but requires an analytical expression for L'. A description of L' was introduced into Level 2 and Level 3, but it is somewhat complicated. In BSIM2, a two-dimensional analysis is once again used to determine a simpler analytical expression. By definition, $L' = 0$ when $V_{ds} \leq V_{dsat}$. For $V_{ds} > V_{dsat}$, two-dimensional analysis [8], [9] shows that L' increases logarithmically with V_{ds}; the effect on β_0 is depicted in Figure 10.8. In BSIM2, a mathematical function is chosen to approximate β_0', with emphasis on the behavior when $V_{ds} = V_{dsat}$ (Figure 10.7):

$$\beta_0' = \beta_0 + \beta_1 \cdot tanh \left(\frac{\beta_2 \cdot V_{ds}}{V_{dsat}} \right) + \beta_3 \cdot V_{ds} - \beta_4 \cdot V_{ds}^2, \tag{10.70}$$

where β_2, β_3, and β_4 are model parameters that will be described in more detail below. The term β_1 is described by

$$\beta_1 = \beta_s - (\beta_0 + \beta_3 \cdot V_{dd} - \beta_4 \cdot V_{dd}^2), \tag{10.71}$$

where β_s is the saturation region equivalent of β_0, defined as

$$\beta_s = \frac{\mu_o\big|_{V_{ds}=\mathbf{VDD}} C_{ox} W_{eff}}{L_{eff}}, \tag{10.72}$$

where the mobility is defined by (10.13). For completeness, (10.71) should be rewritten with the parameter names boldfaced:

$$\beta_1 = \beta_s - (\beta_0 + \beta_3 \cdot \mathbf{VDD} - \beta_4 \cdot \mathbf{VDD}^2). \tag{10.73}$$

The model parameters β_2, β_3, and β_4 are composite parameters, which, by an unfortunate choice of terminology, use *MU* in their names and can be confused with the mobility parameters:

$$\beta_2 = MU20 + MU2B \cdot V_{bs} + MU2G \cdot V_{gs}, \tag{10.74}$$

$$\beta_3 = MU30 + MU3B \cdot V_{bs} + MU3G \cdot V_{gs}, \tag{10.75}$$

$$\beta_4 = MU40 + MU4B \cdot V_{bs} + MU4G \cdot V_{gs}. \tag{10.76}$$

Again, it should be noted that despite the appearance of *MU* in the parameter names of (10.74)–(10.76), the composite parameters β_2, β_3, and β_4 describe the output conductance and are loosely related to the mobility.

The relationship between the output conductance and the mobility is subtle, but is very important to understanding the structure of BSIM2. Recall from (10.67) that the composite term β_0 is defined by

$$\beta_0 \equiv \frac{\mu_o \big|_{V_{ds}=0} C_{ox} W_{eff}}{L_{eff}}. \tag{10.77}$$

The output conductance model is introduced by defining β_0' as a modified version of β_0:

$$\beta_0' = \beta_0 + \beta_1 \cdot tanh\left(\frac{\beta_2 \cdot V_{ds}}{V_{dsat}}\right) + \beta_3 \cdot V_{ds} - \beta_4 \cdot V_{ds}^2, \tag{10.78}$$

where β_1, β_2, β_3, and β_4 provide the terms for the output conductance model. These modifications are all composite terms, defined by (10.73)–(10.76). When (10.78) is used, β_0' replaces β_0 in (10.77). When the output conductance model is neglected, β_1, β_2, β_3, and β_4 are all zero, and β_0' reduces to β_0.

Now reconsider (10.77) by rearranging that expression to an equation for the low drain bias mobility:

$$\mu_o \big|_{V_{ds}=0} = \frac{\beta_0}{C_{ox} \cdot \frac{W_{eff}}{L_{eff}}}. \tag{10.79}$$

This expression describes the low drain bias mobility in the absence of the output conductance terms. When the output conductance terms are added to the model, β_0 is replaced by β_0', and (10.79) becomes

$$\mu_o\bigg|_{V_{ds}=0} = \frac{\beta'_0}{C_{ox} \cdot \frac{W_{eff}}{L_{eff}}}. \tag{10.80}$$

The introduction of the output conductance terms changes the computed value of the low drain bias mobility $\mu_o\big|_{V_{ds}=0}$. Thus, BSIM2 describes the output conductance behavior entirely through a modification to the low drain bias mobility.

For larger drain biases, the output conductance/resistance is largely dominated by a hot electron induced change in the effective substrate potential. Consider a MOSFET operating with a high drain voltage. The high field near the drain accelerates carriers to high energies; this energy is dissipated by impact ionization, generating an electron and a hole. In an nFET, the holes thus generated are swept into the substrate, and have the effect of slightly forward biasing the source-substrate junction. This small bias can be defined as

$$V'_{bs} = I_{sub} \cdot R_{sub}, \tag{10.81}$$

where I_{sub} is the substrate current and R_{sub} is the substrate resistance. This added bias reduces the effective threshold voltage, which increases the drain current I_{ds} above its expected value. (This mechanism is considered in more detail in BSIM3, where it is called the *substrate current induced body effect* (SCBE).)

This change to the drain current can be described by

$$I'_{ds} = I_{ds} \cdot (1 + C \cdot I_{sub} \cdot R_{sub}), \tag{10.82}$$

where C is an arbitrary constant. To simplify (10.82) to a useful model expression, an equation for I_{sub} is taken from a popular model [10]:

$$I_{sub} = I_{ds} \left(\frac{\alpha_i}{\beta_i}\right) \left(\frac{E_{max}}{l}\right) e^{-\beta_i/E_{max}}, \tag{10.83}$$

where α_i and β_i are model coefficients (the subscript i represents *ionization*, and is **not** a counting variable), E_{max} is the maximum field near the drain, defined as

$$E_{max} = \frac{V_{ds} - V_{dsat}}{l}, \tag{10.84}$$

and l is a characteristic length given by

$$l \equiv 0.22 t_{ox}^{\frac{1}{3}} x_j^{\frac{1}{2}}; \tag{10.85}$$

x_j is the drain junction depth. By using (10.83) in (10.82) and performing some algebraic manipulations, (10.82) can be transformed into a workable model equation:

$$I'_{ds} = I_{ds} \left[1 + A_i e^{-B_i/(V_{ds} - V_{dsat})}\right]. \tag{10.86}$$

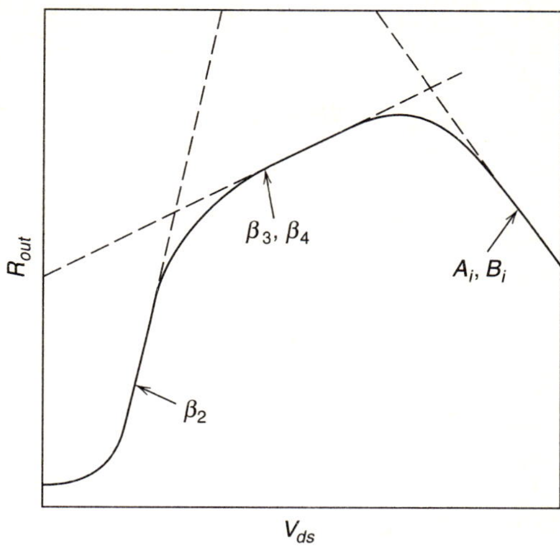

Figure 10.9 The output resistance versus the drain bias, showing where the various parameters dominate the characteristic. From [1]. © 1990. The Regents of the University of California. Used by permission.

The terms A_i and B_i are composite model parameters, given by

$$A_i = AI0 + AIB \cdot V_{bs}, \tag{10.87}$$

$$B_i = BI0 + BIB \cdot V_{bs}. \tag{10.88}$$

(Again, the subscript i represents *ionization*, and is **not** a counting variable.) Under the typical conditions of device operation,

$$A_i e^{-B_i/(V_{ds}-V_{dsat})} \ll 1, \tag{10.89}$$

and this hot carrier induced effect causes little change to the magnitude of the drain current I_{ds}. However, as shown in Figure 10.9, the effect on the output conductance/resistance can be quite important.

Temperature Dependence

As noted throughout this chapter, the original BSIM2 formulation contains no parameters to describe the temperature dependent behavior of the model. The HSPICE implementation of BSIM2 (HSPICE Level 39) adds several parameters to account for temperature variations. The approach to the temperature dependence of the mobility was briefly described in Section 10.4; however, it was noted there that the mobility structure of BSIM2 would not be complete until the output conductance was discussed.

The temperature dependence of the mobility in HSPICE Level 39 is described by

$$\mu_o\Big|_{V_{ds}=0}(T) = \mu_o\Big|_{V_{ds}=0}(T_{nom}) \cdot \left(\frac{T}{T_{nom}}\right)^{\textbf{BEX}}, \tag{10.90}$$

where T and T_{nom} are in **degrees Kelvin**. Thus, the temperature dependence of the mobility is addressed entirely through the low drain bias mobility, defined in (10.80) as

$$\mu_o\Big|_{V_{ds}=0} = \frac{\beta_0'}{C_{ox} \cdot \frac{W_{eff}}{L_{eff}}}, \qquad (10.91)$$

where β_0' contains the terms that constitute the output conductance model. As in the earlier cases, **BEX** is easily extracted from measured data.

10.7   The Charge Model

The original formulation of BSIM2 [1] contained no new model for the device charge. As implemented in SPICE2, BSIM2 used the BSIM charge model [11], which is described in Chapter 8. Later a charge model based on the BSIM2 formulation was developed, and introduced into both University of California/Berkeley SPICE (SPICE3E2) and HSPICE. In SPICE3E2, only this new model can be used when BSIM2 is employed. In HSPICE, the user has the choice of using either this new charge model or the older BSIM model.

Unlike BSIM, the BSIM2 charge model does not offer the user any choice of source/drain charge partitioning schemes. Instead, partitioning is (drain/source) 50%/50% in the linear region and 40%/60% in the saturation region.

Unfortunately, this model was not well documented, and other than by extracting their form directly from the SPICE3E2 code, the only reference for these equations ("on paper") is the HSPICE manual [3]. Thus, the BSIM2 charge model equations are presented here in a very terse form.

10.7.1   The Accumulation Region

In the accumulation region ($V_{gs} < V_{bs} + VFB$), the node charge equations are

$$Q_{GATE} = C_{ox} \cdot W_{eff} \cdot L_{eff} \cdot (V_{gs} - V_{bs} - VFB), \qquad (10.92)$$

$$Q_{BULK} = -Q_{GATE}, \qquad (10.93)$$

and

$$Q_{INV} = 0. \qquad (10.94)$$

10.7.2   The Subthreshold Region

In the subthreshold region ($V_{bs} + VFB < V_{gs} < V_t + VGLOW$), the charge equations become more complicated. Here,

$$Q_{GATE} = C_{ox} \cdot W_{eff} \cdot L_{eff} \cdot (V_{gs} - V_{bs} - VFB) \cdot \left\{ 1 - \frac{V_{gs} - V_{fb} - VFB}{V_{gs} - V_{bs} - VFB - (V_{gs} - V_t)} \right.$$

$$\left. + \frac{1}{3} \cdot \left[\frac{V_{gs} - V_{bs} - VFB}{V_{gs} - V_{bs} - VFB - (V_{gs} - V_t)} \right]^2 \right\}, \tag{10.95}$$

$$Q_{DEPL} = -Q_{GATE}, \tag{10.96}$$

and

$$Q_{INV} = 0. \tag{10.97}$$

10.7.3   The Linear Region

In the linear region ($V_{ds} < V_{dsat}$), the inversion charge is partitioned equally (50%/50%) between the drain and the source. The charge equations are

$$Q_{GATE} = \frac{2}{3} \cdot C_{ox} \cdot W_{eff} \cdot L_{eff} \cdot (V_{gs} - V_t) \cdot \left[\frac{3 \cdot \left(1 - \frac{V_{ds}}{V_{dsat}}\right) + \left(\frac{V_{ds}}{V_{dsat}}\right)^2}{2 - \frac{V_{ds}}{V_{dsat}}} \right]$$

$$- Q_{DEPL}, \tag{10.98}$$

$$Q_{DEPL} = -\frac{1}{3} \cdot C_{ox} \cdot W_{eff} \cdot L_{eff} \cdot (V_t - V_{bs} - VFB), \tag{10.99}$$

$$Q_D = -\frac{1}{3} \cdot C_{ox} \cdot W_{eff} \cdot L_{eff} \cdot (V_{gs} - V_t) \cdot \left[\frac{3 \cdot \left(1 - \frac{V_{ds}}{V_{dsat}}\right) + \left(\frac{V_{ds}}{V_{dsat}}\right)^2}{2 - \frac{V_{ds}}{V_{dsat}}} \right.$$

$$\left. + \frac{\frac{V_{ds}}{V_{dsat}} \cdot \left(1 - \frac{V_{ds}}{V_{dsat}}\right) + \frac{1}{5} \cdot \left(\frac{V_{ds}}{V_{dsat}}\right)^2}{\left(2 - \frac{V_{ds}}{V_{dsat}}\right)^2} \right] - Q_{DEPL}, \tag{10.100}$$

and

$$Q_S = -(Q_{GATE} + Q_{DEPL} + Q_D). \tag{10.101}$$

10.7.4   The Saturation Region

In the saturation region ($V_{ds} \geq V_{dsat}$), the inversion charge is partitioned 40%/60% between the drain and the source. The charge equations here are

$$Q_{GATE} = \frac{2}{3} \cdot C_{ox} \cdot W_{eff} \cdot L_{eff} \cdot (V_{gs} - V_t) - Q_{DEPL}, \tag{10.102}$$

$$Q_{DEPL} = -\frac{1}{3} \cdot C_{ox} \cdot W_{eff} \cdot L_{eff} \cdot (V_t - V_{bs} - VFB), \qquad (10.103)$$

$$Q_D = -\frac{4}{15} \cdot C_{ox} \cdot W_{eff} \cdot L_{eff} \cdot (V_{gs} - V_t), \qquad (10.104)$$

and

$$Q_S = -(Q_{GATE} + Q_{DEPL} + Q_D). \qquad (10.105)$$

10.8   The Extrinsic Model Structure

As described earlier in this chapter, the extrinsic structure of BSIM2 is identical to that of BSIM; this structure was extensively described in Section 8.8. When parameter extraction was actually carried out, it was noted that this simple extrinsic structure, in which the extracted independent device parameters are supposed to vary as $\frac{1}{L_{eff}}$ and $\frac{1}{W_{eff}}$, leads to poor results in submicron devices, as the extracted parameters do not follow this relationship. For proper implementation, some type of binning scheme must be imposed on this extrinsic structure. Since BSIM is now rarely used, binning was not discussed in Chapter 8. Instead, as BSIM2 is at present a commonly used model, a binning scheme will be discussed here.

As in BSIM, BSIM2 model parameters come in triplets, as described by (10.1), and are used to compute the composite parameters which appear in the model equations. As in Chapters 8 and 9, let Z be some composite model parameter which is computed from a triplet set of model parameters in the form

$$Z = \mathbf{Z} + \frac{\mathbf{LZ}}{L_{eff}} + \frac{\mathbf{WZ}}{W_{eff}}. \qquad (10.106)$$

Six devices will be used for parameter extraction, as shown in Figure 10.10; there are four devices along the length axis and three devices along the width axis. An independent set of intrinsic model parameters will be extracted for each of the six devices. For the discussion here, the particular value of Z extracted at each geometry will be subscripted with the corresponding length and width; for example, Z_{43} indicates the extracted value of Z at $L_{eff} = L_4$ and $W_{eff} = W_3$. If a full matrix of twelve devices was available, it would be possible to extract twelve sets of independent intrinsic parameters, as was done in Chapter 9. However, as will be discussed below, the basic extrinsic structure of BSIM2 (and, by implication, BSIM as well) prevents these additional geometries from being of use for model building. However, as shown in Figure 10.10, the approach to be described here results in twelve model bins, as was the case for HSPICE Level 28. The model parameters for each bin will be subscripted for identification. For example, $\mathbf{Z}_9$, $\mathbf{LZ}_9$, and $\mathbf{WZ}_9$ will indicate the value of the final triplet parameter set for bin **9**.

Consider the set of four wide devices with varying channel lengths. By treating the effective channel width W_3 as very large, the **WZ** term in (10.106) vanishes, simplifying that expression to

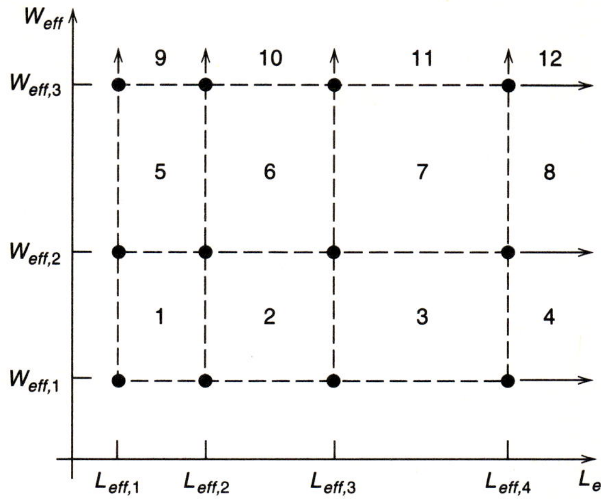

Figure 10.10   The devices used to construct a binned BSIM2 model, and the bins which result.

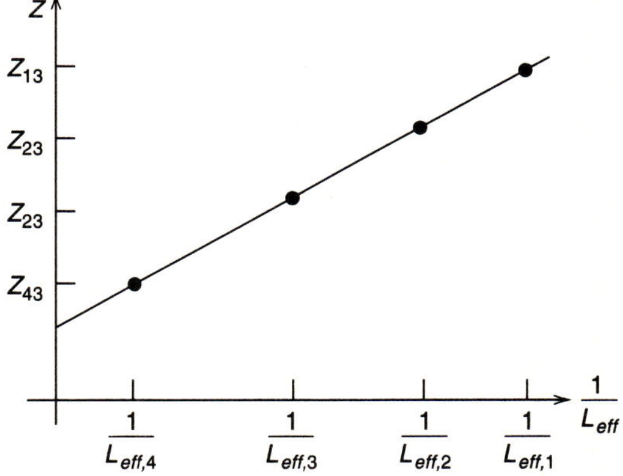

Figure 10.11   Ideal extrinsic model result, in which a plot of the values of Z versus $\frac{1}{L_{eff}}$ follows a straight line.

$$Z = \mathbf{Z} + \frac{\mathbf{LZ}}{L_{eff}}. \tag{10.107}$$

As described in Chapter 8, plotting the extracted values of Z versus $\frac{1}{L_{eff}}$ is intended to produce a straight line with a slope of $\mathbf{LZ}$ and an intercept of $\mathbf{Z}$, as shown in Figure 10.11. However, in submicron devices, the actual result is generally like that shown in Figure 10.12; the data are not amenable to a simple linear fit.

Any binning scheme must find some way to use the linear structure of (10.107) to describe the nonlinear form of the data depicted in Figure 10.12. The basic concept is to cope with the data curve by using several line segments. This implies that a larger number of devices will allow more line segments and better accuracy. Each of the three

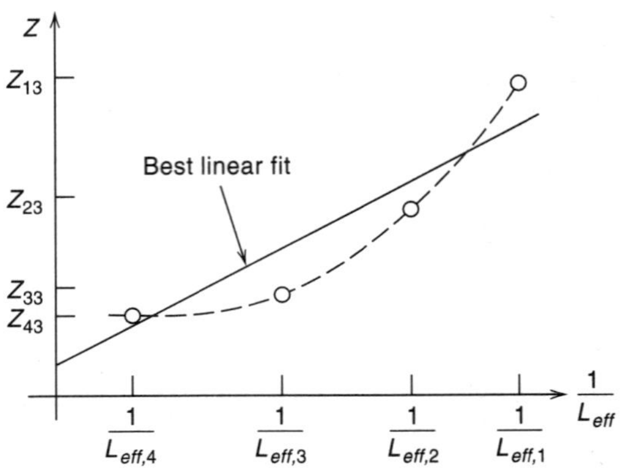

Figure 10.12   A plot of Z versus $\frac{1}{L_{eff}}$ for a submicron technology. The data do not follow a straight line.

shorter channel devices will be regarded as independent, and (10.107) will be used to perform a two-point fit between the short device and the long, wide base device. (This method is very similar to the approach described for Level 3, and the same deficiencies that resulted there will also appear here.)

Due to the extrinsic model structure, the reference point for each two-point fit is the long, wide base device rather than the next device. Unlike the scheme described in Chapter 9, the extrinsic model structure prevents the bins from communicating. For example, consider the situation where bins **9** and **10** meet. This allows two expressions to be written for the extracted model parameter Z_{23}. From bin **9**,

$$Z_{23} = \mathbf{Z}_9 + \frac{\mathbf{LZ}_9}{L_2},\qquad(10.108)$$

while from bin **10**,

$$Z_{23} = \mathbf{Z}_{10} + \frac{\mathbf{LZ}_{10}}{L_2}.\qquad(10.109)$$

Setting these two expressions equal to each other leads to

$$\mathbf{Z}_9 + \frac{\mathbf{LZ}_9}{L_2} = \mathbf{Z}_{10} + \frac{\mathbf{LZ}_{10}}{L_2}.\qquad(10.110)$$

Even if $\mathbf{Z}_9$ and $\mathbf{Z}_{10}$ are known, the equation still contains two unknowns, which prevents unique values of $\mathbf{LZ}_9$ and $\mathbf{LZ}_{10}$ from being determined.

Instead, the determination of parameters for each bin must refer to the longest device, as depicted graphically in Figure 10.13. The solid line in each bin defines **LZ** for that bin, while the dashed lines are traced back to the y-axis to define the intercept **Z**. Note that since the reference is to the long, wide device rather than the next bin (as discussed above), when the parameter Z is reconstructed, it will be discontinuous across

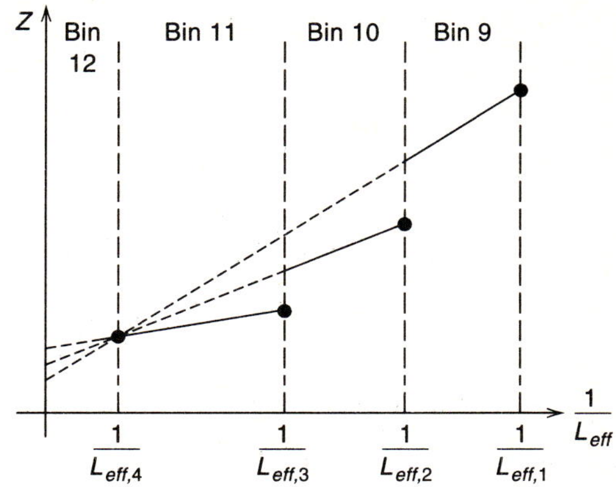

Figure 10.13   Depiction of how extrinsic model parameters are extracted for each bin. The slope of the solid line in each bin defines **LZ**, while the dashed line indicates the reference back to the long, wide base device.

the bin boundaries. Since the parameter set is discontinuous across the bin boundaries, the device characteristics, such as the drain current, will suffer from the same problem. This will be demonstrated after a full model is constructed in Section 10.9.

A similar method is used to determine the various **WZ** values. Here the three devices with a length of L_4 and various widths are employed. As in the case of the length parameters, the long, wide base device is used as the reference point, which will introduce discontinuities across the bin boundaries.

As was noted in Section 8.9, a problem with using the y-intercept to extract any **Z** is that the separate length and width extractions will produce different values of **Z**. The simplest solution is to define the largest device (here, $L_{eff} = L_4$ and $W_{eff} = W_3$) as infinitely long and wide, and thus to set the value of **Z** in every bin to Z_{43}. This is the approach that is generally used. This also indicates how bin **12** is described, as depicted in Figure 10.13. Treating the largest device as infinitely long and wide implies that $\mathbf{LZ}_{12} \rightarrow 0$; the solid line segment in this bin is treated as horizontal with a slope of zero, and no length parameters need be defined.

These results also indicate why the method used in Chapter 9 of employing twelve devices to define the twelve bins cannot be used here. Consider the boundary between bins **1** and **2** along the line where $W_{eff} = W_1$. From each bin, two equations can be written for the extracted parameter Z_{21}:

$$Z_{21} = \mathbf{Z}_1 + \frac{\mathbf{LZ}_1}{L_2} + \frac{\mathbf{WZ}_1}{W_1}, \tag{10.111}$$

and

$$Z_{21} = \mathbf{Z}_2 + \frac{\mathbf{LZ}_2}{L_2} + \frac{\mathbf{WZ}_2}{W_1}. \tag{10.112}$$

Even allowing for $\mathbf{Z}_1$ and $\mathbf{Z}_2$ to be included as Z_{43}, when these two expressions are set equal to each other, there are four unknowns. Instead, the simple expedient is taken of

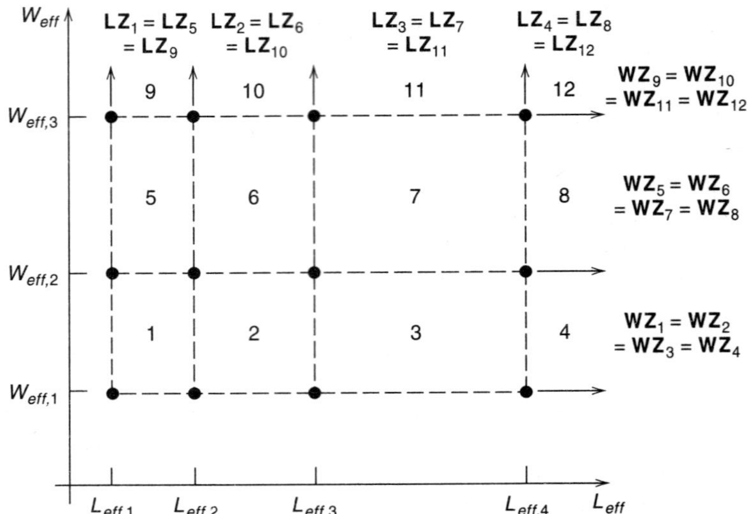

Figure 10.14   The model bins, indicating how length and width parameters are the same across different bins.

defining the length dependence entirely from the set of wide devices with varying channel lengths, and the width dependence from the set of long devices with various widths. In this form, the length dependence will be the same in bins **1**, **5**, and **9** (see Figure 10.14); that is, $\mathbf{LZ_1 = LZ_5 = LZ_9}$. A similar statement could be made about bins **2**, **6**, and **10**, etc. The width dependence will also be the same across various groups of bins.

The most important weakness of this method is that the model can lose accuracy for devices that are *both* short and narrow, as combined small geometry effects are not considered. It is possible that some scheme could be devised using twelve devices in a grid and employing an iterative (virtually finite element) method of solving (10.106) for **Z**, **LZ**, and **WZ** in each bin; this would be very complex and apparently has never been tried.

It should be noted that the HSPICE implementations of both BSIM and BSIM2 (HSPICE Levels 13 and 39, respectively) make available the $L_{ref,eff}$ and $W_{ref,eff}$ method introduced with HSPICE Level 28, which is described in Chapter 9. This approach allows the use of a device matrix grid for model binning. The Level 39 implementation (but **not** the Level 13 implementation) also makes available the product terms, which are also described in Chapter 9.

10.9   Parameter Extraction and Model Development

Like the other second-generation models, BSIM2 is very empirical; thus, the model formulation is a shell that is heavily dependent on parameter extraction for proper final results. Therefore, detailed parameter extraction is once again required to demonstrate the strengths and weaknesses of BSIM2, and also to clearly separate the effects of the intrinsic and extrinsic parts of the model structure. As with BSIM and HSPICE Level 28, due to the empirical nature of the model, many possible parameter extraction routes are possible; the method used here is a reasonable choice.

10.9.1   Basic Parameter Extraction

As was the case with BSIM and HSPICE Level 28, BSIM2 parameter extraction proceeds in three distinct phases. A long, wide device is once again used as a base, to which small geometry effects are added as corrections.

Here parameter extraction will proceed as follows:

- The oxide thickness and the differences between the drawn and effective channel dimensions are provided as process input.
- In the first phase of parameter extraction, a long, wide device is used to determine some base parameters, which are used as the starting point for each individual device extraction of the second phase. As with BSIM and HSPICE Level 28, the user can choose to define certain parameters as global, and determine them in this phase. In the original formulation of BSIM2, the low field mobility is treated as a global parameter, with no length and width dependence (this was also the case in BSIM).
- In the second phase, a set of parameters is extracted independently for each device, using the parameter results of the first phase as the starting point for each particular device. Once again, this phase represents the fitting of the data for each independent device to the intrinsic equation structure of the model.
- In the third phase, the compiled parameters from the second phase are used to determine the geometry parameters; this represents the imposition of the extrinsic structure onto the model.

The basic approach to parameter extraction is identical to that used for BSIM. Five devices are used, in the form detailed in Figure 10.15.

The First Phase

In this phase of parameter extraction, a long, wide device is used to determine a set of base parameters, which is used as the starting point for the second-phase extractions. In the first step, fitting is done to the linear characteristic with no substrate biases applied. As Figure 10.16 depicts, very good results are obtained, showing that BSIM2 works well for high vertical channel fields. In this step, the low field mobility **MU0** is determined. In the original formulation of BSIM2, **MU0** had no geometry parameters, and remained unchanged throughout the remainder of model construction. The HSPICE implementation of BSIM2 (HSPICE Level 39) introduced geometry dependence for **MU0** with the parameters **LMU0** and **WMU0**. In order to allow a more meaningful comparison of HSPICE Level 28 and BSIM2, the HSPICE implementation, with its geometry dependence, was chosen for use here.

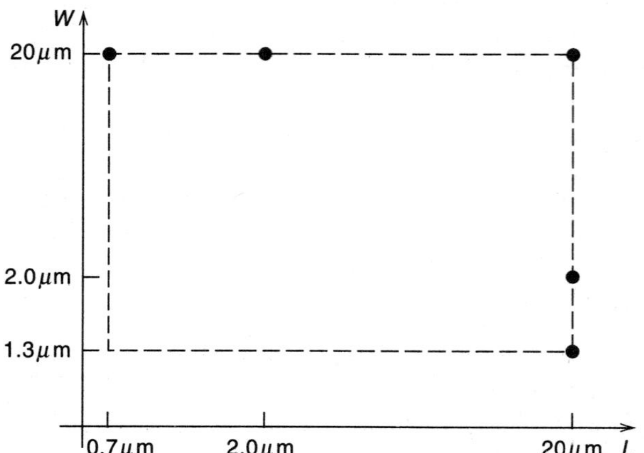

Figure 10.15 The specific geometry scheme used to extract the parameters for the BSIM model developed here.

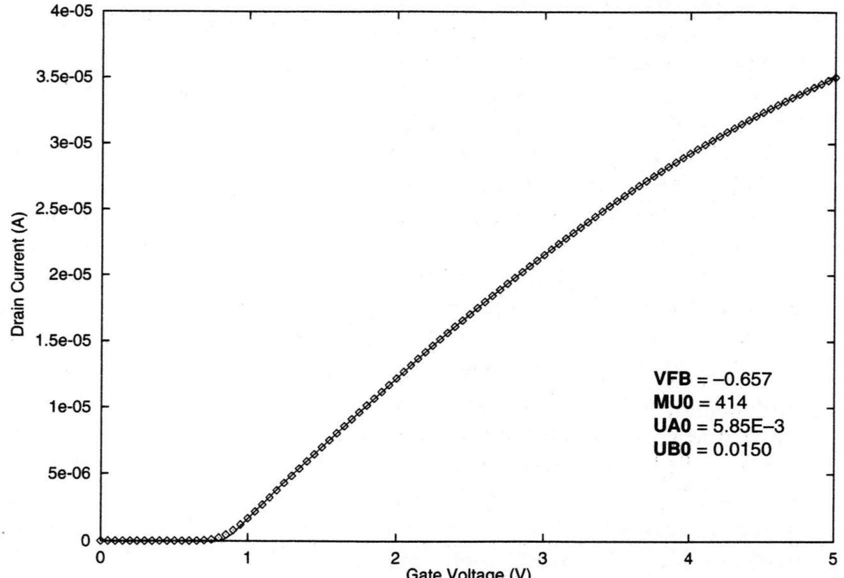

Figure 10.16 Phase I: The extraction of the base device parameters **VFB**, **MU0**, **UA0**, and **UB0**, $W/L = 20$ μm/20 μm.

In the second step, the linear characteristic is employed with substrate biases applied. As shown in Figure 10.17, very good results are obtained; these results are much better than the first-phase results of BSIM and HSPICE Level 28. Note, though, that once again these results represent a starting point for the second phase, which is where detailed accuracy is very important.

For the third step, the linear characteristic is once again used with substrate biases applied, to extract the parameter **PHI**. The results, depicted in Figure 10.18, are once again excellent. As with BSIM and HSPICE Level 28, **PHI** comes with length and width parameters; however, experience has shown that **PHI** varies little with geometry. Thus, the choice is made here, as was the case with BSIM and HSPICE Level 28, to fix **PHI** for good here, and make it a global parameter with no geometry dependence.

The Second Phase

The second phase of parameter extraction involves the independent fitting of *each* device to the intrinsic model structure (i.e., the basic set of model equations). The procedure described here is repeated for each device that is being used.

For the first step, fitting is done for the linear characteristic with substrate biases applied. As shown in Figure 10.19, excellent results are obtained for a short (20/0.7) device (Figure 10.19a), a longer (20/2.0) device (Figure 10.19b), and a narrow (1.3/20) device (Figure 10.19c). This represents an improvement over the small geometry results obtained with BSIM. Note once again that the option is used here of extracting a value of the low field mobility **MU0** for each device.

In the second step, the saturation characteristic is used with no substrate biases applied. Once again, the results represent an improvement over the situation with BSIM; however, some minor fitting difficulties are still present in the short (20/0.7) device (Figure 10.20a) and the narrow (1.3/20) device (Figure 10.20c).

The third step makes use of the saturation characteristic with substrate biases applied. In a submicron channel length (20/0.7) device, the result (in contrast to BSIM) is qualitatively correct (Figure 10.21a); these results are adequate for most applications. In a longer channel (20/2.0) device, the result is quantitatively better (Figure 10.21b).

The subthreshold characteristics are used in the fourth step, with results depicted in Figure 10.22. These results are good, as they properly describe the current for several decades (of current) below the weak-strong inversion transition.

The final step involves the fitting of the output conductance characteristics. Unlike HSPICE Level 28, BSIM2 contains several parameters that are included specifically for the modeling of the output conductance. There should thus be less corruption of the current result by the output conductance extraction. With no output conductance fitting, the results of Figure 10.23a are obtained. This result is basically adequate for most situations. A slight improvement can be obtained by extracting the output conductance fitting parameters, with the result shown in Figure 10.23b. This causes some loss of accuracy in the drain current results, as shown in Figure 10.24. As was noted with HSPICE Level 28, the proper fitting of the output conductance is a very sophisticated process, involving various methods of weighting the drain current and the output conductance during parameter extraction; this is certainly the case with BSIM2. However, note that due to the inclusion of separate output conductance parameters in the model, the potential

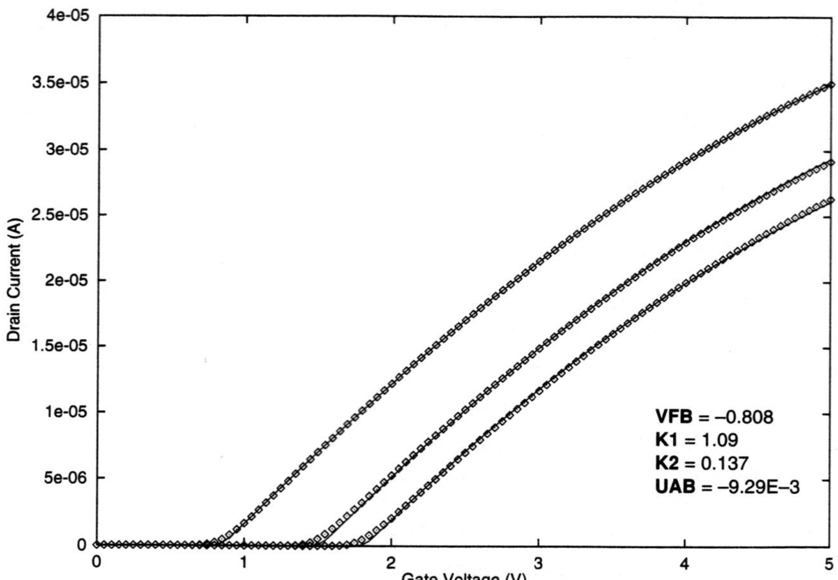

Figure 10.17   Phase I: The extraction of the base device parameters **VFB**, **K1**, **K2**, and **UAB**, $W/L = 20$ μm/20 μm.

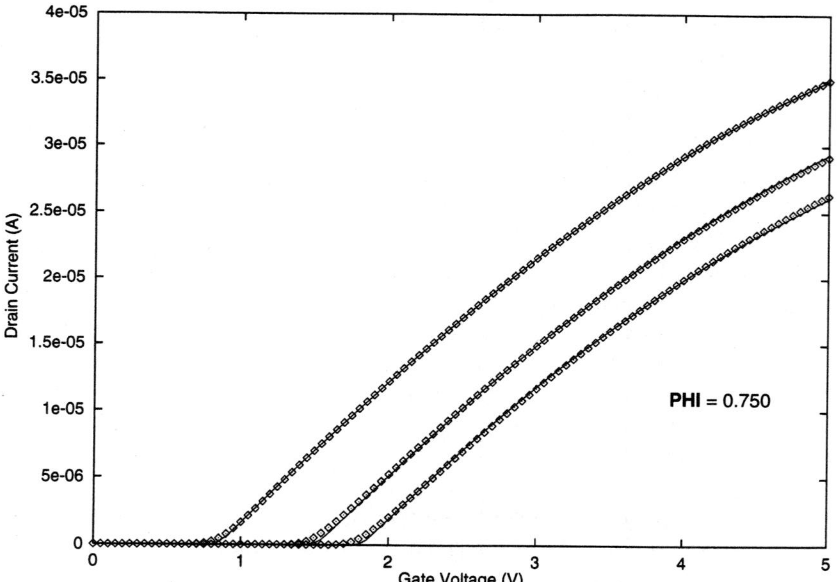

Figure 10.18   Phase I: The extraction of the base device parameter **PHI**, $W/L = 20$ μm/20 μm.

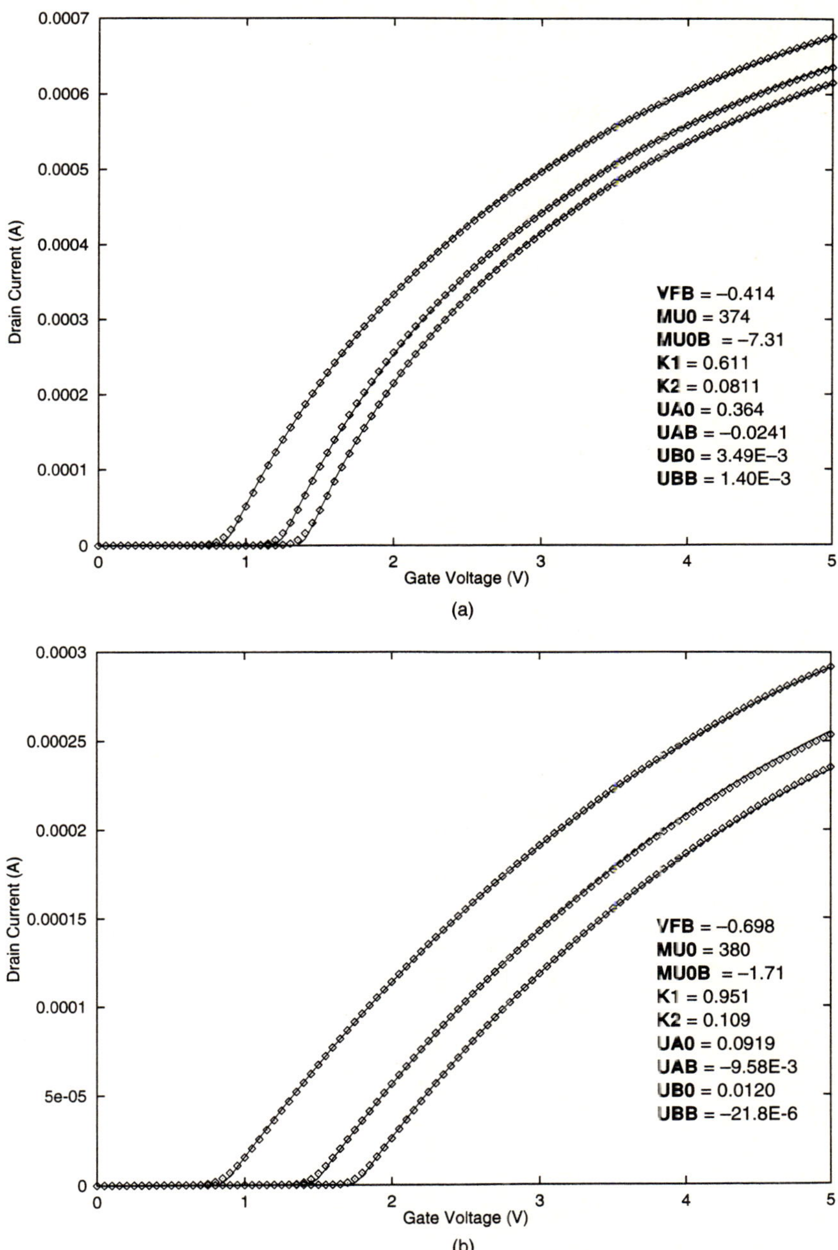

Figure 10.19 Phase II: The extraction of the device parameters **VFB**, **MU0**, **MU0B**, **K1**, **K2**, **UA0**, **UAB**, **UB0**, and **UBB**; (a) $W/L = 20\,\mu\text{m}/0.7\,\mu\text{m}$; (b) $W/L = 20\,\mu\text{m}/2.0\,\mu\text{m}$; (c) $W/L = 1.3\,\mu\text{m}/20\,\mu\text{m}$.

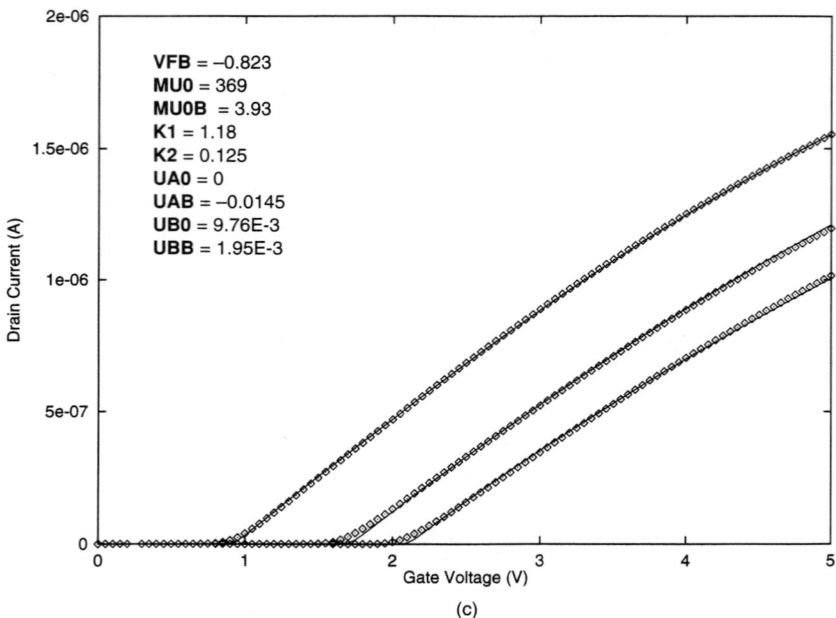

Figure 10.19   *(cont.)*

for corruption of the drain current results is much smaller in BSIM2 than in HSPICE Level 28.

Using this procedure, an independent set of model parameters is compiled for each device. This set of parameters is collected in Table 10.4 (see pg. 364).

The Third Phase

The third phase of parameter extraction involves using the compiled list of independent device parameters, determined during the second phase, as data for the extraction of the length and width parameters that form the extrinsic part of the model structure. As with BSIM, in BSIM2, the extrinsic model structure expects that the individual device parameters will be able to fit the equation

$$Z = \mathbf{Z} + \frac{\mathbf{LZ}}{L_{eff}} + \frac{\mathbf{WZ}}{W_{eff}}. \tag{10.113}$$

However, once again, the individual device parameters do not vary linearly with the inverse of the effective channel dimensions, as demonstrated with **VFB** data in Figure 10.25. Therefore, as was the case with BSIM, a linear regression is carried out on the individual device data; this produces the best-fit straight line of Figure 10.25, and in this case determines the extrinsic model parameter **LVFB**.

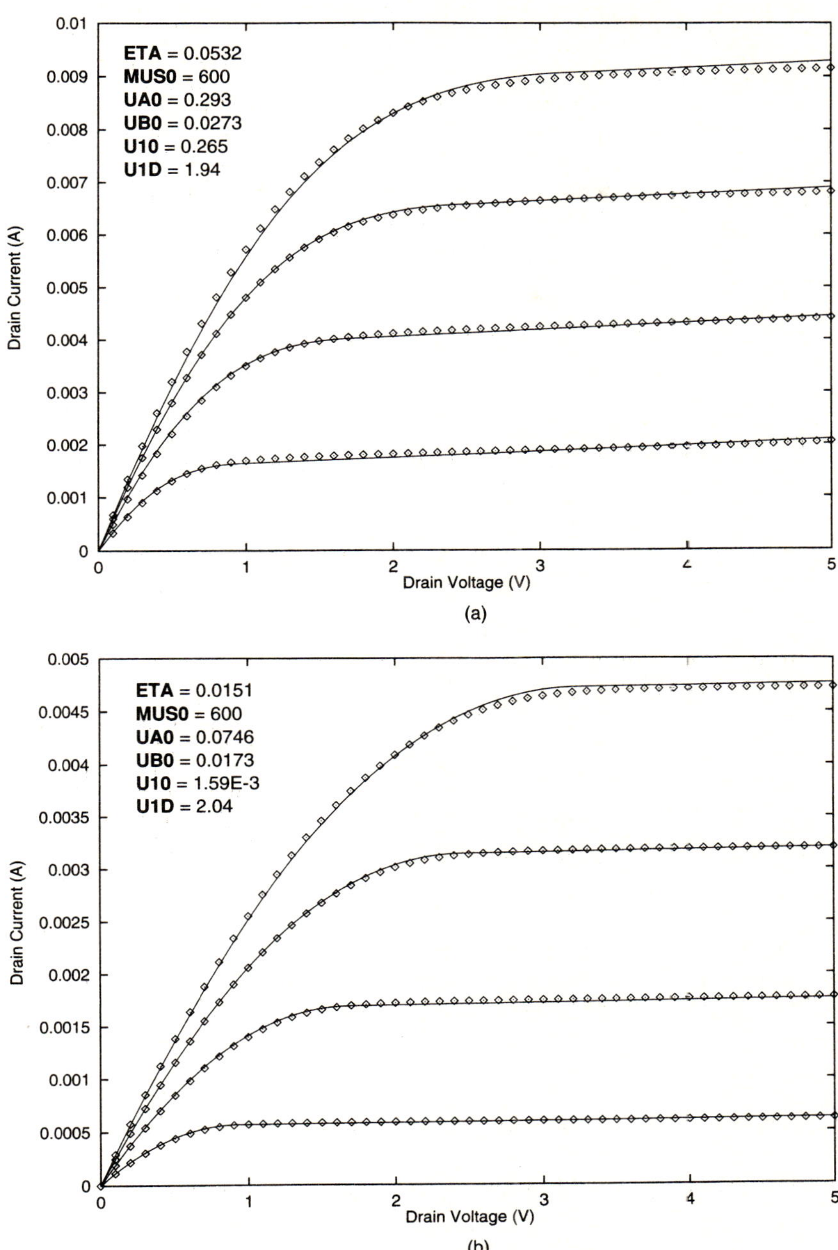

Figure 10.20   Phase II: The extraction of the device parameters **ETA**, **MUS0**, **UA0**, **UB0**, **U10**, and **U1D**; (a) $W/L = 20\,\mu\text{m}/0.7\,\mu\text{m}$; (b) $W/L = 20\,\mu\text{m}/2.0\,\mu\text{m}$; (c) $W/L = 1.3\,\mu\text{m}/20\,\mu\text{m}$.

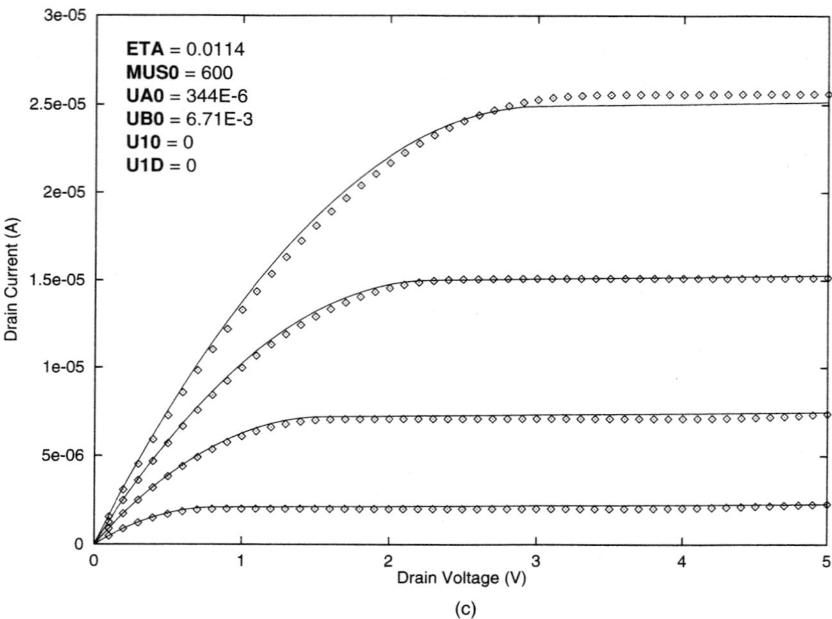

Figure 10.20　*(cont.)*

The final model thus constructed is compiled in Table 10.5 (see pag. 366). Although the basic method is the same, it should be noted that the overall results here tend to be better than those obtained with BSIM; this is detailed below.

Temperature Dependence

As noted at the beginning of the chapter, the original BSIM2 structure makes no provision for any variation of the temperature from that used for data collection. The HSPICE implementation of BSIM2 (HSPICE Level 39) includes a temperature-dependence scheme that is essentially identical to that used for HSPICE Level 28 (see Chapter 9).

10.9.2　Model "Playback"

Once the model has been completed, it is useful to compare it to original data, to evaluate its predictive behavior. In addition, this is necessary in the second-generation models if the effect of the imposition of the extrinsic model structure is to be detailed.

Figure 10.26 compares the final model with data for the 20/0.7 device. As in the other second-generation models, the linear characteristic (Figure 10.26a) suffers some loss of accuracy due to the later model extraction steps of the second phase. The saturation characteristic of Figure 10.26b can be compared with the second phase result of Figure 10.24. This comparison clearly indicates that the extrinsic model structure is

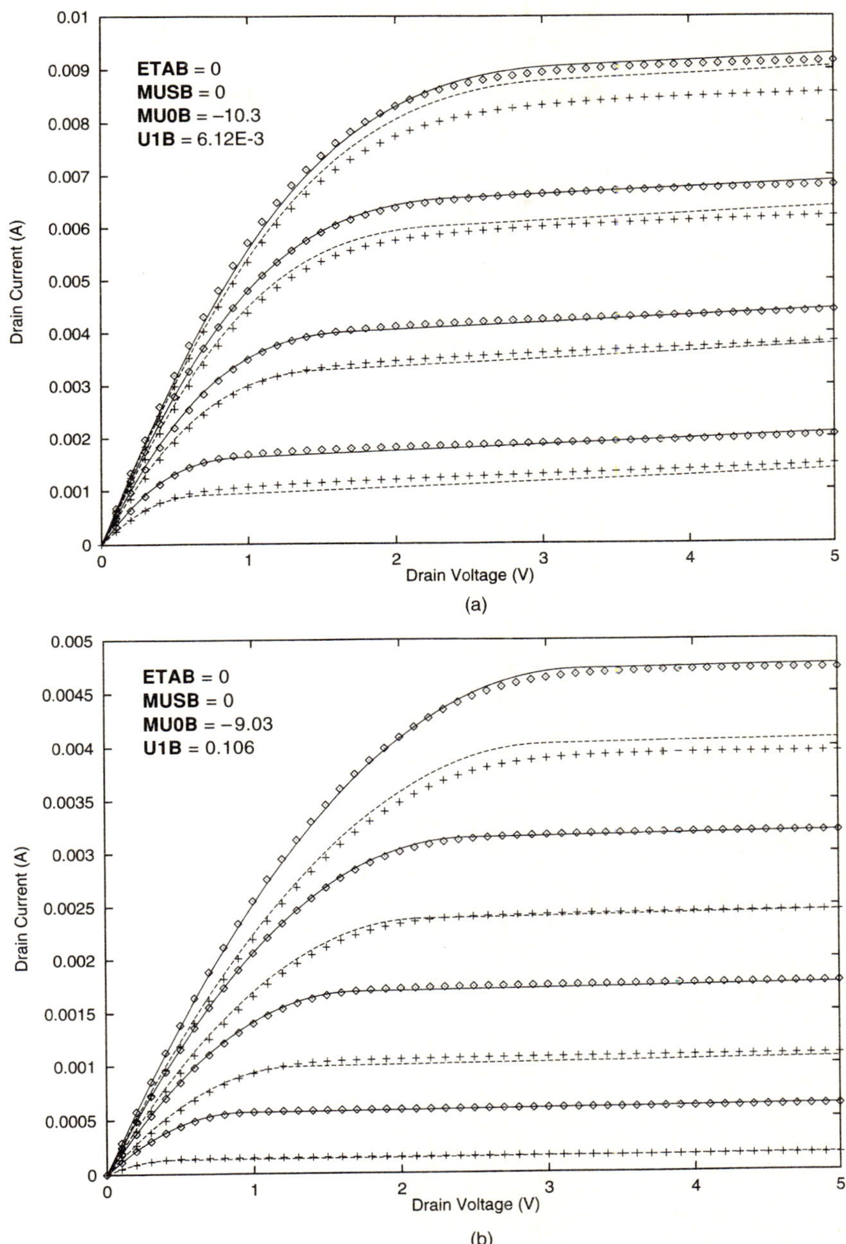

Figure 10.21   Phase II: The extraction of the device parameters **ETAB, MUSB, MU0B, U1B**; (a) $W/L = 20$ μm/0.7 μm; (b) $W/L = 20$ μm/2.0 μm.

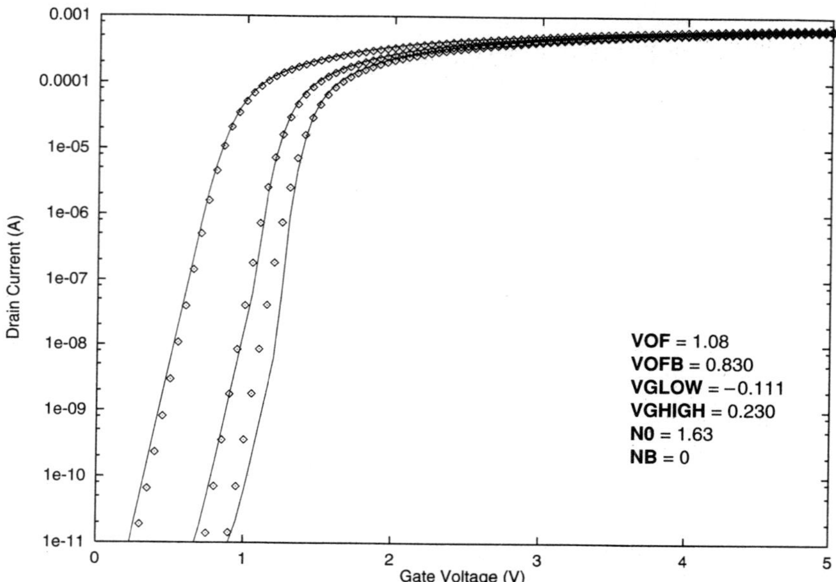

Figure 10.22   Phase II: The extraction of the device parameters **VOF**, **VOFB**, **VGLOW**, **VGHIGH**, **N0**, and **NB**, $W/L = 20$ μm/0.7 μm.

responsible for a large change in the characteristic, with a considerable reduction of the final model accuracy. Figure 10.27 examines a 20/2.0 device. As in the previous case, the linear characteristic (Figure 10.27a) shows a loss of accuracy due to later extraction steps. It is worthwhile to compare the final saturation characteristic (Figure 10.27b) with the result in the second phase (Figure 10.20b). It can once again be seen that the use of the second-generation extrinsic structure has caused a reduction in the accuracy of the final model. However, the results are much improved from those obtained with the final BSIM model. In general, BSIM2 provides much better model results for intermediate channel lengths. Figure 10.28 examines a 20/0.9 device; note that this device geometry was *not* used during parameter extraction. The results are very similar to those just examined for the 20/2.0 device. Again, it should be noted that BSIM2 has an improved ability to describe intermediate channel lengths when compared with BSIM.

A narrow (1.3/20) device is examined in Figure 10.29. When compared with the second phase result (Figure 10.20c), the saturation characteristic (Figure 10.29b) shows a significant loss of accuracy due to the use of the extrinsic model structure. In the case of a 2.0/20 device (Figure 10.30), similar problems are noted. BSIM2 does not do as well with the imposition of the extrinsic width model as it does with the extrinsic length model.

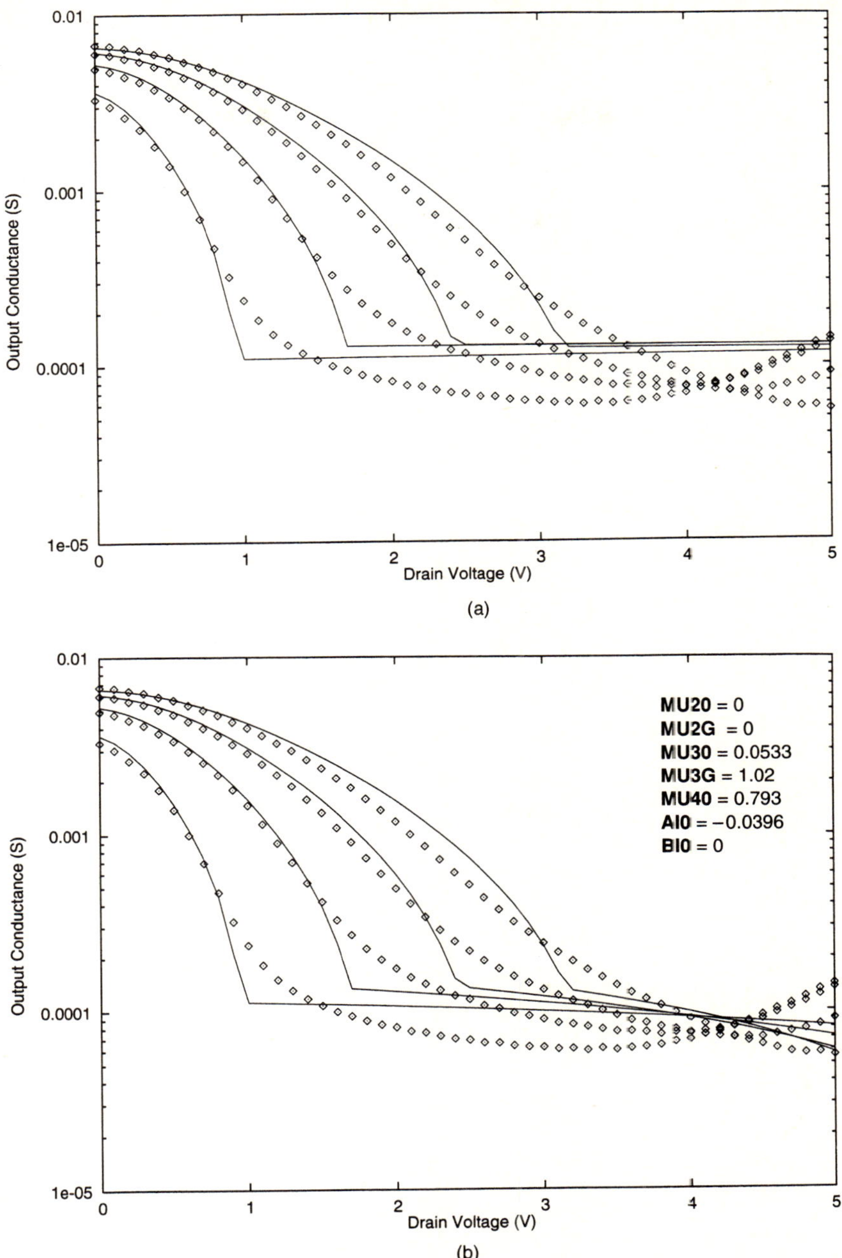

Figure 10.23   Phase II: The extraction of **MU20, MU2G, MU30, MU3G, MU40, AI0, AIB, BI0** and **BIB** for output conductance fitting, $W/L = 20$ μm/0.7 μm; (a) before the extraction; (b) after the extraction.

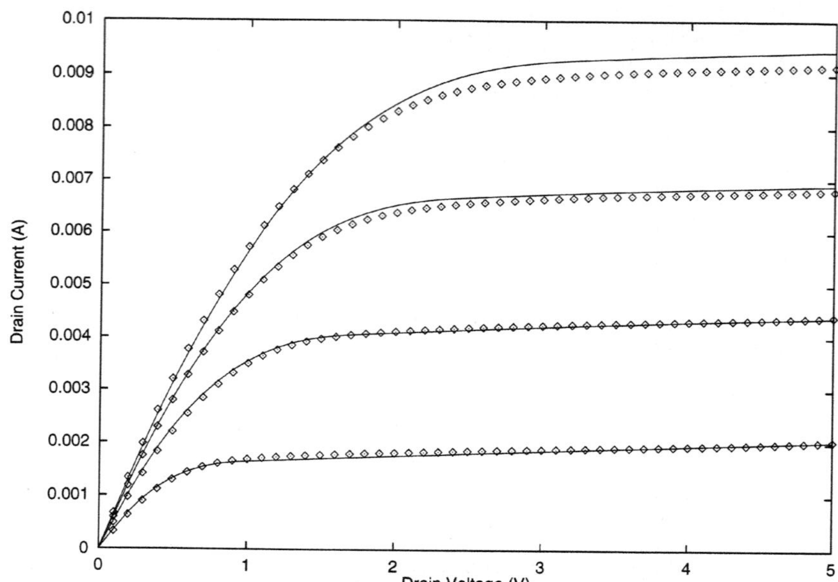

Figure 10.24   The drain current-drain voltage characteristics of a $W/L = 20\ \mu\text{m}/0.7\ \mu\text{m}$ device after the output conductance extraction depicted in Figure 10.23b.

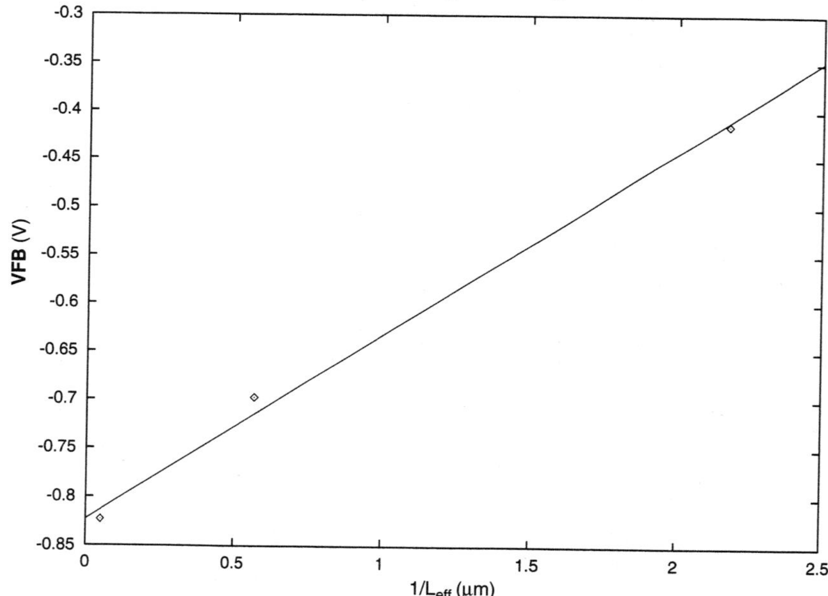

Figure 10.25   The individual device results for the model parameter **VFB** versus $\frac{1}{L_{eff}}$, and the linear regression fit to the extrinsic model equation for the determination of **LVFB**. Note how the data points do not follow a straight line.

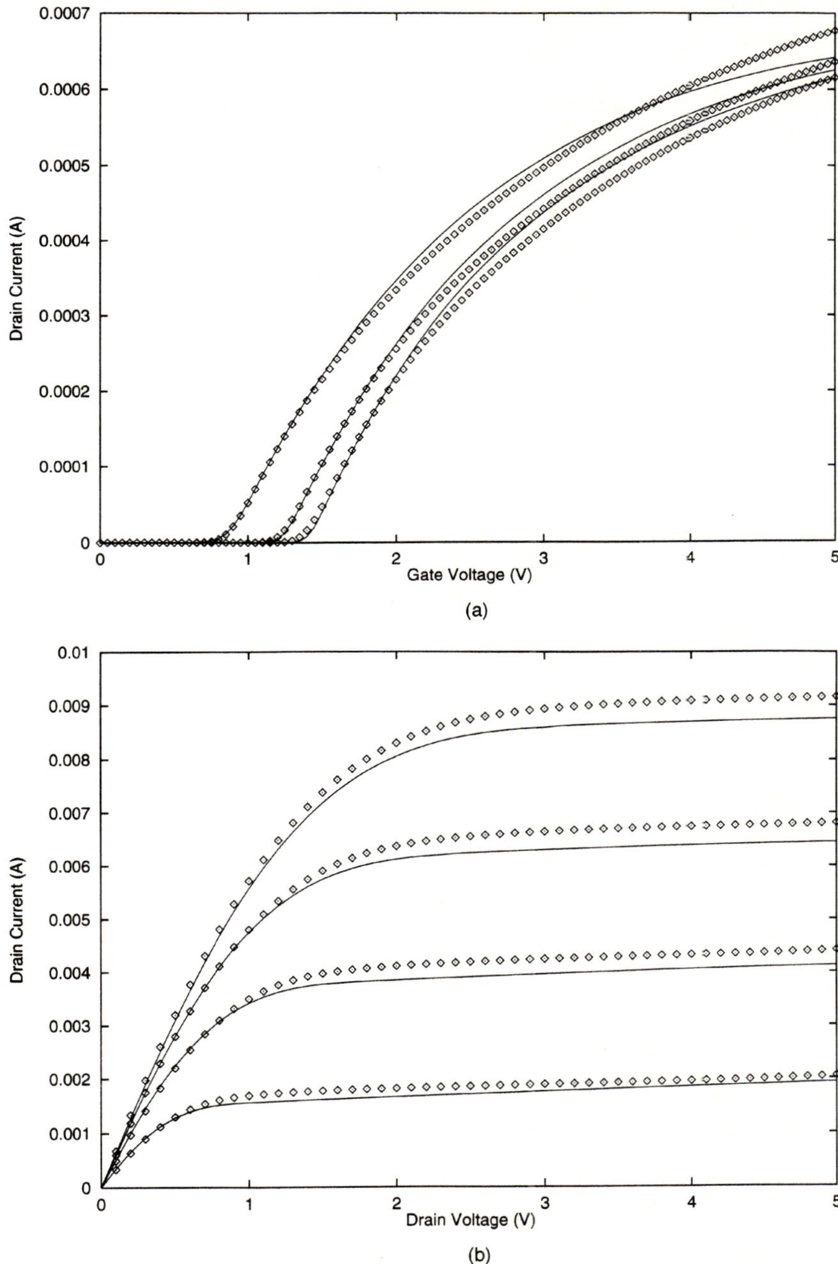

Figure 10.26   Comparison of measured data and the final model for the 20/0.7 device; (a) linear characteristic; (b) saturation characteristic.

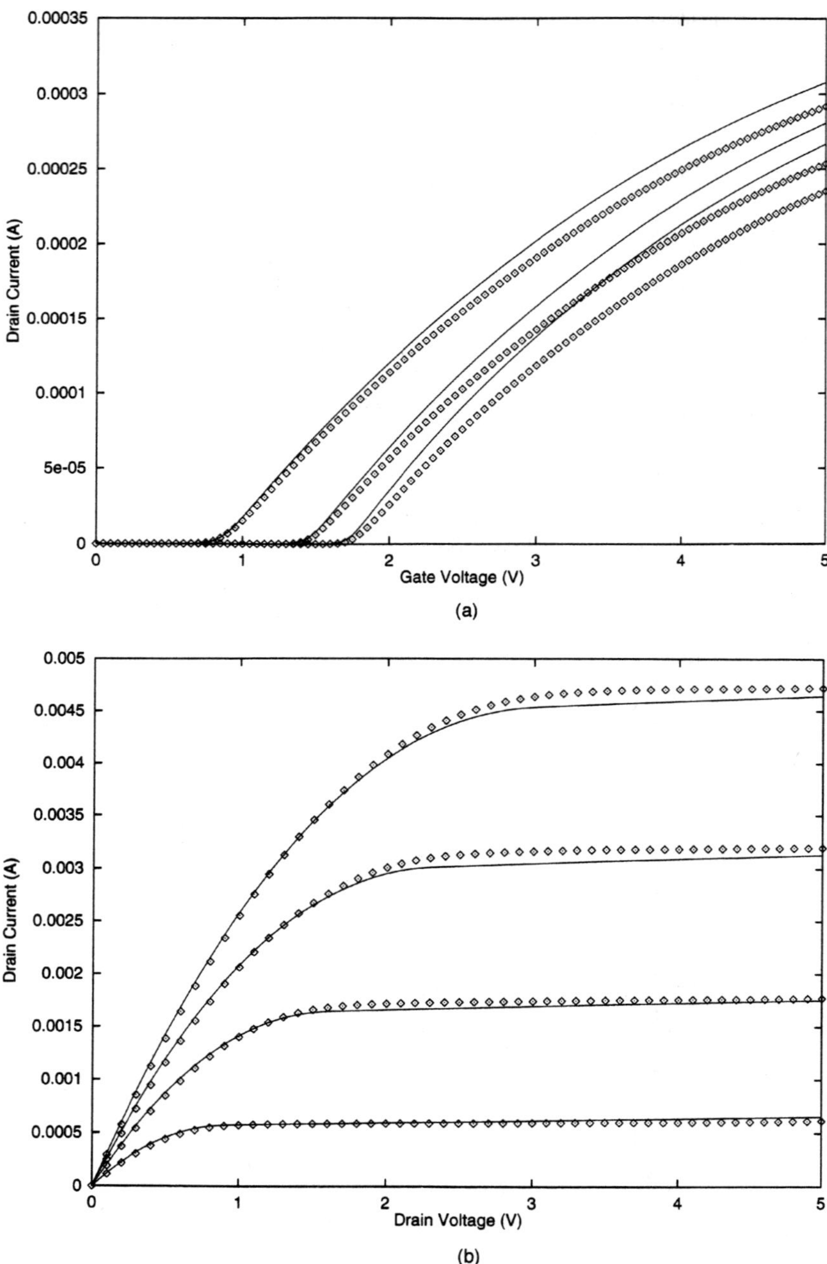

Figure 10.27 Comparison of measured data and the final model for the 20/2.0 device; (a) linear characteristic; (b) saturation characteristic.

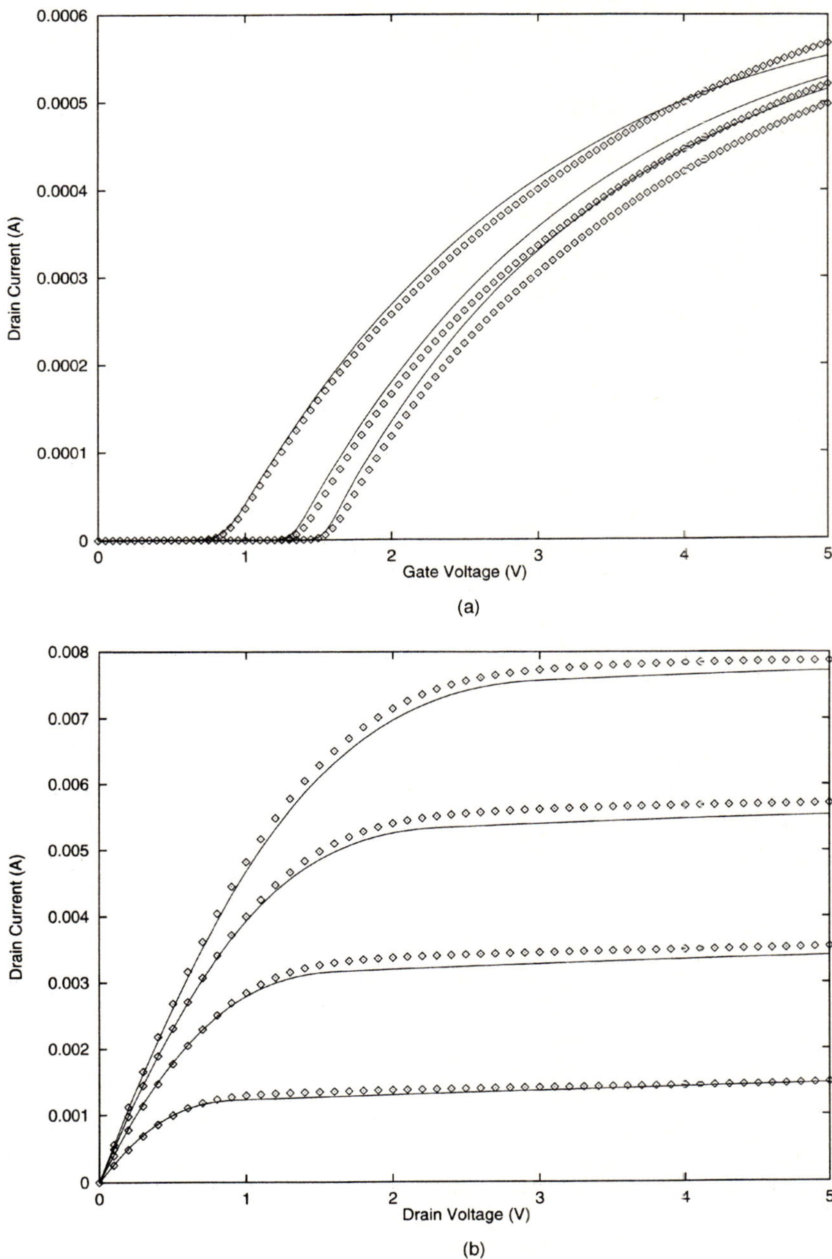

Figure 10.28   Comparison of measured data and the final model for the 20/0.9 device; (a) linear characteristic; (b) saturation characteristic.

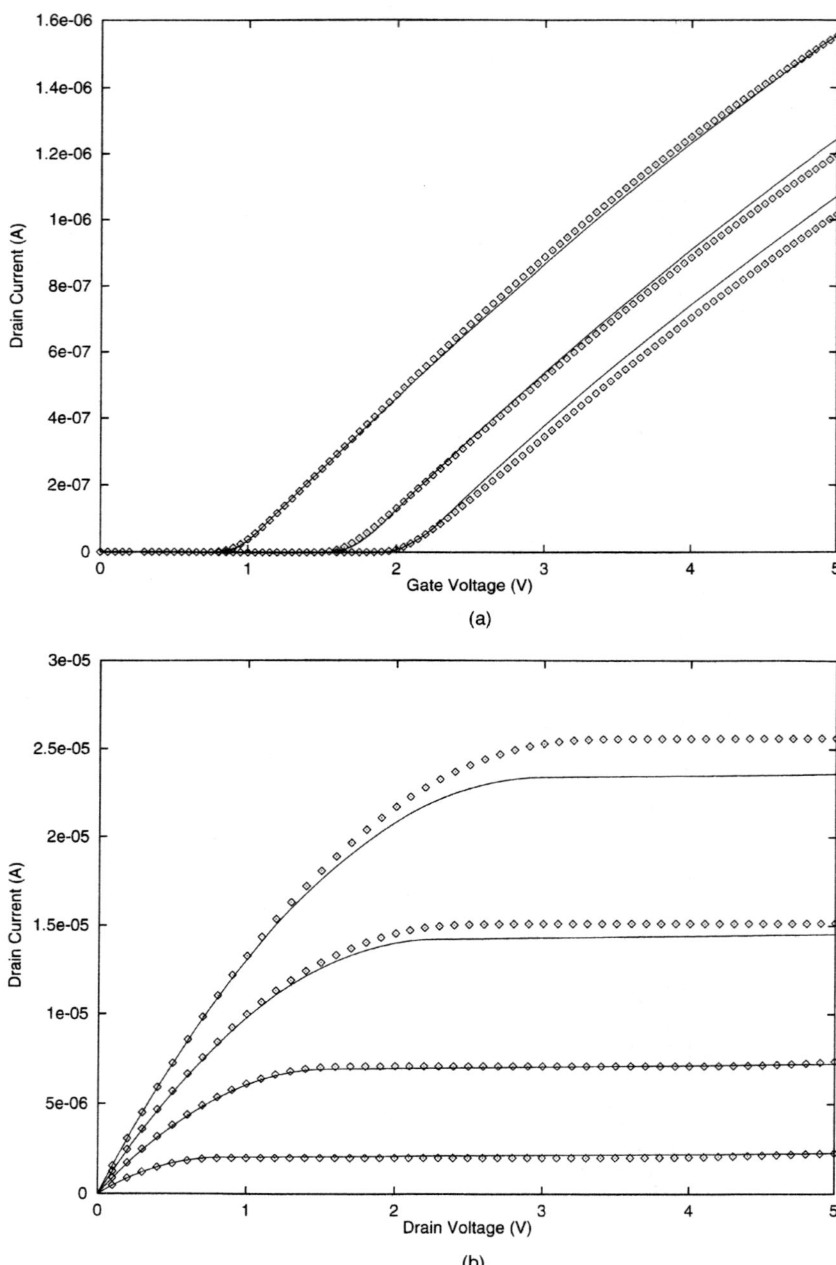

Figure 10.29 Comparison of measured data and the final model for the 1.3/20 device;
(a) linear characteristic; (b) saturation characteristic.

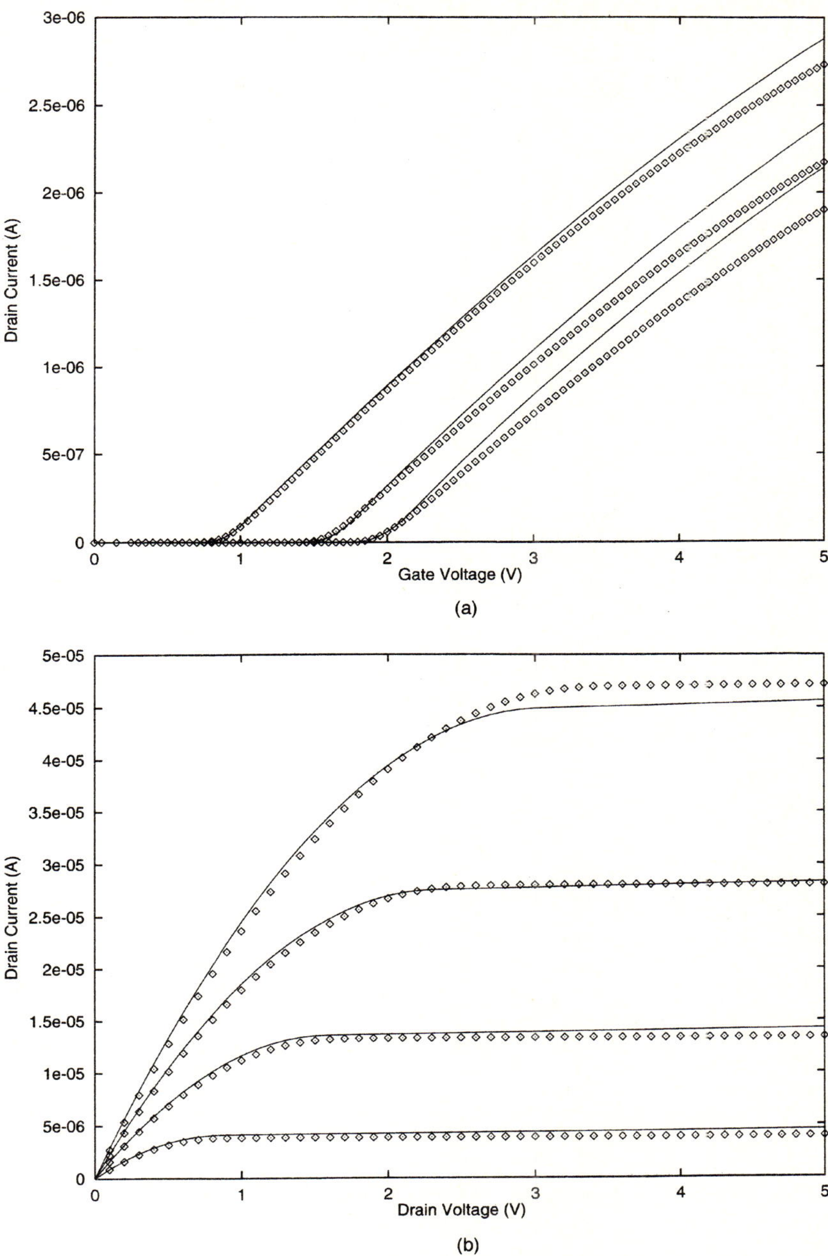

Figure 10.30 Comparison of measured data and the final model for the 2.0/20 device; (a) linear characteristic; (b) saturation characteristic.

Table 10.4 The individual parameters extracted independently for each device during the second phase of parameter extraction.

Parameter	20/20	20/2.0	20/0.9	20/0.7	2.0/20	1.3/20
TOX	13.0E-9	13.0E-9	13.0E-9	13.0E-9	13.0E-9	13.0E-9
LD	0.240E-6	0.240E-6	0.240E-6	0.240E-6	0.240E-6	0.240E-6
WD	0.400E-6	0.400E-6	0.400E-6	0.400E-6	0.400E-6	0.400E-6
VDD	5.0	5.0	5.0	5.0	5.0	5.0
MU0	406	380	384	374	370	369
MU0B	−8.79	−9.03	−12.2	−10.3	−4.47	−1.74
MUS0	600	600	600	600	600	600
MUSB	0	0	0	0	0	0
UA0	4.61E-6	0.0750	0.199	0.293	468E-6	344E-6
UAB	−8.49E-3	−9.58E-3	−0.0218	−0.0241	−0.0116	−0.0145
UB0	9.43E-3	0.0173	0.0263	0.0273	6.84E-3	6.71E-3
UBB	66.2E-6	−21.8E-6	1.06E-3	1.40E-3	992E-6	1.95E-3
VFB	−0.823	−0.698	−0.562	−0.414	−0.825	−0.823
PHI	0.750	0.750	0.750	0.750	0.750	0.750
K1	1.09	0.951	0.819	0.611	1.16	1.18
K2	0.138	0.109	0.109	0.0811	0.136	0.125
ETA	5.60E-3	0.0151	0.0395	0.0532	0.0158	0.0114
ETAB	0	0	0	0	0	0
U10	0	1.59E-3	0.146	0.265	0	0
U1B	0	0.106	0.0228	6.12E-3	0	0
U1D	0	2.04	2.22	1.94	0	0
VOF	0.695	0.770	0.720	1.08	0.855	1.15
VOFB	−0.0943	0.0645	0.128	0.830	−0.158	−0.188
VOFD	−0.0700	−0.0700	−0.0700	−0.0700	−0.0700	−0.0700
VGLOW	−8.75E-3	−0.0115	−0.0114	−0.111	−0.0103	−0.0402
VGHIGH	0.193	0.204	0.196	0.230	0.224	0.245
N0	1.63	1.67	1.61	1.63	1.87	1.62
NB	0	0	0.333	0	0	0
ND	0	0	0	0	0	
MU20	0.0240	9.40E-3	6.99E-3	0	0	0
MU2B	0	0	0	0	0	0
MU2G	0	0	0	0	0	0
MU30	0.876	86.9E-6	0	0.0533	0	0
MU3B	0	0	0	0	0	0
MU3G	1.04	0.0535	0.148	1.02	0	0
MU40	0.410	0.0787	0.358	0.793	0	0
MU4B	0	0	0	0	0	0
MU4G	0	0	0	0	0	0
AI0	−0.230	−0.0391	−0.0726	−0.0396	−0.209	0
AIB	0	0	0	0	0	0
BI0	0	0	0	0	0	0.200
BIB	0	0	0	0	0	0

A short, narrow (1.3/0.7) device is detailed in Figure 10.31. The linear characteristic (Figure 10.31a) is very poor, as short and narrow effects are modeled independently but not together in the general second-generation extrinsic model structure. The saturation characteristic is extremely poor, showing a kink in the high drain bias region. Note carefully that *this is despite the fact that all of the devices actually used in parameter*

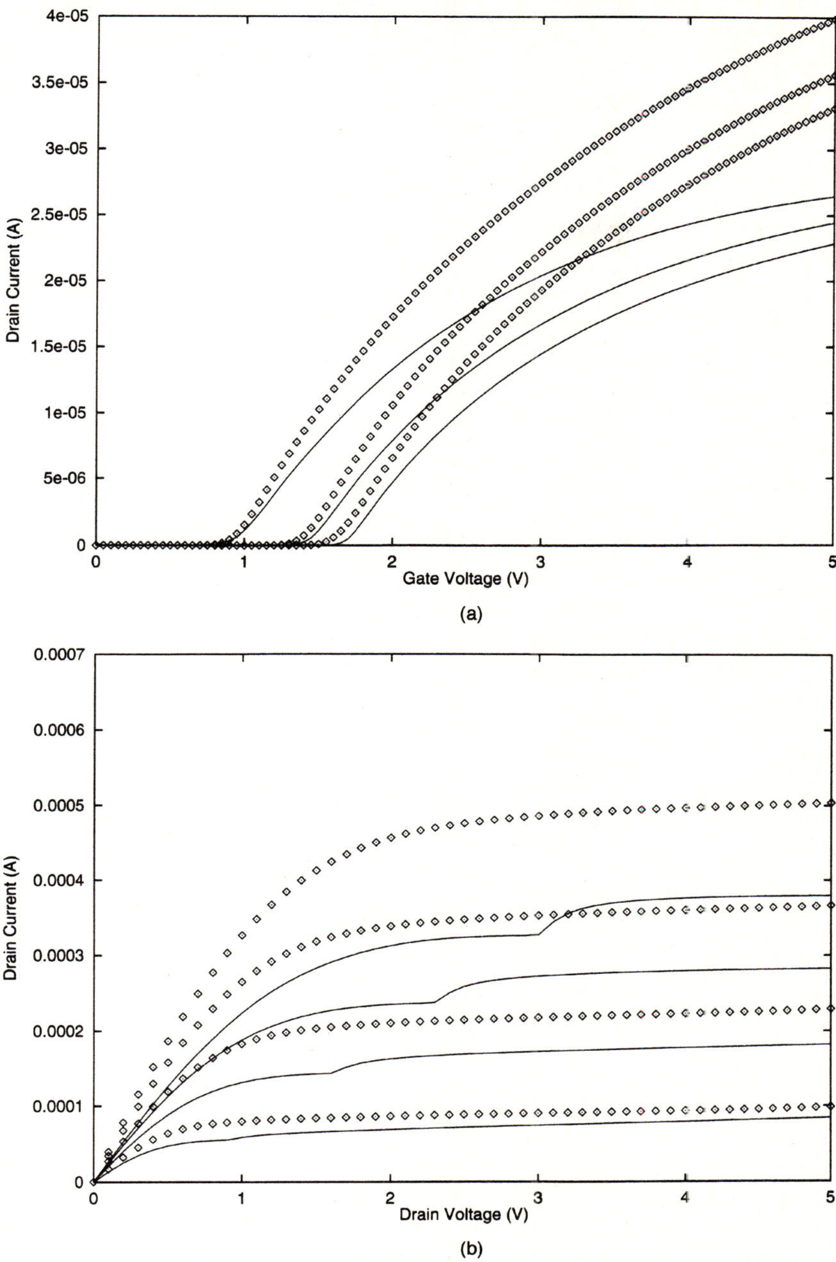

Figure 10.31 Comparison of measured data and the final model for the 1.3/0.7 device; (a) linear characteristic; (b) saturation characteristic.

Table 10.5   The final (unbinned) BSIM2 model parameter set.

Base Parameter	Length Parameter	Width Parameter
TOX = 13.0E-9	–	–
LD = 0.240E-6	–	–
WD = 0.400E-6	–	–
VDD = 5.0	–	–
MU0 = 406	**LMU0** = −16.7	**WMU0** = −39.1
MU0B = −8.79	**LMU0B** = −0.677	**WMU0B** = 6.47
MUS0 = 600	**LMUS0** = 0	**WMUS0** = 0
MUSB = 0	**LMUSB** = 0	**WMUSB** = 0
UA0 = 4.61E-6	**LUA0** = 0.135	**WUA0** = 3.65E-4
UAB = −8.49E-3	**LUAB** = −6.84E-3	**WUAB** = −5.30E-3
UB0 = 9.43E-3	**LUB0** = 8.58E-3	**WUB0** = −2.85E-3
UBB = 66.2E-6	**LUBB** = 5.64E-4	**WUBB** = 1.64E-3
VFB = −0.823	**LVFB** = 0.190	**WVFB** = −7.68E-4
PHI = 0.750	**LPHI** = 0	**WPHI** = 0
K1 = 1.09	**LK1** = −0.222	**WK1** = 8.83E-2
K2 = 0.138	**LK2** = −0.0277	**WK2** = −9.64E-3
ETA = 5.60E-3	**LETA** = 0.0216	**WETA** = 7.88E-3
ETAB = 0	**LETAB** = 0	**WETAB** = 0
U10 = 0	**LU10** = 0.114	**WU10** = 0
U1B = 0	**LU1B** = 0.0146	**WU1B** = 0
U1D = 0	**LU1D** = 1.06	**WU1D** = 0
VOF = 0.695	**LVOF** = 0.174	**WVOF** = 0.372
VOFB = −0.0943	**LVOFB** = 0.416	**WVOFB** = −8.84E-2
VOFD = −0.0700	**LVOFD** = 0	**WVOFD** = 0
VGLOW = −8.75E-3	**LVGLOW** = −0.0443	**WVGLOW** = −2.21E-2
VGHIGH = 0.193	**LVGHIGH** = 0.0172	**WVGHIGH** = 4.74E-2
N0 = 1.63	**LN0** = 1.54E-2	**WN0** = 8.53E-2
NB = 0	**LNB** = 0.143	**WNB** = 0.275
ND = 0	**LND** = 0	**WND** = 0
MU20 = 0.0240	**LMU20** = −0.0120	**WMU20** = −2.56E-2
MU2B = 0	**LMU2B** = 0	**WMU2B** = 0
MU2G = 0	**LMU2G** = 0	**WMU2G** = 0
MU30 = 0.876	**LMU30** = −0.453	**WMU30** = −0.934
MU3B = 0	**LMU3B** = 0	**WMU3B** = 0
MU3G = 1.04	**LMU3G** = −0.120	**WMU3G** = −1.11
MU40 = 0.410	**LMU40** = 0.128	**WMU40** = −0.437
MU4B = 0	**LMU4B** = 0	**WMU4B** = 0
MU4G = 0	**LMU4G** = 0	**WMU4G** = 0
AI0 = −0.230	**LAI0** = 0.103	**WAI0** = 0.165
AIB = 0	**LAIB** = 0	**WAIB** = 0
BI0 = 0	**LBI0** = 0	**WBI0** = 0
BIB = 0	**LBIB** = 0	**WBIB** = 0

extraction showed smooth simulated saturation characteristics; a result like this could constitute a nasty surprise for a circuit designer. It must again be stressed that the basic second-generation extrinsic model structure is built to accommodate a geometric scheme like that shown in Figure 10.15, where short and narrow effects are described independently, and there is no real provision for the specific analysis of short, narrow devices. A unique feature of the extrinsic model structure introduced into HSPICE Level

28 (with the reference channel geometries $L_{eff,ref}$ and $W_{eff,ref}$) is that their use makes possible a description of short, narrow devices. If a need was present, further work could be done to try to repair the kink in Figure 10.31b. However, this result will be left to stand here, as a warning of a potential major pitfall for the circuit design user of BSIM2.

In Chapter 8, it was noted that in BSIM, in a submicron channel length device with a gate bias slightly larger than the threshold voltage, a negative output conductance can occur for high drain biases. In Chapter 9, it was noted that this problem does not occur in HSPICE Level 28. As Figure 10.32 shows, this problem has also been eliminated in BSIM2. Another problem in BSIM, which occurs in submicron channel length devices with a low drain bias, is a jump in the current at the weak-strong inversion transition; this problem is eliminated in HSPICE Level 28. BSIM2 also repairs this problem, producing a smooth current result at the weak-strong inversion transition, as shown in Figure 10.33.

10.9.3   A Binning Scheme for BSIM2

As detailed earlier, the second-generation extrinsic model structure forces the use of a linear regression fit of individual device parameters as data to $\frac{1}{L_{eff}}$ and $\frac{1}{W_{eff}}$. If the fit is not good, a loss of final model accuracy will occur at every geometry. In HSPICE Level 28, a special feature was introduced, using reference geometries; this allows for the construction of models with very detailed binning. This type of approach is not generally available for BSIM and BSIM2; a binning method must be imposed on the model structure in a manner similar to the one described for use with Level 3 (see Chapter 7).

The binning scheme that will be used here was described in Section 10.8. Each bin will be constructed using a local (small geometry device) and the long, wide device; the latter serves the role of a reference device. The device geometries to be used are detailed in Figure 10.34, where it is shown that 12 model bins will result. Note that like Level 3 but unlike HSPICE Level 28, the bins will not be totally independent, since the second-generation structure does not permit combined analysis of short and narrow effects at the same time. For example, the length-related bins **1**, **2**, **3**, and **4** will all have the same width parameters, while the width-related bins **1**, **5**, and **9** will all have the same length parameters. It should also be noted again that the HSPICE implementation of BSIM2 (HSPICE Level 39) contains the reference geometries of HSPICE Level 28; detailed binning is possible there.

The basic method of binning involves the extrinsic model structure, as detailed in Section 10.8; this is best illustrated using the example of Figure 10.35 for the parameter **VFB**. As noted earlier (Figure 10.25), the individual device data do not fall on a straight line when plotted against the inverse channel length; this is clearly visible in Figure 10.35. Here the longest channel device (L_{eff} = 19.76 μm) is used as an infinitely long ($\frac{1}{L_{eff}}$ = 0) reference. The geometry parameter **LVFB** is determined for each bin by plotting a straight line from each small geometry data point back to the reference point; the slope of the line is **LVFB**. For the region of L_{eff} > 19.76 μm, **LVFB** is set to zero.

Using this method, an entire binned model is constructed. Due to its large size, this completed model is contained in Appendix E.

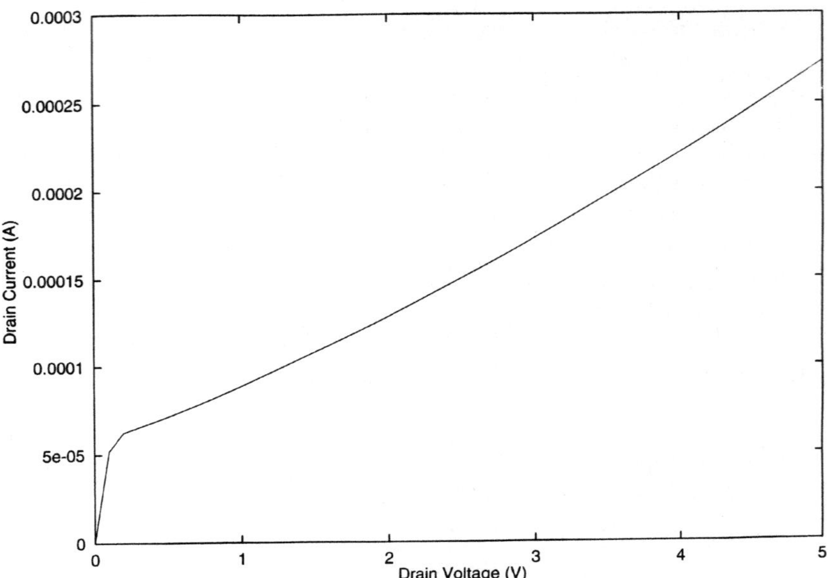

Figure 10.32 Final model result for the 20/0.7 device with a gate bias of $V_{gs} = 1.0\,\text{V}$. Note that in contrast to the BSIM result (Figure 8.24), there is no negative output conductance for high drain biases.

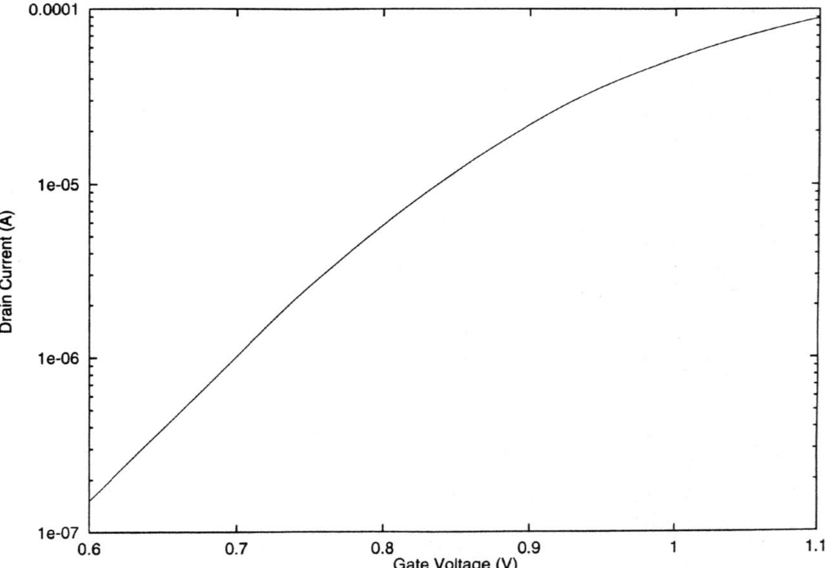

Figure 10.33 Detail of the final model result for the 20/0.7 device in the weak-strong inversion transition region, $V_{ds} = 0.1\,\text{V}$. Note that the transition region is smooth, while in the BSIM result (Figure 8.25), there is a jump in the characteristic near $V_{gs} = 0.9\,\text{V}$.

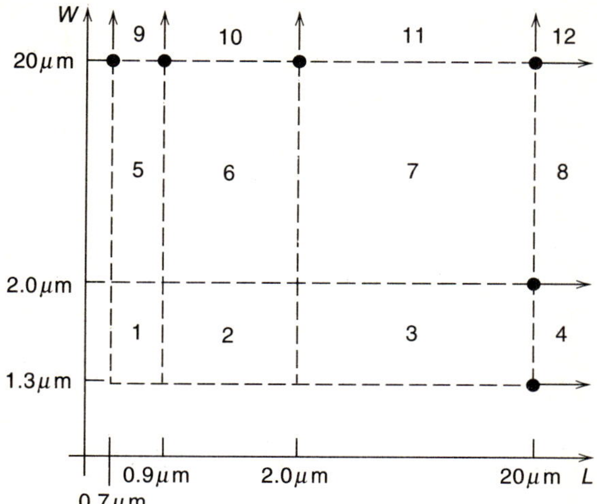

Figure 10.34   The binning scheme used here, resulting in twelve model bins.

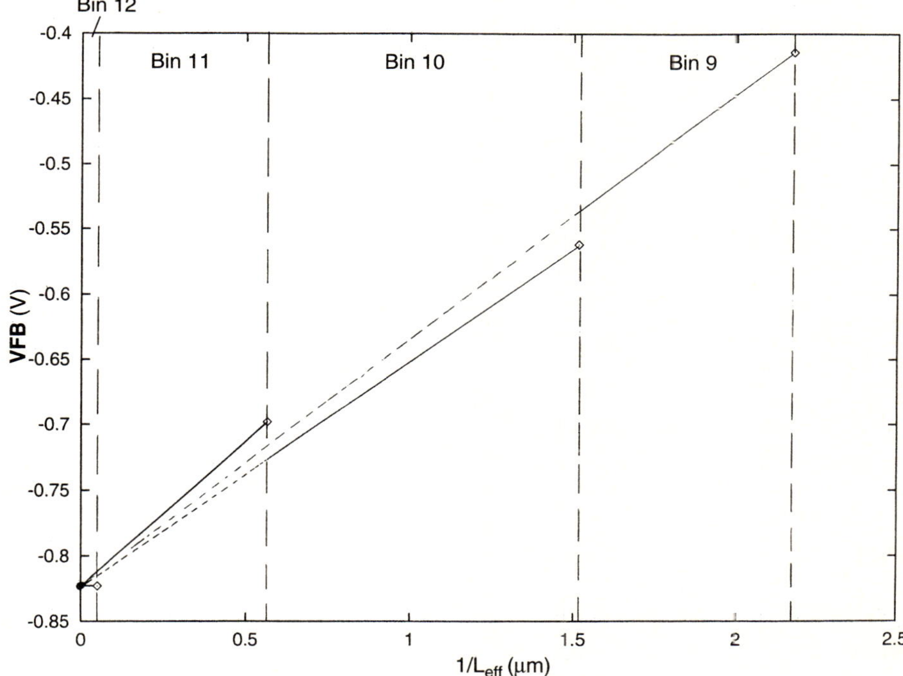

Figure 10.35   The individual device results for the model parameter **VFB** versus $\frac{1}{L_{eff}}$, depicting how the geometry parameter **LVFB** is determined for each bin.

10.9.4   "Playback" of the Binned Model

For the analysis of the binned model, the same set of devices used for playback of the simple model (Figures 10.26–10.31) will be used here. These results are assembled in Figures 10.36–10.41; the interested reader is encouraged to browse these results in detail. It is most important to note that the binned results are much closer to the individual device results of the second phase of parameter extraction than are the results of the unbinned model. This would be expected, as the binning scheme anchors the results at the given geometry data points, while the unbinned scheme does its best to reach a global approximation. For example, consider the saturation characteristics of the 20/0.7 device at the end of the second phase (Figure 10.24), in the unbinned model (Figure 10.26b), and in the binned model (Figure 10.36b). The unbinned model underpredicts the current, while the binned model is more accurate, with a slight underprediction of the current; note also that the second phase result (Figure 10.24) and the binned result (Figure 10.36b) are virtually identical.

The intermediate geometry results are also improved. Note once again that the saturation characteristics are now virtually identical to the second phase results. For example, the final fit for the 20/0.9 device (Figure 10.38b) is now very good. The narrow channel results are also improved, as a comparison of Figure 10.29b and Figure 10.39b (the 1.3/20 device) clearly shows. However, the short, narrow (1.3/0.7) results (Figure 10.41) are still very poor, and the saturation characteristic still shows the kink which occurred earlier. With no method of properly describing short, narrow devices, both the second-generation extrinsic structure, and any binning scheme used with it, are severely limited. Circuit designers attempting to use short, narrow devices should proceed with great caution.

Finally, as with the first-generation binning scheme described in Chapter 7 for Level 3, the binning method described here is imposed after the fact on the model structure. Since this model structure is not designed for binning, it cannot properly accommodate the process; the bins are constructed in isolation from each other, which causes model discontinuities at the bin boundaries. For example, the drain and gate biases can be set at 5.0 V, the channel width set at 20 μm, and the drain current plotted against the channel length; the results are depicted in Figure 10.42. A large discontinuity in the drain current occurs at the bin boundary of $L_{drawn} = 0.9$ μm. As noted in Chapter 7, this occurrence can have unpleasant consequences. For example, if statistical simulation runs involving process variations are used in the neighborhood of this bin boundary, very poor results will be obtained. Also, if a channel length is near this boundary and is optimized in a circuit design, the results will be unpredictable.

10.10   Final Comments

BSIM2 is, in essence, an improved version of BSIM, and so shows the basic second-generation behavior. As with the other second-generation models, extensive mathematical conditioning is introduced, resulting in a model formulation that is highly empirical.

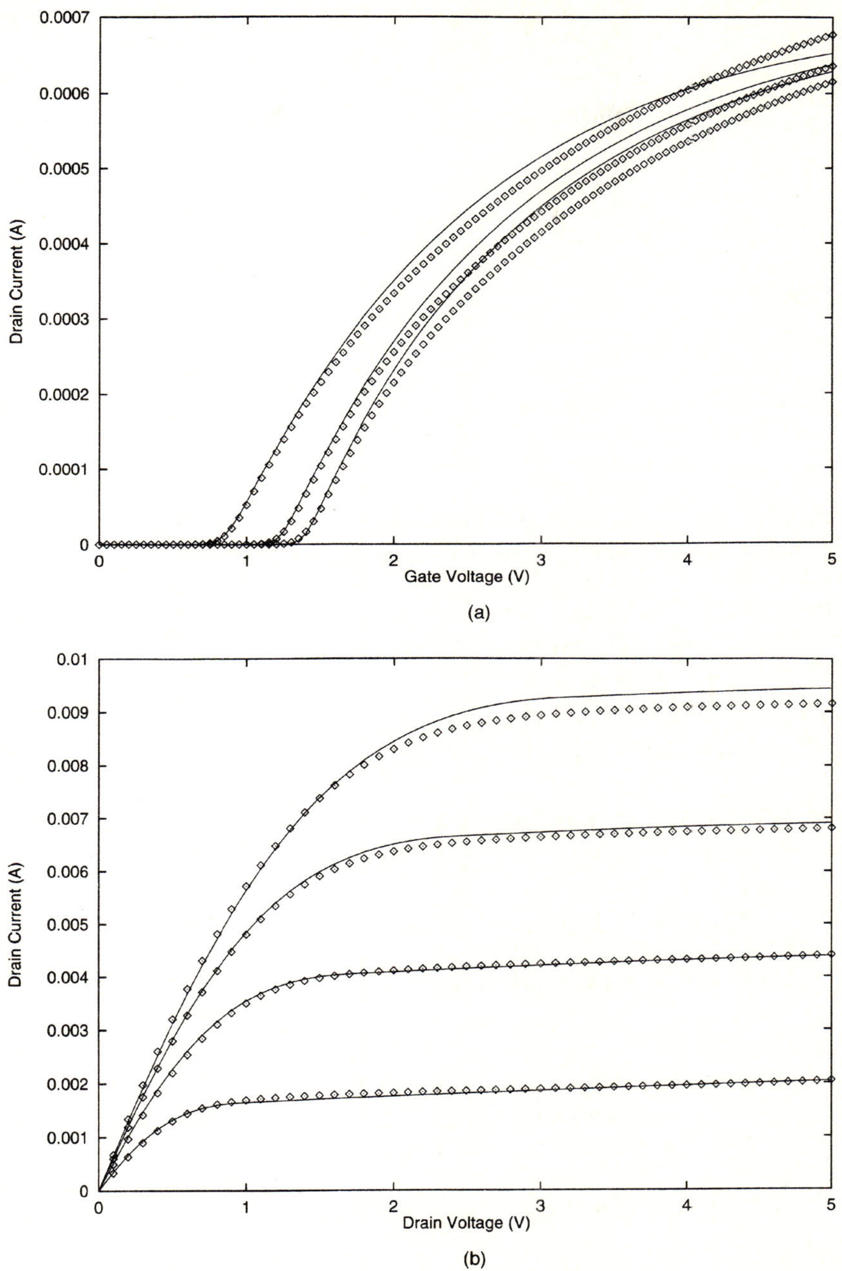

Figure 10.36 Comparison of measured data and the binned model for the 20/0.7 device; (a) linear characteristic; (b) saturation characteristic.

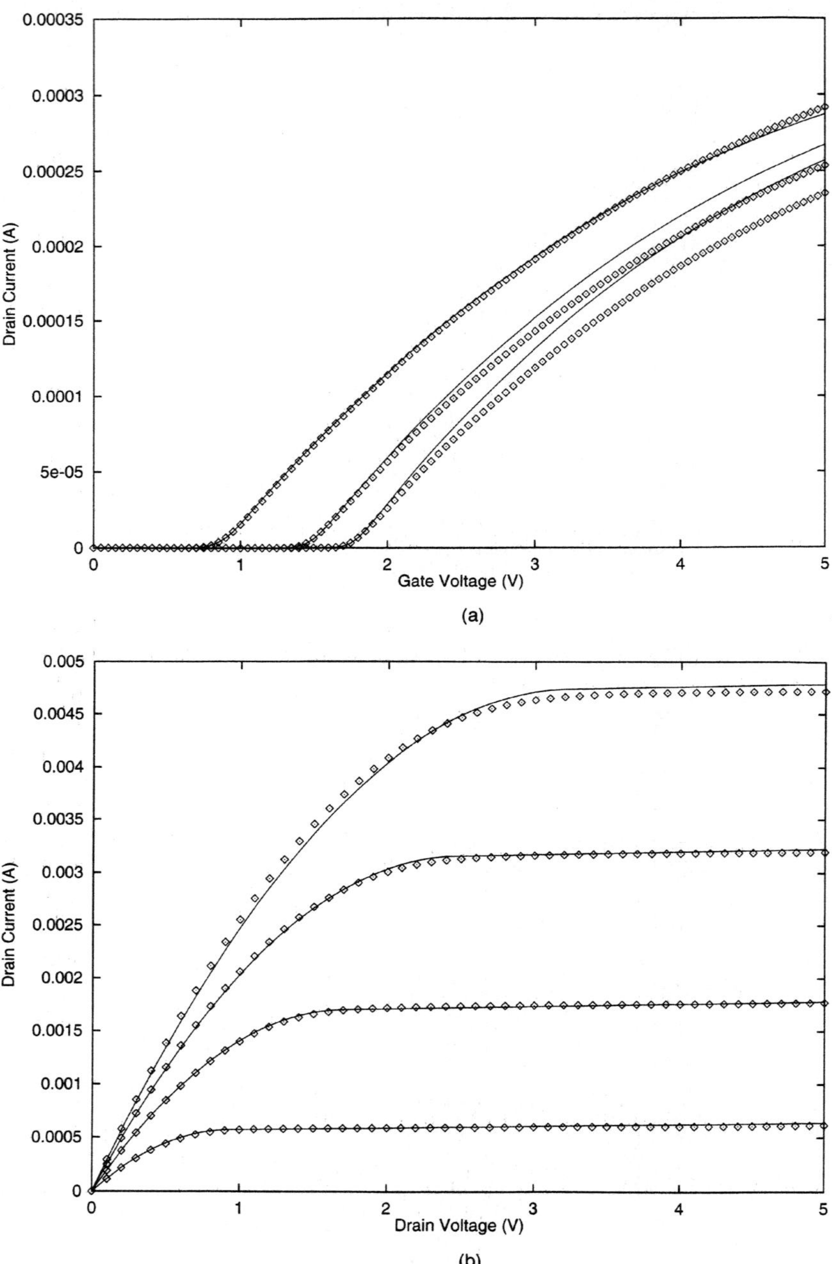

Figure 10.37 Comparison of measured data and the binned model for the 20/2.0 device; (a) linear characteristic; (b) saturation characteristic.

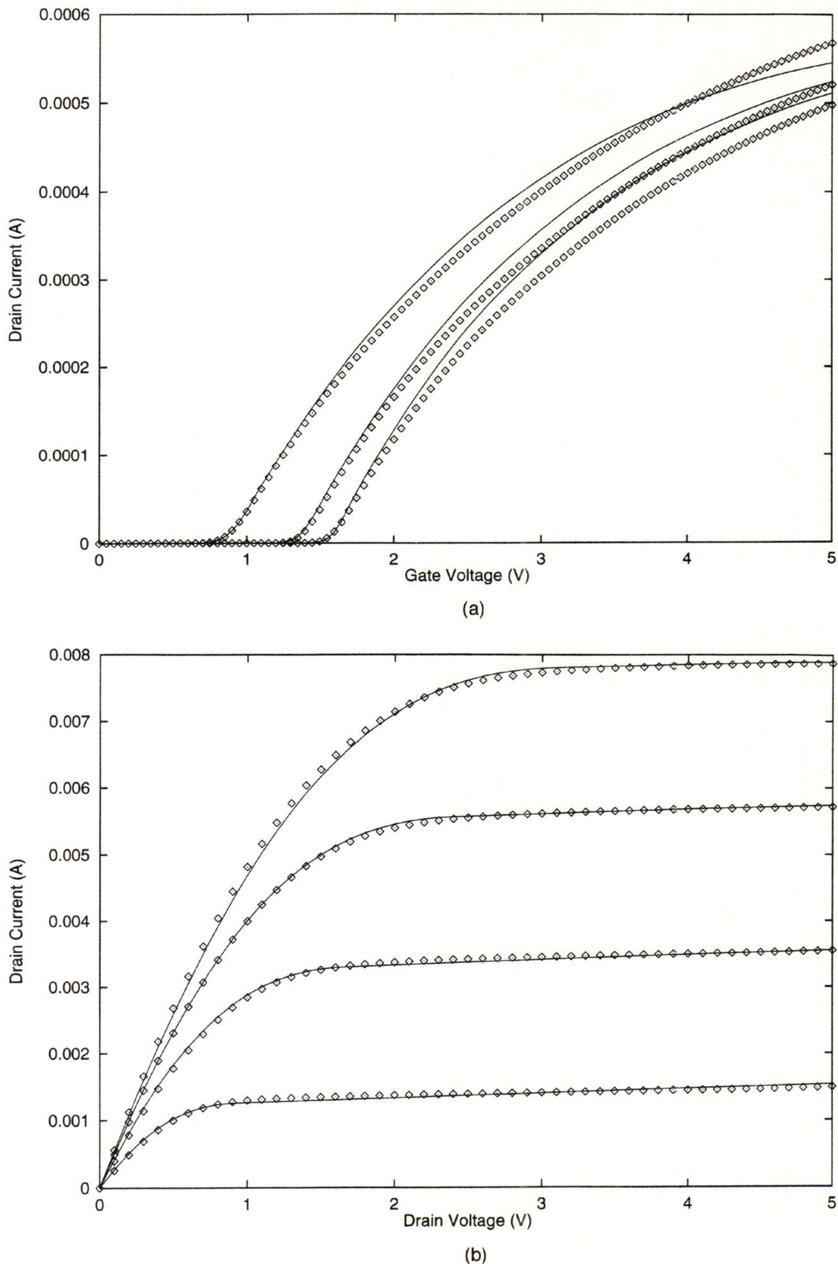

Figure 10.38 Comparison of measured data and the binned model for the 20/0.9 device; (a) linear characteristic; (b) saturation characteristic.

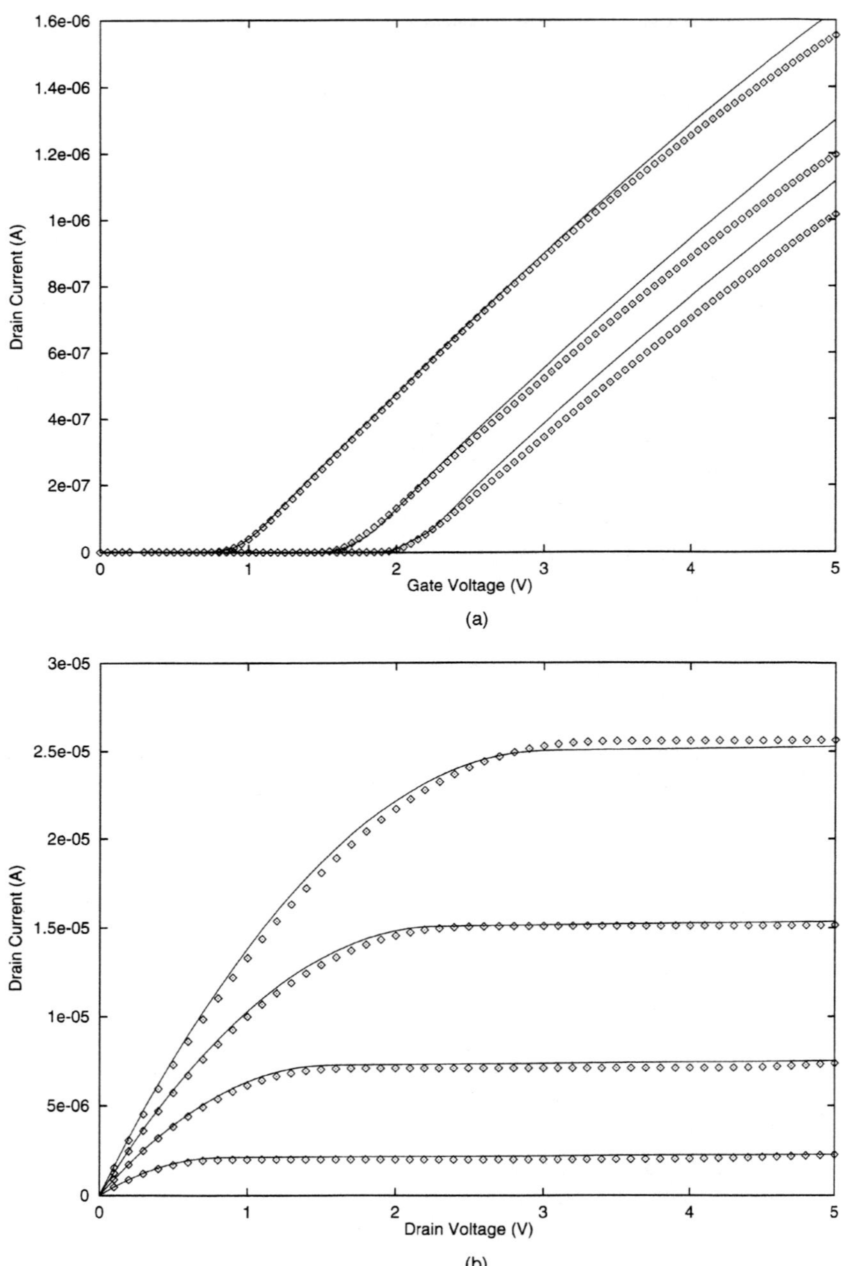

Figure 10.39 Comparison of measured data and the binned model for the 1.3/20 device; (a) linear characteristic; (b) saturation characteristic.

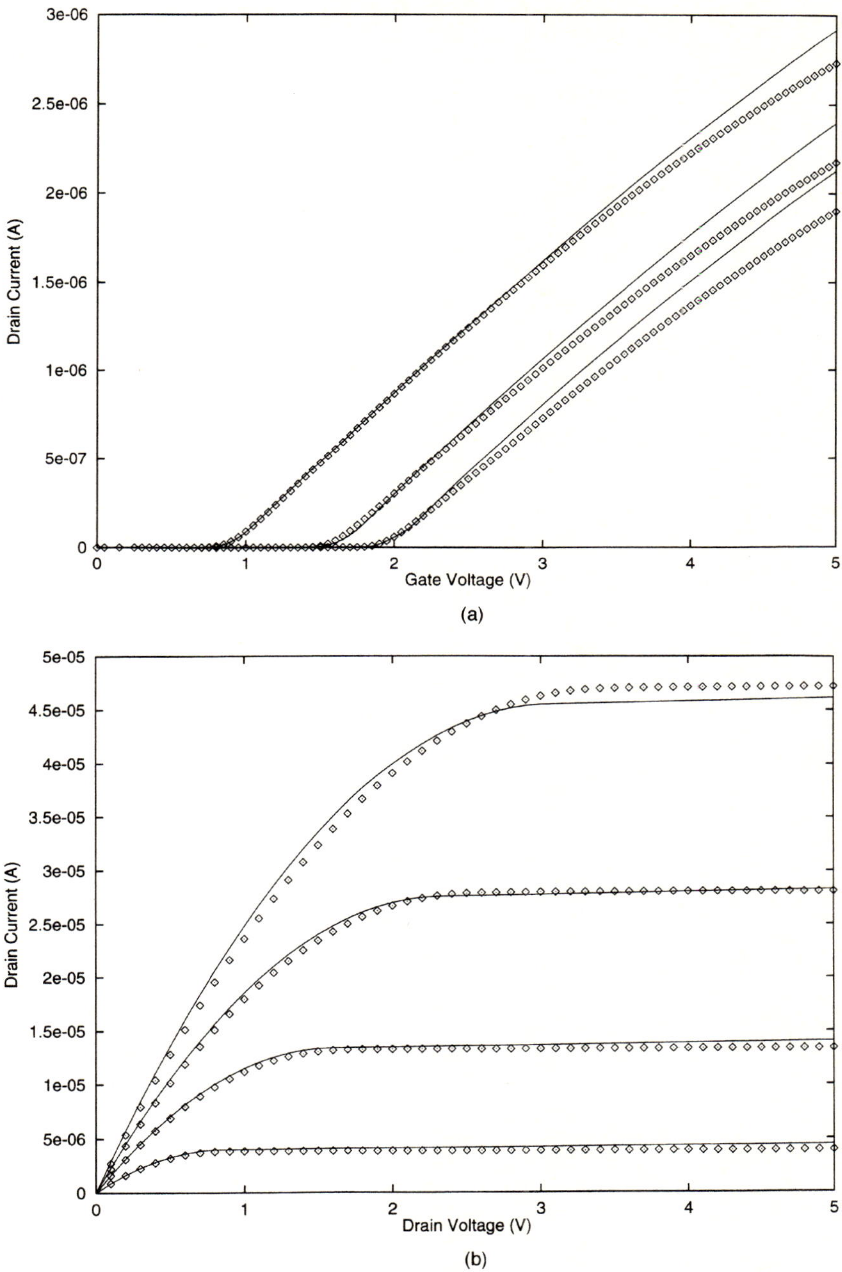

Figure 10.40 Comparison of measured data and the binned model for the 2.0/20 device; (a) linear characteristic; (b) saturation characteristic.

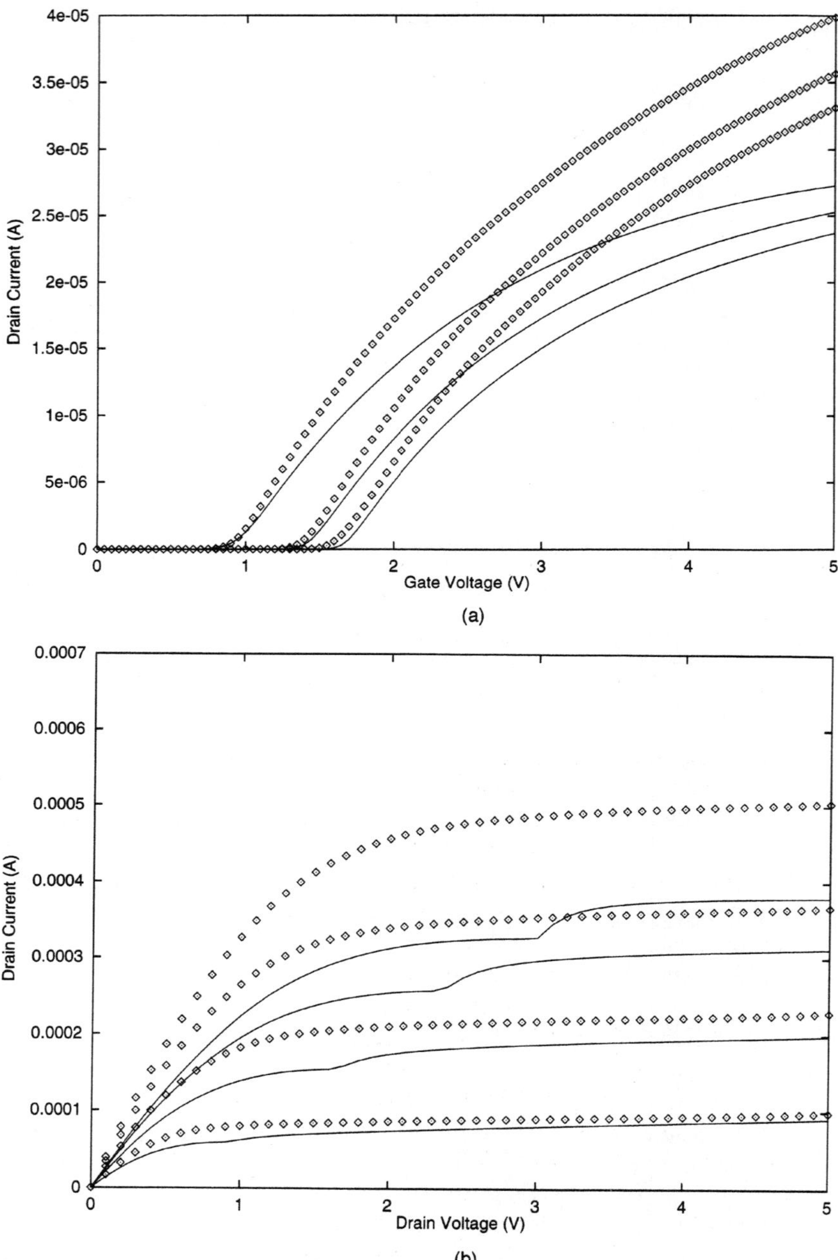

Figure 10.41   Comparison of measured data and the binned model for the 1.3/0.7 device; (a) linear characteristic; (b) saturation characteristic.

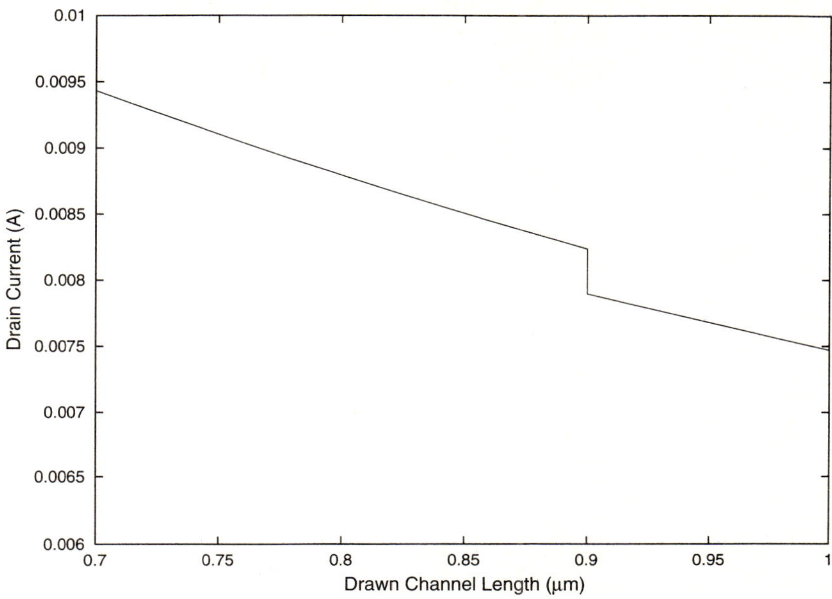

Figure 10.42   The modeled drain current versus channel length, $V_{gs} = V_{ds} = 5.0$ V, showing the discontinuity at the $L = 0.9$ μm bin boundary.

Once again, the model is basically a shell, which places the burden of final model quality quite firmly on parameter extraction.

The improvements that are contained in BSIM2 are quite substantial. The difficulties with polynomial functions in BSIM are repaired here, which allows BSIM2 to be useful for submicron channel lengths. More accurate results are possible in the linear-saturation transition region of the saturation characteristics; also, the behavior of the saturation characteristics with a substrate bias applied is qualitatively correct in BSIM2. The negative output conductance which can occur in BSIM for low gate biases and high drain biases does not occur in BSIM2, and the weak-strong inversion transition, which can show a jump for low drain biases in BSIM, is smooth in BSIM2.

In addition, the extrinsic structure, although unchanged from BSIM, provides better fitting results in BSIM2, especially for intermediate geometries. The intrinsic model formulation contains additional model parameters for the description of the output conductance; this makes BSIM2 suitable for use in analog circuit design. However, since the second-generation extrinsic structure is used, short and narrow channel effects are treated separately but not in combination; therefore, poor results are obtained for short, narrow devices. BSIM2 (as well as BSIM) can be binned; however, the basic second-generation structure prevents the binning scheme from dealing with short, narrow devices. Binning can improve the model results; however, the extrinsic model structure was not designed to accommodate binning; model discontinuities will occur at bin boundaries. For example, the drain current will be discontinuous across bin boundries, which can have unpleasant consequences.

Perhaps the most important drawback of BSIM2 is its complexity; this complexity is the price that is paid for improved accuracy. The complicated model structure renders BSIM2 rather slow and inefficient during circuit simulation. Also, there is a large number of model parameters, which can be tedious to supervise. This complexity provided interest in finding approaches for model simplification, which became the original intent of the third generation of analytical FET models. These models are discussed in the next section.

BIBLIOGRAPHY

1. M. Jeng, "Design and Modeling of Deep Submicrometer MOSFETs," University of California/Berkeley, Electronics Research Laboratory Memorandum No. UCB/ERL M90/90 (1990).

2. C. Wan and B. Sheu, "Temperature Dependence Modeling for MOS VLSI Circuit Simulation," *IEEE Trans. Comp. Aid. Des.* vol. CAD-8, pp. 1065–1072 (1989).

3. *HSPICE User's Manual*, Meta-Software, Inc., Campbell, California, 1993.

4. S. Sze, *Physics of Semiconductor Devices* (2nd ed.), John Wiley & Sons, 1981.

5. D. Caughey and R. Thomas, "Carrier Mobilities in Silicon Empirically Related to Doping and Field," *Proc. IEEE* vol. 55, pp. 2192–2193 (1967).

6. C. Sodini, P. Ko, and J. Moll, "The Effect of High Fields on MOS Device and Circuit Performance," *IEEE Trans. Elec. Dev.* vol. ED-31, pp. 1386–1393 (1984).

7. J. Brews, "A Charge Sheet Model of the MOSFET," *Sol. St. Elec.* vol. 21, pp. 345–355 (1978).

8. Y. El-Mansy and A. Boothroyd, "A Simple Two Dimensional Model for IGFET Operation in the Saturation Region," *IEEE Trans. Elec. Dev.* vol. ED-24, pp. 254–262 (1977).

9. P. Ko, R. Muller, and C. Hu, "A Unified Model for Hot Electron Currents in MOSFETs," *1981 IEDM Tech. Dig.*, pp. 600–603.

10. C. Hu *et al.*, "Hot Electron Induced MOSFET Degradation - Model, Monitor, and Improvement," *IEEE Trans. Elec. Dev.* vol. ED-32, pp. 375–385 (1985).

11. B. Sheu, "MOS Transistor Modeling and Characterization for Circuit Simulation," University of California/Berkeley, Electronics Research Laboratory Memorandum No. UCB/ERL M85/85 (1985).

11
BSIM3

11.1   Introduction

As detailed in Chapter 8, the Level 1, Level 2, and Level 3 models comprise the first generation of SPICE FET models. These models focus on detailed analytical descriptions of device behavior, which leads to a small number of parameters that are relatively easy to extract. The successors to these models are the second-generation models BSIM, HSPICE Level 28, and BSIM2. These models place more emphasis on mathematical conditioning of the model equations for robust and efficient circuit simulation. However, these equations are constructed on a semi-empirical and even empirical basis; thus, the model parameters are largely empirical in both structure and numerical values, providing little physical information about the process technology which they describe. The number of parameters is very large; the real information content of the second-generation models (versus the first-generation models) has shifted from the model equations to the model parameter set.

BSIM3 [1, 2] and some other new models represent another major change; they represent the emergence of the third generation of SPICE FET models. The most important general aspect of these models is an attempt to reintroduce a physical basis into the model form and its associated parameters, while maintaining the mathematical fitness of the model equations. The goal is to allow for a more concrete connection between the parameter set and the underlying process technology, while avoiding the complicated and often ill-behaved (but rather commonly used) polynomial expressions of the second-generation models.

BSIM3 itself has evolved through three different versions. The first version forms the original basis for the model, but had some severe mathematical problems. The second version was largely a correction of these mathematical difficulties, and several new parameters were introduced. The third version significantly changed the form of the model to guarantee continuous and smooth model equations; a number of empirical fitting expressions were also introduced, along with many additional parameters. This

379

third version is logically receiving the most attention, and is the version that is discussed here.

A careful examination of BSIM3 reveals that the form of its equations and terms is very much based on the BSIM and BSIM2 structures. Many of the BSIM3 equations are identical to the earlier models, or differ only in the content of one or two composite terms. The electrostatic charge and threshold voltage models include a number of additional effects, such as a more detailed description of nonuniform channel doping in both the vertical and lateral directions, while the mobility model includes a more realistic description of the effects of vertical and lateral fields. More importantly, a number of auxiliary functions are introduced, which allow for the construction of a single drain current equation that is valid for all regions of operation; the drain current expression is also continous and smooth across all of the transition regions. A very detailed model for the output conductance is developed; somewhat surprisingly, it is very closely based on the "λ-model" for the saturation region which was used in Level 1 and Level 2, but not employed in subsequent models. Finally, the charge model structure is virtually identical to that of BSIM, but also includes auxiliary functions; this allows for a continuous and smooth expression for each node charge equation for all regions of device operation, which represents a major improvement over earlier models.

A list of the model parameters used in BSIM3 may be found in Table 11.1.

Before discussing BSIM3 in detail, it is appropriate to first consider the general structure of the emerging third-generation models.

11.2   The Philosophy of the Third-Generation Models

As described in Section 8.2, the second-generation models shifted the focus of device modeling away from a description of the FET structure and toward the problems of circuit simulation usage of the FET model. The main emphasis was placed on developing a robust and efficient set of equations, which would provide good convergence behavior with a minimal amount of computational effort. As a result, the models took on a great deal of empirical character, with a very large number of model parameters. It is thus very difficult to make connections between a model parameter set and the process technology which that model describes. The second-generation models also constructed a separate extrinsic model structure on top of the conventional intrinsic structure to allow a more accurate description of the geometry dependence of device behavior.

The very large number of model parameters and the essentially empirical character of the second-generation models have caused many problems. Since the model structure is very complex, parameter extraction is difficult and tedious, and extensive care is required when the same extraction methodology is applied to different process technologies. The lack of a physical basis for the parameter set is not a concern for the description of individual device characteristics; however, the very weak connection to the underlying process technology makes it difficult to achieve a reasonable and efficient description of process variations (see Chapter 14). Finally, the extensive use of polynomial equations and other mathematical functions has led to occasional surprises, as these expressions can behave badly under certain circumstances.

Table 11.1   The BSIM3 model parameter set.

Parameter	Units	Description
Process Parameters		
TOX	m	Gate Oxide Thickness
LINT	m	Source/Drain Underdiffusion of Gate
WINT	m	Isolation Reduction of Channel Width
XJ	m	Junction Depth
NGATE	cm^{-3}	Gate Doping Concentration
Electrical Parameters		
LL	m^{LLN}	Coefficient for Length Dependence of Length
LLN		Power Coefficient for Length Dependence of Length
LW	m^{LWN}	Coefficient for Width Dependence of Length
LWN		Power Coefficient for Width Dependence of Length
LWL	$m^{(LLN+LWN)}$	Coefficient for $L \cdot W$ Product Dependence of Length
DWG	m/V	Gate Bias Dependence of Effective Channel Width
DWB	$m/V^{\frac{1}{2}}$	Substrate Bias Dependence of Effective Channel Width
WL	m^{WLN}	Coefficient for Length Dependence of Width
WLN		Power Coefficient for Length Dependence of Width
WW	m^{WWN}	Coefficient for Width Dependence of Width
WWN		Power Coefficient for Width Dependence of Width
WWL	$m^{(WLN+WWN)}$	Coefficient for $L \cdot W$ Product Dependence of Width
A0		Short Channel Bulk Charge Coefficient
AGS	V^{-1}	Gate Bias Effect on A_{bulk}
B0	m	Narrow Channel Bulk Charge Coefficient
B1	m	Offset to **B0**
KETA	V^{-1}	Substrate Bias Effect on the Bulk Charge
VOFF	V	Offset Voltage
VTH0	V	Threshold Voltage, Long, Wide Device, Zero Substrate Bias
NCH	cm^{-3}	Channel Doping Concentration
NSUB	cm^{-3}	Substrate Doping Concentration
NLX	m	Nonuniform Lateral Doping Parameter
K1	$V^{\frac{1}{2}}$	Body Effect on Threshold Voltage (First-Order Term)
K2		Body Effect on Threshold Voltage (Second-Order Term)
K3		Narrow Width Coefficient
K3B	V^{-1}	Substrate Bias Dependence of **K3**
W0	m	Narrow Width Parameter
DVT0		Short Channel Effect Coefficient 1
DVT1		Short Channel Effect Coefficient 2
DVT2	V^{-1}	Substrate Bias Dependence of Short Channel Effects
DVT0W	m^{-1}	Width Effect on Short Channel Effect Coefficient 1
DVT1W	m^{-1}	Width Effect on Short Channel Effect Coefficient 2
DVT2W	V^{-1}	Substrate Bias Influence of Width on Short Channel Effects
DSUB		Substrate Bias Effect on DIBL
ETA0		Subthreshold DIBL Coefficient
ETAB		Substrate Bias Effect on Subthreshold DIBL
MOBMOD		Mobility Selection Parameter
UO	$cm^2/V{\cdot}s$	Low Field Mobility
UA	m/V	Gate Field Induced Mobility Reduction Parameter (First Order)
UB	m^2/V^2	Gate Field Induced Mobility Reduction Parameter (Second Order)

Table 11.1 *(cont.)*

Parameter	Units	Description
Electrical Parameters (cont.)		
UC	m/V^2	Substrate Bias Dependence of Mobility Reduction
VSAT	m/s	Carrier Saturation Velocity, $T = T_{nom}$
RDSW	$\Omega\text{-}\mu m^{\mathbf{WR}}$	Series Resistance Per Unit Width
PRWG	V^{-1}	Gate Bias Dependence of the Series Resistance
PRWB	$\Omega\text{-}\mu m/V^2$	Substrate Bias Dependence of the Series Resistance
WR	m	Narrow Channel Effect on the Series Resistance
A1	V^{-1}	Saturation Voltage Fitting Parameter 1
A2		Saturation Voltage Fitting Parameter 2
DELTA	V	V_{dsx} Smoothing Parameter
PCLM		Channel Length Modulation Parameter
PDIBLC1		Output Conductance DIBL Parameter 1
PDIBLC2		Output Conductance DIBL Parameter 2
PDIBLCB	V^{-1}	Substrate Bias Dependence of DIBL
DROUT		Short Channel Correction to DIBL Effect on Output Conductance
PVAG		Gate Bias Dependence of the Early Voltage
PSCBE1	V/m	SCBE Parameter 1
PSCBE2	V/m	SCBE Parameter 2
NFACTOR		Subthreshold slope coefficient
CDSC	F/m^2	Drain/Source—Channel Coupling Capacitance
CDSCB	$F/V{\cdot}m^2$	Substrate Bias Dependence of **CDSC**
CDSCD	$F/V{\cdot}m^2$	Drain Bias Dependence of **CDSC**
CIT	F/m^2	Interface Trap Capacitance
ALPHA0	m/V	Substrate Current Fitting Parameter 1
BETA0	V	Substrate Current Fitting Parameter 2
DLC	m	Capacitive Channel Length Reduction
DWC	m	Capacitive Channel Width Reduction
CLC	m	Short Channel Charge Coefficient 1
CLE		Short Channel Charge Coefficient 2
CGS0	F/m	Zero Bias Gate-Source Overlap Capacitance
CGS1	F/m	Correction to **CGS0**
CGD0	F/m	Zero Bias Gate-Drain Overlap Capacitance
CGD1	F/m	Correction to **CGD0**
CKAPPA	F/m	Fringing Field Coefficient
CGB0	F/m	Zero Bias Gate-Bulk Capacitance
Temperature Parameters		
KT1	V	Temperature Effect on Short Channel Threshold Voltage
KT1L	V·m	Channel Length Dependence of Temperature Sensitivity
KT2		Temperature Effect on Substrate Bias Effect on Threshold Voltage
UTE		Temperature Effect on the Low Drain Bias Mobility
UA1	m/V	Temperature Coefficient for $U_a(T)$
UB1	m^2/V^2	Temperature Coefficient for $U_b(T)$
UC1	m/V^2	Temperature Coefficient for $U_b(T)$
AT	m/s	Temperature Effect on the Carrier Saturation Velocity
PRT	$\Omega\text{-}\mu m$	Temperature Dependence of **RDSW**

The third generation of FET models attempts to address these concerns. Since these models are very new, the situation is quite fluid, and the structure of these models cannot yet be considered final; also, it is not yet clear which models will find widespread acceptance. To date, this generation of models includes BSIM3 from the University of California/Berkeley [1, 2] and MOS Model 9 from Philips Electronics [3]. These models are presently being implemented in commonly used circuit simulators, and so are discussed in detail here. In addition, other candidate third-generation models are discussed in Chapter 16. It is not known if any of these particular models will find their way into the simulation mainstream; however, they also provide examples of the future directions that studies of MOS modeling for circuit simulation are taking. The reader should also note that additional analytical FET models are regularly published in the technical literature.

With the available information, the general characteristics of the third-generation FET models can be described. Their basic structure is actually very similar to the first-generation models, in that all geometry dependence is in the intrinsic model structure, while the extrinsic geometry structure of the second-generation models (associated with the L and W parameters) is discarded. The model equations are taken to be valid for all geometries, and thus the models are not designed to accommodate binning. The exception to this description is MOS Model 9, which actually extends the extrinsic structure of the second-generation models.

The models contain a small (compared to the second-generation models) number of physically based parameters, with the intention of providing a better description of both the underlying process technology and the variations associated with that process. The exception to this statement is BSIM3, which started out with this structure, but has since evolved into a considerably more empirical form with a large number of model parameters.

The most definite common characteristic of all the third generation models is the extensive use of *smoothing functions*, which guarantee continuous and smooth behavior of the model equations across all regions of device operation. Separate equations are derived to describe the different regions of device operation; these equations are then linked using one or more of these smoothing functions. The smoothing functions serve to reduce the expressions to the separate-region expressions in the appropriate limit, while providing a well-behaved intermediate transition region across the operational boundaries. In this way, a single equation is made to be valid for all regions of device operation; for example, a model contains one drain current equation, one equation for each node charge, etc.

Finally, most of these models include descriptions of some small geometry effects that have been ignored in the past. These small geometry effects are described in Section 3.5.

As noted earlier, the third-generation models are very new. As yet, there is virtually no data base of experience for either the extraction of the model parameters or the behavior of these models when employed in a circuit simulator. This kind of usage also provides invaluable insight into the quality of the model structure, both for the description of various process technologies and for the simulation of different types of circuits. At this point, it is only possible to comment on the model structure, rather than its behavior when actually used for modeling and simulation; for example, it is not yet possible to make a definitive statement about any possible need for model binning.

11.3 The Electrostatic Model of the MOSFET Structure

11.3.1 The Inversion Charge Density in Strong Inversion

In a simple two-terminal MOS capacitor (see Section 3.3) or an FET with a very small drain bias, the inversion charge density can, to a reasonable approximation, be written as

$$Q_{inv} = -C_{ox} \cdot (V_{gs} - V_t). \tag{11.1}$$

In the development of BSIM [4], this expression was modified to account for the influence of a larger drain bias. When a drain bias is applied, the depletion charge density is no longer uniform along the channel; the drain bias increases the size of the depletion region near the drain diffusion. As a result, the depletion charge density increases there, so that the inversion charge density decreases in the same region. In BSIM, (11.1) is modified semi-empirically, leading to an expression for the average inversion charge density along the channel:

$$Q_{inv} = -C_{ox} \cdot (V_{gs} - V_t - a \cdot \phi_n), \tag{11.2}$$

where ϕ_n is the electron quasi-Fermi level due to the applied drain bias and a is a composite term which was derived in Section 8.7. When no drain bias is applied, $\phi_n \to 0$, and (11.2) reduces to (11.1).

BSIM3 uses the same approach, except that the BSIM term a is replaced with a much more complex expression:

$$A_{bulk} = \left(1 + \frac{\mathbf{K1}}{2 \cdot (\phi_s - V_{bsx})^{\frac{1}{2}}} \cdot \left\{ \frac{\mathbf{A0} \cdot L_{eff}}{L_{eff} + 2 \cdot (\mathbf{XJ} + x_d)} \cdot \left[1 - \mathbf{AGS} \cdot V_{gsx} \right. \right. \right.$$
$$\left. \left. \left. \times \left(\frac{L_{eff}}{L_{eff} + 2 \cdot (\mathbf{XJ} + x_d)} \right)^2 \right] + \frac{\mathbf{B0}}{W_{eff} + \mathbf{B1}} \right\} \right) \cdot \frac{1}{1 + \mathbf{KETA} \cdot V_{bsx}}, \tag{11.3}$$

where x_d is the gate-induced depletion depth derived in Section 3.3:

$$x_d = \left[\frac{2\epsilon_{Si} \cdot (\phi_s - V_{bsx})}{q \cdot \mathbf{NCH}} \right]^{\frac{1}{2}}; \tag{11.4}$$

NCH is the *surface* doping (rather than the *substrate* doping), as is explained in greater detail in Section 11.4, while ϕ_s is the surface potential. In BSIM3, **XJ** is intended to be an input parameter representing the physical depth of the source/drain junctions in the silicon; however, in all earlier models, **XJ** ends up being an extracted parameter whose value has no resemblance to the actual junction depth. It remains to be seen what will happen in BSIM3.

In the first-generation models, the substrate doping is a parameter (**NSUB**), and the surface potential is computed from it. In the second-generation models, the substrate doping parameter is discarded, and the surface potential appears as an extracted parameter (**PHI**) which serves the same purpose. In BSIM3, as just noted, the doping reappears as a parameter; at the surface, the doping is **NCH**. Although ϕ_s can be specified in BSIM3 as a parameter, it is redundant with **NCH**. The approach to be recommended here is to use the method of the first-generation models, and to treat **NCH** as a model parameter and compute ϕ_s from it. Using the basic approach discussed in Section 3.2, the surface potential is computed from

$$\phi_s = 2 \cdot \frac{k_b T}{q} \cdot ln\left(\frac{\mathbf{NCH}}{n_i(T)}\right). \tag{11.5}$$

The term $n_i(T)$ is the intrinsic carrier concentration, computed in BSIM3 using

$$n_i(T) = 1.45 \times 10^{10} \cdot \left(\frac{T}{300.15}\right)^{\frac{3}{2}} \cdot exp\left(21.5565981 - \frac{q}{2k_b T} \cdot E_g(T)\right), \tag{11.6}$$

where $E_g(T)$ is the silicon band gap energy value, computed here from

$$E_g(T) = 1.16 - \frac{7.02 \times 10^{-4} \cdot T^2}{T + 1108}. \tag{11.7}$$

In both (11.6) and (11.7), the temperature is specified in **degrees Kelvin** rather degrees Celsius.

The term V_{bsx} is an auxiliary substrate bias expression, which serves to prevent the applied substrate bias from becoming too large:

$$V_{bsx} = V_{BC} + \frac{1}{2}\left\{V_{bs} - V_{BC} - \delta_1 + \left[(V_{bs} - V_{BC} - \delta_1)^2 - 4 \cdot \delta_1 \cdot V_{BC}\right]^{\frac{1}{2}}\right\}, \tag{11.8}$$

where V_{BC} is given by

$$V_{BC} = 0.9 \cdot \left(\phi_s - \frac{\mathbf{K1}^2}{4 \cdot \mathbf{K2}^2}\right); \tag{11.9}$$

the model parameters **K1** and **K2** are discussed in Section 11.4. The term δ_1 is a constant, which is set to $\delta_1 = 0.02$. Although (11.8) is quadratic, the origin of this expression is unclear. Thus, in BSIM3, (11.2) becomes

$$Q_{inv}(y) = -C_{ox} \cdot (V_{gs} - V_t - A_{bulk} \cdot \phi_n(y)). \tag{11.10}$$

11.3.2   The Combined Description of the Inversion Charge Density

In weak inversion, a completely different description of the inversion charge density is required. This description must be connected with the strong inversion expression, so that a smooth result is produced across the weak inversion-strong inversion transition. BSIM3 uses the basic third-generation approach of deriving the weak and strong inversion expressions separately, then defining a smoothing function to mediate between the two regions in a manner which leads to a single equation which is valid over the entire range of operation.

With a small drain bias applied, the inversion charge density in weak inversion can be written as [5]

$$Q_{inv,w,V_{ds}=0} = Q_o \cdot exp\left(\frac{q}{k_b T} \cdot \frac{V_{gs} - V_t}{n}\right), \tag{11.11}$$

where n is the *ideality factor* for the subthreshold slope, given approximately by [6]

$$n = 1 + \frac{C_{depl}}{C_{inv}}, \tag{11.12}$$

and detailed below, while

$$Q_o = \left(\frac{q\epsilon_{Si} \cdot \mathbf{NCH}}{2\phi_s}\right)^{\frac{1}{2}} \cdot \frac{k_b T}{q} \cdot exp\left(-\frac{q}{k_b T} \cdot \frac{\mathbf{VOFF}}{n}\right); \tag{11.13}$$

the term **VOFF** is an offset voltage, which was introduced in BSIM2 (see Section 10.5) to smooth the weak inversion-strong inversion transition region.

In strong inversion, as noted above in (11.1),

$$Q_{inv,s,V_{ds}=0} = -C_{ox} \cdot (V_{gs} - V_t). \tag{11.14}$$

To smooth the transition region, and to arrive at a single equation for the entire operating range, BSIM3 introduces an auxiliary gate bias expression, given by

$$V_{gsx} = \frac{2 \cdot \frac{k_b T}{q} \cdot ln\left[1 + exp\left(\frac{q}{k_b T} \cdot \frac{V_{gs} - V_t(T)}{2 \cdot n}\right)\right]}{1 + 2 \cdot n \cdot C_{ox} \cdot \left(\frac{2\phi_s}{q\epsilon_{Si} \cdot \mathbf{NCH}}\right)^{\frac{1}{2}} \cdot exp\left(-\frac{q}{k_b T} \cdot \frac{V_{gs} - V_t(T) - 2 \cdot \mathbf{VOFF}}{2 \cdot n}\right)}, \tag{11.15}$$

where n, the subthreshold ideality factor, is described by an expression more detailed than (11.12):

$$n = 1 + \mathbf{NFACTOR} \cdot \frac{C_{depl}}{C_{ox}} + \frac{1}{C_{ox}} \cdot (\mathbf{CDSC} + \mathbf{CDSCB} \cdot V_{bsx} + \mathbf{CDSCD} \cdot V_{ds})$$

$$\times \left[exp\left(-\frac{\mathbf{DVT1} \cdot L_{eff}}{2 \cdot L_t}\right) + 2 \cdot exp\left(-\frac{\mathbf{DVT1} \cdot L_{eff}}{L_t}\right)\right] + \frac{C_{it}}{C_{ox}}, \tag{11.16}$$

where C_{depl} is found from

$$C_{depl} = \frac{\epsilon_{Si}}{x_d}. \tag{11.17}$$

The term x_d is the gate-induced depletion depth computed in Section 3.3,

$$x_d = \left(\frac{2\epsilon_{Si} \cdot (\phi_s - V_{bsx})}{q \cdot \mathbf{NCH}}\right)^{\frac{1}{2}}, \tag{11.18}$$

while L_t is a *characteristic length*, given by

$$L_t = \left(\frac{\epsilon_{Si} \cdot x_d}{C_{ox}}\right)^{\frac{1}{2}} \cdot (1 + \mathbf{DVT2} \cdot V_{bsx}). \tag{11.19}$$

The term C_{it} is the capacitance due to interface traps. In modern MOS technology, the interface trap density is usually very low ($< 10^{10}/\text{cm}^2$), so C_{it} can be neglected in most situations.

The inversion charge density for the entire range of operation is thus found from

$$Q_{inv,V_{ds}=0} = C_{ox} \cdot V_{gsx}. \tag{11.20}$$

When $V_{gs} \gg V_t$,

$$V_{gsx} \rightarrow (V_{gs} - V_t), \tag{11.21}$$

and the simple strong inversion expression (11.14) is recovered. When $V_{gs} \ll V_t$, V_{gsx} reduces to the weak inversion region expression (11.11). The use of V_{gsx} ensures a continuous and smooth expression across the transition region.

Now consider the effect of the drain bias. In strong inversion, the inversion charge density is given by (11.10):

$$Q_{inv}(y) = -C_{ox} \cdot (V_{gs} - V_t - A_{bulk} \cdot \phi_n(y)). \tag{11.22}$$

The drain bias has the effect of changing the inversion charge density by an amount $\Delta Q_{inv,s}$, which is easily found by subtracting (11.14) from (11.22):

$$\Delta Q_{inv,s} = -C_{ox} \cdot A_{bulk} \cdot \phi_n(y). \tag{11.23}$$

In the subthreshold region, when a drain bias is applied, (11.11) becomes

$$Q_{inv,w}(y) = Q_o \cdot exp\left(\frac{q}{k_bT} \cdot \frac{V_{gs} - V_t - A_{bulk} \cdot \phi_n(y)}{n}\right)$$

$$= Q_{inv,w,V_{ds}=0} \cdot exp\left(-\frac{q}{k_bT} \cdot \frac{A_{bulk} \cdot \phi_n(y)}{n}\right). \tag{11.24}$$

To simplify this expression, a first-order Taylor expansion is performed on the exponential term, which leads to

$$Q_{inv,w}(y) = Q_{inv,w,V_{ds}=0} \cdot \left(1 - \frac{q}{k_b T} \cdot \frac{A_{bulk} \cdot \phi_n(y)}{n}\right). \tag{11.25}$$

In a manner identical to that used to develop (11.23), the change in the weak inversion charge density due to the drain bias can be found by subtracting (11.11) from (11.25):

$$\Delta Q_{inv,w} = -\frac{q}{k_b T} \cdot \frac{A_{bulk} \cdot \phi_n(y)}{n} \cdot Q_{inv,w,V_{ds}=0}. \tag{11.26}$$

Some method of linking the weak inversion and strong inversion charge expressions must be found. In BSIM3, the changes in the inversion charge due to the application of a drain bias (11.23) and (11.26) are connected using a familiar expression:

$$\Delta Q_{inv}(y) = \frac{\Delta Q_{inv,s}(y) \cdot \Delta Q_{inv,w}(y)}{\Delta Q_{inv,s}(y) + \Delta Q_{inv,w}(y)}. \tag{11.27}$$

The form of this equation is identical to that used to smooth the connection between the weak inversion and strong inversion currents in BSIM (see (8.146)). The limiting behavior of (11.27) is easy to find. When $\Delta Q_{inv,s}(y) \gg \Delta Q_{inv,w}(y)$,

$$\Delta Q_{inv}(y) \rightarrow \Delta Q_{inv,w}(y); \tag{11.28}$$

when $\Delta Q_{inv,s}(y) \ll \Delta Q_{inv,w}(y)$,

$$\Delta Q_{inv}(y) \rightarrow \Delta Q_{inv,s}(y). \tag{11.29}$$

The use of (11.27) ensures a smooth and continuous expression for the inversion charge density along the channel for all regions of operation.

Substituting (11.23) and (11.26) into (11.27) leads to

$$\Delta Q_{inv}(y) = \frac{\phi_n(y)}{V_b} \cdot Q_{inv,s,V_{ds}=0}, \tag{11.30}$$

where

$$V_b = \frac{V_{gsx} + n \cdot \frac{k_b T}{q}}{A_{bulk}}. \tag{11.31}$$

To improve the results, n in (11.31) is replaced with 2:

$$V_b = \frac{V_{gsx} + 2 \cdot \frac{k_b T}{q}}{A_{bulk}}. \tag{11.32}$$

Therefore, the final expression for the inversion charge density along the channel for all regions of device operation is

$$Q_{inv}(y) = Q_{inv, V_{ds}=0} \left(1 - \frac{\Phi_n(y)}{V_b}\right),$$ (11.33)

where $Q_{inv, V_{ds}=0}$ is computed from (11.20).

11.3.3   The Effective Channel Length and Width

In most models, the effective channel length and width are computed using

$$L_{eff} = L_{drawn} - \Delta L = L_{drawn} - 2 \cdot \mathbf{LINT},$$ (11.34)

and

$$W_{eff} = W_{drawn} - \Delta W = W_{drawn} - 2 \cdot \mathbf{WINT}.$$ (11.35)

The terms **LINT** and **WINT** describe the difference between the layout and on-wafer dimensions when the diffusion of dopants into the channel region is taken into account (see Appendix B); they are usually measured and taken to be valid for all geometries for any particular MOS process technology.

BSIM3 includes the somewhat unusual feature of allowing ΔL and ΔW to take on geometry dependence. Along with (11.34) and (11.35), different composite expressions are also made available.

Besides (11.34), BSIM3 also offers

$$L_{eff} = L_{drawn} - 2 \cdot L_D,$$ (11.36)

where L_D is given by

$$L_D = \mathbf{LINT} + \frac{\mathbf{LL}}{L_{drawn}^{\mathbf{LLN}}} + \frac{\mathbf{LW}}{W_{drawn}^{\mathbf{LWN}}} + \frac{\mathbf{LWL}}{L_{drawn}^{\mathbf{LLN}} \cdot W_{drawn}^{\mathbf{LWN}}};$$ (11.37)

the term **LINT** represents the value provided by the standard description of the channel length, as in (11.34). In the long channel limit, (11.37) reduces to **LINT**. Note the similarity between (11.37) and the HSPICE Level 28 composite parameter expression (9.1), which includes a product term similar to the last term in (11.37).

The description of the effective channel width is even more complicated. Along with (11.35), BSIM3 also offers

$$W_{eff} = W_{drawn} - 2 \cdot W_D$$ (11.38)

and

$$W'_{eff} = W_{drawn} - 2 \cdot W'_D, \tag{11.39}$$

where

$$W_D = W'_D + \mathbf{DWG} \cdot V_{gsx} + \mathbf{DWB} \cdot \left[(\phi_s - V_{gsx})^{\frac{1}{2}} - \phi_s^{\frac{1}{2}} \right], \tag{11.40}$$

and

$$W'_D = \mathbf{WINT} + \frac{\mathbf{WL}}{L_{drawn}^{\mathbf{WLN}}} + \frac{\mathbf{WW}}{W_{drawn}^{\mathbf{WWN}}} + \frac{\mathbf{WWL}}{L_{drawn}^{\mathbf{WLN}} \cdot W_{drawn}^{\mathbf{WWN}}}. \tag{11.41}$$

Equation (11.40) serves to add gate and substrate bias dependence to the effective channel width calculation, while (11.41) adds geometry dependence in a manner identical to that used in (11.37). Note again the resemblance of (11.41) to the composite parameter expression of HSPICE Level 28.

The reader should be cautioned that ΔL and ΔW are usually physically meaningful, well defined, and constant with geometry for a given MOS process technology. This addition to BSIM3 is basically empirical, and should improve the results in small geometry devices. However, ΔL and ΔW are key variables when a description of the effect of process variations is required; they are used directly from measurements of the variations observed in manufactured wafers. The use of the method described here could weaken this connection, and make it more difficult to properly model process variations in small geometry FETs. Best results are usually obtained when ΔL and ΔW, like t_{ox}, are determined before SPICE parameter extraction, and then provided as input to that process. Also, the use of noninteger parameters as exponents indicates that these expressions will be computationally inefficient.

11.3.4 Polysilicon Gate Depletion

In an ideal situation, the polysilicon gate is heavily doped all the way down to its interface with the gate oxide, so that it behaves like a metal. In reality, since the polysilicon is a semiconductor material and not a metal, there is some voltage drop across the gate; this causes some degree of band bending at the gate/gate oxide interface (as described in Section 3.5). Thus, a depletion region is able to form at the bottom of the gate; this region is not very thick (values on the order of 1 nm being typical), but this small depletion region is important in modern thin gate oxide FETs, where the gate oxide thickness is commonly 15 nm or less. This depletion region, being devoid of free carriers, behaves like a dielectric; there is a voltage drop in the gate that serves to increase the apparent gate oxide thickness above its physical value.

A second (and more important) cause of polysilicon gate depletion is the difficulty encountered when attempting to completely dope the gate material all the way down to its interface with the gate oxide without driving dopant through the gate oxide and into the substrate. This is particularly a problem when p^+ polysilicon is used [7].

BSIM3 adds a model to account for polysilicon gate depletion due to the first cause (simple band bending), but not the second (incomplete polysilicon doping). This model

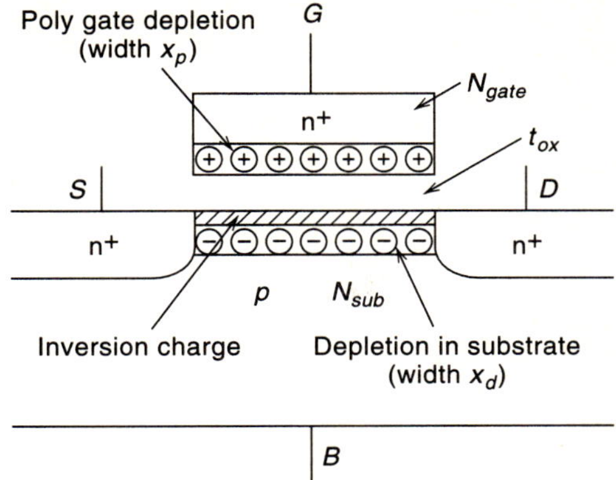

Figure 11.1   The physical polysilicon gate depletion model used in BSIM3. From [2]. © 1995. The Regents of the University of California. Used by permission.

treats the gate as another MOS capacitor, above the substrate. As is the case throughout BSIM3, the model attempts to be physical; however, as in many physical situations, it may be difficult to precisely define the values in question. A potentially simpler method of accounting for polysilicon depletion is discussed at the end of this section.

The basis of the model developed for use in BSIM3 is shown in Figure 11.1, where a depletion layer of ionized impurities forms at the bottom of the gate material layer. Using the one-dimensional depletion layer thickness model developed in Section 3.4, from (3.87) the depletion depth in the polysilicon is

$$x_p = \left(\frac{2 \cdot \epsilon_{Si} \cdot V_p}{q \cdot N_g}\right)^{\frac{1}{2}} = \left(\frac{2 \cdot \epsilon_{Si} \cdot V_p}{q \cdot \mathbf{NGATE}}\right)^{\frac{1}{2}}, \tag{11.42}$$

where **NGATE** is the gate doping (which is assumed to be uniform) and V_p is the voltage drop across the polysilicon gate (i.e., the voltage drop across the depletion region). Equation (11.42) can be rewritten as an expression for V_p:

$$V_p = \frac{q \cdot \mathbf{NGATE} \cdot x_p^2}{2 \cdot \epsilon_{Si}}. \tag{11.43}$$

At the polysilicon/oxide interface, a continuity condition must be satisfied:

$$\epsilon_{ox} \cdot E_{ox} = \epsilon_{Si} \cdot E_p, \tag{11.44}$$

where E_p is the field in the polysilicon depletion region. Using (3.82), the field in this depletion region is

$$E_p = \frac{q \cdot \mathbf{NGATE} \cdot x_p}{\epsilon_{Si}}. \tag{11.45}$$

Treating the oxide as a perfect dielectric leads to

$$E_{ox} = \frac{V_{ox}}{t_{ox}}.$$ (11.46)

Substituting (11.45) and (11.46) into (11.44) produces

$$\frac{\epsilon_{ox} \cdot V_{ox}}{t_{ox}} = \left(2\epsilon_{Si}qN_gV_p\right)^{\frac{1}{2}}.$$ (11.47)

The total voltage drop from the gate to the substrate is

$$V_{gb} = V_{fb} + V_{ox} + \phi_s + V_p,$$ (11.48)

where V_{fb} is the flatband voltage and ϕ_s is the total voltage drop in the substrate (across the gate-induced depletion region). Adjusting the bias reference point for conventional FET terminology using

$$V_{gb} = V_{gs} + V_{sb} = V_{gs} - V_{bs},$$ (11.49)

(11.48) becomes

$$V_{gs} = V_{fb} + V_{bs} + V_{ox} + \phi_s + V_p.$$ (11.50)

In the BSIM3 derivation, it is assumed that $V_{bs} = 0$, so (11.50) becomes

$$V_{gs} = V_{fb} + V_{ox} + \phi_s + V_p.$$ (11.51)

Using (11.43) and (11.47) in (11.51) produces (after a long algebraic exercise)

$$\zeta \cdot \left(V_{gs} - V_{fb} - \phi_s - V_p\right)^2 - V_p = 0,$$ (11.52)

where

$$\zeta = \frac{\epsilon_{ox}^2}{2q\epsilon_{Si}N_g t_{ox}^2}.$$ (11.53)

Equation (11.52) can be manipulated into the form of an effective gate bias, reflecting the decrease in the gate field at the channel due to the voltage drop across the depletion region at the bottom of the gate:

$$V_{gs,eff} = V_{fb} + \phi_s + \frac{q\epsilon_{Si} \cdot \mathbf{NGATE} \cdot t_{ox}^2}{\epsilon_{ox}^2} \cdot \left[1 + \frac{2\epsilon_{ox}^2 \cdot (V_{gs} - V_{fb} - \phi_s)}{q\epsilon_{Si} \cdot \mathbf{NGATE} \cdot t_{ox}^2}\right]^{\frac{1}{2}}.$$ (11.54)

This effective gate voltage is used to describe the reduction of the drain current in the following form:

$$\frac{I_{ds}(V_{gs,eff})}{I_{ds}(V_{gs})} = \frac{V_{gs,eff} - V_t}{V_{gs} - V_t}.$$

(11.55)

Experiments confirm that polysilicon depletion does reduce the drain current, albeit by only a few percent. BSIM3 can account for this effect using the description developed here. However, as noted earlier, the most common root cause of polysilicon depletion is incomplete doping due to the need to prevent gate dopant atoms from being driven through the gate oxide into the substrate. As a result, there is often a very steep doping gradient near the bottom of the gate material, making it difficult to define a value of the gate doping N_g.

A simpler alternative, which is employed by necessity in other models due to the lack of a specific description of polysilicon depletion, is to account for this behavior using the gate oxide thickness t_{ox}. If measured in the laboratory via a physical method (such as ellipsometry), the actual (or physical) value of the oxide thickness is found. The oxide thickness can also be measured from the behavior of a low frequency capacitance-voltage curve in the strong inversion region, yielding an electrical value for the oxide thickness. The difference between these two measurements is the thickness of the depletion region in the polysilicon gate. For example, if the physical value of the oxide thickness is 9.0 nm, and the electrical value is 9.7 nm, the depletion region thickness is 0.7 nm. The electrical value of 9.7 nm (rather than the physical value of 9.0 nm) is used as the oxide thickness t_{ox} in the model parameter set. To use this approach, **NGATE** should be set to a very large value (e.g., 10^{20}/cm^3), so that it does not enter the calculations.

This method is able to produce good results. It should be noted that it does neglect any gate bias dependence of the depth of the polysilicon depletion region. This could be a concern in analog and low power digital design, where low values of the gate bias are more commonly used. However, the error introduced is usually negligible (rarely exceeding about 1%) for oxide thicknesses greater than 5 nm, and becomes only slightly larger (2–3%) for thinner gate oxides.

11.4   The Threshold Voltage Model

11.4.1   The Base Model

The base threshold voltage model for a long, wide device is the same as that of a simple two-terminal MOS capacitor, derived in Section 3.3:

$$V_t = V_{fb} + \phi_s + \gamma \cdot (\phi_s - V_{bsx})^{\frac{1}{2}},$$

(11.56)

where γ is the conventional body effect term, given by

$$\gamma = \frac{(2\epsilon_{Si}q N_{sub})^{\frac{1}{2}}}{C_{ox}},$$

(11.57)

and V_{bsx} is the auxiliary substrate bias expression (11.8), detailed in Section 11.3. Assuming that the substrate is uniformly doped, the surface potential ϕ_s is defined very simply from

$$\phi_s = 2\phi_f = 2 \cdot \frac{k_b T}{q} \cdot ln\left[\frac{N_{sub}}{n_i}(T)\right],$$

(11.58)

where $n_i(T)$ is the intrinsic carrier concentration.

For notational convenience, (11.56) can be rearranged to

$$V_t = V_{to} + \gamma \cdot \left[(\phi_s - V_{bsx})^{\frac{1}{2}} - \phi_s^{\frac{1}{2}}\right],$$

(11.59)

where V_{to} is the threshold voltage of a long wide device with no substrate bias applied:

$$V_{to} = V_{fb} + \phi_s + \gamma \cdot \phi_s^{\frac{1}{2}}.$$

(11.60)

Rather than compute V_{to} in a composite manner using (11.60), this term is replaced with the model parameter **VTH0**, which is extracted. This transforms (11.59) to

$$V_t = \mathbf{VTH0} + \gamma \cdot \left[(\phi_s - V_{bsx})^{\frac{1}{2}} - \phi_s^{\frac{1}{2}}\right].$$

(11.61)

11.4.2 Nonuniform Channel Doping

If the substrate is not uniformly doped, the simple description of the effect of the substrate bias on the threshold voltage (11.57) is insufficient. The surface doping profile is usually of a form similar to that shown in Figure 11.2; due to the use of ion implantation and thermal annealing cycles, the doping profile takes on an approximately Gaussian shape. However, if this doping profile is used, as the gate-induced depletion region grows with increasing gate bias, it will be expanding into a region where the doping varies with

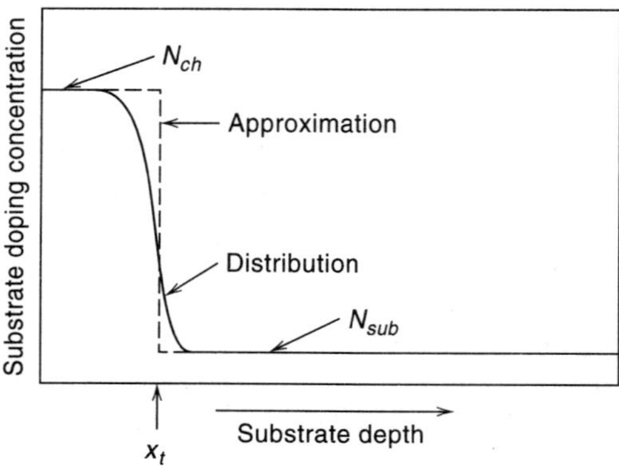

Figure 11.2 The MOSFET channel doping profile; the solid line represents a typical Gaussian profile produced by ion implantation, while the dashed line depicts the simplified "block" doping approximation used in BSIM3. From [2]. © 1995. The Regents of the University of California. Used by permission.

position; the body effect parameter γ will itself have substrate bias dependence. To fully solve the equations, a computationally inefficient iterative scheme would be required.

The BSIM3 solution to this problem is also shown in Figure 11.2. The Gaussian profile is replaced with a surface block of a higher doping concentration N_{ch}, which is found from the surface down to a depth of x_t. Below x_t, the substrate doping is the base value N_{sub}. This approach leads to two separate γ terms [5], defined by

$$\gamma_1 = \frac{(2\epsilon_{Si}qN_{ch})^{\frac{1}{2}}}{C_{ox}} = \frac{(2\epsilon_{Si}q \cdot \mathbf{NCH})^{\frac{1}{2}}}{C_{ox}}, \tag{11.62}$$

and

$$\gamma_2 = \frac{(2\epsilon_{Si}qN_{sub})^{\frac{1}{2}}}{C_{ox}} = \frac{(2\epsilon_{Si}q \cdot \mathbf{NSUB})^{\frac{1}{2}}}{C_{ox}}. \tag{11.63}$$

As the gate-induced depletion region expands in the region with a doping concentration of **NCH**, the effect of the substrate doping is described by γ_1. As this depletion region continues to grow, it eventually expands into the main substrate, and γ_2 is used.

While simple, this approach to constructing γ_1 and γ_2 requires that three physical parameters (**NCH**, **NSUB**, and x_t) be known. A variety of simulations and measurements can allow values of **NCH** and **NSUB** to be found with relative ease, but the assignment of x_t to a curved doping profile is somewhat arbitrary. The compromise method used in BSIM3 is to modify the threshold voltage expression (11.61) to the form

$$V_t = \mathbf{VTH0} + \mathbf{K1} \cdot \left[(\phi_s - V_{bsx})^{\frac{1}{2}} - \phi_s^{\frac{1}{2}} \right] - \mathbf{K2} \cdot V_{bsx}, \tag{11.64}$$

where **K1** and **K2** are model parameters. Note that (11.64) is very similar to the threshold voltage expression used in BSIM and BSIM2:

$$V_t = V_{fb} + \mathbf{K1} \cdot (\phi_s - V_{bs})^{\frac{1}{2}} - \mathbf{K2} \cdot (\phi_s - V_{bs}); \tag{11.65}$$

the BSIM3 expression (11.64) differs only in the specific arrangement of the surface potential ϕ_s in the equation. As was noted in Chapters 8 and 10, the use of **K1** and **K2** is intended to allow for the description of nonuniform substrate doping. However, the introduction of **K2** is essentially the inclusion of another term into a power series in $(\phi_s - V_{bs})^{\frac{1}{2}}$; this should (and does) improve the model fitting results.

As was noted in the introduction to this chapter, the BSIM3 structure makes a strong effort to allow for model parameters to be computed from physical variables. This is the case for **K1** and **K2**; expressions are derived to allow these parameters to be computed from the physical variables **NCH**, **NSUB**, and x_t, if they are known. However, in most situations, **K1** and **K2** will be extracted directly from data in the usual manner; they will thus be treated here as electrical parameters rather than as composite physical terms.

11.4.3 Lateral Nonuniform Doping

As described in Section 3.5, in some process technologies, the oxidation of the sidewall of the etched polysilicon gate can induce the redistribution of impurities near the silicon surface. Under the edges of the gate, in the vicinity of what will eventually become the ends of the channel, point defects are injected during sidewall oxidation [8]; these point defects gather doping impurities from the substrate, thereby increasing the doping concentration near the source and drain diffusions. As the channel length decreases, these more heavily doped regions consume a larger fraction of the total channel; this causes the threshold voltage to increase as the channel length is shortened.

This effect is modeled in BSIM3 by computing the value of this increased threshold voltage, V_{LND}, and adding it to the threshold voltage computation. The simplified method of describing the doping along the channel that is used in BSIM3 is shown in Figure 11.3. In the middle of the channel, the surface doping remains at a value of N_{ch}, while near the ends of the channel, the doping concentration is N_{ds}. To allow for a reasonable calculation, the transition between the regions is taken to be abrupt, and the regions near the source and drain are assigned an arbitrary length L_x. To compute the effect on the threshold voltage, the lateral doping profile shown in Figure 11.3 is used to compute an average effective doping along the channel:

$$N_{eff} = \frac{\mathbf{NCH} \cdot (L_{eff} - 2 \cdot L_x) + N_{ds} \cdot (2 \cdot L_x)}{L_{eff}} = \mathbf{NCH} \cdot \left(1 + \frac{2 \cdot L_x}{L_{eff}} \cdot \frac{N_{ds} - N_{ch}}{N_{ch}}\right)$$

$$= \mathbf{NCH}\left(1 + \frac{N_{LX}}{L_{eff}}\right), \tag{11.66}$$

where

$$N_{LX} = 2 \cdot L_x \cdot \frac{N_{ds} - \mathbf{NCH}}{\mathbf{NCH}}. \tag{11.67}$$

Examining (11.67), N_{LX} could be computed from knowledge of the physical terms L_x, N_{ds}, and **NCH**. However, the values of L_x and N_{ds} are difficult to determine; instead,

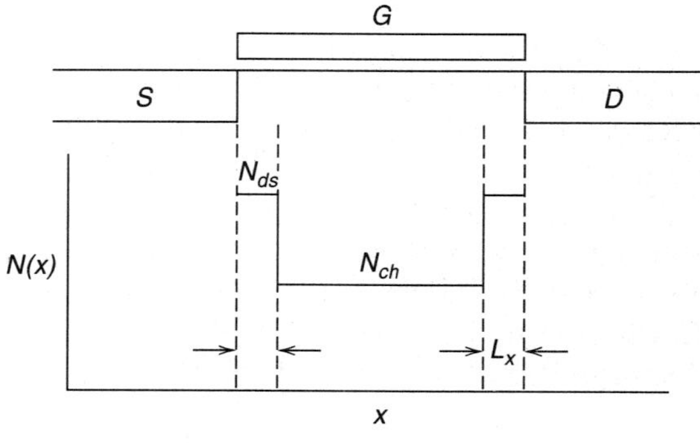

Figure 11.3 The simplified approximation for nonuniform lateral doping which is used in BSIM3. From [2]. © 1995. The Regents of the University of California. Used by permission.

N_{LX} is treated as a model parameter (**NLX**). By setting $V_{bs} = 0$ and noting that **K1**, like γ, is proportional to $N_{eff}^{\frac{1}{2}}$, the final expression is reached:

$$V_{LND} = \mathbf{K1} \cdot \left[\left(1 + \frac{\mathbf{NLX}}{L_{eff}} \right)^{\frac{1}{2}} - 1 \right] \cdot \phi_s^{\frac{1}{2}}. \tag{11.68}$$

11.4.4   Short Channel Effects

In BSIM3, as in the other models, short channel effects are included in the threshold voltage model by computing the additional effect of a short channel on the threshold voltage (V_{SCE}) and subtracting that term from the computed threshold voltage. Due to their importance, many models have been proposed to describe these effects. BSIM3 chooses to employ the model proposed by Liu *et al.* [9], which is based on a quasi-two dimensional solution to the Poisson equation along the channel:

$$V_{SCE} = \Theta_t \cdot \left[2 \cdot (V_{bi} - \phi_s) + V_{ds} \right], \tag{11.69}$$

where V_{bi} is the built-in potential of the source-substrate and drain-substrate junctions. The term Θ_t is a function of the channel length, and is given by

$$\Theta_t = exp \left(-\frac{L_{eff}}{2 \cdot L_{t,SCE}} \right) + 2 \cdot exp \left(-\frac{L_{eff}}{L_{t,SCE}} \right), \tag{11.70}$$

where $L_{t,SCE}$ is a characteristic length, defined as

$$L_{t,SCE} = \left(\frac{\epsilon_{ox} \cdot t_{ox}}{\epsilon_{Si}} \cdot \frac{x_d}{\xi} \right). \tag{11.71}$$

The term x_d is the depth of the gate-induced depletion region in the substrate, as derived in Section 3.3:

$$x_d = \left(\frac{2\epsilon_{Si} \cdot (\phi_s - V_{bs})}{q \cdot \mathbf{NCH}} \right)^{\frac{1}{2}}. \tag{11.72}$$

When a drain bias is applied, x_d will vary along the channel, and will be larger at the drain due to the larger depletion region induced by the drain bias; the term ξ is introduced into (11.71) to account for this behavior by averaging the value of x_d along the channel.

The model described by (11.69) includes all of the basic physics underlying short channel effects; in this manner, both source/drain charge sharing and the effect of drain-induced barrier lowering (DIBL) are contained in the same expression. However, improved accuracy is achieved by modifying this model, so that charge-sharing effects and DIBL are treated separately, while a number of empirical model parameters are introduced.

The change in the threshold voltage due to short channel effects is separated into a charge-sharing component and a DIBL component:

$$V_{SCE} = V_{CS} + V_{DIBL}. \tag{11.73}$$

This implies a similar splitting of the term Θ_t into a charge-sharing component (Θ_{CS}) and a DIBL component (Θ_{DIBL}).

Charge Sharing

The charge-sharing term Θ_{CS} is given by

$$
\begin{aligned}
\Theta_{CS} = \mathbf{DVT0} \cdot & \left[exp\left(-\frac{\mathbf{DVT1} \cdot L_{eff}}{2 \cdot L_{t,CS}} \right) + 2 \cdot exp\left(-\frac{\mathbf{DVT1} \cdot L_{eff}}{L_{t,CS}} \right) \right] \\
\mathbf{DVT0W} \cdot & \left[exp\left(-\frac{\mathbf{DVT1W} \cdot W_{eff} \cdot L_{eff}}{2 \cdot L_{t,CS,W}} \right) \right. \\
& \left. + 2 \cdot exp\left(-\frac{\mathbf{DVT1W} \cdot W_{eff} \cdot L_{eff}}{L_{t,CS,W}} \right) \right],
\end{aligned}
\tag{11.74}
$$

where $L_{t,CS}$ is identical to the characteristic length L_t described by (11.19)

$$L_{t,CS} = \left(\frac{\epsilon_{Si} \cdot t_{ox}}{\epsilon_{ox}} \cdot x_d \right)^{\frac{1}{2}} \cdot (1 + \mathbf{DVT2} \cdot V_{bsx}), \tag{11.75}$$

and $L_{t,CS,W}$ is another characteristic length for the added width expression:

$$L_{t,CS,W} = \left(\frac{\epsilon_{Si} \cdot t_{ox}}{\epsilon_{ox}} \cdot x_d \right)^{\frac{1}{2}} \cdot (1 + \mathbf{DVT2W} \cdot V_{bsx}). \tag{11.76}$$

Finally,

$$V_{CS} = \Theta_{CS} \cdot (V_{bi} - \phi_s). \tag{11.77}$$

Drain-Induced Barrier Lowering

The drain-induced barrier lowering term Θ_{DIBL} is given by

$$\Theta_{DIBL} = exp\left(-\frac{\mathbf{DSUB} \cdot L_{eff}}{2 \cdot L_{t,DIBL}} \right) + 2 \cdot exp\left(-\frac{\mathbf{DSUB} \cdot L_{eff}}{L_{t,DIBL}} \right), \tag{11.78}$$

where $L_{t,DIBL}$ is the characteristic length $L_{t,CS}$ with a zero substrate bias:

$$L_{t,DIBL} = \left(\frac{\epsilon_{Si} \cdot t_{ox}}{\epsilon_{ox}} \cdot x_d \right)^{\frac{1}{2}} . \tag{11.79}$$

The final expression for V_{DIBL} is

$$V_{DIBL} = \Theta_{DIBL} \cdot (\mathbf{ETA0} + \mathbf{ETAB} \cdot V_{bsx}) \cdot V_{ds}. \tag{11.80}$$

Note how the composite DIBL parameter in (11.80)

$$\eta = \mathbf{ETA0} + \mathbf{ETAB} \cdot V_{bsx} \tag{11.81}$$

includes substrate bias dependence in a form identical to that found in the second generation models. Also note that as in most previous models, the change in the threshold voltage due to DIBL is taken to be linearly proportional to the drain bias [10].

11.4.5 Narrow Channel Effects

As described in Section 6.2 and Section 7.2, the gate-induced depletion region spreads outside the width of the channel (as defined by the drawn polysilicon gate). This increases the total depletion charge above its otherwise expected value; in narrow devices, this can have a significant effect upon the threshold voltage.

In BSIM3, a narrow channel correction voltage V_{NCE} is computed and added to the total threshold voltage expression. The expression chosen is originally due to Muller and Kamins [11]:

$$V_{NCE} = \frac{\pi \cdot q \cdot N_{sub} \cdot x_{d,max}^2}{2 \cdot C_{ox} \cdot W_{eff}} = 3\pi \cdot \frac{t_{ox}}{W_{eff}} \cdot \phi_s. \tag{11.82}$$

To improve the results, this expression is modified with the addition of several empirical model parameters, so that

$$V_{NCE} = (\mathbf{K3} + \mathbf{K3B} \cdot V_{bsx}) \cdot \frac{t_{ox}}{(W_{eff} + \mathbf{WO})} \cdot \phi_s. \tag{11.83}$$

Note how $\mathbf{K3}$ and $\mathbf{K3B}$ are introduced to account for the effect of the substrate bias in the same manner as that found in the second generation models.

11.4.6 The Final Threshold Voltage Expression

By combining all the corrections to the base threshold voltage model for a long, wide device with a uniformly doped substrate, the complete expression used in BSIM3 to describe the threshold voltage is reached:

$$V_t = \mathbf{VTH0} + \mathbf{K1} \cdot \left[(\phi_s - V_{bsx})^{\frac{1}{2}} - \phi_s^{\frac{1}{2}} \right] - \mathbf{K2} \cdot V_{bsx} + V_{LND} - V_{SCE} + V_{NCE}$$

$$= \mathbf{VTH0} + \mathbf{K1} \cdot \left[(\phi_s - V_{bsx})^{\frac{1}{2}} - \phi_s^{\frac{1}{2}} \right] - \mathbf{K2} \cdot V_{bsx}$$

$$+ \mathbf{K1} \cdot \left[\left(1 + \frac{\mathbf{NLX}}{L_{eff}} \right)^{\frac{1}{2}} - 1 \right] \cdot \phi_s^{\frac{1}{2}}$$

$$- \Theta_{CS} \cdot (V_{bi} - \phi_s) - \Theta_{DIBL} \cdot (\mathbf{ETA0} + \mathbf{ETAB} \cdot V_{bsx}) \cdot V_{ds}$$

$$+ (\mathbf{K3} + \mathbf{K3B} \cdot V_{bsx}) \cdot \frac{t_{ox}}{(W_{eff} + \mathbf{WO})} \cdot \phi_s. \tag{11.84}$$

Temperature Dependence

As in other models, in BSIM3, parameters are extracted at a temperature of T_{nom} (usually roughly 25^oC), which must be specified with the model; in most SPICE simulators, this information is entered at the top of the SPICE circuit simulation input file. Other parameters are then included to account for the device behavior when the operating temperature differs from T_{nom}.

It has been observed that the threshold voltage, to a very good approximation, varies linearly with the temperature. Thus, in BSIM3, the temperature dependence of the threshold voltage is described by

$$V_t(T) = V_t(T_{nom}) + \left(\mathbf{KT1} + \frac{\mathbf{KT1L}}{L_{eff}} + \mathbf{KT2} \cdot V_{bsx} \right) \cdot \left(\frac{T}{T_{nom}} - 1 \right), \tag{11.85}$$

where $V_t(T_{nom})$ is computed from (11.84) with $T = T_{nom}$. Note that in (11.85), the temperature must be specified in **degrees Kelvin** rather than degrees Celsius. The composite expression involving **KT1** and **KT2** is similar in intent to the parameter **TCV**, which appears in HSPICE Level 28 and the HSPICE implementation of several other models.

11.5   The Mobility Model

11.5.1   The Effect of the Vertical Field

Strong Inversion

In BSIM3, three choices are offered for the description of the reduction of the mobility by the vertical field. One model is virtually identical to the BSIM2 description; while it is intended for use with depletion devices, there is no restriction on its more general use. The other two models are based on a more physically realistic evaluation of the mobility. The specific model to be employed is chosen using the flag parameter **MOBMOD**, as discussed below.

As just noted, the new offering in BSIM3 is a more physically realistic model for the reduction of the channel mobility by the vertical field. This model is based on an *average field* description [12], such that the carriers in the inversion layer, regardless of their depth below the interface, are influenced by an *effective field* [13]:

$$E_{eff} = \frac{Q_{depl} + \frac{Q_{inv}}{2}}{\epsilon_{Si}}. \qquad (11.86)$$

It has been shown [12] that this expression can be rewritten as

$$E_{eff} \cong \frac{V_{gs} + V_t}{6 \cdot t_{ox}}. \qquad (11.87)$$

It has also been shown that the relationship between the mobility and the effective field is well described by [14]

$$\mu_v = \frac{\mu_o}{1 + \left(\frac{E_{eff}}{E_o}\right)^\nu}, \qquad (11.88)$$

where E_o is a *critical field* value for the onset of the reduction of the mobility by the vertical field and ν is a parameter which varies between 1 and 2. If ν takes on any value besides 1, (11.88) is in the form of a power series, which will be computationally inefficient particularly if ν takes on a noninteger value. As an alternative, as was done in BSIM2 (see Section 10.4), the denominator of (11.88) is expanded in a power series out to the quadratic term, leading to

$$\mu_v = \frac{\mu_o}{1 + A \cdot \left(\frac{E_{eff}}{E_o}\right) + B \cdot \left(\frac{E_{eff}}{E_o}\right)^2}. \qquad (11.89)$$

Substituting (11.87) into (11.89) produces

$$\mu_v = \frac{\mu_o}{1 + A \cdot \left(\frac{V_{gs}+V_t}{6 \cdot t_{ox} \cdot E_o}\right) + B \cdot \left(\frac{V_{gs}+V_t}{6 \cdot t_{ox} \cdot E_o}\right)^2}. \qquad (11.90)$$

Note that in contrast to the earlier models, in which $(V_{gs} - V_t)$ (rather than $(V_{gs} + V_t)$) appeared, the form of (11.90) provides a more intuitively correct result. For example, if the substrate doping is increased, the mobility μ_v should decrease. If the substrate doping is increased, the threshold voltage will increase; this will increase the value of the denominator of (11.90), which will decrease the value of the mobility (as expected).

From here, the implementation of (11.90) becomes somewhat arbitrary, as the introduction of constants and parameters has the effect of combining various terms. BSIM3 offers two variations on (11.90). One expression is

$$\mu_v = \frac{\mathbf{UO}}{1 + (\mathbf{UA} + \mathbf{UC} \cdot V_{bsx}) \cdot \left(\frac{V_{gs}+2 \cdot V_t(T)}{t_{ox}}\right) + \mathbf{UB} \cdot \left(\frac{V_{gs}+2 \cdot V_t(T)}{t_{ox}}\right)^2}, \qquad (11.91)$$

where $\mathbf{UO}$ is the model parameter which represents the low field mobility. The reason for the introduction of $2 \cdot V_t$ will become clear below. The second expression is a minor rearrangement of (11.91):

$$\mu_v = \cfrac{\textbf{UO}}{1 + \left[\textbf{UA} \cdot \left(\frac{V_{gs} + 2 \cdot V_t(T)}{t_{ox}} \right) + \textbf{UB} \cdot \left(\frac{V_{gs} + 2 \cdot V_t(T)}{t_{ox}} \right)^2 \right] \cdot (1 + \textbf{UC} \cdot V_{bsx})}. \qquad (11.92)$$

BSIM3 also makes available a mobility model which is very similar to the BSIM2 description. This model is

$$\mu_v = \cfrac{\textbf{UO}}{1 + (\textbf{UA} + \textbf{UC} \cdot V_{bsx}) \cdot \left(\frac{V_{gs} - V_t(T)}{t_{ox}} \right) + \textbf{UB} \cdot \left(\frac{V_{gs} - V_t(T)}{t_{ox}} \right)^2}, \qquad (11.93)$$

Equation (11.93) is quite similar to the BSIM2 mobility expression (10.15).

Connecting Weak and Strong Inversion

As was the case for the depletion charge model, it is advantageous to arrange for the mobility model to be smooth and continuous at the weak inversion-strong inversion transition. In BSIM3, this is actually done quite simply; the auxiliary gate bias expression V_{gsx}, introduced in Section 11.3, replaces V_{gs} in (11.91), (11.92), and (11.93), yielding the final basic mobility model equations. With this replacement, (11.91) becomes

$$\mu_v = \cfrac{\textbf{UO}}{1 + (\textbf{UA} + \textbf{UC} \cdot V_{bsx}) \cdot \left(\frac{V_{gsx} + 2 \cdot V_t(T)}{t_{ox}} \right) + \textbf{UB} \cdot \left(\frac{V_{gsx} + 2 \cdot V_t(T)}{t_{ox}} \right)^2}; \qquad (11.94)$$

this model is selected when **MOBMOD** = 1. Note that when $V_{gs} \gg V_t(T)$, $V_{gsx} \rightarrow (V_{gs} - V_t(T))$, and thus

$$(V_{gsx} + 2 \cdot V_t(T)) \rightarrow (V_{gs} + V_t(T)), \qquad (11.95)$$

properly reflecting the relationship described by (11.87). When $V_{gs} \ll V_t(T)$,

$$V_{gsx} \propto exp(V_{gs} - V_t(T)); \qquad (11.96)$$

the V_{gsx} terms in the denominator become very small, and $\mu_v \rightarrow \mu_o$ in the subthreshold region. This is the expected behavior, given that the low gate bias should not influence the channel mobility.

In the same manner, (11.92) becomes

$$\mu_v = \cfrac{\textbf{UO}}{1 + \left[\textbf{UA} \cdot \left(\frac{V_{gsx} + 2 \cdot V_t(T)}{t_{ox}} \right) + \textbf{UB} \cdot \left(\frac{V_{gsx} + 2 \cdot V_t(T)}{t_{ox}} \right)^2 \right] \cdot (1 + \textbf{UC} \cdot V_{bsx})}; \qquad (11.97)$$

this model is selected when **MOBMOD** $= 3$. Finally, (11.93) is transformed to

$$\mu_v = \cfrac{\textbf{UO}}{1 + (\textbf{UA} + \textbf{UC} \cdot V_{bsx}) \cdot \left(\frac{V_{gsx}}{t_{ox}}\right) + \textbf{UB} \cdot \left(\frac{V_{gsx}}{t_{ox}}\right)^2} : \qquad (11.98)$$

this model is selected when **MOBMOD** $= 2$. Note that the discussion of the limiting behavior in (11.95) and (11.96) also applies to (11.97) and (11.98).

Mobility Model Selection

In the earlier models and elsewhere in BSIM3, selection parameters (such as **XPART** for charge partitioning) are usually chosen at the circuit simulation level, as they offer the circuit simulation user a choice among several possible options within the model. However, the mobility model selection parameter **MOBMOD** is somewhat different. A mobility model must be selected *before* parameter extraction is performed; thus, the extracted parameters will only provide a proper description for that particular choice of mobility model. Therefore, the mobility selection flag parameter should be provided as an FET model parameter.

Temperature Dependence

The temperature dependence of the mobility (as reduced by the vertical field) in BSIM3 is more complicated than similar expressions in earlier models. For **MOBMOD** $= 1$, (11.94) becomes

$$\mu_v(T) = \cfrac{\textbf{UO} \cdot \left(\frac{T}{T_{nom}}\right)^{\textbf{UTE}}}{1 + (U_a(T) + U_c(T) \cdot V_{bsx}) \cdot \left(\frac{V_{gsx}+2\cdot V_t}{t_{ox}}\right) + U_b(T) \cdot \left(\frac{V_{gsx}+2\cdot V_t}{t_{ox}}\right)^2} , \qquad (11.99)$$

where

$$U_a(T) = \textbf{UA} + \textbf{UA1} \cdot \left(\frac{T}{T_{nom}} - 1\right), \qquad (11.100)$$

$$U_b(T) = \textbf{UB} + \textbf{UB1} \cdot \left(\frac{T}{T_{nom}} - 1\right), \qquad (11.101)$$

and

$$U_c(T) = \textbf{UC} + \textbf{UC1} \cdot \left(\frac{T}{T_{nom}} - 1\right). \qquad (11.102)$$

The parameter **UTE** serves the same role that **BEX** fulfills in HSPICE Level 28 and in the HSPICE implementations of several other models. The parameters **UA**, **UB**, and **UC** are determined during the extraction of parameters at $T = T_{nom}$, while **UA1**, **UB1**, and **UC1** are extracted later to account for temperature dependence.

For **MOBMOD** $= 3$, (11.97) becomes

$$\mu_v(T) = \frac{\mathbf{UO} \cdot \left(\frac{T}{T_{nom}}\right)^{\mathbf{UTE}}}{1 + \left[U_a(T) \cdot \left(\frac{V_{gsx} + 2 \cdot V_t(T)}{t_{ox}}\right) + U_b(T) \cdot \left(\frac{V_{gsx} + 2 \cdot V_t(T)}{t_{ox}}\right)^2\right] \cdot (1 + U_c(T) \cdot V_{bsx})}. \tag{11.103}$$

For **MOBMOD** $= 2$, (11.98) becomes transformed to

$$\mu_v = \frac{\mathbf{UO} \cdot \left(\frac{T}{T_{nom}}\right)^{\mathbf{UTE}}}{1 + (U_a(T) + U_c(T) \cdot V_{bsx}) \cdot \left(\frac{V_{gsx}}{t_{ox}}\right) + U_b(T) \cdot \left(\frac{V_{gsx}}{t_{ox}}\right)^2}. \tag{11.104}$$

11.5.2   The Effect of the Lateral Field

In contrast to some models, BSIM3 uses a very simple description of the effect of the lateral field on the mobility. This is done through the model for the carrier velocity, so that the result does not actually appear in the mobility model. To describe the carrier velocity-lateral field relationship, a simple two-part piecewise-continuous model is used [15]:

$$\begin{aligned}
v(y) &= \frac{\mu_v(T) \cdot E_y(y)}{1 + \frac{E_y(y)}{E_{sat}}} \quad \cdots \quad E_y(y) < E_{sat} \\
&= v_{SAT} \quad\quad\quad \cdots \quad E_y(y) > E_{sat}
\end{aligned} \tag{11.105}$$

where E_{sat} is the field at the pinch-off point (where $V_{ds} = V_{dsat}$); more generically, E_{sat} is the lateral field at the point in the channel where the carriers reach the saturation velocity. To guarantee the continuity of (11.105) at $E_y(y) = E_{sat}$, substitution of this constraint into (11.105) leads to the definition

$$E_{sat} = \frac{2 \cdot v_{SAT}}{\mu_v(T)} = \frac{2 \cdot \mathbf{VSAT}}{\mu_v(T)}; \tag{11.106}$$

The saturation velocity **VSAT** is extracted as a model parameter. Note that in this model, as the gate bias increases and μ_v decreases, E_{sat} increases.

Temperature Dependence

The saturation velocity is weakly affected by changes in temperature; however, when the temperature varies by a large amount from T_{nom}, this small change is very important to the saturation region characteristics. In BSIM3, the temperature dependence of the saturation velocity is computed using

$$v_{SAT}(T) = \mathbf{VSAT} - \mathbf{AT} \cdot \left(\frac{T}{T_{nom}} - 1 \right), \tag{11.107}$$

where **VSAT** is a model parameter which is extracted at $T = T_{nom}$. The BSIM3 model parameter **AT** is similar in intent to the parameter **FEX**, which appears in HSPICE Level 28 and the HSPICE implementations of several other models.

11.6   The Drain Current Model

11.6.1   Introduction

As described in Section 11.2, the third-generation models introduce the use of smoothing functions to permit continuous and smooth descriptions of the device behavior across the two major operating transition regions: the weak inversion-strong inversion transition and the linear region-saturation region transition. In Section 11.3, the BSIM3 depletion charge model, which is smooth and continuous across both of these transitions, was developed, while Section 11.5 considered the BSIM3 mobility model, which is smooth and continuous across the weak inversion-strong inversion transition region and piecewise continuous at the linear region-saturation region transition.

Here the drain current model will be developed in a similar fashion, making use of the inversion charge and mobility models which were already described. The discussion proceeds rather logically, beginning with the basic equation and developing the linear region model. A saturation voltage model is introduced to describe the transition point, and an additional mathematical function is introduced to ensure a continuous and smooth current expression in that region. The development of a drain current model for the saturation region is then considered in the form of an output conductance model.

11.6.2   The Linear Region

The Basic Model Without Series Resistance

Neglecting the effect of source/drain series resistance, a general expression for the drain current is [11]

$$I_{ds,R_{ds}=0}(y) = W_{eff} \cdot Q_{inv}(y) \cdot \mu_{eff}(y) \cdot \frac{d\phi_n(y)}{dy}. \tag{11.108}$$

As noted above, expressions have already been derived for $Q_{inv}(y)$ and $\mu_{eff}(y)$ that are valid for all regions of device operation. Here only the linear region of operation will be considered. Substituting (11.33) for $Q_{inv}(y)$ and the linear region ($E_y(y) < E_{sat}$) portion of (11.105) into (11.108) leads to

$$I_{ds,lin,R_{ds}=0}(y) = W_{eff} \cdot \left[Q_{inv,V_{ds}=0} \cdot \left(1 - \frac{\phi_n(y)}{V_b} \right) \right] \cdot \left(\frac{\mu_v(T)}{1 + \frac{E_y(y)}{E_{sat}}} \right) \cdot \frac{d\phi_n(y)}{dy}, \tag{11.109}$$

where E_{sat} is given by (11.106). Integrating (11.109) from the source to the drain in the usual manner produces the basic linear region current expression:

$$I_{ds,lin,R_{ds}=0} = \frac{\mu_v(T) \cdot W_{eff}}{L_{eff}} \cdot \frac{Q_{inv,V_{ds}=0} \cdot V_{ds} \cdot \left(1 - \frac{V_{ds}}{2 \cdot V_b}\right)}{1 + \frac{V_{ds}}{E_{sat} \cdot L_{eff}}}. \tag{11.110}$$

The Effect of Source/Drain Series Resistance

In the first-generation models, the additional resistance in the current path due to the source and drain was included by adding lumped resistors to that current path. In the second-generation models, the problem was simplified by absorbing this extrinsic resistance into the mobility model and its associated parameters. BSIM3 returns to a first generation approach, by including the extrinsic source/drain series resistance as a lumped resistance, rather than in the mobility model.

BSIM3 introduces the lumped resistance term R_{ds} by applying Ohm's law to the linear region current expression:

$$I_{ds,lin} = \frac{V_{ds}}{R_{tot}} = \frac{V_{ds}}{R_{chan} + R_{ds}}, \tag{11.111}$$

where R_{chan} is the channel resistance

$$R_{chan} = \frac{V_{ds}}{I_{ds}\Big|_{R_{ds}=0}}, \tag{11.112}$$

while $I_{ds}\Big|_{R_{ds}=0}$ is the linear region current expression (11.110). Using (11.110) in (11.112) leads to the expression for the linear region current with the series resistance included:

$$I_{ds,lin} = \frac{\mu_v(T) \cdot W_{eff}}{L_{eff}} \cdot \frac{1}{1 + \frac{V_{ds}}{E_{sat} \cdot L_{eff}}} \cdot \frac{Q_{inv,V_{ds}=0} \cdot V_{ds} \cdot \left(1 - \frac{V_{ds}}{2 \cdot V_b}\right)}{1 + \frac{R_{ds}}{V_{ds}} \cdot \frac{\mu_v(T) \cdot W_{eff}}{L_{eff}} \cdot \frac{Q_{inv,V_{ds}=0} \cdot V_{ds} \cdot \left(1 - \frac{V_{ds}}{2 \cdot V_b}\right)}{1 + \frac{V_{ds}}{E_{sat} \cdot L_{eff}}}}$$

$$= \frac{I_{ds,lin,R_{ds}=0}}{1 + \frac{R_{ds} \cdot I_{ds,lin,R_{ds}=0}}{V_{ds}}}. \tag{11.113}$$

The term R_{ds} is not introduced as a simple parameter, but is instead included as a composite term, computed using

$$R_{ds} = \frac{\mathbf{RDSW} \cdot \left\{1 + \mathbf{PRWG} \cdot V_{gsx} + \mathbf{PRWB} \cdot \left[(\phi_s - V_{bsx})^{\frac{1}{2}} - \phi_s^{\frac{1}{2}}\right]\right\}}{\left(10^{-6} \cdot W'_{eff}\right)^{\mathbf{WR}}}, \tag{11.114}$$

where W'_{eff} is given by (11.39), **RDSW** represents the resistance per unit channel width, and **PRWB** accounts for the effect of the substrate bias. Note that the use of a width dependent parameter (**RDSW**) is a much better approach than using a simple lumped resistance, as it allows a model to be valid for an arbitrary effective channel width; this is in contrast to the use of lumped terms such as **RS** and **RD**, which must be determined separately for each particular effective channel width. Also, note that (11.114) contains a parameter as an exponent; since this parameter is likely to take on noninteger values, this expression may be computationally inefficient.

Equation (11.113) will be considered to be valid for all regions of operation as long as $V_{ds} \leq V_{dsat}$. For $V_{ds} > V_{dsat}$, the drain current I_{ds} becomes virtually constant; thus, the description provided by (11.113) is incomplete, and must be further improved; these modifications are discussed in Section 11.6.3. Also, note that the mobility appears in (11.113) as $\mu_v(T)$; this expression only takes account of the effect of the vertical field on the mobility. As discussed earlier, lateral field effects enter the drain current description through the carrier velocity model, and so are contained in other parts of the drain current model.

Temperature Dependence

Like the saturation velocity, the series resistance is weakly influenced by changes in temperature. However, for temperatures quite different than T_{nom}, the change can be significant. In BSIM3, the temperature dependence of the series resistance is described by

$$R_{dsw}(T) = \textbf{RDSW} + \textbf{PRT} \cdot \left(\frac{T}{T_{nom}} - 1 \right), \tag{11.115}$$

where **RDSW** is extracted at $T = T_{nom}$.

11.6.3   The Saturation Voltage

The Simple Expression With No Series Resistance

To begin the evaluation of the saturation voltage V_{dsat}, an expression for the lateral field at any point along the channel $E_y(y)$ is required. This is done using the general current expression (11.108), which is integrated to some arbitrary point along the channel; this yields

$$E_y(y) = \frac{I_{ds,lin,R_{ds}=0}}{\left[\left(W_{eff} \cdot Q_{inv,V_{ds}=0} \cdot \mu_v(T) - \frac{I_{ds,lin,R_{ds}=0}}{E_{sat}} \right)^2 - \frac{2 \cdot I_{ds,lin,R_{ds}=0} \cdot W_{eff} \cdot \mu_v(T) \cdot y \cdot Q_{inv,V_{ds}=0}}{V_b} \right]^{\frac{1}{2}}}. \tag{11.116}$$

The drift velocity will saturate in the channel at the point y at which $E_y(y) = E_{sat}$, and the drain current will reach its saturated value. Thus, $I_{ds,lin,R_{ds}=0}$ becomes $I_{dsat,R_{ds}=0}$. Solving (11.116) with these constraints leads to

$$I_{dsat,R_{ds}=0} = \frac{\mu_v(T) \cdot W_{eff}}{2 \cdot y} \cdot \frac{Q_{inv,V_{ds}=0} \cdot E_{sat} \cdot y \cdot V_b}{E_{sat} \cdot y + V_b}. \qquad (11.117)$$

Since at the simplest level the drain current does not increase above $I_{dsat,R_{ds}=0}$, the choice of y is somewhat arbitrary. This allows a choice of $y = L_{eff}$, when the pinch-off point is located at the drain diffusion. This transforms (11.117) into

$$I_{dsat,R_{ds}=0} = \frac{\mu_v(T) \cdot W_{eff}}{2 \cdot L_{eff}} \cdot \frac{Q_{inv,V_{ds}=0} \cdot E_{sat} \cdot L_{eff} \cdot V_b}{E_{sat} \cdot L_{eff} + V_b}. \qquad (11.118)$$

A second expression for $I_{dsat,R_{ds}=0}$ may be found by setting $V_{ds} = V_{dsat}$ in the linear region expression (11.110):

$$I_{dsat,lin,R_{ds}=0} = \frac{\mu_v(T) \cdot W_{eff}}{L_{eff}} \cdot \frac{Q_{inv,V_{ds}=0} \cdot V_{dsat} \cdot \left(1 - \frac{V_{dsat}}{2 \cdot V_b}\right)}{1 + \frac{V_{dsat}}{E_{sat} \cdot L_{eff}}}. \qquad (11.119)$$

Equations (11.118) and (11.119) can be solved together, leading to a solution for the saturation voltage V_{dsat}:

$$V_{dsat,R_{ds}=0} = \frac{E_{sat} \cdot L_{eff} \cdot \left(V_{gsx} + 2 \cdot \frac{k_b T}{q}\right)}{A_{bulk} \cdot E_{sat} \cdot L_{eff} + \left(V_{gsx} + 2 \cdot \frac{k_b T}{q}\right)}. \qquad (11.120)$$

Using this simplified but workable expression for the saturation voltage, some observations about the long and short channel limits of the strong inversion drain current model can be made. In long channel devices,

$$E_{sat} \cdot L_{eff} \gg V_{gsx}, \qquad (11.121)$$

so (11.120) becomes

$$V_{dsat} \doteq \frac{V_{gs} - V_t(T)}{A_{bulk}}; \qquad (11.122)$$

recalling the correspondence between A_{bulk} in BSIM3 and a in BSIM, (11.122) is identical to the long channel limit of the BSIM saturation voltage expression (8.136). Using (11.122) in (11.119), it can be shown that

$$I_{ds} \propto \frac{1}{L_{eff}} \cdot (V_{gs} - V_t(T))^2; \qquad (11.123)$$

the long channel "square law" expression for the saturation region current is recovered. In short devices,

$$E_{sat} \cdot L_{eff} \ll V_{gsx},\tag{11.124}$$

and (11.120) becomes

$$V_{dsat} = E_{sat} \cdot L_{eff}.\tag{11.125}$$

Substituting (11.125) into (11.119), it can be shown that

$$I_{ds} \propto (V_{gs} - V_t(T));\tag{11.126}$$

the drain current is independent of the channel length and is only linearly proportional to $(V_{gs} - V_t(T))$. Thus, the basic structure of the BSIM3 drain current model is able to reproduce these important features of the short channel limit.

The Complete Saturation Voltage Model

To accurately describe the physical situation, the model for the saturation voltage, like the model for the linear region drain current, must be upgraded to include the effect of the source/drain series resistance. The introduction of this added complication leads to a large amount of tedious and unpleasant algebra, most of which is eliminated here in the interest of brevity. As in the case above, where the series resistance is neglected, an expression is found for $E_y(y)$, the field along the channel; however, the full linear expression (11.113) is used rather than the simpler equation (11.110); this in turn leads to a very complicated expression for I_{dsat}. The second expression for I_{dsat} is found quite simply, by replacing V_{ds} in the linear region equation (11.113) with the saturation voltage V_{dsat}:

$$I_{ds,lin,V_{ds}=V_{dsat}} = \frac{\mu_v(T) \cdot W_{eff}}{L_{eff}} \cdot \frac{1}{1 + \frac{V_{dsat}}{E_{sat} \cdot L_{eff}}}$$

$$\times \frac{Q_{inv,V_{ds}=0} \cdot V_{dsat} \cdot \left(1 - \frac{V_{dsat}}{2 \cdot V_b}\right)}{1 + \frac{R_{ds}}{V_{dsat}} \cdot \frac{\mu_v(T) \cdot W_{eff}}{L_{eff}} \cdot \frac{Q_{inv,V_{dsat}=0} \cdot V_{dsat} \cdot \left(1 - \frac{V_{dsat}}{2 \cdot V_b}\right)}{1 + \frac{V_{dsat}}{E_{sat} \cdot L_{eff}}}}$$

$$= \frac{I_{ds,lin,R_{ds}=0,V_{ds}=V_{dsat}}}{1 + \frac{R_{ds} \cdot I_{ds,lin,R_{ds}=0,V_{ds}=V_{dsat}}}{V_{dsat}}}.\tag{11.127}$$

In contrast to the relatively simple form of (11.120), this system of two equations leads to a complicated quadratic expression:

$$V_{dsat} = \frac{1}{2a}\left[-b \pm \left(b^2 - 4 \cdot a \cdot c\right)^{\frac{1}{2}}\right],\tag{11.128}$$

where

$$a = A_{bulk}^2 \cdot R_{ds} \cdot C_{ox} \cdot W_{eff} \cdot v_{SAT}(T) + \left(\frac{1}{\lambda} - 1\right) \cdot A_{bulk}, \tag{11.129}$$

$$b = -\left[\left(V_{gsx} - 2 \cdot \frac{k_b T}{q}\right) \cdot \left(\frac{2}{\lambda} - 1\right) + A_{bulk} \cdot E_{sat} \cdot L_{eff}\right.$$

$$\left. + 3 \cdot A_{bulk} \cdot R_{ds} \cdot C_{ox} \cdot W_{eff} \cdot v_{SAT}(T) \cdot \left(V_{gsx} + 2 \cdot \frac{k_b T}{q}\right)\right], \tag{11.130}$$

$$c = E_{sat} \cdot L_{eff} \cdot \left(V_{gsx} + 2 \cdot \frac{k_b T}{q}\right)$$

$$+ 2 \cdot R_{ds} \cdot C_{ox} \cdot W_{eff} \cdot v_{SAT}(T) \cdot \left(V_{gsx} + 2 \cdot \frac{k_b T}{q}\right)^2, \tag{11.131}$$

and

$$\lambda = \mathbf{A1} \cdot (V_{gs} - V_t(T)) + \mathbf{A2}. \tag{11.132}$$

The quadratic expression (11.128) has two possible solutions, only one of which is correct; some method of choosing between the two possibilities must be found. In the limit of $R_{ds} \to 0$, (11.128) should reduce to the simpler expression (11.120). This occurs for the "$-$" choice, so the correct saturation voltage expression is

$$V_{dsat} = \frac{1}{2a}\left[-b - \left(b^2 - 4 \cdot a \cdot c\right)^{\frac{1}{2}}\right]. \tag{11.133}$$

Linking the Linear and Saturation Regions

As described in Section 11.2, one of the key goals of the third generation models is to guarantee smooth and continuous expressions for the drain current and its first derivative for all regions of device operation. This is done by introducing special mathematical smoothing functions into the model, which reduce to the simpler limiting cases away from the transition regions which they are describing.

In Section 11.3, the auxiliary gate bias V_{gsx} was introduced; this function guarantees a smooth and continuous transition between the weak inversion region and the strong inversion region. To guarantee a smooth and continuous transition between the linear region and the saturation region, an auxiliary drain bias V_{dsx} is introduced:

$$V_{dsx} = V_{dsat} - \frac{1}{2} \cdot \left\{ V_{dsat} - V_{ds} - \mathbf{DELTA}\right.$$

$$\left. + \left[(V_{dsat} - V_{ds} - \mathbf{DELTA})^2 + 4 \cdot \mathbf{DELTA} \cdot V_{dsat}\right]^{\frac{1}{2}} \right\}, \tag{11.134}$$

where **DELTA** is an empirical smoothing parameter and V_{dsat} is computed from (11.133). (Note that this **DELTA** is unrelated to narrow channel behavior, which is described by a parameter of this name in Level 2 and Level 3.) This auxiliary function is very similar to a smoothing function introduced into MOS Model 9 [3] for the same purpose (see Section 12.5).

The role of V_{dsx} can easily be defined by examining its limiting behavior. For this discussion, let **DELTA** $= 0$, so that the limiting behavior can be reached without the complication of dealing with this added empirical parameter. When $V_{ds} = V_{dsat}$, $V_{dsx} \rightarrow V_{dsat}$; the saturation voltage is recovered. When $V_{ds} \gg V_{dsat}$, $V_{dsx} \rightarrow V_{dsat}$; in the absence of the output conductance model (discussed in Section 11.6.4), the drain bias is pinned at V_{dsat}, and the drain current maintains the fixed value of I_{dsat}. When $V_{ds} \ll V_{dsat}$, $V_{dsx} \rightarrow V_{ds}$; the auxiliary function reduces to the simple linear region drain bias.

Thus, the role of V_{dsx} is rather clear. In the saturation region of operation, V_{dsx} will replace V_{dsat} in the development of the output conductance model, which is discussed below. In the linear region of operation, V_{dsx} replaces V_{ds}. Therefore, the current equations (11.110) and (11.113) must be rewritten as

$$I_{ds,lin,R_{ds}=0} = \frac{\mu_v(T) \cdot W_{eff}}{L_{eff}} \cdot \frac{Q_{inv,V_{ds}=0} \cdot V_{dsx} \cdot \left(1 - \frac{V_{dsx}}{2 \cdot V_b}\right)}{1 + \frac{V_{dsx}}{E_{sat} \cdot L_{eff}}} \tag{11.135}$$

and

$$I_{ds,lin} = \frac{\mu_v(T) \cdot W_{eff}}{L_{eff}} \cdot \frac{1}{1 + \frac{V_{dsx}}{E_{sat} \cdot L_{eff}}} \cdot \frac{Q_{inv,V_{ds}=0} \cdot V_{dsz} \cdot \left(1 - \frac{V_{dsx}}{2 \cdot V_b}\right)}{1 + \frac{R_{ds}}{V_{dsx}} \cdot \frac{\mu_v(T) \cdot W_{eff}}{L_{eff}} \cdot \frac{Q_{inv,V_{cs}=0} \cdot V_{dsx} \cdot \left(1 - \frac{V_{dsx}}{2 \cdot V_b}\right)}{1 + \frac{V_{dsx}}{E_{sat} \cdot L_{eff}}}}$$

$$= \frac{I_{ds,lin,R_{ds}=0}}{1 + \frac{R_{ds} \cdot I_{ds,lin,R_{ds}=0}}{V_{dsx}}} \cdot \tag{11.136}$$

In (11.136), $I_{ds,lin,R_{ds}=0}$ is now provided by (11.135) (rather than by (11.110)).

11.6.4 The Output Conductance Model

If (11.136) is used to describe the saturation region current, for a given gate bias V_{gs}, the drain current will maintain a constant value for $V_{ds} \geq V_{dsat}$. Since this is not in fact what happens, additional modifications to the current expression are required.

BSIM3 describes this behavior by developing an output conductance model for the saturation region. The basis of this approach is identical to that developed in BSIM2 [16]; as shown in Figure 11.4, the output conductance is separated into regions that are deemed to be separately dominated by different physical effects. Individual models for the output conductance behavior are then developed for each of these regions, and these submodels are added to the general current model.

The BSIM3 approach begins by noting that, in general, in the saturation region, the drain current is a weak function of the drain bias (i.e., it is a weak function of

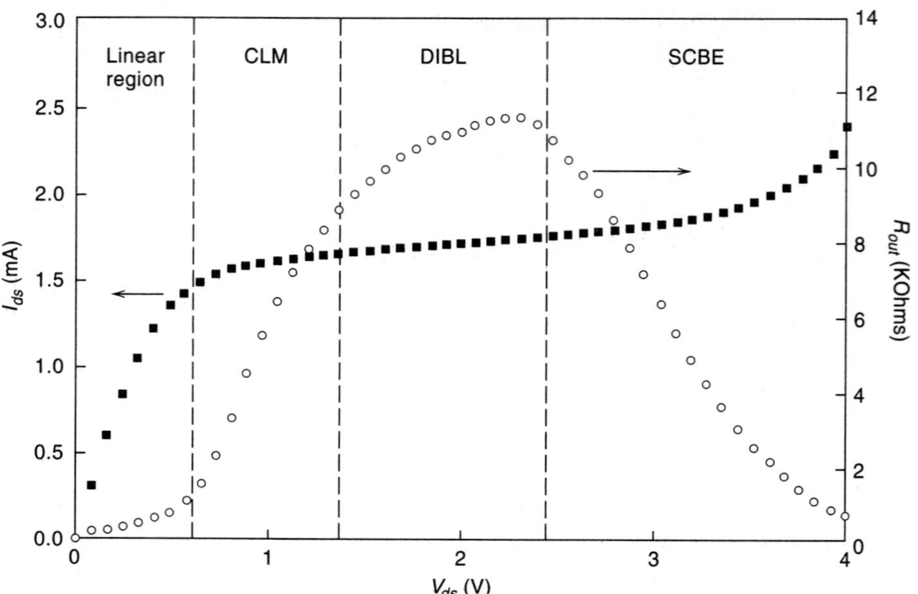

Figure 11.4   The different regions of the output conductance; a separate model is developed for each region. From [2]. © 1995. The Regents of the University of California. Used by permission.

$(V_{ds} - V_{dsat})$). Treating the drain current expression as a general function of the gate and drain biases, a first order Taylor expansion for the saturation region yields

$$I_{ds}(V_{gs}, V_{ds}) = I_{ds}(V_{gs}, V_{dsat}) + \frac{\partial I_{ds}(V_{gs}, V_{ds})}{\partial V_{ds}} \cdot (V_{ds} - V_{dsat}). \tag{11.137}$$

However, since by definition

$$I_{ds}(V_{gs}, V_{dsat}) = I_{dsat}, \tag{11.138}$$

(11.137) can be simplified to

$$I_{ds}(V_{gs}, V_{ds}) = I_{dsat} + \frac{\partial I_{ds}(V_{gs}, V_{ds})}{\partial V_{ds}} \cdot (V_{ds} - V_{dsat}). \tag{11.139}$$

Note that the drain current in (11.139) is linear in $(V_{ds} - V_{dsat})$, which is equivalent to being linear in V_{ds}. To describe this situation, BSIM3 uses the concept of the Early voltage in bipolar transistors [5, 17], as shown in Figure 11.5. However, this approach is identical to the λ-method (with the parameter **LAMBDA**), which is used in Level 1 and Level 2; as shown in Figure 11.5,

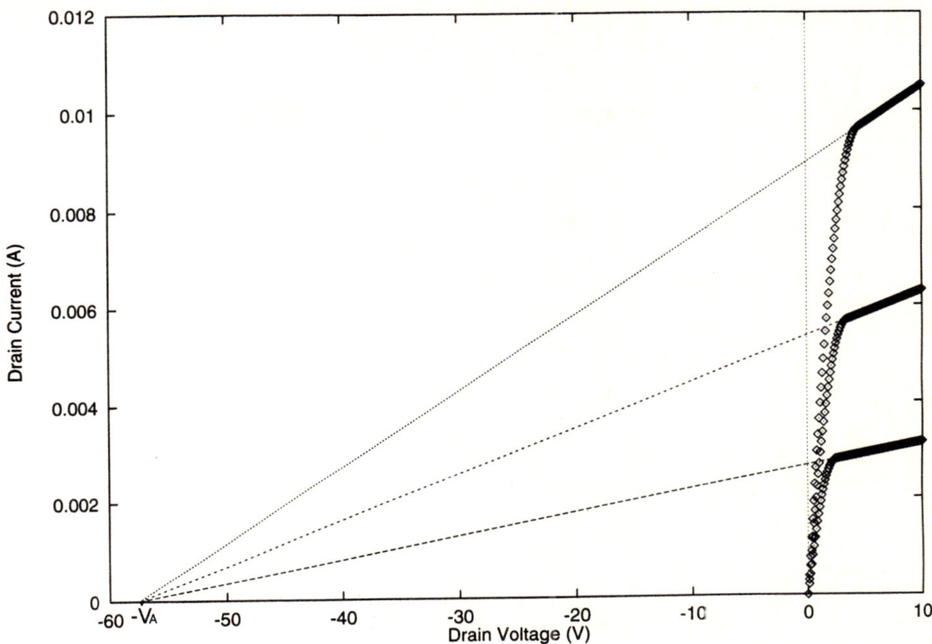

Figure 11.5   The basic Early voltage definition in a MOSFET.

$$V_A = \frac{1}{\textbf{LAMBDA}}, \tag{11.140}$$

where V_A is the Early voltage.

The model implemented in BSIM3 is best illustrated using the graph in Figure 11.6. In the saturation region, the drain current can be expressed as

$$I_{ds} = I_{dsat} + (I_{ds} - I_{dsat}). \tag{11.141}$$

The slope of the current in the saturation region is

$$\alpha = \frac{I_{ds} - I_{dsat}}{V_{ds} - V_{dsat}}, \tag{11.142}$$

so (11.141) can be rewritten as

$$I_{ds} = I_{dsat} + \alpha \cdot (V_{ds} - V_{dsat}). \tag{11.143}$$

In contrast to the development used in Level 1 (see Section 5.2), the slope α is defined here as a derivative:

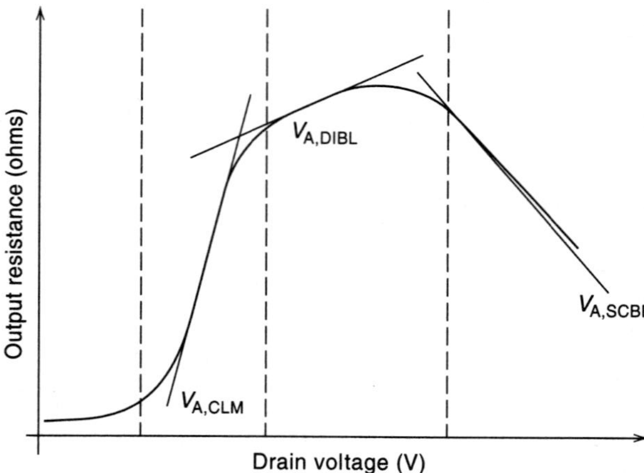

Figure 11.6   Graphical depiction of the method used in BSIM3 to describe the output conductance.

$$\alpha = \frac{\partial I_{ds}}{\partial V_{ds}}. \tag{11.144}$$

Substituting this expression into (11.143) leads to

$$I_{ds} = I_{dsat} + \frac{\partial I_{ds}}{\partial V_{ds}} \cdot (V_{ds} - V_{dsat}); \tag{11.145}$$

the Taylor expansion (11.139) is recovered.

Equation (11.145) is in slope-intercept form ($y = mx + b$), with the zero of x defined at $V_{ds} = V_{dsat}$. Thus, the x-intercept ($y = 0$) is found at

$$V_{ds} = -V_A + V_{dsat}. \tag{11.146}$$

Substituting this expression into (11.145) yields

$$\left(\frac{\partial I_{ds}}{\partial V_{ds}}\right) \cdot (-V_A) + I_{dsat} = 0, \tag{11.147}$$

which can be solved for the Early voltage V_A:

$$V_A = \frac{I_{dsat}}{\left(\frac{\partial I_{ds}}{\partial V_{ds}}\right)}. \tag{11.148}$$

Using (11.148) in (11.145) produces the general saturation region current expression:

$$I_{ds} = I_{dsat} + \left[\left(\frac{I_{dsat}}{V_A}\right) \cdot (V_{ds} - V_{dsat})\right] = I_{dsat} \cdot \left(1 + \frac{V_{ds} - V_{dsat}}{V_A}\right). \tag{11.149}$$

Introducing V_{dsx} as defined by (11.134) transforms (11.149) into

$$I_{ds} = I_{dsat} + \left[\left(\frac{I_{dsat}}{V_A} \right) \cdot (V_{ds} - V_{dsx}) \right] = I_{dsat} \cdot \left(1 + \frac{V_{ds} - V_{dsx}}{V_A} \right). \qquad (11.150)$$

Note that these equations are essentially identical to the Level 1 expression

$$I_{ds,sat} = I_{dsat} \cdot (1 + \textbf{LAMBDA} \cdot V_{ds}). \qquad (11.151)$$

In BSIM3, it is assumed that a specific Early voltage can be computed independently for each of the three saturation region mechanisms considered in Figure 11.4 (channel length modulation, drain-induced barrier lowering, and substrate current induced body effect). In each region of the curve of Figure 11.4, an Early voltage is computed and used in (11.149). For all intents and purposes, this is an extension of the old λ-method, except that three different versions of λ are introduced to replace the original single parameter.

Channel Length Modulation

Using the chain rule, channel length modulation is described in (11.148) as

$$V_{A,CLM} = \frac{I_{dsat}}{\frac{\partial I_{ds}}{\partial V_{ds}}} = \frac{I_{dsat}}{\frac{\partial I_{ds}}{\partial L_{eff}} \cdot \frac{\partial L_{eff}}{\partial V_{ds}}}. \qquad (11.152)$$

A very complex derivation follows, which is based on a two-dimensional analysis of the space charge region [1] and uses the drain current expression (11.135). This results in

$$V_{A,CLM} = \frac{A_{bulk} \cdot E_{sat} \cdot L_{eff} + V_{gsx}}{A_{bulk} \cdot E_{sat} \cdot l} \cdot (V_{ds} - V_{dsx}), \qquad (11.153)$$

where l is a composite length term, defined by

$$l = \left(\frac{\epsilon_{Si}}{\epsilon_{ox}} \cdot t_{ox} \cdot \textbf{XJ} \right)^{\frac{1}{2}} \qquad (11.154)$$

and **XJ** is the junction depth. As noted earlier, use of the model will be required to determine if **XJ** will actually represent the physical junction depth, or instead serve as an empirical parameter (e.g., as in Level 3).

For relative simplicity during development, (11.153) was derived using the simple drain current expression (11.135), where it was assumed that the source/drain series resistance R_{ds} is negligible. However, as subsequently noted, this is not a good approximation in short channel FETs, and the more detailed expression (11.136) was developed. However, in BSIM3, rather than use this more complicated expression in the derivation, the simpler result (11.153) is modified by the addition of an empirical parameter (**PCLM**) for improved accuracy. This leads to the final expression for the Early voltage due to channel length modulation:

$$V_{A,CLM} = \frac{1}{\textbf{PCLM}} \cdot \frac{A_{bulk} \cdot E_{sat} \cdot L_{eff} + V_{gsx}}{A_{bulk} \cdot E_{sat} \cdot l} \cdot (V_{ds} - V_{dsx}). \tag{11.155}$$

Drain-Induced Barrier Lowering

The BSIM3 expression for the effect of drain-induced barrier lowering (DIBL) on the output characteristics is developed (as described above for channel length modulation) by applying the chain rule to (11.148):

$$V_{A,DIBL} = \frac{I_{dsat}}{\frac{\partial I_{ds}}{\partial V_{ds}}} = \frac{I_{dsat}}{\frac{\partial I_{ds}}{\partial V_t} \cdot \frac{\partial V_t}{\partial V_{ds}}}. \tag{11.156}$$

In the same manner as that used to reach (11.153), a very complicated derivation follows, which produces

$$V_{A,DIBL} = \frac{1}{\Theta_t} \cdot \left[V_{gsx} - \left(\frac{1}{\frac{1}{A_{bulk} \cdot V_{dsx}} + \frac{1}{V_{gsx}}} \right) \right], \tag{11.157}$$

where Θ_t is the short channel threshold voltage coefficient defined by (11.70). However, the same problem discussed above occurs again—the simple drain current expression (11.135), which assumes that the source/drain series resistance is negligible, has been used in the derivation rather than the more accurate expression (11.136). Once again, this problem is dealt with empirically, by making modifications to (11.157). The threshold voltage coefficient Θ_t is replaced with another expression **only in the description of the effect of DIBL on the output conductance behavior**:

$$\Theta_{rout} = \textbf{PDIBLC1} \cdot \left[exp \left(-\frac{\textbf{DROUT} \cdot L_{eff}}{2 \cdot L_{t,DIBL}} \right) + 2 \cdot exp \left(-\frac{\textbf{DROUT} \cdot L_{eff}}{L_{t,DIBL}} \right) \right]$$
$$+ \textbf{PDIBLC2}, \tag{11.158}$$

where $L_{t,DIBL}$ is the characteristic length defined by (11.79). There is extensive reasoning behind these modifications [2], but the overall result is empirical. Thus, the final expression for the Early voltage due to drain-induced barrier lowering is

$$V_{A,DIBL} = \frac{1}{\Theta_{rout}} \cdot \left[V_{gsx} - \left(\frac{1}{\frac{1}{A_{bulk} \cdot V_{dsx}} + \frac{1}{V_{gsx}}} \right) \right]. \tag{11.159}$$

Ensuring Continuity

When the final drain current expression is assembled, the continuity of both the drain current equation and its first derivative (the output conductance) must be ensured at the linear-saturation transition point. In BSIM3, an expression is introduced for the Early voltage at $V_{ds} = V_{dsat}$:

$$V_{A,sat} = \frac{E_{sat} \cdot L_{eff} + V_{dsx} + 2 \cdot R_{ds} \cdot v_{SAT}(T) \cdot C_{ox} \cdot W_{eff} \cdot \left(V_{gsx} - \frac{A_{bulk}}{2} \cdot V_{dsx}\right)}{1 + A_{bulk} \cdot R_{ds} \cdot v_{SAT}(T) \cdot C_{ox} \cdot W_{eff}}. \tag{11.160}$$

At the linear-saturation transition point, both channel length modulation and drain-induced barrier lowering are important, and so in BSIM3 their effects are included together in a composite Early voltage expression:

$$V_{A2} = V_{A,sat} + \frac{1}{\left(\frac{1}{V_{A,CLM}} + \frac{1}{V_{A,DIBL}}\right)}. \tag{11.161}$$

However, to improve the final model accuracy, BSIM3 once again makes empirical modifications, so that (11.161) becomes

$$V_{A2} = V_{A,sat} + \frac{\left(1 + \frac{\mathbf{PVAG} \cdot V_{gs}}{E_{sat} \cdot L_{eff}}\right)}{\left(\frac{1}{V_{A,CLM}} + \frac{1}{V_{A,DIBL,C}}\right)}, \tag{11.162}$$

where $V_{A,DIBL,C}$ is a corrected DIBL Early voltage, described by

$$V_{A,DIBL,C} = \frac{V_{gsx} + 2 \cdot \frac{k_b T}{q}}{\Theta_{rout} \cdot (1 + \mathbf{PDIBLCB} \cdot V_{bsx})} \cdot \left(1 - \frac{A_{bulk} \cdot V_{dsat}}{A_{bulk} \cdot V_{dsat} + V_{gsx} + 2 \cdot \frac{k_b T}{q}}\right). \tag{11.163}$$

Using (11.150), the expression for the drain current is

$$I_{ds,cd} = \frac{I_{ds,lin,R_{ds}=0,V_{ds}=V_{dsx}}}{1 + \frac{R_{ds} \cdot I_{ds,lin,R_{ds}=0,V_{ds}=V_{dsx}}}{V_{dsx}}} \cdot \left(1 + \frac{V_{ds} - V_{dsx}}{V_{A2}}\right). \tag{11.164}$$

Note that due to the use of V_{gsx} and V_{dsx}, this expression is valid for all regions of device operation.

Substrate Current Induced Body Effect

For high drain biases (corresponding to high drain fields), impact ionization will cause a significant substrate current to flow in the device. As described in Section 10.6, the presence of a finite substrate current implies the existence of a substrate resistance; this causes the substrate bias to be altered to

$$V_{bs}' = V_{bs} + I_{sub} \cdot R_{sub}. \tag{11.165}$$

This modified substrate bias slightly forward biases the source-substrate junction, which causes a decrease in the effective threshold voltage and a corresponding increase in the drain current. This behavior is known as the substrate current induced body effect (SCBE) [18]; in BSIM2 [16], the identical process is referred to as the *hot electron induced output conductance shift.*

In BSIM3, it is assumed that this behavior is significant only at drain biases well above the saturation voltage; thus, the continuity concerns with the linear region, described above, are not considered. To describe the substrate current, a popular model [19] is used:

$$I_{sub} = \frac{A_i}{B_i} \cdot I_{ds,sat} \cdot (V_{ds} - V_{dsx}) \cdot exp\left(-\frac{B_i \cdot l}{V_{ds} - V_{dsx}}\right), \tag{11.166}$$

where l is given by (11.154). The additional current is simply added to the drain current that is already present:

$$I_{ds} = I_{ds,cd} + I_{sub}, \tag{11.167}$$

where $I_{ds,cd}$ is given by (11.164). Using (11.166), (11.167) becomes

$$I_{ds} = I_{ds,cd}\left[1 + \frac{A_i}{B_i} \cdot I_{ds,lin,V_{ds}=V_{dsx}} \cdot (V_{ds} - V_{dsx}) \cdot exp\left(-\frac{B_i \cdot l}{V_{ds} - V_{dsx}}\right)\right]$$

$$= I_{ds,cd} \cdot \left[1 + \frac{V_{ds} - V_{dsx}}{\frac{B_i}{A_i} \cdot exp\left(\frac{B_i \cdot l}{V_{ds} - V_{dsx}}\right)}\right], \tag{11.168}$$

where $I_{ds,lin,V_{ds}=V_{dsx}}$ is given by (11.136) with V_{ds} set to V_{dsx}. This allows for the definition of the Early voltage due to the substrate current induced body effect:

$$V_{A,SCBE} = \frac{B_i}{A_i} \cdot exp\left(\frac{B_i \cdot l}{V_{ds} - V_{dsx}}\right). \tag{11.169}$$

BSIM3 also makes modifications to the Early voltage expression. Once again reasoning is provided [2], but the essence is empirical:

$$V_{A,SCBE} = \frac{L_{eff}}{\textbf{PSCBE2}} \cdot exp\left(\frac{\textbf{PSCBE1} \cdot l}{V_{ds} - V_{dsat}}\right). \tag{11.170}$$

The Final Expression for the Drain Current

Combining (11.168), (11.169), and (11.170) leads to the final BSIM3 expression for the drain current in the saturation region:

$$I_{ds} = I_{ds,lin,V_{ds}=V_{dsx}} \cdot \left(1 + \frac{V_{ds} - V_{dsx}}{V_{A2}}\right) \cdot \left(1 + \frac{V_{ds} - V_{dsx}}{V_{A,SCBE}}\right). \tag{11.171}$$

This equation is valid for all regions of device operation. Due to the use of the smoothing functions V_{gsx} and V_{dsx}, continuous and smooth behavior across both transition regions is guaranteed.

11.6.5   The Substrate Current Model

In addition to the drain current model, BSIM3 also offers a model for the substrate current:

$$I_{sub} = \frac{\textbf{ALPHA0}}{L_{eff}} \cdot (V_{ds} - V_{dsx}) \cdot exp\left(-\frac{\textbf{BETA0}}{V_{ds} - V_{dsx}}\right)$$
$$\cdot I_{ds,lin,V_{ds}=V_{dsx}} \cdot \left(1 + \frac{V_{ds} - V_{dsx}}{V_{A2}}\right). \qquad (11.172)$$

The specific origins of this expression are unclear. The substrate current is generally not used in circuit design, but can be employed as a monitor of hot carrier degradation [19], if this is critical to a particular circuit design.

11.7   The Charge Model

The BSIM3 charge model employs the development used for the charge models of BSIM and BSIM2 as its base. As in the current model, the equations that represent the linear region are developed first. Then, rather than develop separate descriptions for the saturation and subthreshold regions, auxiliary smoothing functions are introduced to allow these linear region expressions to be generalized for all regions of device operation. Some of these auxiliary functions were already introduced into the current model, while others are developed specifically for use in the charge model.

This process results in node charge expressions which are continuous and smooth for all regions of device operation. Of particular note, the glaring discontinuity in the charge expressions at the threshold voltage (the point of the weak inversion-strong inversion transition) which occurs in the first- and second-generation models is now eliminated; this improves both the accuracy of the FET model, and the circuit simulation convergence properties. This is particularly important in low voltage digital circuit designs, for which the subthreshold region represents a larger fraction of the voltage operating range, and in analog circuit designs, which often operate statically at voltages near the threshold voltage.

11.7.1   The Active Channel Length and Width

In BSIM3, separate values of the effective channel length and width must be computed for use in the charge calculations. As described in Appendix B, the current and charge models treat the effective channel dimensions differently. For example, the current depends on the distance between the source and drain junctions (the effective channel length), while the charge behavior of the gate depends on the physical size of the polysilicon gate material;

these differ from the drawn dimensions in different ways. As detailed in Appendix B, a careful accounting for these differences can greatly simplify the charge calculations.

The channel length and width that are used in the BSIM3 charge computations are defined by

$$L_{active} = L_{drawn} - 2 \cdot \Delta L_c \tag{11.173}$$

and

$$W_{active} = W_{drawn} - 2 \cdot \Delta W_c. \tag{11.174}$$

The terms ΔL_c and ΔW_c are similar to the composite terms (11.37) and (11.38) in Section 11.3.2:

$$\Delta L_c = \mathbf{DLC} + \frac{\mathbf{LL}}{L_{drawn}^{\mathbf{LLN}}} + \frac{\mathbf{LW}}{W_{drawn}^{\mathbf{LWN}}} + \frac{\mathbf{LWL}}{L_{drawn}^{\mathbf{LLN}} \cdot W_{drawn}^{\mathbf{LWN}}} \tag{11.175}$$

and

$$\Delta W_c = \mathbf{DWC} + \frac{\mathbf{WL}}{L_{drawn}^{\mathbf{WLN}}} + \frac{\mathbf{WW}}{W_{drawn}^{\mathbf{WWN}}} + \frac{\mathbf{WWL}}{L_{drawn}^{\mathbf{WLN}} \cdot W_{drawn}^{\mathbf{WWN}}}; \tag{11.176}$$

DLC and **DWC** serve the same role as **LINT** in (11.34) and **WINT** in (11.35). The intent here is identical to that described in Section 11.3.2; a large number of terms are added to improve the accuracy, but at the cost of causing ΔL_c and ΔW_c to have geometry dependence. As noted in the earlier discussion, if these terms are determined properly, they are normally well-defined and unchanging for a given process technology. While this method can improve the fitting accuracy, it runs the risk of making it more difficult to properly describe the effect of process variations, as normal statistical variations in ΔL_c and ΔW_c provide very important contributions to the description of those variations.

11.7.2   Auxiliary Functions

As in the current model, the charge model uses smoothing functions to guarantee continuous and smooth node charge equations across all regions of device operation. Some of these auxiliary functions are taken directly from the current model, while others use the same basic approach, but are developed specifically for the charge model. These auxiliary functions will be described in detail here, and then introduced in passing to the node charge equations as those expressions are developed.

Two auxiliary functions are introduced in the same manner in which they appear in the current model. The auxiliary function V_{gsx}, defined in (11.15), will replace ($V_{gs} - V_t$) where this expression occurs. Also, the auxiliary function V_{bsx}, defined by (11.8), will replace the substrate bias V_{bsx}.

To smooth the linear-saturation transition, an auxiliary drain bias function V_{dsq}, which is similar in intent to the auxiliary drain bias expression V_{dsx}, is introduced:

$$V_{dsq} = V_{dsat,q} - \frac{1}{2} \cdot \left[V_4 + \left(V_4^2 + 4 \cdot \delta_4 \cdot V_{dsat,q} \right) \right], \tag{11.177}$$

where

$$V_4 = V_{dsat,q} - V_{ds} - \delta_4 \tag{11.178}$$

and δ_4 is set to $\delta_4 = 0.02$. The term $V_{dsat,q}$ is an auxiliary saturation voltage expression for use in the charge model:

$$V_{dsat,q} = \frac{V_{gsx}}{A_{bulk} \cdot \left[1 + \left(\frac{\mathbf{CLC}}{L_{active}} \right)^{\mathbf{CLE}} \right]} = \frac{V_{gsx}}{A_{bulk,q}}, \tag{11.179}$$

which also defines $A_{bulk,q}$, the charge model version of the bulk charge term A_{bulk}; $A_{bulk,q}$ is introduced to account for short channel effects. When $\mathbf{CLC} \to 0$, $A_{bulk,q}$ reduces to A_{bulk}, and (11.179) reverts to the long channel limit of the saturation voltage expression (11.122). Note the presence in (11.179) of a parameter as an exponential term, which indicates that the expected noninteger value of $\mathbf{CLE}$ will harm the computational efficiency of the expression.

The structure of (11.179) indicates another important consideration. In contrast to the earlier charge models, the BSIM3 charge model requires the extraction of many parameters specifically for use in the charge model. This places additional demands on test site design, and also introduces requirements for AC parameter extraction.

Although it is less important and sees more limited use, a smoothing function is also introduced for the accumulation-depletion transition region, which occurs at the flatband voltage ($V_{gs} = V_{fb}$):

$$V_{fbx} = V_{fb} - \frac{1}{2} \cdot \left[V_3 + \left(V_3^2 + 4 \cdot \delta_3 \cdot V_{fb} \right)^{\frac{1}{2}} \right], \tag{11.180}$$

where

$$V_3 = V_{fb} - V_{gb} - \delta_3 \tag{11.181}$$

and δ_3 is set to $\delta_3 = 0.02$.

11.7.3 The Basic Charge Equations

The basic device charge equations employ the description developed by Yang, Epler, and Chatterjee [20] as implemented in BSIM (see Section 8.8), with a few minor notational adjustments and the introduction of the various auxiliary functions discussed above. The core of the formulation uses the linear region equations; then, as is done with the current model, auxiliary functions are introduced to make these expressions valid for the saturation region as well. Separate expressions are introduced for the accumulation and subthreshold regions, but these are also modified to allow for continuous and smooth transitions across the region boundaries.

The Accumulation and Subthreshold Regions

In the accumulation region, the inversion charge is zero, and the gate charge is mirrored in the substrate as accumulation charge near the silicon surface. The structure behaves as a two dimensional MOS capacitor, so that the accumulation charge is described by [6]

$$Q_{ACC} = -W_{active} \cdot L_{active} \cdot C_{ox} \cdot (V_{gb} - V_{fb}). \qquad (11.182)$$

In the subthreshold region, the inversion charge is again negligible, and the gate charge is mirrored in the substrate as depletion charge. This depletion charge is computed from [4, 20]

$$Q_{DEPL, V_{ds}=0} = -W_{active} \cdot L_{active} \cdot C_{ox} \cdot \frac{\mathbf{K1}^2}{2} \cdot \left\{ -1 + \left[1 + \frac{4 \cdot (V_{gb} - V_{fb})^{\frac{1}{2}}}{\mathbf{K1}^2} \right] \right\}, \qquad (11.183)$$

which is identical to the BSIM subthreshold region depletion charge expression (8.154).

    Both (11.182) and (11.183) represent the bulk charge in the substrate, and so should be linked in a continuous and smooth manner across the boundary at $V_{gs} = V_{fb}$. Using auxiliary functions, (11.182) becomes

$$Q_{ACC} = -W_{active} \cdot L_{active} \cdot C_{ox} \cdot (V_{fbx} - V_{fb}), \qquad (11.184)$$

where V_{fbx} is the auxiliary flatband voltage expression given by (11.180). Again using auxiliary functions, (11.183) becomes

$$Q_{DEPL, V_{ds}=0} = -W_{active} \cdot L_{active} \cdot C_{ox} \cdot \frac{\mathbf{K1}^2}{2}$$

$$\cdot \left\{ -1 + \left[1 + \frac{4 \cdot (V_{gs} - V_{fbx} - V_{gsx} - V_{bsx})^{\frac{1}{2}}}{\mathbf{K1}^2} \right] \right\}, \qquad (11.185)$$

where V_{gsx} is the auxiliary gate bias given by (11.15) and V_{bsx} is the auxiliary substrate bias given by (11.8). The use of V_{gsx} also guarantees a continuous and smooth transition between the subthreshold region and the strong inversion region.

    The gate charge is simply the mirror value of the charge in the substrate. In the accumulation region,

$$Q_{GATE} = -Q_{ACC}, \qquad (11.186)$$

while in the subthreshold region,

$$Q_{GATE} = -Q_{DEPL, V_{ds}=0}. \qquad (11.187)$$

The Strong Inversion Region

As noted earlier, BSIM3 uses the BSIM charge model development, which is closely based on the description provided by Yang, Epler, and Chatterjee [20]. The linear region expressions developed there are employed; then, instead of deriving a separate set of charge expressions for the saturation region, auxiliary functions are introduced to make the model generally valid for the linear and saturation regions.

As in all the charge model descriptions, charge neutrality must be satisfied for both the charge per unit area and the total charge. Thus,

$$Q_{gate}(y) + Q_{inv}(y) + Q_{depl}(y) = 0 \qquad (11.188)$$

(where the lower case subscripts indicate charge per unit area), and

$$Q_{GATE} + Q_{INV} + Q_{DEPL} = 0 \qquad (11.189)$$

(where the capitalized subscripts indicate total charge). In the linear region [20], the gate charge per unit area is

$$Q_{gate}(y) = C_{ox} \cdot \left[V_{gs} - V_{fb} - \phi_s - \phi_n(y) \right], \qquad (11.190)$$

while the inversion charge per unit area is

$$Q_{inv}(y) = -C_{ox} \cdot \left[V_{gs} - V_t - A_{bulk} \cdot \phi_n(y) \right]. \qquad (11.191)$$

The depletion charge per unit area is easily found from (11.188), by using (11.190) and (11.191):

$$Q_{depl}(y) = -\left[Q_{gate}(y) + Q_{inv}(y) \right] = -C_{ox} \cdot \left[V_t - V_{fb} - \phi_s - (1 - A_{bulk}) \cdot \phi_n(y) \right]. \qquad (11.192)$$

Each total charge component is found by integrating (11.190), (11.191), and (11.192) across the width and along the length of the FET channel. In contrast to BSIM, BSIM3 uses the special charge model expressions W_{active} and L_{active} (in contrast to W_{eff} and L_{eff}), as discussed above (see also Appendix B). As in the earlier models, it is assumed that the FET is uniform across the channel, so that the integral across the channel width reduces to W_{active}. Thus, the total inversion charge is found from

$$Q_{INV} = -W_{active} \cdot \int_0^{L_{active}} Q_{inv}(y) \cdot dy. \qquad (11.193)$$

Substituting (11.190) into (11.193), changing variables from dy to $d\phi_n(y)$, and carrying out the integration as described in Section 8.8 leads to

$$Q_{INV} = -W_{active} \cdot L_{active} \cdot C_{ox} \cdot \left[V_{gs} - V_t - \frac{A_{bulk}}{2} \cdot V_{ds} + \frac{1}{12} \cdot \frac{A_{bulk}^2 \cdot V_{ds}^2}{V_{gs} - V_t - \frac{A_{bulk}}{2} \cdot V_{ds}} \right], \qquad (11.194)$$

which is identical to the BSIM expression (8.165). For use in BSIM3, (11.194) is generalized by introducing the auxiliary functions V_{gsx} for $(V_{gs} - V_t)$, $A_{bulk,q}$ for A_{bulk}, and V_{dsq} for V_{ds}. With these changes to (11.194), the final expression for the total inversion charge is

$$Q_{INV} = -W_{active} \cdot L_{active} \cdot C_{ox} \cdot \left[V_{gsx} - \frac{A_{bulk,q}}{2} \cdot V_{dsq} + \frac{1}{12} \cdot \frac{A_{bulk,q}^2 \cdot V_{dsq}^2}{V_{gsx} - \frac{A_{bulk,q}}{2} \cdot V_{dsq}} \right]. \quad (11.195)$$

The BSIM3 expressions for the total depletion charge and the total inversion charge are the same as those developed for use in BSIM. However, some minor notational changes are introduced which alter the appearance of the final expressions. The total depletion charge is found from

$$Q_{DEPL} = W_{active} \cdot \int_0^{L_{active}} Q_{depl}(y) \cdot dy. \quad (11.196)$$

Substituting (11.192) into (11.196) produces

$$Q_{DEPL} = -W_{active} \cdot C_{ox} \cdot \int_0^{L_{active}} \left[V_t - V_{fb} - \phi_s - (1 - A_{bulk}) \cdot \phi_n(y) \right] \cdot dy. \quad (11.197)$$

Introducing the long channel (i.e., two-dimensional MOS capacitor) threshold voltage expression (3.96) (which is also valid for very low drain biases)

$$V_t = V_{fb} + \phi_s + \gamma \cdot (\phi_s - V_{bs})^{\frac{1}{2}} = V_{fb} + \phi_s + \frac{(2\epsilon_{Si} q N_{sub})^{\frac{1}{2}}}{C_{ox}} \cdot (\phi_s - V_{bs})^{\frac{1}{2}} \quad (11.198)$$

into (11.197) transforms that expression to

$$Q_{DEPL} = -W_{active} \cdot C_{ox} \cdot \int_0^{L_{active}} \left[(2\epsilon_{Si} q N_{sub})^{\frac{1}{2}} \cdot (\phi_s - V_{bs})^{\frac{1}{2}} - (1 - A_{bulk}) \cdot \phi_n(y) \right] \cdot dy$$

$$= -W_{active} \cdot L_{active} \cdot C_{ox} \cdot (2\epsilon_{Si} q N_{sub})^{\frac{1}{2}} \cdot (\phi_s - V_{bs})^{\frac{1}{2}}$$

$$+ W_{active} \cdot C_{ox} \cdot \int_0^{L_{active}} (1 - A_{bulk}) \cdot \phi_n(y) \cdot dy. \quad (11.199)$$

The integral for Q_{DEPL} has been split into two parts. The first part is a simple expression for the depletion charge when no drain bias is applied, while the second part includes A_{bulk}, and thus provides a description of the change in the depletion charge due to the drain bias. Solving the integral in (11.199) as in Section 8.8 leads to

$$Q_{DEPL} = -W_{active} \cdot L_{active} \cdot C_{ox} \cdot (2\epsilon_{Si} q N_{sub})^{\frac{1}{2}} \cdot (\phi_s - V_{bs})^{\frac{1}{2}}$$

$$+ W_{active} \cdot L_{active} \cdot C_{ox} \cdot \left(\frac{1 - A_{bulk}}{2} \cdot V_{ds} - \frac{1}{12} \cdot \frac{(1 - A_{bulk}) \cdot A_{bulk} \cdot V_{ds}^2}{V_{gs} - V_t - \frac{A_{bulk}}{2} \cdot V_{ds}} \right)$$

$$= Q_{DEPL, V_{ds}=0} + \Delta Q_{DEPL}, \quad (11.200)$$

where ΔQ_{DEPL} represents the change in the drain depletion charge due to the application of a drain bias. Equation (11.200) is actually identical to the BSIM expression (8.164), although the form has a different appearance. Once again, auxiliary functions are introduced to make this expression have general applicability. Substituting V_{bsx} for V_{bs}, V_{gsx} for $(V_{gs} - V_t)$, $A_{bulk,q}$ for A_{bulk}, and V_{dsq} for V_{ds} transforms (11.200) into the final expression for the total depletion charge:

$$
\begin{aligned}
Q_{DEPL} &= -W_{active} \cdot L_{active} \cdot C_{ox} \cdot (2\epsilon_{Si}q \cdot \mathbf{NSUB})^{\frac{1}{2}} \cdot (\phi_s - V_{bsx})^{\frac{1}{2}} \\
&\quad + W_{active} \cdot L_{active} \cdot C_{ox} \cdot \left(\frac{1 - A_{bulk,q}}{2} \cdot V_{dsq} - \frac{1}{12} \cdot \frac{(1 - A_{bulk,q}) \cdot A_{bulk,q} \cdot V_{dsq}^2}{V_{gsx} - \frac{A_{bulk,q}}{2} \cdot V_{dsq}} \right) \\
&= Q_{DEPL,V_{ds}=0} + \Delta Q_{DEPL}.
\end{aligned}
\tag{11.201}
$$

Note that the simple substrate doping $\mathbf{NSUB}$ is used, rather than some more complicated expression involving both $\mathbf{NCH}$ and $\mathbf{NSUB}$.

The total gate charge expression is found from the integral

$$
Q_{GATE} = W_{active} \cdot \int_0^{L_{active}} Q_{gate}(y) \cdot dy.
\tag{11.202}
$$

Substituting (11.190) into (11.202) leads to

$$
Q_{GATE} = W_{active} \cdot C_{ox} \cdot \int_0^{L_{active}} \left[V_{gs} - V_{fb} - \phi_s - \phi_n(y) \right] \cdot dy.
\tag{11.203}
$$

This expression is modified in a manner similar to that used on the depletion charge equation. The long channel/zero drain bias threshold voltage expression (11.198) is manipulated into the form

$$
V_{fb} + \phi_s = V_t - \frac{(2\epsilon_{Si}qN_{sub})^{\frac{1}{2}}}{C_{ox}} \cdot (\phi_s - V_{bs})^{\frac{1}{2}}.
\tag{11.204}
$$

Substituting (11.204) into (11.203) yields

$$
\begin{aligned}
Q_{GATE} &= W_{active} \cdot C_{ox} \int_0^{L_{active}} \left[V_{gs} - V_t + \frac{(2\epsilon_{Si}qN_{sub})^{\frac{1}{2}}}{C_{ox}} \cdot (\phi_s - V_{bs})^{\frac{1}{2}} - \phi_n(y) \right] \cdot dy \\
&= W_{active} \cdot L_{active} \cdot C_{ox} \cdot (2\epsilon_{Si}qN_{sub})^{\frac{1}{2}} \cdot (\phi_s - V_{bs})^{\frac{1}{2}} \\
&\quad + W_{active} \cdot C_{ox} \int_0^{L_{active}} \left[V_{gs} - V_t - \phi_n(y) \right] \cdot dy.
\end{aligned}
\tag{11.205}
$$

In this form, the expression for the total gate charge has been separated into the mirror of the zero bias depletion charge and an additional component. By changing variables from

dy to $d\phi_n(y)$, (11.205) can be solved as in Section 8.8:

$$Q_{GATE} = -Q_{DEPL,V_{ds}=0}$$

$$+ W_{active} \cdot L_{active} \cdot C_{ox} \cdot \left(V_{gs} - V_t - \frac{V_{ds}}{2} + \frac{1}{12} \cdot \frac{A_{bulk} \cdot V_{ds}^2}{V_{gs} - V_t - \frac{A_{bulk}}{2} \cdot V_{ds}} \right). \quad (11.206)$$

This expression, like the other charge expressions derived here, is also identical to its BSIM forebearer, with all visible differences being due to notational modifications. Once again, the four auxiliary functions are introduced, which converts (11.206) into the final expression for the gate charge:

$$Q_{GATE} = -Q_{DEPL,V_{ds}=0} + W_{active} \cdot L_{active} \cdot C_{ox} \cdot \left(V_{gsx} - \frac{V_{dsq}}{2} + \frac{1}{12} \cdot \frac{A_{bulk,q} \cdot V_{dsq}^2}{V_{gsx} - \frac{A_{bulk,q}}{2} \cdot V_{dsq}} \right)$$

$$= W_{active} \cdot L_{active} \cdot C_{ox} \cdot (2\epsilon_{Si}q \cdot \mathbf{NSUB})^{\frac{1}{2}} \cdot (\phi_s - V_{bsx})^{\frac{1}{2}}$$

$$+ W_{active} \cdot L_{active} \cdot C_{ox} \cdot \left(V_{gsx} - \frac{V_{dsq}}{2} + \frac{1}{12} \cdot \frac{A_{bulk,q} \cdot V_{dsq}^2}{V_{gsx} - \frac{A_{bulk,q}}{2} \cdot V_{dsq}} \right). \quad (11.207)$$

Note once again that the simple substrate doping **NSUB** is used, rather than some more complicated expression involving both **NCH** and **NSUB**.

11.7.4   Charge Partitioning

To properly create a node charge description, the inversion charge must be partitioned (in other words, divided) between the source and the drain. The simplest approach is to divide the inversion charge in half, and assign equal halves as the node charges Q_S and Q_D. However, when comparing their model results with numerical simulations, Ward and Dutton [21] found the best agreement when they assigned 60% of the inversion charge to the source and 40% to the drain; this division of the inversion charge has become the most popular choice. In addition, the simple physical argument can be made [20] that in saturation, the drain is electrically disconnected from the inversion layer, so that all the inversion charge should be assigned to the source. This method also has advantages in certain situations, where the numerical results are improved by eliminating the drain charge [2], as will be discussed below.

Like BSIM and BSIM2, BSIM3 offers the user the choice of all three of these suggested partitioning methods. Partitioning is carried out using the integral approach suggested by Ward and Dutton [21]:

$$Q_S = W_{active} \cdot \int_0^{L_{active}} \left(1 - \frac{y}{L_{active}} \right) \cdot q_{inv}(y) \cdot dy \quad (11.208)$$

and

$$Q_D = W_{active} \cdot \int_0^{L_{active}} \frac{y}{L_{active}} \cdot q_{inv}(y) \cdot dy, \qquad (11.209)$$

where y is the point in the channel which defines the partitioning fraction (e.g., for equal charge partitioning, $y = \frac{L_{active}}{2}$). In all cases, in the limit of $V_{ds} \to 0$ (or, more specifically for BSIM3, $V_{dsq} \to 0$), $Q_S = Q_D$; with no drain bias applied, the inversion charge is partitioned equally between the source and the drain, which is the expected result.

As in BSIM, the choice among the three possible partitioning methods is made by setting the flag parameter **XPART** to 0 for the 40%/60% method, 1 for the 0%/100% method, and 0.5 for the 50%/50% method. This choice is usually made at the circuit simulation level.

The 40%/60% Charge-Partitioning Method

As noted above, the 40%/60% charge-partitioning method was suggested by Ward and Dutton [21] as providing the best agreement between their model and more detailed numerical simulations. Setting y in (11.208) and (11.209) to 0.4, changing variables from dy to $d\phi_n(y)$, integrating, and introducing the auxiliary functions leads to

$$Q_S = -\frac{1}{2} \cdot \frac{W_{active} \cdot L_{active} \cdot C_{ox}}{\left(V_{gsx} - \frac{A_{bulk,q}}{2} \cdot V_{dsq}\right)^2} \cdot \left(V_{gsx}^3 - \frac{4}{3} \cdot V_{gsx}^2 \cdot A_{bulk,q} \cdot V_{dsq}\right.$$

$$\left. + \frac{2}{3} V_{gsx} \cdot A_{bulk,q}^2 \cdot V_{dsq}^2 - \frac{2}{15} \cdot A_{bulk,q}^3 \cdot V_{dsq}^3\right) \qquad (11.210)$$

and

$$Q_D = -\frac{1}{2} \cdot \frac{W_{active} \cdot L_{active} \cdot C_{ox}}{\left(V_{gsx} - \frac{A_{bulk,q}}{2} \cdot V_{dsq}\right)^2} \cdot \left(V_{gsx}^3 - \frac{5}{3} \cdot V_{gsx}^2 \cdot A_{bulk,q} \cdot V_{dsq}\right.$$

$$\left. + V_{gsx} \cdot A_{bulk,q}^2 \cdot V_{dsq}^2 - \frac{1}{5} \cdot A_{bulk,q}^3 \cdot V_{dsq}^3\right). \qquad (11.211)$$

In the low drain bias limit of $V_{dsq} \to 0$, (11.210) and (11.211) reduce to

$$Q_S = Q_D = -\frac{1}{2} \cdot \frac{W_{active} \cdot L_{active} \cdot C_{ox}}{V_{gsx}} = \frac{1}{2} \cdot Q_{INV}. \qquad (11.212)$$

The 50%/50% Charge-Partitioning Method

The 50%/50% charge-partitioning method is the simplest possible approach, as it divides the inversion charge equally between the source node and the drain node in all situations.

Setting y in (11.208) and (11.209) to 0.5, changing variables from dy to $d\phi_n(y)$, integrating, and introducing the auxiliary functions yields

$$Q_S = Q_D = -\frac{1}{2} \cdot W_{active} \cdot L_{active} \cdot C_{ox} \cdot \left(V_{gsx} - \frac{A_{bulk,q}}{2} \cdot V_{dsq} + \frac{1}{12} \cdot \frac{A_{bulk,q}^2 \cdot V_{dsq}^2}{V_{gsx} - \frac{A_{bulk,q}}{2} \cdot V_{dsq}} \right)$$

$$= \frac{1}{2} \cdot Q_{INV}. \tag{11.213}$$

Note that this expression is essentially equal to the BSIM 50%/50% charge-partitioning result (8.182), with the only differences being due to the introduction of the BSIM3 auxiliary functions. In the low drain bias limit of $V_{dsq} \to 0$, (11.213) reduces to

$$Q_S = Q_D = -\frac{1}{2} \cdot \frac{W_{active} \cdot L_{active} \cdot C_{ox}}{V_{gsx}} = \frac{1}{2} \cdot Q_{INV}, \tag{11.214}$$

which is identical to the limiting result in the 40%/60% charge-partitioning method (11.212).

The 0%/100% Charge-Partitioning Method

The 0%/100% charge-partitioning method was suggested by Yang, Epler, and Chatterjee [20], based on the physical argument that in saturation, the drain is electrically disconnected from the inversion layer, so that all the inversion charge should be assigned to the source. It has also been noted [2] that in the simulation of very rapid transients in certain circuits, a drain charge spike can occur as a numerical artifact, and that the 0%/100% charge-partitioning method by definition eliminates this phenomenon.

Setting y in (11.208) and (11.209) to 0, changing variables from dy to $d\phi_n(y)$, integrating, and introducing the auxiliary functions produces

$$Q_S = -W_{active} \cdot L_{active} \cdot C_{ox} \cdot \left(\frac{V_{gsx}}{2} + \frac{1}{4} \cdot A_{bulk,q} \cdot V_{dsq} - \frac{1}{24} \cdot \frac{A_{bulk,q}^2 \cdot V_{dsq}^2}{V_{gsx} - \frac{A_{bulk,q}}{2} \cdot V_{dsq}} \right) \tag{11.215}$$

and

$$Q_D = -W_{active} \cdot L_{active} \cdot C_{ox} \cdot \left(\frac{V_{gsx}}{2} - \frac{3}{4} \cdot A_{bulk,q} \cdot V_{dsq} + \frac{1}{8} \cdot \frac{A_{bulk,q}^2 \cdot V_{dsq}^2}{V_{gsx} - \frac{A_{bulk,q}}{2} \cdot V_{dsq}} \right). \tag{11.216}$$

As was the case for 50%/50% charge partitioning, these results are essentially identical to the BSIM results (in this case, (8.177) and (8.178)), differing only in the use of the auxiliary functions in the BSIM3 expressions. In the low drain bias limit of $V_{dsq} \to 0$, (11.215) and (11.216) reduce to

$$Q_S = Q_D = -\frac{1}{2} \cdot \frac{W_{active} \cdot L_{active} \cdot C_{ox}}{V_{gsx}} = \frac{1}{2} \cdot Q_{INV}, \tag{11.217}$$

which is identical to the limiting results in the 40%/60% charge partitioning method (11.212) and in the 50%/50% charge-partitioning method (11.214).

Additional Charge Contributions

In addition to a model for the inversion charge (partitioned between the source and drain nodes), BSIM3 also includes models for the extra charge contributed by the "outer-fringe" portion of the gate, and the region of overlap between the gate and the diffusion (see Figure 11.7). The charge values involved are negligible in large devices, but since they do not scale as the channel length is reduced, their contribution is more significant in short channel devices. The approach used in BSIM3 is more sophisticated than the simple zero bias capacitance method of earlier models (see Chapter 13 for a detailed discussion), but also introduces a requirement for AC parameter extraction.

The outer fringe capacitance between the gate and a diffusion is a simple, bias independent, reciprocal capacitance. As discussed in Chapter 13, the use of a simple dielectric model for the capacitor can be employed, which leads to a capacitance value of [22, 23]

$$C_{of} = \frac{2}{\pi} \cdot \epsilon_{ox} \cdot ln\left(1 + \frac{t_{poly}}{t_{ox}}\right),$$
(11.218)

where t_{poly} is the thickness of the gate polysilicon. (Note that this expression assumes that all of the dielectric material between the gate and the diffusion is silicon dioxide, which is not always the case.) The outer fringe capacitance can also be found using numerical device simulation; a result virtually identical to that found from (11.218) is obtained. More directly, if the appropriate test structures are available, C_{of} can be measured directly to determine a value. As in the earlier charge computations, it is assumed that this behavior is uniform across the width of the device; as a result, C_{of} is in units of F/m.

Once C_{of} has been determined, the source and drain charges due to the outer fringe capacitance are computed using the very simple linear capacitance relationships

$$Q_{S,of} = -W_{active} \cdot C_{of} \cdot V_{gs}$$
(11.219)

and

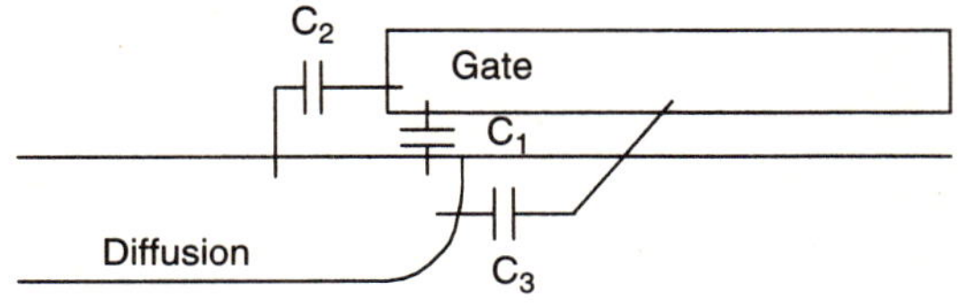

Figure 11.7   The different components of the gate-diffusion capacitance.

$$Q_{D,of} = -W_{active} \cdot C_{of} \cdot V_{gd}. \tag{11.220}$$

Note that the gate-source bias V_{gs} is used as–is in (11.219), and is **not** replaced with the auxiliary gate bias function V_{gsx}. The gate charge due to the outer-fringe contribution is simply the mirror image of the source and drain charges:

$$Q_{GATE,of} = -\left(Q_{S,of} + Q_{D,of}\right). \tag{11.221}$$

The overlap capacitance is also often treated as a simple capacitor, but this contribution actually does have bias dependence. Under the appropriate bias conditions, the diffusion will contain a depletion region, which will affect the charge calculation; this is particularly apparent if a lightly doped drain (LDD) region is present.

The BSIM3 model for the overlap charge is semi-empirical, and includes several parameters which must be extracted. As noted earlier, this is a change from earlier models, and introduces a requirement for the extraction of AC parameters. Once again, it is assumed that the charge behavior will be uniform across the width of the device. The overlap charge in the source node is computed from

$$Q_{S,ov} = W_{active} \cdot \left(\mathbf{CGS0} \cdot V_{gs} + \mathbf{CGS1} \cdot \left\{ V_{gs} - V_{gs,ov} \right.\right.$$
$$\left.\left. + \frac{\mathbf{CKAPPA}}{2} \cdot \left[\left(1 + \frac{4 \cdot V_{gs,ov}}{\mathbf{CKAPPA}}\right)^{\frac{1}{2}} - 1 \right] \right\}\right), \tag{11.222}$$

where $V_{gs,ov}$ is an auxiliary function given by

$$V_{gs,ov} = \frac{1}{2} \cdot \left\{ V_{gs} + \delta_1 + \left[(V_{gs} + \delta_1)^2 + 4 \cdot \delta_1 \right]^{\frac{1}{2}} \right\} \tag{11.223}$$

and δ_1 is a constant which is set to $\delta_1 = 0.02$. Note once again that the gate bias V_{gs} is used as–is here, and is **not** replaced with the auxiliary gate bias function V_{gsx}. An essentially identical expression is used for the drain overlap charge:

$$Q_{D,ov} = W_{active} \cdot \left(\mathbf{CGD0} \cdot V_{gd} + \mathbf{CGD1} \cdot \left\{ V_{gd} - V_{gd,ov} \right.\right.$$
$$\left.\left. + \frac{\mathbf{CKAPPA}}{2} \cdot \left[\left(1 + \frac{4 \cdot V_{gd,ov}}{\mathbf{CKAPPA}}\right)^{\frac{1}{2}} - 1 \right] \right\}\right), \tag{11.224}$$

where $V_{gd,ov}$ is an auxiliary function which is in essence identical to $V_{gs,ov}$:

$$V_{gd,ov} = \frac{1}{2} \cdot \left\{ V_{gd} + \delta_1 + \left[(V_{gd} + \delta_1)^2 + 4 \cdot \delta_1 \right]^{\frac{1}{2}} \right\}. \tag{11.225}$$

The gate overlap charge is merely the mirror image of the source and drain overlap charges, and is thus computed from

$$Q_{GATE,ov} = -\left(Q_{S,ov} + Q_{D,ov}\right). \qquad (11.226)$$

The charge contributions due to the outer fringe capacitance and the overlap region are added on to the already computed values of the gate, source, and drain charges, which were described earlier. The total node charge expressions for these three device terminals thus become

$$Q_{GATE,total} = Q_{GATE} + Q_{GATE,of} + Q_{GATE,ov}, \qquad (11.227)$$

$$Q_{S,total} = Q_S + Q_{S,of} + Q_{S,ov}, \qquad (11.228)$$

and

$$Q_{D,total} = Q_D + Q_{D,of} + Q_{D,ov}. \qquad (11.229)$$

Note that like the earlier models, this scheme does not provide a description of the bias dependent "inner-fringe" contribution (see Figure 11.7). This situation is considered in more detail in Chapter 13.

The Gate to Bulk Capacitance

BSIM3 also includes the parameter **CGB0** to represent the zero bias gate-bulk capacitance. This term is identical in function to the parameter **CGBO**, which appears in the earlier models. In modern MOSFETs, the gate-diffusion capacitance is usually much larger than the gate-bulk capacitance, and thus **CGB0** can usually be neglected. These zero bias capacitance terms are discussed more fully in Chapter 13.

11.7.5   Non-Quasi-Static Model

In addition to the traditional quasi-static charge equations just described, BSIM3 also includes a non-quasi-static model for the node charges. As discussed in detail in Chapter 13, a non-quasi-static charge model will only become necessary at very high frequencies that are much larger than those currently in use in MOSFETs, or in very long channel devices. Since the mainstream user is unlikely to encounter such a situation at this time, a discussion of the BSIM3 non-quasi-static model will be omitted.

11.8   Final Comments

As noted at the beginning of this chapter, BSIM3, like the other third generation models, has only recently been introduced, so a detailed commentary based on experience cannot yet be offered. However, some general observations can be made.

The original intent of restoring a more sound physical basis to the model parameters is a welcome objective. In addition, discarding the extrinsic model structure (based on the use of the L and W parameters) considerably simplifies the basic form of the model. The elimination of the polynomial expressions which were introduced into the second generation models also simplifies the mathematical formulation.

Rather than use these polynomial terms, BSIM3 employs a more sophisticated set of smoothing functions that shows much improved mathematical behavior. These smoothing functions also serve in a much cleaner fashion; they allow the model current and charge equations to reduce to the expected limiting behavior, while providing a continuous and smooth transition between the different regions of device operation. This is a major improvement over the situation encountered in the earlier models. This approach also guarantees continuous and smooth behavior of the current and charge equations and their first derivatives; this is required for the accurate simulation of analog circuits, and is a major aid to the simulation of low voltage digital circuits.

The original intent of BSIM3 was to achieve a relatively simple model with a small number of parameters. However, as BSIM3 has evolved, many empirical expressions have been introduced, which in turn has led to the introduction of a large number of (empirical) parameters. It is not yet known if all of these empirical expressions will be well behaved when they are employed in circuit simulation. Also the geometry dependence of the terms ΔL and ΔW in the model is a bit unusual; these terms, like the oxide thickness t_{ox}, should have a physical basis. As noted earlier in the text, adding geometry dependence to ΔL and ΔW could seriously impair the ability to model process variations.

However, final judgments on BSIM3 (and the other third-generation models) will have to wait until the model is extensively exercised in both parameter extraction and circuit simulation. For example, the usual questions regarding the ease of parameter extraction, the quality of the model's physical basis, the possible need for binning, convergence and efficiency during circuit simulation, etc., can then be answered.

BIBLIOGRAPHY

1. J. Huang et al., "BSIM3 Manual (Version 2.0)," University of California/Berkeley, Electronics Research Laboratory (1994).

2. Y. Cheng et al., "BSIM3 Version 3.0 Manual," University of California/Berkeley, Electronics Research Laboratory (1995).

3. R. Velghe, D. Klaassen, and F. Klaassen, "MOS Model 9," Unclassified Report NL-UR 003/94, Philips Electronics N.V. (1994).

4. B. Sheu, "MOS Transistor Modeling and Characterization for Circuit Simulation," University of California/Berkeley, Electronics Research Laboratory Memorandum No. UCB/ERL M85/85 (1985).

5. H. deGraaff and F. Klaassen, *Compact Transistor Modeling for Circuit Design*, Springer-Verlag, 1990.

6. S. Sze, *Physics of Semiconductor Devices* (2nd ed.), John Wiley & Sons, 1981.

7. M. Chen et al., "Constraints in p-Channel Device Engineering for Submicron CMOS Technologies," *1988 IEDM Tech. Dig.*, pp. 390–393.

8. M. Orlowski, C. Mazure, and F. Lau, "Submicron Short Channel Effects Due to Gate Reoxidation Induced Lateral Interstitial Diffusion," *1987 IEDM Tech. Dig.*, pp. 632–635.

9. Z. Liu et al., "Threshold Voltage Model for Deep Submicrometer MOSFETs," *IEEE Trans. Elec. Dev.* vol. ED-40, pp. 86–95 (1993).

10. H. Masuda, M. Nakai, and M. Kubo, "Characteristics and Limitation of Scaled-Down MOSFETs Due to Two-Dimensional Field Effect," *IEEE Trans. Elec. Dev.* vol. ED-26, pp. 980–986 (1979).

11. R. Muller and T. Kamins, *Device Electronics for Integrated Circuits* (2nd edition), John Wiley & Sons, 1986.

12. G. Gildenblat, "VLSI Electronics: Microstructure Science," vol. 18, Academic Press, 1989.

13. A. Sabnis and J. Clemens, "Characterization of Electron Velocity in the Inverted <100> Si Surface," *1979 IEDM Tech. Dig.*, pp. 18–21.

14. M. Liang, J. Choi, P. Ko, and C. Hu, "Inversion Layer Capacitance and Mobility of Very Thin Gate Oxide MOSFETs," *IEEE Trans. Elec. Dev.* vol. ED-33, pp. 409–413 (1986).

15. E. Talkhan, I. Manour, and A. Barboor, "Investigation of the Effect of Drift Field Dependent Mobility on MOSFET Characteristics," *IEEE Trans. Elec. Dev.* vol. ED-19, pp. 899–916 (1972).

16. M. Jeng, "Design and Modeling of Deep Submicrometer MOSFETs," University of California/Berkeley, Electronics Research Laboratory Memorandum No. UCB/ERL M90/90 (1990).

17. J. Early, "Effects of Space Charge Layer Widening in Junction Transistors," *Proc. IRE* vol. 40, pp. 1401–1406 (1952).

18. F. Hsu et al., "An Analytical Breakdown Model for Short Channel MOSFETs," *IEEE Trans. Elec. Dev.* vol. ED-29, pp. 1735–1740 (1982).

19. C. Hu et al., "Hot Electron Induced MOSFET Degradation—Model, Monitor, and Improvement," *IEEE Trans. Elec. Dev.* vol. ED-32, pp. 375–385 (1985).

20. P. Yang, B. Epler, and P. Chatterjee, "An Investigation of the Charge Conservation Problem for MOSFET Circuit Simulation," *IEEE J. Sol. St. Circ.* SC-18, pp. 128–138 (1983).

21. D. Ward and R. Dutton, "A Charge-Oriented Model for MOS Transistor Capacitances," *IEEE J. Sol. St. Circ.* vol. SC-13, pp. 703–708 (1978).

22. R. Shrivastava and K. Fitzpatrick, "A Simple Model for the Overlap Capacitance of a VLSI MOS Device," *IEEE Trans. Elec. Dev.* vol. ED-29, pp. 1870–1875 (1982).

23. N. Arora, *MOSFET Models for VLSI Circuit Simulation*, Springer-Verlag, 1993.

12

MOS Model 9

12.1 Introduction

MOS Model 9 is another third-generation model, developed at Philips Laboratories [1, 2]. The structure of the model is based on the ideas developed in a popular textbook written by members of the same group [3]. Although MOS Model 9 was developed internally at Philips Electronics, it has been made generally available; with support from its authors, it is being included in many widely used general-purpose circuit simulators, and thus is drawing a significant amount of attention.

MOS Model 9 follows the general third-generation approach (see Section 11.2) of introducing numerical smoothing functions into the model structure; these smoothing functions serve two related purposes. First, these functions permit continuous and smooth equations across the various transition points (such as the saturation voltage V_{dsat}) of FET operation. Second, the appropriate use of smoothing functions allows for the development of a single model equation (e.g., a drain current equation) that is valid in all regions of device operation. Taken together, these improvements allow third-generation models to provide improved descriptions of various device phenomena in a manner that is efficient during circuit simulation.

However, in contrast to BSIM3 and some other potential third-generation candidate models (see Chapter 16), MOS Model 9 retains the second-generation approach to describing the geometry dependence of the model characteristics. While the basic method of the second-generation models is used, the method is extensively modified to improve the results.

In addition, the form of the model shows the clear influence of its origins in industrial device modeling and circuit simulation. A relatively small number of model equations are developed, and these equations are very "clean" and mathematically simple. The number of parameters is also relatively small when compared with many other models. Since the model is new, it is not yet possible to give a detailed evaluation of its behavior in actual use; however, the simplicity of the model equations and the small number of parameters both suggest that the model will be efficient in circuit simulation and will offer comparatively simple parameter extraction.

MOS Model 9 is also somewhat unique, as it is the first mainstream SPICE FET model that does not have its origins at the University of California/Berkeley. (HSPICE Level 28, described in Chapter 9, is a modified version of BSIM). This suggests that the field of SPICE FET modeling is undergoing an important change, as a wider variety of FET models will now be considered. Other possible candidate models, which are not yet widely used in commonly available circuit simulators, are briefly described in Chapter 16; these and other models may also begin to achieve widespread acceptance.

12.2   The Structure of MOS Model 9

While it is clearly a third-generation model, MOS Model 9 incorporates many of the features of the second-generation models to describe the geometry dependence of the transistor characteristics.

Like BSIM3 and the candidate models discussed in Chapter 16, numerical smoothing functions are added to the model to allow for the use of a single expression for a device characteristic (e.g., the drain current or a node charge) over the entire operating range of the FET.

The geometry dependence is included with a second-generation-like approach; however, the method is applied selectively (to only certain parameters), and is used in different ways for different parameters. The use of reference transistors, developed in HSPICE, is also included in MOS Model 9; this makes the model more amenable to binning. In addition, many more parameters are included to describe temperature dependence; these additions are very much intertwined with the geometry dependence.

Finally, MOS Model 9 introduces a more detailed description of the channel length and width which separates the differences between drawn and on-wafer dimensions from the usual underdiffusion effects. This level of detail can be very important for the computation of node charge, and is of significance for the modeling of process variations.

12.2.1   Smoothing Functions

One of the hallmarks of the third generation models is the use of numerical smoothing functions to improve the mathematical behavior of the models. These smoothing functions serve two interrelated purposes. First, they eliminate the need to change equations at particular operating points (e.g., when the drain bias V_{ds} crosses the saturation voltage V_{dsat}), which allows for the use of a single equation for all regions of device operation (e.g., one drain current equation rather than separate equations for the subthreshold, linear, and saturation regions). Furthermore, the use of these functions permits continuous and smooth model equations across all the transition points. This allows smooth and continuous first derivatives to also be guaranteed where this is appropriate (e.g., in the development of the expressions for the device conductances).

MOS Model 9 employs a collection of hyperbolic functions to serve this role; these functions are described as they are introduced into the model. These functions meet the requirements discussed above. In addition, these functions are well understood, and are mathematically well behaved over their entire range of use; this is in contrast to the difficulties that are sometimes encountered with the polynomial functions introduced into the second-generation models.

12.2.2 The Extrinsic Geometry Structure

As noted above, MOS Model 9 employs the basic second-generation approach to describe the geometry dependence of the device characteristics. However, extensive modifications are made to this basic structure. Therefore, before discussing the MOS Model 9 implementation, it will be worthwhile to review the methods used in the second generation models.

The BSIM Extrinsic Geometry Structure

In BSIM (and BSIM2), a single uniform extrinsic model structure is imposed on all the parameters (see Section 8.3). All the extrinsic model parameters are computed from a triplet of intrinsic model parameters using the general form

$$X3MS = \mathbf{X3MS} + \frac{\mathbf{LX3MS}}{L_{eff}} + \frac{\mathbf{WX3MS}}{W_{eff}},\tag{12.1}$$

where **X3MS**, **LX3MS**, and **WX3MS** are the intrinsic parameters which appear in the FET model parameter set, and $X3MS$ is the composite extrinsic model parameter which is computed and used in the model equations. In the limit of L_{eff}, $W_{eff} \rightarrow \infty$, $X3MS \rightarrow$ **X3MS**. The strengths and weaknesses of this approach are discussed in detail in Chapter 8 and Chapter 10. A key concern is that all parameters are assumed to follow the geometry dependent behavior as described by (12.1); when this is not the case, the structure of (12.1) does not readily accommodate model binning, as discussed in Section 10.8.

HSPICE Modifications

The proprietary model HSPICE Level 28, along with the HSPICE implementations of BSIM (HSPICE Level 13) and BSIM2 (HSPICE Level 39) modify the form of (12.1) to include reference lengths and widths. In this way, (12.1) is recast to

$$X3MS = \mathbf{X3MS} + \mathbf{LX3MS} \cdot \left(\frac{1}{L_{eff}} - \frac{1}{L_{eff,ref}} \right) + \mathbf{WX3MS} \cdot \left(\frac{1}{W_{eff}} - \frac{1}{W_{eff,ref}} \right).\tag{12.2}$$

If L_{eff}, $W_{eff} \rightarrow \infty$, (12.2) reduces to (12.1), and nothing has changed. As an alternative, a long, wide transistor (e.g., $W_{eff}/L_{eff} = 10\mu m/10~\mu m$) can be chosen as a reference transistor. However, this is a minor modification, which has little effect on the situation. If binning is introduced, the discontinuity problems described in Section 10.8 will occur.

However, as described in detail in Section 9.8, the availability of these reference dimensions can be used to greatly improve the continuity of the device characteristics

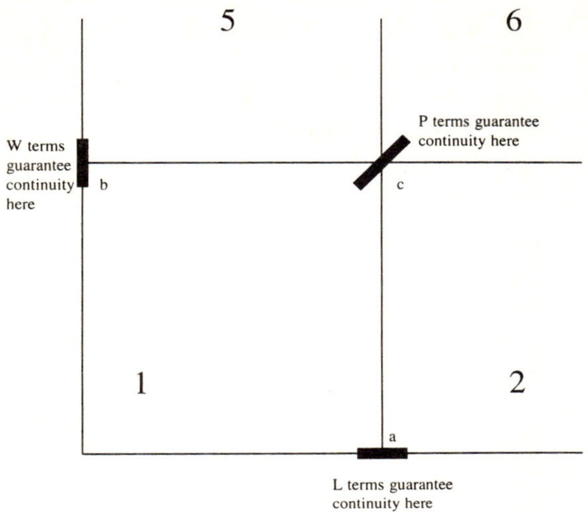

Figure 12.1   A detail of a model bin, indicating where the use of reference terms allows parameter continuity across the bin boundaries.

across bin boundaries. Consider the detail of a binned model in Figure 12.1. If the reference dimensions $L_{eff,ref,1}$ and $W_{eff,ref,1}$ in bin **1** are defined as, respectively, the upper limits of L_{eff} and W_{eff} for that bin, all model parameters (and thus all model characteristics) will be continous across these two particular corners of the bin. While this does not guarantee parameter continuity across the rest of the bin boundary, the introduction of two fixed reference points, which *do* guarantee continuity, greatly improves the situation.

In the HSPICE Level 28 structure (and in HSPICE Level 39), the use of the reference dimensions is carried even further, and (12.2) is modified to

$$X3MS = \mathbf{X3MS} + \mathbf{LX3MS} \cdot \left(\frac{1}{L_{eff}} - \frac{1}{L_{eff,ref}} \right) + \mathbf{WX3MS} \cdot \left(\frac{1}{W_{eff}} - \frac{1}{W_{eff,ref}} \right)$$

$$+ \mathbf{PX3MS} \cdot \left(\frac{1}{L_{eff}} - \frac{1}{L_{eff,ref}} \right) \cdot \left(\frac{1}{W_{eff}} - \frac{1}{W_{eff,ref}} \right). \tag{12.3}$$

As shown in Figure 12.1, this addition to the extrinsic structure also guarantees model continuity at another corner of the bin. This has the total effect of guaranteeing parameter and model continuity across all the bin faces.

MOS Model 9 Implementation

MOS Model 9 implements its extrinsic geometry structure in the general modified form of (12.2); product terms, as used in (12.3), do not appear. However, in contrast to the second-generation approach, geometry dependence is introduced in a selective manner, as only certain parameters are deemed to have geometry dependence. In addition, that geometry dependence is not introduced in a uniform manner; some parameters have only length dependence. Furthermore, for certain parameters, the geometry dependence is further modified to include L_{eff} and $L_{eff,ref}$ as power functions; quadratics, square roots, and extracted powers (as model parameters) are used. Finally, the introduction of temperature dependent parameters is greatly expanded; many of these parameters are deeply intertwined with the geometry dependence.

It has been claimed that MOS Model 9 does not require model binning. If this is the case, $L_{eff,ref}$ and $W_{eff,ref}$ can be set to values chosen for a long, wide reference transistor and fixed for the entire model; this is assumed in the MOS Model 9 documentation [1], and the derivations which appear there follow that line of reasoning. However, with the general availability of the reference parameters, MOS Model 9 can readily accommodate binning in the manner described above. To date, claims that models do not require binning have yet to hold up in every possible situation. Therefore, in the derivations contained in the remainder of this chapter, it is assumed that the user will be forced to create binned models. If binning is not used (or, more hopefully, is not needed), the nonbinned case is easily determined.

However, it should be noted that in contrast to HSPICE Level 28, the inclusion of binning in MOS Model 9 must be approached with caution. Only certain parameters have geometry dependence (and thus include $L_{eff,ref}$ and $W_{eff,ref}$); therefore, if continuity is to be maintained at the boundary points depicted in Figure 12.1, only those parameters can be allowed to have different values in different bins. Furthermore, since several parameters have only length dependence, parameter continuity can be maintained along the length axis of Figure 12.1, but not along the corresponding width axis.

The various parameters employed in MOS Model 9 are listed in Tables 12.1–12.3.

12.2.3 The Channel Length and Width

In the basic original SPICE structure, the effective channel length and width were defined very simply; the differences between the drawn (layout) dimensions and the effective channel dimensions were lumped together. However, the actual situation is somewhat more complicated, as shown in Figure 12.2 and described in Appendix B. Consider the polysilicon gate shape which defines the drawn channel length L_{drawn}. Due to processing, a different value of the polysilicon channel length will actually be printed on the wafer; this is one part of the difference between L_{drawn} and L_{eff}. In addition, the source and drain will underdiffuse the gate, reducing L_{eff} to a value less than the printed polysilicon gate length; this is the second part of the difference between L_{drawn} and L_{eff}. While the simplest solution is to describe this difference with a single parameter, this introduces practical problems, as described in Appendix B.

MOS Model 9 copes with this situation in the manner described in Figure 12.2. In the usual manner, L_{drawn} and W_{drawn} are the layout dimensions of these shapes. The differences between these drawn dimensions and the printed dimensions on a wafer are described, respectively, by the parameters **LVAR** and **WVAR**. The reduction of the channel dimensions by the underdiffusion of dopants is accounted for by **LAP** and **WOT**.

The effective channel length and width are then described by

$$L_{eff} = L_{drawn} + \textbf{LVAR} - 2 \cdot \textbf{LAP} \tag{12.4}$$

Table 12.1   The basic MOS Model 9 parameter set.

Parameter	Units	Description
Process Parameters		
TOX	m	Gate Oxide Thickness
LVAR	m	Process Bias on the Polysilicon Channel Length
WVAR	m	Process Bias on the Channel Width
LAP	m	Source/Drain Underdiffusion of Gate
WOT	m	Isolation Reduction of Channel Width
Electrical Parameters		
VTO	V	Threshold Voltage With No Substrate or Drain Bias
PHIB	V	Surface Potential for Strong Inversion
KO	$V^{\frac{1}{2}}$	Substrate Sensitivity When Depleting Surface Doping
K	$V^{\frac{1}{2}}$	Substrate Sensitivity When Depleting Bulk Doping
VSBX	V	Voltage of Transition Between **KO** and **K**
GAMOO		DIBL Coefficient
ETAGAM		Substrate Bias Dependence of DIBL
VSBT	V	Transition Voltage for Substrate Bias Dependence of DIBL
GAM1	$V^{(1-\textbf{ETADS})}$	Static Feedback Coefficient
ETADS		Substrate Bias Dependence of Static Feedback
THE1	V^{-1}	Gate Field Mobility Reduction Coefficient
THE2	$V^{-\frac{1}{2}}$	Substrate Bias Mobility Reduction Coefficient
THE3	V^{-1}	Drain Bias Mobility Reduction Coefficient
MO		Subthreshold Slope Ideality Factor
ETAM		Substrate Bias Dependence of the Subthreshold Slope
ALP		Channel Length Modulation Coefficient
VP	V	Characteristic Voltage of Channel Length Modulation
ZET1		Weak Inversion Correction Factor
BET	A/V^2	Composite Gain Factor
A1		Weak Avalanche Current Coefficient
A2	V	Weak Avalanche Current Exponent
A3		Weak Avalanche Turn-On Coefficient
COL	F/m	Zero Bias Gate-Diffusion Capacitance

and

$$W_{eff} = W_{drawn} + \textbf{WVAR} - 2 \cdot \textbf{WOT}. \tag{12.5}$$

In this form, the MOS Model 9 parameters **LVAR**, **LAP**, **WVAR**, and **WOT** are identical, respectively, to the HSPICE parameters **XL**, **LD**, **XW**, and **WD**.

This description of the channel geometry provides a more physically realistic description of the situation. In addition, the underdiffusion parameters (here, **LAP** and **WOT**) tend not to vary, while the drawn/print offset parameters **LVAR** and **WVAR** are subject to random process variations (see Chapter 14). This situation is described in more detail in Appendix B.

Table 12.2　The MOS Model 9 geometry and temperature parameters.

Parameter	Units	Description
Geometry Parameters		
SLVTO	V·m	Length Dependence of *VTO* 1
SL2VTO	V·m^2	Length Dependence of *VTO* 2
SWVTO	V·m	Width Dependence of *VTO*
SLKO	V$^{\frac{1}{2}}$·m	Length Dependence of *KO*
SWKO	V$^{\frac{1}{2}}$·m	Width Dependence of *KO*
SLK	V$^{\frac{1}{2}}$·m	Length Dependence of *K*
SWK	V$^{\frac{1}{2}}$·m	Width Dependence of *K*
SLVSBX	V·m	Length Dependence of *VSBX*
SWVSBX	V·m	Width Dependence of *VSBX*
SLVSBT	V·m	Length Dependence of *VSBT*
SLGAM1	V$^{(1-\mathbf{ETADS})}$	Length Dependence of *GAM1*
SWGAM1	V$^{(1-\mathbf{ETADS})}$	Width Dependence of *GAM1*
SLTHE1	m/V	Length Dependence of *THE1*
SWTHE1	m/V	Width Dependence of *THE1*
SLTHE2	m/V	Length Dependence of *THE2*
SWTHE2	m/V	Width Dependence of *THE2*
SLTHE3	m/V	Length Dependence of *THE3*
SWTHE3	m/V	Width Dependence of *THE3*
SLMO	m$^{-\frac{1}{2}}$	Length Dependence of *MO*
SLALP	m$^{\mathbf{ETAALP}}$	Length Dependence of *ALP*
ETAALP		Exponent of the Length Dependence of *ALP*
SWALP	m	Width Dependence of *ALP*
SLZET1	m$^{\mathbf{ETAZET}}$	Length Dependence of *ZET1*
ETAZET		Exponent of the Length Dependence of *ZET1*
SLA1	m	Length Dependence of *A1*
SWA1	m	Width Dependence of *A1*
SLA2	m	Length Dependence of *A2*
SWA2	m	Width Dependence of *A2*
SLA3	m	Length Dependence of *A3*
SWA3	m	Width Dependence of *A3*
Temperature Parameters		
STVTO	V/K	Temperature Dependence of **VTO**
STTHE1	V^{-1}·K^{-1}	Temperature Dependence of **THE1**
STLTHE1	m/V^{-1}·K^{-1}	Temperature Dependence of Length Dependence of *THE1*
STTHE2	V^{-1}·K^{-1}	Temperature Dependence of **THE2**
STLTHE2	m/V^{-1}·K^{-1}	Temperature Dependence of Length Dependence of *THE1*
STTHE3	V^{-1}·K^{-1}	Temperature Dependence of **THE3**
STLTHE3	m/V^{-1}·K^{-1}	Temperature Dependence of Length Dependence of *THE1*
ETABET		Temperature Dependence of **BET**
STMO	K^{-1}	Temperature Dependence of **MO**
STA1	K^{-1}	Temperature Dependence of **A1**

12.3　The Threshold Voltage Model

One of the more interesting features of MOS Model 9 is the detail of its threshold voltage model. When compared to the other analytical FET models, it is clear that more of the model structure is "tied up" in the threshold voltage model.

Table 12.3   A summary of the composite parameters used in MOS Model 9 and the parameters which contribute to their computation.

Parameter	Units	Contributing Model Parameters
		Composite Parameters
VTO	V	**VTO, STVTO, SLVTO, SL2VTO, SWVTO**
PHIB	V	**PHIB**
KO	$V^{\frac{1}{2}}$	**KO, SLKO, SWKO**
K	$V^{\frac{1}{2}}$	**K, SLK, SWK**
VSBX	V	**VSBX, SLVSBX, SWVSBX**
GAMOO		**GAMOO, SLGAMOO**
VSBT	V	**VSBT, SLVSBT**
GAM1	$V^{(1-\text{ETADS})}$	**GAM1, SLGAM1, SWGAM1**
THE1	V^{-1}	**THE1, STTHE1, SLTHE1, STLTHE1, SWTHE1**
THE2	$V^{-\frac{1}{2}}$	**THE2, STTHE2, SLTHE2, STLTHE2, SWTHE2**
THE3	V^{-1}	**THE3, STTHE3, SLTHE3, STLTHE3, SWTHE3**
MO		**MO, STMO, SLMO**
ALP		**ALP, SLALP, ETAALP, SWALP**
VP	V	**VP**
ZET1		**ZET1, SLZET1, ETAZET**
BET	A/V^2	**BET, ETABET**
A1		**A1, STA1, SLA1, SWA1**
A2	V	**A2, SLA2, SWA2**
A3		**A3, SLA3, SWA3**

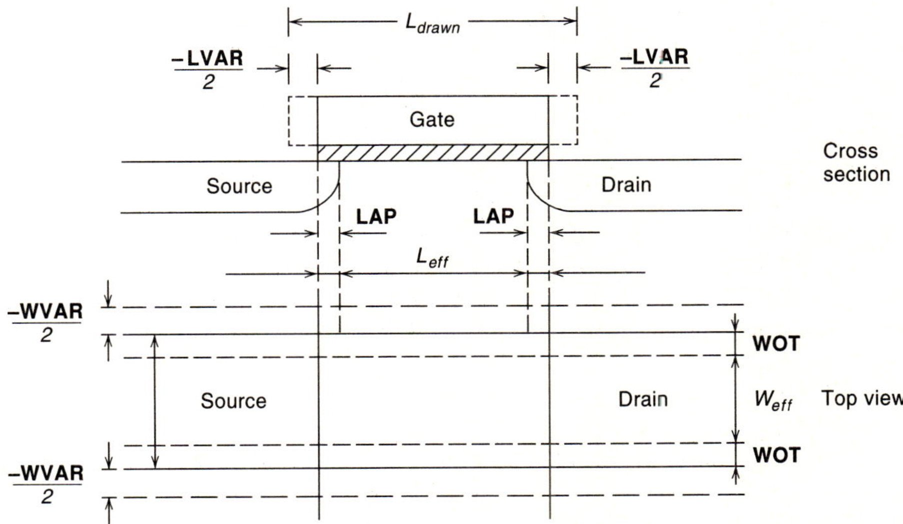

Figure 12.2   The MOS Model 9 description of the channel length and width. The different physical effects are carefully separated in this model. From [1]. © 1994. Philips Electronics N.V. Used by permission.

For a particular FET, a basic capacitor-like zero substrate bias threshold voltage is defined as a model parameter. The model then adds corrections to that value to account for the application of substrate and drain biases. This results in a rather straightforward model for the threshold voltage.

12.3.1   The Basic Model

The development begins with the simple three terminal MOS capacitor threshold voltage model developed in Section 3.4 (see (3.113)):

$$V_t = V_{fb} + \phi_s + \gamma \cdot (\phi_s - V_{bs})^{\frac{1}{2}}, \tag{12.6}$$

where from first principles

$$\gamma = \frac{1}{C_{ox}} \cdot (2\epsilon_{Si} q N_{sub})^{\frac{1}{2}}; \tag{12.7}$$

in the general case,

$$\phi_s = 2\phi_f, \tag{12.8}$$

where ϕ_f is the bulk potential of a uniformly doped substrate. For $V_{bs} = 0$, (12.6) becomes

$$V_{t,V_{bs}=0} = V_{to} = V_{fb} + \phi_s + \gamma \cdot \phi_s^{\frac{1}{2}}. \tag{12.9}$$

Solving this expression for $(V_{fb} + \phi_s)$ yields

$$V_{fb} + \phi_s = V_{to} - \gamma \cdot \phi_s^{\frac{1}{2}}. \tag{12.10}$$

Substituting (12.10) into (12.6) allows for the creation of a threshold voltage expression in which the nonzero substrate bias effects are isolated in a separate term:

$$V_t = V_{to} - \gamma \cdot \phi_s^{\frac{1}{2}} + \gamma \cdot (\phi_s - V_{bs})^{\frac{1}{2}} = V_{to} + \gamma \cdot \left[(\phi_s - V_{bs})^{\frac{1}{2}} - \phi_s^{\frac{1}{2}} \right]; \tag{12.11}$$

this expression is very commonly employed in analysis of the FET threshold voltage, and is used in Level 1 (see Section 5.2).

MOS Model 9 uses (12.11), but as in many other models, the zero substrate bias threshold voltage V_{to} is treated as a model parameter. Thus, (12.11) becomes

$$V_t = VTO + \gamma \cdot \left[(\phi_s - V_{bs})^{\frac{1}{2}} - \phi_s^{\frac{1}{2}} \right]. \tag{12.12}$$

12.3.2 Substrate Bias Effects

As noted above, (12.12) was derived with the assumption that the substrate is uniformly doped. Modern FETs generally include a surface "tailoring" implant to adjust and control the device threshold voltage, so that a designed value is achieved. In the first and second generation models, this situation was not considered; either the substrate doping **NSUB** or the inversion potential **PHI** is treated as a simple model parameter, and reasonable results are usually obtained. In the third-generation models, a major goal is to provide a more physically based description of MOSFET behavior. MOS Model 9 follows this trend by including a model for nonuniformly doped substrates.

Like BSIM3, MOS Model 9 simplifies the typical Gaussian surface doping profile to a rectangular surface layer, as shown in Figure 12.3. However, unlike BSIM3, MOS Model 9 does not bother to define these two doping concentrations. Instead, two different versions of the substrate sensitivity term γ are introduced to account for the two different doping regions.

Consider the gate-induced depletion region in a MOSFET when there is no drain bias applied. For small gate biases, this depletion region expands into the surface doping region, and one particular value of γ will describe the device behavior. For larger gate biases, the depletion region will expand in the substrate doping region, and a different value of γ is required.

Examining (12.12), it can be seen that if V_t is plotted against $\left[(\phi_s - V_{bs})^{\frac{1}{2}} - \phi_s^{\frac{1}{2}}\right]$ for a uniformly doped substrate, a straight line is produced, as shown in Figure 12.4a. The slope of this line is γ, and the y-intercept is the zero substrate bias threshold voltage *VTO*. If a similar plot is created for a nonuniformly doped substrate, a plot similar to Figure 12.4b is found. As expected, two different regions are noted in the plot. For small values of $\left[(\phi_s - V_{bs})^{\frac{1}{2}} - \phi_s^{\frac{1}{2}}\right]$, where the gate-induced depletion region is expanding in the surface doping layer, there is a distinct slope for the line. For larger values of $\left[(\phi_s - V_{bs})^{\frac{1}{2}} - \phi_s^{\frac{1}{2}}\right]$, as the depletion region expands in the substrate, a different slope is noted.

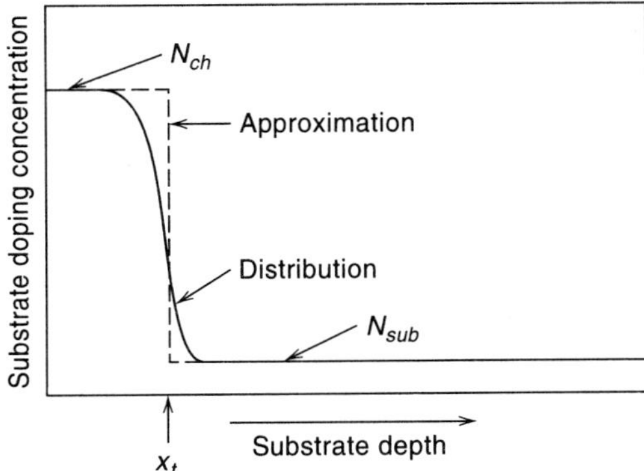

Figure 12.3 A typical surface doping profile in a modern FET, showing the Gaussian doping profile produced by ion implantation (solid line), and the "box" approximation (dashed line) that is used by MOS Model 9. From [6]. © 1995. The Regents of the University of California. Used by permission.

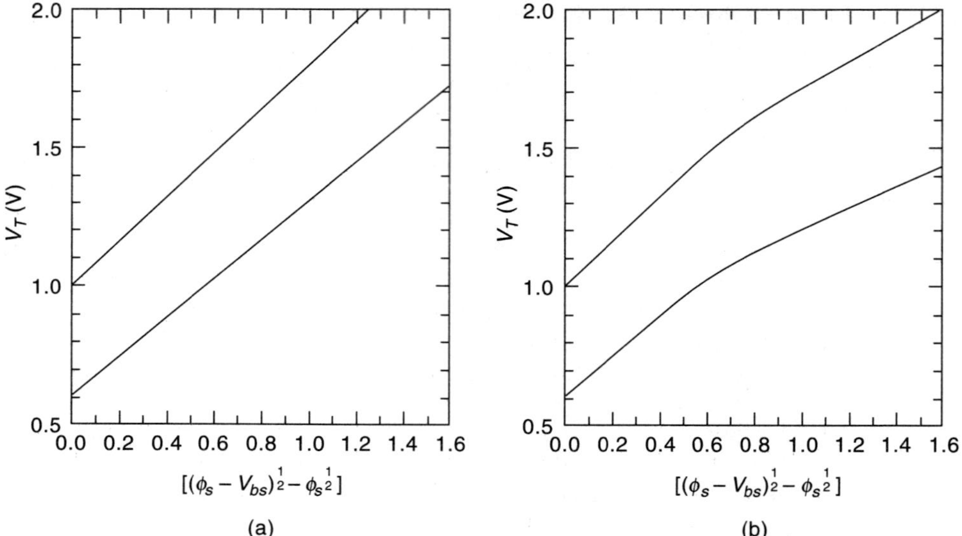

Figure 12.4   Plots of V_t versus $\left[(\phi_s - V_{bs})^{\frac{1}{2}} - \phi_s^{\frac{1}{2}}\right]$; (a) uniformly doped substrate, which produces a single straight line; (b) nonuniformly doped substrate, showing the change in behavior when the gate-induced depletion region has finished expanding into the surface doping region and then expands into the main substrate. From [1]. © 1994. Philips Electronics N.V. Used by permission.

A problem with this discussion is that the form of Figure 12.4a and Figure 12.4b assumes that the surface doping is known, so that the surface potential can be computed from

$$\phi_s = 2\phi_f = 2 \cdot \frac{k_b T}{q} \cdot ln\left(\frac{N_{surf}}{n_i(T)}\right), \tag{12.13}$$

where $n_i(T)$ is the intrinsic carrier concentration. However, like the second-generation models, MOS Model 9 chooses to avoid complexity by defining ϕ_s as a model parameter (here, *PHIB*). This parameter is extracted along with the other threshold voltage parameters.

Figure 12.4b contains the basis of the model implemented in MOS Model 9, where two different values of γ are defined. When the gate-induced depletion region is expanding into the surface depletion layer, the device behavior is described by $\gamma = KO$. When this depletion region is expanding into the main substrate, the device behavior is instead accounted for with $\gamma = K$.

To prepare for the discussion to follow, redefine (12.12) as

$$V_{t1} = VTO + \Delta V_{to}, \tag{12.14}$$

where V_{t1} is the threshold voltage including substrate bias effects (but not drain bias effects), while ΔV_{to} is the specific computed value of those effects.

As shown in Figure 12.4b, there is a *breakpoint* between the two regions of device behavior. MOS Model 9 uses a model parameter ($VSBX$) as the voltage at which this transition takes place. For $-V_{bs} < VSBX$, the gate-induced depletion region is contained entirely in the surface doping layer, and the threshold voltage change is well described by

$$\Delta V_{to} = KO \cdot \left[(PHIB - V_{bs})^{\frac{1}{2}} - PHIB^{\frac{1}{2}} \right]. \tag{12.15}$$

For $-V_{bs} > VSBX$, the situation becomes more complicated, as the gate-induced depletion region now spans two different doping regions. MOS Model 9 employs an expression which is derived by integration over the doping profile as depicted in Figure 12.3; this results in

$$\Delta V_{to} = \left[1 - \left(\frac{K}{KO} \right)^2 \right] \cdot KO \cdot (VSBX + PHIB)^{\frac{1}{2}} - KO \cdot PHIB^{\frac{1}{2}}$$

$$+ K \cdot \left\{ (PHIB - V_{bs}) - \left[1 - \left(\frac{K}{KO} \right)^2 \right] \cdot (PHIB + VSBX) \right\}^{\frac{1}{2}}. \tag{12.16}$$

Two problems occur in the description of the effect of the substrate bias on the threshold voltage as provided by (12.15) and (12.16). First, in the spirit of the third-generation models, it is desirable to guarantee a continuous and smooth transition at the breakpoint of $-V_{bs} = VSBX$. Second, with the equations in NMOS form, during normal circuit usage, $V_{bs} \leq 0$; however, due to numerical transients during circuit simulation iterations, V_{bs} can temporarily take on values greater than zero. As noted in Section 12.2, MOS Model 9 creates continuous and smooth transitions by introducing hyperbolic functions into the model formulation. Here, (12.15) and (12.16) are replaced with a single expression for ΔV_{to}:

$$\Delta V_{to} = K \cdot \left\{ \left[hyp4(-V_{bs}, VSBX, \epsilon_2) + \left(\frac{K}{KO} \right)^2 \cdot (PHIB + VSBX) \right]^{\frac{1}{2}} \right.$$

$$\left. - \left(\frac{K}{KO} \right) \cdot (PHIB + VSBX)^{\frac{1}{2}} \right\}$$

$$+ KO \left\{ [h_1 - hyp4(-V_{bs}, VSBX, \epsilon_2)]^{\frac{1}{2}} - PHIB^{\frac{1}{2}} \right\}, \tag{12.17}$$

where

$$\epsilon_2 = 0.1, \tag{12.18}$$

$hyp4$ is a hyperbolic function given by

$$hyp4(x, x_o, \epsilon) = hyp1(x - x_o, \epsilon) - hyp1(-x_o, \epsilon), \tag{12.19}$$

h_1 is a composite function

$$h_1 = hyp1(-V_{bs} + \frac{PHIB}{2}, \epsilon_1) + \frac{PHIB}{2}, \tag{12.20}$$

with

$$\epsilon_1 = 0.01, \tag{12.21}$$

and $hyp1$ is another hyperbolic function, given by

$$hyp1(x, \epsilon) = \frac{1}{2} \cdot \left[x + (x^2 + 4 \cdot \epsilon)^{\frac{1}{2}} \right]. \tag{12.22}$$

For $-V_{bs} \gg VSBX$, (12.17) reduces to (12.16). In the limit of $-V_{bs} \ll VSBX$, (12.17) becomes identical to (12.15), while any transient problems which can occur when $V_{bs} > 0$ during circuit simulation are eliminated. The term ΔV_{to}, as described by (12.17), is used in (12.14) to compute V_{t1}; as noted earlier, V_{t1} represents the threshold voltage as modified by the application of a substrate bias.

12.3.3 Drain Bias Effects

As the drain bias V_{ds} increases, the depletion layer expands around the drain diffusion; this causes two effects on the threshold voltage. As this depletion region expands, the overlap between the gate-induced depletion region and the drain depletion region increases. This leads to a decrease in the total depletion charge, which has the effect of increasing the inversion charge and thus decreasing the threshold voltage; this *static feedback* effect is described in more detail in Section 6.2 and Section 7.2. In addition, in short channel FETs, the application of the drain bias reduces the channel potential barrier, which decreases the value of the threshold voltage. This mechanism is known as drain-induced barrier lowering (DIBL). MOS Model 9 develops separate expressions to describe each contribution.

Drain-Induced Barrier Lowering

Drain-induced barrier lowering (DIBL) is particularly important in the subthreshold region, where a small change in the threshold voltage can lead to a significant change in the magnitude of the drain current. The DIBL mechanism also has considerable substrate bias dependence, and very different behavior is noted for low vs. high substrate biases.

MOS Model 9 describes the threshold voltage lowering effect of DIBL using

$$\Delta V_{t,DIBL} = -\gamma_o \cdot \frac{V_{gtx}^2}{V_{gtx}^2 + V_{gt1}^2} \cdot V_{ds}, \tag{12.23}$$

where V_{gtx} is a smoothing term given by

$$V_{gtx} = \frac{1}{2} \cdot (2)^{\frac{1}{2}} \tag{12.24}$$

and V_{gt1} is a function designed to prevent the driving voltage from becoming negative in the subthreshold region:

$$\begin{aligned} V_{gt1} &= V_{gs} - V_{t1}, \quad \cdots \quad V_{gs} \geq V_{t1} \\ &= \quad 0 \quad\quad \cdots \quad V_{gs} < V_{t1} \end{aligned} \tag{12.25}$$

the term V_{t1} is the threshold voltage as modified by an applied substrate bias, as described by (12.14). The term γ_o details the effect of the substrate bias on DIBL, and is given by

$$\gamma_o = GAMOO \cdot \left(\frac{u_{s1}}{PHIB^{\frac{1}{2}}} \right)^{\mathbf{ETAGAM}}, \tag{12.26}$$

where u_{s1} defines two separate regions of the effect of the substrate bias on DIBL:

$$\begin{aligned} u_{s1} &= \quad (PHIB - V_{bs})^{\frac{1}{2}} \quad \cdots \quad -V_{bs} \leq VSBT \\ &= (PHIB + VSBT)^{\frac{1}{2}} \quad \cdots \quad -V_{bs} > VSBT. \end{aligned} \tag{12.27}$$

Thus, *VSBT* defines a transition voltage, above and below which DIBL is affected differently by the substrate bias.

For a uniformly doped substrate, a value of **ETAGAM** $= 1$ provides a proper description for use in (12.26). When the substrate is non-uniformly doped, good results have been found empirically for **ETAGAM** $= 2$. Of course, **ETAGAM** can simply be extracted as a model parameter; however, note that if **ETAGAM** takes on a noninteger value, the computational efficiency of (12.26) is considerably reduced.

Once again, the problem of guaranteeing continuous and smooth transitions across the defined boundaries arises; here, this is the case in (12.25) at $V_{gs} = V_{t1}$, and in (12.27) at $-V_{bs} = VSBT$. Therefore, (12.25) is replaced by

$$V_{gt1} = hyp1(V_{gs} - V_{t1}, \epsilon_4), \tag{12.28}$$

where

$$\epsilon_4 = 5.0 \times 10^{-4} \tag{12.29}$$

and the hyperbolic function *hyp*1 is given by (12.22). In the appropriate limits, (12.28) reduces to the results of (12.25). Equation (12.27) is superceded by

$$u_{s1} = hyp2(h_1^{\frac{1}{2}}, (PHIB + VSBT), \epsilon_3), \tag{12.30}$$

where

$$\epsilon_3 = 0.01 \tag{12.31}$$

and h_1 is determined using (12.20). The hyperbolic function $hyp2$ is computed using

$$hyp2(x, x_o, \epsilon) = x - hyp1(x - x_o, \epsilon), \tag{12.32}$$

and the hyperbolic function $hyp1$ is described in (12.22).

DIBL is the dominant drain bias effect on the threshold voltage in the subthreshold region of operation; here, this region is defined by $V_{gt1} < V_{gtx}$. Note that as in most of the other models, in (12.23) the effect of DIBL on the threshold voltage is taken to be linearly proportional to V_{ds} [4].

Static Feedback

The effect of static feedback on the threshold voltage is particularly important at large drain biases, when the drain depletion region is a major portion of the total device depletion region. Static feedback has negligible direct substrate bias dependence; the substrate bias has only a secondary effect, entering the situation only through its effect on the (long channel) threshold voltage.

In MOS Model 9, the effect of static feedback is included in the threshold voltage model using

$$\Delta V_{t,SF} = -GAM1 \cdot \frac{V_{gt1}^2}{V_{gtx}^2 + V_{gt1}^2} \cdot V_{ds}^{\textbf{ETADS}}, \tag{12.33}$$

where V_{gt1} is described by (12.28) and V_{gtx} is given by (12.24). Empirical observation [1] suggests a value of **ETADS** $= 0.6$; however, this model parameter can be extracted. Note the possible computational inefficiency due to the use of a non-integer exponent.

Static feedback dominates the effect of the drain bias on the threshold voltage in the strong inversion region. Here, that region is defined by $V_{gt1} > V_{gtx}$.

Combining DIBL and Static Feedback

The effect of the drain bias on the threshold voltage is described by

$$V_{t2} = V_{t1} + \Delta V_{t1}, \tag{12.34}$$

where ΔV_{t1} is the collected description of the effect of the drain bias on the threshold voltage and V_{t1} is the threshold voltage as modified by the substrate bias (as given by (12.14)). Thus, V_{t2} is the **complete** model for the threshold voltage of a device, comprised of a simple capacitor-like threshold voltage VTO and corrections due to the application of substrate and drain biases.

The term ΔV_{t1} is the linear combination of the effect of DIBL (12.23) and the effect of static feedback (12.33):

$$\Delta V_{t1} = -\gamma_o \cdot \frac{V_{gtx}^2}{V_{gtx}^2 + V_{gt1}^2} \cdot V_{ds} - GAM1 \cdot \frac{V_{gt1}^2}{V_{gtx}^2 + V_{gt1}^2} \cdot V_{ds}^{\textbf{ETADS}}. \tag{12.35}$$

However, the first and second derivatives of this expression are not well behaved at $V_{ds} = 0$ due to the presence of **ETADS** as a power exponent; **ETADS** usually takes on a noninteger value of about 0.6. To solve this problem, (12.35) is superceded by

$$\Delta V_{t1} = \left\{ -\gamma_o - \left[GAM1 \cdot (V_{ds} + \lambda_2)^{\textbf{ETADS}-1} - \gamma_o \right] \cdot \frac{V_{gt1}^2}{V_{gtx}^2 + V_{gt1}^2} \right\} \cdot \frac{V_{ds}^2}{V_{ds} + 1}, \tag{12.36}$$

where

$$\lambda_1 = 0.1 \tag{12.37}$$

and

$$\lambda_2 = 1.0 \times 10^{-4}. \tag{12.38}$$

The term ΔV_{t1} from (12.36) is then used in (12.34) to compute the final threshold voltage V_{t2} for a particular device, which includes the corrections to *VTO* due to both substrate bias and drain bias effects.

For convenience, the total drive voltage can be written as

$$V_{gt2} = V_{gs} - V_{t2}. \tag{12.39}$$

This is the MOS Model 9 implementation of the common expression $(V_{gs} - V_t)$ which appears throughout the various analytical FET models.

12.3.4   Temperature and Geometry Dependence

As described in Section 12.2, MOS Model 9 includes geometry dependence in the model in a manner based on the approach used in the second generation models. However, rather than set up a general scheme and then force all parameters to conform to that scheme, the geometry dependence is introduced in a very selective manner. In contrast to the second-generation models, the temperature dependence is very much intertwined with the geometry dependence. Therefore, throughout this chapter, the geometry dependence and temperature dependence will be considered in each section, rather than in one all-encompassing discussion at the end of the chapter.

Temperature Dependence

The temperature dependence of the surface potential is computed in two parts. First, an auxiliary function is defined:

$$S_{T,\phi_s} = \frac{\textbf{PHIB} - 1.13 - 2.5 \times 10^{-4} \cdot T_{nom}}{300}, \tag{12.40}$$

where T_{nom} is the temperature at which parameter extraction was carried out (usually around 300K). The final temperature dependence of ϕ_s is then described by

$$PHIB = \textbf{PHIB} + S_{T,\phi_s} \cdot (T - T_{nom}). \tag{12.41}$$

Note that in (12.42), T and T_{nom} must be specified in **degrees Kelvin** rather than in degrees Celsius.

Next, MOS Model 9 makes use of the well-known observation that to a reasonable approximation, the threshold voltage varies linearly with the temperature. All of the temperature dependence is assigned to the zero bias threshold voltage, yielding

$$V_{to}(T) = \textbf{VTO} + \textbf{STVTO} \cdot (T - T_{nom}), \tag{12.42}$$

Note once again that T and T_{nom} must be specified in degrees Kelvin.

As originally intended [1], **VTO** is supposed to be specified for use in (12.42) as the zero substrate bias threshold voltage for the so-called reference transistor. As noted in Section 12.2, this is supposed to be a long, wide device which is free from small geometry effects; all geometry dependence is then accounted for by the extrinsic model structure. However, by **not** placing this restriction on the use of **VTO** in (12.42), MOS Model 9 can more readily accommodate model binning.

Geometry Dependence

Once the temperature dependence of the zero bias threshold voltage has been computed using (12.42), that result is used in the geometry dependence calculation. MOS Model 9 uses the basic second-generation approach; however, to improve the accuracy, a term that is quadratic in the effective channel length L_{eff} is added to the model. The final computed value of the zero bias threshold voltage is

$$VTO = V_{to}(T) + \textbf{SLVTO} \cdot \left(\frac{1}{L_{eff}} - \frac{1}{L_{eff,ref}} \right) + \textbf{SL2VTO} \cdot \left(\frac{1}{L_{eff}^2} - \frac{1}{L_{eff,ref}^2} \right)$$

$$+ \textbf{SWVTO} \cdot \left(\frac{1}{W_{eff}} - \frac{1}{W_{eff,ref}} \right), \tag{12.43}$$

where $V_{to}(T)$ is computed from (12.42); the geometries subscripted with *ref* indicate the reference length and width, as discussed in Section 12.2.

Several of the parameters which appear in the model for the effect of the substrate bias on the threshold voltage also have geometry dependence. These are

$$KO = \textbf{KO} + \textbf{SLKO} \cdot \left(\frac{1}{L_{eff}} - \frac{1}{L_{eff,ref}} \right) + \textbf{SWKO} \cdot \left(\frac{1}{W_{eff}} - \frac{1}{W_{eff,ref}} \right), \tag{12.44}$$

$$K = \mathbf{K} + \mathbf{SLK} \cdot \left(\frac{1}{L_{eff}} - \frac{1}{L_{eff,ref}} \right) + \mathbf{SWK} \cdot \left(\frac{1}{W_{eff}} - \frac{1}{W_{eff,ref}} \right), \tag{12.45}$$

and

$$VSBX = \mathbf{VSBX} + \mathbf{SLVSBX} \cdot \left(\frac{1}{L_{eff}} - \frac{1}{L_{eff,ref}} \right) + \mathbf{SWVBSX} \cdot \left(\frac{1}{W_{eff}} - \frac{1}{W_{eff,ref}} \right). \tag{12.46}$$

The geometry dependence of the effect of the drain bias on the threshold voltage is accounted for using

$$VSBT = \mathbf{VSBT} + \mathbf{SLVSBT} \cdot \left(\frac{1}{L_{eff}} - \frac{1}{L_{eff,ref}} \right) \tag{12.47}$$

and

$$GAM1 = \mathbf{GAM1} + \mathbf{SLGAM1} \cdot \left(\frac{1}{L_{eff}} - \frac{1}{L_{eff,ref}} \right) + \mathbf{SWGAM1} \cdot \left(\frac{1}{W_{eff}} - \frac{1}{W_{eff,ref}} \right). \tag{12.48}$$

Note that $VSBT$ has length dependence but does **not** have width dependence.

The remaining parameters are treated as having neither temperature dependence nor geometry dependence; they are allowed to have the same value over the entire bin in question. If binning is not employed, these parameters maintain their values for the entire geometry range.

12.4   The Mobility Model

MOS Model 9 employs a relatively simple model for the mobility. As in the second-generation models, the source/drain series resistance is absorbed into this model.

12.4.1   The Effect of the Vertical Field

The description of the reduction of the mobility by the vertical field begins with the basic expression developed for use in earlier models, such as Level 3 (see Section 7.4):

$$\mu_v = \frac{\mu_o}{1 + \theta_{10} \cdot (V_{gs} - V_{t1})}, \tag{12.49}$$

where μ_v is the mobility as affected by the vertical field, μ_o is the low field mobility, θ_{10} is the standard mobility reduction parameter (identical to **THETA** in Level 3), and V_{t1} is the threshold voltage including substrate bias effects, as given by (12.14). In MOS Model 9, this basic expression is augmented with an additional substrate bias dependence [3], which transforms (12.49) to

$$\mu_v = \frac{\mu_o}{1 + \theta_{10} \cdot (V_{gs} - V_{t1}) + \theta_2 \cdot \left[(PHIB - V_{bs})^{\frac{1}{2}} - PHIB^{\frac{1}{2}} \right]}. \tag{12.50}$$

Note that when $V_{bs} = 0$, the term associated with θ_2 vanishes, and V_{t1} simplifies to the zero substrate bias threshold voltage parameter *VTO*; the simplest possible expression for the mobility as reduced by the vertical field is recovered, which is identical to the Level 3 expression (7.43).

The drive voltage $(V_{gs} - V_{t1})$ cannot be allowed to become negative in the subthreshold region; this must be prevented, while also providing a continuous and smooth transition at $V_{gs} = V_{t1}$. The auxiliary term V_{gt1}, defined by (12.28), is used, which modifies (12.50) to

$$\mu_v = \frac{\mu_o}{1 + \theta_{10} \cdot V_{gt1} + \theta_2 \cdot \left[(PHIB - V_{bs})^{\frac{1}{2}} - PHIB^{\frac{1}{2}} \right]}. \tag{12.51}$$

12.4.2   The Effect of the Lateral Field

A choice from a number of possible models can be made for the description of the effect of the lateral field on the mobility. MOS Model 9 chooses to employ a modification to (12.51) of the form [3]

$$\mu_{eff} = \frac{\mu_o}{\left\{ 1 + \theta_{10} \cdot V_{gt1} + \theta_2 \cdot \left[(PHIB - V_{bs})^{\frac{1}{2}} - PHIB^{\frac{1}{2}} \right] \right\} \cdot (1 + \theta_{30} \cdot V_{ds})}, \tag{12.52}$$

where μ_{eff} is the mobility as affected by both the vertical and lateral fields; with the use of V_{ds}, this expression is valid in the linear region of operation. As discussed in Section 12.5, MOS Model 9 introduces an auxiliary function V_{ds1} to replace V_{ds}; V_{ds1} reduces to V_{ds} in the linear region, to V_{dsat} in the saturation region, and provides a continuous and smooth transition between the two regions. Thus, (12.52) becomes

$$\mu_{eff} = \frac{\mu_o}{\left\{ 1 + \theta_{10} \cdot V_{gt1} + \theta_2 \cdot \left[(PHIB - V_{bs})^{\frac{1}{2}} - PHIB^{\frac{1}{2}} \right] \right\} \cdot (1 + \theta_{30} \cdot V_{ds1})}. \tag{12.53}$$

In this way, the influence of carrier velocity saturation on the mobility is subsumed into the expression for the saturation voltage; this occurs as part of the limiting behavior of V_{ds1}.

12.4.3   The Source/Drain Series Resistance

In the first-generation models, the source/drain series resistance is included in the form of lumped elements (**RS** and **RD**, or as a similar expression which more readily accommodates geometry dependence); the same approach is used in BSIM3 (see Section 11.6). In contrast, the second-generation models take the expedient approach of absorbing the series resistance into the mobility. This method is quite convenient, as it bypasses any need for a complicated description of the bias dependencies of the series resistance; this is in contrast to the complicated expressions that are developed in BSIM3 to describe the series resistance (see Section 11.6).

MOS Model 9 uses the second-generation approach of absorbing the series resistance into the mobility model. This is done using [3]

$$\theta_1 = \theta_{10} + (R_S + R_D) \cdot \frac{\mu_o \cdot C_{ox} \cdot W_{eff}}{L_{eff}} \qquad (12.54)$$

and

$$\theta_3 = \theta_{30} - R_D \cdot \frac{\mu_o \cdot C_{ox} \cdot W_{eff}}{L_{eff}} - \frac{\theta_{10}}{2}. \qquad (12.55)$$

Note that the vertical field description includes the series resistance as a single lumped sum of R_S and R_D; the drain bias expression only includes the drain resistance R_D while also factoring in gate bias dependence through θ_{10}. The terms θ_1 and θ_3 replace θ_{10} and θ_{30} in (12.53), which becomes

$$\mu_{eff} = \frac{\mu_o}{\left\{ 1 + \theta_1 \cdot V_{gt1} + \theta_2 \cdot \left[(PHIB - V_{bs})^{\frac{1}{2}} - PHIB^{\frac{1}{2}} \right] \right\} \cdot (1 + \theta_3 \cdot V_{ds1})}. \qquad (12.56)$$

12.4.4   The Final Mobility Expression

MOS Model 9 treats θ_1, θ_2, and θ_3 as model parameters. Making this modification to (12.56) leads to the final expression for the mobility:

$$\mu_{eff} = \frac{\mu_o}{\left\{ 1 + THE1 \cdot V_{gt1} + THE2 \cdot \left[(PHIB - V_{bs})^{\frac{1}{2}} - PHIB^{\frac{1}{2}} \right] \right\} \cdot (1 - THE3 \cdot V_{ds1})}. \qquad (12.57)$$

Note that in contrast to most other models, the low field mobility μ_o is **not** specified as a model parameter; instead, it is absorbed into the model parameter *BET*, which is described in Section 12.5.

12.4.5   Temperature and Geometry Dependence

The temperature dependence and geometry dependence of the mobility model are rather complex, as these two adjustments are intertwined to a very large degree. Each θ-like parameter is taken to have its own temperature dependence, in a manner very much like that developed for the threshold voltage parameter *VTO*, as described by (12.42). Then, a combined length and temperature parameter is determined. Finally, these two computations are unified to produce the final expression.

Consider the parameter *THE1*. Temperature dependence involving this parameter is introduced in the form of (12.42):

$$\theta_1(T) = \mathbf{THE1} + \mathbf{STTHE1} \cdot (T - T_{nom}), \qquad (12.58)$$

where T and T_{nom} are specified in **degrees Kelvin**. Next a composite model parameter which accounts for both the length and the temperature dependence is included:

$$S_{L,\theta_1} = \mathbf{SLTHE1} + \mathbf{STLTHE1} \cdot (T - T_{nom}), \tag{12.59}$$

Finally, the complete geometry and temperature dependence expression is reached by combining the results of (12.58) and (12.59) into a more familiar form:

$$THE1 = \theta_1(T) + S_{L,\theta_1} \cdot \left(\frac{1}{L_{eff}} - \frac{1}{L_{eff,ref}} \right) + \mathbf{SWTHE1} \cdot \left(\frac{1}{W_{eff}} - \frac{1}{W_{eff,ref}} \right). \tag{12.60}$$

The width dependence of the final composite parameter *THE1* is found in the direct and usual manner, while the length dependence takes the slightly more complicated path through (12.59).

The same approach is applied to the parameters *THE2* and *THE3*. This leads to the expressions

$$\theta_2(T) = \mathbf{THE2} + \mathbf{STTHE2} \cdot (T - T_{nom}), \tag{12.61}$$

$$S_{L,\theta_2} = \mathbf{SLTHE2} + \mathbf{STLTHE2} \cdot (T - T_{nom}), \tag{12.62}$$

$$THE2 = \theta_2(T) + S_{L,\theta_2} \cdot \left(\frac{1}{L_{eff}} - \frac{1}{L_{eff,ref}} \right) + \mathbf{SWTHE2} \cdot \left(\frac{1}{W_{eff}} - \frac{1}{W_{eff,ref}} \right), \tag{12.63}$$

$$\theta_3(T) = \mathbf{THE3} + \mathbf{STTHE3} \cdot (T - T_{nom}), \tag{12.64}$$

$$S_{L,\theta_3} = \mathbf{SLTHE3} + \mathbf{STLTHE3} \cdot (T - T_{nom}), \tag{12.65}$$

and

$$THE3 = \theta_3(T) + S_{L,\theta_3} \cdot \left(\frac{1}{L_{eff}} - \frac{1}{L_{eff,ref}} \right) + \mathbf{SWTHE3} \cdot \left(\frac{1}{W_{eff}} - \frac{1}{W_{eff,ref}} \right). \tag{12.66}$$

Note once again that all temperatures are specified in degrees Kelvin.

12.5   The Drain Current Model

MOS Model 9 develops a drain current model which is semi-empirical and includes a number of physical effects in a relatively simple manner. As in the threshold voltage model, a number of auxiliary and hyperbolic functions are introduced to guarantee continuous and smooth transitions between the various regions of device operation.

Among the various physical considerations included in the drain current model are the effect of the substrate bias on the subthreshold slope, an auxiliary function to smooth the weak inversion-strong inversion transition, a saturation voltage expression which is folded into an auxiliary drain bias to guarantee a smooth linear region-saturation region transition and smooth output conductance results, and a more involved channel length modulation description to improve the output conductance results. This leads to a single current equation for all regions of device operation, which is a trademark of the third-generation models. Furthermore, an expression for the weak avalanche current is added; this is the same physical effect that is referred to as the *hot electron induced change in the effective substrate potential* in BSIM2 and the *substrate current induced body effect* in BSIM3. This addition is required for a proper description of the output conductance behavior at a combination of a low gate bias (slightly larger than the threshold voltage) and a high drain bias. The resulting drain current expression is very clean, and is easy to implement in a circuit simulator.

12.5.1 Substrate Bias Effect on the Subthreshold Slope

When a substrate bias is applied to an FET, the extent of the source-substrate junction depletion region is altered, which changes the depletion charge; this will affect the inversion charge and thus the subthreshold slope of the device. MOS Model 9 uses a description of the departure of the subthreshold slope from ideality, embodied in [3]

$$m = 1 + MO \cdot \left(\frac{PHIB^{\frac{1}{2}}}{u_{s1}} \right)^{ETAM} , \tag{12.67}$$

where u_{s1} is an auxiliary function provided by (12.30). In the ideal situation, $PHIB^{\frac{1}{2}} \ll u_{s1}$, and $m \to 1$; this corresponds to an ideal subthreshold slope of 60 mV/decade at $T = 300K$. For a uniformly doped substrate, **ETAM** $= 1$, and all departure from ideal behavior is accounted for by the model parameter MO. For a nonuniformly doped substrate, it has been empirically observed [1] that a value of **ETAM** $= 2$ provides good results. Of course, **ETAM** can simply be extracted; however, if **ETAM** takes on a noninteger value, the computational efficiency of (12.67) is seriously reduced.

12.5.2 The Weak Inversion-Strong Inversion Transition

As noted in the descriptions of the other models, diffusion conduction dominates in the weak inversion region, while drift conduction is the most important transport mechanism in the strong inversion region. There is a transition region where both mechanisms are important; this complicates the description of the physical behavior. This situation arises in the neighborhood of the threshold voltage, where a good description of the device behavior is particularly important.

To guarantee a continuous and smooth transition through this region, MOS Model 9 once again employs auxiliary functions. The basic function is

$$G_1 = exp \left(\frac{q}{k_b T} \cdot \frac{V_{gt2}}{2 \cdot m} \right), \tag{12.68}$$

where V_{gt2} is the total drive voltage as defined by (12.39) and m is given by (12.67). This expression is used in a composite auxiliary gate bias function:

$$V_{gt3} = 2 \cdot m \cdot \frac{k_b T}{q} \cdot ln(1 + G_1). \tag{12.69}$$

In the subthreshold region, $V_{gt2} < 0$; with a negative argument in the exponential term of (12.68), G_1 takes on a very small value. An expansion of the logarithmic term in (12.69) converts that equation to

$$V_{gt3} = 2 \cdot m \cdot \frac{k_b T}{q} \cdot [ln(1) + G_1] = 2 \cdot m \cdot \frac{k_b T}{q} \cdot exp\left(\frac{q}{k_b T} \cdot \frac{V_{gt2}}{2 \cdot m}\right); \qquad (12.70)$$

the drain current will depend exponentially on the gate bias, as is the case in the subthreshold region. In the strong inversion region, $V_{gt2} > 0$, and thus $G_1 \gg 1$. In this case, (12.69) becomes

$$V_{gt3} = 2 \cdot m \cdot \frac{k_b T}{q} \cdot ln(G_1) = V_{gt2}; \qquad (12.71)$$

the drain current now depends on the gate bias in the usual form of $(V_{gs} - V_{t2})$, as described in Section 12.3. The structure of (12.69) allows for a continuous and smooth transition between the two regions; this approach is very similar to one originally suggested by Oguey and Cserveny [5].

However, the form of (12.69) can lead to numerical problems. In standard double precision computation and storage, the magnitude of a number is limited to maximum and minimum values of $1 \times 10^{\pm 308}$; attempts to compute larger or smaller numbers lead to, respectively, numerical overflow and numerical underflow. The term G_1 as described by (12.68) is vulnerable to this problem, and the consequences for (12.69) can be serious due to the computation of a logarithmic expression from an exponential function. For any particular value of x, a result of

$$ln\,[exp(x)] = x \qquad (12.72)$$

should be produced. However, this will not be the case if exponential overflow or underflow occurs during the computation.

The simplest method of avoiding the possibility that the results of the exponential computation will fall outside of the range $1 \times 10^{\pm 308}$ is to limit the range of the magnitude of the argument in the exponential, so that it will always be less than ± 37. This is done in MOS Model 9 through a series of steps. An auxiliary limiting function is defined:

$$V_{gta} = 2 \cdot m \cdot \frac{k_b T}{q} \cdot \lambda_7, \qquad (12.73)$$

where

$$\lambda_7 = 37. \qquad (12.74)$$

When $V_{gt2} < V_{gta}$, the argument in the exponential of (12.68) will not exceed 37, and no numerical problems are encountered; G_1 is computed using (12.68). The composite auxiliary gate bias expression V_{gt3} is then computed using a modified form of (12.69):

$$V_{gt3} = 2 \cdot m \cdot \frac{k_b T}{q} \cdot ln(1 + G_1) + \lambda_3, \tag{12.75}$$

where

$$\lambda_3 = 1 \times 10^{-8}. \tag{12.76}$$

The reason for the introduction of the small constant λ_3 will be discussed in Section 12.6.

When $V_{gt2} \geq V_{gta}$, the problem of exponential overflow can occur, so the computation of G_1 is dispensed with entirely. Instead, the limiting case of V_{gt3}, as computed in (12.71), is used directly, with one minor modification:

$$V_{gt3} = V_{gt2} + \lambda_3, \tag{12.77}$$

where λ_3 is given by (12.76). By inspection, it can be seen that λ_3 is added to (12.77) to guarantee a continuous and smooth transition with (12.75).

12.5.3   The Saturation Voltage

Since the saturation voltage defines the linear region-saturation transition point, an accurate model which produces continuous and smooth results for the drain current and the output conductance across this transition is required. This description must account for the effect of both the substrate bias and the drain bias; since both of these biases will change the FET depletion region, they will affect the depletion charge and thus will alter the inversion charge.

The approach used in MOS Model 9 is to develop a description in which

$$\frac{\partial I_{ds}}{\partial V_{ds}} = 0 \tag{12.78}$$

at $V_{ds} = V_{dsat}$, and then to add in separate expressions to describe the nonzero output conductance in the saturation region. Many of the models discussed in earlier chapters have also used this method.

The saturation voltage expression employed in MOS Model 9 [3] was originally developed for a digital model, in which the detailed behavior of the first derivative of the drain current was not considered:

$$V_{dsat} = \frac{V_{gt3}}{1 + \delta_1} \cdot \frac{2}{1 + \left(1 + \frac{2 \cdot THE3 \cdot V_{gt3}}{1 + \delta_1}\right)^{\frac{1}{2}}}, \tag{12.79}$$

where V_{gt3} is determined using either (12.75) or (12.77); δ_1 is given by

$$\delta_1 = \frac{\lambda_4}{(PHIB - V_{bs})^{\frac{1}{2}}} \cdot \left[K + \frac{(KO - K) \cdot VSBX^2}{VSBX^2 + (\lambda_5 \cdot V_{gt1} - V_{bs})^2}\right], \tag{12.80}$$

where

$$\lambda_4 = 0.3, \tag{12.81}$$

$$\lambda_5 = 0.1, \tag{12.82}$$

and V_{gt1} is computed using (12.28).

12.5.4   The Output Conductance Behavior

Second Derivative Continuity at $V_{ds} = V_{dsat}$

Equation (12.79) was derived such that the continuity of both the drain current and its first derivative at $V_{ds} = V_{dsat}$ are guaranteed. However, the continuity of the second derivative was not enforced. As a result, as depicted in Figure 12.5, the output conductance is continuous, but is not smooth at $V_{ds} = V_{dsat}$.

For more accurate results and improved circuit simulation behavior, the continuity of the second derivative at $V_{ds} = V_{dsat}$ must also be guaranteed. This is done in MOS Model 9 using a smoothing function in the form of an auxiliary drain bias; the same method is used in BSIM3 [6]. MOS Model 9 employs [1]

$$V_{ds1} = hyp5(V_{ds}, V_{dsat}, \epsilon_5), \tag{12.83}$$

where

$$\epsilon_5 = \lambda_6 \cdot \frac{V_{dsat}}{1 + V_{dsat}} \tag{12.84}$$

and

$$\lambda_6 = 0.3; \tag{12.85}$$

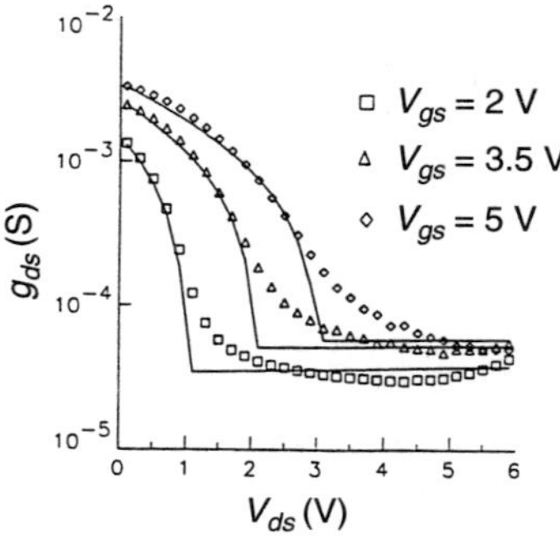

Figure 12.5   The output conductance computed using the digital saturation voltage model (12.79); the results are continuous, but not smooth at $V_{ds} = V_{dsat}$. From [1]. © 1994. Philips Electronics N.V. Used by permission.

hyp5 is a hyperbolic smoothing function given by

$$hyp5(x, x_o, \epsilon) = x_o - hyp1\left(x_o - x - \frac{\epsilon^2}{x_o}, \epsilon\right), \qquad (12.86)$$

and *hyp1* is another hyperbolic smoothing function, found using (12.22).

Channel Length Modulation and DIBL

As noted above, MOS Model 9 takes a commonly-used approach of first assuring that the first derivative of the drain current vanishes at $V_{ds} = V_{dsat}$, then developing separate expressions to account for the saturation region. Like BSIM2 [7] and BSIM3 [6, 8], these additions include the effects of channel length modulation, drain induced barrier lowering (DIBL), and a *weak avalanche* current. Under most circumstances, channel length modulation and DIBL are the key factors determining the output conductance behavior.

When the drain bias is larger than the saturation voltage, channel length modulation and DIBL interact [3]; channel length modulation changes the charge profile in the device (reducing the inversion charge near the drain), which changes how DIBL depends on the gate bias. MOS Model 9 includes an auxiliary function [3] to account for this behavior:

$$G_2 = 1 + ALP \cdot ln\left(1 + \frac{V_{ds} - V_{ds1}}{VP}\right); \qquad (12.87)$$

this expression will be included in the final drain current equation.

In addition, the saturation of the current in the subthreshold region must be more carefully described. In strong inversion, the saturation voltage V_{dsat} increases with increasing gate bias V_{gs}. In the subthreshold region, this is not the case. Instead, the saturation of the current is described by [1]

$$I_{ds} \propto 1 - exp\left(-\frac{q}{k_b T} \cdot V_{ds}\right). \qquad (12.88)$$

To guarantee continuous and smooth behavior at the weak inversion-strong inversion boundary, MOS Model 9 implements this expression through an auxiliary function:

$$G_3 = \frac{ZET1 \cdot \left[1 - exp\left(-\frac{q}{k_b T} \cdot V_{ds}\right)\right] + G_1 \cdot G_2}{\frac{1}{ZET1} + G_1}, \qquad (12.89)$$

where G_1 is given by (12.68),

$$G_1 = exp\left(\frac{q}{k_b T} \cdot \frac{V_{gt2}}{2 \cdot m}\right), \qquad (12.90)$$

V_{gt2} is described by (12.39),

$$V_{gt2} = V_{gs} - V_{t2},\tag{12.91}$$

and V_{t2}, the final threshold voltage expression including the effect of both the substrate bias and the drain bias, is computed using (12.34).

In the subthreshold region, V_{gt2} is negative, and thus G_1 becomes very small. In this case, (12.89) simplifies to

$$G_3 = ZET1^2 \cdot \left[1 - exp\left(-\frac{q}{k_b T} \cdot V_{ds} \right) \right].\tag{12.92}$$

In strong inversion (when $V_{gs} \gg V_{t2}$), G_1 becomes very large, and (12.89) reduces to

$$G_3 = G_2.\tag{12.93}$$

However, with G_1 appearing in (12.89), the numerical overflow/underflow problems due to the exponential term in (12.89), discussed earlier, will arise here as well. As described above, due to the limits of double precision computation and storage, numbers with magnitudes larger or smaller than $1 \times 10^{\pm 308}$ cannot be dealt with in numerical solutions. This forces the use of a numerical filter, such that the magnitude of any argument inside the exponential term is bounded by ± 37. Once again, a limiting value is computed:

$$V_{gta} = 2 \cdot m \cdot \frac{k_b T}{q} \cdot \lambda_7,\tag{12.94}$$

where

$$\lambda_7 = 37.\tag{12.95}$$

For $V_{gt2} < V_{gta}$, no numerical problems are encountered, and G_3 is computed using (12.89). For $V_{gt2} \geq V_{gta}$, exponential overflow will occur; however, since this numerical situation will only arise well into strong inversion, G_1 does not need to be calculated, and G_3 is found directly from the strong inversion limit expression (12.93).

The Drain Current Equation Without Weak Avalanche

Combining the components of the above discussion, the MOS Model 9 drain current equation without the inclusion of the weak avalanche contribution is [1, 3]

$$I_{dso} = BET \cdot G_3$$

$$\times \frac{V_{gt3} \cdot V_{ds1} - \left(\frac{1+\delta_1}{2}\right) \cdot V_{ds1}^2}{\left\{ 1 + THE1 \cdot V_{gt1} + THE2 \cdot \left[(PHIB - V_{bs})^{\frac{1}{2}} - PHIB^{\frac{1}{2}} \right] \right\} \cdot (1 + THE3 \cdot V_{ds1})}.\tag{12.96}$$

Note how, as mentioned in Section 12.4, the low field mobility has been absorbed into the model parameter *BET*. In the usual FET description,

$$\beta = \frac{\mu_o \cdot C_{ox} \cdot W_{eff}}{L_{eff}}. \tag{12.97}$$

However, in MOS Model 9, all geometry dependence is removed from the basic equations and is instead placed in the extrinsic geometry structure; this is discussed below. Thus, in MOS Model 9, β is converted to the model parameter **BET**, which will apply specifically to one particular device geometry.

The Weak Avalanche Current

For high drain biases, hot carriers lead to impact ionization and a substrate current. The presence of a substrate current, even for $V_{bs} = 0$, implies the existence of an induced change in the effective substrate bias:

$$V'_{bs} = V_{bs} + I_{sub} \cdot R_{sub}, \tag{12.98}$$

where R_{sub} is the substrate resistance. This change in the substrate potential slightly forward biases the source-substrate junction; a current flows, which must be added to the total drain current of the device. This behavior is also included in BSIM2 and BSIM3; in BSIM2 [7], this phenomenon is referred to as the *hot electron induced change in the effective substrate potential*, while in BSIM3 [6, 8] it is called the *substrate current induced body effect*.

The MOS Model 9 description of the weak avalanche current is

$$\begin{aligned} I_{avl} &= \quad\quad\quad 0 \quad\quad\quad\quad\quad\quad V_{ds} \leq V_{dsa} \\ &= I_{dso} \cdot A1 \cdot exp\left(-\frac{A2}{V_{ds} - V_{dsa}}\right), \quad V_{ds} > V_{dsa} \end{aligned} \tag{12.99}$$

where I_{dso} is given by (12.97) and V_{dsa} is the drain bias for the onset of the weak avalanche current, which is computed using

$$V_{dsa} = A3 \cdot V_{dsat}. \tag{12.100}$$

In bulk MOSFETs, this current is not large (usually no more than about 50 nA), so its effect on the total drain current is negligible. However, its effect on the output conductance behavior can be significant, as is shown below.

The Final Drain Current Expression

The final MOS Model 9 drain current expression is

$$I_{ds} = I_{dso} + I_{avl}, \tag{12.101}$$

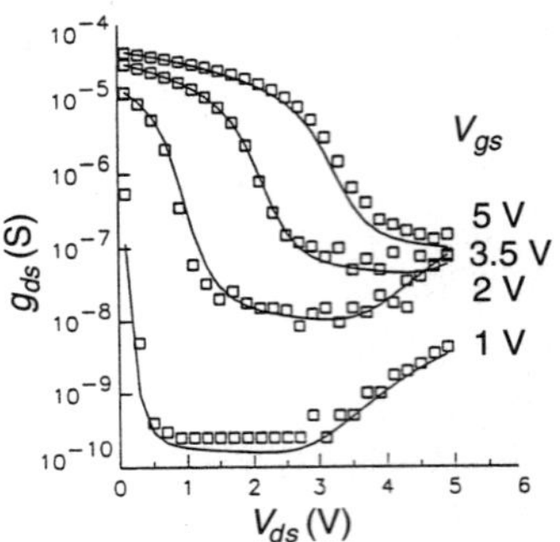

Figure 12.6 The output conductance including the weak avalanche contribution to the drain current. For high drain biases with a low gate bias, the output conductance begins to increase with an increasing drain bias. From [1]. © 1994. Philips Electronics N.V. Used by permission.

where I_{dso} is computed using (12.97) and I_{avl} is given by (12.99). The weak avalanche current is not large, but can be important to the output conductance for high drain biases with a gate bias that is only slightly larger than the threshold voltage. This is shown in Figure 12.6, where the output conductance actually begins to increase for an increasing drain bias when the gate bias is low.

12.5.5 Temperature and Geometry Dependence

Temperature and geometry dependence are introduced into MOS Model 9 in a manner consistent with the second-generation approach, as discussed in Section 12.2. However, the application of this method to the individual parameters is carried out in a very specific fashion. A variety of geometry dependencies is introduced, as is discussed below.

Earlier, it was noted that the current prefactor term **BET** was converted into a model parameter with no geometry dependence, so that all geometry dependence could be removed from the core model equations and instead isolated in the extrinsic model structure. Here, that geometry dependence is re-introduced. However, temperature dependence is first introduced in the form

$$\beta(T) = \mathbf{BET} \cdot \left(\frac{T_{nom}}{T}\right)^{\mathbf{ETABET}}. \tag{12.102}$$

Careful examination of this expression indicates that this is a description of the temperature dependence of the low field mobility, and that **ETABET** is the exponent which describes that temperature dependence; thus **ETABET** is identical in function to the HSPICE parameter **BEX** and the BSIM3 parameter **UTE**. The geometry dependence is then introduced in a very simple fashion:

$$BET = \beta(T) \cdot \frac{W_{eff}}{L_{eff}}; \tag{12.103}$$

this is consistent with the form of the current prefactor in the generic drain current equation discussed throughout this text.

The channel length modulation description has geometry dependence, which is introduced in an unusual form:

$$ALP = \textbf{ALP} + \textbf{SLALP} \cdot \left(\frac{1}{L_{eff}^{\textbf{ETAALP}}} - \frac{1}{L_{eff,ref}^{\textbf{ETAALP}}} \right) + \textbf{SWALP} \cdot \left(\frac{1}{W_{eff}} - \frac{1}{W_{eff,ref}} \right). \qquad (12.104)$$

The parameter **ETAALP** is introduced to allow ALP to have different geometry dependence for nFETs and pFETs. It is suggested [1] that a value of **ETAALP** $= 0$ be used for nFETs and that a value of **ETAALP** $= 1$ be used for pFETs; note that this implies that ALP will have length dependence for pFETs but not for nFETs. It is also possible to extract **ETAALP**; however, it has been observed that this can lead to strange model results [1], and is not recommended. Also the extraction of an arbitrary (noninteger) value for **ETAALP** will cause (12.104) to be less efficient during circuit simulation.

In addition,

$$VP = \textbf{VP} \cdot \left(\frac{L_{eff}}{L_{eff,ref}} \right). \qquad (12.105)$$

The DIBL parameter $GAMOO$ is allowed to have length dependence but not width dependence; this length dependence is taken to be quadratic. Thus,

$$GAMOO = \textbf{GAMOO} + \textbf{SLGAMOO} \cdot \left(\frac{1}{L_{eff}^{2}} - \frac{1}{L_{eff,ref}^{2}} \right). \qquad (12.106)$$

The subthreshold slope parameter MO, not unexpectedly, has both temperature and geometry dependence. The temperature dependence is introduced in a form consistent with that for the threshold voltage, as described by (12.42):

$$m_o(T) = \textbf{MO} \cdot \textbf{STMO} \cdot (T - T_{nom}), \qquad (12.107)$$

where the temperatures must be specified in **degrees Kelvin**. The geometry dependence of $m_o(T)$ is taken to be affected only by the channel length; this is not a surprising assumption, as the width of a device has very little effect on the subthreshold slope. This behavior is deemed to have square root dependence:

$$MO = m_o(T) + \textbf{SLMO} \cdot \left(\frac{1}{L_{eff}^{\frac{1}{2}}} - \frac{1}{L_{eff,ref}^{\frac{1}{2}}} \right). \qquad (12.108)$$

The weak inversion correction parameter $ZET1$ has length dependence in a form very similar to that of the channel length modulation parameter ALP:

$$ZET1 = \mathbf{ZET1} + \mathbf{SLZET1} \cdot \left(\frac{1}{L_{eff}^{\mathbf{ETAZET}}} - \frac{1}{L_{eff,ref}^{\mathbf{ETAZET}}} \right) \cdot \tag{12.109}$$

Unlike ALP, $ZET1$ has no width dependence. Like $\mathbf{ETAALP}$, the parameter $\mathbf{ETAZET}$ is introduced to allow $ZET1$ to have different geometry dependence for nFETs and pFETs. Suggested values [1] are $\mathbf{ETAZET} = 0.5$ for nFETs and $\mathbf{ETAZET} = 1$ for pFETs. It is also possible to extract $\mathbf{ETAZET}$ from data; however, this is not recommended [1], as it can lead to strange model results. The extraction of an arbitrary (noninteger) value for $\mathbf{ETAZET}$ will also cause (12.109) to be less efficient during circuit simulation.

Finally, the weak avalanche current parameters can be considered. Of these three parameters ($A1$, $A2$, and $A3$), only $A1$ is deemed to contain temperature dependence. This is introduced in the standard MOS Model 9 form:

$$a_1(T) = \mathbf{A1} + \mathbf{STA1} \cdot (T - T_{nom}), \tag{12.110}$$

where the temperatures are specified in degrees Kelvin. The geometry dependence is then introduced in the basic second-generation form:

$$A1 = a_1(T) + \mathbf{SLA1} \cdot \left(\frac{1}{L_{eff}} - \frac{1}{L_{eff,ref}} \right) + \mathbf{SWA1} \cdot \left(\frac{1}{W_{eff}} - \frac{1}{W_{eff,ref}} \right). \tag{12.111}$$

The remaining two parameters ($A2$ and $A3$) do not have temperature dependence, but have geometry dependence in the standard second-generation form:

$$A2 = \mathbf{A2} + \mathbf{SLA2} \cdot \left(\frac{1}{L_{eff}} - \frac{1}{L_{eff,ref}} \right) + \mathbf{SWA2} \cdot \left(\frac{1}{W_{eff}} - \frac{1}{W_{eff,ref}} \right), \tag{12.112}$$

and

$$A3 = \mathbf{A3} + \mathbf{SLA3} \cdot \left(\frac{1}{L_{eff}} - \frac{1}{L_{eff,ref}} \right) + \mathbf{SWA3} \cdot \left(\frac{1}{W_{eff}} - \frac{1}{W_{eff,ref}} \right). \tag{12.113}$$

12.6   The Charge Model

Like the rest of the model, the MOS Model 9 node charge description is quite simple and produces clean expressions for the node charges. The development also recognizes a proper approach to the node charge/active gate capacitance problem, which is considered in detail in Chapter 13. Only charge expressions are derived, reflecting the proper level of detail for communication with the circuit simulator (see Chapter 13). The derivation is also carried out in a manner which ensures that continuous and smooth expressions for the node charges are produced; this improves both the charge model accuracy and the behavior of circuit simulation using these expressions. Finally, only a simple quasi-static model for the node charges is developed; as discussed in Chapter 13, this type of model will be valid for the vast majority of MOSFET simulation situations.

12.6.1   The Basic Method

Like all the other charge models, the MOS Model 9 development is based on charge conservation in the form

$$Q_{GATE} + Q_{DEPL} + Q_{INV} = 0, \qquad (12.114)$$

where the capitalized subscripts indicate *total* charge (rather than charge per unit area). The depletion charge Q_{DEPL} and the inversion charge Q_{INV} are computed from the model equations; the gate charge Q_{GATE} is then found by solving

$$Q_{GATE} = -(Q_{DEPL} - Q_{INV}). \qquad (12.115)$$

In addition, it will be necessary to split the inversion charge into node charges for the source and the drain:

$$Q_{INV} = Q_S + Q_D; \qquad (12.116)$$

this situation is considered in detail in Section 12.6.4.

12.6.2   Additional Expressions

The development of the charge model introduces a number of additional expressions into the structure of MOS Model 9. For convenience, these will be discussed here, before the main parts of the charge model are considered. These additions are due to the need to use the drain terminal as a reference in the charge model development, and by a need to modify the saturation voltage expression that was originally derived with the current model.

The Drain Terminal as a Reference

The drain terminal must be used as a reference to describe the various charge/voltage relationships involving the drain. These equations are easily created by replacing the source with the drain in the expressions derived earlier with the source as the reference.
    The *drain version* of V_{bs} is

$$V_{bd} = -V_{ds} + V_{bs}. \qquad (12.117)$$

There is also a need for a drain version of the auxiliary function h_1 (12.20):

$$h_2 = hyp1\left(-V_{bd} + \frac{PHIB}{2}, \epsilon_1\right) + \frac{PHIB}{2}, \qquad (12.118)$$

where ϵ_1 is given by (12.21) and $hyp1$ is a hyperbolic smoothing function computed using (12.22). In addition, there is a drain version of ΔV_{to} ((12.17)), the shift in the threshold voltage due to the substrate bias:

$$\Delta V_{to,d} = K \cdot \left\{ \left[hyp4(-V_{bd}, VSBX, \epsilon_2) + \left(\frac{K}{KO}\right)^2 \cdot (PHIB + VSBX) \right]^{\frac{1}{2}} \right.$$

$$\left. - \left(\frac{K}{KO}\right) \cdot (PHIB + VSBX)^{\frac{1}{2}} \right\}$$

$$+ KO\left\{ [h_1 - hyp4(-V_{bd}, VSBX, \epsilon_2)]^{\frac{1}{2}} - PHIB^{\frac{1}{2}} \right\}, \tag{12.119}$$

where ϵ_2 is given by (12.18) and $hyp4$ is the hyperbolic smoothing function described by (12.19). Finally, there is a drain version of V_{t1} ((12.14)), the threshold voltage as affected by the applied substrate bias:

$$V_{t1,d} = VTO + \Delta V_{to,d}, \tag{12.120}$$

where VTO is a model parameter representing the threshold voltage for very low substrate and drain biases.

The Saturation Voltage

The term δ_1, as defined by (12.80), appears in the saturation voltage expression (12.79) and in the equation for I_{dso} (12.96). However, it has been found [1] that the derivatives $\frac{dQ_S}{dV_D}$ and $\frac{dQ_D}{dV_S}$ will contain a discontinuity at $V_{bs} = V_{bd}$ if δ_1 is used in the charge model. To solve this problem, δ_1 is replaced in the charge model with

$$\delta_2 = -\frac{\partial V_{t2}}{\partial V_{bs}} - \frac{\partial V_{t2}}{\partial V_{gs}} - \frac{\partial V_{t2}}{\partial V_{ds}}, \tag{12.121}$$

where V_{t2}, given by (12.34), is the final threshold voltage, including the effects of both the substrate bias and the drain bias. Replacing δ_1 in the saturation voltage expression (12.79) produces the version that is used in the charge model:

$$V_{dsat,q} = \frac{V_{gt3}}{1 + \delta_2} \cdot \frac{2}{1 + \left(1 + \frac{2 \cdot THE3 \cdot V_{gt3}}{1 + \delta_2}\right)^{\frac{1}{2}}}. \tag{12.122}$$

As an additional result of this change, it becomes necessary to compute a charge model version of the auxiliary drain bias function V_{ds1}, computed using (12.83); this is the function which guarantees both a continuous and smooth linear region-saturation region transition and continuous and smooth output conductance results. Replacing V_{dsat} in (12.83) with $V_{dsat,q}$ leads to the auxiliary drain bias function that is used in the charge model:

$$V_{ds2} = hyp5(V_{ds}, V_{dsat,q}, \epsilon_7), \tag{12.123}$$

where ϵ_7 is simply the earlier expression ϵ_5 with $V_{dsat,q}$ replacing V_{dsat}:

$$\epsilon_7 = \lambda_8 \cdot \frac{V_{dsat,q}}{1 + V_{dsat,q}}, \tag{12.124}$$

with

$$\lambda_8 = 0.1; \tag{12.125}$$

note that this value of λ_8 is different from the value of the equivalent term λ_6, which appears in (12.85). The term $hyp5$ is the hyperbolic smoothing function given by (12.86).

12.6.3   The Depletion Charge

The depletion charge is computed in two parts, using both the source and the drain as separate references; the total depletion charge is computed from an average of the two values found. Hyperbolic smoothing functions are also used here to produce continuous and smooth results for all regions of device operation.

Additional Terms

To begin, some terms and expressions must be defined. The gate-substrate bias is simply

$$V_{gb} = V_{gs} - V_{bs}. \tag{12.126}$$

An expression for the flatband voltage will also be required, to separate the different operating regions of the device. This is found beginning with the three-terminal MOS threshold voltage expression (3.113):

$$V_t = V_{fb} + PHIB + \gamma \cdot (PHIB - V_{bs})^{\frac{1}{2}}. \tag{12.127}$$

As in Section 12.3, V_{to} is defined as the threshold voltage when no substrate bias is applied; thus, (12.127) becomes

$$V_{t,V_{bs}=0} = V_{to} = V_{fb} + PHIB + \gamma \cdot PHIB^{\frac{1}{2}}. \tag{12.128}$$

The flatband voltage is then determined directly from (12.128):

$$V_{fb} = V_{to} - PHIB - \gamma \cdot PHIB^{\frac{1}{2}}. \tag{12.129}$$

Substituting the appropriate model parameters into this expression produces the final expression:

$$V_{fb} = VTO - PHIB - KO \cdot PHIB^{\frac{1}{2}}. \tag{12.130}$$

Note that only the surface doping needs to be considered to determine the flatband voltage; hence, KO (rather than K or some hybrid combination) replaces γ in (12.130).

The Depletion Charge with the Source as the Reference

The depletion charge with the source as the reference is computed using a basic solution to the Poisson equation [1, 3]. In accumulation, as defined by $V_{gb} < V_{fb}$,

$$Q_{DEPL,s} = -C_{ox} \cdot (V_{gb} - V_{fb}). \tag{12.131}$$

In weak inversion, defined by $V_{fb} \leq V_{gb} \leq -V_{bs} + V_{t1}$, the inversion charge is negligible compared to the depletion charge; the Poisson equation solution yields

$$Q_{DEPL,s} = -C_{ox} \cdot KO \cdot \left\{ -\frac{KO}{2} + \left[\frac{KO^2}{4} + (V_{gb} - V_{fb}) \right]^{\frac{1}{2}} \right\}. \tag{12.132}$$

Note that at the boundary $V_{gb} = V_{fb}$, $Q_{DEPL,s} = 0$ in both (12.131) and (12.132); thus, the continuity of $Q_{DEPL,s}$ at this boundary is assured. Also note that since the gate bias is low, it is assumed that the depletion region expands only into the surface doping region; only the surface doping parameter KO needs to be included.

In strong inversion, defined by $V_{gb} > -V_{bs} + V_{t1}$, the depletion charge is screened by the inversion layer and no longer changes with a changing gate-substrate bias V_{gb}. In this circumstance, the solution to the Poisson equation results in

$$Q_{DEPL,s} = -C_{ox} \cdot KO \cdot \left\{ -\frac{KO}{2} + \left[\frac{KO^2}{4} + (-V_{bs} + V_{t1} - V_{fb}) \right]^{\frac{1}{2}} \right\}. \tag{12.133}$$

Note that at the boundary $V_{gb} = -V_{bs} + V_{t1}$, (12.133) becomes identical to (12.132), which ensures the continuity of $Q_{DEPL,s}$ across that boundary. However, the derivative $\frac{\partial Q_{DEPL,s}}{\partial V_{gb}}$ is **not** continuous there. Therefore, it is necessary to replace (12.132) and (12.133) with a modified expression, using hyperbolic smoothing functions, which guarantees a continuous and smooth transition. In addition, (12.131) must be replaced with a similarly compatible expression.

For $V_{gb} < V_{fb}$, (12.131) is replaced with

$$Q_{DEPL,s} = -C_{ox} \cdot hyp3(V_{gb} - V_{fb}, -V_{bs} + V_{t1}, \epsilon_6), \tag{12.134}$$

where

$$\epsilon_6 = 0.03, \tag{12.135}$$

$hyp3$ is a hyperbolic smoothing function, computed using

$$hyp3 = hyp2(x, x_o, \epsilon) - hyp2(0, x_o, \epsilon), \tag{12.136}$$

and $hyp2$ is another hyperbolic smoothing function, given by (12.32).

For $V_{gb} \geq V_{fb}$, (12.132) and (12.133) are replaced by a single equation:

$$Q_{DEPL,s} = -C_{ox} \cdot KO$$

$$\cdot \left\{ -\frac{KO}{2} + \left[\left(\frac{KO}{2} \right)^2 + hyp3(V_{gb} - V_{fb}, -V_{bs} + V_{t1} - V_{fb}, \epsilon_6) \right]^{\frac{1}{2}} \right\}. \quad (12.137)$$

The use of (12.134) and (12.137) guarantees that $Q_{DEPL,s}$ will be continuous and smooth across all operating regions, and that all the first derivatives will be continuous at the defined boundaries.

The Depletion Charge with the Drain as the Reference

Like the source development, the computation of the depletion charge with the drain as the reference employs a basic solution to the Poisson equation [1, 3]. In accumulation ($V_{gb} < V_{fb}$), the drain bias is unimportant, and a solution identical to the source expression (12.131) is reached:

$$Q_{DEPL,d} = -C_{ox} \cdot (V_{gb} - V_{fb}). \quad (12.138)$$

In a manner identical to that used for the source solution, weak and strong inversion regions are defined, and an expression for $Q_{DEPL,d}$ is determined for each region. However, the simple source expression V_{bs} is replaced with the more complicated expression for V_{bd}, as described in (12.117). For use in the charge model, it is appropriate to replace V_{ds} in (12.117) with the auxiliary drain bias function V_{ds2}, which is described by (12.123). Thus, (12.117) becomes

$$V_{bd} = -V_{ds2} + V_{bs}; \quad (12.139)$$

this expression will be used in the drain-related computations.

In weak inversion ($V_{fb} \leq V_{gb} \leq -V_{bd} + V_{t1,d}$), the inversion charge is vanishingly small; since the drain bias has no effect on the depletion charge, a solution identical to the source expression (12.132) is found:

$$Q_{DEPL,d} = -C_{ox} \cdot KO \cdot \left\{ -\frac{KO}{2} + \left[\frac{KO^2}{4} + (V_{gb} - V_{fb}) \right]^{\frac{1}{2}} \right\}. \quad (12.140)$$

Note that (12.138) and (12.140) are continuous at $V_{fb} = V_{gb}$, as both expressions reduce to $Q_{DEPL,d} = 0$ at that point.

In strong inversion ($V_{gb} > -V_{bd} + V_{t1,d}$), the depletion charge is screened by the inversion layer and is no longer affected by the gate-substrate bias. This results in a solution which is a slightly modified form of the source expression (12.133):

$$Q_{DEPL,d} = -C_{ox} \cdot KO \cdot \left\{ -\frac{KO}{2} + \left[\frac{KO^2}{4} + (-V_{bd} + V_{t1,d} - V_{fb}) \right]^{\frac{1}{2}} \right\}. \quad (12.141)$$

Note that (12.140) and (12.141) are continuous across the boundary $V_{gb} = -V_{bd} + V_{t1,d}$. However, the derivative $\frac{\partial Q_{DEPL,d}}{\partial V_{gb}}$ is **not** continuous there. Thus, as in the source derivation, (12.140) and (12.141) must be replaced with an expression which solves this problem. It is also necessary to replace (12.138) with a compatible expression.

The resulting expressions are virtually identical to those developed using the source as the reference. For $V_{gb} < V_{fb}$, (12.138) is replaced with

$$Q_{DEPL,d} = -C_{ox} \cdot hyp3(V_{gb} - V_{fb}, -V_{bd} + V_{t1,d}, \epsilon_6), \quad (12.142)$$

where ϵ_6 is given by (12.135) and $hyp3$ is the hyperbolic smoothing function described by (12.136). For $V_{gb} \geq V_{fb}$, (12.140) and (12.141) are replaced by a single equation:

$$Q_{DEPL,d} = -C_{ox} \cdot KO$$

$$\cdot \left\{ -\frac{KO}{2} + \left[\left(\frac{KO}{2}\right)^2 + hyp3(V_{gb} - V_{fb}, -V_{bd} + V_{t1,d} - V_{fb}, \epsilon_6) \right]^{\frac{1}{2}} \right\}. \quad (12.143)$$

Together, (12.142) and (12.143) produce continuous and smooth results for $Q_{DEPL,d}$, and ensure that all the first derivatives are continuous at the region boundaries.

The Total Depletion Charge

As noted earlier, the total depletion charge is computed by averaging the computed values of $Q_{DEPL,s}$ and $Q_{DEPL,d}$:

$$Q_{DEPL} = \frac{1}{2} \cdot (Q_{DEPL,s} + Q_{DEPL,d}), \quad (12.144)$$

where $Q_{DEPL,s}$ is found from either (12.134) or (12.137) and $Q_{DEPL,d}$ is found from either (12.142) or (12.143).

12.6.4   The Inversion Charge

The other part of the charge in the device is the inversion charge Q_{INV}. This charge must be split into a source charge Q_S and a drain charge Q_D, as described in (12.116):

$$Q_{INV} = Q_S + Q_D. \quad (12.145)$$

MOS Model 9 avoids the explicit assignment of a partition fraction, commonly employed in other models [9, 10], by the use of a physical method.

The inversion charge is computed from

$$Q_{INV} = -W_{eff} \cdot \int_0^{L_{eff}} q \cdot Q_{inv}(y) \cdot dy, \tag{12.146}$$

where $Q_{inv}(y)$ is the inversion charge per unit area along the channel; this common approach assumes that $Q_{inv}(y)$, to a reasonable approximation, is uniform across the width of the channel. To describe Q_S and Q_D, (12.146) is divided into two parts:

$$Q_S = -q \cdot W_{eff} \cdot \int_0^{L_{eff}} Q_{inv}(y) \cdot \frac{y}{L_{eff}} \cdot dy \tag{12.147}$$

and

$$Q_D = -q \cdot W_{eff} \cdot \int_0^{L_{eff}} Q_{inv}(y) \cdot \left(1 - \frac{y}{L_{eff}}\right) \cdot dy, \tag{12.148}$$

where y is an arbitrary *partition point*, located somewhere along the channel. MOS Model 9 eliminates y by changing the variable of integration from y to the lateral potential V, and by employing the current equation [3]

$$I_{ds} \cdot y = -q \cdot W_{eff} \cdot \int_0^V \mu_{eff} \cdot Q_{inv}(V) \cdot dV. \tag{12.149}$$

The solutions to (12.147) and (12.148) are

$$Q_S = -C_{ox} \left[\frac{1}{2} \cdot V_{gt3} + (1 + \delta_2) \cdot V_{ds2} \cdot \left(\frac{1}{12} \cdot F_j - \frac{1}{60} \cdot F_j^2 - \frac{1}{6} \right) \right] \tag{12.150}$$

and

$$Q_D = -C_{ox} \left[\frac{1}{2} \cdot V_{gt3} + (1 + \delta_2) \cdot V_{ds2} \cdot \left(\frac{1}{12} \cdot F_j + \frac{1}{60} \cdot F_j^2 - \frac{1}{3} \right) \right], \tag{12.151}$$

where F_j is an auxiliary function computed using

$$F_j = \frac{(1 + \delta_2) \cdot (1 + THE3 \cdot V_{ds2}) \cdot V_{ds2}}{2 \cdot V_{gt3} - (1 + \delta_2) \cdot V_{ds2}}, \tag{12.152}$$

and V_{gt3} is the auxiliary gate bias function given by (12.75) and (12.77). Note that when $V_{ds} = 0$ (and thus $V_{ds2} = 0$), (12.150) and (12.151) reduce to

$$Q_S = Q_D = -\frac{1}{2} \cdot C_{ox} \cdot V_{gt3}. \tag{12.153}$$

Here the reason for the addition of the small constant λ_3 to the expressions for V_{gt3}, discussed briefly in Section 12.5, becomes clear. Without the small constant, when

$V_{gt3} \to 0$, the denominator of F_j also goes to zero; this leads to division by zero. A small constant could be added directly to the denominator of F_j; however, it has been shown [1] that this will lead to discontinuities in the first derivatives of Q_S and Q_D. As an alternative, a small constant is added to V_{gt3}; that modification to V_{gt3} is used throughout the model. Therefore, as described in Section 12.5, V_{gt3} is computed using

$$
\begin{aligned}
V_{gt3} &= 2 \cdot m \cdot \frac{k_b T}{q} \cdot ln(1 + G_1) + \lambda_3 \qquad V_{gt2} < V_{gta} \\
&= V_{gt2} + \lambda_3 \qquad V_{gt2} \geq V_{gta},
\end{aligned}
\tag{12.154}
$$

where

$$
\lambda_3 = 1 \times 10^{-8};
\tag{12.155}
$$

the remaining terms and their origins are discussed in Section 12.5.

As derived, with the auxiliary gate bias function V_{gt3} present, (12.150) and (12.151) provide expressions for Q_S and Q_D which are valid for all regions of device operation. However, as $V_{ds} \to 0$, there is a discontinuous transition between the derivatives $\frac{dQ_S}{dV_D}$ and $\frac{dQ_D}{dV_S}$ [1]. This problem is solved by replacing the term $(1 + \delta_2)$ in (12.150) and (12.151) with

$$
\Delta_2 = \frac{\partial V_{gt3}}{\partial V_{bs}} + \frac{\partial V_{gt3}}{\partial V_{gs}} + \frac{\partial V_{gt3}}{\partial V_{ds}}.
\tag{12.156}
$$

With this substitution, the final MOS Model 9 expressions for the source and drain charges are

$$
Q_S = -C_{ox} \left[\frac{1}{2} \cdot V_{gt3} + \Delta_2 \cdot V_{ds2} \cdot \left(\frac{1}{12} \cdot F_j - \frac{1}{60} \cdot F_j^2 - \frac{1}{6} \right) \right]
\tag{12.157}
$$

and

$$
Q_D = -C_{ox} \left[\frac{1}{2} \cdot V_{gt3} + \Delta_2 \cdot V_{ds2} \cdot \left(\frac{1}{12} \cdot F_j + \frac{1}{60} \cdot F_j^2 - \frac{1}{3} \right) \right].
\tag{12.158}
$$

Once again, in the limit of $V_{ds} = 0(V_{ds2} = 0)$, (12.157) and (12.158) simplify to

$$
Q_S = Q_D = -\frac{1}{2} \cdot C_{ox} \cdot V_{gt3}.
\tag{12.159}
$$

12.6.5   The Zero Bias Gate Capacitance

Like the other models, MOS Model 9 includes an offset to accommodate the zero bias gate-drain and gate-source capacitances. However, MOS Model 9 assumes that the source and drain are identical, and that separate expressions (embodied in the earlier models as **CGDO** and **CGSO**) are not necessary; this assumption is quite reasonable in modern self-aligned source/drain technology. The zero bias gate-source and gate-drain capacitance are computed from

$$C_{gd,z} = C_{gs,z} = W_{eff} \cdot \mathbf{COL}, \tag{12.160}$$

where **COL** is the gate-diffusion capacitance per unit width of the polysilicon gate. A shortcoming of this simple (and commonly used) approach is the neglect of the screening of the inner-fringe capacitance by the inversion layer; this issue becomes important in short channel devices, as discussed in detail in Chapter 13.

12.7   Final Comments

Like BSIM3, MOS Model 9 is a very new model which has yet to be extensively exercised in both parameter extraction and circuit simulation. It is therefore not yet possible to give a detailed analysis of its behavior in active use.

However, MOS Model 9 has a number of important positive points in its favor. The clean and simple formulation of the equations suggests that it will be very efficient in circuit simulation. In addition, the relatively small number of parameters indicates that parameter extraction will be quite simple. Finally, the model is unique among mainstream models at this time in having an industrial (versus an academic) heritage. This indicates that MOS Model 9 has been used extensively in an industrial circuit simulation environment, and has proven itself to be tractable for that purpose.

Finally, the industrial heritage of MOS Model 9 may be a sign that a new era in SPICE FET models is about to begin. It appears that FET models will enter common usage from a wide variety of sources; the door is now open to a number of new models (see Chapter 16 for a brief discussion of several possible candidate models). It remains to be seen which models will gain widespread acceptance and find common usage. However, it is important to note that any particular CMOS process technology may require a growing portfolio of FET model types.

BIBLIOGRAPHY

1. R. Velghe, D. Klaassen, and F. Klaassen, "MOS Model 9," Unclassified Report NL-UR 003/94, Philips Electronics N.V. (1994).

2. R. Velghe, D. Klaassen, and F. Klaassen, "Compact MOS Modeling for Analog Circuit Simulation," *1993 IEDM Tech. Dig.*, pp. 485–488.

3. H. deGraaff and F. Klaassen, *Compact Transistor Modeling for Circuit Design*, Springer-Verlag, 1990.

4. H. Masuda, M. Nakai, and M. Kubo, "Characteristics and Limitation of Scaled-Down MOSFETs Due to Two-Dimensional Field Effect," *IEEE Trans. Elec. Dev.* vol. ED-26, pp. 980–986 (1979).

5. H. Oguey and S. Cserveny, "Modhle du Transistor MOS Valable dans un Grand Domaine de Courants," *Bull. SEV/VSE*, February 1982.

6. Y. Cheng et al., "BSIM3 Version 3.0 Manual," University of California/Berkeley, Electronics Research Laboratory (1995).

7. M. Jeng, "Design and Modeling of Deep Submicrometer MOSFETs," University of California/Berkeley, Electronics Research Laboratory Memorandum No. UCB/ERL M90/90 (1990).

8. J. Huang et al., "BSIM3 Manual (Version 2.0)," University of California/Berkeley, Electronics Research Laboratory (1994).

9. D. Ward and R. Dutton, "A Charge-Oriented Model for MOS Transistor Capacitances," *IEEE J. Sol. St. Circ.* vol. SC-13, pp. 703–708 (1978).

10. B. Sheu, "MOS Transistor Modeling and Characterization for Circuit Simulation," University of California/Berkeley, Electronics Research Laboratory Memorandum No. UCB/ERL M85/85 (1985).

13

The Active Device Capacitance

13.1 Introduction

As noted during the descriptions of the SPICE FET models, each FET model suggests a set of equations which describes the total charge found at the four MOSFET nodes. Using these charges to compute capacitances, in particular the capacitance between the gate and the other device terminals, was noted to be a twofold problem, involving the charge model associated with each FET model, and the implementation of those charge models at the circuit simulation level for the actual capacitance calculation.

This latter problem, the computation of the gate capacitance by the circuit simulator, will be considered here. As hinted at earlier and discussed more fully in this chapter, an accurate FET charge model is required for an accurate capacitance computation; however, this represents a necessary but not sufficient condition. In particular, if the circuit node voltage (rather than the circuit node charge) is used as the state variable in the circuit matrix solution, the node charges are computed by an integration that imparts path dependence to the solution for the charge. This approximation can cause FET charge neutrality to be violated, a problem that is commonly referred to as the failure to conserve charge.

The gate capacitance models and the various difficulties with charge conservation will be considered here. In addition, all of the models currently used in SPICE assume that the charge can be computed from the voltage in a quasi-static manner; that is, the charge is taken to be an instantaneous function of the voltage. This assumption has its limits, which will also be considered here. The objective of this chapter is not to examine the circuit simulation implementation of gate capacitance models in great detail. Instead, the intention is to allow the FET model user to develop a clear view of what requirements must be imposed on the circuit simulation methods if accurate results are to be achieved.

13.2 The Zero Bias Gate Capacitance

Before examining the active device capacitance models in full, the zero bias capacitance must be described. In very large FETs, those used when SPICE FET models were first introduced, the active gate capacitance is much larger than any zero bias capacitance; thus, the zero bias gate capacitance could be neglected. As noted in Chapter 5, this assumption was used when the original Level 1 model was developed. As FET sizes decreased, the zero bias capacitance was no longer negligible, and had to be included in the FET models.

Consider the gated FET diffusion depicted in Figure 13.1. With no biases applied, there is no inversion layer present. However, there is a capacitance between the gate and the diffusion, which is divided here into three subcomponents. The term C_1 is the capacitance between the gate and the diffusion through the gate oxide, and is quite simply expressed using the oxide capacitance. The term C_2 is a fringe capacitance between the edge of the gate and the diffusion; some more sophisticated treatments even include the capacitance between the *top* of the gate and the diffusion in C_2 [1]. The term C_3 is another fringing capacitance, between the gate and the inner side of the diffusion.

A derivation of C_2 and C_3 is somewhat complex and approximate, due to the nonplanar geometries associated with those terms. A workable expression for the capacitance between the gate and the diffusion under the zero bias condition is [2]

$$C_{gdo} = C_1 + C_2 + C_3 = C_{ox} \cdot \mathbf{LD} + \epsilon_{ox} \cdot ln\left(1 + \frac{t_{poly}}{t_{ox}}\right) + \frac{2\epsilon_{Si}}{\pi} \cdot ln\left(1 + \frac{x_j}{t_{ox}}\right), \quad (13.1)$$

where C_{gdo} is the capacitance per unit length of gate *width* (the axis normal to the page in Figure 13.1), $\mathbf{LD}$ is the SPICE model parameter representing the length of underdiffusion, t_{poly} is the thickness of the gate polysilicon, and x_j is the junction depth. The form of (13.1) assumes that the gate sidewalls were etched vertically with no sloping. In modern self-aligned CMOS technology, the structure of the gate-diffusion region is identical for the source and the drain, so this discussion applies equally at both ends of the device.

For a modern FET, some typical values of the physical terms in (13.1) are $t_{ox} = 10$ nm, $\mathbf{LD} = 0.05$ μm, $x_j = 0.1$ μm, and $t_{poly} = 300$ nm. Substituting these values into (13.1) leads to

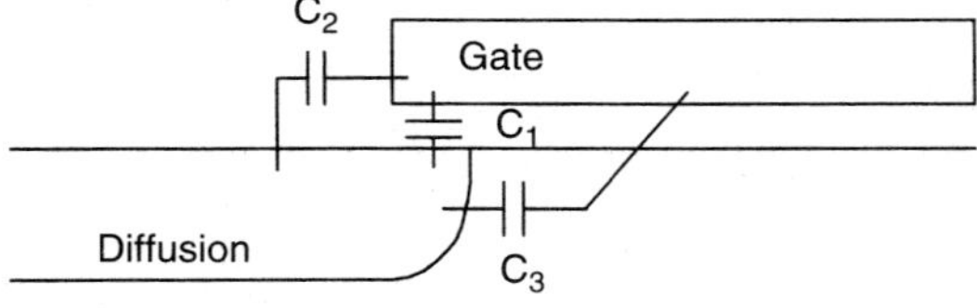

Figure 13.1 The gated FET diffusion structure, showing the direct gate-diffusion capacitance (C_1), the outer-fringe capacitance (C_2), and the inner-fringe capacitance (C_3).

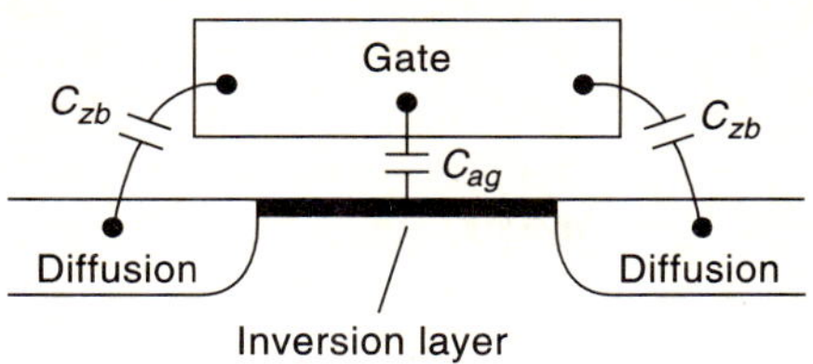

Figure 13.2 The total active gate capacitance, comprised of the sum of the active gate capacitance and the zero bias capacitance.

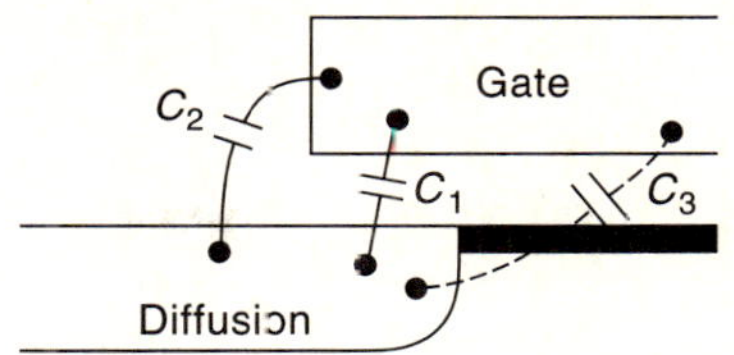

Figure 13.3 The screening out of the inner-fringe capacitance (C_3) by the inversion layer.

$$C_{gdo} = C_{gso} = 0.17\,f\text{F}/\mu\text{m} + 0.11\,f\text{F}/\mu\text{m} + 0.16\,f\text{F}/\mu\text{m} = 0.44\,f\text{F}/\mu\text{m}, \qquad (13.2)$$

where, as noted above, the final result is expressed as the capacitance per unit width of the gate (the axis pointing out of the page in Figure 13.1). These terms appear in SPICE as the zero bias gate-diffusion capacitance parameters **CGDO** and **CGSO**, which are implemented in units of F/m. Thus,

$$\textbf{CGDO} = \textbf{CGSO} = 4.4 \times 10^{-10}(\text{F/m}). \qquad (13.3)$$

For the SPICE implementation (Figure 13.2), the zero bias capacitance is treated as a constant over the range of applied biases, and is merely added to the active gate capacitance, which does vary with the applied biases (as discussed below). Again assuming that the source and the drain regions of the device are identical, the total gate capacitance as computed in SPICE is

$$C_{GATE,total} = C_{ag} + C_{zb} = C_{ag} + 2 \cdot (C_{gdo} \cdot W_{eff}), \qquad (13.4)$$

where C_{ag} is the active gate capacitance for some particular combination of applied voltages and C_{zb} is the zero bias gate capacitance; the gate-diffusion capacitance has been multiplied by the effective channel width to yield the *total* zero bias gate capacitance.

There is a major problem with this approach of computing a zero bias capacitance and adding it onto the active gate capacitance as a constant. As shown in Figure 13.3, the fringing term C_3 is screened out by the inversion layer, so that $C_3 = 0$ when the inversion layer is present. When this is taken into account, the computation of the gate-diffusion capacitance (13.2) with an inversion layer present should be

$$C_{gdo} = C_1 + C_2 = 0.17\,f\text{F}/\mu\text{m} + 0.11\,f\text{F}/\mu\text{m} = 0.28\,f\text{F}/\mu\text{m}. \qquad (13.5)$$

When compared with the result of (13.2), it is found that retaining the C_3 term when an inversion layer is present introduces an error of about 36% into C_{gdo}. Although this percentage may vary somewhat for various fabrication technologies, the overall effect, the introduction of a large error, is clear.

The effect of this omission in the gate-diffusion capacitance model must be examined more closely. Consider a rather large FET, with channel width and length of 100 μm; these are typical dimensions for an FET of the mid 1960s, when SPICE FET models were first introduced. To a very reasonable approximation, when an inversion layer is present, the active gate capacitance of an FET is described by

$$C_{ag} = C_{ox} \cdot W_{eff} \cdot L_{eff}. \tag{13.6}$$

For $W/L = 100$ μm/100 μm, (13.6) produces $C_{ag} = 34.5$ pF, while the zero bias gate capacitance from (13.4) is $2 \cdot C_{gdo} \cdot W_{eff} = 88$ fF. The zero bias capacitance represents a correction to the active gate capacitance of less than 0.25% and can be entirely neglected, as noted as the beginning of this section.

Now consider a smaller device, with $W_{eff}/L_{eff} = 10$ μm/2 μm. In this case, $C_{ag} = 69$ fF, while the zero bias gate capacitance is $C_{zb} = 8.8$ fF. If C_{zb} is added to C_{ag}, as in (13.4), the result is $C_{GATE,total} = 77.8$ fF. The inclusion of C_{zb} now represents a 12% correction to C_{ag}; this is a significant change, and C_{zb} now cannot be neglected. If C_{zb} is computed without the C_3 term, the result is $C_{zb} = 5.6$ fF and $C_{GATE,total} = 74.6$ fF; thus, the neglect of C_3 introduces an error of about 4%, which is still not large enough to be of concern.

Finally, consider a very small present-day device, with $W_{eff}/L_{eff} = 10$ μm/0.25 μm. In this case, $C_{ag} = 8.63$ fF, which is comparable to the value (including C_3) of $C_{zb} = 8.8$ fF; the total capacitance is $C_{GATE,total} = 17.43$ fF. In this "modern" situation, it is clear that the zero bias gate capacitance is relatively large, and must be included in any description of the gate capacitance. In addition, if C_3 is properly discarded, $C_{zb} = 5.6$ fF, and $C_{GATE,total} = 14.23$ fF. This value is very different from that found by the erroneous inclusion of C_3 in the computation of C_{zb}; in fact, this inclusion of C_3 introduces an error of 22% into the calculation of the total gate capacitance.

This creates a quandry for the FET model building effort. The term C_{gdo} (implemented in SPICE as the parameters **CGDO** and **CGSO**) is supposed to represent the zero bias capacitance; however, if those values are included, a significant overprediction of the total gate capacitance at high gate biases occurs. In digital circuit design, $V_{dd} \gg V_t$, and in any transient from V_{dd} to ground (or vice versa), during most of that transient, an inversion layer will be present. (While still used, this assumption becomes questionable in newer low-power CMOS technologies, where $V_{dd} \approx 2.5$ V and $V_t \approx 0.7$ V.) An empirical solution, as shown in Figure 13.4, is to adjust **CGDO** and **CGSO** to give the best possible results for high gate biases; as a result, these parameters lose their physical definitions as zero bias gate values, and the modeled gate capacitance characteristic becomes inaccurate for low gate biases. This latter problem, the loss of accuracy for low gate biases, has an obvious origin; since C_3 was factored out by hand for high gate biases, it is also missing for low gate biases, and the model now underestimates the low gate capacitance values.

This situation often becomes so complicated in practice that all pretense of maintaining any connection to the device physics is abandoned. Instead, **CGDO** and **CGSO** are treated as purely empirical AC constants, and are adjusted to provide the best

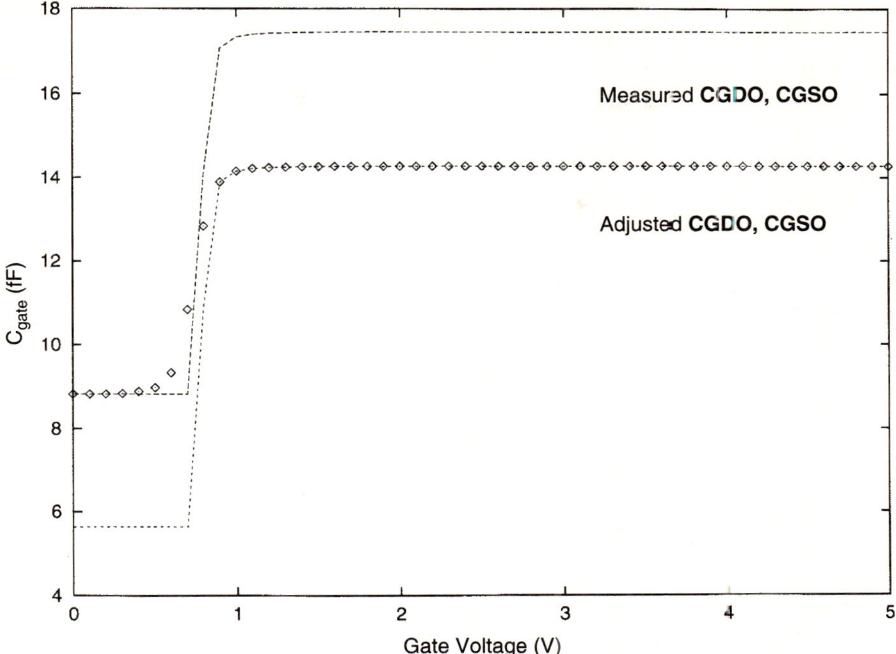

Figure 13.4   The result of adjusting the zero bias capacitance parameters **CGDO** and **CGSO** to compensate for the inability to account for the screening out of the inner-fringe capacitance (C_3) at high gate biases. If this approach is adopted, **CGDO** and **CGSO** will **not** agree with measurements of the zero bias gate-diffusion capacitance.

possible agreement with some chosen sample of circuits, such as ring oscillators. This common method is considered in more detail in Chapter 15. This problem is even more serious in analog designs, where accurate gate capacitance values are required over the entire range of gate biases. As Figure 13.4 clearly shows, with no ability to turn C_3 "on" at low gate biases and "off" at high gate biases, accuracy over that entire range is not possible.

Although not as drastic a concern as the problems associated with the implementation of the C_3 terms, it has also been shown that the C_1 and C_2 terms have some gate bias dependence [3], [4]. This can be an important concern for LDD MOS-FETs.

From this discussion, it is clear that the state of the zero bias gate capacitance model in SPICE is questionable. The present approach, using a fixed zero bias capacitance as a constant to be added to the active gate capacitance, is physically incorrect and leads to incorrect results in small geometry FETs. Clumsy, ad hoc solutions can be found, but the methods are empirical and are not widely extendable beyond a narrow range of validity. A new approach to modeling the zero bias gate capacitance is needed, particularly for low-voltage analog circuit design.

13.3 The Quasi-Static Approximation

Consider a MOSFET operating under DC bias conditions. A current flows from the source to the drain; this current can be generically described by

$$I_{ds} = M(V_d, V_g, V_s, V_b), \qquad (13.7)$$

where $M(V_d, V_g, V_s, V_b)$ is some mathematical function that describes the drain-source current in terms of the four-terminal voltages. Here these voltages are assigned specifically to each terminal, while, for example, the drain-source voltage would be written as

$$V_{ds} = V_d - V_s. \qquad (13.8)$$

Note that under DC bias conditions, the inversion layer charge density Q_{inv} is a constant, even though carriers are entering and leaving the inversion layer as drain-source current [5].

At a more detailed level, the current passing through each device terminal can be described by the four terms (i_d, i_g, i_s, i_b). With only DC biases applied, and neglecting any current due to forward biasing of the source by hot carrier generation, i_d and i_s are both described by the drain-source current I_{ds}. For this discussion, the DC values of the gate and substrate current (i_g and i_s) are set to zero and ignored. (These terms have nonzero DC values, but the currents are several orders of magnitude smaller than the drain-source current. Measurements and models of i_g and i_b are used for detailed descriptions of FET hot carrier behavior, but these effects are only considered for the detailed modeling of the output conductance.)

In a similar fashion, the charge on each FET node is a function of the applied biases:

$$Q_G = f(V_d, V_g, V_s, V_b), \qquad (13.9)$$

where $f(V_d, V_g, V_s, V_b)$ is a mathematical function describing that relationship and Q_G is the gate charge; similar expressions can be written for the other nodes (Q_D, Q_S, and Q_B).

With DC biases applied, all voltages are constant with time. Since their charge is constant under these conditions, all capacitors can be treated as open circuit elements. In fact, many circuit simulators actually do this to capacitor elements when a DC solution is being computed; this reduces the number of nonzero circuit matrix elements, and eliminates unnecessary computations. However, when terminal voltages vary with time, the various capacitances must be computed and included in the circuit model. Thus, a separate description of the FET is required.

Under transient conditions, there will still be currents flowing through the four device terminals, as described by i_d, i_g, i_s, and i_b. Although the values will be different than those for the DC solution, the terminal currents must still satisfy the Kirchoff current law and sum to zero [2]:

$$i_d + i_g + i_s + i_b = 0. \tag{13.10}$$

The transient solution must also satisfy the basic electrical requirements, but is superimposed on top of the DC solution. Similarly, under transient conditions, the charge neutrality of the device is still required:

$$Q_D + Q_G + Q_S + Q_B = 0. \tag{13.11}$$

It is with this basis that the quasi-static approximation is made. It is assumed that in any transient condition, the voltages vary slowly. The expression *slowly* implies that the current and charge at each device terminal can be described as instantaneous functions of the terminal voltages; that is, (13.7) and (13.9) can be used to compute (i_d, i_g, i_s, i_b) and (Q_G, Q_D, Q_S, Q_B), with the instantaneous transient voltages for any given time treated as DC bias voltages for the solution of (13.7) and (13.9). Since the DC expressions already exist, no additional work is required to derive expressions for the AC conditions.

All of the major gate capacitance models used for mainstream FET simulation in SPICE employ the quasi-static approximation. Of course, this approximation is based on slow voltage variations; if voltages vary quickly, the approximation obviously breaks down. Details on the limits of the quasi-static approximation are described in Section 13.8.

Under the quasi-static approximation, the transient terminal currents can be written as

$$i_d = I_{ds} + \frac{dQ_D}{dt}, \tag{13.12}$$

$$i_s = -I_{ds} + \frac{dQ_S}{dt}, \tag{13.13}$$

$$i_g = \frac{dQ_G}{dt}, \tag{13.14}$$

and

$$i_b = \frac{dQ_B}{dt}, \tag{13.15}$$

where I_{ds} is described by (13.7). An important difficulty with (13.12) and (13.13) is that the inversion layer is electrically connected to the source and drain, making it difficult to separately define Q_D and Q_S [2]. Instead, since

$$Q_{INV} = Q_D + Q_S, \tag{13.16}$$

(13.12) and (13.13) are combined into the more general expression

$$i_d + i_s = I_{ds} + \frac{dQ_{INV}}{dt}.$$

(13.17)

Equation (13.17) is used in modeling the device behavior, leading to a value of Q_{INV}. Once Q_{INV} has been computed, some method must be used to partition it between Q_D and Q_S; these methods are specific to particular gate capacitance modeling schemes, and will be considered there. In this form, Q_D and Q_S enter the device equations through Q_{INV}, while the gate charge Q_G is electrically isolated from the rest of the device by the gate oxide. This leaves only the depletion charge Q_{DEPL}, which is electrically connected to the substrate terminal. Therefore,

$$Q_B = Q_{DEPL}.$$

(13.18)

These two expressions (Q_B and Q_{DEPL}) are often used interchangably, which is appropriate in view of (13.18). However, when discussed here, Q_{DEPL} will refer to the depletion region *inside* the device, while Q_B will refer to the charge associated with the substrate terminal.

Now consider only the AC component of each terminal. Examining (13.12)–(13.15), a generic expression can be written as

$$i_z = \frac{dQ_z}{dt},$$

(13.19)

where z is the particular device terminal in question. Using the chain rule [5], (13.19) is expanded; for example, for the drain charge Q_D,

$$\frac{dQ_D}{dt} = \frac{\partial Q_D}{\partial V_d} \cdot \frac{dV_d}{dt} + \frac{\partial Q_D}{\partial V_g} \cdot \frac{dV_g}{dt} + \frac{\partial Q_D}{\partial V_s} \cdot \frac{dV_s}{dt} + \frac{\partial Q_D}{\partial V_b} \cdot \frac{dV_b}{dt}.$$

(13.20)

Similar expressions can be written down for the other three device terminal charges (Q_G, Q_S, and Q_B). To evaluate (13.20) (and the transient equation (13.19)), expressions are required for the terminal charges as functions of the terminal voltages. These expressions were derived from each FET current model, as described in Chapters 5–12. Thus, the charge model associated with each FET current model is indeed independent of the particular gate capacitance model chosen. Instead, the charge model is used as input to the gate capacitance model.

13.4   The Meyer Model

The oldest and simplest model for the gate capacitance was developed by Meyer [6]. The original derivation used the Level 1 charge expressions; however, any properly derived set of charge expressions can be employed. This simple model is commonly used with the first generation of FET models, but is not employed with the later FET models, as more sophisticated gate capacitance derivations have become available.

In the Meyer model, the gate capacitance is attributed entirely to changes in the gate charge; it is also assumed that all capacitances are reciprocal (e.g., $C_{gd} = C_{dg}$). The active gate capacitance terms are computed by taking the derivative of the gate charge against each specific terminal voltage:

$$C_{gz} = \frac{\partial Q_{GATE}}{\partial V_{gz}}, \tag{13.21}$$

where C_{gz} represents the capacitance between the gate and terminal z, while V_{gz} is the voltage between the gate and terminal z.

The Linear Region

In the linear region, the gate capacitance terms are found using (5.40):

$$C_{gs} = \frac{\partial Q_{GATE}}{\partial V_{gs}} = \frac{2}{3} W_{eff} L_{eff} C_{ox} \left[1 - \frac{(V_{gd} - V_t)^2}{(V_{gs} - V_t + V_{gd} - V_t)^2} \right], \tag{13.22}$$

and

$$C_{gd} = \frac{\partial Q_{GATE}}{\partial V_{gs}} = \frac{2}{3} W_{eff} L_{eff} C_{ox} \left[1 - \frac{(V_{gs} - V_t)^2}{(V_{gs} - V_t + V_{gd} - V_t)^2} \right]. \tag{13.23}$$

Note also that

$$C_{gb} = \frac{\partial Q_{GATE}}{\partial V_{gb}} = 0. \tag{13.24}$$

For the development presented here, this is a reasonable conclusion, as in strong inversion the inversion layer screens the silicon bulk from the gate charge.

The Saturation Region

In the saturation region, (5.43) is used in (13.21). The capacitances are then

$$C_{gs} = \frac{\partial Q_{GATE}}{\partial V_{gs}} = \frac{2}{3} W_{eff} L_{eff} C_{ox}, \tag{13.25}$$

$$C_{gd} = \frac{\partial Q_{GATE}}{\partial V_{gd}} = 0, \tag{13.26}$$

and

$$C_{gb} = \frac{\partial Q_{GATE}}{\partial V_{gb}} = 0. \tag{13.27}$$

Note that all the capacitances are now constants. This is also a reasonable outcome, as, due to channel pinch-off, the drain bias no longer affects the device behavior.

The Subthreshold Region

The original Meyer treatment of the gate capacitance was only concerned with the strong inversion region of operation ($V_{gs} > V_t$); values were computed for C_{gs} and C_{gd}, and C_{gb} was effectively set to zero. Most implementations of the Meyer model [7] expand the original development to include the capacitance conditions for $V_{gs} < V_t$. In this regime, $Q_{inv} \ll Q_{depl}$, so in contrast to the case of strong inversion, the gate capacitance characteristics will be determined by the depletion charge. Since the inversion charge is negligible, using (5.44) it can be readily seen that

$$C_{gs} = 0, \tag{13.28}$$

and

$$C_{gd} = 0. \tag{13.29}$$

This leads to discontinuities in C_{gs} and C_{gd} at V_t; in most implementations of the Meyer model, some sort of numerical interpolation is used to eliminate this problem.

From here, since this computation was not included in the original Meyer model, a variety of solutions for Q_{DEPL} is implemented in various simulators. Whichever solution is found, the gate to bulk capacitance is then computed from

$$C_{gb} = \frac{\partial Q_{GATE}}{\partial V_{gb}}. \tag{13.30}$$

The resulting expression is forced to zero at V_t to maintain continuity with the results for C_{gb} described above.

The Accumulation Region

For $V_{gs} < V_{fb}$, the MOS structure is in accumulation, and behaves like a simple two plate capacitor. Thus,

$$C_{gb} = C_{ox}. \tag{13.31}$$

At $V_{gs} = V_{fb}$, the weak inversion expression for C_{gb} is forced to equal C_{ox} to guarantee continuity. As noted in (13.28) and (13.29), the Meyer model sets both C_{gs} and C_{gd} to zero when $V_{gs} < V_t$.

The Complete Meyer Model

As derived, the Meyer model neglects any zero bias capacitance between the gate and the other device terminals. As described in Section 13.2, this is a poor approximation in small geometry FETs; zero bias capacitance terms are added to the active gate capacitance to account for this behavior. With these modifications, the Meyer model is implemented in most simulators as follows. In the linear region, (13.22)–(13.24) become

$$C_{gs} = \frac{2}{3} W_{eff} L_{eff} C_{ox} \left[1 - \frac{(V_{gd} - V_t)^2}{(V_{gs} - V_t + V_{gd} - V_t)^2} \right] + \textbf{CGSO} \cdot W_{eff}, \quad (13.32)$$

$$C_{gd} = \frac{2}{3} W_{eff} L_{eff} C_{ox} \left[1 - \frac{(V_{gs} - V_t)^2}{(V_{gs} - V_t + V_{gd} - V_t)^2} \right] + \textbf{CGDO} \cdot W_{eff}, \quad (13.33)$$

and

$$C_{gb} = 0. \quad (13.34)$$

In the saturation region, (13.25)–(13.27) become

$$C_{gs} = \frac{2}{3} W_{eff} L_{eff} C_{ox} + \textbf{CGSO} \cdot W_{eff}, \quad (13.35)$$

$$C_{gd} = \textbf{CGDO} \cdot W_{eff}, \quad (13.36)$$

and

$$C_{gb} = 0. \quad (13.37)$$

Otherwise, for $V_{gs} < V_t$, (13.28) and (13.29) become

$$C_{gs} = \textbf{CGSO} \cdot W_{eff}, \quad (13.38)$$

and

$$C_{gd} = \textbf{CGDO} \cdot W_{eff}. \quad (13.39)$$

For $V_{gb} = 0$, C_{gb} takes on the value

$$C_{gb} = \textbf{CGBO} \cdot L_{eff}, \quad (13.40)$$

to go along with whatever model is derived from (13.30) for $V_{gb} \neq 0$.

(Unless the device will be operated in the regime of $V_{gs} < V_{fb}$, the parameter **CGBO** can be neglected and set to zero. However, it is advisable to include a value of

CGBO in the FET model parameter file, set to zero, as some simulators will attempt to compute a value for **CGBO** which can be misleading.)

Although originally derived for use with Level 1, any charge model can be used in the derivation of the Meyer model. In Berkeley SPICE, the Meyer model was replaced with the Ward-Dutton model when the Level 2 FET model was introduced. In other simulators, the Meyer model is available for use with the other first-generation models (Level 2 and Level 3). The Level 2 FET model was considered to be so complex that a separate charge model is not derived; the Level 1 charge model is used instead. In contrast, since Level 3 is much simpler than Level 2, a charge model is derived from the current model, leading to a modified form of the Meyer model specific to Level 3. The results are nearly identical to those obtained from the Level 1 charge model.

Implementation of the Meyer Model

As originally derived, the Meyer model [6] has no specific parameters. The charge values are computed using the parameterized current model, and these values are used to compute the gate capacitance. Some implementations of the Meyer model [7] include additional parameters to improve the fit to data.

Due to its simplicity, the Meyer model faces a number of problems. As shown in Figure 13.5, the Meyer gate capacitance model is well defined in the strong inversion region and the off condition, with some form of numerical interpolation required to mediate between the two regimes. This has historically not been a problem in digital circuit simulation; since V_{dd} has tended to be much larger than V_t, node voltages spend a very small fraction of their time in the regime of $0 < V_{node} < V_t$, and the interpolation is not very important. However, as V_{dd} continues to decrease while V_t remains relatively constant, the fractional importance of the region where $0 < V_{node} < V_t$ becomes larger. In addition, analog circuits make use of this region in both dynamic and static modes. The inherently poor quality of the simple interpolation "fix" is becoming a danger to accurate simulation results.

Beyond these difficulties in the basic implementation of the Meyer model for digital simulations with large supply voltages, there are more serious problems. As has been noted [8], these problems are generally centered on three shortcomings. First, the Meyer model assumes that the depletion charge Q_{depl} is constant along the channel, is influenced only by the gate-bulk bias V_{gb}, and is not affected by either the drain or source voltages [2]. That is to say,

$$\frac{\partial Q_{depl}}{\partial V_d} = \frac{\partial Q_{depl}}{\partial V_s} = 0. \tag{13.41}$$

Since, as noted above, the depletion charge and the charge at the bulk terminal Q_B are used interchangably, and since all capacitive elements are treated as reciprocal, this implies that

$$C_{bd} = C_{bs} = C_{db} = C_{sb} = 0. \tag{13.42}$$

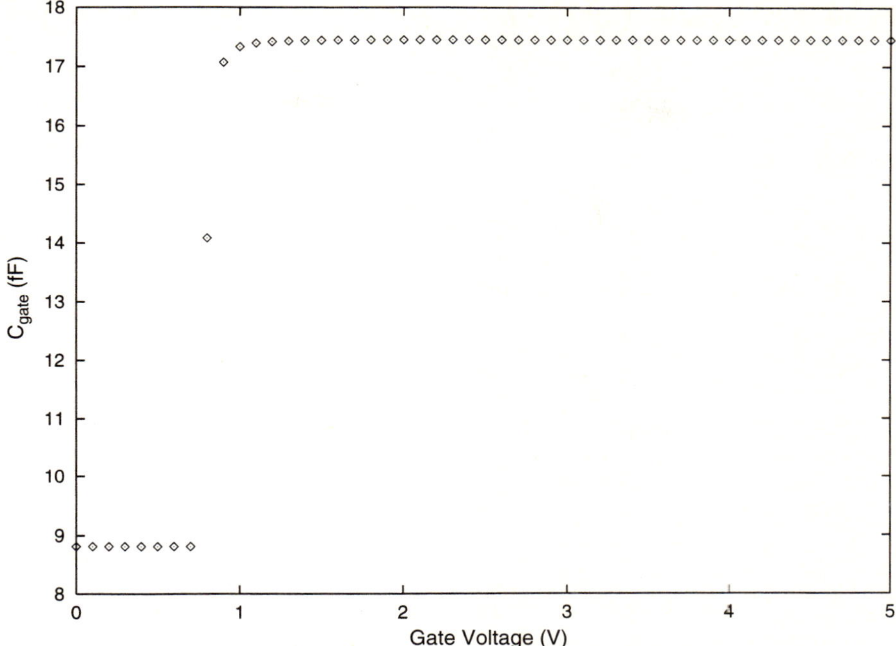

Figure 13.5   The computed Meyer gate capacitance in a $W_{eff}/L_{eff} = 10\,\mu m/0.25\,\mu m$ device with a low drain bias, showing the very sharp transition between the off condition and strong inversion.

(Note: The term Q_{depl} has been used in (13.41) to clearly indicate that the capacitances in (13.42) are the capacitances between the *depletion charge* and the *inversion charge*; the latter was noted earlier to be partitioned between the source and the drain terminals. The capacitances in (13.42) are **not** the bulk p-n junction capacitances, which are regarded as being extrinsic rather than intrinsic to the device [2].) While these capacitances are easily added to the Meyer description [8, 9], since the inversion layer screens the bulk charge, their contribution to the active device capacitance is rather small.

The second and third problems are related and much more important. The Meyer model assumes that all capacitances are reciprocal; that is, for any two device terminals y and z,

$$C_{yz} = C_{zy}. \tag{13.43}$$

This assumption may seem rather straightforward, but it is not valid for the four-terminal MOSFET structure, as is described in Section 13.5. Finally, it can be shown that the Meyer model does not obey charge conservation during circuit transients; that is, the charge neutrality relationship (13.11) is violated. As noted above, these two problems are actually related. In addition, the difficulties with charge conservation in the Meyer model involve not only the model formulation, but larger issues which must be dealt with at the circuit simulation level. This entire situation is considered in detail in Section 13.5.

Despite the various shortcomings, the Meyer model is still very useful. It is the only gate capacitance model available in SPICE which offers a direct, *analytical* solution for the gate capacitance terms, rather than the *iterative numerical* techniques that more sophisticated models require. This, of course, makes the Meyer model the most computationally efficient gate capacitance model available. In circuits where charge conservation is not important (which is the case in most digital circuits), the Meyer model is a good choice; this is particularly true for many very large digital circuit simulations, where simulation efficiency is more important than detailed physical accuracy.

13.5 The Charge Conservation Problem

The problem of charge conservation is very important to FET modeling and CMOS circuit simulation, but its origins are very subtle and require careful consideration. The expression *charge conservation* is simply a statement that the FET charge neutrality condition

$$Q_G + Q_D + Q_S + Q_B = 0 \qquad (13.44)$$

must hold for all applied biases and all times. Although (13.44) is built into the derivation of the Meyer active gate capacitance model [6], this condition can actually be violated under certain circumstances. While this failure is not always a serious problem, charge conservation is very important in certain types of CMOS circuits, such as dynamic RAMs and switched capacitor filters. In such vulnerable circuits, the failure of charge conservation manifests itself by the continual charging of some nodes, so that the simulated voltage there increases continually with time.

As noted in Section 13.4, the charge conservation problem is intertwined with the issue of whether or not the capacitive elements can be treated as reciprocal (e.g., does $C_{gd} = C_{dg}$?). This illustrates the first consideration in discussing charge conservation— the description of the charges and capacitances of the FET. A second and more global concern is the implementation of the FET models in the circuit simulator; an accurate FET model is insufficient to guarantee charge conservation. These items will be considered in turn.

13.5.1 The Nonreciprocity of the Capacitive Elements

In the Meyer model, it is assumed that all the capacitive elements are reciprocal; that is,

$$C_{yz} = C_{zy}, \qquad (13.45)$$

where y and z are any two device terminals. This implies that

$$\frac{\partial Q_y}{\partial V_z} = \frac{\partial Q_z}{\partial V_y}; \qquad (13.46)$$

the change in the charge on terminal y due to a change in the voltage of terminal z is equal to the change in the charge on terminal z due to a change in the voltage of terminal y. In addition, (13.45) implies that

$$\frac{\partial Q_y}{\partial V_{yz}} = -\frac{\partial Q_z}{\partial V_{yz}}; \tag{13.47}$$

for a change in the potential difference between terminals y and z, the change in the charge induced in terminal y is exactly mirrored by a change in the charge of terminal z. This is the standard description of a capacitor formed by two metal plates separated by a dielectric, as found in a standard electrodynamics textbook [10].

This description would be adequate if each FET capacitive element involved only two terminals, and could be described linearly in the form [8]

$$\Delta i_{yz} = \alpha_o \cdot C_{yz} \cdot \Delta V_{yz}, \tag{13.48}$$

in accordance with (13.45) and (13.46). However, the charge on each capacitor is *not* a simple function of the voltage on the two terminals; as described by (13.9), the charge on any one terminal is in fact a function of the voltages on *all four* device terminals; e.g.,

$$Q_G = f(V_d, V_g, V_s, V_b). \tag{13.49}$$

Since these individual voltage contributions cannot simply be added together [9], these capacitors are nonlinear, and cannot be described by the linear relationship (13.48), or by the simple two-plate capacitor equations (13.45)–(13.47). There is thus no particular reason for the reciprocity expression (13.45) to be valid.

13.5.2   Reciprocity and Charge Conservation

Given the nonlinear nature of the MOSFET capacitances, which form the basis for their nonreciprocal nature, the relationship between that nonreciprocity and the problem of charge conservation can be examined in more detail. This sets constraints on the nature of any method of computing the active gate capacitance terms.

The charge neutrality requirement in the FET was described by (13.44):

$$Q_G + Q_D + Q_S + Q_B = 0. \tag{13.50}$$

In the Meyer model, the substrate bias V_{bs} is taken to be a constant, and therefore does not enter into the capacitance calculations. Taking the derivative of (13.50) with respect to V_{gs} [9] thus leads to

$$\frac{\partial Q_G}{\partial V_{gs}} + \frac{\partial Q_D}{\partial V_{gs}} + \frac{\partial Q_S}{\partial V_{gs}} = 0. \tag{13.51}$$

If reciprocity is assumed, the general expression for reciprocal capacitive elements (13.47) can be invoked, so that

$$\frac{\partial Q_G}{\partial V_{gs}} = -\frac{\partial Q_S}{\partial V_{gs}}. \tag{13.52}$$

Substituting (13.52) into (13.51) results in

$$\frac{\partial Q_D}{\partial V_{gs}} = 0. \tag{13.53}$$

This is an interesting result, as it indicates that for reciprocity to be valid, the charge on the drain cannot be affected by the gate-source voltage [9]. If considered more generally [2], and even including the substrate bias V_{bs} in the derivation, it can be shown that the assumption of reciprocal capacitance elements will lead to the general requirement that

$$\frac{\partial Q_x}{\partial V_{yz}} = 0, \tag{13.54}$$

where y and z are any two device terminals, and x is either of the two *remaining* device terminals. However, an examination of any of the charge models developed in Chapters 5–12 clearly shows that (13.54) is not correct; if (13.54) *is* used, it will lead to incorrect values of the charge. This leads to discrepancies in the computation of the charge neutrality condition (13.50), which is another way of saying that charge is not conserved. If reciprocity is discarded, the charge neutrality condition (13.50) can be satisfied.

The final conclusion is that reciprocal capacitance elements and charge neutrality are incompatible; a model which chooses one cannot have the other. For example, the Meyer model employs reciprocal capacitance elements, and therefore cannot guarantee the conservation of charge. If charge conservation is required from a model, the various capacitance elements must be treated as nonreciprocal.

Once this incompatibility was realized, a few attempts were made to "fix" the Meyer model so that it would conserve charge. One approach [11] modifies the Meyer model so that each capacitor is instead a composite structure of a capacitor, a resistor, and a current source. This method does achieve charge conservation, but is forced to introduce the nonreciprocity of the capacitive elements. An extension of this approach [8] uses a composite structure of a capacitor, a resistor, and *two* current sources; charge conservation is again achieved, but this time the reciprocity of the capacitive elements is maintained. A third approach [9] attacks the charge conservation problem by independently specifying Q_D and Q_S and forcing their sum to equal Q_{INV}. Some method of partitioning the inversion charge between the source and the drain must be included, which is a situation much like that found in the Ward-Dutton model, described in Section 13.6. Charge conservation is guaranteed, but the nonreciprocity of the capacitive elements must be introduced. All of these fixes to the Meyer model are interesting, but none have yet found their way into any of the mainstream SPICE simulators.

13.5.3 Charge Conservation and Circuit Simulation

In addition to the issue of how the node charges are simulated at the device level, there are a number of concerns at the circuit simulation level. These difficulties are quite subtle, and may at first glance appear to actually be device level issues. However, as will be shown here, implementation at the circuit simulation level is actually the most important constraint on the ability to conserve charge.

As originally implemented in SPICE, the voltage is specified as the *state variable* at each circuit node; that is, at each node, if the voltage is not fixed by connection to an external node (such as the power supply or ground), it is the item which the matrix circuit solution produces. Once the voltage is determined at each node, the circuit behavior is computed from knowledge of all the node voltages.

In the most general case, either the voltage or the charge could be specified as the state variable at each node. However, it has been shown [12] that due to the highly nonlinear relationship between the voltage and the charge in MOSFETs, charge conservation can only be ensured by using the charge as the state variable in MOS simulation. If the voltage is specified as the state variable, the charge must be computed by integrating the voltage, which will lead to incorrect results.

Consider the computation of a node charge during SPICE circuit simulation. The Meyer model computes capacitance values, which are made available at the circuit simulation level; what follows is the case for *any* model which provides capacitance (rather than charge) to the circuit simulator. The charge on any particular node is found by integrating the transient current [2]:

$$Q_z(t_1) - Q_z(t_0) = \int_{t_0}^{t_1} i_z dt = \int_{t_0}^{t_1} C(V)dV, \tag{13.55}$$

where z is an arbitrary node. Careful examination of (13.55) reveals a serious problem. The node capacitance, as provided from the Meyer model or any other model, is specified for the endpoints of the integration (at the voltages V_{t_0} and V_{t_1}), but not in the interval between t_0 and t_1. Since $C(V)$ is not specified as a function in this region, the result of the integral in (13.55) (used to compute the charge) is path dependent. This is a mathematically sophisticated way of saying that since the function $C(V)$ is not completely specified in the interval between t_1 and t_2, the integration can follow any path; as a result, the computation of the area under the curve (in this case, the charge) is arbitrary, and does not lead to a well defined result. The charge neutrality requirement (13.50), however, demands well-defined results for the node charges. Therefore, (13.55) contains no method of guaranteeing charge neutrality; it is thus easily violated, and charge is not conserved.

The best possible solution to this problem is to attempt to specify the capacitance in the interval from t_0 to t_1 as the average of the capacitance at those two ends of the integral. With this approach, (13.55) simplifies to [2]

$$Q(t_1) - Q(t_0) = \overline{C} \cdot (V_{t_2} - V_{t_1}). \tag{13.56}$$

For example, the computation of the gate charge is

$$Q_G(t_1) - Q_G(t_0) = \overline{C}_{gs} \cdot \Delta V_{gs} + \overline{C}_{gd} \cdot \Delta V_{gd} + \overline{C}_{gb} \cdot \Delta V_{gb}, \qquad (13.57)$$

which is obviously different from the exact result

$$Q_G(t_1) - Q_G(t_0) = \int_{t_0}^{t_1} C_{gs} \cdot dV_{gs} + \int_{t_0}^{t_1} C_{gd} \cdot dV_{gd} + \int_{t_0}^{t_1} C_{gb} \cdot dV_{gb}. \qquad (13.58)$$

The incomplete knowledge of the capacitance leads to an inaccurate computation of the charge.

The most important item here is to note again that the above discussion applies to the *circuit simulation* level, rather than the device model level. In any FET charge model derivation, and in the Meyer capacitance model itself, the charge neutrality condition (13.50) is built in to the derivation; it thus may seem strange that charge is not conserved. However, as this discussion shows, if the voltage is specified as the state variable, incorrect values of the node charges are computed, and charge is not conserved. Instead, the charge must be chosen as the state variable for circuit simulation. Thus, the accuracy of the device model and the problem of charge conservation during circuit simulation are actually two independent issues [12].

Between the device model and the circuit simulation, an important link must be made. The device model must be able to completely communicate the charge neutrality requirement to activities at the circuit simulation level. Some implementations of SPICE [12] used charge as the state variable, but did not make the charge expressions from the FET models available at the circuit simulation level. If this is the case, the charges, although now state variables, are computed by integrating over the transient voltage points, as in (13.55). The same problems occur, and charge is not conserved.

It is required that the FET model *charge* expressions, rather than the *capacitance* expressions, be available at the circuit simulation level. Only by using analytical expressions for the charge as a function of the applied voltages can charge conservation be guaranteed. In fact, with charge as a state variable, and the FET terminal charges expressed as analytical functions of the voltage at the circuit simulation level, the accuracy of the capacitance expressions is not critical to the accuracy of transient circuit simulations [12]. As demonstrated in Sections 13.6 and 13.7, in these circumstances, the capacitance expressions appear only in the Newton-Raphson iterations for computing the node charges.

13.5.4   A Brief Summary of the Requirements

Since the issue of charge conservation is subtle yet so important, the three basic requirements for ensuring charge conservation are briefly summarized here:

1. The FET model must make available node charge expressions that observe charge neutrality, while making no requirement that the capacitance elements be

reciprocal. In addition, some method must be available to partition the inversion charge between the source and the drain terminals. More detailed aspects of the accuracy requirements are intertwined with the features and failings of each FET model.

2. Charge (rather than voltage) must be chosen as the state variable for circuit simulation.

3. The FET model charge expressions (rather than the capacitance expressions) must be made available at the circuit simulation level.

Note that of the three requirements, two apply at the circuit simulation level. The requirements on the FET model are rather straightforward, and are for the most part imposed as a matter of course. The second and third requirements have forced major revisions in the structure and implementation of circuit simulators. This is the justification for the earlier statement that charge conservation is more of an issue at the circuit simulation level than at the device model level.

As a final note, it must be stressed that charge conservation is a *mathematical* problem rather than a *physical* problem. A model can be made to conserve charge and at the same time provide totally incorrect results for the device node charges.

13.6   The Ward-Dutton Model

To achieve charge conservation, Ward and Dutton [13] introduced a new gate capacitance model. In an important departure from the Meyer model, the charge at each node is iteratively computed as an analytical function of the terminal voltages in a self-consistent manner; this forces the inversion charge to be explicitly partitioned between the source and drain terminals. The transient current into and out of each node is also considered in more detail. These modifications form the basis for all succeeding charge and gate capacitance models, where charge conservation is a paramount issue.

The Ward-Dutton model begins by noting that the transient current into or out of any arbitrarily defined region of space is simply the instantaneous change in the charge with time:

$$i = \frac{dQ}{dt}.$$

(13.59)

In the usual fashion, the charge is divided into the gate charge Q_{GATE}, the bulk charge Q_{DEPL}, and the inversion charge Q_{INV}. The inversion charge is found from the charge neutrality condition

$$Q_{GATE} + Q_{DEPL} + Q_{INV} = 0,$$

(13.60)

so that

$$Q_{INV} = -(Q_{GATE} + Q_{DEPL}).$$

(13.61)

To describe the capacitance behavior, the inversion charge must be related to accessible device terminals. Since it is electrically connected to the source and the drain, the inversion charge can be expressed as

$$Q_{INV} = Q_S + Q_D. \tag{13.62}$$

Partitioning the inversion charge between the source and the drain is a somewhat arbitrary process. The simplest approach is to divide the charge equally:

$$Q_S = Q_D = \frac{1}{2} Q_{INV}. \tag{13.63}$$

A more generic approach is to add a parameter to the model to allow the degree of charge partitioning to be selected; that is

$$Q_D = b \cdot Q_{INV}, \tag{13.64}$$

and

$$Q_S = (1 - b) \cdot Q_{INV}, \tag{13.65}$$

where $0 \leq b \leq 1$. This model is implemented in all versions of SPICE with the charge partitioning parameter **XQC**, which replaces b in (13.64) and (13.65):

$$Q_D = \mathbf{XQC} \cdot Q_{INV}, \tag{13.66}$$

$$Q_S = (1 - \mathbf{XQC}) \cdot Q_{INV}. \tag{13.67}$$

In all cases, if **XQC** is not specified, it defaults to 0.5 (equal charge partitioning). However, **XQC** is often specified, usually with a value of less than 0.5; a value of 0.4 is common, as Ward and Dutton [13] noted that this value of the charge partitioning parameter gave results closest to those of a detailed numerical simulation. However, the reader is strongly cautioned that the choice is entirely arbitrary, as the physical situation is not necessarily as clear-cut as this description indicates, and there is no good method of extracting **XQC** from data.

As (13.59) indicates, the current into or out of a node is the instantaneous change in the charge with time:

$$i_z = \frac{dQ_z}{dt}, \tag{13.68}$$

where z is the node under consideration. Note that due to (13.62), the channel charge is treated as a combined source and drain charge:

$$i_S + i_D = \frac{d}{dt}(Q_S + Q_D). \tag{13.69}$$

The basic objective of this approach is to specify the node charges as analytical functions of the terminal voltages; as described in Section 13.5, this guarantees that charge is conserved. For a given time interval, the change in the node charge is found by integrating (13.68):

$$\Delta Q_x = \int_{t_0}^{t_1} i_x dt = Q_x(t_1) - Q_x(t_0), \tag{13.70}$$

where $(t_1 - t_0)$ is the particular time interval under consideration. Let t_0 be a time point for which the charges have all been computed, while t_1 is a future time point at which it is desired to compute the charges. Some nodes are connected directly to external terminals, such as ground, DC biases, and (most importantly) signal waveforms; these nodes provide the input voltages for computation, and constraints (similar to boundary conditions) on those computations. (The situation is similar to that found in numerical device simulation, where the potential of contact nodes is determined solely by the applied bias there, while the potential at the other nodes must be computed.) At the time point t_1, the externally connected nodes have specified voltages; the voltages at all the other nodes are then computed. Once the terminal voltages on any particular FET have been determined, the charges on the various terminals are computed. Although this requires integration of the transient current, as in (13.55), the capacitance is eliminated from the expression [2]:

$$Q = \int_{t_1}^{t_2} i \cdot dt = Q_{t_2} - Q_{t_1} = Q(V_{t_2}) - Q(V_{t_1}); \tag{13.71}$$

since the charge is always specified as an analytic function of the terminal voltages (rather than as an integral of the averaged capacitance), charge conservation is guaranteed. Once the charges have been computed, the capacitances are easily determined. Of particular note is the difference between the "chain of computation" in the Meyer model (voltage, capacitance, charge) and in the Ward-Dutton model (voltage, charge, capacitance). The time point t_1 is then treated as the new t_0, and the process is repeated for another t_1.

The specifics of the method will now be considered. As described above, let the time point t_0 represent a time at which all node voltages, charges, and transient currents are known and computed. This time point could be a DC bias point at the start of a transient, or a fully solved point along the transient before the next time point is computed. A trapezoidal numerical integration of (13.70) leads to

$$\frac{(t_1 - t_0)}{2} \left(i_{t_1} - i_{t_0} \right) = Q_{t_1} - Q_{t_0}; \tag{13.72}$$

an equation of this form applies to every node. Note that (13.72) contains two unknowns, i_{t_1} and Q_{t_1}, and thus cannot be solved directly for either unknown. Instead, an iterative Newton-Raphson procedure is employed. For any given node, the voltage at time point t_0 is known, and will be written here as V_{t_0}. The iterative procedure begins with an initial guess for the node voltage at time point t_1, written here as $V_{t_1}^{(0)}$. An updated node voltage,

$V_{t_1}^{(1)}$, is then computed; $V_{t_1}^{(1)}$ replaces $V_{t_1}^{(0)}$ in the next iteration, where it is treated exactly as $V_{t_1}^{(0)}$. This process continues until the difference between $V_{t_1}^{(1)}$ and $V_{t_1}^{(0)}$ becomes small, at which point the final node voltage is deemed to be found.

For the Newton-Raphson computation, the charge on the node in question at time point t_1 is written as a Taylor expansion, with only the linear term retained:

$$Q_{t_1}^{(1)} = Q_{t_1}^{(0)} + \sum_x \left. \frac{\partial Q}{\partial V_x} \right|_{V_x = V_{x,t_1}^{(0)}} \left(V_{x,t_1}^{(1)} - V_{x,t_1}^{(0)} \right). \tag{13.73}$$

The summation over x represents a summation over all four terminals of the device. The charge Q inside the partial derivative represents the charge on the node under consideration, and is found directly by inserting the voltages computed by the latest Newton-Raphson iteration into the analytical model for the node charge as a function of the terminal voltages. As noted earlier, this voltage dependent charge can be computed from any MOS transistor model; however, the Ward-Dutton model, as originally implemented, uses the simple Level 1 charge expressions described in Chapter 5. Using (13.61) and a choice of either (13.63) or (13.66) and (13.67), the derivatives can be solved either analytically or numerically. Using (13.73), the trapezoidal integration (13.72) becomes

$$\frac{(t_1 - t_0)}{2} \left(i_{t_1}^{(1)} - i_{t_0} \right) = Q_{t_1}^{(0)} - Q_{t_0} + \sum_x \left. \frac{\partial Q}{\partial V_x} \right|_{V_x = V_{x,t_1}^{(0)}} \left(V_{x,t_1}^{(1)} - V_{x,t_1}^{(0)} \right). \tag{13.74}$$

Equation (13.74) can then be solved for $i_{t_1}^{(1)}$.

To proceed further, consider what will be done with $i_{t_1}^{(1)}$ once it is found; the individual device capacitance terms must be reconsidered. The capacitance between any two device terminals is described by

$$C_{yz} = \frac{\partial Q_y}{\partial V_z}. \tag{13.75}$$

Since the transient current into node y, from (13.68), is

$$i_y = \frac{dQ_y}{dt}, \tag{13.76}$$

the current flowing into node y due to a change in the voltage on node z is, in its most generic form,

$$i_y = C_{yz} \frac{dV_z}{dt}. \tag{13.77}$$

However, a change in a particular terminal voltage can cause current to flow into or out of the node; to describe the current flow for all four node voltages, positive and negative signs must be added to (13.77). For this discussion, consider the gate terminal and let $y = G$. If an increasingly positive bias is applied to the gate ($z = G$), electrons will flow into the gate; for historical reasons [2, 5], this is defined as a positive current. Equation (13.77) for this specific situation is

$$i_G = C_{GG}\frac{dV_G}{dt};\qquad(13.78)$$

the positive sign indicates electron flow *into* the gate. Now consider the situation when an increasingly positive bias is applied to the drain ($y = D$). Electrons will flow into the drain; this negative charge is imaged as positive charge in the gate, and electrons will flow out of the gate. Using the just described historical definition, the gate current for this situation is written as

$$i_G = -C_{GD}\frac{dV_D}{dt};\qquad(13.79)$$

the negative sign indicates that electrons flow *out of* the gate. Continuing with this reasoning, an expression for the gate current is

$$i_G = C_{GG}\frac{dV_G}{dt} - \sum_z C_{Gz}\frac{dV_z}{dt},\qquad(13.80)$$

where the subscript z denotes the other three device terminals. Using these concepts, (13.80) can be rewritten into an even more general form for any particular device terminal y:

$$i_y = C_{yy}\frac{dV_y}{dt} - \sum_z C_{yz}\frac{dV_z}{dt}.\qquad(13.81)$$

The structure of (13.81) is that of a system of linear equations, which can be expressed in matrix form as

$$\begin{bmatrix} i_G \\ i_D \\ i_S \\ i_B \end{bmatrix} = \begin{bmatrix} C_{GG} & -C_{GD} & -C_{GS} & -C_{GB} \\ -C_{DG} & C_{DD} & -C_{DS} & -C_{DB} \\ -C_{SG} & -C_{SD} & C_{SS} & -C_{SB} \\ -C_{BG} & -C_{BD} & -C_{BS} & C_{BB} \end{bmatrix} \begin{bmatrix} \frac{dV_G}{dt} \\ \frac{dV_D}{dt} \\ \frac{dV_S}{dt} \\ \frac{dV_B}{dt} \end{bmatrix}.\qquad(13.82)$$

The 4×4 matrix of capacitance terms is usually referred to as the admittance matrix.

Consideration can now return to the solutions for $i_{t_1}^{(1)}$ found from (13.74). For each particular terminal of the FET, the result for $i_{t_1}^{(1)}$ is substituted into the appropriate term in the current column vector of (13.82); each current term includes the voltage for this iteration, $V_{x,t_1}^{(1)}$, which is the desired final result. Each capacitance term is found from use

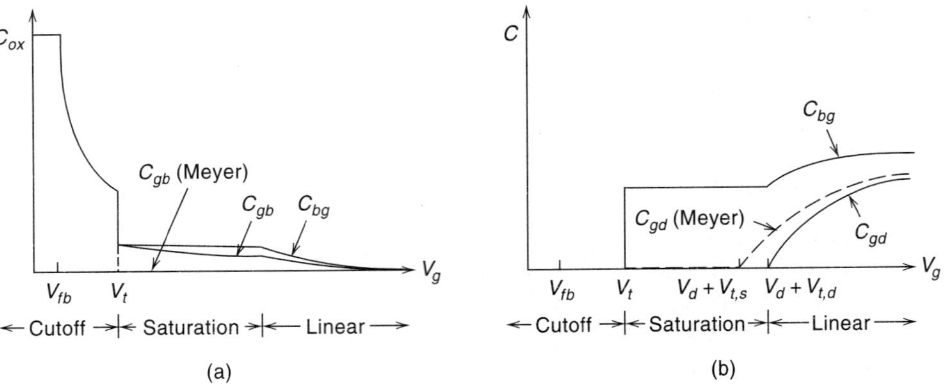

Figure 13.6 Results from the Ward-Dutton gate capacitance model; (a) gate-bulk results; (b) gate-drain results. From [13]. © 1978. IEEE. Used by permission.

of (13.75) with the terminal voltages and node charges of the latest Newton-Raphson iteration. Using the appropriate matrix operations, (13.82) is then solved for the set of four terminal voltages $V_{x,t_1}^{(I)}$. These results are then used as new initial guesses for the node voltages in the next Newton-Raphson iteration. This procedure continues until the change in the node voltage with each iteration becomes very small, at which point the node voltages are considered to be found.

Once the FET terminal voltages have been determined, the node charges are computed, and the capacitance values are found using (13.75). As described earlier, any FET charge model can be used, but Ward and Dutton [13] chose to use the very simple Level 1 charge model described in Chapter 5 [14] to reduce the complexity and computation time. Sample results are shown in Figure 13.6. Differences between the Meyer model and the more accurate Ward-Dutton model are clearly evident. Also, the nonreciprocity of the individual capacitance terms (e.g., $C_{GD} \neq C_{DG}$) is readily apparent.

Implementation of the Ward-Dutton Model

Like the Meyer model, the Ward-Dutton model is a self-contained active gate capacitance model, without specific model parameters (other than the charge-partitioning parameter **XQC**). As is the case for the Meyer model, all implementations of the Ward-Dutton model include the zero bias capacitance parameters **CGSO**, **CGDO**, and **CGBO**; these parameters represent, respectively, the zero bias gate-source, gate-drain, and gate-bulk capacitances. These parameters are described in detail in Section 13.2.

When the Ward-Dutton model is used, the total gate capacitances are

$$C_{GS,total} = C_{GS,Ward-Dutton} + \textbf{CGSO} \cdot W_{eff}, \tag{13.83}$$

$$C_{GD,total} = C_{GD,Ward-Dutton} + \textbf{CGDO} \cdot W_{eff}, \tag{13.84}$$

$$C_{GB,total} = C_{GB,Ward-Dutton} + \textbf{CGBO} \cdot L_{eff}. \tag{13.85}$$

Similar relationships hold for the capacitances C_{SG}, C_{DG}, and C_{BG}; the zero bias capacitances are treated as reciprocal.

Comments on the Ward-Dutton Model

When compared to the Meyer model, the Ward-Dutton model has a more realistic physical basis, and provides more accurate results. A major part of this improvement is the guarantee that charge neutrality is always preserved (charge is conserved). In addition, the discontinuities in the Meyer model, which are either simply tolerated or worked around with some form of numerical interpolation, are not present.

Although it represents an improvement, the Ward-Dutton model does have some shortcomings of its own. As originally implemented, the model used the simpler Level 1 expressions for charge in the device; the Level 2 charge expressions were judged to be too complicated to be computationally efficient. By using the simpler charge expressions, small geometry effects, which Level 2 was created to describe, are not considered in the gate capacitance model. However, the charge expressions of Level 3, being less complicated than those of Level 2, have been implemented in the Ward-Dutton model. Also, although the discontinuities of the Meyer model are not present, the Ward-Dutton model is nearly discontinuous at the threshold voltage. While this is not a significant problem for digital circuit design, a poor description in this region can be a major concern in certain analog circuit designs which operate near and around the threshold voltage. The introduction of charge partitioning in any form, including the added parameter **XQC**, is rather arbitrary, and thus questionable.

Finally, it should be noted that the Ward-Dutton model was designed for the era when SPICE employed the voltage rather than the charge as the state variable for circuit simulation. Due to this limitation, the Ward-Dutton model must compute the node voltages in order to eventually compute the charges. This results in quite a bit of complexity, which would be eliminated if charge were the state variable in SPICE. Newer models take advantage of this more recent improvement in the structure of SPICE.

13.7   The Yang-Epler-Chatterjee Model

With the improvements to the device current equations of the second-generation models, a need arose for an improved device charge model. This model was provided by Yang, Epler, and Chatterjee [12] in 1983. This device charge model is used in BSIM, HSPICE Level 28, BSIM2, and BSIM3, and is described in detail in Chapter 8. Yang *et al.* also described a method of implementing this new charge model in SPICE; when the three conditions for charge conservation (described in Section 13.5) are met, charge conservation is guaranteed. This method became the standard SPICE implementation at the time of the second-generation models.

As noted several times in this chapter, the current through a node and the charge on that node are described by

$$i_y = \frac{dQ_y}{dt}. \tag{13.86}$$

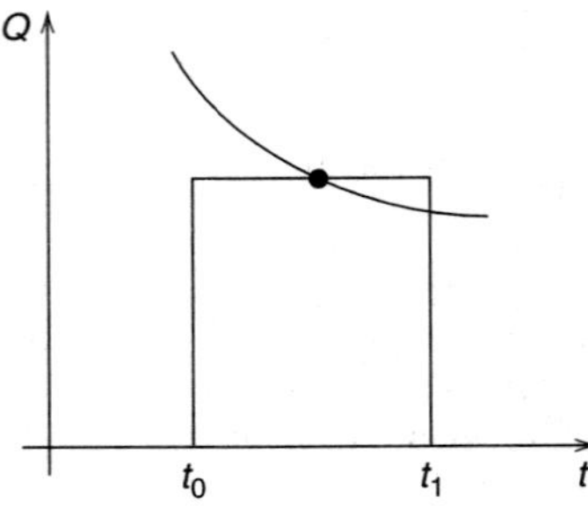

Figure 13.7 Determination of the change in charge between time t_0 and t_1, via the numerical integration of the current versus time. The trapezoidal integration method uses the midpoint of the interval to build a rectangle that is an approximation for the area under the curve.

This equation is easily rewritten in integral form as

$$\int_{t_0}^{t_1} dQ_y = \int_{t_0}^{t_1} i_y dt. \tag{13.87}$$

The meaning of (13.87) is shown in Figure 13.7. The current i_y varies over time; the focus here will be on the arbitrary interval from t_0 to t_1. The left-hand side of (13.87) is easily integrated:

$$\int_{t_0}^{t_1} dQ_y = Q_y(t_1) - Q_y(t_0); \tag{13.88}$$

the solution of (13.87) is the change in charge on node y that occurs as the time goes from t_0 to t_1. The right-hand side of (13.87) can be solved using the trapezoidal integration method [15]:

$$\int_{t_0}^{t_1} i_y dt = \left(\frac{t_1 - t_0}{2}\right)\left[i_y(t_1) + i_y(t_0)\right]. \tag{13.89}$$

This method is described in Figure 13.7, where the midpoint of the time scale is used to define one side of a rectangle, while the current values at each end of the time interval are summed to form the other side of the rectangle.

Using (13.88) and (13.89) in (13.87) produces

$$Q_y(t_1) - Q_y(t_0) = \left(\frac{\Delta t}{2}\right)\left[i_y(t_1) + i_y(t_0)\right], \tag{13.90}$$

where Δt is the time interval

$$\Delta t = \frac{t_1 - t_0}{2}. \tag{13.91}$$

Equation (13.90) is then rearranged to a form which describes the current through node y at time t_1:

$$i_y(t_1) = -i_y(t_0) + \frac{2}{\Delta t}\left[Q_y(t_1) - Q_y(t_0)\right]. \tag{13.92}$$

Examination of (13.92) indicates that if the state of the device at the time t_0 is assumed to be computed and known, there are still two unknowns, $i_y(t_1)$ and $Q_y(t_1)$; (13.92) cannot be solved analytically. However, a solution can be found using an iterative numerical Newton-Raphson method. For some general function $f(x)$, a first-order Taylor expansion yields

$$f(x) = f(x_o) + (x - x_o) \cdot \left.\frac{\partial f}{\partial x}\right|_{x_o}, \tag{13.93}$$

where x_o is fixed and x varies. The Newton-Raphson scheme treats x_o as the value for a solved iteration, and x as the value for the next iteration; when $x = x_o$, $f(x) = f(x_o)$, and the solution has been found. Here,

$$f(x) = i(V), \tag{13.94}$$

so

$$x_o = V^k(t_1) \tag{13.95}$$

and

$$x_o = V^{k+1}(t_1), \tag{13.96}$$

where the superscripts indicate Newton-Raphson iteration numbers. In this approach, the iteration denoted by k is treated as completed (from an initial guess to start, then from successive iterations), while $k + 1$ is the succeeding iteration. Thus, in its most generic form, (13.93) becomes

$$i(V^{k+1}) = i(V^k) + \left[V^{k+1}(t_1) - V^k(t_1)\right] \cdot \frac{\partial i(V^k(t_1))}{\partial V^k(t_1)}. \tag{13.97}$$

The iterations continue until the difference between $V^{k+1}(t_1)$ and $V^k(t_0)$ becomes small, at which point

$$i(V^{k+1}) = i(V^k), \tag{13.98}$$

and the terminal voltage is deemed to be computed.

In the four terminal MOSFET, the current i is written as i_y, the current through node y. The voltages must account for the voltage difference between node y and the three other terminals. Thus, for the full MOSFET, (13.97) is expanded to

$$i_y^{k+1}(t_1) = i_y^k(t_1) + \sum_{y \neq z} \frac{\partial i(V_{yz}^k(t_1))}{\partial V_{yz}^k(t_1)} \left[V_{yz}^{k+1}(t_1) - V_{yz}^k(t_1) \right]. \tag{13.99}$$

From the trapezoidal integration (13.90), it can be seen that

$$dQ_y = \left(\frac{\Delta t}{2} \right) \cdot di_y, \tag{13.100}$$

so that

$$\frac{\partial i(V_{yz}^k(t_1))}{\partial V_{yz}^k(t_1)} = \frac{2}{\Delta t} \cdot \frac{\partial Q_y(V_{yz}^k(t_1))}{\partial V_{yz}^k(t_1)}. \tag{13.101}$$

Substituting this expression and (13.92) into (13.99) leads to

$$i_y^{k+1} = -i_y(t_0) + \frac{2}{\Delta t} \left[Q_y(t_1) - Q_y(t_0) \right] + \sum_{y \neq z} \frac{2}{\Delta t} \cdot \frac{\partial Q_y(t_1)}{\partial V_{yz}^k(t_1)} \left[V_{yz}^{k+1}(t_1) - V_{yz}^k(t_1) \right]; \tag{13.102}$$

here, the voltage dependence of i_y and Q_y is dropped for notational convenience. Note that the node voltages are changed as the iteration proceeds, and the node charges and currents are continually recomputed as those voltages change from iteration to iteration.

To simplify the form of (13.102), define the capacitance between any two nodes after the completion of iteration k as

$$C_{yz}^k = \frac{\partial Q_y}{\partial V_{yz}^k}. \tag{13.103}$$

This allows (13.102) to be rewritten in the somewhat more manageable form of

$$i_y^{k+1} = -i_y(t_0) + \frac{2}{\Delta t} \left[Q_y(t_1) - Q_y(t_0) \right] + \sum_{y \neq z} \frac{2}{\Delta t} \cdot \left\{ C_{yz}^k \left[V_{yz}^{k+1}(t_1) - V_{yz}^k(t_1) \right] \right\}. \tag{13.104}$$

It is important to note that the capacitance C_{yz} appears as a by-product of the solution of the charge state of the FET. Here the capacitance is computed numerically (and properly) from (13.103); when the Newton-Raphson iterative solution has converged, the capacitance has been determined. However, it is the charge state of the FET that is important for the description of the transient behavior, not the (incidental) capacitance. Note that here, the capacitance is computed from the charge solution (the proper approach); this is in constrast to the older (and improper) method of computing the charge state from the capacitance.

For additional convenience, all possible iteration k terms in (13.104) can be pulled together into an equivalent current:

$$i_{y,eq}^{k} = -i_y(t_0) + \frac{2}{\Delta t}\left[Q_y(t_1) - Q_y(t_0)\right] - \sum_{y\neq z}\frac{2}{\Delta t}\cdot C_{yz}^{k}\cdot V_{yz}^{k}(t_1). \tag{13.105}$$

Substituting this expression into (13.104) leads to

$$i_y^{k+1}(t_1) = i_{y,eq}^{k} + \sum_{y\neq z}\frac{2}{\Delta t}\cdot C_{yz}^{k}\cdot V_{yz}^{k+1}(t_1). \tag{13.106}$$

To proceed, consider once again the Kirchoff current law

$$i_d + i_g + i_s + i_b = 0, \tag{13.107}$$

and the conservation of charge

$$Q_D + Q_G + Q_S + Q_B = 0. \tag{13.108}$$

There are four node currents in the device. However, only three are independent; constrained by (13.107), once three of the currents have been determined, the fourth is by definition also determined. Similarly, between the four device terminals, twelve capacitances are defined. Conservation of charge imposes the constraint that only nine of them are independent; once nine of the capacitances are determined independently, the remaining three are by definition also determined.

In addition, the current-charge relationship (13.86) can be expanded using the chain rule, as shown in Section 13.3. For example, for the gate current,

$$i_g = \frac{dQ_G}{dt} = \frac{\partial Q_G}{\partial V_{gd}}\cdot\frac{dV_{gd}}{dt} + \frac{\partial Q_G}{\partial V_{gb}}\cdot\frac{dV_{gb}}{dt} + \frac{\partial Q_G}{\partial V_{gs}}\cdot\frac{dV_{gs}}{dt}; \tag{13.109}$$

similar expressions can be written down for i_d, i_s and i_b. Using the charge conservation requirement (13.108) and the chain rule expressions suggested by (13.109), it can be shown [12] that the capacitance terms are governed by

$$\sum_{y\neq z}C_{yz} = \sum_{y\neq z}C_{zy}. \tag{13.110}$$

Now, note that (13.106) is an expression for the current through node y as computed during Newton-Raphson iteration $k + 1$. The Kirchoff current law (13.107) applies here as well; thus,

$$\sum_y i_y^{k+1}(t_1) = 0. \tag{13.111}$$

As a consequence, from (13.106),

$$\sum_y i_y^{k+1}(t_1) = \sum_y i_{y,eq}^k + \sum_{y \neq z} \frac{2}{\Delta t} \cdot C_{yz}^k \cdot V_{yz}^{k+1}(t_1) = 0. \tag{13.112}$$

From (13.110), it can also be shown that

$$\sum_y \sum_{y \neq z} \frac{2}{\Delta t} \cdot C_{yz}^k \cdot V_{yz}^{k+1}(t_1) = 0. \tag{13.113}$$

The combination of (13.112) and (13.113) yields the unsurprising result that

$$\sum_y i_{y,eq}^k = 0. \tag{13.114}$$

What was true for the simple node current equations is also true for the equivalent current—only three of the four values can be computed independently. Once three of the four values have been determined, the fourth is by definition also determined.

Some sample capacitance values computed using the Yang-Epler-Chatterjee model are shown in Figure 13.8. The results are similar to those produced by the Ward-Dutton model, but there are significant differences. For example, the results for C_{gd} are quite different in the two models.

Taking full advantage of the implementation of charge as the state variable in SPICE, the Yang-Epler-Chatterjee model has been a most successful charge and capacitance model. The smooth transitions between the various regions of device operation also set this model apart from earlier charge models.

13.8 The Limits of the Quasi-Static Approximation

In Section 13.3, the quasi-static approximation for MOS charge models was stated to be valid for terminal voltages that varied sufficiently slowly. Of course, to determine the limits of usefulness of the quasi-static approximation, some method of quantifying *sufficiently slowly* would be useful.

Not unexpectedly, there are no hard and fast rules for determining a particular limit, outside of which non-quasi-static effects will have to be considered. A general rule of thumb for the usefulness of the quasi-static approximation has been found to be [5]

$$\tau_r > 20\tau_t, \tag{13.115}$$

where τ_r is the *rise time* of a waveform, and τ_t is the *transit time* for carriers which leave the source and arrive at the drain. The use of 20 is somewhat arbitrary, and cases can be made for values from 15–25 [5]. If the carriers traverse the channel at their maximum possible value of the saturation velocity v_{SAT}, then the transit time is easily defined by

$$\tau_t = \frac{L_{eff}}{v_{SAT}}. \tag{13.116}$$

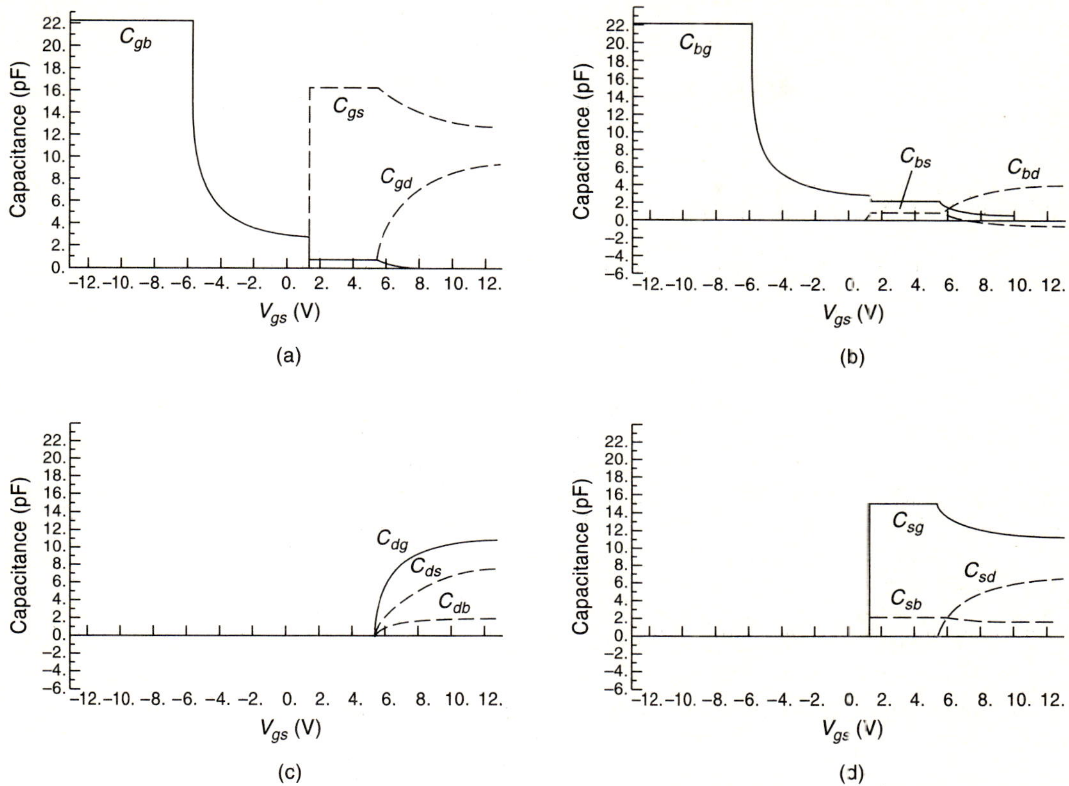

Figure 13.8   Results from the Yang-Epler-Chatterjee gate capacitance model; (a) gate capacitances; (b) bulk capacitances; (c) drain capacitances; (d) source capacitances. From [12]. © 1983. IEEE. Used by permission.

In an n-channel MOSFET, $v_{SAT} \approx 10^7$ cm/s. For $L_{eff} = 1.0$ μm, the transit time is $\tau_t = 10$ ps; as long as waveform rise times are larger than about 200 ps, the quasi-static approximation should be valid. A reported stage delay for a ring oscillator with this channel dimension is about 10 ns [2]; this indicates that the quasi-static approximation is reasonable in this geometry range. For $L_{eff} = 0.25$ μm, $\tau_t = 2.5$ ps, so the quasi-static approximation is valid for rise times τ_r greater than about 50 ps. This indicates that for a given frequency, the greatest vulnerability to non-quasi-static effects will be found in *longer* channel devices.

Of course, a decreasing channel length implies a higher operating frequency, so it is more instructive to estimate the frequency where the quasi-static approximation begins to break down for a given channel length. For the discussion of $L_{eff} = 0.25$ μm above, a value of $\tau_r = 25$ ps implies a minimum total waveform size of 50 ps, which translates into an operating frequency of 20 GHz; this is two orders of magnitude larger than the fastest silicon CMOS digital integrated circuits reported for this geometry, indicating that the quasi-static approximation is a good one. This result also indicates that non-quasi-static effects are most likely to be encountered in microwave devices, as has been noted [16].

This discussion applies quite well to digital circuit design, where the shortest available channel lengths are generally used. The greatest vulnerability to non-quasi-static effects occurs at high frequencies and long channel lengths. These conditions are found in some analog and signal processing circuits [17]. Circuits of this type are likely to be the first to require a non-quasi-static charge model.

There are additional special circumstances that may require non-quasi-static charge models for silicon MOS technology. A number of models have been reported [17–21]; however, none of these models have appeared in a generally available SPICE simulation package.

13.9   Final Comments

As noted in Chapters 5–12, the computation of the active device capacitance is an issue at the circuit simulation level, rather than at the device level. In those chapters, where the individual FET models were derived, only node charge equations were produced. The proper simulation of the active device capacitance is actually a matter of properly simulating the node charges in the entire circuit.

The most demanding aspect of simulating the node charges is the enforcement of the requirement that charge be conserved; FET charge neutrality should hold true for all bias conditions at all times. This seemingly simple issue is subtle but quite complex. However, three conditions can be listed which must be met if charge conservation is to be assured during circuit simulation. Only one of the three applies to the FET model itself; the node charges must always sum to zero, and some method of partitioning the inversion charge between the source and the drain must be available. The other two issues require proper implementation at the circuit simulation level. First, the node charge, rather than the node voltage, must be used as the state variable for circuit simulation. Second, the charge equations of the FET model (rather than subsequently derived capacitance equations) must be available at the circuit simulation level.

However, care must be used to note that charge conservation is a matter of model *implementation* rather than model *accuracy*. A model can conserve charge and still provide completely inaccurate node charge results.

It is interesting to note that if the approach is implemented properly, the active gate capacitance essentially disappears from the basic level of circuit simulation. These capacitances can only be computed numerically, and only as a by-product of the proper simulation of the charges in the circuit. It is perhaps best to only consider charge in MOS simulation, and not impose capacitance artificially on the system.

One key element of all these derivations is the assumption that the charge at any node can be expressed as an instantaneous DC function of the node voltages (the quasi-static approximation). This approximation is a good one under virtually all circumstances, and can in general be used without question. However, the quasi-static approximation is weakest under the combination of high frequencies and *long* channel devices. It may encounter difficulty in some high-speed analog and signal processing circuits.

A final concern is the treatment of the zero bias gate-diffusion capacitance as an additive constant for all biases. The *inner-fringe* term should be screened out by the inversion layer, and thus should disappear at high gate biases, but be present at low gate biases. Its global inclusion causes unrealistically high values of the total gate capacitance at high gate biases, which is the main region of interest in digital circuits. As a result, the zero bias capacitance parameters are often treated as empirical terms, to be adjusted to give the best results when compared with a collection of circuits, such as ring oscillators.

BIBLIOGRAPHY

1. E. Greeneich, "An Analytical Model for the Gate Capacitance of Small Geometry MOS Structures," *IEEE Trans. Elec. Dev.* vol. ED-30, pp. 1838–1839 (1983).

2. N. Arora, *MOSFET Models for VLSI Circuit Simulation*, Springer-Verlag, 1993.

3. T. Smedes and F. Klaassen, "Effects of the Lightly Doped Drain Configuration on Capacitance Characteristics of Submicron MOSFETs," *1990 IEDM Tech. Dig.*, pp. 197–200.

4. N. Arora, D. Bell, and L. Bair, "An Accurate Method of Determining MOSFET Gate Overlap Capacitance," *Sol. St. Elec.* vol. 35, pp. 1817–1822 (1992).

5. Y. Tsividis, *The MOS Transistor*, McGraw-Hill, 1987.

6. J. Meyer, "MOS Models and Circuit Simulation," *RCA Review* vol. 32, pp. 42–63 (1971).

7. *HSPICE User's Manual*, Meta-Software, Inc., Campbell, California, 1993.

8. M. Cirit, "The Meyer Model Revisited: Why is Charge Not Conserved?" *IEEE Trans. Comp.-Aid. Des.* vol. 8, pp. 1033–1037 (1989).

9. K. Sakallah, Y. Yen, and S. Greenberg, "A First-Order Charge Conserving MOS Capacitance Model," *IEEE Trans. Comp.-Aid. Des.* vol. 9, pp. 99–108 (1990).

10. D. Griffiths, *Introduction to Electrodynamics*, Prentice-Hall, 1981.

11. C. Turchetti, P. Prioretti, G. Masetti, E. Profumo, and M. Vanzi, "A Meyer-like Approach for the Transient Analysis of Digital MOS ICs," *IEEE Trans. Comp.-Aid. Des.* vol. 5, pp. 499–506 (1986).

12. P. Yang, B. Epler, and P. Chatterjee, "An Investigation of the Charge Conservation Problem for MOSFET Circuit Simulation," *IEEE J. Sol. St. Circ.* SC-18, pp. 128–138 (1983).

13. D. Ward and R. Dutton, "A Charge-Oriented Model for MOS Transistor Capacitances," *IEEE J. Sol. St. Circ.* vol. SC-13, pp. 703–708 (1978).

14. H. Ihantola and J. Moll, "Design Theory of a Surface Field-Effect Transistor," *Sol. St. Elec.* vol. 7, pp. 423–430 (1964).

15. *Handbook of Applied Mathematics* (2nd ed.) (ed. by C. Pearson), Van Nostrand Reinhold, 1990.

16. P. Roblin, S. Kang, and W. Liou, "Improved Small Signal Equivalent Circuit Model and Large Signal State Equations for the MOSFET/MODFET Wave Equation," *IEEE Trans. Elec. Dev.* vol. ED-38, pp. 1706–1718 (1991).

17. P. Vandeloo and W. Sansen, "Modeling of the MOS Transistor for High Frequency Analog Design," *IEEE Trans. Comp.-Aid. Des.* vol. 8, pp. 713–723 (1989).

18. M. Bagheri and Y. Tsividis, "A Small Signal DC to High Frequency Nonquasistatic Model for the Four Terminal MOSFET Valid in All Regions of Operation," *IEEE Trans. Elec. Dev.* vol. ED-32, pp. 2383–2391 (1985).

19. P. Mancini, C. Turchetti, and G. Masetti, "A Non-Quasi-Static Analysis of the Transient Behavior of the Long Channel MOST Valid in All Regions of Operation," *IEEE Trans. Elec. Dev.* vol. ED-34, pp. 325–335 (1987).

20. L. Pu and Y. Tsividis, "Small Signal Parameters and Thermal Noise of the Four Terminal MOSFET in Non-Quasistatic Operation," *Sol. St. Elec.* vol. 33, pp. 513–521 (1990).

21. H. Park, P. Ko, and C. Hu, "A Charge Conserving Non-Quasistatic (NQS) MOSFET Model for SPICE Transient Analysis," *IEEE Trans. Comp.-Aid. Des.* vol. 10, pp. 629–642 (1991).

14

Accounting for Systematic Process Variations

14.1 Introduction

As noted in Chapter 2, in an ideal world, absolutely identical FETs would be produced during every fabrication run, on every die and on every wafer. In reality, even when a process is stable and well controlled, systematic statistical variations around the center of that process will exist. These variations will affect the device characteristics and circuit behavior, and so must be accounted for in the SPICE FET models.

Providing a useful description of that variability can be a daunting task. Model distributions must be wide enough to encompass the variations to be expected from the process. However, if the variation of the model is too wide, designed circuits will be much less aggressive than they might otherwise be; the process is not pushed to its maximum possible usefulness.

Unfortunately, the practical modeling of CMOS process variations in SPICE has only recently begun to attract significant attention. This is somewhat surprising, given the long industrial use of SPICE. Since the problem rarely offers a clean solution, various attempts have generally lacked a sense of completeness. However, the proper ability to model process variations can be put to competitive advantage, as is now being realized. Without changing either the manufacturing process or the basic structure of the designed circuits, an accurate understanding of process variations allows circuits to be designed to the (correct) edge of their capabilities, without crossing the line into failure.

Of course, any attempt to discuss process variations will rely on a certain amount of statistical analysis; this will be discussed here. It is important to note that the application of such statistical analysis implies that the manufacturing process is stable, and is subject only to random variations around the mean. The statistical analysis techniques can be applied to an unstable process; however, this is an incorrect use of statistics, and the designer should be on the lookout for such occurrences.

The actual effort for describing statistical variations can range from the relatively simple to the very complex. The standard historical approach has been to (somewhat arbitrarily) designate a few parameters as important for the description of process variations, and provide the manufacturing data for the statistical variation of those parameters. Recently, the more sophisticated technique of principal component analysis has been applied to this problem; improved results are obtained at the cost of greatly expanded requirements for repetitive modeling to compute the statistical variations of the parameters.

The actual application to CMOS technology is complicated by the need for separate statistical models for nMOSFETs and pMOSFETs. In theory, the variability of nFETs and pFETs could be treated as separate and independent. However, since they are fabricated in close proximity on the same wafer, this is not the case. Some method of correlating the nFET and pFET variations should be developed. Also, in some circuits, the overall process variation is less of an issue than is the ability to describe how two FETs fabricated in close proximity will differ from each other.

These issues are addressed in this chapter. Since this subject is not well developed, as many questions as answers may be raised. However, enough has been done to allow useful approaches to be developed.

14.2   Stable Processes and Process Variations

Before proceeding, it is worthwhile to briefly consider the meaning and implications of a *stable process*. This simply indicates that fabrication produces consistent and repeatable results; the mean values of key process parameters are well defined and do not vary over time. The process is also well controlled, which implies that the variations which occur are random variations around the mean. If this is the case, the data collected for any parameter should form a simple Gaussian distribution, as shown in Figure 14.1. When a process consistently produces data with a stable mean and random variations about that mean, it is said to be operating in statistical control. Improvements in process control will tighten up the distribution (i.e., reduce the value of the standard deviation σ), but

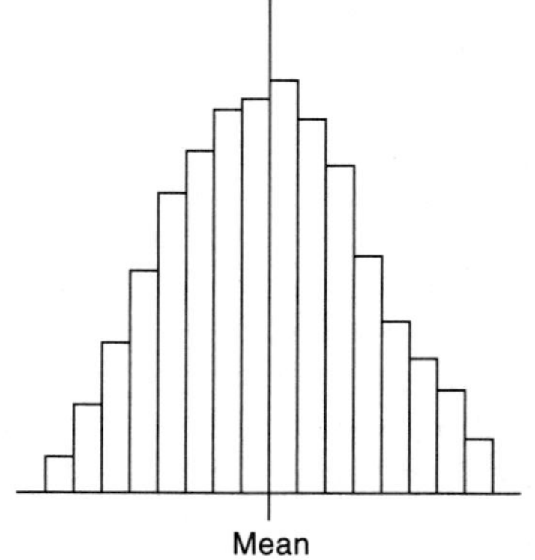

Figure 14.1   A histogram depicting the distribution of a large number of measured parameter values. The Gaussian shape of the distribution indicates that the process is in statistical control, and that the deviations from the mean value are due to statistically random variations.

will not alter the mean value. If results fall outside the expected distribution, they are due to special causes, which are not part of the normal expected process; these special causes should be identified and eliminated, as they indicate the intrusion of nonsystematic error(s). These points are fundamental to the theory of statistical process control, as developed by Shewhart [1] and Deming [2].

From the designer's point of view, the goal of process development is to eliminate all nonsystematic errors, and produce a stable process which is in statistical control. A stable process is required if the models are to describe the FET behavior when that circuit is actually fabricated. Once the process is stable, it can be improved by tightening up the parameter distributions, which reduces the amount of variation that must be accounted for in the design. While this is often referred to as *tightening up the specifications* (which is another way of saying that the σ values have been reduced), a more enlightened description involves the reduction of the Taguchi loss function [3].

A stable process can be intentionally changed for two reasons. First, it may be deemed desirable to alter a well-defined physical value; reduction of the gate oxide thickness would be a good example. The process is changed to accommodate the new gate oxide thickness, then must be brought back into statistical control before it can be made available for fabrication. Second, a process might be modified to improve the behavior, cost, and/or yield of one part of that process; this might unintentionally change certain parameters. This is generally undesirable, as the benefit of improving the process might be outweighed by the unforeseen need to redesign various circuits fabricated in that technology.

For best circuit design results, a designer expects to be working with a stable process that is in statistical control. This is the only way that confidence can exist that the fabrication process is consistent and well behaved over time. In the discussion which follows, it is assumed that this is the case. Drift of the process over time and continual updates of process means and/or variations are warning signs that a process is not properly under statistical control.

14.3   Simple Skewing with Fabrication Data

To describe the effect of process variations on the behavior of designed circuits, some method of communicating the variability of the process must exist between the fabrication facility and the circuit designer. If all the model parameters were physically meaningful and completely independent, it would be a simple matter to collect data on these parameters, and perform statistical analysis to determine all the means and variations. By providing means and standard deviations to the circuit designer (via the FET models), the process would be completely and adequately described.

However, as detailed in Chapters 5–12, some parameters have clear physical meaning, while others are electrical parameters which are determined by parameter extraction. Among the electrical parameters, some have partial physical meaning, while others are completely empirical. Many of these parameters are also correlated rather than independent; a description of their variations is clouded by the fact that they shift together rather than separately.

It is possible to cope with this situation with very complicated forms of numerical analysis, as described in Section 14.4. However, a greatly simplified approach is commonly used. In this method, a subset of physical parameters (parameters that have quantifiable physical meaning) is selected as responsible for the variations in the process. Fabrication data are collected on these designated parameters, and means and standard deviations are computed. The results are deemed to represent the variations in the process. Note that this method is identical to the ideal approach described above, except that an acknowledged subset of the full parameter set is employed.

For example, a common set of chosen parameters is the oxide thickness (**TOX**), the reduction of the channel length from its drawn value (2·**LD**), the reduction of the channel length from its drawn value (2·**WD**), and the variation in the model's computed threshold voltage (ΔV_t). Sample results are shown in Table 14.1a, where the mean and 3σ values are described. Using the data in Table 14.1a, the 3σ-fast and 3σ-slow skew parameters of Table 14.1b are produced. (Note that a fast skew corresponds to a larger (than the mean) value of **LD** and smaller values of **TOX**, **WD**, and ΔV_t; a shorter, wider channel with a thinner gate oxide and a lower threshold voltage will produce more current, and thus be

Table 14.1 Statistical data for four process parameters; (a) the mean and $\pm 3\sigma$ deviations of the data; (b) the conversion of the data of (a) into fast and slow process skew parameter sets.

(a)

Parameter	Mean	$+3\sigma$	-3σ
nFET			
TOX (nm)	10.0	10.6	9.4
LD (μm)	0.10	0.13	0.07
WD (μm)	0.80	1.10	0.50
ΔV_t (V)	0.000	0.060	−0.060
pFET			
TOX (nm)	10.0	10.6	9.4
LD (μm)	0.12	0.16	0.06
WD (μm)	0.90	1.30	0.50
ΔV_t (V)	0.000	0.080	−0.080

(b)

Parameter	Mean	3σ-fast	3σ-slow
nFET			
TOX (nm)	10.0	9.4	10.6
LD (μm)	0.10	0.13	0.07
WD (μm)	0.80	0.50	1.10
ΔV_t (V)	0.000	−0.060	0.060
pFET			
TOX (nm)	10.0	9.4	10.6
LD (μm)	0.12	0.16	0.08
WD (μm)	0.90	0.50	1.30
ΔV_t (V)	0.000	0.080	−0.080

faster. Also note that the pFET threshold voltage offset is described in terms of the real (negative) value of the threshold voltage.)

By plugging the skew values into the FET models, descriptions of the 3σ-slow and 3σ-fast FET behavior can be reached. This method also allows the selection of other σ values, if these are deemed to be more representative of device and circuit behavior. This last issue is considered in detail in Chapter 15.

This approach does have a number of advantages. The parameters selected have physical meaning, so their values actually describe the behavior of the process. These parameters are usually compiled as part of routine monitoring of the fabrication process, so their description is usually available as a matter of course. This obviates any need to engage in repetitive extraction of electrical parameters to determine means and variations; this is very time consuming, particularly for the more complicated FET models. Finally, this simple subset of parameters occurs in all the different FET models; using these parameters, a description of process variability can be constructed which applies equally well to all the different FET models. It is common for design requirements to force there to be more than one type of model for the same process technology (e.g., Level 3 and BSIM2 models might be constructed for large-scale digital designs and analog circuits, respectively). Using this simple method, only one set of parameters for describing process variations is developed for all the different model types. In addition, if a binned model is used, these parameters are independent of the device geometry (the channel length and width) and can be applied transparently for all model bins.

However, this approach does have a number of disadvantages. Clearly, the selection of a designated subset of parameters is a somewhat arbitrary exercise, and the resulting description is by definition incomplete. Any physical meaning that is hidden in the electrical parameters is neglected. As a result, further work must be done to correlate the model of process variability with specific circuit level results (rather than FET model results). Often, different results must be imposed on different types of circuits. This problem is considered in detail in Chapter 15.

14.4 Principal Component Analysis

As noted in Section 14.3, the selection of a few parameters for the description of process variations is a somewhat arbitrary exercise, and the exclusion of electrical parameters that have hidden physical meaning indicates that any such description is, by its very nature, incomplete. On the other hand, if each parameter is treated independently and its variations are included directly, the predicted variations will be larger than is really the case; some parameters are correlated, and vary together.

It would be desirable to find some quantitative (and hopefully rigorous) method of analyzing all the parameters and their variations which allows the correlations to be taken into account; this would permit a more proper description of the *total* variation of the process, rather than an erroneous summation of the variability of various parameters.

A method of statistics know as principal component analysis can be used for this purpose. This approach has been known for some time, but it has only recently been applied to the problem of describing the variability of a CMOS fabrication process using the FET model parameter set. There is presently a paucity of results, but the work to date has been promising. Here, the application of principal component analysis to the problem of process variations will be considered. Details of the origin and fundamentals of this method may be found in a textbook on multivariate statistical methods [4].

14.4.1 Principal Component Analysis and FET Models

To apply principal component analysis to process variations and SPICE FET models, a list of the model parameters to be considered must be drawn up. From first principles, this list would include *all* the parameters (both DC and AC) associated with the particular FET model to be used. This can be quite daunting in some of the more complex models, which contain dozens of parameters. The absolute best results are obtained by including all the model parameters; however, an experienced user may be able to carefully screen the parameter list. The repeated use of principal component analysis should also allow certain parameters to be deemed as not involved in process variations, and thus discarded from the analysis.

Next, the sources of data must be identified. As will be discussed below, principal component analysis is a statistical method, so a large amount of data is required. For analysis, data can be collected to compare within-die variations, die-to-die variations, and/or wafer-to-wafer variations; or, a huge bulk data base of manufacturing line information can be continually collected. The methods used are the same, regardless of how the data are assembled.

For purposes here, consider a test structure containing all the FETs required for parameter extraction; let this pattern be repeated once in each die. The data from each die are collected independently, and used for extraction of the appropriate model parameters. From the extraction, a set of model parameters specific to each die is assembled. There should be enough data for meaningful values of the mean and standard deviation of each parameter to be found. For this exercise, let there be n total parameters. As a working example here, a fictitious model with 6 parameters will be employed. These parameters are named X_1, X_2, ... X_6. The mean values of these parameters are listed in Table 14.2. For the later purposes of determining which parameter skewing direction corresponds to the fast or slow side of the process, the parameter X_5 is the low field channel mobility. Note that in the following description of principal component analysis,

Table 14.2   The mean values of the six parameters used in the fictitious model of this example.

Parameter	Mean
X_1	6.907
X_2	14.23
X_3	27.37
X_4	2.429
X_5	645.56
X_6	17.56

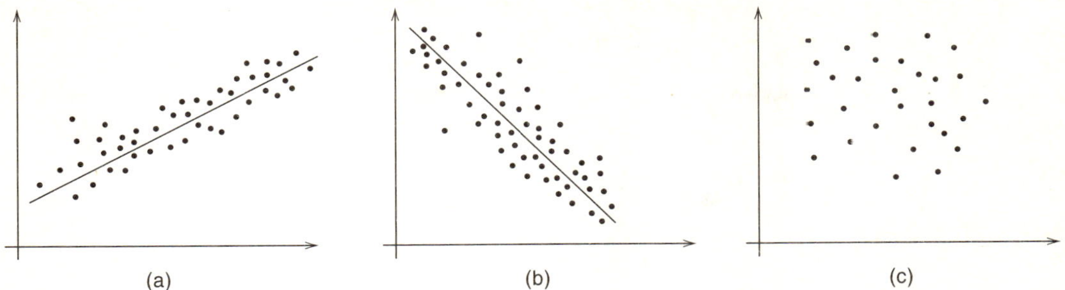

Figure 14.2   The correlation of two parameters from a large collection of data; (a) strong positive correlation; (b) strong negative correlation; (c) weak correlation.

the mean parameter values are employed in the computations, but the standard deviations are not. The standard deviations enter the calculations through the matrix of correlation coefficients, to be described below.

The next step involves determining the correlation coefficient between each pair of parameters. This is equivalent to treating a pair of parameters from each die as an (x,y) point in an x-y plot; when all the points for all the dies are assembled, an attempt is made to carry out a linear regression fit of the data to a straight line. If the two parameters correlate well, the data will be well fit by a straight line, and the computed correlation coefficient will be close to ± 1 (Figures 14.2a and 14.2b). If the parameters are less well correlated, a correlation coefficient lower in magnitude will be found (Figure 14.2c).

Since there are n parameters, the correlation between each parameter and every other parameter can be used to the fill the $n \times n$ correlation matrix. Since each parameter correlates perfectly with itself, each element on the matrix diagonal is 1. The matrix will also be symmetric, so only the elements on one side of the diagonal need to be filled. Any particular matrix element can be described as C_{ij}, the correlation coefficient between parameter i and parameter j. The correlation matrix values to be used here are collected in Table 14.3.

Once the correlation matrix has been filled, it is diagonalized. The theoretical details of matrix diagonalization may be found in a linear algebra textbook [5], while numerical

Table 14.3   The matrix of correlation coefficients for the six parameters. Since the matrix is symmetric, only the elements on one side of the diagonal need to be filled.

	X_1	X_2	X_3	X_4	X_5	X_6
X_1	1					
X_2	0.7817	1				
X_3	0.5755	0.6721	1			
X_4	−0.3795	−0.5385	−0.5523	1		
X_5	0.2477	−0.0894	0.1793	0.3241	1	
X_6	0.1167	0.0828	0.2394	0.3751	0.0920	1

Table 14.4 The results of diagonalizing the matrix
of correlation coefficients; (a) the eigenvalues;
(b) the eigenvectors associated with each particular eigenvalue.

(a)

Λ_1	0.0818
Λ_2	0.183
Λ_3	0.515
Λ_4	0.943
Λ_5	1.51
Λ_6	2.77

(b)

Λ_1	Λ_2	Λ_3	Λ_4	Λ_5	Λ_6
-0.382	0.491	0.539	-0.160	0.204	0.505
0.436	-0.593	0.377	0.126	-0.0129	0.548
-0.495	-0.279	-0.622	0.0468	0.156	0.513
-0.447	-0.492	0.357	0.0601	0.498	-0.424
0.372	0.00933	-0.191	-0.698	0.581	0.0195
0.284	0.297	-0.127	0.682	0.591	0.0396

routines for carrying out diagonalization can be obtained from the appropriate references
[6]. Diagonalization produces an $n \times n$ matrix of eigenvectors $\mathbf{U}$ and their n associated
eigenvalues; each eigenvector has a particular eigenvalue associated with it. Each column
of the eigenvector matrix $\mathbf{U}$ (e.g., U_{11}, U_{21}, ... U_{n1}) is a specific eigenvector, with
its associated eigenvalue (for the example here, Λ_1). The eigenvalues and eigenvectors
produced by diagonalizing the correlation matrix are compiled in Table 14.4. Note that the
eigenvalues are arranged from least to greatest, and that each eigenvector is normalized;
that is, for eigenvector i,

$$\sum_{j=1}^{n} U_{ji}^2 = 1. \tag{14.1}$$

Examining the eigenvector matrix, each eigenvector column contains the coefficients
that describe a principal component. Like the parameters, each principal component is
a number. As there are n parameters under consideration, there are also n principal
components.

The connection between the parameters and the principal components can now be
described. Let X_i be the numbered parameter values and P_i the principal component
values, where $i = 1, 2, ...n$. The principal components are expressed as a linear
combination of the parameters using

$$P_i = \sum_{j=1}^{n} U_{ji} \cdot X_j, \tag{14.2}$$

where U_{ji} is the jth component of the column eigenvector i. For the time being, the mean value of each parameter X_j is used in the summation. The parameters can also be computed from the principal components, using

$$X_i = \sum_{j=1}^{n} U_{ij} \cdot P_j, \tag{14.3}$$

where U_{ij} is the jth element of each column eigenvector i. How (14.2) and (14.3) are used will be discussed below.

Returning to the eigenvalues, as noted above, each eigenvalue is associated with a particular eigenvector, and thus with one specific principal component. The eigenvalue represents the variance of the principal component; the standard deviation of the principal component may be found from

$$\sigma_i = \sqrt{\Lambda_i}, \tag{14.4}$$

where σ_i is the standard deviation of principal component i. A larger eigenvalue implies a larger variance and a larger standard deviation; by implication, that particular principal component has a larger effect on the variability of the parameters.

The matrix computation of the eigenvalues (Λ_i) can produce them in any order; this is due to the arbitrary selection of the order of the input model parameters. As noted earlier, the numerical routine used here has chosen to order the eigenvalues (after their computation) from least to greatest. To determine the relative importance of each principal component, the eigenvalues can be arranged in order from greatest to least. These variance values can then be converted to a percentage value using

$$\Pi_i = \left(\frac{\Lambda_i}{n}\right) \times 100, \tag{14.5}$$

where n is the number of parameters and principal components. The percentage value reflects the importance of each principal component to the measured variations in the process parameters. The percentage variance values for the example in use here may be found in Table 14.5.

Table 14.5   The percentage variance and cumulative percentage variance of the six eigenvalues.

Eigenvalue	Pct. Var.	Cum. Pct. Var.
Λ_6	46.13	46.13
Λ_5	25.14	71.28
Λ_4	15.72	86.99
Λ_3	8.589	95.58
Λ_2	3.054	98.64
Λ_1	1.363	100.0

Furthermore, by adding up the percentage values, the contribution of the combined principal components can be determined. As above, let the percentage variances computed from (14.5) be ordered from greatest to least, where Π_6 is the largest percentage variance and Π_1 is the smallest. By itself, Π_6 represents the percentage of parameter variability that is accounted for by principal component 6. Next, the sum $(\Pi_6 + \Pi_5)$ can be computed; the result represents the percentage of parameter variability that is accounted for by principal components 6 and 5. This can be continued, with $(\Pi_6 + \Pi_5 + \Pi_4)$ representing the percentage of parameter variability accounted for by principal components 6, 5, and 4. This successive summation process is continued for all the percentage variances; these results are also found in Table 14.5. By definition,

$$\sum_{i=1}^{n} \Pi_i = 100\%, \tag{14.6}$$

since all the variability of the parameters is described by all the principal components.

This process can be useful for simplifying the number of principal components which are later used to compute the actual variations of the model parameters. Note that the four most important principal components account for more than 95% of the variability, while the first five principal components account for nearly 99% of the variability. To reduce the magnitude of succeeding computations, principal components which make small contributions to the overall variability are sometimes discarded.

The use of the principal components to compute the actual skew of the model parameters can now be considered. The first step is to determine if adding the standard deviation σ_i to its associated principal component P_i will cause the model to be skewed toward the slower or the faster end of the process. This is best illustrated by considering the model parameters themselves; a lower value of the oxide thickness represents a skew toward the faster end of the process, while a lower value of the mobility represents a skew toward the slower end of the process. Each principal component must be considered individually. This is done by computing the parameters from the principal components as in (14.3), except that one principal component has three times its standard deviation added to it. Selecting principal component k as the one under scrutiny at one time,

$$X_i = \left[\sum_{j \neq k}^{n} U_{ij} \cdot P_j \right] + [U_{ik} \cdot (P_k + 3 \cdot \sigma_k)]. \tag{14.7}$$

This will isolate the effect of skewing each individual principal component, independently of all the other principal components.

Whether the addition of σ_k to P_k represents fast or slow skewing may be obvious from examination of the resulting parameter set. If this is not the case, DC or even AC SPICE simulations may have to be performed to clearly determine the answer. Here the identification of X_5 as the low field channel mobility is sufficient to allow the determination of this information. The effect of skewing each individual principal component on X_5 is compiled in Table 14.6. By recalling the mean value of X_5 from

Table 14.6   The effect of skewing each principal component
on the value of parameter X_5, the low field channel mobility;
this allows for the determination of the sign sensitivity
of each principal component, as represented by the parameter ζ_i.

Prin. Comp. Skewed	X_5	ζ_i
P_1	645.88	+1
P_2	645.57	+1
P_3	645.15	−1
P_4	643.53	−1
P_5	647.70	+1
P_6	645.66	+1

Table 14.7   The computed mean, 3σ-fast, and 3σ-slow skew
parameter sets, as determined by principal component analysis.

Parameter	Mean	3σ-fast	3σ-slow
X_1	6.907	9.789	4.025
X_2	14.23	15.35	13.11
X_3	27.37	30.92	23.81
X_4	2.429	0.1892	4.669
X_5	645.56	650.57	640.55
X_6	17.56	18.84	16.28

Table 14.3, it can be seen that addition represents fast skewing for principal components
1, 2, 5, and 6, while addition represents slow skewing for principal components 3 and
4. This allows for the assignment of the sign sensitivity term ζ_i; as shown in Table
14.6, ζ_i is assigned a value of +1 or −1 based on the sign sensitivity of each principal
component. The introduction of ζ_i is a mathematical convenience that will be put to use
below.

After all the sign sensitivities are determined for all the principal components,
the skewed parameter sets can finally be determined. For the example here, let the first
skewed parameter set to be determined represent the 3σ-fast process. For this case, (14.3)
is modified to

$$X_i^{3\sigma,f} = \sum_{j=1}^{n} U_{ij} \cdot \left\{ P_j + \left[\zeta_j \cdot (3 \cdot \sigma_j) \right] \right\}, \tag{14.8}$$

where $X_i^{3\sigma,f}$ is the 3σ-fast version of parameter X_i; the introduction of ζ_j serves to
simplify the introduction of the sign sensitivity into the computations. The resulting 3σ-
fast skewed parameter set may be found in Table 14.7.

The 3σ-slow parameter set is determined in the same manner, by merely reversing
the sign of the sign sensitivity term ζ_j. The resulting equation is very similar to (14.8):

$$X_i^{3\sigma,s} = \sum_{j=1}^{n} U_{ij} \cdot \left\{ P_j + \left[-\zeta_j \cdot (3 \cdot \sigma_j) \right] \right\}, \tag{14.9}$$

where $X_i^{3\sigma,s}$ is the 3σ-slow version of parameter X_i. The 3σ-slow skewed parameter set that results may also be found in Table 14.7. Note that for five of the parameters (X_1, X_2, X_3, X_5, and X_6), a larger parameter value corresponds to fast skewing, while a larger value of X_4 corresponds to slow skewing.

The choice here of 3σ to represent the fast and slow regions of the process is completely arbitrary; any choice for the number of standard deviations can be employed. The number of standard deviations chosen is often determined by semi-empirical observation of the available DC and AC data. These issues are considered in some detail in, respectively, Section 14.5 and Chapter 15.

14.4.2 Discussion

Principal component analysis has a number of advantages. It provides a more complete and rigorous description of the effect of process variations on the model parameter set. This is achieved by accounting for the variability of all the parameters in a consistent manner. By developing a more accurate description of the variabilities of the process, designs can stay within the expected process results without being too cautious and thus less aggressive.

This improved rigor and accuracy does have its disadvantages. The method is very complex and computationally intensive. Since simple routine fabrication data is not used for describing the process variations, a massive amount of parameter extraction must be carried out to create a large data base of model parameter sets for analysis. If more than one type of model (e.g, Level 3, BSIM2, etc.) is employed for a particular process technology, the entire analysis must be carried out for *each* type of FET model. In the more complex models, with dozens of parameters, the analysis becomes very bulky. In addition, if a model is binned, the results of principal component analysis do not apply across the entire geometry space, but must be carried out independently for each model bin. To further add to the situation, if some method is employed to guarantee the continuity of the model across bin boundaries (as in HSPICE Level 28, described in Chapter 9), for each set of skewed model parameters, the length, width, and (if available) product terms must be recomputed and included for each specific skewed set of model parameters. The full application of principal component analysis in such a situation can rapidly lead to a modeling nightmare!

As noted at the outset of this section, the use of principal component analysis to describe process variability is a recent development, so as yet there are insufficient results to firmly judge the effectiveness of the method. The available results are encouraging. Principal component analysis has been successfully used to describe the slow corner of a digital model [7], and the mismatch of devices on the same die [8]; the latter is an important consideration in analog circuit design. Hopefully, as the method is employed more frequently, the trade-offs between its rigor and its concomitant complexity can be more fully evaluated.

14.5 Describing Fast and Slow Process Corners

The analysis described in Sections 14.3 and 14.4 was limited; the goal was to describe the means and standard deviations of the nFET technology and the pFET technology separately. Naturally, this subject must be extended to describe the combined behavior of nFETs and pFETs. Surprisingly, little rigorous effort in this area has been reported. Generally, an empirical or semi-empirical approach is used; results are usually adequate, but as will be noted in the descriptions, the development of more rigorous methods would greatly improve the ability to account for process variations.

For the time being, let there be available a description of both the nFET and the pFET that describes the statistically typical devices and the standard deviations around that typical point; due to its simplicity, the skew method using a subset of four parameters, as described in Section 14.3, is employed here (see Table 14.1). Consider the plot of pFET current versus nFET current in Figure 14.3a. By noting the typical, 3-σ slow and 3-σ fast points as pairs, a straight line is produced with those points noted along it. The fast-fast pair denotes the fastest corner of the process, while the slow-slow pair denotes the slowest corner of the process. If the nFETs and pFETs always varied around the typical point by exactly the same σ value, Figure 14.3a would properly describe the possible variations.

However, there will be some offset between the nFETs and the pFETs. A common way of coping with this situation is depicted in Figure 14.3b. Here, the fast-fast and slow-slow process corners still exist. In addition, the alternate combinations of fast and slow are included to more completely describe the process. New corners, representing slow nFET-fast pFET and slow pFET-fast nFET are added, leading to the rectangle of Figure 14.3b. This approach completely covers any nFET/pFET offset from the diagonal line of Figure 14.3a.

While this method is all-encompassing, it assumes that the deviations from the typical point of the nFET and the pFET are completely independent of each other. While not as perfect as shown in Figure 14.3a, there is obviously some correlation between the nFET and the pFET. This is confirmed by the collection of data points in Figure 14.3c. The data follow the general line described in Figure 14.3a, but do spread out around it. Note that no data points come even close to populating either fast-slow corner.

Using the data of Figure 14.3c, an alternative and more accurate method can be developed; this is shown in Figure 14.3d. The diagonal line of Figure 14.3a and the rectangle of Figure 14.3b form the basis of outlining the area to be considered. However, two diagonal lines are drawn to (empirically) bound the data; this is in essence a method of more carefully defining a subset of the rectangle approach. A stretched hexagon is defined which contains all the measured data, but without the addition of the two process corners which are devoid of data. The two fast-slow corners of the rectangle method are replaced by four new process corners, offset in pairs from the slow-slow and fast-fast corners. Toward the slow corner of the process, there are now corners representing a slow pFET with a slightly faster nFET, and a slow nFET with a slightly faster pFET. Toward

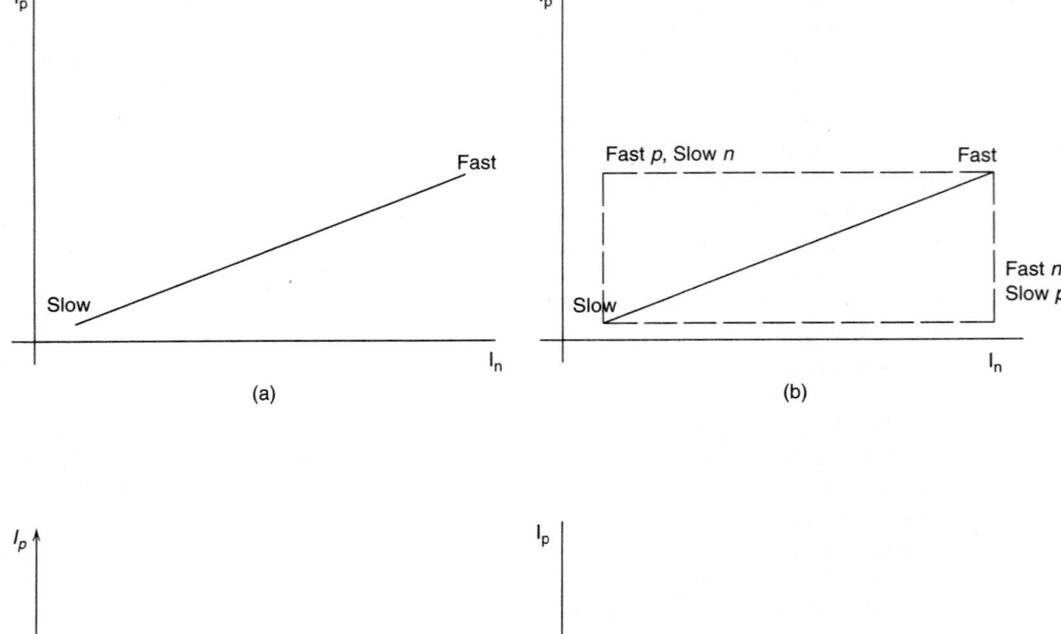

Figure 14.3 Combined nFET versus pFET current for some CMOS technology, for fixed values of drawn channel length and width at some given temperature; (a) perfect correlation—the model follows a diagonal line; (b) no correlation—the model describes a rectangle encompassing all possible nFET and pFET current values; (c) typical data—the correlation between the nFET and the pFET is not perfect, but is fairly good; (d) the construction of a compromise method, based on the rectangle method of (b), but which eliminates the wasteful fast-slow corners; note that this method leads to six process corners.

the fast corner of the process, there are now corners representing a fast nFET with a slightly slower pFET, and a fast pFET with a slightly slower nFET.

This approach has the advantage of factoring out the fast-slow corners, which are not reached by the process. This in turn leads to more realistic circuit design, as those unpopulated corners do not have to be considered when simulating the effect of process variations on the behavior of circuits. A proper determination of these four new process corners requires an examination of device mismatch, which is described in Section 14.6.

As noted in Chapter 2, if the data of Figure 14.3d can be modeled, the results can actually be used for more aggressive designs. If the region is well described, it is possible to push the design toward the fast corner of the process (instead of targeting the typical point). This method is often employed in high-performance digital circuit design, when there is an overwhelming need to produce faster parts. In this manner, the number of faster parts can be increased; this is achieved at the cost of lowering the overall yield, as the process distribution will spread outside the normal window. FETs which fall beyond the fast-fast corner will likely fail due to extreme short channel effects.

14.6   Device Mismatch

In Section 14.5, the two extremes of the combined description of nFETs and pFETs with process variations were considered. In one case (Figure 14.3a), nFETs and pFETs correlate perfectly, and the combined variations follow a straight line. At the other extreme (Figure 14.3b), the variations are completely independent of each other, which leads to a rectangle defined by the ends of the straight line of Figure 14.3a. Examination of data (Figure 14.3c) shows that the real situation is somewhere between the two extremes, leading to the compromise solution of Figure 14.3d. A quantified description of the process requires consideration of the degree of correlation between nFETs and pFETs, which is neither perfect nor nonexistent.

Despite the obvious importance of this topic, it has received little attention in the literature, and there are few references to cite [8, 9]. Without several rigorous studies, only a limited discussion can be carried out; this suggests some approaches to be used, but the reader is cautioned that only a sketchy outline is possible at this time.

Device mismatch generally needs to be considered on two separate levels. One level involves producing a more accurate description of the correlation between nFETs and pFETs as discussed above; this is an important consideration in digital circuit design. A second level concerns the mismatch between different transistors in the same circuit; this situation is of particular concern in analog circuit design.

14.6.1   Global Mismatch

For this discussion, consider once again the method employed in Section 14.3 (using the skew parameter sets of Table 14.1), where a small number of parameters were selected as being physically based and assigned as process skew components. Also assume that data are repeatedly gathered from many sample dies on many different wafers. In addition to measuring the nFET and pFET skew parameters on each die, the difference between them should be collected as well. With sufficient data, statistical analysis can be performed, and a mean and standard deviation produced; typical results are compiled in Table 14.8.

Reexamining Figure 14.3d, with its six process corners, it is clear that the two most extreme corners correspond to fast-fast and slow-slow models for the nFET and pFET. The four offset corners must also be described. Consider the fast-fast process corner. One

Table 14.8   The computed 3σ mismatch between nFET and pFET process parameters.

TOX (nm)	0.000
LD (μm)	0.010
WD (μm)	0.040
ΔV_t (V)	0.010

Table 14.9   The computed offset corner parameters, which describe all six corners of the stretched hexagon of Figure 14.3d; (a) the fast end of the hexagon; (b) the slow end of the hexagon.

(a)

	Fast-Fast		Fast-n/Mod.-p		Fast-p/Mod.-n	
	n	p	n	p	n	p
TOX (nm)	9.4	9.4	9.4	9.4	9.4	9.4
LD (μm)	0.13	0.16	0.13	0.15	0.12	0.16
WD (μm)	0.50	0.50	0.50	0.54	0.54	0.50
ΔV_t (V)	−0.060	0.080	−0.060	0.070	−0.050	0.080

(b)

	Slow-Slow		Slow-n/Mod.-p		Slow-p/Mod.-n	
	n	p	n	p	n	p
TOX (nm)	10.6	10.6	10.6	10.6	10.6	10.6
LD (μm)	0.07	0.08	0.07	0.09	0.08	0.08
WD (μm)	1.10	1.30	1.10	1.26	1.06	1.30
ΔV_t (V)	0.060	−0.080	0.060	−0.070	0.050	−0.080

offset corner involves a fast nFET and a slightly slower pFET. The model for the slightly slower pFET is constructed by starting with the fast pFET model, then changing the parameters to slightly slower values using the standard deviations from the *correlation* data described in Table 14.8. The results are found in Table 14.9a, as the skew model for a fast nFET and a modified pFET; the modified pFET values are reached by starting with the fast pFET values, and using the mismatch values of Table 14.8 to slow the fast pFET model slightly. By the same procedure, a model for the other offset corner, a fast pFET and a slightly slower nFET, is also constructed (as contained in Table 14.9a).

At the slow-slow end, the same general approach is used, except that the correlation offsets are skewed toward slightly *faster* parameter sets. This produces the offset corners for, respectively, a slow nFET with a slightly faster pFET, and a slow pFET with a slightly faster nFET. Skew parameters for the three corners at the slow end of the process are compiled in Table 14.9b.

This method is essentially semi-empirical, so the choice of the number of standard deviations to be used to quantify the offset corners is somewhat arbitrary. Here 3σ was used, since this is the number of standard deviations of the fast-fast and slow-slow corners. However, the actual number of standard deviations to be used is up to the discretion of the model builder; 3σ is a common choice for skew model construction, with a different multiple of σ being used at the circuit simulation level, as discussed in Chapter 15.

Although the method is more complicated, principal component analysis can also be used to determine the offset corners. This will more accurately determine these values; the method of implementation is identical to that just described.

Whatever method is employed, the goal is to develop a model for the variability of the process; this model should allow the circuit designer to verify that the circuit will function anywhere within the defined polygon of operation, as described in the form of Figure 14.3d. In addition, a model of this sort allows construction of quantitative predictions of circuit behavior for the most extreme expected variations in the fabrication process.

14.6.2 Local Mismatch

The offsets considered in Section 14.6.1 involved finding a more accurate description of an allowed region of FET behavior; a bounded area is produced which includes all possible FET current values while excluding regions where no combined nFET-pFET current value pairs exist. This type of description works well for digital circuit design.

Analog circuit design imposes additional requirements. It is necessary to describe the largest amount of offset that can occur between two specific transistors [8–11]; the distance between two such transistors becomes a consideration.

As noted at the beginning of this section, the modeling of device mismatch is an important topic, but it has received little attention, and there are few archival references to cite. Rather than attempt to produce a detailed description from limited sources, a simple example will be given to illustrate the basic requirements. More details may be found in an excellent monograph on the subject [12].

Consider the current mirror circuit shown in Figure 14.4. If the two FETs are identical, then the currents i_1 and i_2 will also be identical. However, it is (statistically) likely that the two nFETs will be somewhat different, so that i_1 and i_2 will differ by some amount. If the structures of the two nFETs were completely independent of each other, one nFET could be described by the 3σ-fast model, and the other nFET by the 3σ-slow model. However, this is not the case; some degree of correlation exists.

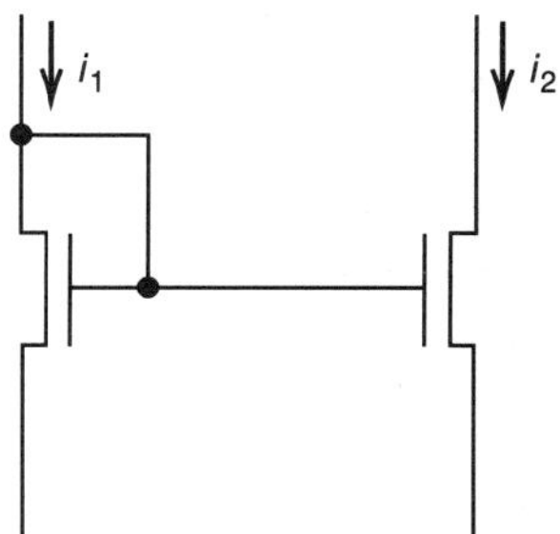

Figure 14.4   A two-nFET current mirror circuit.

Table 14.10   The effect of nFET mismatch on the skew parameter set, which is used to describe the mismatch between the nFETs of the current mirror circuit of Figure 14.4; (a) 3σ skew parameter mismatch versus distance between the two nFETs; (b) nFET mismatch around the typical process point for a distance d_1 between the two nFETs.

(a)

	d_1	d_2	d_3
TOX (nm)	0.00	0.00	0.00
LD (μm)	± 0.010	± 0.020	± 0.023
WD (μm)	± 0.010	± 0.021	± 0.037
ΔV_t (V)	± 0.0023	± 0.0036	± 0.0042

(b)

	Center	More Current	Less Current
TOX (nm)	10.0	10.0	10.0
LD (μm)	0.10	0.11	0.09
WD (μm)	0.80	0.79	0.81
ΔV_t (V)	0.000	-0.0023	0.0023

For this example, the methods of Section 14.6.1 will be employed, but with two significant differences. First, sets of nFET parameters will be correlated with other sets of nFET parameters. Second, the distance between individual nFETs appears as a new variable; the smaller the distance between two FETs, the greater the degree of parameter correlation. Sample results are collected in Table 14.10a for three different distances between two individual nFETs.

Using the data of Table 14.10a, skew models for mismatch are easily constructed. As an example, mismatch models for two nFETs separated by a distance d_1 are compiled in Table 14.10b. One nFET (n_1) is skewed to produce larger current values, while the other (n_2) is skewed to produce smaller current values. By plugging these skew parameter sets into the main FET model, the predicted current values are found.

This example used the nFET typical point as its base; however, this selection was completely arbitrary. Using some multiple of the standard deviation σ, any point along the line of possible nFET current values can be selected, and the same mismatch analysis performed from there. In addition to the effect of process variations on the overall situation, the effect of mismatch variations is thus considered.

Note that this aspect of analog circuit design increases the extent of the FET modeling requirements. In particular, the need for data collection and modeling of process variability expands significantly. Furthermore, the constraints faced by a circuit designer are also increased. A proper design can only be carried out to the extent that models for mismatch (over various FET separation distances) are available.

The method described here is simple, using only the subset approach of selecting a few model parameters for skew analysis. Attention is now being paid to the more sophisticated method of principal component analysis for dealing with this problem [8, 12].

14.7   Final Comments

As noted at the beginning of this chapter, the subject of accounting for systematic process variations in the SPICE FET models is not well developed. Historically, the models have been designed for the description of individual transistor characteristics; the inclusion of the effect of process variations has been added as an afterthought as part of model postprocessing.

The basic approach is to find some method of defining boundaries on the device behavior, so that designed circuits remain within those boundaries. Defining useful boundaries is not difficult; however, the simple approaches which are commonly employed often lack rigor. Because the circuit designer is inadvertently forced to deal with process results which in fact do not occur, design aggressiveness is limited.

In an ideal FET model, all parameters would *in toto* be physically based and completely independent of each other. By compiling statistical data on the process, it would be a simple matter to transfer that information to the SPICE FET model parameter set; an accurate description of skewing would be readily available.

However, since the models contain a number of electrical and empirical parameters, all of which have a murky physical basis, there is no clean route to such a description. A common solution to this problem is to select a subset of parameters and designate them as physically based and responsible for all the process variations. Since this designation is somewhat arbitrary, the method is clumsy and incomplete, and is usually forced to work at the circuit simulation level; this is described in Chapter 15.

More recently, principal component analysis has been employed to describe the effect of process variations on the model parameter set. This method is more complete and rigorous, and provides a much more accurate representation of process variability. On the down side, principal component analysis requires a tremendous amount of parameter extraction to build a statistical data base; the data and parameter extraction needs become excessive in binned models.

There are thus many shortcomings in any of the available methods of accounting for process variations at the SPICE FET model level. This transfers much of the problem to the circuit simulation level, as discussed in Chapter 15.

BIBLIOGRAPHY

1. W. Shewhart, *Statistical Methods from the Viewpoint of Quality Control*, Department of Agriculture, The Graduate School, Washington University, 1939.

2. W. Deming, *Statistical Adjustment of Data*, Dover, 1964.

3. P. Ross, *Taguchi Techniques for Quality Engineering*, McGraw-Hill, 1988.

4. W. Dillon and M. Goldstein, *Multivariate Analysis Methods and Applications*, John Wiley & Sons, 1984.

5. S. Isaak and M. Manougian, *Basic Concepts of Linear Algebra*, W. W. Norton, 1976.

6. W. Press, S. Teukolsky, W. Vetterling, and B. Flannery, *Numerical Recipes in Fortran* (2nd ed.), Cambridge University Press, 1992.

7. K. Burke et al., "Worst-Case MOSFET Parameter Extraction for a 2 μm CMOS Process," *Proc. 1994 Int. Conf. on Microelec. Test Struc.*, pp. 119–125.

8. C. Michael and M. Ismail, "Statistical Modeling of Device Mismatch for Analog MOS Integrated Circuits," *IEEE J. Sol. St. Circ.* vol. SC-27, pp. 154–166 (1992).

9. M. Pelgrom, A. Duinmaiger, and A. Welbers, "Matching Properties of MOS Transistors," *IEEE J. Sol. St. Circ.* vol. SC-24, pp. 1433–1439 (1989).

10. J. Shyu, G. Temes, and F. Krummenacher, "Random Error Effects in Matched MOS Capacitors and Current Sources," *IEEE J. Sol. St. Circ.* vol. SC-19, pp. 948–955 (1984).

11. K. Lakshmikumar, R. Hadaway, and M. Copeland, "Characterization and Modeling of Mismatch in MOS Transistors for Precision Analog Design," *IEEE J. Sol. St. Circ.* vol. SC-21, pp. 1057–1066 (1986).

12. C. Michael and M. Ismail, *Statistical Modeling for Computer-Aided Design of MOS VLSI Circuits*, Kluwer, 1993.

15

Circuit Level Correlation of Models and Hardware

15.1   Introduction

In the bulk of this text, the discussion has centered on the relationship between models and data at the FET level. In Chapters 5–12, the individual model formulations were examined; with each particular formulation representing the upper limit of a model's capability, methods for the extraction of model parameters that allow the best possible fit to measured DC transistor characteristics were considered. In Chapter 13, the discussion expanded to include the active device capacitance/charge models, while Chapter 14 discussed methods of accounting for the effect of systematic process variations on the FET characteristics.

For some circuits, a good description of the DC characteristics of the devices is sufficient for accurate circuit simulation. However, in other circuits, the AC and frequency response characteristics are the most important results. In most digital circuits, an accurate model for the delay time of a signal propagating through that circuit is required.

In this chapter, some methods of matching circuit results and SPICE FET models are considered. As was the case for approaches to describing systematic process variations (discussed in Chapter 14), this subject is rather poorly developed; results are "made to work" in a fashion that is largely empirical. It is at this level that the various model shortcomings described in Chapters 5–14 are most clearly seen; these problems must be dealt with for reasonable AC simulation results.

Here the problem will be considered at two levels. First, as discussed in Chapter 13, there are structural problems with the SPICE FET gate capacitance models which must be remedied by semi-empirical or empirical corrections; this is necessary to reach agreement with delay results from simple circuits. Second, the process skew descriptions of Chapter 14 translate directly into a predicted distribution of circuit delay results. As

529

these predictions usually differ from measured data, some method of bringing the model and the data together must be developed.

15.2   Circuit Correlation with the Gate Capacitance Parameters

As described in Chapter 13, the SPICE FET gate capacitance models that are currently available suffer from some shortcomings that become critical in short channel devices. The zero bias gate capacitance is introduced with the gate-drain and gate-source capacitance parameters **CGDO** and **CGSO**; these capacitances are added as constants to the total active device capacitance, which is computed for various terminal biases using the FET model charge equations.

As shown in Figure 15.1, one component of the zero bias gate capacitance, the inner-fringe contribution, is screened out by the inversion layer when $V_{gs} > V_t$. This contribution to **CGDO** and **CGSO** should be removed for higher gate biases. However, this is not done in the available gate capacitance models. In large devices, this oversight is negligible, and does not affect the results. In short devices, as detailed in Chapter 13, the erroneous inclusion of the inner-fringe capacitance for high gate biases has a significant effect.

15.2.1   Correcting the Capacitance for High Gate Biases

In most models, the only parameters which are available to affect the device capacitance are **CGDO** and **CGSO**; these are adjusted in an attempt to properly describe the device capacitance. Consider the results shown in Figure 15.2. It is assumed here that a very large FET structure has been used to measure the active device capacitance at high gate biases, when an inversion layer is present and the inner-fringe capacitance should be screened out; it is also assumed that the zero bias capacitance parameters **CGDO** and **CGSO** have been measured from an appropriate structure, as described in Chapter 13. The results show that the zero bias capacitance results are good (as expected), but that a larger than expected capacitance is predicted for high gate biases; this is the error caused by the failure to remove the inner-fringe contribution to **CGDO** and **CGSO**. Since $V_{dd} \gg V_t$, this error occurs in the region of greatest importance for accurate digital circuit simulation.

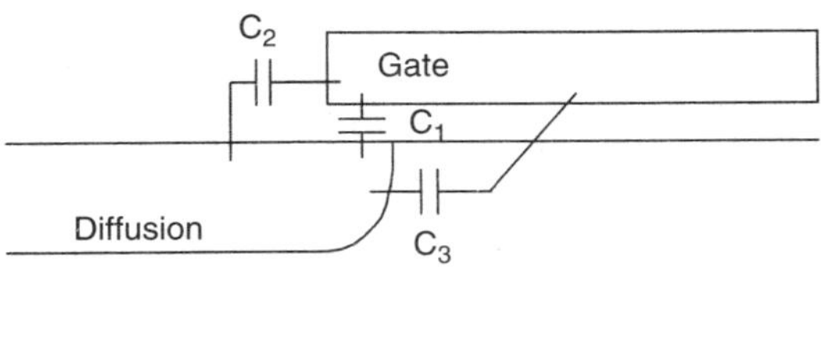

Figure 15.1   The zero bias components of the gate-diffusion capacitance. The direct overlap capacitance (C_1) and the outer-fringe capacitance (C_2) are unaffected by the presence of an inversion layer; however, the inner-fringe capacitance (C_3) is screened by the inversion layer and should not be included when an inversion layer is present.

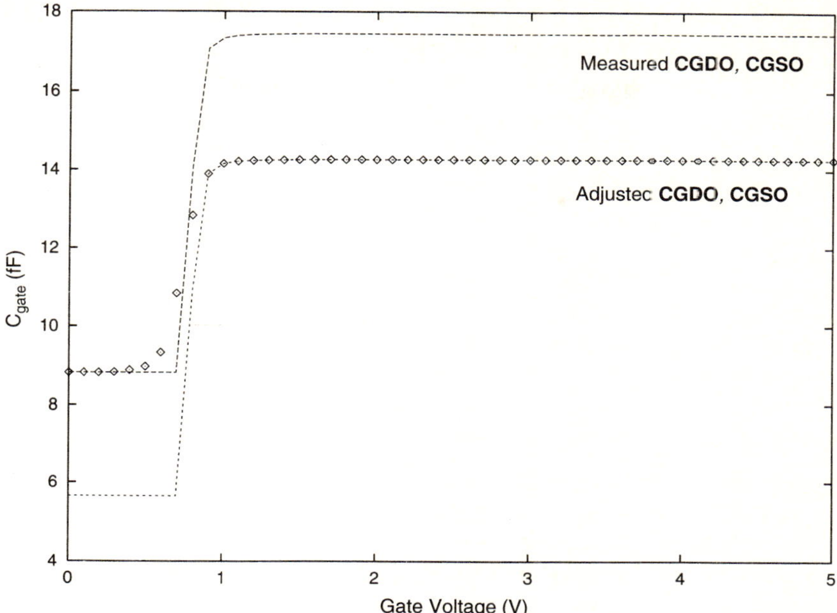

Figure 15.2 The active device capacitance as a function of the gate bias with a low drain bias. The points represent data, while the lines describe the active device capacitance computed with, respectively, measured and adjusted values of **CGDO** and **CGSO**.

Since the only capacitance model parameters are **CGDO** and **CGSO**, these terms provide the only method of improving the results. By decreasing the values of **CGDO** and **CGSO**, improved results are obtained, as shown in Figure 15.2. Essentially, the contribution of the inner-fringe term is subtracted from **CGDO** and **CGSO**, allowing for a better correlation with the data at high gate bias values. However, this is achieved at the cost of damaging the correlation for low gate bias values. Note also that the accuracy in the neighborhood of the threshold voltage is questionable; this is a concern in analog circuit design.

This is an unfortunate but unavoidable method of forcing the model to reach its best possible fit to the data. For digital circuits in which $V_{gs} \gg V_t$, this approach actually works quite well; since the load capacitance will be some value averaged over the entire range of the switching (gate) bias, the accuracy over the largest region of device operation $(V_{gs} > V_t)$ usually leads to good results. However, in low-voltage digital circuits, the supply voltage V_{dd} is greatly decreased while the threshold voltage V_t is not decreased [1, 2]. The region of inferior accuracy $(V_{gs} < V_t)$ becomes a larger percentage of the total operating range, and simulation accuracy consequently suffers. Furthermore, analog circuits can operate statically for virtually any value of V_{gs}. Since the structure of the

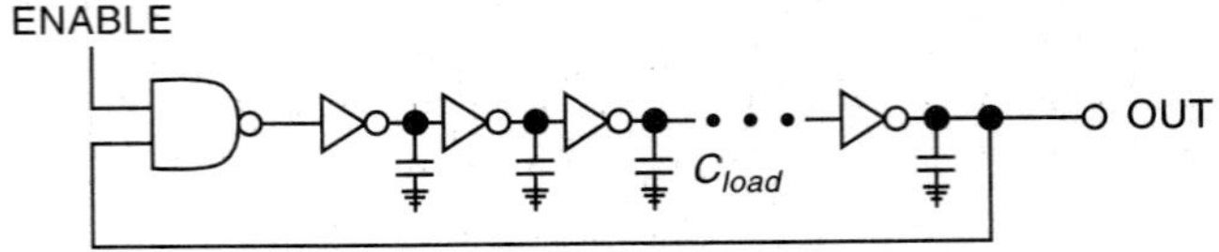

Figure 15.3   A ring oscillator circuit, with the output driving a load capacitor.

gate capacitance model forces the results to be inaccurate for some part of the operating range, an analog circuit will be forced to deal with an inaccurate device capacitance model during some part of its operation. This is a serious issue which has yet to be dealt with in SPICE FET modeling.

15.2.2   Direct Correlation with Circuits

The clumsy but workable method described by Figure 15.2 relies on the ability to measure the active device capacitance of some large FET structure. This requires very careful and delicate measurements which are difficult to perform [3]. As an alternative, the semi-empirical approach of Section 15.2.1 can be bypassed entirely in favor of an empirical method of correlating the gate capacitance parameters directly with circuit results.

Consider the ring oscillator circuit depicted in Figure 15.3. This ring oscillator drives a load capacitor; for the approach to be described here, the oscillator must be fabricated with several different values of that load capacitor (here, 0.1 pF, 1.0 pF, 10 pF, and 100 pF). To begin, the values of the zero bias gate capacitance parameters **CGDO** and **CGSO** are measured (**CGDO** = **CGSO** = 0.248 fF/μm) and inserted into the FET model; the resulting comparison of model and simulation results is shown in Table 15.1. For a large load capacitance, the correlation between the simulated and the measured stage delay is quite good; the large load capacitance is dominant, and the active gate capacitance is not important. The results deteriorate for smaller load capacitances, reaching a discrepancy of nearly 25% for a very light load of 0.1 pF. For the lightly loaded oscillator, the active device capacitance is dominant; the error introduced by the inclusion of the inner-fringe capacitance at high gate biases causes the substantial error which is observed.

Table 15.1   A comparison of measured and simulated ring oscillator delays, with **CGDO** and **CGSO** at both measured and adjusted values.

	Data	Model,**CGDO**=**CGSO**=0.248 fF/μm		Model,**CGDO**=**CGSO**=0.220 fF/μm	
C_{load}	Delay	Delay	Pct. Error	Delay	Pct. Error
0.1 pF	6.09 ps	7.56 ps	+24.3%	6.12 ps	+0.49%
1.0 pF	63.7 ps	71.6 ps	+12.4%	70.7 ps	−1.3%
10 pF	629 ps	681 ps	+8.3%	639 ps	+1.5%
100 pF	6.43 ns	6.37 ns	−0.93%	6.36 ns	−1.1%

By trial and error (or, if desired, by a somewhat more sophisticated computation), the values of **CGDO** and **CGSO** can be adjusted empirically to allow the best possible agreement with the measured results. Here, the optimum value of the gate capacitance was found to be **CGDO** = **CGSO** = 0.220 fF/μm. Results using this parameter value are also compiled in Table 15.1. Good results are now obtained, with discrepancies between the simulations and the data of no more than ±2%.

15.2.3   Summary

All of the approaches described in this section are caused by a basic shortcoming of the available active gate capacitance models; these models fail to properly describe the inner-fringe contribution. Due to this deficiency, semi-empirical and/or empirical approaches are employed, in which **CGDO** and **CGSO** are used as variable parameters for matching measured delay data and simulations. As shown here, results are "made to work" for a particular set of circuits. Since the approach is essentially empirical, it might be necessary to revisit the determination of **CGDO** and **CGSO** for different types of circuits. It should also be noted that this method is at greater risk in low voltage digital circuits and analog circuits, where the accuracy of the active gate capacitance over the entire bias range is more important.

15.3   Circuit Correlation with Process Variations

In Sections 14.3, 14.5, and 14.6, various methods of accounting for systematic process variations in SPICE FET models were considered. This involved some approach to describing the distribution of nFET and pFET drain current values, and the correlation between them. A similar approach to examining the distribution of circuit delay results must now be undertaken.

In Section 14.3, a simple method of describing FET process variations was developed; several parameters were selected, and their statistical variability was used to compute the resulting predicted variation in the FET drain current. There, it was noted that the statistical ±3σ definition should encompass 99.6% of the available data; however, in reality, that fraction of the data is often contained within roughly ±2σ. Some method of connecting the *predicted* 3σ distribution and the *measured* 3σ distribution must be found. Although this could be done for the current values, such a determination is usually carried out on circuit results.

To model the distribution of circuit delays, a large amount of data must be collected. This allows for the construction of the histogram distribution of ring oscillator stage delay data depicted in Figure 15.4. If the fabrication process is stable, and variations about the mean are due only to statistically random events (see Section 14.2), the results should form a Gaussian distribution, like that found in Figure 15.4. The mean value of the distribution should correlate with the simulated mean; this is likely done using the methods described in Section 15.2. Next, the ±3σ delay values predicted by the process skew model are computed and noted on the distribution of data. As expected from the results of Section 14.6, the predicted delay values of ±3σ are well outside the points

Figure 15.4   The distribution of delay values for a ring oscillator.

Table 15.2   The modified skew parameter set, showing both the original 3σ values, and the modified values used to represent ±3σ circuit behavior.

Parameter	Mean	3σ-fast	2.7σ-fast	3σ-slow	2.8σ-slow
nFET					
TOX (nm)	10.00	9.40	9.46	10.60	10.56
LD (μm)	0.100	0.130	0.127	0.070	0.072
WD (μm)	0.800	0.500	0.530	1.10	1.08
ΔV_t (V)	0.000	−0.060	−0.054	0.060	0.056
pFET					
TOX (nm)	10.00	9.40	9.46	10.60	10.56
LD (μm)	0.120	0.160	0.156	0.080	0.083
WD (μm)	0.900	0.500	0.540	1.30	1.27
ΔV_t (V)	0.000	0.080	0.072	−0.080	−0.075

within which 99.6% of the data are contained. Using the data, the points within which 99.6% of the data *are* contained can be found, as noted in Figure 15.4. These points correspond to some other multiple of σ; in this case, 2.8σ on the slow side, and 2.7σ on the fast side.

These artificial multiples of σ represent the de facto ±3σ values. Instead of the 3σ values of the process skew parameters (e.g., **TOX, LD**, etc.), values of 2.8σ (slow skew) and 2.7σ (fast skew) are employed. This is best illustrated by example in Table 15.2, where the simple process skew data developed in Section 14.3 (see Table 14.1b) is considered.

Also note, for example, that on the slow side of the distribution, *circuit* standard deviations of σ, 2σ, and 3σ correspond to *process* standard deviations of 0.9σ, 1.8σ, and 2.7σ. These artificial σ values are employed to compute the individual process skew parameter values for use in both direct worst-case analysis and in statistical simulations.

In digital circuits, this approach is not elegant, but it tends to work quite well. A large weakness of the method is that different results tend to be obtained for different groups of circuits (e.g., ring oscillator results will likely differ from those of logic gates, etc.). Different artificial multiples of σ must be determined for each circuit family. Another noteable drawback of this approach is that the model is forced empirically to agree with particular sets of measured data. Without a firm basis to build on, the extension of the models to groups of circuits for which data are not available must be undertaken with caution.

Finally, it should be noted that the selection of σ values specific to various circuit families transfers some of the modeling responsibility from the foundry to the design user. The foundry can provide generic information on the variability of the process. However, the design user must collect data and determine the artificial σ multiple values which are appropriate for each group of designed circuits.

15.4   Final Comments

As noted at the beginning of this chapter, the correlation of models and hardware is not firmly rooted in a physical description of the FET technology, but is instead a matter of forcing results together in a rather ad hoc manner. Simulations are "made to work"; this situation calls for considerable improvement. In particular, the shortcoming of the active gate capacitance model, in which the inner-fringe term is not removed when an inversion layer exists, requires correction. The glaring discrepancy between the predicted $\pm 3\sigma$ distribution of circuit delays and the actual values shows where a physically based model would be most valuable for circuit simulation; a physically based model is not necessary to describe individual transistor characteristics, but would be invaluable for a proper description of the effect of process variations.

The key feature of the methods described here is that they force simulated and measured results to agree for specific instances, rather than for the general case. The extendability of such models is thus limited, and care must be taken when using those models to predict the behavior of circuits that have not been fabricated.

The transfer of responsibility for the final circuit results away from the basic foundry modeling effort puts additional responsibility on the shoulders of the circuit design users. With this diffusion of effort, close cooperation among the fabrication facility, the model building effort, and the circuit design users becomes an important asset, and even a competitive advantage.

BIBLIOGRAPHY

1. E. Nowak, "Ultimate CMOS ULSI Performance," *1993 IEDM Tech. Dig.*, pp. 115–118.
2. D. Foty and E. Nowak, "MOSFET Technology for Low-Voltage/Low-Power Applications," *IEEE Micro*, July 1994, pp. 68–77.
3. N. Arora, D. Bell, and L. Bair, "An Accurate Method of Determining MOSFET Gate Overlap Capacitance," *Sol. St. Elec.* vol. 35, pp. 1817–1822 (1992).

16

New Model Candidates

16.1 Introduction

In Chapters 5–12, the FET models which are widely available for use in various general-purpose SPICE circuit simulators were reviewed in detail. For historical reasons, most of these models have originated at the University of California/Berkeley, with the UCB-developed versions of SPICE serving as the vehicle for distribution and acceptance of each model. Two significant exceptions were also included. HSPICE Level 28, discussed in Chapter 9, is a modified version of BSIM which has become widely employed due to the common use of the Meta-Software's HSPICE circuit simulator. More recently, MOS Model 9, developed by Philips Laboratories and discussed in Chapter 12, has been made widely available and is thus becoming an important model.

These models form the core FET models that the general user of publicly available versions of SPICE will encounter. There also exist a large number of proprietary FET models which are implemented in various proprietary versions of SPICE; since no details of these models are available, they cannot be discussed here.

In addition, any reader who has browsed the technical literature will note that new analytical FET models for use in circuit simulation are continually being published. While these models are interesting, as was noted in Chapter 1, few of these models find their way into general use; due to both physical and mathematical requirements, any model requires a well-defined vehicle (an accepted circuit simulator) if it is to attain widespread acceptance.

However, a few new models stand out as possible candidates for common use in SPICE. Examining the criteria discussed in Chapter 11, these models have the characteristics of third-generation FET models. However, they are not as yet used in generally available SPICE simulators, and so are presented here (rather than along with the other third-generation models). Only time will determine if these models *will* find general use, or if other models will appear that will fill that role. These models are included in this text for completeness, and to show in which directions efforts to build new analytical FET models are headed.

16.2   The Power-Lane Model

The Power-Lane model [1] uses the original Level 3 formulation as its base. Modifications are then made to correct the difficulties that prevented Level 3 from being used for analog circuit design. Since the resulting model is not much more complicated than its Level 3 forebearer, it is relatively simple. It thus can offer a reasonable model for analog circuit design, and may find some use.

The most interesting aspect of the Power-Lane model is that it was the first FET model to properly introduce smoothing functions for the various transition regions in the operation of the device. This made it possible to assemble a single equation (such as a drain current equation) which is valid for all regions of device operation. This technique is now widely used, and has become one of the key features of the third-generation models and various other candidate models.

16.2.1   The Threshold Voltage Model

The Power-Lane threshold voltage expression modifies the original Level 3 expression by adding another term which empirically accounts for nonuniform substrate doping [2]. This term is very similar to the expression associated with the parameter **K2** in the second-generation models; that is, the added expression essentially represents the addition of another term of a power series in $V_{bs}^{\frac{1}{2}}$. Also the Power-Lane model discards the short and narrow channel effects on the depletion region, which are represented in Level 3 with the parameters f_s and f_n; the implications of this approach will be considered after the entire model has been fully developed.

The Level 3 threshold voltage expression is

$$V_t = V_{fb} + 2\phi_f + f_s \cdot \gamma \cdot (2\phi_f - V_{bs})^{\frac{1}{2}} + f_n \cdot (2\phi_f - V_{bs}) - \sigma \cdot V_{ds}, \qquad (16.1)$$

where γ is the composite body effect term given by

$$\gamma = \frac{(2\epsilon_{Si} q N_A)^{\frac{1}{2}}}{C_{ox}}, \qquad (16.2)$$

and σ is the composite drain-induced barrier lowering term

$$\sigma = \mathbf{ETA} \cdot \frac{8.15 \times 10^{-22}}{C_{ox} \cdot L_{eff}^3}. \qquad (16.3)$$

The Power-Lane threshold voltage expression is

$$V_t = V_{fb} + 2\phi_f + \gamma_1 \cdot (2\phi_f - V_{bs})^{\frac{1}{2}} + \gamma_2 \cdot V_{bs} - \sigma \cdot V_{ds}, \qquad (16.4)$$

where γ_1 is identical to the Level 3 expression γ, as defined by (16.2), while σ is defined differently:

$$\sigma = \mathbf{ETA} \cdot \frac{8.15 \times 10^{-10}}{C_{ox} \cdot L_{\mathit{eff}}}. \tag{16.5}$$

As described above, note that the Level 3 terms which describe small geometry effects on the depletion region (f_s and f_n) have been omitted. The term γ_2 could be defined in the form of (16.2), with an additional doping parameter of the form **NSUB2**. However, here, γ_2 is introduced as a strictly empirical model parameter. Thus, for convenience, (16.4) can be rewritten as

$$V_t = V_{fb} + 2\phi_f + \gamma_1 \cdot (2\phi_f - V_{bs})^{\frac{1}{2}} + \mathbf{GAMMA2} \cdot V_{bs} - \sigma \cdot V_{ds}. \tag{16.6}$$

Examining (16.6), it can be seen that γ_1 is a composite term based on the parameter **NSUB**, while **GAMMA2** is a model parameter.

16.2.2　The Mobility Model

The Vertical Field

The Power-Lane model corrects a major deficiency of the Level 3 model by introducing substrate bias dependence into the model for the reduction of the mobility by the vertical field. The Level 3 expression is

$$\mu_v = \frac{\mathbf{UO}}{1 + \mathbf{THETA} \cdot (V_{gs} - \mathbf{VTO})}. \tag{16.7}$$

The Power-Lane expression is

$$\mu_v = \frac{\mu_o}{1 + \theta \cdot (V_{gsx} - V_t) + 2 \cdot \zeta \cdot \gamma_1 \cdot \left[(2\phi_f - V_{bs})^{\frac{1}{2}} - (2\phi_f)^{\frac{1}{2}} \right]}. \tag{16.8}$$

The term V_{gsx} is a modified expression for the gate bias which is introduced for the modeling of the subthreshold region; it will thus be considered when subthreshold conduction is discussed below. In strong inversion, V_{gsx} reduces to the simple gate bias value V_{gs}. The term ζ is a model parameter for the substrate bias portion of the mobility expression [3], so (16.8) should be rewritten as

$$\mu_v = \frac{\mu_o}{1 + \theta(V_{gsx} - V_t) + 2 \cdot \mathbf{ZETA} \cdot \gamma_1 \cdot \left[(2\phi_f - V_{bs})^{\frac{1}{2}} - (2\phi_f)^{\frac{1}{2}} \right]}. \tag{16.9}$$

The Lateral Field

The Power-Lane description of the effect of the lateral field on the mobility is virtually identical to the Level 3 expression. In Level 3, the total effective mobility is described by

$$\mu_{eff} = \frac{\mu_v}{1 + \frac{\mu_v \cdot V'_{ds}}{\textbf{VMAX} \cdot L_{eff}}}, \qquad (16.10)$$

where the mobility computed with just the vertical field (μ_v) is taken from (16.8), and V'_{ds} is defined from

$$V'_{ds} = min(V_{ds}, V_{dsat}). \qquad (16.11)$$

The Power-Lane expression is

$$\mu_{eff} = \frac{\mu_v}{1 + \frac{\mu_v \cdot V_{dsx1}}{\textbf{VMAX} \cdot L_{eff}}}, \qquad (16.12)$$

where μ_v is now included from (16.9), and V_{dsx1} is a modified drain bias expression that is introduced to allow for more accurate modeling of the linear-saturation transition region, which is discussed below.

16.2.3   The Drain Current Model

In the Level 3 formulation, there are two drain current equations. The main equation describes the strong inversion region (the linear and saturation regions), while a subthreshold current equation is added seemingly as an afterthought. The Power-Lane model improves the situation by using one drain current equation for all regions of operation. A modified expression for channel length modulation is also introduced, which includes gate bias effects. In addition, a variety of smoothing functions for the linear-saturation transition region, similar in style to those employed in HSPICE Level 28 and the third-generation models, are introduced.

The Strong Inversion Region

The Level 3 drain current expression for the strong inversion region is

$$I_{ds} = \frac{\mu_{eff} \cdot W_{eff} \cdot C_{ox}}{L_{eff} - L'} \left[V_{gs} - V_t - \frac{1 + f_b}{2} \cdot V'_{ds} \right] \cdot V'_{ds}, \qquad (16.13)$$

where V'_{ds} is defined by (16.11), L' is the reduction in the lateral length of the inversion layer by channel length modulation, the effective channel length and width are

$$L_{eff} = L - 2 \cdot \textbf{LD} \qquad (16.14)$$

and

$$W_{eff} = W - 2 \cdot \textbf{WD}, \qquad (16.15)$$

and f_b is defined by

$$f_b = \frac{\gamma \cdot f_s}{4 \cdot (2\phi_f - V_{bs})^{\frac{1}{2}}} + f_n. \tag{16.16}$$

The Power-Lane drain current expression, which will be used for all regions of operation, is

$$I_{ds} = \frac{\mu_{eff} \cdot W_{eff} \cdot C_{ox}}{L_{eff} - L'} \left[V_{gsx} - V_t - \frac{1 + f_b}{2} \cdot V_{dsx1} \right] \cdot V_{dsx2}, \tag{16.17}$$

where now f_b is defined as

$$f_b = \frac{\gamma_1}{4 \cdot (2\phi_f - V_{bs})^{\frac{1}{2}}} - \textbf{GAMMA2}, \tag{16.18}$$

while V_{dsx1} and V_{dsx2} are effective model drain biases which are used to more effectively describe the linear-saturation transition (discussed below). Note once again that the Level 3 small geometry depletion region terms f_s and f_n have been discarded, while the additional parameter **GAMMA2** has been included.

As noted earlier, the terms V_{dsx1} and V_{dsx2} are effective drain bias expressions which are used to allow for improved modeling of the linear-saturation transition region. These terms are defined by

$$V_{dsx1} = \frac{V_{ds}}{\left[1 + \left(\frac{V_{ds}}{V_{dsat}} \right)^{\frac{1}{\textbf{XA}}} \right]^{\textbf{XA}}}, \tag{16.19}$$

and

$$V_{dsx2} = \frac{V_{dsat1} \cdot V_{ds}}{V_{dsat2} \cdot \left[1 + \left(\frac{V_{ds}}{V_{dsat}} \right)^{\frac{1}{\textbf{XA}}} \right]^{\textbf{XA}}}; \tag{16.20}$$

the saturation voltage terms V_{dsat}, V_{dsat1}, and V_{dsat2} will be discussed below. The parameter **XA** is a smoothing term; in their original paper, Power and Lane [1] recommend the use of **XA** = 0.1, although it is clear that this parameter could also be extracted from data. **XA** serves to describe the rate at which V_{dsx1} goes to V_{ds} or V_{dsat} on opposite sides of the linear-saturation transition region. The use of **XA** allows for well-defined linear, saturation, and transition regions, in a manner virtually identical to that employed in HSPICE Level 28 and the third-generation models. In this manner, the continuity of the drain current expression and its derivatives is ensured for all regions of device operation.

It is instructive to examine the limiting behavior of V_{dsx1} and V_{dsx2}. In the generically defined linear region ($V_{ds} < V_{dsat}$), $V_{dsx1} \rightarrow V_{ds}$, while in the generically defined saturation region ($V_{ds} > V_{dsat}$), $V_{dsx1} \rightarrow V_{dsat}$. In weak inversion ($V_{gs} \ll V_t$),

$V_{dsx1} \to 0$. The limiting behavior of V_{dsx2} is somewhat different. When $V_{dsx2} < V_{dsat2}$, $V_{dsx2} \to V_{ds}$, while when $V_{dsx2} > V_{dsat2}$, $V_{dsx2} \to V_{dsat1}$. In contrast to the situation for V_{dsx1}, in weak inversion, V_{dsx2} does **not** go to zero.

16.2.4   The Saturation Voltage

The Level 3 saturation voltage expression is

$$V_{dsat} = \frac{V_{gs} - V_t}{1 + f_b} + \frac{L_{eff} \cdot \mathbf{VMAX}}{\mu_{eff}} - \left[\left(\frac{V_{gs} - V_t}{1 + f_b} \right)^2 + \left(\frac{L_{eff} \cdot \mathbf{VMAX}}{\mu_{eff}} \right)^2 \right]^{\frac{1}{2}}. \tag{16.21}$$

The Power-Lane model uses the same expression, except that the gate bias V_{gs} is replaced with the modified gate bias expression V_{gsx}:

$$V_{dsat} = \frac{V_{gsx} - V_t}{1 + f_b} + \frac{L_{eff} \cdot \mathbf{VMAX}}{\mu_{eff}} - \left[\left(\frac{V_{gsx} - V_t}{1 + f_b} \right)^2 + \left(\frac{L_{eff} \cdot \mathbf{VMAX}}{\mu_{eff}} \right)^2 \right]^{\frac{1}{2}}. \tag{16.22}$$

The auxiliary saturation voltage terms V_{dsat1} and V_{dsat2} are defined by

$$V_{dsat1} = \frac{k_b T}{q} \cdot \mathbf{XMU} + \frac{k_b T}{q} \cdot ln \left[1 + exp \left(\frac{V_{dsat} - \frac{k_b T}{q} \cdot \mathbf{XMU}}{\frac{k_b T}{q}} \right) \right], \tag{16.23}$$

and

$$V_{dsat2} = \frac{k_b T}{q} + \frac{k_b T}{q} \cdot ln \left[1 + exp \left(\frac{V_{dsat} - \frac{k_b T}{q}}{\frac{k_b T}{q}} \right) \right], \tag{16.24}$$

where $\mathbf{XMU}$ is a parameter which improves the ability of the model to fit data in the subthreshold region. Note that when $\mathbf{XMU} = 1$, $V_{dsat2} = V_{dsat1}$.

Channel Length Modulation

To describe channel length modulation, the Level 3 formulation made slight modifications to an expression originally due to Baum and Beneking [4], leading to

$$L' = \left[\left(\frac{V_{dsat}}{2 \cdot a \cdot L_{eff}} \right)^2 + \frac{\mathbf{KAPPA} \cdot (V_{ds} - V_{dsat})}{a} \right]^{\frac{1}{2}} - \frac{V_{dsat}}{2 \cdot a \cdot L_{eff}}, \tag{16.25}$$

where the term a is described by

$$a = \frac{q \cdot \mathbf{NSUB}}{2\epsilon_s}. \tag{16.26}$$

The Power-Lane formulation uses a channel length modulation model that was originally developed by Reddi and Sah [5], and later modified to include the effect of the gate bias [6]:

$$L' = \left[\frac{2\epsilon_s \cdot \mathbf{KAPPA} \cdot (V_{ds} - V_{dsx3})}{q \cdot \mathbf{NSUB} \cdot (1 + \mathbf{KAPPAG} \cdot (V_{gsx} - V_t))} \right]^{\frac{1}{2}}, \tag{16.27}$$

where the auxiliary drain bias V_{dsx3} is

$$V_{dsx3} = V_{dsx2} \cdot \frac{V_{dsat2}}{V_{dsat1}}. \tag{16.28}$$

This adds the parameter **KAPPAG**, which describes the gate bias dependence, to the model parameter set.

The Subthreshold Region

In Level 3, a different drain current equation is used to describe the subthreshold region, which is defined as the region in which $V_{gs} < V_{on}$; the subthreshold and superthreshold expressions are forced to match at $V_{gs} = V_{on}$, but the first derivative is discontinuous at this point. The model equation is

$$I_{ds} = I_{on} e^{q(V_{gs} - V_{on})/nk_b T}, \tag{16.29}$$

where

$$V_{on} = V_t + n \cdot \frac{k_b T}{q}, \tag{16.30}$$

$$I_{on} = I_{ds} \bigg|_{V_{gs} = V_{on}}, \tag{16.31}$$

and

$$n = 1 + \frac{q \cdot \mathbf{NFS}}{C_{ox}} + \frac{C_B}{C_{ox}} = 1 + \frac{q \cdot \mathbf{NFS}}{C_{ox}} + \frac{1}{C_{ox}} \cdot \left[\frac{f_s \cdot \gamma \cdot (2\phi_f - V_{bs})^{\frac{1}{2}} - f_n \cdot (2\phi_f - V_{bs})}{2 \cdot (2\phi_f - V_{bs})} \right]. \tag{16.32}$$

The Power-Lane model uses the single drain current equation (16.17) for all regions of operation. The modified gate bias expression V_{gsx}, mentioned above, becomes important here, and is given by

$$V_{gsx} = V_A \cdot \ln \left[1 + exp \left(\frac{V_{gs} - V_t}{V_A} \right) \right] + V_t; \tag{16.33}$$

the term V_A is

$$V_A = \frac{k_b T}{q} \cdot \left[1 + \frac{q \cdot \mathbf{NFS}}{C_{ox}} + \frac{\gamma_1 \cdot (2\phi_f - V_{bs})^{\frac{1}{2}} - \mathbf{GAMMA2} \cdot (2\phi_f - V_{bs})}{2 \cdot (2\phi_f - V_{bs})} \right]. \quad (16.34)$$

As noted earlier, in strong inversion, $V_{gsx} \to V_{gs}$. In weak inversion ($V_{gs} \ll V_t$), so (16.33) can be rewritten using an approximation for a small value of z:

$$ln(1 + z) \doteq z. \quad (16.35)$$

Using (16.35), (16.33) becomes

$$V_{gsx} = V_A \cdot exp\left(\frac{V_{gs} - V_t}{V_A} \right) + V_t. \quad (16.36)$$

In addition, in strong inversion, V_{dsx1}, V_{dsx2}, and V_{dsx3} all go to V_{ds} in the linear region. Also under strong inversion, both V_{dsat1} and V_{dsat2} go to V_{dsat}. In these circumstances, the Power-Lane drain current model equation (16.17) reduces to the Level 3 strong inversion region drain current expression (16.13):

$$I_{ds} = \frac{\mu_{eff} \cdot W_{eff} \cdot C_{ox}}{L_{eff} - L'} \left[V_{gs} - V_t - \frac{1 + f_b}{2} \cdot V'_{ds} \right] \cdot V'_{ds}. \quad (16.37)$$

Thus, the Level 3 and Power-Lane expressions for the drain current are the same in the linear and saturation regions. However, the Power-Lane model differs from Level 3 by also containing a separate *transition* region between the linear and saturation regions; in that region, the Power-Lane drain current model equation does *not* reduce to the Level 3 expression given by (16.37).

In weak inversion, for $V_{ds} > \frac{k_b T}{q}$,

$$V_{dsx1} = V_{dsat} = \frac{V_A}{1 + f_b} \cdot exp\left(\frac{V_{gs} - V_t}{V_A} \right) \cdot V_{dsx2}, \quad (16.38)$$

while

$$V_{dsx2} \to \frac{k_b T}{q} \cdot \mathbf{XMU}. \quad (16.39)$$

For $V_{ds} < \frac{k_b T}{q}$,

$$V_{dsx2} \to V_{ds}. \quad (16.40)$$

Examining (16.39) and (16.40), it can be seen that V_{dsx2} behaves in a manner described by

$$V_{dsx2} = \left[1 - exp\left(-\frac{V_{ds}}{\frac{k_b T}{q}} \right) \right].$$

$$(16.41)$$

Its net effect (and that of **XMU**) is similar to the empirical smoothing term $e^{1.8}$, which was introduced into the subthreshold current expression of BSIM; this accounts for the slight offset between the threshold definitions of the subthreshold and strong inversion regions.

16.2.5   The Charge Model

The node charge equations, which can be derived from the current model equations, were not derived in the original Power-Lane paper. In the Power-Lane model, the Yang-Epler-Chatterjee [7] node charge and gate capacitance model, discussed in detail in Chapter 8, was employed. As long as the circuit simulator uses charge (rather than voltage) as the state variable (as discussed in Chapter 13), use of this model guarantees that charge in conserved.

16.2.6   Comments

The Power-Lane model is a Level 3-like model that has been made suitable for use in analog circuit design, without the introduction of a great deal of additional complexity. Due to this simplicity, this model could find some use in general purpose SPICE simulation.

There are some shortcomings, both in the model formulation and in some unavoidable difficulties that are inherited from Level 3. The short and narrow geometry corrections to the depletion charge model (f_s and f_n) have been ignored; in fact, it appears that the intention of the model is to describe *individual* transistor characteristics, rather than to create a *general* model for an entire process technology. The original Power-Lane paper [1] quotes different values of the zero substrate bias threshold voltage parameter **VTO** for different channel lengths; as originally intended, **VTO** is supposed to describe a long, wide device for a given process technology. While this is not a concern for the description of individual device characteristics, it does indicate that the model could be vulnerable to problems when process variations of the channel length and width occur around that individual description; the model formulation assumes that the variations are small enough that the changes are described entirely by changing L_{eff} and W_{eff}. If desired, it would not be difficult to include f_s and f_n in the model formulation; however, the difficulties with binning and the discontinuity of the drain current at the bin boundaries, as discussed in Chapter 7, would still be present.

The formulation of the subthreshold region model represents a significant improvement over the "afterthought" approach of the original Level 3 development. However, the Power-Lane subthreshold model still employs only one parameter, **NFS**, to describe the subthreshold region. As noted in Chapter 7, since this parameter has no geometry, bias, or temperature dependence, an accurate subthreshold model must be chosen for only one value of each of these terms. Finally, temperature dependence is not discussed in the development of this model, and it is unclear if any temperature dependence is included. If desired, it would not be difficult to add temperature dependence to the mobility, in a manner consistent with the two possible approaches discussed in Chapter 7.

Overall, the Power-Lane model represents a useful extension of the basic Level 3 formulation, so that a model suitable for analog circuit design is produced. However, due to its heritage, the model will be subject to many of the shortcomings inherent in Level 3.

16.3   The PCIM Model

PCIM is the acronym for the **P**hysically-Based, **C**ontinuous **I**GFET **M**odel, produced by the device modeling group at the Digital Equipment Corporation [8]. The goal of the model development is to attain smooth and continuous current, conductance, charge, and capacitance expressions over the entire range of device operation. As with all the third-generation models, a major part of this effort involves the derivation of a single drain current equation for use in all regions of device operation. In contrast to other third-generation models, PCIM includes a number of semi-empirical "from experience" expressions in several places. Smoothing functions similar to those used in the Power-Lane model (see Section 16.2) are frequently employed. Many of the equations which appear during the derivation bear a strong resemblance to expressions found in BSIM; differences often occur only in the form of the composite terms. Perhaps uniquely, PCIM attempts to introduce a model for the carrier velocity-electric field relationship which has a better physical basis than those employed in other models.

The final model formulation is mathematically robust, with a small (about 12) number of model parameters requiring extraction. The inclusion of a smooth and continuous charge model is of particular note for analog circuit design.

16.3.1   The Electrostatic Model of the MOSFET Structure

In strong inversion, the depletion charge along the channel of a small geometry MOSFET can be written as [9]

$$Q_{depl}(y) = C_{ox} \cdot F_l \cdot F_w \cdot \gamma \cdot \left(2\phi_f - V_{bs} + V(y)\right)^{\frac{1}{2}} = C_{ox} \cdot \gamma_e \cdot \left(2\phi_f - V_{bs} + V(y)\right)^{\frac{1}{2}}, \quad (16.42)$$

where γ_e is a small geometry version of the body effect parameter γ, which is defined as

$$\gamma = \frac{(2\epsilon_{Si}q \cdot \mathbf{NSUB})^{\frac{1}{2}}}{C_{ox}}. \quad (16.43)$$

The terms F_l and F_w represent, respectively, short and narrow channel corrections to the depletion charge, and are thus similar to the terms f_s and f_n used in Level 2 (see Section 6.2) and Level 3 (see Section 7.2). In the PCIM development [8], no specific expressions for F_l and F_w are given; the reader is referred to [9] for a compilation of suggestions.

In a manner similar to that used in BSIM (see Section 8.4), a Taylor series expansion is carried out on (16.42), leading to

$$Q_{depl}(y) = C_{ox} \cdot \gamma_e \cdot \left[(2\phi_f - V_{bs})^{\frac{1}{2}} + \delta \cdot V(y) \right]. \tag{16.44}$$

A number of expressions have been proposed for δ; PCIM chooses to employ [9]

$$\delta = \frac{1}{2(2\phi_f - V_{bs})^{\frac{1}{2}}} \cdot \left[1 - \frac{1}{1.43 + 1.41 \cdot (2\phi_f - V_{bs})^{\frac{1}{2}}} \right]. \tag{16.45}$$

Note that this expression is very similar to (8.69) in BSIM. The inversion charge is then easily computed from the charge neutrality requirement, leading to

$$Q_{inv}(y) = C_{ox} \cdot \left[V_{gs} - V_t - \alpha \cdot V(y) \right], \tag{16.46}$$

which is identical to the BSIM inversion charge expression (8.94). However, in contrast to BSIM, the term α is defined as

$$\alpha = 1 + \delta \cdot \gamma_e. \tag{16.47}$$

16.3.2   The Threshold Voltage Model

The threshold voltage model used in PCIM is very similar to the expression used in Level 3. It is particularly noteworthy that in contrast to many other models, PCIM does not include any auxiliary body effect terms, which were added in an attempt to describe the effect of nonuniform substrate doping.

The basic threshold voltage expression is

$$V_t = V_{to} - \sigma \cdot V_{ds}, \tag{16.48}$$

where V_{to} is the low drain bias threshold voltage (**not** the long, wide device threshold voltage; V_{to} includes geometry effects) given by

$$V_{to} = V_{fb} + 2\phi_f + \gamma_e \cdot (2\phi_f - V_{bs})^{\frac{1}{2}} + V_{bump}. \tag{16.49}$$

Small geometry effects on the depletion charge are included through the use of γ_e. The term V_{bump} is included to describe threshold voltage roll-up (an *increasing* threshold voltage with decreasing channel length), described in Section 3.5. The term σ is the drain-induced barrier lowering (DIBL) parameter, which has commonly appeared in FET models, beginning with its introduction in Level 3. From experience, the suggested expression for σ used here is

$$\sigma = \frac{\epsilon_{Si} \cdot (\sigma_0 - \sigma_1 \cdot V_{bs})}{C_{ox} \cdot L_{eff}^m}, \tag{16.50}$$

where σ_0, σ_1, and m are all model parameters; (16.50) can thus be recast as

$$\sigma = \frac{\epsilon_{Si} \cdot (\mathbf{SIGMA0} - \mathbf{SIGMA1} \cdot V_{bs})}{C_{ox} \cdot L_{eff}^{\mathbf{M}}}.$$

(16.51)

Note that the substrate bias dependence of DIBL is included in a manner like that employed to describe the substrate bias dependence of *any* parameter in BSIM. Note also that for $V_{bs} = 0$, (16.51) essentially reduces to the Level 3 DIBL expression (except for the fixing of $\mathbf{M} = 3$ and the multiplicative constant which appears in Level 3).

16.3.3 The Mobility Model

PCIM uses a very simple model for the reduction of the mobility by the vertical field. The surface mobility is described by [10–12]

$$\mu_v = \frac{\mu_o}{1 + \alpha_\theta \cdot E_{eff}},$$

(16.52)

where μ_o is the low field mobility and α_θ is a scattering constant which describes the interaction of the carriers with the silicon-silicon dioxide interface. Since both μ_o and α_θ are model parameters, (16.52) can be rewritten as

$$\mu_v = \frac{\mathbf{UO}}{1 + \mathbf{ALPHATH} \cdot E_{eff}}.$$

(16.53)

The term E_{eff} is an average effective vertical field, defined as [10]

$$E_{eff} = \frac{1}{\epsilon_{Si}} \cdot \left(Q_{depl} + \zeta \cdot Q_{inv} \right),$$

(16.54)

where ζ is a constant, taken to be $\zeta = 0.50$ for electrons and $\zeta = 0.25 - 0.30$ for holes. Treating E_{eff} as a constant along the channel leads to the expression used in PCIM:

$$E_{eff} = \frac{\zeta \cdot C_{ox}}{\epsilon_{Si}} \cdot \left[V_{gs} - V_t + \frac{1}{\zeta} \cdot \gamma \cdot (2\phi_f - V_{bs})^{\frac{1}{2}} - \frac{V_{ds}}{2} \cdot \left(\frac{1}{\zeta} - \frac{1}{\zeta} \cdot \alpha + \alpha \right) \right].$$

(16.55)

16.3.4 The Carrier Velocity–Lateral Field Model

It has been known for some time that the relationship between the carrier velocity and the lateral electric field is well described by [13]

$$v = \frac{\mu_v \cdot E_y}{\left[1 + \left(\frac{E_y}{E_c} \right)^v \right]^{\frac{1}{v}}},$$

(16.56)

where $\nu = 2$ for electrons and $\nu = 1$ for holes. As was shown in Section 7.4, the critical field for the onset of velocity saturation can be defined using the basic Ohm's law expression, so that

$$E_c = \frac{\upsilon_{SAT}}{\mu_\upsilon}. \tag{16.57}$$

Reported measurements of υ_{SAT} vary considerably [13, 14]; for electrons, $\upsilon_{SAT} = 6 - 9 \times 10^6$ cm/s, while for holes, $\upsilon_{SAT} = 4 - 8 \times 10^6$ cm/s. Since this makes it difficult to specify a value for υ_{SAT}, this term is instead taken to be a model parameter (**VSAT**).

If (16.56) is used to describe the velocity-field relationship for electrons, a general solution for the drain current I_{ds} cannot be reached. A number of approximate solutions have been reported [15–17], but the drain current expressions produced are very complicated and thus ill-suited for use in circuit simulation. In addition, these equations are forced to use an iterative method of solving for the saturation voltage V_{dsat}; iterative approaches are computationally inefficient, and are generally avoided in FET models for circuit simulation. Alternatively, it is common to simply set $\nu = 1$, which reduces (16.56) to the Caughey-Thomas velocity-field model [18], which was used in Level 3; this causes the critical field E_c to be treated as a parameter. However, since the saturation velocity **VSAT** and the surface mobility μ_υ are well defined, the critical field E_c should also be well defined.

The development of PCIM uses a different approximation for the velocity-field relationship:

$$\upsilon \doteq \frac{\mu_\upsilon \cdot E_y}{1 + \delta_o \cdot \left(\frac{E_y}{E_c}\right)}, \tag{16.58}$$

where the lateral field E_y can be approximated as

$$E_y \doteq \frac{V_{ds}}{L_{eff}}. \tag{16.59}$$

Using (16.59), the term δ_o is written as

$$\delta_o = \frac{\left(\frac{E_y}{E_c}\right)}{\left[1.5 + \frac{E_y}{E_c}\right]} \doteq \frac{\left(\frac{V_{ds}}{L_{eff} \cdot E_c}\right)}{\left[1.5 + \frac{V_{ds}}{L_{eff} \cdot E_c}\right]}. \tag{16.60}$$

If $\delta_o \rightarrow 1$, (16.60) reduces to the Caughey-Thomas velocity-field model. However, if this approach is used, the velocity υ is underpredicted; consequently, the saturation voltage V_{dsat} is underpredicted, which leads to an overprediction of the drain current I_{ds}.

16.3.5 The Drain Current Model

The Basic Equation

The derivation of the PCIM drain current model begins with the simple velocity-charge expression for the drain current anywhere along the channel [19]:

$$I_{ds}(y) = W_{eff} \cdot v(y) \cdot Q_{inv}(y). \tag{16.61}$$

Substituting (16.46) for the inversion charge $Q_{inv}(y)$ and (16.58) for the carrier velocity $v(y)$ produces

$$I_{ds} = \frac{\mu_v \cdot W_{eff} \cdot C_{ox}}{L_{eff} \cdot \left(1 + \delta_o \cdot \frac{V_{ds}}{L_{eff} \cdot E_c}\right)} (V_{gs} - V_t - \frac{\alpha}{2} \cdot V_{ds}) \cdot V_{ds}. \tag{16.62}$$

Note that this equation is very similar to the BSIM linear region expression (8.75). The critical field E_c is not a parameter, but is computed from (15.57), while the term α accounts for small geometry effects.

The form of (16.62) will lead to an iterative expression for the saturation voltage V_{dsat}. Therefore, it is necessary to modify the term δ_o, which is defined by (16.60).

The Saturation Voltage

To avoid using an iterative solution for V_{dsat} in PCIM, the saturation voltage model is modified as follows. At $V_{ds} = V_{dsat}$, the generic expression V_{dsct} is replaced in δ_o ((16.60)) with a long channel saturation voltage expression:

$$V_p = \left(\frac{V_{gs} - V_t}{\alpha}\right), \tag{16.63}$$

where V_p is the long channel saturation voltage; the same expression was developed in BSIM, resulting in (8.136). This leads to the PCIM saturation voltage expression

$$V_{dsat} = \frac{2 \cdot V_c \cdot V_p}{V_c + (1 - \delta_o) \cdot V_p + \left[V_c^2 + (1 - \delta_o)^2 \cdot V_p^2 + 2 \cdot \delta_o \cdot V_c \cdot V_p\right]}, \tag{16.64}$$

where V_c is the critical voltage, defined simply as

$$V_c = L_{eff} \cdot E_c = \frac{L_{eff} \cdot \textbf{VSAT}}{\mu_v}. \tag{16.65}$$

In the limit of $\delta_o \to 1$, V_{dsat} approaches the same result that would have been obtained if (16.56) were used with $\nu = 1$.

To improve the quality of the linear-saturation transition, a smoothing function [20, 21], similar to the one used in the Power-Lane model [1] (see Section 16.2) is employed:

$$F(V_{ds}, V_{dsat}) = 1 - \frac{ln\left[1 + exp\left(\xi \cdot \left(1 - \frac{V_{ds}}{V_{dsat}}\right)\right)\right]}{ln\left[1 + exp(\xi)\right]}. \tag{16.66}$$

Large values of ξ produce a steep transition, while small values of ξ lead to a smooth transition. The equivalent of the Power-Lane suggestion [1] of **XA** $= 0.1$ occurs here as $\xi = 10$. However, it was found that the best results for PCIM were reached with $\xi = 6$; this value is used in the model. This smoothing function takes the form

$$V_{dsx} = F(V_{ds}, V_{dsat}) \cdot V_{dsat}. \tag{16.67}$$

Channel Length Modulation

As the drain bias V_{ds} increases above the saturation voltage V_{dsat}, the pinch-off point moves toward the source; a voltage of $(V_{ds} - V_{dsat})$ is dropped across the pinch-off region, which has a length (in the channel direction) of L'. PCIM employs [22]

$$L' = l \cdot ln\left[\frac{V_{ds} - V_{dsat}}{l \cdot E_c} + \frac{E_m}{E_c}\right], \tag{16.68}$$

where l is a *characteristic length*:

$$l = \left(\frac{\epsilon_{Si}}{\epsilon_{ox}}\right)^{\frac{1}{2}} \cdot t_{ox} \cdot x_j. \tag{16.69}$$

The term E_m is a *maximum field*, given by

$$E_m = E_c \cdot \left[1 + \left(\frac{V_{ds} - V_{dsat}}{l \cdot E_c}\right)^2\right]^{\frac{1}{2}}. \tag{16.70}$$

Examining (16.69), the term l could be defined from a set of physical parameters. However, a problem similar to one which occurs in Level 3 is encountered: x_j, which is supposed to represent the junction depth, takes on an electrical meaning that is different from its physical definition. As a result, in PCIM, l is taken to be a model parameter (**EL**).

Next, a Taylor series expansion (similar to that used on (16.42) to produce (16.44)) is carried out on (16.70), leading to

$$\frac{E_m}{E_c} \cong 1 + \delta' \cdot \left(\frac{V_{ds} - V_{dsat}}{\mathbf{EL} \cdot E_c}\right). \tag{16.71}$$

Substituting (16.71) into (16.68) produces

$$L' = \mathbf{EL} \cdot ln\left[1 + \frac{V_{ds} - V_{dsat}}{V_{pp}}\right], \tag{16.72}$$

where

$$V_{pp} = \frac{\mathbf{EL} \cdot E_c}{\delta'}.$$ (16.73)

The critical field E_c is well defined in this model, and $\mathbf{EL}$ is a parameter, but the term δ' is not well defined; the solution used in PCIM is to treat V_{pp} as a model parameter (**VPP**).

With channel length modulation included, the drain current model (16.62) becomes

$$I_{ds} = \frac{\mu_v \cdot W_{eff} \cdot C_{ox}}{(L_{eff} - L') \cdot \left(1 + \delta_o \cdot \frac{\mu_v \cdot V_{dsx}}{(L_{eff}-L') \cdot \mathbf{VSAT}}\right)} \left(V_{gs} - V_t - \frac{\alpha}{2} \cdot V_{dsx}\right) \cdot V_{dsx},$$ (16.74)

where V_{dsx} is given by (16.67). To guarantee continuity, the pinch-off region length L' is used in both the linear and saturation regions; examination of (16.72) shows that L' becomes negligible in the linear region. To ensure a smooth linear-saturation transition, V_{dsx} is used in place of V_{ds} in (16.72), which thus becomes

$$L' = \mathbf{EL} \cdot ln\left[1 + \frac{V_{dsx} - V_{dsat}}{V_{pp}}\right].$$ (16.75)

The Source-Drain Series Resistance

In PCIM, the source and drain series resistances are included as a single lumped resistance R_t. As in the second-generation models, this added resistance is folded into the mobility model [9, 23]; thus, (16.53) becomes

$$\mu_v = \frac{\mathbf{UO}}{1 + \mathbf{ALPHATH} \cdot E_{eff} + \left(\frac{\mu_v \cdot C_{ox} \cdot R_t}{L_{eff}}\right) \cdot \left[V_{gs} - V_t - \left(\alpha - \frac{1}{2}\right) \cdot V_{dsx}\right]}.$$ (16.76)

The lumped resistance R_t is treated as a model parameter (**RT**), so (16.76) is rewritten as

$$\mu_v = \frac{\mathbf{UO}}{1 + \mathbf{ALPHATH} \cdot E_{eff} + \left(\frac{\mu_v \cdot C_{ox} \cdot \mathbf{RT}}{L_{eff}}\right) \cdot \left[V_{gs} - V_t - \left(\alpha - \frac{1}{2}\right) \cdot V_{dsx}\right]}.$$ (16.77)

It was found that this expression works well as long as the lumped resistance does not become too large (as long as $\mathbf{RT} < 150 - 200 \ \Omega$).

The Subthreshold Current Model

In most of the subthreshold region, it is necessary to model only the diffusion component of the current. Usually, this involves the use of an equation of the form [9, 19, 23]

$$I_{ds} = I_{pf} \cdot exp\left(\frac{q}{k_b T} \cdot \frac{V_{gs} - V_t}{n}\right)\left[1 - exp\left(-\frac{q}{k_b T} \cdot V_{ds}\right)\right],$$ (16.78)

where I_{pf} is a current prefactor; different models tend to vary only in this prefactor. The term n represents the increase in the subthreshold slope from its ideal value (60 mV/decade at $T = 300K$). A simple physical description of n is [24]

$$n = \left(1 + \frac{C_{depl}}{C_{inv}}\right), \tag{16.79}$$

which reaches its ideal value of $n = 1$ when $C_{depl} \ll C_{inv}$. Other models use various approaches to compute n; here, n is treated as a model parameter (**N**).

As noted above, in the subthreshold region conduction is dominated by diffusion, while in strong inversion conduction is dominated by drift. In the transition region between subthreshold and strong inversion both mechanisms are important; the current equations must match in this region. In PCIM, an auxiliary gate bias, similar to that used in the Power-Lane model, is employed as a smoothing function [17]:

$$V_{gsx} = 2 \cdot \mathbf{N} \cdot \frac{k_b T}{q} \cdot ln\left[1 + exp\left(\frac{q}{k_b T} \cdot \frac{V_{gs} - V_t}{2 \cdot \mathbf{N}}\right)\right] + V_t. \tag{16.80}$$

By replacing V_{gs} in the strong inversion current equation with V_{gsx}, a smooth transition from weak to strong inversion is produced. It can also be seen by examining (16.80) that when $V_{gs} \gg V_t$, $V_{gsx} \to V_{gs}$.

When the gate bias V_{gs} is several thermal voltages $\left(\frac{k_b T}{q}\right)$ below the threshold voltage V_t, an approximation for small z,

$$ln(1 + z) \doteq z \tag{16.81}$$

can be used in (16.80); this produces

$$V_{gsx} - V_t \equiv V_{gtx} = 2 \cdot \mathbf{N} \cdot \frac{k_b T}{q} \cdot exp\left(\frac{q}{k_b T} \cdot \frac{V_{gs} - V_t}{2 \cdot \mathbf{N}}\right), \tag{16.82}$$

which defines the auxiliary voltage V_{gtx}. In addition, modifying the long channel pinch-off voltage expression (16.63) leads to

$$V_{dsat} \approx \frac{V_{gtx}}{\alpha}. \tag{16.83}$$

Since (16.78) must match with (16.74) at the weak-strong transition, the prefactor I_{pf} is

$$I_{pf} = \frac{2 \cdot \mu_v \cdot C_{ox}}{\alpha} \cdot \left(\frac{W_{eff}}{L_{eff}}\right) \cdot \left(\mathbf{N} \cdot \frac{k_b T}{q}\right)^2. \tag{16.84}$$

The Final Drain Current Equation

Combining (16.74), (16.78), and (16.84) produces the final drain current equation, which is valid for all regions of device operation:

$$I_{ds} = \frac{\mu_v \cdot W_{eff} \cdot C_{ox}}{L_{eff} \cdot \left(1 - \frac{L'}{L_{eff}} + \delta_o \cdot \frac{\mu_v \cdot V_{dsx}}{L_{eff} \cdot \text{VSAT}}\right)} \cdot \left(V_{gsx} - V_t - \frac{\alpha}{2} \cdot V_{dsx}\right)$$

$$\times V_{gsx} \cdot \left[1 - \frac{exp\left(-\frac{q}{k_b T} \cdot V_{ds}\right)}{1 + exp\left(\frac{q}{k_b T} \cdot \frac{V_{gtx}}{\mathbf{N}}\right)}\right]. \tag{16.85}$$

16.3.6 The Charge Model

In contrast to some other models, PCIM develops its own charge model from its drain
current equation. Using (16.85) and the usual procedures for arriving at a description of
the inversion charge [25, 26] produces

$$Q_{INV} = C_{ox} \cdot W_{eff} \cdot (L_{eff} - L') \cdot \left(V_{gtx} - \frac{\alpha}{2} \cdot V_{dsx} + A\right) + Q_{LD}, \tag{16.86}$$

where the capitalized subscripts in Q_{INV} and Q_{LD} indicate the *total* charge (rather than
the charge per unit area). The term Q_{LD} represents the inversion charge in the pinch-off
region. This charge is small compared with Q_{INV}, but is not zero; in most models, this
charge is neglected (and taken to be zero). The expression used in PCIM for this charge
is

$$Q_{LD} = L' \cdot \left(\frac{I_d}{v}\right) = C_{ox} \cdot W_{eff} \cdot L' \cdot \left(V_{gtx} - \alpha \cdot V_{dsx}\right), \tag{16.87}$$

where the term A is

$$A = \left[1 + \frac{\delta_o \cdot V_{dsx}}{(L_{eff} - L') \cdot E_c}\right] \cdot \frac{\alpha^2 \cdot V_{dsx}^2}{12\left(V_{gtx} - \frac{\alpha}{2} \cdot V_{dsx}\right)}. \tag{16.88}$$

To complete the node charge model, the inversion charge must be partitioned
between the source and the drain. PCIM uses a method [9] which leads to

$$Q_S = C_{ox} \cdot W_{eff} \cdot (L_{eff} - L') \cdot \left[\frac{1}{2} \cdot V_{gtx} - \frac{\alpha}{6} \cdot V_{dsx} + A \cdot (1 + B)\right] + \frac{1}{2} \cdot Q_{LD} \tag{16.89}$$

and

$$Q_D = C_{ox} \cdot W_{eff} \cdot (L_{eff} - L') \cdot \left[\frac{1}{2} \cdot V_{gtx} - \frac{\alpha}{3} \cdot V_{dsx} + A \cdot B\right] + \frac{1}{2} \cdot Q_{LD}, \tag{16.90}$$

where the capitalized subscripts once again indicate *total* charge. The term B is given by

$$B = \frac{\delta_o \cdot V_{dsx}}{2 \cdot (L_{eff} - L') \cdot E_c} - \left(1 + \frac{\delta' \cdot V_{dsx}}{(L_{eff} - L') \cdot E_c}\right) \cdot \frac{5 \cdot V_{gtx} - 2 \cdot \alpha \cdot V_{dsx}}{10 \cdot (V_{gtx} - 5 \cdot \alpha \cdot V_{dsx})}. \tag{16.91}$$

Examination of (16.89) and (16.90) shows that Q_S and Q_D are continuous across the weak inversion-strong inversion transition, and are nonzero in the weak inversion region; taking Q_S and Q_D to be zero in the weak inversion region is not a bad approximation, but this causes a discontinuity at the transition.

A similar approach can be used to calculate the depletion charge Q_{DEPL} [9], after which the gate charge Q_{GATE} is easily computed from the charge neutrality requirement

$$Q_{GATE} + Q_{DEPL} + Q_{INV} = 0. \tag{16.92}$$

The individual terminal capacitances are then found from

$$C_{ij} = \Delta \cdot \frac{\partial Q_i}{\partial V_j}, \tag{16.93}$$

where i and j represent two particular device terminals, and

$$\Delta \equiv +1 (i = j), -1 (i \neq j). \tag{16.94}$$

The capacitance values could be computed numerically using (16.93), but this is inefficient in a circuit simulation environment. Instead, (16.93) is used to produce analytical expressions for the terminal capacitances. These expressions are not reported in the original PCIM paper, but may be found elsewhere [9].

16.3.7 Comments

In total, PCIM is an interesting model, with good mathematical conditioning and a small number of model parameters. It is particularly noteworthy that an effort is made to ensure smooth and continuous behavior of the *charge* expressions, in addition to the currents and conductances; this is a unique and important improvement that is particularly significant for analog circuit design. In contrast to the Power-Lane model, PCIM is clearly intended to describe a particular process technology over its entire geometric range. In fact, it is claimed that one set of model parameters is valid over the entire range of geometries. This is a significant claim; however, extensive exercising of the model for a wide variety of process technologies would be required to confirm its validity.

At present, there are no known plans to incorporate PCIM into any publicly available SPICE circuit simulator. However, if PCIM is able to meet its claims, it could draw interest, particularly if other models show shortcomings.

16.4　The Enz-Krummenacher-Vittoz (EKV) Model

The Enz-Krummenacher-Vittoz (EKV) model [27–32] is a very new candidate model. This model is clearly oriented toward use in low power analog circuit design, which places perhaps the greatest demands on the capabilities of the FET model. While it contains many third-generation features, the EKV model stands somewhat apart from other models, in that it employs a new and fresh approach to the study of the analytical modeling of the FET. Rather than use the source node as the voltage reference point, the substrate node serves this role; this allows the source and drain to be treated symmetrically, with separate voltages, which is particularly useful in circuits where the FET is used bi-directionally (a common practice in analog circuits). From this, a very interesting and new approach to modeling the transition regions of the device follows: by developing a pinch-off voltage which applies independently to the source and drain terminals, a weak inversion model can be used to describe both the subthreshold and pinch-off/saturation behavior of the FET. In this manner, the weak inversion-strong inversion transition and the linear region-saturation region transition, which are treated separately in other models, are both described by the same weak inversion-strong inversion model.

The EKV model develops a somewhat more involved and physically detailed description of the inversion charge. Like the third-generation models and other candidate models, it also uses one drain current equation and a number of smoothing functions, to ensure continuity of that drain current equation and its first derivatives.

In contrast to the Power-Lane model and PCIM, which appear to be finished models, the EKV model is a work in progress, and continues to be improved and extended. Although the EKV model includes only two parameters for the description of short channel effects, it has demonstrated good results using a single parameter set for all geometries for channel lengths as short as 0.7 μm [30, 31]. Further work is in progress to extend the model to shorter channel lengths. In addition, the original formulation of the EKV model [27] supplied mathematically conditioned node capacitance expressions to the circuit simulator, while the node charge expressions were not conditioned; as described in Chapter 13, the FET model should communicate node charges to the circuit simulator to assure charge conservation. More recently [29], improvements have been introduced to correct this problem.

The most important aspect of the EKV model might be the basic approach which it employs. As just noted above, all other models discussed in this text have been based very directly on an original analytical FET model [33]. The EKV model rethinks that basic approach, and thus may represent the first manifestation of a new approach to analytical FET modeling.

The original paper describing the EKV model [27] is unusually well detailed. As a result, the discussion which follows will be somewhat terse in comparison; the reader is referred to the original paper for full details.

16.4.1   The Modes of Device Operation

As noted above, the EKV model defines the substrate contact as the voltage reference, so that the singly-subscripted terms V_s, V_d, and V_g are the voltages with respect to the substrate. In Section 16.4.3, the pinch-off voltage V_p is defined as a transition voltage; on one side of V_p, there will exist an inversion layer with a virtually constant density (the strong inversion condition), while on the other side of V_p, the inversion layer disappears exponentially with the changing voltage (the weak inversion condition). A unique feature of the EKV model is that V_p is used to describe both the subthreshold behavior and the channel pinch-off/current saturation characteristics.

In addition, with the substrate as the voltage reference, the source and drain voltage are independent of each other. This allows the FET to be treated as symmetric, with both forward and reverse modes; the direction of current flow is based on the sign of $(V_d - V_s)$.

The different modes of FET operation are illustrated in Figure 16.1. When both the source voltage V_s and the drain voltage V_d are less than the pinch-off voltage V_p, the entire channel is in strong inversion. This is referred to as *conduction mode*, where the source and drain are completely linked by an inversion layer. This also corresponds to what is more commonly known as the linear region of operation.

When $V_s < V_p$ and $V_d > V_p$, the drain end of the channel pinches off, and the device operates in saturation. This is referred to as *forward saturation mode*. When $V_s > V_p$ and $V_d < V_p$, the source end of the channel pinches off. This is referred to as *reverse saturation mode*. Note that these two modes are symmetric representations of each other; the only difference between the two modes is the direction of current flow.

When V_s and V_d are both larger than V_p, the entire channel is pinched off. If either V_s or V_d is only slightly larger than V_p, the device is in *weak inversion mode*, more commonly known as the subthreshold region of operation. If neither V_s or V_d is close to V_p, then there is no conduction, and the device is said to be in *blocked mode*.

If either V_s or V_d is greater than the junction potential V_j, the junction(s) will be forward biased. This activates a *bipolar mode* in the device, which must be superimposed on the already present FET mode [34].

In Figure 16.1, the various modes of device operation are sharply defined. However, in reality, the transitions between these regions will be smooth. For example, between

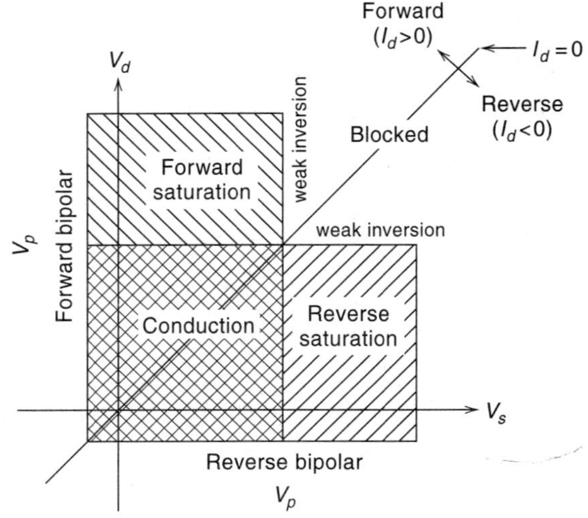

Figure 16.1 A graphical definition of the different FET operating modes. From [27]. © 1995. Kluwer Academic Publishers. Used by permission.

weak and strong inversion, there is an intermediate region which is commonly referred to as moderate inversion [19].

16.4.2   The Electrostatic Model of the Device

Compared to other models, the electrostatic device description of the EKV model is very involved and detailed. The objective is the development of a more physically based model for the inversion charge, to allow for a more accurate and robust model of the device behavior.

   The discussion which follows refers entirely to an nFET; it is easily extended to apply to a pFET. Also, it is assumed that the substrate doping is uniform. The authors [27] note that the model could be modified to describe nonuniform substrate doping; more recently, this addition has been made to the model [32].

Strong Inversion

The EKV model uses a somewhat detailed solution to the Poisson equation to describe the inversion charge Q_{inv}. Consider once again the two-terminal MOS capacitor structure described in Section 3.3 (see Figure 3.18), where ψ_s is the surface potential and ϕ_f is the substrate Fermi potential. In a steady state, no current flows, and the electron and hole quasi-Fermi levels (ϕ_n and ϕ_p, respectively) do not differ from the substrate Fermi potential:

$$\phi_n = \phi_p = \phi_f. \tag{16.95}$$

Next, if the source and drain are included, the application of voltages there will lead to a non-steady-state situation, in which current flows:

$$\phi_n, \phi_p \neq \phi_f. \tag{16.96}$$

This allows for the definition of the channel potential as the difference between the two quasi-Fermi potentials:

$$V_{ch} = \phi_n - \phi_p. \tag{16.97}$$

In an nFET, the current flow is due to electrons rather than holes; this allows the simplification

$$\phi_p = \phi_f, \tag{16.98}$$

so that (16.97) becomes

$$V_{ch} = \phi_n - \phi_f. \tag{16.99}$$

   Integrating the Poisson equation and allowing for the use of the strong inversion condition [19]

$$\phi_s \gg \frac{k_b T}{q} \tag{16.100}$$

produces

$$Q_{inv} = -\gamma \cdot C_{ox} \cdot \left(\frac{k_b T}{q}\right)^{\frac{1}{2}} \cdot \left\{ \left[\frac{q}{k_b T} \cdot \psi_s \right.\right.$$
$$\left.\left. + exp\left(\frac{q}{k_b T} \cdot (\psi_s - 2\phi_f - V_{ch})\right)\right]^{\frac{1}{2}} - \left[\frac{q}{k_b T} \cdot \psi_s\right]^{\frac{1}{2}} \right\}, \tag{16.101}$$

where γ is the body effect coefficient, given by

$$\gamma = \mathbf{GAMMA} = \frac{(2q\epsilon_{Si}N_{sub})^{\frac{1}{2}}}{C_{ox}}. \tag{16.102}$$

Note also that ϕ_f is not computed from N_{sub}, but is a model parameter in the form

$$\mathbf{PHI} = \psi_o - m \cdot \frac{k_b T}{q}. \tag{16.103}$$

Equation (16.101) expresses Q_{inv} in terms of the surface potential ψ_s; however, for use in circuit simulation, Q_{inv} must be expressed in terms of the gate voltage V_g. The total gate voltage is easily found by summing up the various contributions to the voltage drop between the gate and the substrate contact. Using simple electrostatic arguments, and treating the inversion layer as a vanishingly thin charge layer, it can be shown that [23]

$$V_g = V_{fb} + \psi_s + \mathbf{GAMMA} \cdot \psi_s^{\frac{1}{2}} - \frac{Q_{inv}}{C_{ox}}. \tag{16.104}$$

In strong inversion, the surface potential ψ_s is essentially independent of V_g, and so can be approximated as a constant:

$$\psi_s \doteq \psi_o + V_{ch}, \tag{16.105}$$

where

$$\psi_o = \mathbf{PHI} + m \cdot \frac{k_b T}{q}. \tag{16.106}$$

The term m is an integer, representing several thermal voltages; a typical value is $m = 4$.

Solving (16.104) for Q_{inv} and replacing ψ_s with the strong inversion expression (16.105) yields

$$Q_{inv} = -C_{ox} \cdot \left[V_g - V_{tb}(V_{ch}) \right], \tag{16.107}$$

where

$$V_{tb} = V_{fb} + (\psi_o + V_{ch}) + \textbf{GAMMA} \cdot (\psi_o + V_{ch})^{\frac{1}{2}}. \tag{16.108}$$

If source and drain effects are neglected (so that $V_{ch} = 0$) and (16.106) is reduced to $\psi_o = \textbf{PHI}$, then the simple two-terminal MOS capacitor threshold voltage expression, derived in Section 3.3, is recovered:

$$V_t = V_{fb} + \textbf{PHI} + \textbf{GAMMA} \cdot (\textbf{PHI})^{\frac{1}{2}}. \tag{16.109}$$

For convenience, (16.108) can be rewritten as

$$V_{tb} = V_{to} + V_{ch} + \textbf{GAMMA} \cdot \left[(\psi_o + V_{ch})^{\frac{1}{2}} - \psi_o^{\frac{1}{2}} \right], \tag{16.110}$$

where V_{to} is the base threshold voltage, defined as the gate voltage when $Q_{inv} = 0$ and $V_{ch} = 0$:

$$V_{to} = \textbf{VTO} = V_{tb} \Big|_{V_{ch}=0} = V_{fb} + \psi_o + \textbf{GAMMA} \cdot \psi_o^{\frac{1}{2}}. \tag{16.111}$$

As above, if (16.106) is reduced to $\psi_o = \textbf{PHI}$, the two-terminal MOS capacitor threshold voltage expression (16.109) is recovered.

The flatband voltage V_{fb} and the potential ψ_o are determined by the gate material and the substrate doping of a particular technology. A plot of V_{tb} versus V_{ch} (Figure 16.2) is therefore unique to a particular process technology [35]. This graphical method of describing the MOSFET also shows how the pinch-off voltage V_p is used to describe both the weak inversion-strong inversion and linear region-saturation region transitions.

By setting $Q_{inv} = 0$ and $V_{ch} = V_p$ and using (16.110), (16.107) becomes

$$V_g \Big|_{V_{ch}=V_p, Q_{inv}=0} = V_{tb} \Big|_{V_{ch}=V_p} = \textbf{VTO} + V_p + \textbf{GAMMA} \cdot \left[(\psi_o - V_p)^{\frac{1}{2}} - \psi_o^{\frac{1}{2}} \right]. \tag{16.112}$$

This expression can be rearranged to provide a definition of the pinch-off voltage V_p:

$$V_p = V_g - \textbf{VTO} - \textbf{GAMMA}$$

$$\times \left\{ \left[V_g - \textbf{VTO} + \left(\psi_o^{\frac{1}{2}} + \frac{\textbf{GAMMA}}{2} \right)^2 \right]^{\frac{1}{2}} - \left[\psi_o^{\frac{1}{2}} + \frac{\textbf{GAMMA}}{2} \right] \right\}. \tag{16.113}$$

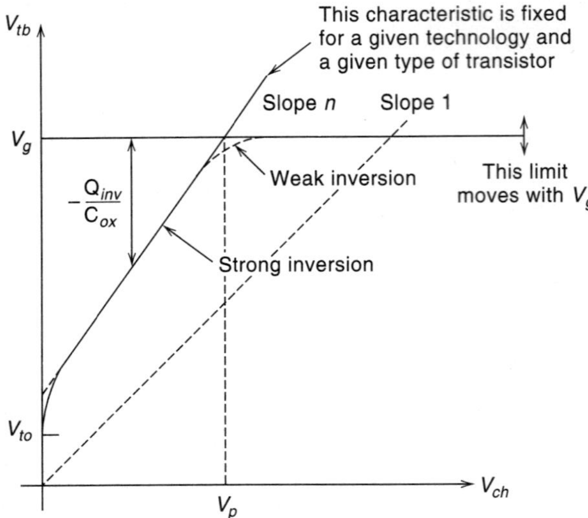

Figure 16.2   A plot of the strong inversion threshold voltage expression V_{tb} versus the channel voltage V_{ch} (solid line); this plot is unique to each particular process technology. The various regions of device operation (dashed lines) are superimposed onto the plot. From [27]. © 1995. Kluwer Academic Publishers. Used by permission.

Equation (16.112) can also be used to describe the slope factor n:

$$n \equiv \frac{dV_g}{dV_p} = 1 + \frac{\textbf{GAMMA}}{2 \cdot \left(\psi_o + V_p\right)^{\frac{1}{2}}}. \tag{16.114}$$

This can be expressed in a different form to allow n to be computed as a function of the gate bias:

$$\frac{1}{n} = \frac{dV_p}{dV_g} = 1 - \frac{\textbf{GAMMA}}{2 \cdot \left[V_g - \textbf{VTO} + \left(\frac{\textbf{GAMMA}}{2} + \psi_o^{\frac{1}{2}}\right)^2\right]^{\frac{1}{2}}}. \tag{16.115}$$

As was shown in the development of BSIM2 (see Section 10.5), it turns out that V_p is essentially linear with V_g. Thus, an approximation virtually identical to the one derived for use in BSIM (see Section 8.7) can be developed here:

$$V_p \cong \frac{V_g - \textbf{VTO}}{n}. \tag{16.116}$$

Equation (16.116) can be used for quick "hand" calculations of FET behavior; however, in the full implementation of the EKV model, the more detailed and accurate expression (16.113) is employed.

By introducing the full expression for V_p (16.113) into the simplified equation for Q_{inv} (16.107),

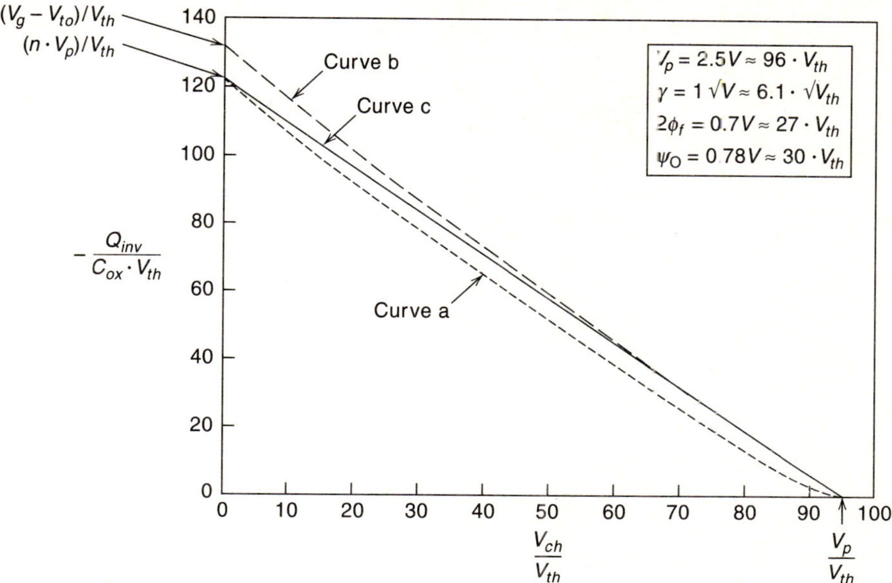

Figure 16.3   The inversion charge Q_{inv} versus the channel voltage V_{ch}. Note how each simplified expression (dashed lines) differs from the more rigorous result (solid line). From [27]. © 1995. Kluwer Academic Publishers. Used by permission.

$$Q_{inv} = -C_{ox} \cdot \left\{ V_p - V_{ch} + \textbf{GAMMA} \cdot \left[(\psi_o + V_p)^{\frac{1}{2}} - (\psi_o + V_{ch})^{\frac{1}{2}} \right] \right\}. \quad (16.117)$$

Plotting Q_{inv} versus V_{ch} (Figure 16.3) shows that (16.117) predicts higher than expected values of the inversion charge when compared with the more rigorous expression given by (16.101) and (16.104); this will lead to a larger than expected prediction of the current. A first-order Taylor expansion of (16.117)

$$Q_{inv} \doteq (V_{ch} - V_p) \cdot \left. \frac{\partial Q_{inv}}{\partial V_{ch}} \right|_{V_{ch}=V_p} = -C_{ox} \cdot n \cdot (V_p - V_{ch}) \quad (16.118)$$

leads to an improved result, but the computed value of Q_{inv} is still slightly higher than expected. The solution to this problem which is employed in the EKV model is to treat **VTO**, **GAMMA**, and ψ_o as independent model parameters (rather than computing them from physical values, such as V_{fb} and N_{sub}). This improves the final result, but has the effect of increasing **VTO** above the value that would be reached with a computation using physical variables.

Weak Inversion

In weak inversion ($V_{ch} \geq V_p$), the inversion charge Q_{inv} does not abruptly vanish, but decreases smoothly as V_p decreases below V_{ch}. In addition, in this region of operation, $Q_{inv} \ll Q_{depl}$, but is nonzero. However, a reasonable first-pass description of the weak inversion region can be found by setting $Q_{inv} = 0$, and employing the gate bias definition (16.104) and the definition of **VTO** (16.111):

$$V_g = \mathbf{VTO} + (\psi_s - \psi_o) + \mathbf{GAMMA} \cdot \left(\psi_s^{\frac{1}{2}} - \psi_o^{\frac{1}{2}} \right). \qquad (16.119)$$

Here, the definition of ψ_s (16.105) has been employed with $V_{ch} = V_p$:

$$\psi_s = \psi_o + V_p. \qquad (16.120)$$

Rather than plot ψ_s versus ($V_g - V_{fb}$), as was done in the development of BSIM2 (see Section 10.5 and Figure 10.2), instead ψ_s is plotted versus V_p for different values of V_{ch} (Figure 16.4). From the plot, it can be seen that in the subthreshold region ($V_p < V_{ch}$), ψ_s is linear with V_p, while in strong inversion ($V_p \geq V_{ch}$), ψ_s is virtually a constant (as noted earlier). Thus, a simplified set of expressions for the surface potential is

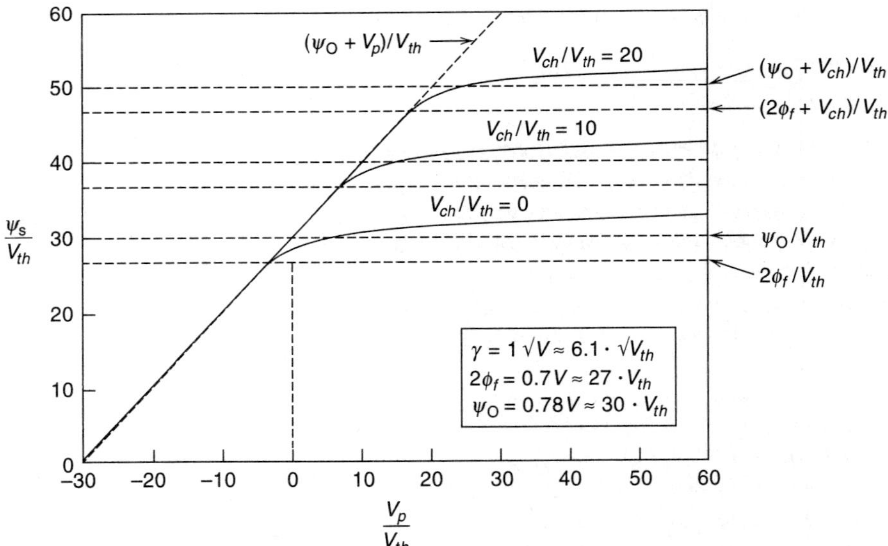

Figure 16.4   The surface potential ψ_s versus the pinch-off voltage V_p. From [27]. © 1995. Kluwer Academic Publishers. Used by permission.

$$\psi_s = \psi_o + V_p \qquad V_p < V_{ch} \quad \text{(weak inversion)}$$
$$\quad = \psi_o + V_{ch} \qquad V_p \geq V_{ch} \quad \text{(strong inversion)}. \tag{16.121}$$

In weak inversion, $\psi_s \ll (\mathbf{PHI} + V_{ch})$, so the exponential term in (16.101) becomes

$$exp\left[\frac{q}{k_b T} \cdot (\psi_s - \mathbf{PHI} - V_{ch})\right] \ll \frac{q}{k_b T} \cdot \psi_s. \tag{16.122}$$

Using this expression and a first-order Taylor series expansion of the first bracketed square root term in (16.101) leads to an expression for the inversion charge Q_{inv} in the weak inversion region:

$$Q_{inv} \cong -C_{ox} \cdot \frac{\mathbf{GAMMA}}{2 \cdot \psi_s^{\frac{1}{2}}} \cdot \frac{k_b T}{q} \cdot exp\left[\frac{q}{k_b T} \cdot (\psi_s - \mathbf{PHI} - V_{ch})\right]$$
$$= -K_w \cdot C_{ox} \cdot \frac{k_b T}{q} \cdot exp\left[\frac{q}{k_b T} \cdot (V_p - V_{ch})\right], \tag{16.123}$$

where

$$K_w = (n - 1) \cdot exp\left[\frac{q}{k_b T} \cdot (\psi_o - \mathbf{PHI})\right]. \tag{16.124}$$

16.4.3   The Current Equations

The Basic Equation

A generalized expression for the drain current that accounts for both drift and diffusion transport while using a charge sheet description of the inversion layer (which neglects carrier velocity saturation) is [19, 36]

$$I_d = -\mu_o \cdot W_{eff} \cdot Q_{inv} \cdot \frac{dV_{ch}}{dy}, \tag{16.125}$$

where μ_o is the low field mobility; μ_o is essentially a parameter, but in the structure of the EKV model, it is subsumed into

$$\mathbf{KP} = \mu_o \cdot C_{ox}. \tag{16.126}$$

Equation (16.125) assumes that the mobility is constant along the channel. Integrating (16.125) from the source (where $V_{ch} = V_s$) to the drain (where $V_{ch} = V_d$) produces

$$I_d = \frac{\mu_o \cdot C_{ox} \cdot W_{eff}}{L_{eff}} \cdot \int_{V_s}^{V_d} -\frac{Q_{inv}(V_{ch})}{C_{ox}} \cdot dV_{ch}. \tag{16.127}$$

Equation (16.127) can be split into a forward current component, which depends only on $(V_p - V_s)$, and a reverse current component, which depends only on $(V_p - V_d)$:

$$I_d = \frac{\mu_o \cdot C_{ox} \cdot W_{eff}}{L_{eff}} \cdot \int_{V_s}^{\infty} -\frac{Q_{inv}(V_{ch})}{C_{ox}} \cdot dV_{ch} - \frac{\mu_o \cdot C_{ox} \cdot W_{eff}}{L_{eff}} \cdot \int_{V_d}^{\infty} -\frac{Q_{inv}(V_{ch})}{C_{ox}} \cdot dV_{ch}$$

$$= I_f - I_r. \tag{16.128}$$

This expression is general, and thus is valid for all regions of device operation.

Strong Inversion

Substituting the strong inversion charge equation (16.118) into (16.128) leads to expressions for the currents in strong inversion:

$$\begin{aligned} I_f &= \frac{n}{2} \cdot \frac{\mu_o \cdot C_{ox} \cdot W_{eff}}{L_{eff}} \cdot (V_p - V_s)^2 & V_s < V_p \\ &= 0 & V_s \geq V_p, \end{aligned} \tag{16.129}$$

and

$$\begin{aligned} I_r &= \frac{n}{2} \cdot \frac{\mu_o \cdot C_{ox} \cdot W_{eff}}{L_{eff}} \cdot (V_p - V_d)^2 & V_d < V_p \\ &= 0 & V_d \geq V_p. \end{aligned} \tag{16.130}$$

Note that these expressions are essentially identical to the square law current equation for long channel devices, which was derived in Section 5.2. However, note also that in these forms, the square law applies to V_s and V_d, but not to V_g.

Weak Inversion

Substituting the weak inversion charge equation (16.123) into (16.128) and integrating yields the weak inversion current expressions:

$$I_f = K_w \cdot \frac{\mu_o \cdot C_{ox} \cdot W_{eff}}{L_{eff}} \cdot \left(\frac{k_b T}{q}\right)^2 \cdot exp\left[\frac{q}{k_b T} \cdot (V_p - V_s)\right], \tag{16.131}$$

and

$$I_r = K_w \cdot \frac{\mu_o \cdot C_{ox} \cdot W_{eff}}{L_{eff}} \cdot \left(\frac{k_b T}{q}\right)^2 \cdot exp\left[\frac{q}{k_b T} \cdot (V_p - V_d)\right]. \tag{16.132}$$

Connecting the Strong and Weak Inversion Regions

The strong inversion and weak inversion current expressions must be connected so that a smooth transition between them is produced. The EKV model uses a semi-empirical approach as a compromise between physical rigor and computational efficiency. The final goal is to ensure the continuity of the current equations and their derivatives for all regions of device operation.

This is done by introducing a normalization scheme. First, define a *specific current*:

$$I_{sp} = 2 \cdot n \cdot \frac{\mu_o \cdot C_{ox} \cdot W_{eff}}{L_{eff}} \cdot \left(\frac{k_b T}{q}\right)^2; \tag{16.133}$$

this implies that

$$K_w = 2 \cdot n. \tag{16.134}$$

Next, the voltages are all normalized to the thermal voltage, using

$$v_p = \frac{V_p}{\left(\frac{k_b T}{q}\right)}, \tag{16.135}$$

$$v_s = \frac{V_s}{\left(\frac{k_b T}{q}\right)}, \tag{16.136}$$

and

$$v_d = \frac{V_d}{\left(\frac{k_b T}{q}\right)}. \tag{16.137}$$

A generalized expression for the normalized current in both strong inversion and weak inversion is

$$i_d \equiv \frac{I_d}{I_s} = i_f - i_r = F(v_p - v_s) - F(v_p - v_d), \tag{16.138}$$

where

$$i_f = \frac{I_f}{I_s} \tag{16.139}$$

and

$$i_r = \frac{I_r}{I_s}. \tag{16.140}$$

The term $F(v)$ is the interpolation function which will be used to smooth the transition region. Examining (16.129)–(16.132), the limiting behavior of this interpolation function can be determined to be

$$F(v) = \left(\frac{v}{2}\right)^2 \qquad v \gg 0$$
$$\hphantom{F(v)} = exp(v) \qquad v \ll 0. \tag{16.141}$$

A simplified version of a mathematical function suggested by Oguey and Cserveny [37] which meets these limits is

$$F(v) = \left\{ ln\left[1 + exp\left(\frac{v}{2}\right)\right] \right\}^2. \tag{16.142}$$

Using this expression, the $F(v)$ terms in (16.138) become

$$i_f = F(v_p - v_s) = \left\{ ln\left[1 + exp\left(\frac{v_p - v_s}{2}\right)\right] \right\}^2, \tag{16.143}$$

and

$$i_r = F(v_p - v_d) = \left\{ ln\left[1 + exp\left(\frac{v_p - v_d}{2}\right)\right] \right\}^2. \tag{16.144}$$

Equations (16.143) and (16.144) can be manipulated to express the voltages in terms of the currents, which is a practice commonly used in analog circuit design:

$$v_p - v_s = 2 \cdot ln\left[exp\left(i_f^{\frac{1}{2}}\right) - 1\right], \tag{16.145}$$

and

$$v_p - v_d = 2 \cdot ln\left[exp\left(i_r^{\frac{1}{2}}\right) - 1\right]. \tag{16.146}$$

The use of the interpolation function $F(v)$ guarantees that the current equations and their derivatives are continuous and smooth. Further details may be found in the original EKV paper [27] .

16.4.4   The Charge Model

In addition to continuous and smooth current equations and derivatives, it is also desirable to have a continuous and smooth charge model for use in the active device capacitance calculations. The EKV model uses an approach similar to the one developed above, by describing the strong inversion and weak inversion regions separately, then developing an interpolation function to smooth the region between them. However, the original form of the EKV model [27] performs this interpolation on the resulting equations for the capacitance terms, rather than on the charge equations. This may introduce problems during the interaction of the model with the circuit simulator, which could cause nonconservation of charge (as detailed in Chapter 13). This problem is currently in the process of being corrected [29].

The Basic Equations

As in all the other models, expressions must be developed for the different charge regions in the device. Assuming that the inversion charge is uniform across the width of the channel, the total inversion charge is found from

$$Q_{INV} = W_{eff} \cdot \int_0^{L_{eff}} Q_{inv}(y) \cdot dy, \qquad (16.147)$$

where the capitalized subscript indicates *total* charge (rather than the areal charge density). By substituting the basic current expression (16.125) and changing variables (as described in detail in Section 5.3), (16.147) becomes

$$Q_{INV} = -\frac{\mu_o \cdot W_{eff}^2}{I_d} \cdot \int_{V_s}^{V_d} [Q_{inv}(V_{ch})]^2 \cdot dV_{ch}. \qquad (16.148)$$

By a similar derivation, the depletion charge is determined to be

$$Q_{DEPL} = W_{eff} \cdot \int_0^{L_{eff}} Q_{depl}(y) \cdot dy = -\frac{\mu_o \cdot W_{eff}^2}{I_d} \cdot \int_{V_s}^{V_d} Q_{depl}(V_{ch}) \cdot Q_{inv}(V_{ch}) \cdot dV_{ch}. \qquad (16.149)$$

The gate charge is then easily found using the charge neutrality condition

$$Q_{GATE} = -Q_{INV} - Q_{DEPL} - Q_{OX}. \qquad (16.150)$$

Note that in modern MOS technology, the areal oxide charge density Q_{ox} is usually less than 10^{-9} C/cm^2, so that Q_{OX} can be safely neglected.

Once the charge expressions have been determined, the capacitance terms are computed using

$$C_{ij} = \xi \cdot \frac{\Delta Q_i}{\Delta V_j}\bigg|_{V_k, k \neq j} = \xi \cdot \frac{\partial Q_i}{\partial V_j}\bigg|_{V_k, k \neq j}, \qquad (16.151)$$

where

$$\begin{aligned} \xi &= -1 \quad i \neq j \\ &= +1 \quad i = j. \end{aligned} \qquad (16.152)$$

Strong Inversion

Substituting the strong inversion charge expression (16.107) into (16.148) leads to

$$Q_{INV} = -\frac{4}{3} \cdot W_{eff} \cdot L_{eff} \cdot C_{ox} \cdot n \cdot \frac{k_b T}{q} \cdot \frac{i_f + (i_f + i_r)^{\frac{1}{2}} + i_r}{i_f^{\frac{1}{2}} + i_r^{\frac{1}{2}}}, \qquad (16.153)$$

where in strong inversion

$$i_f = \left[\frac{q}{k_b T} \cdot \frac{V_p - V_s}{2} \right]^2 \tag{16.154}$$

and

$$i_r = \left[\frac{q}{k_b T} \cdot \frac{V_p - V_d}{2} \right]^2 . \tag{16.155}$$

The depletion charge is computed from (16.149) in a similar fashion:

$$Q_{DEPL} = -W_{eff} \cdot L_{eff} \cdot C_{ox} \cdot \mathbf{GAMMA} \cdot (\psi_o + V_p)^{\frac{1}{2}} - \frac{n-1}{n} \cdot Q_{INV}. \tag{16.156}$$

The gate charge is then computed using the charge neutrality condition (16.150):

$$Q_{GATE} = W_{eff} \cdot L_{eff} \cdot C_{ox} \cdot \mathbf{GAMMA} \cdot (\psi_o + V_p)^{\frac{1}{2}} - \frac{1}{n} \cdot Q_{INV} - Q_{OX}. \tag{16.157}$$

Note once again that Q_{OX} is usually negligible in modern MOS technology.

Weak Inversion

Substituting the weak inversion charge expression (16.123) into (16.148) yields

$$Q_{INV} = -W_{eff} \cdot L_{eff} \cdot C_{ox} \cdot \frac{K_w}{2} \cdot \frac{k_b T}{q} \cdot (i_f + i_r), \tag{16.158}$$

where in weak inversion

$$i_f = exp \left[\frac{q}{k_b T} \cdot (V_p - V_s) \right] \tag{16.159}$$

and

$$i_r = exp \left[\frac{q}{k_b T} \cdot (V_p - V_d) \right]. \tag{16.160}$$

Equation (16.158) can be further simplified by including the definition of K_w (16.134); this leads to

$$Q_{INV} = -W_{eff} \cdot L_{eff} \cdot C_{ox} \cdot n \cdot \frac{k_b T}{q} \cdot (i_f + i_r). \tag{16.161}$$

The depletion charge is computed by replacing ψ_s with $(\psi_o + V_p)$, as defined in (16.121) for weak inversion; this produces

$$Q_{DEPL} = -W_{eff} \cdot L_{eff} \cdot C_{ox} \cdot \mathbf{GAMMA} \cdot (\psi_o + V_p)^{\frac{1}{2}} . \tag{16.162}$$

The gate charge is once again computed from the charge neutrality requirement as expressed in (16.150):

$$Q_{GATE} = W_{eff} \cdot L_{eff} \cdot C_{ox} \cdot \mathbf{GAMMA} \cdot (\psi_o + V_p)^{\frac{1}{2}} - Q_{INV} - Q_{OX}, \tag{16.163}$$

noting that Q_{OX} is usually negligible in modern MOS technology. In weak inversion $Q_{INV} \ll Q_{DEPL}$, so (16.163) can be simplified to

$$Q_{GATE} \cong W_{eff} \cdot L_{eff} \cdot C_{ox} \cdot \mathbf{GAMMA} \cdot (\psi_o + V_p)^{\frac{1}{2}} - Q_{OX}, \tag{16.164}$$

Smoothing the Strong Inversion-Weak Inversion Transition

Once the various charge expressions have been computed, the capacitance equations are derived using (16.151). In the interest of brevity, these expressions are not detailed here. The original form of the EKV model [27] then proceeds to carry out an interpolation and smoothing effort on the capacitance terms, but *not* on the charge terms. This is being corrected in further work on the model [29].

The Non-Quasi-Static Model

As discussed in Chapter 13, it is usually assumed that the charge state of the FET is an instantaneous function of the terminal voltages. This assumption allows the charge-voltage relationships which follow naturally from the current model to be employed in describing the node charge values. As noted there, this quasi-static approach can break down for very high frequencies, or for very long channel lengths. For submicron technologies, the frequency at which the quasi-static approximation breaks down is well into the GHz range, so it is not likely that this situation will be encountered in most digital circuit designs. However, if a very long channel length is used in an analog circuit design, it is possible that a non-quasi-static approach may be required.

The EKV model includes a non-quasi-static model for the active gate capacitance terms. The development of that model is very involved; as it is unlikely that a non-quasi-static model will be required, the details will not be described here. However, the general guideline [27] can be noted that a non-quasi-static model is likely to be required when the operating frequency reaches a limit of roughly

$$\omega > \frac{\omega_o}{10}, \tag{16.165}$$

where

$$\omega_o = \frac{\mu_o}{L_{eff}^2} \cdot (V_p - V_s). \tag{16.166}$$

For example, if the operating frequency is about 200 MHz, a non-quasi-static model may be necessary for channel lengths longer than about 10 μm.

Comments

As noted above, the EKV model carries its development through to a final set of active device capacitance expressions. Using these equations as output from the FET model and as input to the circuit simulator introduces some problems. As discussed in Chapter 13, to guarantee charge conservation, the charge (rather than capacitance) equations should be provided to the circuit simulator, where charge should be the state variable.

In the original description of the EKV model [27], no interpolation was carried out for the charge expressions; they will not be smooth and continuous for all regions of device operation. If these charge models are used in a circuit simulator, the results will not be optimal. On the other hand, if the node capacitance equations (which *are* continuous and smooth) are used, charge may not be conserved. Fortunately, this problem is being addressed in further work on the model [29].

16.4.5   Small Geometry Effects

Surprisingly, small geometry effects are added to the EKV model in a very simple and rudimentary fashion. As noted earlier, despite this simplicity, good model results have been obtained with a single parameter set (for all geometries) down to a channel length of 0.7 μm [30, 31]. Further work is in progress to assess the behavior of the model for shorter channel lengths.

Mobility Reduction by the Vertical Field

The EKV model uses the simple Level 3-like expression for the reduction of the mobility by the vertical field:

$$\mu_v = \frac{\mu_o}{1 + \theta \cdot V_p'} = \frac{\mu_o}{1 + \mathbf{THETA} \cdot V_p'}, \tag{16.167}$$

where V_p' is a smoothed version of

$$V_p' = max(V_p, 0). \tag{16.168}$$

The modified mobility μ_v replaces the low field mobility μ_o in all the previous model equations.

Velocity Saturation

To account for carrier velocity saturation, the computed currents are modified to a form described by [19]

$$I_d' = \frac{I_d}{1 + \frac{V_d' - V_s}{L_{eff} \cdot E_c}} = \frac{I_d}{1 + \frac{V_d' - V_s}{L_{eff} \cdot \mathbf{UCRIT}}}, \tag{16.169}$$

where

$$V_d' = V_d \qquad V_d < V_{dsat}$$
$$= V_{dsat} \qquad V_d \geq V_{dsat}. \tag{16.170}$$

The term I_d' replaces I_d in all the previous equations, where I_d is the difference between the forward current I_f and the reverse current I_r.

Channel Length Modulation

To account for channel length modulation, the EKV model uses an approach [19] in which for $V_d \geq V_p$,

$$\Delta L \cong \zeta \cdot \left(V_d - V_p\right)^{\frac{1}{2}}, \tag{16.171}$$

where

$$\zeta \equiv \frac{2 \cdot \epsilon_{Si}}{\mathbf{GAMMA} \cdot C_{ox}}. \tag{16.172}$$

Replacing **GAMMA** in (16.172) with its definition from (16.102) leads to

$$\zeta = \left(\frac{2\epsilon_{Si}}{q \cdot N_{sub}}\right)^{\frac{1}{2}}. \tag{16.173}$$

Small Channel Geometry Effects

To account for small channel geometries, the pinch-off voltage V_p is modified to a form [19]

$$V_p = V_g - \mathbf{VTO} + \mathbf{GAMMA} \cdot \psi_o^{\frac{1}{2}} - \alpha \cdot \mathbf{GAMMA}$$

$$\times \left\{ \left[V_g - \mathbf{VTO} + \mathbf{GAMMA} \cdot \psi_o^{\frac{1}{2}} - \left(\frac{\alpha \cdot \mathbf{GAMMA}}{2}\right)^2 \right]^{\frac{1}{2}} - \frac{\alpha \cdot \mathbf{GAMMA}}{2} \right\}, \tag{16.174}$$

where

$$\alpha = 1 + \zeta \cdot \left\{ \frac{\eta_W}{W_{eff}} \cdot (\psi_o + v_s)^{\frac{1}{2}} - \frac{\eta_L}{L_{eff}} \cdot \left[(\psi_o + V_s)^{\frac{1}{2}} + (\psi_o + V_d)^{\frac{1}{2}} \right] \right\}$$

$$= 1 + \zeta \cdot \left\{ \frac{\mathbf{WETA}}{W_{eff}} \cdot (\psi_o + v_s)^{\frac{1}{2}} - \frac{\mathbf{LETA}}{L_{eff}} \cdot \left[(\psi_o + V_s)^{\frac{1}{2}} + (\psi_o + V_d)^{\frac{1}{2}} \right] \right\}. \tag{16.175}$$

The terms **WETA** and **LETA** are dimensionless fine-tuning parameters, which must be determined by extraction.

16.4.6 Comments

The EKV model is a very interesting development, and may point out the direction of future activities in analytical MOSFET modeling. This model represents the first significant rethinking of how the FET is modeled since the original analytical FET models were introduced during the 1960s. The use of the substrate as the voltage reference allows the weak inversion-strong inversion transition and the linear region-saturation region transition to be treated as the same physical phenomenon. As a result, only one mathematical function needs to be introduced to ensure that the current equations and their derivatives are continuous and smooth for all regions of device operation. The EKV model also uses an improved mathematical function to carry out this task.

The EKV model also pays more careful attention to the proper use of MOS device physics. This greatly aids in improving the physical basis of the model. In addition, the model is clearly oriented toward low power analog circuit design, which puts great demands on the quality of the FET model.

As noted in this section, in its original form [27], the model does have some shortcomings. Small geometry effects are considered in a very terse fashion after the bulk of the model has been completed; however, very reasonable small geometry results have been obtained [30, 31]. The mathematical core of the model is very elegant and sophisticated. However, as noted at the beginning of this section, this model is a work in progress, and future upgrades will be made available.

In the original model development [27], the active device charge/capacitance description is capacitance oriented rather than charge oriented. As noted in Chapter 13, if charge conservation is to be ensured, one requirement is that the FET model and the circuit simulator must communicate at the charge level (rather than at the capacitance level). The original EKV model mathematically optimized its capacitance expressions for smoothness and continuity, but left the charge equations in their original state. This situation is being addressed [29]; in the near future, smooth and continuous charge expressions will be available for the circuit simulator, so that charge will be conserved and the charge expressions will be smooth and continuous, as well as robust and efficient.

Overall, the EKV model represents a new approach to FET modeling. This rethinking of the approaches used to date will be the subject of further discussion in Chapter 17. However, like the other models examined in this chapter, this model will require extensive exercise in both parameter extraction and circuit simulation if its problems and prospects are to be fully evaluated.

16.5 Final Comments

The models discussed in this chapter are not presently employed in any commonly available SPICE circuit simulator; they are described here because they could become important FET models. All three models are based on the third-generation principles discussed in Section 11.2. In particular, these models put significant emphasis on ensuring the continuity of the model equations and their derivatives. These improvements are required of FET models which are used in analog circuit design; in addition, these mathematically more robust equations improve the efficiency and convergence properties of circuit simulation. By employing one current equation for all regions of device operation, these mathematical goals are more easily met. To ensure that these goals are actually met, the extensive use of smoothing functions is introduced.

The three different models considered here represent different approaches and levels of sophistication in addressing these requirements. The Power-Lane model modifies an existing model. This provides the quickest and simplest solution; however, some of the shortcomings of the original model (in this case, Level 3) will inevitably be brought along to the new model. PCIM uses the original FET model basis developed during the 1960s, but creates a new model formulation. In this manner, the model is intentionally constructed to meet the desired goals. The Enz-Krummenacher-Vittoz (EKV) model represents a brand new approach to the problem of analytical FET modeling. This points out a path for future studies in this area.

Like the mainstream third-generation models (BSIM3 and MOS Model 9), these models are new, so an extensive body of knowledge and experience has yet to be developed. A good deal of practical usage with both parameter extraction and circuit simulation must take place to build up that knowledge base; this must occur before the true strengths and weaknesses of the models can be defined. At present, it is only possible to comment on the structure of each model, rather than the effects of its implementation.

BIBLIOGRAPHY

1. J. Power and W. Lane, "An Enhanced SPICE MOSFET Model Suitable for Analog Applications," *IEEE Trans. Comp.-Aided Des.* vol. CAD-11, pp. 1418–1425 (1992).

2. A. Silburt, R. Foss, and W. Petrie, "An Efficient MOS Transistor Model for Computer-Aided Design," *IEEE Trans. Comp.-Aided Des.* vol. CAD-3, pp. 104–111 (1984).

3. W. Maes, K. de Meyer, and L. Dupas, "DC Characterization of MOS Transistors, SPICE Model Level 3," Internal Report, Dept. Elektrotech., Katholieke University, Leuven, Belgium, 1984.

4. G. Baum and H. Beneking, "Drift Velocity Saturation in MOS Transistors," *IEEE Trans. Elec. Dev.* vol. ED-17, pp. 481–482 (1970).

5. V. Reddi and C. Sah, "Source and Drain Resistance Beyond the Pinch-Off Point in Metal-Oxide-Semiconductor Transistors (MOST)," *IEEE Trans. Elec. Dev.* vol. ED-12, pp. 139–141 (1965).

6. H. Masuda et al., "A Submicrometer MOS Transistor I-V Model for Circuit Simulation," *IEEE Trans. Comp.-Aided Des.* vol. CAD-10, pp. 161–170 (1991).

7. P. Yang, B. Epler, and P. Chatterjee, "An Investigation of the Charge Conservation Problem for MOSFET Circuit Simulation," *IEEE J. Sol. St. Circ.* SC-18, pp. 128–138 (1983).

8. N. Arora, R. Rios, C. Huang, and K. Raol, "PCIM: A Physically Based Continuous Short Channel IGFET Model for Circuit Simulation," *IEEE Trans. Elec. Dev.* vol. ED-41, pp. 988–997 (1994).

9. N. Arora, *MOSFET Models for VLSI Circuit Simulation*, Springer-Verlag, 1993.

10. N. Arora and G. Gildenblat, "A Semi-Empirical Model of the MOSFET Inversion Layer Mobility for Low Temperature Operation," *IEEE Trans. Elec. Dev.* vol. ED-34, pp. 89–93 (1987).

11. S. Lee, "Universality of Mobility-Gate Field Characteristics of Electrons in the Inversion Charge Layer and Its Application in MOSFET Modeling," *IEEE Trans. Comp.-Aided Design* vol. CAD-8, pp. 724–730 (1989).

12. C. Huang and N. Arora, "Characterization and Modeling of the n- and p-Channel MOSFET Inversion Layer Mobility in the Temperature Range 25 to 125°C," *Sol. St. Elec.* vol. 37, pp. 97–104 (1994).

13. R. Coen and R. Muller, "Velocity of Surface Carriers in Inversion Layers of Silicon," *Sol. St. Elec.* vol. 23, pp. 35–40 (1980).

14. T. Chan, S. Lee, and H. Gaw, "Experimental Characterization and Modeling of Electron Saturation Velocity in the MOSFET Inversion Layer from 90 to 350K," *IEEE Elec. Dev. Lett.* vol. EDL-2, pp. 466–468 (1980).

15. G. Taylor, "Velocity-Saturated Characteristics of Short-Channel MOSFETs," *AT&T Bell System Tech. Journal* vol. 63, pp. 1325–1381 (1984).

16. T. Grotjohn and B. Hoefflinger, "A Parametric Short-Channel MOS Transistor Model for Subthreshold and Strong Inversion Current," *IEEE Trans. Elec. Dev.* vol. ED-31, pp. 109–112 (1984).

17. G. Wright, "Simple and Continuous MOSFET Models for the Computer-Aided Design of VLSI," *Proc. Inst. Elec. Eng.* vol. 132, part I, pp. 187–194 (1985).

18. D. Caughey and R. Thomas, "Carrier Mobilities in Silicon Empirically Related to Doping and Field," *Proc. IEEE* vol. 55, pp. 2192–2193 (1967).

19. Y. Tsividis, *The MOS Transistor*, McGraw-Hill, 1987.

20. H. Park, P. Ko, and C. Hu, "A Charge Sheet Capacitance Model of Short Channel MOSFETs for SPICE," *IEEE Trans. Comp.-Aided Des.* vol. CAD-10, pp. 376–389 (1991).

21. C. McAndrew, B. Bhattacharyya, and O. Wing, "A Single Piece C_∞-Continuous MOSFET Model Including Subthreshold Conduction," *IEEE Elec. Dev. Lett.* vol. EDL-12, pp. 565–567 (1991).

22. P. Ko, "Approaches to Scaling," *Advances MOS Device Physics* (ed. by G. Einspruch and G. Gildenblat), *VLSI Electronics: Microstructure Science* (vol. 18), Academic, 1989.

23. H. deGraaff and F. Klaassen, *Compact Transistor Modeling for Circuit Design*, Springer-Verlag, 1990.

24. S. Sze, *Physics of Semiconductor Devices* (2nd ed.), John Wiley & Sons, 1981.

25. J. Meyer, "MOS Models and Circuit Simulation," *RCA Review* vol. 32, pp. 42–63 (1971).

26. D. Ward and R. Dutton, "A Charge-Oriented Model for MOS Transistor Capacitances," *IEEE J. Sol. St. Circ.* vol. SC-13, pp. 703–708 (1978).

27. C. Enz, F. Krummenacher, and E. Vittoz, "An Analytical MOS Transistor Model Valid in All Regions of Operation and Dedicated to Low Voltage and Low Current Applications," *Analog Int. Circ. and Signal Proc.* vol. 8, pp. 83–114 (1995).

28. M. Bucher, C. Lallement, C. Enz, and F. Krummenacher, "The EPFL-EKV MOSFET Model Equations for Simulation, Version 2.3," Electronics Laboratories LEG, Swiss Federal Institute of Technology EPFL, Lausanne, Switzerland, December 1995.

29. C. Enz, "The EKV Model: A MOST Model Dedicated to Low Current and Low Voltage Analogue Circuit Design and Simulation," in *Low Power HF Microelectronics*, IEE Circuits & Systems Series No. 8, IEE Press, 1996.

30. M. Bucher, C. Lallement, and C. Enz, "An Efficient Parameter Extraction Methodology for the EKV MOST Model," 1996 IEEE Int. Conf. on Microelectronic Test Structures (in press).

31. M. Bucher, C. Lallement, C. Enz, and F. Krummenacher, "Accurate MOS Modeling for Analog Circuit Simulation Using the EKV Model," 1996 IEEE Int. Symposium on Circuits and Systems (in press).

32. C. Lallement, C. Enz, and M. Bucher, "Simple Solutions for Modeling Non-Uniform Substrate Doping," 1996 IEEE Int. Symposium on Circuits and Systems (in press).

33. H. Ihantola and J. Moll, "Design Theory of a Surface Field-Effect Transistor," *Sol. St. Elec.* vol. 7, pp. 423–430 (1964).

34. E. Vittoz, "MOS Transistors Operated in the Lateral Bipolar Mode and Their Application in CMOS Technology," *IEEE J. Sol. St. Circ.* vol. SC-18, pp. 273–279 (1983).

35. H. Wallinga and K. Bult, "Design and Analysis of CMOS Analog Signal Processing Circuits by Means of a Graphical MOST Model," *IEEE J. Sol. St. Circ.* vol. SC-24, pp. 672–680 (1989).

36. H. Pao and C. Sah, "Effects of Diffusion Current on the Characteristics of Metal-Oxide(Insulator)-Semiconductor Transistors," *Sol. St. Elec.* vol. 10, pp. 927–937 (1966).

37. H. Oguey and S. Cserveny, "Modèle du Transistor MOS Valable dans un Grand Domaine de Courants," *Bull. SEV/VSE*, February 1982.

17

The Future of Device Models for Circuit Simulation

As was noted in Chapter 2, the typical user of SPICE and its FET element models is on the receiving end of a history of activities. This history has set the constraints for any new developments which have taken place in this field.

Thus, it is reasonable to expect that any further developments will evolve in a logical fashion from what has already been learned and accomplished. Any attempt to sketch the future direction of efforts in analytical FET modeling for circuit simulation must be connected with what has been done in the past, and with the situation in the present state of the art.

17.1   The Past

The roots of the analytical FET models used in SPICE are fairly straightforward. When the field-effect transistor appeared on the scene as an important microelectronic device in the early 1960s, a set of analytical equations were derived [1] to provide an understandable description of FET behavior. There was no explicit intent of using that description in a circuit simulator; in fact, circuit simulators did not exist at that time. Later, when SPICE [2, 3] appeared on the scene, it was quite natural to use this simple description as the FET element model and imbed it in the circuit simulator.

In this form, the intent of the FET model is clearly to provide a physically based description, regardless of any possible concomitant mathematical difficulties. In addition, due to this structure, most of the model parameters have some correspondence to physically meaningful quantities. The original base model was extensively modified by the continual addition of various corrections for small geometries. However, due to the emphasis on a physically satisfying FET model, problems were encountered with mathematical efficiency and convergence behavior during circuit simulation.

As time passed, the reality of the situation, that these element models were being used in a circuit simulation environment, began to creep into developments. The FET model equations began to be reworked in a variety of ways, with the clear goal of providing more efficient and robust circuit simulation behavior. As a consequence, the number of model parameters began to grow, and these parameters began to take on an increasingly empirical character.

17.2   The Present

In the present environment, the FET element models have become heavily empirical, with extensive mathematical conditioning. The number of model parameters has become very large, and most of these parameters are basically empirical in character. The focus is clearly on the circuit simulation use of the FET element model, rather than on a physical description of the FET. As such, mathematical fitness overrides physical understanding in the description of the FET. In this form, the FET element model serves as the communication vehicle between the circuit designer and the silicon fabrication foundry.

With the growth of both the number and empirical character of model parameters, increasing emphasis must be placed on parameter extraction. It is important to note once again that the mathematical formulation of an FET element model represents the upper limit of what is possible from that model. To reach (or at least closely approach) that limit, well-executed parameter extraction is required. This discussion also points out a very important distinction between two separate issues in analytical FET modeling. A model must be analyzed very carefully to determine which aspects of device behavior it is able to describe. Often, a model is capable of describing most important aspects of device behavior; this can be achieved by appropriate parameter extraction methods. On the other hand, it is equally important to understand where some part of the underlying model formulation truly *is* unable to provide a proper description; the neglect of the screening of the inner-fringe capacitance by the inversion layer, described in Chapter 13 and Chapter 15, provides a good example. In these situations, some clever work-around is often found; however, such solutions usually have very limited validity. These differences must be understood, both to maximize model effectiveness, and to avoid potential pitfalls which can be unintentionally introduced.

There is also, at present, a growing need to properly describe what are loosely described as the analog functions of an FET [4]. These arise due to the different use that is made of the FET in analog circuits (compared to digital circuits). For example, in a digital circuit, bias switching is from the power supply voltage (V_{dd}) to ground (large signal approach); however, in an analog circuit, an FET is often biased at intermediate voltages, and small biases are applied around that point (small signal approach). In this latter approach, the finer structure of the FET behavior, described by the device conductances, becomes very important to a proper description of the FET. Models available at this time can (for the most part) meet these requirements for accuracy; however, this continuing need for model accuracy has forced the use of model binning, which introduces its own problems.

A final problem with the present state of FET modeling for circuit design does not affect the circuit designer directly, but presents an increasingly unpleasant problem for the builder of models. Given the proliferation of analytical FET models, the model provider now encounters a portfolio problem, in which one particular CMOS process technology requires a portfolio of FET model types to meet the needs of different circuit design groups. For example, a given CMOS technology will likely require a Level 3 model, for quick analysis of digital circuit performance and for the time-efficient simulation of very large digital ICs. At the same time, the same technology will likely require an HSPICE Level 28 model for analog and mixed signal designs *which employ the same process technology*. Furthermore, designers using simulation packages which cannot use HSPICE Level 28 require a BSIM2 model as a workable alternative. Finally, new model demands (for BSIM3 and/or MOS Model 9) are being made before the need for older models has disappeared.

This problem may seem to be academic in nature, but it is very important to the present industry situation. With a larger number of FET model types to support, model building resources are spread more thinly. As a consequence, the quality of each particular model can suffer, due to the need to put less effort into each individual type of model so that the full menu of model choices can be supported.

17.3   The Future

Exactly what the future holds for analytical FET modeling is, of course, a matter of conjecture. However, certain trends and driving forces can be identified; these can serve as guideposts for what can be expected in the years ahead.

At present, the new model situation (of the third-generation models and the new candidate models) is in a state of flux, with new developments appearing regularly. It is not yet clear which of these models will find common use, and which will fall by the wayside. It is also possible that other new models will emerge to become important.

Regardless of which models are accepted and which are not, trends that are influencing the development and use of FET models can be identified. Perhaps the most important consideration [4] is the growth of analog and mixed signal circuits, particularly as subcomponents of larger digital ICs. In addition, low-voltage/low-power digital circuits are becoming more important; interestingly, many of the problems in FET modeling which are regarded as being analog in nature also begin to appear in low-voltage digital circuits. Finally, the increasing clock frequencies in all ICs force a continuing reevaluation of the methods used to cope with these high switching speeds.

The third-generation models demonstrate a structure which addresses problems of this sort. The use of smoothing functions guarantees that the model is well behaved near transition voltages, such as the saturation voltage (V_{dsat}). This usually leads to a single equation for a particular device characteristic (such as the drain current) which is valid for all regions of device operation; this approach is superior to the development of separate equations which are then pieced together at the transition points. As a result, in the third-generation models, continuous and smooth equations (and, where necessary, derivatives) are much more easily provided. An interesting suggestion [5] is to use an approach common in analog circuit design, and exploit the symmetry of the FET structure in the development of the FET model itself.

The future goals of a good FET model are fairly easy to state. A model should be accurate (or, at least, have the ability to be *made* accurate through parameter extraction) in its description of FET behavior. Its formulation should be mathematically robust and efficient, so that circuit simulation convergence can be reached with little difficulty. Finally, parameter extraction must be a reasonable task, if the potential (structural) capabilities of the model are to be achieved. It is once again important to note that a distinction must be made between what a model is genuinely capable of achieving (through parameter extraction, to reach the full capability of that model) and true physical shortcomings of that model (which *cannot* be repaired by parameter extraction, other than in the form of patches which will have extremely limited validity).

While other features are not strictly necessary, their inclusion would simplify model building and usage; they are thus desirable, and can hopefully be introduced in future models. For example, the ability to use a small number of physically based parameters in an FET model would obviously simplify the description of the device and reduce the complexity of parameter extraction. While this improvement is not necessary for the description of individual device characteristics, it would greatly improve the ability to describe process variations (see Chapter 14). Another "pipedream" is the use of a single set of model parameters for all device geometries, which is tantamount to the elimination of any need for model binning. If this could be done, the side effects of binning, detailed in Chapters 7, 9, and 10, could be eliminated. In addition, binned parameters are rarely (if ever) used for the description of process variations; the elimination of binning would also allow for an improved description of process variations.

Taken together, these two pipedreams (a model with a small number of physically based parameters which are valid for all geometries) could have major implications for the accurate description of process variations. With a small number of parameters and no bins, parameter extraction becomes a much less time consuming process. This opens the door to the tractable use of principal component analysis [6] (see Chapter 14), which requires repetitive data measurement, parameter extraction, and model building. Principal component analysis would provide a fully accurate description of the effect of process variations on FET behavior.

Most importantly, new ideas are under consideration, and are finding their way into new models. For example, the use of smoothing functions is a new development that is becoming widely adopted in various forms. However, it must be noted that the present state of the analysis of FET behavior is still based on the original base model [1], with the continuing inclusion of more and more small geometry corrections in the model formulation. It is now widely acknowledged that much of the trouble in the analytical FET description is caused by the various breakpoints (e.g., V_t, V_{dsat}, etc.) in the model formulation. While these breakpoints were originally developed to simplify the device description, they have introduced mathematical problems which make it difficult to achieve a continuous and smooth description in their vicinity. The effective use of smoothing functions, imposed on the device description, mitigates the problem but also points out the serious level of the situation. It is reasonable to suggest that the time may have arrived for a complete reconsideration of FET modeling, in a modern context and without the baggage of the original model for an FET.

In addition, a model must have a circuit simulator as a vehicle if it is to gain widespread acceptance. The University of California/Berkeley FET models have always had that institution's SPICE simulator as a vehicle, while HSPICE Level 28 has been able to make good use of the widespread industrial implementation of HSPICE to achieve the same purpose. MOS Model 9 is being introduced to the industry with this problem in mind, and a great deal of effort is being made to introduce that model into commonly used circuit simulators. Other analytical FET models are regularly published in the literature; however, unless a strong effort is made to introduce them into the available modeling infrastructure, these models will not be widely used, regardless of their potential capabilities.

Finally, the basic role of the FET element models used in SPICE must be carefully reconsidered. At the most basic level, the circuit designer is the model consumer; it therefore behooves every circuit designer to be an educated model consumer. However, in the present structure, parameter extraction and model building activities are most often associated with the foundry side of the business, and then provided in a "black box" form to the circuit designer. This is (apparently) a relic of the days when model building was in essence an exercise in process characterization, with the results eventually provided to the circuit designer. However, with the growth of the empirical character of the FET models and their parameters, this is no longer the case; parameter extraction and model building provide little or no information to process engineering teams. Instead, process engineering requires a completely independent effort in process characterization, which is beyond the scope of this text. Clearly, in the present situation, there is a need for closer interaction between model building activities and the ultimate users of those activities, the circuit designers. Interactions of this type will have a number of advantages. For example, in such an environment, circuit designers are forced to be aware of (inevitable) situations in which particular types of models have inherent weaknesses. Also, since no analytical FET model is perfect, models can be more carefully built to address specific circuit design needs, while avoiding the wasting of time required to achieve model accuracy where it is not needed.

Perhaps engineering itself can be defined as the art and science of coping with the imperfect. As such, to make the best use of both present and future analytical FET models, a more connected approach must be taken to link model building with its true *raison d'être*, the communication of information to the circuit designer.

BIBLIOGRAPHY

1. H. Ihantola and J. Moll, "Design Theory of a Surface Field-Effect Transistor," *Sol. St. Elec.* vol. 7, pp. 423–430 (1964).

2. L. Nagel and D. Pederson, "Simulation Program with Integrated Circuit Emphasis," University of California/Berkeley, Electronics Research Laboratory Memorandum No. UCB/ERL M352 (1973).

3. L. Nagel, "SPICE2: A Computer Program to Simulate Semiconductor Circuits," University of California/Berkeley, Electronics Research Laboratory Memorandum No. UCB/ERL M520 (1975).

4. Y. Tsividis and K. Suyama, "MOSFET Modeling for Analog Circuit CAD: Problems and Prospects," *IEEE J. Sol. St. Circ.* vol. SC-29, pp. 210–216 (1994).

5. C. Enz, F. Krummenacher, and E. Vittoz, "An Analytical MOS Transistor Model Valid in All Regions of Operation and Dedicated to Low Voltage and Low Current Applications," *Analog Int. Circ. and Signal Proc.* vol. 8, pp. 83–114 (1995).

6. C. Michael and M. Ismail, "Statistical Modeling of Device Mismatch for Analog MOS Integrated Circuits," *IEEE J. Sol. St. Circ.* vol. SC-27, pp. 154–166 (1992).

Appendix A
An Executive Summary of the Various Models

Here a thumbnail sketch of the various FET element models in SPICE is given. The intent is to provide a brief executive summary, which states all of the key points but without much detail. Those details, naturally, may be found in the chapters of the text.

A.1   The FET Model Generations

Initially, an analytical model was developed to provide a simple description of the FET; the intent was to create an understanding of FET behavior, rather than an element model for circuit simulation. Later, when SPICE was developed, this model was used as the element model for the FET.

As device dimensions decreased, this original model remained as the base, and a variety of corrections was added to account for various small geometry effects. With time, device geometries continued to shrink, and more and more corrections were added to describe a wider and wider variety of small geometry effects. However, as the true role of the FET models, as part of circuit simulation, began to become more clear, mathematical fitness became the dominant issue. The FET model focus shifted from a physically based analytical description of device behavior to a structure which would provide the best results for robust and efficient circuit simulation.

For historical reasons, it is useful to divide the SPICE FET models into three evolutionary generations. The first generation models are clearly based on the original intention of a physically based analytical FET model, with all geometry dependence included in the model equations. These models concentrate on the description of the FET rather than the behavior of the model equations in a circuit simulator. This generation of models is comprised of Level 1, Level 2, and Level 3.

The second generation of models represented a major change from the first-generation philosophy. In this set of models, the equations are subject to extensive mathematical conditioning, with a clear focus on their circuit simulation usage. The parameter structure is twofold; there are individual device parameters, and there are also geometry (length and width) parameters which are fit to the individual device parameters. The model structure and its parameters take on considerable empirical character; this has the effect of shifting the action in the model to parameter extraction, which is imposed on a rather simple shell model formulation. This generation of models encompasses BSIM and BSIM2, along with HSPICE Level 28 (which is based on BSIM).

At this time, the third generation of models is beginning to emerge. The basic intent has been to return to a simpler model structure, with a reduced number of parameters; these parameters are also physically based, rather than empirical. Mathematical conditioning is also important in these models; however, in contrast to the polynomial functions which are heavily used in the second-generation models (and which can behave badly in certain circumstances), the third-generation models employ more specialized smoothing functions which are mathematically well behaved. With these smoothing functions, most of the device equations (and, where necessary, their derivatives) are continuous and smooth, and a single equation is used for all regions of device operation. At the present time, BSIM3 and MOS Model 9 are established as third-generation models; however, new models are appearing on the scene which may or may not become important.

A.2   The First-Generation Models

The first generation of FET models is based on a very simple description of the device. The focus of these models is the analytical description of the FET, not circuit simulation usage.

A.2.1   Level 1

The original FET element model is Level 1. The structure of this model is basically identical to that which is developed in an introductory text on solid-state devices, and thus provides a very elementary description of FET behavior. Simplifications such as the gradual channel approximation and the square law for the saturated drain current are employed. The only small geometry effect is the inclusion of a simple λ-model for channel length modulation, which leads to a finite value of the output conductance; however, the output conductance is taken to have a single value throughout the saturation region. Also, no subthreshold conduction model is included. The Level 1 model only has validity in devices which are, by present standards, very large; it can be regarded as obsolete, and useful only as a teaching tool (in that, for example, the few model parameters can be determined using a calculator).

A.2.2   Level 2

The Level 2 model represents the first attempt to include small geometry effects in a MOSFET model. More detailed descriptions of the device depletion region, the threshold voltage, and mobility reduction by the vertical field are included; however, effects such as drain-induced barrier lowering (DIBL) and the lateral field reduction of the channel mobility are neglected, and the vertical field effect on the mobility is described in a mathematically clumsy fashion. The drain current expression fully integrates the description of the depletion charge; this leads to an equation with $\frac{3}{2}$-power terms, which are computationally inefficient.

A model for the reduction of the saturation voltage by velocity saturation is included; unfortunately, the resulting expressions are hideously complicated, with a quartic equation providing the final description. Four possible solutions to the quartic equation exist, and these must be sorted to find the correct one; it is also feasible that all of the quartic solutions will be rejected, forcing the use of another expression to compute the saturation voltage. Despite all this effort, there is still a very serious first derivative discontinuity in the drain current expression in the neighborhood of the saturation voltage.

A subthreshold current model is also introduced for the first time; this model is extremely simple, with only one parameter, so that the subthreshold current can only be modeled for one choice of channel length, drain bias, and temperature. Furthermore, although continuity with the strong inversion drain current expression is guaranteed, the first derivative is badly discontinuous at the transition point; this is known to cause convergence problems during circuit simulation.

The Level 2 model was developed to address small geometry device behavior. However, the model structure was allowed to become very complicated; in circuit simulation, it is slow, inefficient, and frequently encounters convergence problems. As a result, it was quickly replaced with the Level 3 model, to be discussed below. Ironically, one feature of Level 2 has proven to have some use; the λ-model of the output conductance is available in Level 2, so that Level 2 is still used for the description of this important device characteristic in analog circuit design. However, this description is very weak, and should only be used as a quick method of obtaining an initial oversimplified result; for real analog circuit design, a better model, such as HSPICE Level 28 or BSIM2, should be used. Another problem with Level 2 is that a relatively good description can be reached at one particular geometry; however, if the same parameters are used and the simulation geometry is changed (which is what the model structure intends), the results deteriorate. This has led to the development of model binning, which is described below.

A.2.3   Level 3

The Level 3 model was developed to overcome the observed shortcomings of Level 2. While the structure is basically similar to Level 2, Level 3 takes a semi-empirical approach, which places more emphasis on parameter extraction. Some physical effects are added, such as DIBL and the reduction of the mobility by the lateral field. The main difference (versus Level 2) is that the Level 3 model can provide a description which is at least as accurate as the Level 2 result, but with a mathematical structure that is considerably more efficient. Therefore, Level 3 is still in wide use today for the modeling of digital circuits.

The Level 3 structure concentrates on simplicity. The description of the reduction of the mobility by the vertical field is very simple and semi-empirical, although it neglects the influence of the substrate bias. The effect of the lateral field on the mobility is included as well, and is used with the saturation velocity to compute the reduced value of the saturation voltage. In contrast to Level 2, though, a very simple and efficient expression for the saturation voltage is produced. The drain current model makes use of a Taylor series expansion of the depletion charge expression, which eliminates the $\frac{3}{2}$-power terms that appear in Level 2. As a result, the Level 3 drain current expression is much more efficient; this represents the first instance of mathematical conditioning in the development of an FET model. Level 3 employs the same subthreshold current model as that of Level 2, with the associated problems discussed earlier.

Level 3 also suffers from the geometry dependence problem described above for Level 2; model parameters are supposed to be valid for all device geometries. The parameters can be extracted for an optimum result for one particular device geometry; however, if the geometry (such as the channel length) is changed, the model results usually become quite poor. The clumsy solution to this problem has been the introduction of model binning, in which the geometry space (the length and width extent of the technology) is broken into subregions, each with its own model. While this method improves the model results as the channel length and width are varied, the model structure is not designed to accommodate binning, so new problems are created. For example, the model characteristics (such as the drain current) cannot be made continuous across the bin boundaries. This can be a serious problem if the channel length is optimized during circuit design, and during statistical analysis of process variations.

Despite its many shortcomings, the Level 3 model has proven to be relatively simple, robust, and reliable; it is thus very popular as a model for digital circuit design, particularly for very large circuit simulations, where the simplicity of Level 3 allows for a result to be reached in a reasonable amount of time. However, the Level 3 model shows a very unpleasant discontinuity in the first derivative of the drain current at the saturation voltage; this discontinuity grows worse for shorter channel lengths. This leads to very poor output conductance results; combined with a very weak subthreshold current model, this renders Level 3 unsuitable for analog circuit design.

A.3   The Second-Generation Models

In the second-generation models, the device geometry is included in the basic model equations. In addition, an entirely separate parameter structure is created solely to describe the geometry dependence. Independent parameters are extracted for each device; then the geometry parameters are extracted to fit the initial set of independent parameters across the device length and width. The goal is to provide an apparatus in which the original independent device parameters can be reconstituted for any particular choice of channel length and width.

A.3.1   BSIM

BSIM was the first of the second-generation models. In the model structure, the emphasis is clearly on mathematical conditioning for circuit simulation (rather than on a "nice" analytical description of the FET). The approach to the description of small geometry effects is strictly empirical; this weakens the connection between the model parameters and the underlying process technology, and strongly shifts the action to parameter extraction.

BSIM includes improved (although empirical) descriptions of the threshold voltage and the mobility. The drain current model employs a number of polynomial expressions to avoid the problem of the $\frac{3}{2}$-power terms which appears in an "exact" derivation (such as that developed in Level 2), which leads to a relatively simple description of the drain current. A very simple description for the saturation voltage, as reduced by velocity saturation, is also included. In addition, a more detailed subthreshold current model is introduced, which has more widespread applicability. The serious first derivative discontinuity at the transition to the strong inversion region is also repaired, greatly improving the convergence behavior.

BSIM represents a major improvement in the mathematical structure of a model, and thus leads to improved circuit simulation behavior. However, BSIM is known to have difficulties in submicron FETs; for example, although mathematically well-behaved, poor results are obtained in the region of the saturation voltage. In addition, the polynomial expressions contain quadratics which can behave very badly; a common symptom is the appearance of a negative output conductance in certain circumstances. A further problem is that the weak inversion (subthreshold) current is merely added to the strong inversion current; it is assumed that the strong inversion current is very large, and that no problems will occur. However, this is not the case for a gate bias slightly larger than the threshold voltage, and a jump in the current occurs at the transition point.

Overall, BSIM is an improved digital model; however, it isn't truly suitable for analog circuit design. Due to its loss of accuracy at submicron channel lengths, it has largely been replaced by HSPICE Level 28 and BSIM2.

A.3.2   HSPICE Level 28

HSPICE Level 28 is a proprietary model developed by Meta-Software, which has used HSPICE as a vehicle to gain rather widespread acceptance. The model structure is based on BSIM, but has been extensively modified. Through extensive mathematical conditioning, HSPICE Level 28 has been made suitable for analog design; it is thus commonly encountered in the IC industry. However, since the model structure is proprietary, it is somewhat difficult to discuss in detail.

A unique feature of HSPICE Level 28 is that the model structure is designed to accommodate model binning. Additional terms are introduced to guarantee the continuity of the model parameters (and thus the model characteristics) at all four corners of a bin; this also guarantees continuity across the faces of a bin.

In its structure, HSPICE Level 28 conditions the quadratic expressions which appear in BSIM, so that they become less troublesome. The drain current model contains very extensive conditioning of the various transition points; this foreshadows the structure of the third-generation models, and has led to very successful results. The saturation voltage model defines a transition region (in addition to the normal use of the linear region and the saturation region), which produces good results. A similar conditioning approach is used on the subthreshold model, where several subregions are defined. In particular, the problems introduced in BSIM by the simple addition of the weak inversion current to the strong inversion current are eliminated.

HSPICE Level 28 maintains the clear focus of the second-generation models on the circuit design use of the FET description. However, the model parameter set is almost entirely empirical. In addition, due to its roots, HSPICE Level 28 is carrying virtually the entire BSIM mathematical structure; it can thus be slow in the simulation of large circuits. Due to its suitability for analog circuit design, HSPICE Level 28 is commonly encountered in the industrial circuit design environment. However, an important drawback of this model is that it locks the user into employing HSPICE for circuit simulation.

A.3.3   BSIM2

BSIM2 is closely based on BSIM. It employs several expressions developed from two dimensional analysis, and makes extensive modifications to the BSIM description of the mobility and the drain current, including a new subthreshold current model. Most importantly, an output conductance model is added, making BSIM2 suitable for analog circuit design.

The BSIM threshold voltage model is modified, so that a negative output conductance will not occur (as sometimes happens in BSIM). The mobility model is similar to the BSIM expression, but adds a quadratic term to improve the final accuracy at higher vertical fields. A new expression for velocity saturation is developed, and the drain current expression is modified to greatly improve the short channel accuracy. Unlike BSIM, the same current expression is used for both the linear and saturation regions. An improved subthreshold current model is introduced, which includes a parametrically defined transition region and is very similar to the one developed in HSPICE Level 28. The output conductance model is complex, with several additional parameters; however, it provides a tractable result for analog circuit design.

The ability to handle analog design requirements is the major feature of BSIM2. In addition, the drain current model is more accurate, and provides better convergence behavior during circuit simulation. The main problem with BSIM2 is its complexity, as it contains a very large number of parameters. BSIM2 is commonly encountered when there is a need to use certain simulators which by definition cannot contain HSPICE Level 28 (due to that model's proprietary nature).

A.4   The Third-Generation Models

As noted earlier, the third generation of FET models is now beginning to emerge. The original intent was simplification of the FET model formulation, a reduction of the number of model parameters, and the development of parameters which are physically based (rather than empirical, as in the second-generation models). There is also extensive use of well-behaved mathematical smoothing functions, which allows for smooth and continuous expressions for model equations and, where necessary, their derivatives. The use of these smoothing functions usually leads to a single equation for each device characteristic which is valid for all regions of device operation.

With new developments appearing regularly, the third-generation model situation is quite fluid. It remains to be seen which models will be accepted, and if new models will arise and become important.

A.4.1   BSIM3

BSIM3 is the initial third-generation model. At present, there is a bit of confusion, as several different versions have appeared. The original intent of BSIM3 was simplicity, with a simplified model structure and a small number of physically based parameters. However, Versions 1 and 2 showed several shortcomings; an attempt to repair these problems is made in Version 3, but this is done with a large infusion of empirical equations and new (largely empirical) model parameters. The model has evolved into an extremely complex form with a very large number of parameters; this seems to violate the original intent of the third-generation models. In addition, the complexity of the model and the large number of parameters suggest that parameter extraction for BSIM3 will be complicated.

A.4.2   MOS Model 9

MOS Model 9 was developed at Philips Laboratories, and is being made widely available in mainstream circuit simulators. It is significant to note that this is the first model without any University of California/Berkeley origins which is finding its way into mainstream use. An examination of the model structure shows that it has an extensive heritage of industrial use.

The model equations are very clean and simple. This suggests that they will be very efficient during circuit simulation, and that parameter extraction should be a relatively straightforward exercise. Very well-behaved hyperbolic expressions are used as smoothing functions; this should result in good behavior during usage. The number of parameters is also quite small. Finally, MOS Model 9 makes partial use of the second-generation approach to describing the geometry dependence; some of the structure added to HSPICE Level 28 is also included, allowing MOS Model 9 to more properly accommodate model binning.

A.4.3   Other Model Candidates

Several other models of note have also recently appeared on the scene. At this time, it is not clear if any of these models will find their way into the mainstream. The Power-Lane model contains modifications to the Level 3 model which make it suitable for analog circuit design. PCIM, from Digital Equipment Corporation, contains a very clean model formulation, with a small number of model parameters. The Enz-Krummenacher-Vittoz (EKV) model takes a fresh approach to the description of the FET, and leads to a model description which is clean and simple; the number of model parameters is small, and the model is very well suited to the needs of analog circuit design.

A.4.4   A Final Comment

At this time, it is difficult to make extensive comments on the new third-generation models. A full discussion must wait until there has been extensive exercise of these models in parameter extraction and circuit simulation.

Appendix B
Channel Length and Width

When an integrated circuit is "floorplanned" at the graphic design level, the channel length and width are generally defined by two types of shapes. The drawn channel length is defined by the width of a shape representing the polysilicon gate, while the drawn channel width is defined by a shape which represents a region which does not have thick oxide isolation. However, the electrical behavior of the channel length and width is affected by silicon processing, and is actually rather complicated.

B.1 FET Layout

Consider the graphic design layout of an FET, as depicted in Figure B.1. The "dogbone" shape represents a region that will be masked from the formation of the thick isolation oxide; this shape defines the drawn channel width (W_{drawn}). At the same time, the width of the polysilicon gate line defines the drawn channel length (L_{drawn}). The intersection of these two shapes defines the FET channel region. This is a designer's view of an FET.

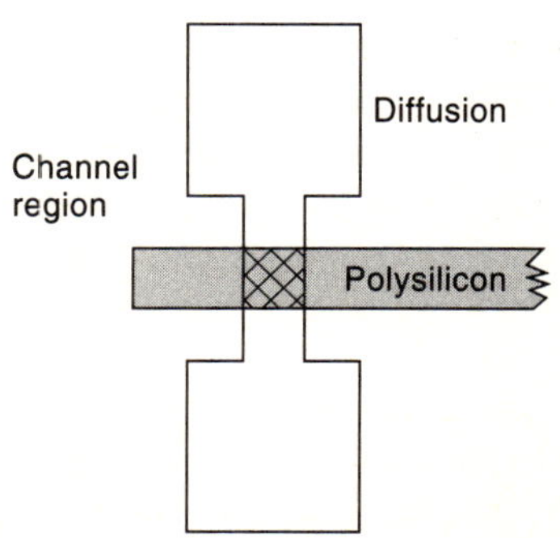

Figure B.1 A graphic layout view of an FET, showing the opening in the isolation oxide, the gate polysilicon, and the intersection of these two shapes which defines the channel region.

B.2   Channel Length

Next consider the cross section of an FET along the channel length, as shown in Figure B.2. This is a device physicist's view of an FET. The effective channel length (L_{eff}) is defined as the distance between the two diffused junctions; the simplest method is to describe the effective channel length as

$$L_{eff} = L_{drawn} - \Delta L. \tag{B.1}$$

However, in this approach, two different effects are combined into ΔL (this was often done in SPICE model building many years ago). As shown in Figure B.2, the drawn channel length does *not* appear on the wafer; in actuality, it is the printed channel length ($L_{printed}$) which is on the wafer. As also shown in Figure B.2, there is an offset between L_{drawn} and $L_{printed}$ due to processing. Using the notation employed in HSPICE to describe this situation, the relationship between the drawn and printed channel length is

$$L_{printed} = L_{drawn} - 2 \cdot \left(-\frac{\textbf{XL}}{2}\right). \tag{B.2}$$

In addition, as also depicted in Figure B.2, the source and drain diffuse under the *printed* gate. Thus, the effective channel length is described by

$$L_{eff} = L_{printed} - 2 \cdot \textbf{LD}. \tag{B.3}$$

If these distinctions are not made, the two effects are combined, and the SPICE FET model is constructed so that

$$L_{eff} = L_{drawn} - 2 \cdot \textbf{LD}; \tag{B.4}$$

this type of approach is commonly used in all varieties of SPICE. However, a more proper physical description is provided by combining (B.2) and (B.3), yielding

$$L_{eff} = L_{drawn} + \textbf{XL} - 2 \cdot \textbf{LD}. \tag{B.5}$$

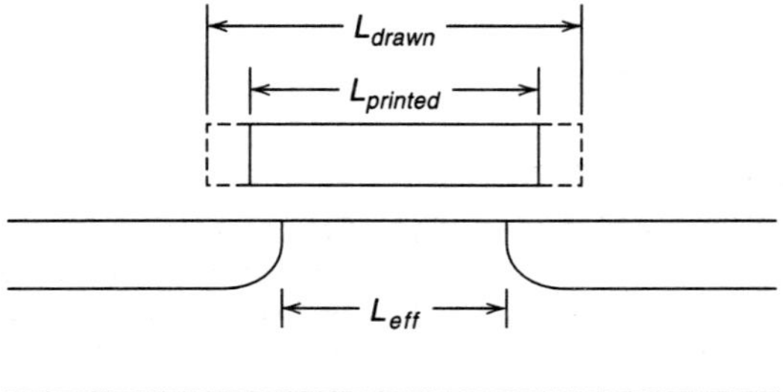

Figure B.2   A cross section along the channel of an FET, showing the drawn channel length, the printed channel length, and the effective channel length.

This expression has employed the notation used in HSPICE; the same type of development appears in MOS Model 9 (see Chapter 12) with slightly different notation, so that (B.5) becomes

$$L_{eff} = L_{drawn} + \textbf{LVAR} - 2 \cdot \textbf{LAP}. \tag{B.6}$$

The need to separate the effect of processing on the printed gate length and the underdiffusion of the printed gate by the source and drain is important for two reasons. First, the offset between the drawn and printed gate lengths provides a correct description of process variations and the device channel length; that is, **XL** (or **LVAR**) varies statistically about a mean value, while **LD** (or **LAP**) is essentially a constant in all samples. Second, a proper description of the true size of the gate electrode is required for a proper description of the total gate charge.

B.3   Channel Width

Finally, consider Figure B.3, a cross section across the channel of an FET, which shows the LOCOS isolation on either side of the device. This situation is virtually identical to that just described for the channel length. The effective channel width could be described quite simply using

$$W_{eff} = W_{drawn} - \Delta W. \tag{B.7}$$

It is interesting to note that until the development of BSIM3, the SPICE FET models developed at the University of California/Berkeley did not even contain a provision for an effective channel width, as described by (B.7). The drawn channel width (W_{drawn}) was taken from the circuit level and used as is; any value for an *effective* channel width had to be inserted by hand at the circuit level. Other varieties of SPICE (such as HSPICE and PSPICE) have added the appropriate parameters to remedy this shortcoming.

However, once again, there are two separate effects which determine the offset between W_{drawn} and W_{eff}. On the wafer, it is not W_{drawn} which appears, but rather $W_{printed}$; $W_{printed}$, as shown in Figure B.3, represents the size of the opening in the isolation oxide which actually appears on the wafer. Using HSPICE notation once again, the printed channel width is

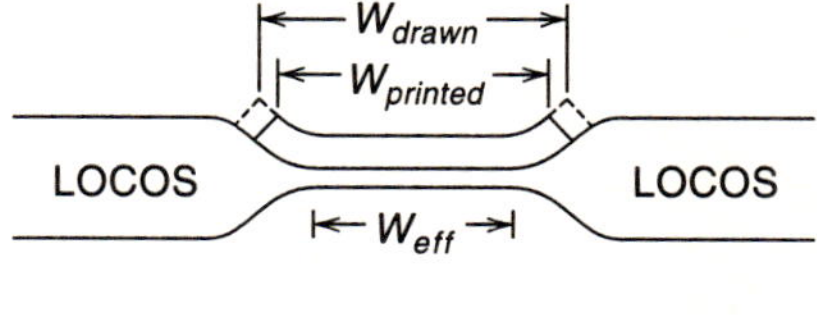

Figure B.3   A cross section across the channel of an FET, showing the drawn channel width, the printed channel width, and the effective channel width.

$$W_{printed} = W_{drawn} - 2 \cdot \left(-\frac{\mathbf{XW}}{2}\right). \tag{B.8}$$

The second part of the problem is slightly more complicated than the situation for the channel length. There is a diffusion of dopants into the channel region from under the field oxide, which reduces the effective channel width to a value less than the printed channel width. In addition, the LOCOS bird's beak encroaches into the channel region, and also reduces the channel width. These two effects are combined, so that

$$W_{eff} = W_{printed} - 2 \cdot \mathbf{WD}. \tag{B.9}$$

If all of these effects are taken together, the effective channel width is found using

$$W_{eff} = W_{drawn} - 2 \cdot \mathbf{WD}; \tag{B.10}$$

this approach was commonly used in the past. However, a more proper physical description is provided by combining (B.8) and (B.9), which produces

$$W_{eff} = W_{drawn} + \mathbf{XW} - 2 \cdot \mathbf{WD}. \tag{B.11}$$

Once again, a slightly different set of notation is used in MOS Model 9, so that (B.11) appears there as

$$W_{eff} = W_{drawn} + \mathbf{WVAR} - 2 \cdot \mathbf{WOT}. \tag{B.12}$$

As was the case for the channel length, the separation of the different physical effects is important for two reasons. First, **XW** (or **WVAR**) provides the correct parameter for the description of the effect of process variations on the channel width; **WD** (or **WOT**) actually varies very little from sample to sample. Second, this approach provides a correct description of the size of the opening in the isolation oxide; this is required for a proper description of the charge behavior of the channel region.

Appendix C
The Final Model Equations

C.1   Level 1

The Drain Current Model

$$I_{ds,lin} = \frac{\mathbf{UO} \cdot C_{ox} \cdot W_{eff}}{L_{eff}} \cdot \left[(V_{gs} - V_t) \cdot V_{ds} - \frac{V_{ds}^2}{2} \right]$$
$$\cdot (1 + \mathbf{LAMBDA} \cdot V_{ds}) \quad V_{ds} < V_{dsat}. \tag{C.1}$$

$$I_{ds,sat} = \frac{\mathbf{UO} \cdot C_{ox} \cdot W_{eff}}{L_{eff}} \cdot (V_{gs} - V_t)^2 \cdot (1 + \mathbf{LAMBDA} \cdot V_{ds}) \quad V_{ds} \geq V_{dsat}. \tag{C.2}$$

$$V_{dsat} = V_{gs} - V_t. \tag{C.3}$$

$$V_t = V_{fb} + 2\phi_f + \gamma \cdot (2\phi_f - V_{bs})^{\frac{1}{2}}. \tag{C.4}$$

The Charge Model

$$Q_{GATE,lin} = \frac{2}{3} \cdot W_{eff} \cdot L_{eff} \cdot C_{ox} \cdot \left[\frac{(V_{gd} - V_t)^3 - (V_{gs} - V_t)^3}{(V_{gd} - V_t)^2 - (V_{gs} - V_t)^2} \right] \quad V_{ds} < V_{dsat}. \tag{C.5}$$

$$Q_{GATE,sat} = \left[\frac{2}{3} \cdot W_{eff} \cdot L_{eff} \cdot C_{ox} \cdot (V_{gs} - V_t) \right] - Q_{DEPL} \quad V_{ds} \geq V_{dsat}. \tag{C.6}$$

$$Q_{DEPL} = W_{eff} \cdot L_{eff} \cdot C_{ox} \cdot \gamma \cdot (2\phi_f - V_{bs})^{\frac{1}{2}}. \tag{C.7}$$

C.2   Level 2

The Threshold Voltage Model

$$V_t = V_{fb} + 2\phi_f + f_s \cdot \gamma \cdot (2\phi_f - V_{bs})^{\frac{1}{2}} + f_n \cdot (2\phi_f - V_{bs}). \tag{C.8}$$

$$f_s = 1 - \frac{\mathbf{XJ}}{2L_{eff}} \left\{ \left[\left(1 + \frac{2W_{ds}}{\mathbf{XJ}}\right)^{\frac{1}{2}} - 1 \right] + \left[\left(1 + \frac{2W_{dd}}{\mathbf{XJ}}\right)^{\frac{1}{2}} - 1 \right] \right\}. \tag{C.9}$$

$$W_{ds} = \left(\frac{2\epsilon_{Si}(2\phi_f - V_{bs})}{q\mathbf{NSUB}} \right)^{\frac{1}{2}}. \tag{C.10}$$

$$W_{dd} = W_{jd} = \left(\frac{2\epsilon_{Si}(2\phi_f - V_{bs} + V_{ds})}{q\mathbf{NSUB}} \right)^{\frac{1}{2}}. \tag{C.11}$$

$$f_n = \mathbf{DELTA} \cdot \frac{\pi\epsilon_s}{4C_{ox}W_{eff}}. \tag{C.12}$$

The Mobility Model

$$\mu_s = \mathbf{UO} \quad E_s \le \mathbf{UCRIT}. \tag{C.13}$$

$$\mu_s = \mathbf{UO} \left(\frac{\epsilon_{Si}}{C_{ox}} \frac{\mathbf{UCRIT}}{(V_{gs} - V_t - \mathbf{UTRA} \cdot V_{ds})} \right)^{\mathbf{UEXP}} \quad E_s > \mathbf{UCRIT}. \tag{C.14}$$

The Drain Current Model

$$\begin{aligned} I_{ds,lin} = \frac{\mu_s W_{eff} C_{ox}}{L_{eff}} &\left\{ (V_{gs} - (V_{fb} + 2\phi_f))V_{ds} - \frac{V_{ds}^2}{2} \right. \\ &- \frac{2}{3} f_s \cdot \gamma \cdot \left[(V_{ds} + 2\phi_f - V_{bs})^{\frac{3}{2}} - (2\phi_f - V_{bs})^{\frac{3}{2}} \right] \\ &\left. - f_n \cdot \left[\frac{V_{ds}^2}{2} + (2\phi_f - V_{bs})V_{ds} \right] \right\} \qquad V_{ds} < V_{dsat}. \end{aligned} \tag{C.15}$$

$$\begin{aligned} I_{ds,sat} = \frac{\mu_s W_{eff} C_{ox}}{L_{eff} - L'} &\left\{ \left[(V_{gs} - (V_{fb} + 2\phi_f)) \right] V'_{dsat} - \frac{V'_{dsat}}{2} \right. \\ &- \frac{2}{3} f_s \cdot \gamma \cdot \left[(V_{ds} + 2\phi_f - V_{bs})^{\frac{3}{2}} - (2\phi_f - V_{bs})^{\frac{3}{2}} \right] \\ &\left. - f_n \cdot \left[\frac{V'_{dsat}}{2} + (2\phi_f - V_{bs})V'_{dsat} \right] \right\} \qquad V_{ds} \ge V_{dsat}. \end{aligned} \tag{C.16}$$

The Saturation Voltage

If neither **ECRIT** nor **VMAX** is specified,

$$
V_{dsat} = \frac{V_{gs} - (V_{fb} + 2\phi_f + f_n \cdot (2\phi_f - V_{bs}))}{(1 + f_n)} + \frac{1}{2}\left[\frac{f_s \cdot \gamma}{1 + f_n}\right]^2 \left\{1 - \left[1 + 4\left(\frac{1 + f_n}{f_s \cdot \gamma}\right)^2\right.\right.
$$

$$
\left.\left.\times \left(\frac{V_{gs} - (V_{fb} + 2\phi_f + f_n \cdot (2\phi_f - V_{bs}))}{1 + f_n} + 2\phi_f - V_{bs}\right)\right]^{\frac{1}{2}}\right\};
\tag{C.17}
$$

V_{dsat} replaces V'_{dsat} in (C.16). If **ECRIT** is specified but **VMAX** is not,

$$
V'_{dsat} = V_{dsat} + V_{crit} - (V_{dsat}^2 + V_{crit}^2)^{\frac{1}{2}}.
\tag{C.18}
$$

$$
V_{crit} = \textbf{ECRIT} \cdot L_{eff}.
\tag{C.19}
$$

If **VMAX** is specified,

$$
x^4 + ax^3 + bx^2 + cx + d = 0,
\tag{C.20}
$$

$$
x \equiv \left(V'_{dsat} + 2\phi_f - V_{bs}\right)^{\frac{1}{2}},
\tag{C.21}
$$

$$
a = \frac{4}{3} f_s \cdot \gamma,
\tag{C.22}
$$

$$
b = -2(G_1 + G_3),
\tag{C.23}
$$

$$
c = -2 f_s \cdot \gamma \cdot G_3,
\tag{C.24}
$$

$$
d = -G_2^2 - \frac{4}{3} f_s \cdot \gamma \cdot G_2^{\frac{3}{2}} + 2G_1 G_2 + 2G_1 G_3,
\tag{C.25}
$$

$$
G_1 = \frac{V_{gs} - \left(V_{fb} + 2\phi_f + f_n \cdot (2\phi_f - V_{bs})\right]}{1 + f_n} + 2\phi_f - V_{bs},
\tag{C.26}
$$

$$
G_2 = 2\phi_f - V_{bs},
\tag{C.27}
$$

$$
G_3 = \frac{\textbf{VMAX} \cdot L_{eff}}{\mu}.
\tag{C.28}
$$

Channel Length Modulation

If **LAMBDA** is specified,

$$
L' = \lambda V_{ds} L_{eff} = \textbf{LAMBDA} \cdot V_{ds} \cdot L_{eff}.
\tag{C.29}
$$

If neither **LAMBDA** nor **VMAX** is specified,

$$L' = x_D \cdot \left\{ \frac{V_{ds} - V_{dsat}}{4} + \left[1 + \left(\frac{V_{ds} - V_{dsat}}{4} \right)^2 \right]^{\frac{1}{2}} \right\}^{\frac{1}{2}}, \qquad \text{(C.30)}$$

$$x_D = \left(\frac{2\epsilon_{Si}}{q \cdot \mathbf{NSUB}} \right)^{\frac{1}{2}}. \qquad \text{(C.31)}$$

If **VMAX** is specified,

$$L' = \left[\left(\frac{E_{sat}}{2a} \right)^2 + \frac{(V_{ds} - V_{dsat})}{a} \right]^{\frac{1}{2}} - \frac{E_{sat}}{2a}, \qquad \text{(C.32)}$$

$$a = \frac{q \cdot \mathbf{NEFF} \cdot \mathbf{NSUB}}{2\epsilon_s}. \qquad \text{(C.33)}$$

The Subthreshold Current Model

$$I_{ds} = I_{on} e^{q(V_{gs} - V_{on})/nk_b T}. \qquad \text{(C.34)}$$

$$
\begin{aligned}
I_{on,lin} = \frac{\mu_s W_{eff} C_{ox}}{L_{eff}} \Bigg\{ & (V_{on} - (V_{fb} + 2\phi_f)) V_{ds} - \frac{V_{ds}^2}{2} \\
& - \frac{2}{3} f_s \cdot \gamma \cdot \left[(V_{ds} + 2\phi_f - V_{bs})^{\frac{3}{2}} - (2\phi_f - V_{bs})^{\frac{3}{2}} \right] \\
& - f_n \cdot \left[\frac{V_{ds}^2}{2} + (2\phi_f - V_{bs}) V_{ds} \right] \Bigg\} \qquad V_{ds} < V_{dsat}.
\end{aligned}
\qquad \text{(C.35)}
$$

$$
\begin{aligned}
I_{on,sat} = \frac{\mu_s W_{eff} C_{ox}}{L_{eff} - L'} \Bigg\{ & \left[(V_{on} - (V_{fb} + 2\phi_f)] V'_{dsat} - \frac{V'_{dsat}}{2} \right. \\
& - \frac{2}{3} f_s \cdot \gamma \cdot \left[(V_{ds} + 2\phi_f - V_{bs})^{\frac{3}{2}} - (2\phi_f - V_{bs})^{\frac{3}{2}} \right] \\
& - f_n \cdot \left[\frac{V'_{dsat}}{2} + (2\phi_f - V_{bs}) V'_{dsat} \right] \Bigg\} \qquad V_{ds} \geq V_{dsat}.
\end{aligned}
\qquad \text{(C.36)}
$$

$$V_{on} = V_t + n \frac{k_b T}{q}. \qquad \text{(C.37)}$$

$$n = 1 + \frac{q \cdot \mathbf{NFS}}{C_{ox}} + \frac{1}{C_{ox}} \left[\frac{f_s \cdot \gamma \cdot (2\phi_f - V_{bs})^{\frac{1}{2}} - f_n \cdot (2\phi_f - V_{bs})}{2(2\phi_f - V_{bs})} \right]. \tag{C.38}$$

The Charge Model

In the strong inversion region,

$$Q_{GATE,lin} = \frac{2}{3} W_{eff} L_{eff} C_{ox} \left[\frac{(V_{gd} - V_t)^3 - (V_{gs} - V_t)^3}{(V_{gd} - V_t)^2 - (V_{gs} - V_t)^2} \right] - Q_{DEPL} \qquad V_{ds} < V_{dsat}, \tag{C.39}$$

For $V_{gs} \geq V_{dsat}$,

$$Q_{GATE,sat} = \left[\frac{2}{3} W_{eff} L_{eff} C_{ox} (V_{gs} - V_t) \right] - Q_{DEPL} \qquad V_{ds} \geq V_{dsat}, \tag{C.40}$$

$$Q_{DEPL} = W_{eff} \cdot L_{eff} \cdot C_{ox} \cdot \gamma \cdot (2\phi_f - V_{bs})^{\frac{1}{2}}. \tag{C.41}$$

Also,

$$Q_{INV} = -(Q_{GATE} + Q_{DEPL}), \tag{C.42}$$

$$Q_D = \mathbf{XQC} \cdot Q_{INV}, \tag{C.43}$$

$$Q_S = (1 - \mathbf{XQC}) \cdot Q_{INV}. \tag{C.44}$$

In the subthreshold region

$$Q_{DEPL} = W_{eff} \cdot L_{eff} \cdot C_{ox} \cdot \gamma \cdot (\phi_s - V_{bs})^{\frac{1}{2}}, \tag{C.45}$$

$$Q_{GATE} = -Q_{DEPL}. \tag{C.46}$$

C.3   Level 3

The Threshold Voltage Model

$$V_t = V_{fb} + 2\phi_f + f_s \cdot \gamma \cdot (2\phi_f - V_{bs})^{\frac{1}{2}} + f_n \cdot (2\phi_f - V_{bs})$$

$$+ \mathbf{ETA} \frac{8.15 \times 10^{-22}}{C_{ox} L_{eff}^{\,3}} V_{ds}. \tag{C.47}$$

$$f_s = 1 - \frac{1}{L_{eff}} \left[(W_C + \mathbf{LD}) \left[1 - \left(\frac{W_P}{\mathbf{XJ} + W_P} \right)^2 \right]^{\frac{1}{2}} - \mathbf{LD} \right]. \tag{C.48}$$

$$W_C = 0.0631353 + 0.8013292 \cdot W_P - 0.01110777 \cdot \frac{W_P^2}{\mathbf{XJ}}. \tag{C.49}$$

$$W_P = \left(\frac{2\epsilon_s V_x}{q \cdot \mathbf{NSUB}}\right)^{\frac{1}{2}}. \tag{C.50}$$

$$f_n = \mathbf{DELTA} \cdot \frac{\pi \epsilon_s}{2 C_{ox} W_{eff}}. \tag{C.51}$$

The Mobility Model

$$\mu_v = \frac{\mathbf{UO}}{1 + \mathbf{THETA} \cdot (V_{gs} - \mathbf{VTO})}. \tag{C.52}$$

$$\mu_{eff} = \frac{\mu_v}{1 + \frac{\mu_v \cdot V'_{ds}}{\mathbf{VMAX} \cdot L_{eff}}}. \tag{C.53}$$

$$V'_{ds} = min(V_{ds}, V_{dsat}). \tag{C.54}$$

In Berkeley SPICE,

$$\mu(T) = \mu(T_{nom}) \cdot \left(\frac{T}{T_{nom}}\right)^{-1.5}, \tag{C.55}$$

while in HSPICE,

$$\mu(T) = \mu(T_{nom}) \cdot \left(\frac{T}{T_{nom}}\right)^{\mathbf{BEX}}. \tag{C.56}$$

The Drain Current Model

$$I_{ds} = \frac{\mu_{eff} C_{ox} W_{eff}}{L_{eff} - L'} \left\{ (V_{gs} - V_t) V'_{ds} - \frac{V'^2_{ds}}{2} \left[1 + \frac{f_s \cdot \gamma}{4(2\phi_f - V_{bs})^{\frac{1}{2}}} + f_n \right] \right\}. \tag{C.57}$$

The Saturation Voltage

If **VMAX** is specified,

$$V_{dsat} = \frac{(V_{gs} - V_t)}{\left(1 + \frac{f_s \cdot \gamma}{4(2\phi_f - V_{bs})^{\frac{1}{2}}} + f_n\right)} + \frac{\mathbf{VMAX} \cdot L_{eff}}{\mu_{eff}}$$

$$- \sqrt{\left[\frac{(V_{gs} - V_t)}{\left(1 + \frac{f_s \cdot \gamma}{4(2\phi_f - V_{bs})^{\frac{1}{2}}} + f_n\right)}\right]^2 + \left[\frac{\mathbf{VMAX} \cdot L_{eff}}{\mu_{eff}}\right]^2}, \tag{C.58}$$

while if **VMAX** is not specified,

$$V_{dsat} = \frac{(V_{gs} - V_t)}{\left(1 + \frac{f_s \cdot \gamma}{4(2\phi_f - V_{bs})^{\frac{1}{2}}} + f_n\right)},$$ (C.59)

Channel Length Modulation

If **VMAX** is specified,

$$L' = \left[\left(\frac{V_{dsat}}{2a \cdot L_{eff}}\right)^2 + \frac{\textbf{KAPPA} \cdot (V_{ds} - V_{dsat})}{a}\right]^{\frac{1}{2}} - \frac{V_{dsat}}{2a \cdot L_{eff}},$$ (C.60)

while if **VMAX** is not specified,

$$L' = \left[\frac{\textbf{KAPPA} \cdot (V_{ds} - V_{dsat})}{a}\right]^{\frac{1}{2}}.$$ (C.61)

The Subthreshold Current Model

$$I_{ds} = I_{on} e^{q(V_{gs} - V_{on})/nk_b T}.$$ (C.62)

$$I_{on} = \frac{\mu_{eff} C_{ox} W_{eff}}{L_{eff} - L'} \left\{(V_{on} - V_t)V'_{ds} - \frac{V_{ds}'^2}{2}\left[1 + \frac{f_s \cdot \gamma}{4(2\phi_f - V_{bs})^{\frac{1}{2}}} + f_n\right]\right\}.$$ (C.63)

$$V_{on} = V_t + n\frac{k_b T}{q}.$$ (C.64)

$$n = 1 + \frac{q \cdot \textbf{NFS}}{C_{ox}} + \frac{1}{C_{ox}} \cdot \left[\frac{f_s \cdot \gamma \cdot (2\phi_f - V_{bs})^{\frac{1}{2}} - f_n \cdot (2\phi_f - V_{bs})}{2(2\phi_f - V_{bs})}\right].$$ (C.65)

The Charge Model

$$Q_{GATE} = W_{eff} \cdot L_{eff} \cdot C_{ox}\left[V_{gs} - (V_{fb} + 2\phi_f - \sigma \cdot V_{ds}) - \frac{V_{ds}}{2}\right.$$
$$\left. + \frac{1 + f_b}{12 \cdot f_i} \cdot V_{ds}^2\right].$$ (C.66)

$$\sigma = \textbf{ETA} \cdot \frac{8.15 \times 10^{-22}}{C_{ox} L_{eff}^3}.$$ (C.67)

$$f_b = \left[\frac{\gamma \cdot f_s}{4(2\phi_f - V_{bs})^{\frac{1}{2}}} + f_n\right].$$ (C.68)

$$f_i = V_{gs} - V_t - \frac{1 + f_b}{2} \cdot V_{ds}. \tag{C.69}$$

$$Q_{DEPL} = -W_{eff} \cdot L_{eff} \cdot C_{ox}$$
$$\times \left[f_s \cdot \gamma \cdot (2\phi_f - V_{bs})^{\frac{1}{2}} + f_n \cdot (2\phi_f - V_{bs}) + \frac{f_b}{2} \cdot V_{ds} - \frac{f_b(1 + f_b)}{12 \cdot f_i} \cdot V_{ds}^2 \right]. \tag{C.70}$$

$$Q_{INV} = -(Q_{GATE} + Q_{DEPL}). \tag{C.71}$$

In the linear region ($V_{ds} < V_{dsat}$)

$$Q_S = Q_D = \frac{1}{2} Q_{INV}, \tag{C.72}$$

while in the saturation region ($V_{ds} \geq V_{dsat}$)

$$Q_D = \mathbf{XQC} \cdot Q_{INV}, \tag{C.73}$$

$$Q_S = (1 - \mathbf{XQC}) \cdot Q_{INV}. \tag{C.74}$$

C.4   BSIM

The Threshold Voltage Model

$$V_t = VFB + PHI + K1 \cdot (PHI - V_{bs})^{\frac{1}{2}} - K2 \cdot (PHI - V_{bs}) - \eta \cdot V_{ds}. \tag{C.75}$$

$$\eta = ETA + X2E \cdot V_{bs} + X3E \cdot (V_{ds} - \mathbf{VDD}). \tag{C.76}$$

The Mobility Model

$$\mu_o \big|_{V_{ds}=0} = \mathbf{MUZ} + X2MZ \cdot V_{bs}. \tag{C.77}$$

$$\mu_o \big|_{V_{ds}=\mathbf{VDD}} = MUS + X2MS \cdot V_{bs}. \tag{C.78}$$

$$\mu_v = \frac{\mu_o}{\left[1 + U_o \cdot (V_{gs} - V_t)\right]}, \tag{C.79}$$

$$U_o = U0 + X2U0 \cdot V_{bs}. \tag{C.80}$$

The Drain Current Model

$$I_{ds,lin} = \frac{\mu_v C_{ox} W_{eff}}{L_{eff}} \left[(V_{gs} - V_t) \cdot V_{ds} - \frac{a}{2} \cdot V_{ds}^2 \right] \qquad V_{ds} < V_{dsat}. \qquad \text{(C.81)}$$

$$a = 1 + \frac{g \cdot V_{ds}^2}{4(PHI - V_{bs})^{\frac{1}{2}}}. \qquad \text{(C.82)}$$

$$g = 1 - \frac{1}{1.744 + 0.8364 \cdot V_A}. \qquad \text{(C.83)}$$

$$I_{ds,sat} = \frac{\mu_v C_{ox} W_{eff}}{L_{eff}} \left(\frac{1}{2 \cdot a \cdot K} \right) (V_{gs} - V_t)^2 \qquad V_{ds} \geq V_{dsat}. \qquad \text{(C.84)}$$

$$K = \frac{1 + V_c + (1 + 2V_c)^{\frac{1}{2}}}{2}. \qquad \text{(C.85)}$$

$$V_c = \left(\frac{U_1}{L_{eff}} \right) \cdot \frac{V_{gs} - V_t}{a}. \qquad \text{(C.86)}$$

$$U_1 = U1 + X2U1 \cdot V_{bs} + X3U1 \cdot (V_{ds} - \mathbf{VDD}). \qquad \text{(C.87)}$$

The Saturation Voltage

$$V_{dsat} = \frac{V_{gs} - V_t}{a \cdot K^{\frac{1}{2}}}. \qquad \text{(C.88)}$$

The Subthreshold Current Model

$$I_{exp} = \frac{\mu_v C_{ox} W_{eff}}{L_{eff}} \cdot \left(\frac{k_b T}{q} \right)^2 e^{1.8} \cdot e^{\frac{q}{k_b T}[(V_{gs} - V_t)/n]} \left(1 - e^{-\frac{q}{k_b T} \cdot V_{ds}} \right). \qquad \text{(C.89)}$$

$$n \equiv N0 + NB \cdot V_{bs} + ND \cdot V_{ds}. \qquad \text{(C.90)}$$

$$I_{weak} = \frac{I_{exp} \cdot I_{limit}}{I_{exp} + I_{limit}}. \qquad \text{(C.91)}$$

$$I_{limit} = \frac{\mu_v C_{ox} W_{eff}}{2L_{eff}} \left(3 \cdot \frac{k_b T}{q} \right)^2. \qquad \text{(C.92)}$$

Combining the Weak and Strong Inversion Currents

$$I_{ds,total} = I_{strong} + I_{weak}. \qquad \text{(C.93)}$$

The Charge Model

In the accumulation region,

$$Q_{GATE} = W_{eff} \cdot L_{eff} \cdot C_{ox} \cdot (V_{gb} - VFB)$$

$$= W_{eff} \cdot L_{eff} \cdot C_{ox} \cdot (V_{gs} - VFB - V_{bs}), \tag{C.94}$$

$$Q_{BULK} = -Q_{GATE}, \tag{C.95}$$

$$Q_{INV} = 0. \tag{C.96}$$

In the subthreshold region,

$$Q_{GATE} = W_{eff} \cdot L_{eff} \cdot C_{ox} \cdot \frac{K1}{2} \left\{ -1 + \left[1 + \frac{4(V_{gs} - VFB - V_{bs})}{K1^2} \right]^{\frac{1}{2}} \right\}, \tag{C.97}$$

$$Q_{DEPL} = -Q_{GATE} \tag{C.98}$$

$$Q_{INV} = 0. \tag{C.99}$$

In the linear region,

$$Q_{GATE} = W_{eff} \cdot L_{eff} \cdot C_{ox}$$
$$\cdot \left[V_{gs} - VFB - PHI - \frac{V_{ds}}{2} + \frac{1}{12} \cdot \frac{\alpha_x \cdot V_{ds}^2}{V_{gs} - V_t - \frac{\alpha_x}{2} \cdot V_{ds}} \right], \tag{C.100}$$

$$Q_{DEPL} = W_{eff} \cdot L_{eff} \cdot C_{ox}$$
$$\cdot \left[-V_t + VFB + PHI + \frac{1 - \alpha_x}{2} \cdot V_{ds} - \frac{1}{12} \cdot \frac{(1 - \alpha_x)\alpha_x V_{ds}^2}{V_{gs} - V_t - \frac{\alpha_x}{2} \cdot V_{ds}} \right], \tag{C.101}$$

$$Q_{INV} = -W_{eff} \cdot L_{eff} \cdot C_{ox}$$
$$\cdot \left[V_{gs} - V_t - \frac{\alpha_x}{2} \cdot V_{ds} + \frac{1}{12} \cdot \frac{(\alpha_x \cdot V_{ds})^2}{V_{gs} - V_t - \frac{\alpha_x}{2} \cdot V_{ds}} \right], \tag{C.102}$$

$$\alpha_x = a \cdot \left[1 + \frac{U_1}{L_{eff}} (V_{gs} - V_t) \right]. \tag{C.103}$$

In the saturation region,

$$Q_{GATE} = W_{eff} \cdot L_{eff} \cdot C_{ox} \cdot \left[V_{gs} - VFB - PHI - \frac{V_{gs} - V_t}{3\alpha_x} \right], \tag{C.104}$$

$$Q_{DEPL} = W_{eff} \cdot L_{eff} \cdot C_{ox} \cdot \left[VFB + PHI - V_t + \frac{(1 - \alpha_x)(V_{gs} - V_t)}{3\alpha_x} \right], \tag{C.105}$$

$$Q_{INV} = -\frac{2}{3} \cdot W_{eff} \cdot L_{eff} \cdot C_{ox} \cdot (V_{gs} - V_t). \tag{C.106}$$

Drain/Source Charge Partitioning

40%/60% Partitioning

In the linear region ($V_{ds} < V_{dsat}$),

$$Q_S = -W_{eff} \cdot L_{eff} \cdot C_{ox}$$

$$\cdot \left\{ \frac{V_{gs} - V_t}{2} + \frac{1}{12} \cdot \frac{(\alpha_x \cdot V_{ds})^2}{V_{gs} - V_t - \frac{\alpha_x}{2} \cdot V_{ds}} - \frac{\alpha_x \cdot V_{ds}}{(V_{gs} - V_t - \frac{\alpha_x}{2} \cdot V_{ds})^2} \right.$$

$$\left. \times \left[\frac{(V_{gs} - V_t)^2}{6} - \frac{\alpha_x \cdot V_{ds} \cdot (V_{gs} - V_t)}{8} + \frac{(\alpha_x \cdot V_{ds})^2}{40} \right] \right\}, \tag{C.107}$$

$$Q_D = -W_{eff} \cdot L_{eff} \cdot C_{ox} \cdot \left\{ \frac{V_{gs} - V_t}{2} - \frac{1}{2} \cdot \alpha_x \cdot V_{ds} + \frac{\alpha_x \cdot V_{ds}}{(V_{gs} - V_t - \frac{\alpha_x}{2} \cdot V_{ds})^2} \right.$$

$$\left. \times \left[\frac{(V_{gs} - V_t)^2}{6} - \frac{\alpha_x \cdot V_{ds} \cdot (V_{gs} - V_t)}{8} + \frac{(\alpha_x \cdot V_{ds})^2}{40} \right] \right\}. \tag{C.108}$$

In the saturation region ($V_{ds} \geq V_{dsat}$),

$$Q_S = -\frac{2}{5} \cdot W_{eff} \cdot L_{eff} \cdot C_{ox} \cdot (V_{gs} - V_t), \tag{C.109}$$

and

$$Q_S = -\frac{4}{15} \cdot W_{eff} \cdot L_{eff} \cdot C_{ox} \cdot (V_{gs} - V_t). \tag{C.110}$$

0%/100% Partitioning

In the linear region ($V_{ds} < V_{dsat}$),

$$Q_S = -W_{eff} \cdot L_{eff} \cdot C_{ox}$$

$$\cdot \left[\frac{V_{gs} - V_t}{2} + \frac{1}{4} \cdot \alpha_x \cdot V_{ds} - \frac{1}{24} \cdot \frac{(\alpha_x \cdot V_{ds})^2}{V_{gs} - V_t - \frac{\alpha_x}{2} \cdot V_{ds}} \right], \tag{C.111}$$

$$Q_D = -W_{eff} \cdot L_{eff} \cdot C_{ox}$$

$$\cdot \left[\frac{V_{gs} - V_t}{2} + \frac{3}{4} \cdot \alpha_x \cdot V_{ds} + \frac{1}{8} \cdot \frac{(\alpha_x \cdot V_{ds})^2}{V_{gs} - V_t - \frac{\alpha_x}{2} \cdot V_{ds}} \right]. \tag{C.112}$$

In the saturation region ($V_{ds} \geq V_{dsat}$),

$$Q_S = Q_{INV} = -\frac{2}{3} \cdot W_{eff} \cdot L_{eff} \cdot C_{ox} \cdot (V_{gs} - V_t), \tag{C.113}$$

$$Q_D = 0. \tag{C.114}$$

50%/50% Partitioning

In the linear region ($V_{ds} < V_{dsat}$),

$$Q_S = Q_D = -\frac{1}{2} W_{eff} \cdot L_{eff} \cdot C_{ox} \cdot \left[V_{gs} - V_t - \frac{\alpha_x}{2} \cdot V_{ds} + \frac{1}{12} \cdot \frac{(\alpha_x \cdot V_{ds})^2}{V_{gs} - V_t - \frac{\alpha_x}{2} \cdot V_{ds}} \right]. \tag{C.115}$$

In the saturation region ($V_{ds} \geq V_{dsat}$),

$$Q_S = Q_D = \frac{Q_{INV}}{2} = -\frac{1}{3} W_{eff} \cdot L_{eff} \cdot C_{ox} \cdot (V_{gs} - V_t). \tag{C.116}$$

C.5 HSPICE Level 28

The Mobility Model

$$\mu_o\big|_{V_{ds}=0} = MUZ + X2M \cdot V_{bs}. \tag{C.117}$$

$$\mu_o\big|_{V_{ds}=\textbf{VDDM}} = (MUZ + X2M \cdot V_{bs})$$
$$\cdot \left\{ 1 + \frac{X3MS}{MUZ + X33M \cdot (V_{gs} - V_t)} \left[(2 - \sqrt{2}) \cdot \textbf{VDDM} \right] \right\}. \tag{C.118}$$

$$\mu_{eff} = \frac{m_{eff}}{1 + U_o \cdot (V_{gs} - V_t)}. \tag{C.119}$$

$$U_o = U0 + X2U0 \cdot V_{bs}. \tag{C.120}$$

$$m_{eff} = (MUZ + X2M \cdot V_{bs})$$
$$\times \left\{ 1 + \frac{X3MS}{MUZ + X33M \cdot (V_{gs} - V_t)} \right.$$
$$\left. \times \left[\textbf{VDDM} + V_{ds} - (\textbf{VDDM}^2 + V_{ds}^2)^{\frac{1}{2}} \right] \right\}. \tag{C.121}$$

$$\mu_{eff}(T) = \mu_{eff}(T_{nom}) \cdot \left(\frac{T}{T_{nom}} \right)^{\textbf{BEX}}, \tag{C.122}$$

The Threshold Voltage Model

$$V_t = VFB + PHI + K1 \cdot (PHI - V_{bs})^{\frac{1}{2}} - K2$$

$$\cdot (PHI - V_{bs}) - \eta \cdot V_{ds}. \tag{C.123}$$

$$\eta = ETA + X2E \cdot V_{bs} + X3E \cdot (V_{ds} - \textbf{VDDM}). \tag{C.124}$$

$$\frac{dV_t}{d\left[(PHI - V_{bs})^{\frac{1}{2}}\right]} = GAMMN. \tag{C.125}$$

$$\frac{dV_t}{dV_{ds}} = -ETAMN. \tag{C.126}$$

$$PHI(T_{nom}) = 2 \cdot \frac{k_b T_{nom}}{q} \cdot \ln\left(\frac{N_A}{n_i}\right). \tag{C.127}$$

$$PHI(T) = 2 \cdot \frac{k_b T}{q} \cdot \ln\left(\frac{N_A}{n_i}\right). \tag{C.128}$$

$$V_t(T) = V_t(T_{nom}) - \textbf{TCV} \cdot (T - T_{nom}). \tag{C.129}$$

The Saturation Voltage

$$V_{dsat} = \frac{2(V_{gs} - V_t)}{(a + R)}. \tag{C.130}$$

$$a = 1 + \frac{g \cdot K1}{2(PHI - V_{bs})^{\frac{1}{2}}}. \tag{C.131}$$

$$g = 1 - \frac{1}{1.744 + 0.8364(PHI - V_{bs})}. \tag{C.132}$$

$$R = \left[a^2 + 2 \cdot U1 \cdot a \cdot (V_{gs} - V_t) + 4 \cdot X3U1 \cdot (V_{gs} - V_t)^2\right]^{\frac{1}{2}}. \tag{C.133}$$

The Transition Voltages

$$V_1 = V_{dsat} - B1 \cdot \frac{V_{dsat}}{1 + V_{dsat}}. \tag{C.134}$$

$$V_2 = V_{dsat} + B2 \cdot (V_{gs} - V_t). \tag{C.135}$$

The Drain Current Model

In strong inversion,

$$I_{ds} = \frac{\mu_{eff} \cdot W_{eff} \cdot C_{ox}}{L_{eff}} \cdot \left[\frac{(V_{gs} - V_t) - a \cdot V_{ds}}{2}\right]$$

$$\cdot \left[\frac{V_{ds}}{1 + (U1 + X3U1 \cdot V_{ds}) \cdot V_{ds}}\right] \cdot \tag{C.136}$$

$$U1(T) = U1(T_{nom}) \cdot \left(\frac{T}{T_{nom}}\right)^{\textbf{FEX}} . \tag{C.137}$$

The weak inversion current is computed in four separately defined subregions. For $\left[\left(\frac{q}{k_bT}\right)\left(\frac{V_{gs}-V_t}{n}\right)\right] < -WFAC + V_{A0}$,

$$I_{weak} = C \cdot exp\left[\left(\frac{q}{k_bT}\right)\left(\frac{V_{gs} - V_t}{n}\right)\right] . \tag{C.138}$$

$$n = N0 + NB \cdot V_{bs} + ND \cdot V_{ds} . \tag{C.139}$$

For $-WFAC + V_{A0} < \left[\left(\frac{q}{k_bT}\right)\left(\frac{V_{gs}-V_t}{n}\right)\right] < 0$,

$$I_{weak} = C \cdot exp\left[\left(\frac{q}{k_bT}\right)\left(\frac{V_{gs} - V_t}{n}\right) - C \cdot W_f\right] . \tag{C.140}$$

W_f is the integral with respect to $\left[\left(\frac{q}{k_bT}\right)\left(\frac{V_{gs}-V_t}{n}\right)\right]$ of

$$dW_f = \left[\left(\frac{q}{k_bT}\right)\left(\frac{V_{gs} - V_t}{n}\right) + WFAC - V_{A0}\right]^2$$

$$\div \left[\left\{1 + \left[\left(\frac{q}{k_bT}\right)\left(\frac{V_{gs} - V_t}{n}\right)\right] + WFAC - V_{A0}\right\}\right.$$

$$\left. \cdot \left\{1 + WFACU \cdot \left[\left(\frac{q}{k_bT}\right)\left(\frac{V_{gs} - V_t}{n}\right)\right] + WFAC - V_{A0}\right\}\right] . \tag{C.141}$$

For $V_{gs} > V_t$ and $V_{gs} < V_{A0}$, $0 < \left[\left(\frac{q}{k_bT}\right)\left(\frac{V_{gs}-V_t}{n}\right)\right] < V_{A0}$,

$$I_{weak} = C \cdot exp\left[\left(\frac{q}{k_bT}\right)\left(\frac{V_{gs} - V_t}{n}\right) - C \cdot W_f\right] - C \cdot \left[\left(\frac{q}{k_bT}\right)\left(\frac{V_{gs} - V_t}{n}\right)\right]^2 . \tag{C.142}$$

For $V_{gs} > V_{A0}$, $\left[\left(\frac{q}{k_bT}\right)\left(\frac{V_{gs}-V_t}{n}\right)\right] > V_{A0}$,

$$I_{weak} = 0 . \tag{C.143}$$

The strong and weak inversion currents are combined using

$$I_{total} = I_{strong} + I_{weak} \cdot \left[1 - exp\left(-\frac{q V_{ds}}{k_b T} \right) \right]. \tag{C.144}$$

C.6 BSIM2

The Threshold Voltage Model

$$V_t = VFB + PHI + K1 \cdot (PHI - V_{bs})^{\frac{1}{2}} - K2 \cdot (PHI - V_{bs}) - \eta \cdot V_{ds}. \tag{C.145}$$

$$\eta = ETA + ETAB \cdot V_{bs}. \tag{C.146}$$

The Mobility Model

$$\mu_o \big|_{V_{ds}=0} = \mathbf{MU0} + MU0B \cdot V_{bs}. \tag{C.147}$$

$$\mu_o \big|_{V_{ds}=\mathbf{VDD}} = MUS0 + MUSB \cdot V_{bs}. \tag{C.148}$$

$$\mu = \frac{\mu_o}{1 + U_a \cdot (V_{gs} - V_t) + U_b \cdot (V_{gs} - V_t)^2}. \tag{C.149}$$

$$U_a = UA0 + UAB \cdot V_{bs}. \tag{C.150}$$

$$U_b = UB0 + UBB \cdot V_{bs}. \tag{C.151}$$

The Drain Current Model

The description of the critical (lateral) field for velocity saturation is

$$E_c = E_{co} \left[1 + U1D \cdot \frac{(V_{ds} - V_{dsat})^2}{V_{dsat}^2} \right] \qquad V_{ds} \le V_{dsat}. \tag{C.152}$$

$$E_c = E_{co} \qquad V_{ds} > V_{dsat}. \tag{C.153}$$

$$E_{co} = U10 + U1B \cdot V_{bs}. \tag{C.154}$$

The drain current in the linear region ($V_{ds} < V_{dsat}$) is given by

$$I_{ds,lin} = \frac{\mu_o C_{ox} W_{eff}}{L_{eff}} \cdot \frac{\left[(V_{gs} - V_t) \cdot V_{ds} - \frac{a}{2} \cdot V_{ds}^2 \right]}{\left[1 + U_a \cdot (V_{gs} - V_t) + U_b \cdot (V_{gs} - V_t)^2 \right] + \frac{U_1}{L_{eff}} \cdot V_{ds}}. \tag{C.155}$$

$$U_1 = (U10 + U1B \cdot V_{bs}) \cdot \left[1 + U1D \cdot \frac{(V_{ds} - V_{dsat})^2}{V_{dsat}^2} \right]. \tag{C.156}$$

$$a = 1 + \frac{g \cdot K1}{2(PHI - V_{bs})^{\frac{1}{2}}}.\tag{C.157}$$

$$g = 1 - \frac{1}{1.744 + 0.8364 \cdot (PHI - V_{bs})}.\tag{C.158}$$

In the saturation region ($V_{ds} \geq V_{dsat}$),

$$I_{ds,sat} = \frac{\mu_o C_{ox} W_{eff}}{L_{eff}} \cdot \frac{\left[(V_{gs} - V_t) \cdot V_{dsat} - \frac{a}{2} \cdot V_{dsat}^2\right]}{\left[1 + U_a \cdot (V_{gs} - V_t) + U_b \cdot (V_{gs} - V_t)^2\right] + \frac{U_1}{L_{eff}} \cdot V_{dsat}}.\tag{C.159}$$

$$V_{dsat} = \frac{(V_{gs} - V_t)}{a \cdot K^{\frac{1}{2}}}.\tag{C.160}$$

$$K = \frac{1}{2}\left[1 + V_c + (1 + 2V_c)^{\frac{1}{2}}\right].\tag{C.161}$$

$$V_c = \frac{U_1}{L_{eff}} \cdot \frac{(V_{gs} - V_t)}{a}.\tag{C.162}$$

$$U_1 = U10 + U1B \cdot V_{bs}.\tag{C.163}$$

The Subthreshold Current Model

$$I_{ds} = \frac{\mu_o C_{ox} W_{eff}}{L_{eff}} \cdot \left(\frac{k_b T}{q}\right) \cdot \frac{e^{V_{gs} - V_t - V_{off}}}{n} \cdot \left[1 - e^{\frac{q}{k_b T} \cdot V_{ds}}\right].\tag{C.164}$$

$$V_{off} = VOF + VOFB \cdot V_{bs} + VOFD \cdot V_{ds}.\tag{C.165}$$

$$n = N0 + \frac{NB}{(PHI - V_{bs})^{\frac{1}{2}}} + ND \cdot V_{ds}.\tag{C.166}$$

In the transition region between *VGLOW* and *VGHIGH*, V_{gs} in (C.164) is replaced by V'_{gs}:

$$V'_{gs} = \alpha_0 + \alpha_1 \cdot V_{gs} + \alpha_2 \cdot V_{gs}^2 + \alpha_3 \cdot V_{gs}^3.\tag{C.167}$$

The coefficients are determined from

$$\alpha_0 = VGHIGH - \alpha1 \cdot VGHIGH - \alpha_2 \cdot VGHIGH^2 - \alpha_3 \cdot VGHIGH^3,\tag{C.168}$$

$$\alpha_1 = \frac{1}{\Delta}\begin{vmatrix} 1 & 2VGHIGH & 3VGHIGH^2 \\ R_d & 2VGLOW & 3VGLOW^2 \\ R_c & VGLOW^2 + VGHIGH^2 & 2VGHIGH^3 + VGLOW^3 \end{vmatrix},\tag{C.169}$$

$$\alpha_2 = \frac{1}{\Delta} \begin{vmatrix} 1 & 1 & 3VGHIGH^2 \\ 1 & R_d & 3VGLOW^2 \\ VGLOW & R_c & 2VGHIGH^3 + VGLOW^3 \end{vmatrix}, \tag{C.170}$$

$$\alpha_3 = \frac{1}{\Delta} \begin{vmatrix} 1 & 2VGHIGH & 1 \\ 1 & 2VGLOW & R_d \\ VGLOW & VGLOW^2 + VGHIGH^2 & R_c \end{vmatrix}, \tag{C.171}$$

$$\Delta = \begin{vmatrix} 1 & 2VGHIGH & 3VGHIGH^2 \\ 1 & 2VGLOW & 3VGLOW^2 \\ VGLOW & VGLOW^2 + VGHIGH^2 & 2VGHIGH^3 + VGLOW^3 \end{vmatrix}. \tag{C.172}$$

The Output Resistance Model

$$\beta_0' = \beta_0 + \beta_1 \cdot tanh\left(\frac{\beta_2 \cdot V_{ds}}{V_{dsat}}\right) + \beta_3 \cdot V_{ds} - \beta_4 \cdot V_{ds}^2. \tag{C.173}$$

$$\beta_0 \equiv \frac{\mu_o\big|_{V_{ds}=0} C_{ox} W_{eff}}{L_{eff}}. \tag{C.174}$$

$$\beta_1 = \beta_s - (\beta_0 + \beta_3 \cdot \mathbf{VDD} - \beta_4 \cdot \mathbf{VDD}^2). \tag{C.175}$$

$$\beta_s = \frac{\mu_o\big|_{V_{ds}=\mathbf{VDD}} C_{ox} W_{eff}}{L_{eff}}. \tag{C.176}$$

$$\beta_2 = MU20 + MU2B \cdot V_{bs} + MU2G \cdot V_{gs}. \tag{C.177}$$

$$\beta_3 = MU30 + MU3B \cdot V_{bs} + MU3G \cdot V_{gs}. \tag{C.178}$$

$$\beta_4 = MU40 + MU4B \cdot V_{bs} + MU4G \cdot V_{gs}. \tag{C.179}$$

$$I_{ds}' = I_{ds}\left[1 + A_i e^{-B_i/(V_{ds}-V_{dsat})}\right]. \tag{C.180}$$

$$A_i = AI0 + AIB \cdot V_{bs}. \tag{C.181}$$

$$B_i = BI0 + BIB \cdot V_{bs}. \tag{C.182}$$

The Charge Model

In SPICE2, BSIM2 makes use of the BSIM charge model, discussed above. In SPICE3, a new charge model was introduced, and is the only charge model available for use in BSIM2. For use with the HSPICE implementation of BSIM2, the user may choose either the BSIM charge model or the BSIM2 charge model.

In the BSIM2 charge model, no charge-partitioning choices are available; the drain/source charge partitioning is 50%/50% in the linear region and 40%/60% in the saturation region.

In the accumulation region ($V_{gs} < V_{bs} + VFB$),

$$Q_{GATE} = C_{ox} \cdot W_{eff} \cdot L_{eff} \cdot (V_{gs} - V_{bs} - VFB), \tag{C.183}$$

$$Q_{BULK} = -Q_{GATE}, \tag{C.184}$$

$$Q_{INV} = 0. \tag{C.185}$$

In the subthreshold region ($V_{bs} + VFB < V_{gs} < V_t + VGLOW$),

$$Q_{GATE} = C_{ox} \cdot W_{eff} \cdot L_{eff} \cdot (V_{gs} - V_{bs} - VFB) \cdot \left\{ 1 - \frac{V_{gs} - V_{fb} - VFB}{V_{gs} - V_{bs} - VFB - (V_{gs} - V_t)} \right.$$

$$\left. + \frac{1}{3} \cdot \left[\frac{V_{gs} - V_{bs} - VFB}{V_{gs} - V_{bs} - VFB - (V_{gs} - V_t)} \right]^2 \right\}, \tag{C.186}$$

$$Q_{DEPL} = -Q_{GATE}, \tag{C.187}$$

$$Q_{INV} = 0. \tag{C.188}$$

In the linear region ($V_{ds} < V_{dsat}$),

$$Q_{GATE} = \frac{2}{3} \cdot C_{ox} \cdot W_{eff} \cdot L_{eff} \cdot (V_{gs} - V_t)$$

$$\cdot \left[\frac{3 \cdot \left(1 - \frac{V_{ds}}{V_{dsat}}\right) + \left(\frac{V_{ds}}{V_{dsat}}\right)^2}{2 - \frac{V_{ds}}{V_{dsat}}} \right] - Q_{DEPL}, \tag{C.189}$$

$$Q_{DEPL} = -\frac{1}{3} \cdot C_{ox} \cdot W_{eff} \cdot L_{eff} \cdot (V_t - V_{bs} - VFB), \tag{C.190}$$

$$Q_D = -\frac{1}{3} \cdot C_{ox} \cdot W_{eff} \cdot L_{eff} \cdot (V_{gs} - V_t) \cdot \left[\frac{3 \cdot \left(1 - \frac{V_{ds}}{V_{dsat}}\right) + \left(\frac{V_{ds}}{V_{dsat}}\right)^2}{2 - \frac{V_{ds}}{V_{dsat}}} \right.$$

$$\left. + \frac{\frac{V_{ds}}{V_{dsat}} \cdot \left(1 - \frac{V_{ds}}{V_{dsat}}\right) + \frac{1}{5} \cdot \left(\frac{V_{ds}}{V_{dsat}}\right)^2}{\left(2 - \frac{V_{ds}}{V_{dsat}}\right)^2} \right] - Q_{DEPL}, \tag{C.191}$$

$$Q_S = -(Q_{GATE} + Q_{DEPL} + Q_D). \tag{C.192}$$

In the saturation region ($V_{ds} \geq V_{dsat}$),

$$Q_{GATE} = \frac{2}{3} \cdot C_{ox} \cdot W_{eff} \cdot L_{eff} \cdot (V_{gs} - V_t) - Q_{DEPL}, \tag{C.193}$$

$$Q_{DEPL} = -\frac{1}{3} \cdot C_{ox} \cdot W_{eff} \cdot L_{eff} \cdot (V_t - V_{bs} - VFB), \tag{C.194}$$

$$Q_D = -\frac{4}{15} \cdot C_{ox} \cdot W_{eff} \cdot L_{eff} \cdot (V_{gs} - V_t),$$
(C.195)

$$Q_S = -(Q_{GATE} + Q_{DEPL} + Q_D).$$
(C.196)

C.7   BSIM3

The Threshold Voltage Model

$$V_t = \mathbf{VTH0} + \mathbf{K1} \cdot \left[(\phi_s - V_{bsx})^{\frac{1}{2}} - \phi_s^{\frac{1}{2}} \right] - \mathbf{K2} \cdot V_{bsz}$$

$$+ \mathbf{K1} \cdot \left[\left(1 + \frac{\mathbf{NLX}}{L_{eff}} \right)^{\frac{1}{2}} - 1 \right] \cdot \phi_s^{\frac{1}{2}}$$

$$- \Theta_{CS} \cdot (V_{bi} - \phi_s) - \Theta_{DIBL} \cdot (\mathbf{ETA0} + \mathbf{ETAB} \cdot V_{bsx}) \cdot V_{ds}$$

$$+ (\mathbf{K3} + \mathbf{K3B} \cdot V_{bsx}) \cdot \frac{t_{ox}}{(W_{eff} + \mathbf{WO})} \cdot \phi_s.$$
(C.197)

$$\Theta_{CS} = \mathbf{DVT0} \cdot \left[exp \left(-\frac{\mathbf{DVT1} \cdot L_{eff}}{2 \cdot L_{t,CS}} \right) + 2 \cdot exp \left(-\frac{\mathbf{DVT1} \cdot L_{eff}}{L_{t,CS}} \right) \right]$$

$$\mathbf{DVT0W} \cdot \left[exp \left(-\frac{\mathbf{DVT1W} \cdot W_{eff} \cdot L_{eff}}{2 \cdot L_{t,CS,W}} \right) \right.$$

$$\left. + 2 \cdot exp \left(-\frac{\mathbf{DVT1W} \cdot W_{eff} \cdot L_{eff}}{L_{t,CS,W}} \right) \right].$$
(C.198)

$$L_{t,CS} = \left(\frac{\epsilon_{Si} \cdot t_{ox}}{\epsilon_{ox}} \cdot x_d \right)^{\frac{1}{2}} \cdot (1 + \mathbf{DVT2} \cdot V_{bsx}).$$
(C.199)

$$L_{t,CS,W} = \left(\frac{\epsilon_{Si} \cdot t_{ox}}{\epsilon_{ox}} \cdot x_d \right)^{\frac{1}{2}} \cdot (1 + \mathbf{DVT2W} \cdot V_{bsx}).$$
(C.200)

$$\Theta_{DIBL} = exp \left(-\frac{\mathbf{DSUB} \cdot L_{eff}}{2 \cdot L_{t,DIBL}} \right) + 2 \cdot exp \left(-\frac{\mathbf{DSUB} \cdot L_{eff}}{L_{t,DIBL}} \right).$$
(C.201)

$$L_{t,DIBL} = \left(\frac{\epsilon_{Si} \cdot t_{ox}}{\epsilon_{ox}} \cdot x_d \right)^{\frac{1}{2}}.$$
(C.202)

$$L_{eff} = L_{drawn} - L_D.$$
(C.203)

$$L_D = \mathbf{LINT} + \frac{\mathbf{LL}}{L_{drawn}^{\mathbf{LLN}}} + \frac{\mathbf{LW}}{W_{drawn}^{\mathbf{LWN}}} + \frac{\mathbf{LWL}}{L_{drawn}^{\mathbf{LLN}} \cdot W_{drawn}^{\mathbf{LWN}}}.$$
(C.204)

$$W_{eff} = W_{drawn} - W_D.$$
(C.205)

$$W'_{eff} = W_{drawn} - W'_D. \tag{C.206}$$

$$W_D = W'_D + \mathbf{DWG} \cdot V_{gsx} + \mathbf{DWB} \cdot \left[(\phi_s - V_{gsx})^{\frac{1}{2}} - \phi_s^{\frac{1}{2}} \right]. \tag{C.207}$$

$$W'_D = \mathbf{WINT} + \frac{\mathbf{WL}}{L_{drawn}^{\mathbf{WLN}}} + \frac{\mathbf{WW}}{W_{drawn}^{\mathbf{WWN}}} + \frac{\mathbf{WWL}}{L_{drawn}^{\mathbf{WLN}} \cdot W_{drawn}^{\mathbf{WWN}}}. \tag{C.208}$$

$$V_{bsx} = V_{BC} + \frac{1}{2} \left\{ V_{bs} - V_{BC} - \delta_1 + \left[(V_{bs} - V_{BC} - \delta_1)^2 - 4 \cdot \delta_1 \cdot V_{BC} \right]^{\frac{1}{2}} \right\}. \tag{C.209}$$

$$V_{BC} = 0.9 \cdot \left(\phi_s - \frac{\mathbf{K1}^2}{4 \cdot \mathbf{K2}^2} \right). \tag{C.210}$$

$$\delta_1 = 0.02. \tag{C.211}$$

$$\phi_s = 2 \cdot \frac{k_b T}{q} \cdot ln \left(\frac{\mathbf{NCH}}{n_i(T)} \right). \tag{C.212}$$

$$n_i(T) = 1.45 \times 10^{10} \cdot \left(\frac{T}{300.15} \right)^{\frac{3}{2}} \cdot exp \left(21.5565981 - \frac{q}{2k_b T} \cdot E_g(T) \right). \tag{C.213}$$

$$E_g(T) = 1.16 - \frac{7.02 \times 10^{-4} \cdot T^2}{T + 1108}. \tag{C.214}$$

$$V_t(T) = V_t(T_{nom}) + \left(\mathbf{KT1} + \frac{\mathbf{KT1L}}{L_{eff}} + \mathbf{KT2} \cdot V_{bsx} \right) \cdot \left(\frac{T}{T_{nom}} - 1 \right). \tag{C.215}$$

The Mobility Model

When $\mathbf{MOBMOD} = 1$,

$$\mu_v = \frac{\mathbf{UO}}{1 + (\mathbf{UA} + \mathbf{UC} \cdot V_{bsx}) \cdot \left(\frac{V_{gsx} + 2 \cdot V_t(T)}{t_{ox}} \right) + \mathbf{UB} \cdot \left(\frac{V_{gsx} + 2 \cdot V_t(T)}{t_{ox}} \right)^2}, \tag{C.216}$$

$$V_{gsx} = \frac{2 \cdot \frac{k_b T}{q} \cdot ln \left[1 + exp \left(\frac{q}{k_b T} \cdot \frac{V_{gs} - V_t(T)}{2 \cdot n} \right) \right]}{1 + 2 \cdot n \cdot C_{ox} \cdot \left(\frac{2\phi_s}{q \epsilon_{Si} \cdot \mathbf{NCH}} \right)^{\frac{1}{2}} \cdot exp \left(-\frac{q}{k_b T} \cdot \frac{V_{gs} - V_t(T) - 2 \cdot \mathbf{VOFF}}{2 \cdot n} \right)}, \tag{C.217}$$

$$n = 1 + \mathbf{NFACTOR} \cdot \frac{C_{depl}}{C_{ox}} + \frac{1}{C_{ox}} \cdot (\mathbf{CDSC} + \mathbf{CDSCB} \cdot V_{bsx} + \mathbf{CDSCD} \cdot V_{ds})$$

$$\times \left[exp \left(-\frac{\mathbf{DVT1} \cdot L_{eff}}{2 \cdot L_t} \right) + 2 \cdot exp \left(-\frac{\mathbf{DVT1} \cdot L_{eff}}{L_t} \right) \right] + \frac{C_{it}}{C_{ox}}, \tag{C.218}$$

$$C_{depl} = \frac{\epsilon_{Si}}{x_d}, \tag{C.219}$$

$$x_d = \left(\frac{2\epsilon_{Si} \cdot (\phi_s - V_{bsx})}{q \cdot \mathbf{NCH}}\right)^{\frac{1}{2}},$$

(C.220)

$$L_t = \left(\frac{\epsilon_{Si} \cdot x_d}{C_{ox}}\right)^{\frac{1}{2}} \cdot (1 + \mathbf{DVT2} \cdot V_{bsx}),$$

(C.221)

$$\mu_v(T) = \frac{\mathbf{UO} \cdot \left(\frac{T}{T_{nom}}\right)^{\mathbf{UTE}}}{1 + (U_a(T) + U_c(T) \cdot V_{bsx}) \cdot \left(\frac{V_{gsx} + 2 \cdot V_t}{t_{ox}}\right) + U_b(T) \cdot \left(\frac{V_{gsx} - 2 \cdot V_t}{t_{ox}}\right)^2},$$

(C.222)

$$U_a(T) = \mathbf{UA} + \mathbf{UA1} \cdot \left(\frac{T}{T_{nom}} - 1\right),$$

(C.223)

$$U_b(T) = \mathbf{UB} + \mathbf{UB1} \cdot \left(\frac{T}{T_{nom}} - 1\right),$$

(C.224)

$$U_c(T) = \mathbf{UC} + \mathbf{UC1} \cdot \left(\frac{T}{T_{nom}} - 1\right).$$

(C.225)

When **MOBMOD** = 3,

$$\mu_v = \frac{\mathbf{UO}}{1 + \left[\mathbf{UA} \cdot \left(\frac{V_{gsx} + 2 \cdot V_t(T)}{t_{ox}}\right) + \mathbf{UB} \cdot \left(\frac{V_{gsx} + 2 \cdot V_t(T)}{t_{ox}}\right)^2\right] \cdot (1 + \mathbf{UC} \cdot V_{bsx})},$$

(C.226)

$$\mu_v(T) = \frac{\mathbf{UO} \cdot \left(\frac{T}{T_{nom}}\right)^{\mathbf{UTE}}}{1 + \left[U_a(T) \cdot \left(\frac{V_{gsx} + 2 \cdot V_t(T)}{t_{ox}}\right) + U_b(T) \cdot \left(\frac{V_{gsx} + 2 \cdot V_t(T)}{t_{ox}}\right)^2\right] \cdot (1 + U_c(T) \cdot V_{bsx})}.$$

(C.227)

When **MOBMOD** = 2,

$$\mu_v = \frac{\mathbf{UO}}{1 + (\mathbf{UA} + \mathbf{UC} \cdot V_{bsx}) \cdot \left(\frac{V_{gsx}}{t_{ox}}\right) + \mathbf{UB} \cdot \left(\frac{V_{gsx}}{t_{ox}}\right)^2},$$

(C.228)

$$\mu_v(T) = \frac{\mathbf{UO} \cdot \left(\frac{T}{T_{nom}}\right)^{\mathbf{UTE}}}{1 + (U_a(T) + U_c(T) \cdot V_{bsx}) \cdot \left(\frac{V_{gsx}}{t_{ox}}\right) + U_b(T) \cdot \left(\frac{V_{gsx}}{t_{ox}}\right)^2}.$$

(C.229)

The Carrier Velocity Model

$$v(y) = \frac{\mu_v(T) \cdot E_y(y)}{1 + \frac{E_y(y)}{E_{sat}}} \qquad E_y(y) < E_{sat}$$

$$= v_{SAT} \qquad E_y(y) > E_{sat}.$$

(C.230)

$$E_{sat} = \frac{2 \cdot v_{SAT}}{\mu_v(T)} = \frac{2 \cdot \mathbf{VSAT}}{\mu_v}. \tag{C.231}$$

$$v_{SAT}(T) = \mathbf{VSAT} - \mathbf{AT} \cdot \left(\frac{T}{T_{nom}} - 1 \right). \tag{C.232}$$

The Drain Current Model

$$I_{ds} = I_{ds,lin,V_{ds}=V_{dsx}} \cdot \left(1 + \frac{V_{ds} - V_{dsx}}{V_{A2}} \right) \cdot \left(1 + \frac{V_{ds} - V_{dsx}}{V_{A,SCBE}} \right). \tag{C.233}$$

$$I_{ds,lin,V_{ds}=V_{dsx}} = \frac{\mu_v(T) \cdot W_{eff}}{L_{eff}} \cdot \frac{1}{1 + \frac{V_{dsx}}{E_{sat} \cdot L_{eff}}}$$

$$\cdot \frac{Q_{inv,V_{ds}=0} \cdot V_{dsx} \cdot \left(1 - \frac{V_{dsx}}{2 \cdot V_b} \right)}{1 + \frac{R_{ds}}{V_{dsx}} \cdot \frac{\mu_v(T) \cdot W_{eff}}{L_{eff}} \cdot \frac{Q_{inv,V_{ds}=0} \cdot V_{dsx} \cdot \left(1 - \frac{V_{dsx}}{2 \cdot V_b} \right)}{1 + \frac{V_{dsx}}{E_{sat} \cdot L_{eff}}}}$$

$$= \frac{I_{ds,lin,R_{ds}=0}}{1 + \frac{R_{ds} \cdot I_{ds,lin,R_{ds}=0}}{V_{dsx}}}. \tag{C.234}$$

$$V_{dsx} = V_{dsat} - \frac{1}{2} \cdot \left\{ V_{dsat} - V_{ds} - \mathbf{DELTA} \right.$$

$$\left. + \left[(V_{dsat} - V_{ds} - \mathbf{DELTA})^2 + 4 \cdot \mathbf{DELTA} \cdot V_{dsat} \right]^{\frac{1}{2}} \right\}. \tag{C.235}$$

$$V_{A2} = V_{A,sat} + \frac{\left(1 + \frac{\mathbf{PVAG} \cdot V_{gs}}{E_{sat} \cdot L_{eff}} \right)}{\left(\frac{1}{V_{A,CLM}} + \frac{1}{V_{A,DIBL,C}} \right)}. \tag{C.236}$$

$$V_{A,SCBE} = \frac{L_{eff}}{\mathbf{PSCBE2}} \cdot exp\left(\frac{\mathbf{PSCBE1} \cdot l}{V_{ds} - V_{dsat}} \right). \tag{C.237}$$

$$Q_{inv,V_{ds}=0} = C_{ox} \cdot V_{gsx}. \tag{C.238}$$

$$V_b = \frac{V_{gsx} + 2 \cdot \frac{k_b T}{q}}{A_{bulk}}. \tag{C.239}$$

$$R_{ds} = \frac{\mathbf{RDSW} \cdot \left\{ 1 + \mathbf{PRWG} \cdot V_{gsx} + \mathbf{PRWB} \cdot \left[(\phi_s - V_{bsx})^{\frac{1}{2}} - \phi_s^{\frac{1}{2}} \right] \right\}}{\left(10^{-6} \cdot W'_{eff} \right)^{\mathbf{WR}}}. \tag{C.240}$$

$$R_{dsw}(T) = \mathbf{RDSW} + \mathbf{PRT} \cdot \left(\frac{T}{T_{nom}} - 1 \right). \tag{C.241}$$

$$I_{ds,lin,R_{ds}=0} = \frac{\mu_v(T) \cdot W_{eff}}{L_{eff}} \cdot \frac{Q_{inv,V_{ds}=0} \cdot V_{dsx} \cdot \left(1 - \frac{V_{dsx}}{2 \cdot V_b}\right)}{1 + \frac{V_{dsx}}{E_{sat} \cdot L_{eff}}}. \tag{C.242}$$

$$V_{dsat} = \frac{1}{2a} \left[-b - \left(b^2 - 4 \cdot a \cdot c\right)^{\frac{1}{2}} \right]. \tag{C.243}$$

$$a = A_{bulk}^2 \cdot R_{ds} \cdot C_{ox} \cdot W_{eff} \cdot v_{SAT}(T) + \left(\frac{1}{\lambda} - 1\right) A_{bulk}. \tag{C.244}$$

$$b = -\left[\left(V_{gsx} - 2 \cdot \frac{k_b T}{q}\right) \cdot \left(\frac{2}{\lambda} - 1\right) + A_{bulk} \cdot E_{sat} \cdot L_{eff} \right.$$
$$\left. + 3 \cdot A_{bulk} \cdot R_{ds} \cdot C_{ox} \cdot W_{eff} \cdot v_{SAT}(T) \cdot \left(V_{gsx} + 2 \cdot \frac{k_b T}{q}\right) \right]. \tag{C.245}$$

$$c = E_{sat} \cdot L_{eff} \cdot \left(V_{gsx} + 2 \cdot \frac{k_b T}{q}\right)$$
$$+ 2 \cdot R_{ds} \cdot C_{ox} \cdot W_{eff} \cdot v_{SAT}(T) \cdot \left(V_{gsx} + 2 \cdot \frac{k_b T}{q}\right)^2. \tag{C.246}$$

$$V_{A,sat} = \frac{E_{sat} \cdot L_{eff} + V_{dsx} + 2 \cdot R_{ds} \cdot v_{SAT}(T) \cdot C_{ox} \cdot W_{eff} \cdot \left(V_{gsx} - \frac{A_{bulk}}{2} \cdot V_{dsx}\right)}{1 + A_{bulk} \cdot R_{ds} \cdot v_{SAT}(T) \cdot C_{ox} \cdot W_{eff}}. \tag{C.247}$$

$$V_{A,CLM} = \frac{1}{\mathbf{PCLM}} \cdot \frac{A_{bulk} \cdot E_{sat} \cdot L_{eff} + V_{gsx}}{A_{bulk} \cdot E_{sat} \cdot l} \cdot (V_{ds} - V_{dsx}). \tag{C.248}$$

$$V_{A,DIBL,C} = \frac{V_{gsx} + 2 \cdot \frac{k_b T}{q}}{\Theta_{rout} \cdot (1 + \mathbf{PDIBLCB} \cdot V_{bsx})} \cdot \left(1 - \frac{A_{bulk} \cdot V_{dsat}}{A_{bulk} \cdot V_{dsat} + V_{gsx} + 2 \cdot \frac{k_b T}{q}}\right). \tag{C.249}$$

$$l = \left(\frac{\epsilon_{Si}}{\epsilon_{ox}} \cdot t_{ox} \cdot \mathbf{XJ}\right)^{\frac{1}{2}}. \tag{C.250}$$

$$A_{bulk} = \left(1 + \frac{\mathbf{K1}}{2 \cdot (\phi_s - V_{bsx})^{\frac{1}{2}}} \cdot \left\{ \frac{\mathbf{A0} \cdot L_{eff}}{L_{eff} + 2 \cdot (\mathbf{XJ} + x_d)} \cdot \left[1 - \mathbf{AGS} \cdot V_{gsx} \right.\right.\right.$$
$$\left.\left.\left. \times \left(\frac{L_{eff}}{L_{eff} + 2 \cdot (\mathbf{XJ} + x_d)}\right)^2 \right] + \frac{\mathbf{B0}}{W_{eff} + \mathbf{B1}} \right\}\right) \cdot \frac{1}{1 + \mathbf{KETA} \cdot V_{bsx}}. \tag{C.251}$$

$$x_d = \left[\frac{2\epsilon_{Si} \cdot (\phi_s - V_{bsx})}{q \cdot \mathbf{NCH}} \right]^{\frac{1}{2}}. \tag{C.252}$$

$$\Theta_{rout} = \textbf{PDIBLC1} \cdot \left[exp \left(-\frac{\textbf{DROUT} \cdot L_{eff}}{2 \cdot L_{t,DIBL}} \right) + 2 \cdot exp \left(-\frac{\textbf{DROUT} \cdot L_{eff}}{L_{t,DIBL}} \right) \right]$$
$$+ \textbf{PDIBLC2}. \tag{C.253}$$

The Charge Model

In the accumulation region,

$$Q_{ACC} = -W_{active} \cdot L_{active} \cdot C_{ox} \cdot (V_{fbx} - V_{fb}), \tag{C.254}$$

$$L_{active} = L_{drawn} - \Delta L_c, \tag{C.255}$$

$$W_{active} = W_{drawn} - \Delta W_c, \tag{C.256}$$

$$\Delta L_c = \textbf{DLC} + \frac{\textbf{LL}}{L_{drawn}^{\textbf{LLN}}} + \frac{\textbf{LW}}{W_{drawn}^{\textbf{LWN}}} + \frac{\textbf{LWL}}{L_{drawn}^{\textbf{LLN}} \cdot W_{drawn}^{\textbf{LWN}}}, \tag{C.257}$$

$$\Delta W_c = \textbf{DWC} + \frac{\textbf{WL}}{L_{drawn}^{\textbf{WLN}}} + \frac{\textbf{WW}}{W_{drawn}^{\textbf{WWN}}} + \frac{\textbf{WWL}}{L_{drawn}^{\textbf{WLN}} \cdot W_{drawn}^{\textbf{WWN}}}, \tag{C.258}$$

$$Q_{DEPL} = -W_{active} \cdot L_{active} \cdot C_{ox} \cdot \frac{\textbf{K1}^2}{2}$$
$$\cdot \left\{ -1 + \left[1 + \frac{4 \cdot (V_{gs} - V_{fbx} - V_{gsx} - V_{bsx})^{\frac{1}{2}}}{\textbf{K1}^2} \right] \right\}, \tag{C.259}$$

$$V_{fbx} = V_{fb} - \frac{1}{2} \cdot \left[V_3 + \left(V_3^2 + 4 \cdot \delta_3 \cdot V_{fb} \right)^{\frac{1}{2}} \right], \tag{C.260}$$

$$V_3 = V_{fb} - V_{gb} - \delta_3, \tag{C.261}$$

$$\delta_3 = 0.02, \tag{C.262}$$

$$Q_{GATE} = -Q_{ACC}. \tag{C.263}$$

In the subthreshold region,

$$Q_{DEPL} = -W_{active} \cdot L_{active} \cdot C_{ox} \cdot \frac{\textbf{K1}^2}{2}$$
$$\cdot \left\{ -1 + \left[1 + \frac{4 \cdot (V_{gs} - V_{fbx} - V_{gsx} - V_{bsx})^{\frac{1}{2}}}{\textbf{K1}^2} \right] \right\}, \tag{C.264}$$

$$Q_{GATE} = -Q_{DEPL}. \tag{C.265}$$

In the strong inversion region,

$$Q_{INV} = -W_{active} \cdot L_{active} \cdot C_{ox}$$

$$\cdot \left[V_{gsx} - \frac{A_{bulk,q}}{2} \cdot V_{dsq} + \frac{1}{12} \cdot \frac{A_{bulk,q}^2 \cdot V_{dsq}^2}{V_{gsx} - \frac{A_{bulk,q}}{2} \cdot V_{dsq}} \right], \tag{C.266}$$

$$Q_{DEPL} = -W_{active} \cdot L_{active} \cdot C_{ox} \cdot (2\epsilon_{Si}q \cdot \mathbf{NSUB})^{\frac{1}{2}} \cdot (\phi_s - V_{bsx})^{\frac{1}{2}} \tag{C.267}$$

$$+ W_{active} \cdot L_{active} \cdot C_{ox}$$

$$\cdot \left(\frac{1 - A_{bulk,q}}{2} \cdot V_{dsq} - \frac{1}{12} \cdot \frac{(1 - A_{bulk,q}) \cdot A_{bulk,q} \cdot V_{dsq}^2}{V_{gsx} - \frac{A_{bulk,q}}{2} \cdot V_{dsq}} \right),$$

$$Q_{GATE} = W_{active} \cdot L_{active} \cdot C_{ox} \cdot (2\epsilon_{Si}q \cdot \mathbf{NSUB})^{\frac{1}{2}} \cdot (\phi_s - V_{bsx})^{\frac{1}{2}}$$

$$+ W_{active} \cdot L_{active} \cdot C_{ox}$$

$$\cdot \left(V_{gsx} - \frac{V_{dsq}}{2} + \frac{1}{12} \cdot \frac{A_{bulk,q} \cdot V_{dsq}^2}{V_{gsx} - \frac{A_{bulk,q}}{2} \cdot V_{dsq}} \right), \tag{C.268}$$

$$V_{dsq} = V_{dsat,q} - \frac{1}{2} \cdot \left[V_4 + (V_4^2 + 4 \cdot \delta_4 \cdot V_{dsat,q}) \right], \tag{C.269}$$

$$V_4 = V_{dsat,q} - V_{ds} - \delta_4, \tag{C.270}$$

$$\delta_4 = 0.02, \tag{C.271}$$

$$V_{dsat,q} = \frac{V_{gsx}}{A_{bulk} \cdot \left[1 + \left(\frac{\mathbf{CLC}}{L_{active}} \right)^{\mathbf{CLE}} \right]} = \frac{V_{gsx}}{A_{bulk,q}}. \tag{C.272}$$

Drain/Source Charge Partitioning

40%/60% Partitioning

$$Q_S = -\frac{1}{2} \cdot \frac{W_{active} \cdot L_{active} \cdot C_{ox}}{\left(V_{gsx} - \frac{A_{bulk,q}}{2} \cdot V_{dsq} \right)^2} \cdot \left(V_{gsx}^3 - \frac{4}{3} \cdot V_{gsx}^2 \cdot A_{bulk,q} \cdot V_{dsq} \right.$$

$$\left. + \frac{2}{3} V_{gsx} \cdot A_{bulk,q}^2 \cdot V_{dsq}^2 - \frac{2}{15} \cdot A_{bulk,q}^3 \cdot V_{dsq}^3 \right). \tag{C.273}$$

$$Q_D = -\frac{1}{2} \cdot \frac{W_{active} \cdot L_{active} \cdot C_{ox}}{\left(V_{gsx} - \frac{A_{bulk,q}}{2} \cdot V_{dsq} \right)^2} \cdot \left(V_{gsx}^3 - \frac{5}{3} \cdot V_{gsx}^2 \cdot A_{bulk,q} \cdot V_{dsq} \right.$$

$$\left. + V_{gsx} \cdot A_{bulk,q}^2 \cdot V_{dsq}^2 - \frac{1}{5} \cdot A_{bulk,q}^3 \cdot V_{dsq}^3 \right). \tag{C.274}$$

50%/50% Partitioning

$$Q_S = Q_D = -\frac{1}{2} \cdot W_{active} \cdot L_{active} \cdot C_{ox}$$

$$\cdot \left(V_{gsx} - \frac{A_{bulk,q}}{2} \cdot V_{dsq} + \frac{1}{12} \cdot \frac{A_{bulk,q}^2 \cdot V_{dsq}^2}{V_{gsx} - \frac{A_{bulk,q}}{2} \cdot V_{dsq}} \right)$$

$$= \frac{1}{2} \cdot Q_{INV}. \tag{C.275}$$

0%/100% Partitioning

$$Q_S = -W_{active} \cdot L_{active} \cdot C_{ox}$$

$$\cdot \left(\frac{V_{gsx}}{2} + \frac{1}{4} \cdot A_{bulk,q} \cdot V_{dsq} - \frac{1}{24} \cdot \frac{A_{bulk,q}^2 \cdot V_{dsq}^2}{V_{gsx} - \frac{A_{bulk,q}}{2} \cdot V_{dsq}} \right). \tag{C.276}$$

$$Q_D = -W_{active} \cdot L_{active} \cdot C_{ox}$$

$$\cdot \left(\frac{V_{gsx}}{2} - \frac{3}{4} \cdot A_{bulk,q} \cdot V_{dsq} + \frac{1}{8} \cdot \frac{A_{bulk,q}^2 \cdot V_{dsq}^2}{V_{gsx} - \frac{A_{bulk,q}}{2} \cdot V_{dsq}} \right). \tag{C.277}$$

The Outer-Fringe Charge

$$C_{of} = \frac{2}{\pi} \cdot \epsilon_{ox} \cdot ln \left(1 + \frac{t_{poly}}{t_{ox}} \right). \tag{C.278}$$

$$Q_{S,of} = -W_{active} \cdot C_{of} \cdot V_{gs}. \tag{C.279}$$

$$Q_{D,of} = -W_{active} \cdot C_{of} \cdot V_{gd}. \tag{C.280}$$

$$Q_{GATE,of} = -\left(Q_{S,of} + Q_{D,of} \right). \tag{C.281}$$

The Overlap-Charge

$$Q_{S,ov} = W_{active} \cdot \left(\mathbf{CGS0} \cdot V_{gs} + \mathbf{CGS1} \cdot \left\{ V_{gs} - V_{gs,ov} \right. \right.$$

$$\left. \left. + \frac{\mathbf{CKAPPA}}{2} \cdot \left[\left(1 + \frac{4 \cdot V_{gs,ov}}{\mathbf{CKAPPA}} \right)^{\frac{1}{2}} - 1 \right] \right\} \right). \tag{C.282}$$

$$V_{gs,ov} = \frac{1}{2} \cdot \left\{ V_{gs} + \delta_1 + \left[\left(V_{gs} + \delta_1 \right)^2 + 4 \cdot \delta_1 \right]^{\frac{1}{2}} \right\}. \tag{C.283}$$

$$\delta_1 = 0.02. \tag{C.284}$$

$$Q_{D,ov} = W_{active} \cdot \left(\mathbf{CGD0} \cdot V_{gd} + \mathbf{CGD1} \cdot \left\{ V_{gd} - V_{gd,ov} \right. \right.$$

$$\left. \left. + \frac{\mathbf{CKAPPA}}{2} \cdot \left[\left(1 + \frac{4 \cdot V_{gd,ov}}{\mathbf{CKAPPA}} \right)^{\frac{1}{2}} - 1 \right] \right\} \right). \tag{C.285}$$

$$V_{gd,ov} = \frac{1}{2} \cdot \left\{ V_{gd} + \delta_1 + \left[(V_{gd} + \delta_1)^2 + 4 \cdot \delta_1 \right]^{\frac{1}{2}} \right\}. \tag{C.286}$$

$$Q_{GATE,ov} = - \left(Q_{S,ov} + Q_{D,ov} \right). \tag{C.287}$$

The Total Node Charges

$$Q_{GATE,total} = Q_{GATE} + Q_{GATE,of} + Q_{GATE,ov}. \tag{C.288}$$

$$Q_{S,total} = Q_S + Q_{S,of} + Q_{S,ov}. \tag{C.289}$$

$$Q_{D,total} = Q_D + Q_{D,of} + Q_{D,ov}. \tag{C.290}$$

The Polysilicon Depletion Model

$$V_{gs,eff} = V_{fb} + \phi_s + \frac{q \epsilon_{Si} \cdot \mathbf{NGATE} \cdot t_{ox}^2}{\epsilon_{ox}^2}$$

$$\cdot \left[1 + \frac{2 \epsilon_{ox}^2 \cdot (V_{gs} - V_{fb} - \phi_s)}{q \epsilon_{Si} \cdot \mathbf{NGATE} \cdot t_{ox}^2} \right]^{\frac{1}{2}}. \tag{C.291}$$

$$\frac{I_{ds}(V_{gs,eff})}{I_{ds}(V_{gs})} = \frac{V_{gs,eff} - V_t}{V_{gs} - V_t}. \tag{C.292}$$

C.8   MOS Model 9

The Threshold Voltage Model

$$V_{t2} = V_{t1} + \Delta V_{t1}. \tag{C.293}$$

$$V_{t1} = VTO + \Delta V_{to}. \tag{C.294}$$

$$\Delta V_{t1} = \left\{ -\gamma_o - \left[GAM1 \cdot (V_{ds} + \lambda_2)^{\mathbf{ETADS}-1} - \gamma_o \right] \cdot \frac{V_{gt1}^2}{V_{gtx}^2 + V_{gt1}^2} \right\}$$

$$\cdot \frac{V_{ds}^2}{V_{ds} + 1}. \tag{C.295}$$

$$\lambda_1 = 0.1. \tag{C.296}$$

$$\lambda_2 = 1.0 \times 10^{-4}. \tag{C.297}$$

$$\Delta V_{to} = K \cdot \left\{ \left[hyp4(-V_{bs}, VSBX, \epsilon_2) + \left(\frac{K}{KO}\right)^2 \cdot (PHIB + VSBX) \right]^{\frac{1}{2}} \right.$$
$$\left. - \left(\frac{K}{KO}\right) \cdot (PHIB + VSBX)^{\frac{1}{2}} \right\}$$
$$+ KO \left\{ [h_1 - hyp4(-V_{bs}, VSBX, \epsilon_2)]^{\frac{1}{2}} - PHIB^{\frac{1}{2}} \right\}. \tag{C.298}$$

$$\epsilon_2 = 0.1. \tag{C.299}$$

$$hyp4(x, x_o, \epsilon) = hyp1(x - x_o, \epsilon) - hyp1(-x_o, \epsilon). \tag{C.300}$$

$$h_1 = hyp1\left(-V_{bs} + \frac{PHIB}{2}, \epsilon_1\right) + \frac{PHIB}{2}. \tag{C.301}$$

$$\epsilon_1 = 0.01. \tag{C.302}$$

$$hyp1(x, \epsilon) = \frac{1}{2} \cdot \left[x + \left(x^2 + 4 \cdot \epsilon\right)^{\frac{1}{2}} \right]. \tag{C.303}$$

$$\gamma_o = GAMOO \cdot \left(\frac{u_{s1}}{PHIB^{\frac{1}{2}}}\right)^{\textbf{ETAGAM}}. \tag{C.304}$$

$$u_{s1} = hyp2(h_1^{\frac{1}{2}}, (PHIB + VSBT), \epsilon_3). \tag{C.305}$$

$$\epsilon_3 = 0.01. \tag{C.306}$$

$$hyp2(x, x_o, \epsilon) = x - hyp1(x - x_o, \epsilon). \tag{C.307}$$

$$V_{gt1} = hyp1(V_{gs} - V_{t1}, \epsilon_4). \tag{C.308}$$

$$\epsilon_4 = 5.0 \times 10^{-4}. \tag{C.309}$$

$$V_{gtx} = \frac{1}{2} \cdot (2)^{\frac{1}{2}}. \tag{C.310}$$

$$S_{T,\phi_s} = \frac{\textbf{PHIB} - 1.13 - 2.5 \times 10^{-4} \cdot T_{nom}}{300}. \tag{C.311}$$

$$PHIB = \textbf{PHIB} + S_{T,\phi_s} \cdot (T - T_{nom}). \tag{C.312}$$

$$V_{to}(T) = \textbf{VTO} + \textbf{STVTO} \cdot (T - T_{nom}). \tag{C.313}$$

$$VTO = V_{to}(T) + \mathbf{SLVTO} \cdot \left(\frac{1}{L_{eff}} - \frac{1}{L_{eff,ref}} \right)$$

$$+ \mathbf{SL2VTO} \cdot \left(\frac{1}{L_{eff}^2} - \frac{1}{L_{eff,ref}^2} \right)$$

$$+ \mathbf{SWVTO} \cdot \left(\frac{1}{W_{eff}} - \frac{1}{W_{eff,ref}} \right). \tag{C.314}$$

$$KO = \mathbf{KO} + \mathbf{SLKO} \cdot \left(\frac{1}{L_{eff}} - \frac{1}{L_{eff,ref}} \right)$$

$$+ \mathbf{SWKO} \cdot \left(\frac{1}{W_{eff}} - \frac{1}{W_{eff,ref}} \right). \tag{C.315}$$

$$K = \mathbf{K} + \mathbf{SLK} \cdot \left(\frac{1}{L_{eff}} - \frac{1}{L_{eff,ref}} \right)$$

$$+ \mathbf{SWK} \cdot \left(\frac{1}{W_{eff}} - \frac{1}{W_{eff,ref}} \right). \tag{C.316}$$

$$VSBX = \mathbf{VSBX} + \mathbf{SLVSBX} \cdot \left(\frac{1}{L_{eff}} - \frac{1}{L_{eff,ref}} \right)$$

$$+ \mathbf{SWVBSX} \cdot \left(\frac{1}{W_{eff}} - \frac{1}{W_{eff,ref}} \right). \tag{C.317}$$

$$VSBT = \mathbf{VSBT} + \mathbf{SLVSBT} \cdot \left(\frac{1}{L_{eff}} - \frac{1}{L_{eff,ref}} \right). \tag{C.318}$$

$$GAM1 = \mathbf{GAM1} + \mathbf{SLGAM1} \cdot \left(\frac{1}{L_{eff}} - \frac{1}{L_{eff,ref}} \right)$$

$$+ \mathbf{SWGAM1} \cdot \left(\frac{1}{W_{eff}} - \frac{1}{W_{eff,ref}} \right). \tag{C.319}$$

The Mobility Model

$$\mu_{eff} = \frac{\mu_o}{\left\{ 1 + THE1 \cdot V_{gt1} + THE2 \cdot \left[(PHIB - V_{bs})^{\frac{1}{2}} - PHIB^{\frac{1}{2}} \right] \right\} \cdot (1 + THE3 \cdot V_{ds1})}. \tag{C.320}$$

$$THE1 = \theta_1(T) + S_{L,\theta_1} \cdot \left(\frac{1}{L_{eff}} - \frac{1}{L_{eff,ref}} \right)$$

$$+ \mathbf{SWTHE1} \cdot \left(\frac{1}{W_{eff}} - \frac{1}{W_{eff,ref}} \right). \tag{C.321}$$

$$\theta_1(T) = \mathbf{THE1} + \mathbf{STTHE1} \cdot (T - T_{nom}). \tag{C.322}$$

$$S_{L,\theta_1} = \mathbf{SLTHE1} + \mathbf{STLTHE1} \cdot (T - T_{nom}). \tag{C.323}$$

$$THE2 = \theta_2(T) + S_{L,\theta_2} \cdot \left(\frac{1}{L_{eff}} - \frac{1}{L_{eff,ref}} \right)$$

$$+ \mathbf{SWTHE2} \cdot \left(\frac{1}{W_{eff}} - \frac{1}{W_{eff,ref}} \right). \tag{C.324}$$

$$\theta_2(T) = \mathbf{THE2} + \mathbf{STTHE2} \cdot (T - T_{nom}). \tag{C.325}$$

$$S_{L,\theta_2} = \mathbf{SLTHE2} + \mathbf{STLTHE2} \cdot (T - T_{nom}). \tag{C.326}$$

$$THE3 = \theta_3(T) + S_{L,\theta_3} \cdot \left(\frac{1}{L_{eff}} - \frac{1}{L_{eff,ref}} \right)$$

$$+ \mathbf{SWTHE3} \cdot \left(\frac{1}{W_{eff}} - \frac{1}{W_{eff,ref}} \right). \tag{C.327}$$

$$\theta_3(T) = \mathbf{THE3} + \mathbf{STTHE3} \cdot (T - T_{nom}). \tag{C.328}$$

$$S_{L,\theta_3} = \mathbf{SLTHE3} + \mathbf{STLTHE3} \cdot (T - T_{nom}). \tag{C.329}$$

The Drain Current Model

$$I_{ds} = I_{dso} + I_{avl}. \tag{C.330}$$

$$I_{dso} = BET \cdot G_3 \tag{C.331}$$

$$\times \frac{V_{gt3} \cdot V_{ds1} - \left(\frac{1+\delta_1}{2} \right) \cdot V_{ds1}^2}{\left\{ 1 + THE1 \cdot V_{gt1} + THE2 \cdot \left[(PHIB - V_{bs})^{\frac{1}{2}} - PHIB^{\frac{1}{2}} \right] \right\} \cdot (1 + THE3 \cdot V_{ds1})}.$$

$$G_3 = \frac{ZET1 \cdot \left[1 - exp \left(-\frac{q}{k_b T} \cdot V_{ds} \right) \right] + G_1 \cdot G_2}{\frac{1}{ZET1} + G_1}. \tag{C.332}$$

$$G_1 = exp \left(\frac{q}{k_b T} \cdot \frac{V_{gt2}}{2 \cdot m} \right). \tag{C.333}$$

$$G_2 = 1 + ALP \cdot ln \left(1 + \frac{V_{ds} - V_{ds1}}{VP} \right). \tag{C.334}$$

$$V_{gt3} = 2 \cdot m \cdot \frac{k_b T}{q} \cdot ln(1 + G_1) + \lambda_3 \qquad V_{gt2} < V_{gta}$$

$$= V_{gt2} + \lambda_3 \qquad V_{gt2} \geq V_{gta}. \tag{C.335}$$

$$V_{gta} = 2 \cdot m \cdot \frac{k_b T}{q} \cdot \lambda_7. \tag{C.336}$$

$$\lambda_7 = 37. \tag{C.337}$$

$$\lambda_3 = 1 \times 10^{-8}. \tag{C.338}$$

$$V_{ds1} = hyp5(V_{ds}, V_{dsat}, \epsilon_5). \tag{C.339}$$

$$\epsilon_5 = \lambda_6 \cdot \frac{V_{dsat}}{1 + V_{dsat}}. \tag{C.340}$$

$$\lambda_6 = 0.3. \tag{C.341}$$

$$hyp5(x, x_o, \epsilon) = x_o - hyp1\left(x_o - x - \frac{\epsilon^2}{x_o}, \epsilon\right). \tag{C.342}$$

$$V_{dsat} = \frac{V_{gt3}}{1 + \delta_1} \cdot \frac{2}{1 + \left(1 + \frac{2 \cdot THE3 \cdot V_{gt3}}{1 + \delta_1}\right)^{\frac{1}{2}}}. \tag{C.343}$$

$$\delta_1 = \frac{\lambda_4}{(PHIB - V_{bs})^{\frac{1}{2}}} \cdot \left[K + \frac{(KO - K) \cdot VSBX^2}{VSBX^2 + (\lambda_5 \cdot V_{gt1} - V_{bs})^2}\right]. \tag{C.344}$$

$$\lambda_4 = 0.3. \tag{C.345}$$

$$\lambda_5 = 0.1. \tag{C.346}$$

$$I_{avl} = 0 \qquad V_{ds} \le V_{dsa}$$

$$= I_{dso} \cdot A1 \cdot exp\left(-\frac{A2}{V_{ds} - V_{dsa}}\right), \qquad V_{ds} > V_{dsa}. \tag{C.347}$$

$$V_{dsa} = A3 \cdot V_{dsat}. \tag{C.348}$$

$$\beta(T) = \mathbf{BET} \cdot \left(\frac{T_{nom}}{T}\right)^{\mathbf{ETABET}}. \tag{C.349}$$

$$BET = \beta(T) \cdot \frac{W_{eff}}{L_{eff}}. \tag{C.350}$$

$$ALP = \mathbf{ALP} + \mathbf{SLALP} \cdot \left(\frac{1}{L_{eff}^{\mathbf{ETAALP}}} - \frac{1}{L_{eff,ref}^{\mathbf{ETAALP}}}\right)$$

$$+ \mathbf{SWALP} \cdot \left(\frac{1}{W_{eff}} - \frac{1}{W_{eff,ref}}\right). \tag{C.351}$$

$$VP = \mathbf{VP} \cdot \left(\frac{L_{eff}}{L_{eff,ref}}\right). \tag{C.352}$$

$$GAMOO = \mathbf{GAMOO} + \mathbf{SLGAMOO} \cdot \left(\frac{1}{L_{eff}^2} - \frac{1}{L_{eff,ref}^2}\right). \tag{C.353}$$

$$m_o(T) = \mathbf{MO} \cdot \mathbf{STMO} \cdot (T - T_{nom}). \tag{C.354}$$

$$MO = m_o(T) + \textbf{SLMO} \cdot \left(\frac{1}{L_{eff}^{\frac{1}{2}}} - \frac{1}{L_{eff,ref}^{\frac{1}{2}}} \right). \tag{C.355}$$

$$ZET1 = \textbf{ZET1} + \textbf{SLZET1} \cdot \left(\frac{1}{L_{eff}^{\textbf{ETAZET}}} - \frac{1}{L_{eff,ref}^{\textbf{ETAZET}}} \right). \tag{C.356}$$

$$a_1(T) = \textbf{A1} + \textbf{STA1} \cdot (T - T_{nom}). \tag{C.357}$$

$$A1 = a_1(T) + \textbf{SLA1} \cdot \left(\frac{1}{L_{eff}} - \frac{1}{L_{eff,ref}} \right)$$
$$+ \textbf{SWA1} \cdot \left(\frac{1}{W_{eff}} - \frac{1}{W_{eff,ref}} \right). \tag{C.358}$$

$$A2 = \textbf{A2} + \textbf{SLA2} \cdot \left(\frac{1}{L_{eff}} - \frac{1}{L_{eff,ref}} \right)$$
$$+ \textbf{SWA2} \cdot \left(\frac{1}{W_{eff}} - \frac{1}{W_{eff,ref}} \right). \tag{C.359}$$

$$A3 = \textbf{A3} + \textbf{SLA3} \cdot \left(\frac{1}{L_{eff}} - \frac{1}{L_{eff,ref}} \right)$$
$$+ \textbf{SWA3} \cdot \left(\frac{1}{W_{eff}} - \frac{1}{W_{eff,ref}} \right). \tag{C.360}$$

The Charge Model

The Depletion Charge

$$Q_{DEPL} = \frac{1}{2} \cdot (Q_{DEPL,s} + Q_{DEPL,d}). \tag{C.361}$$

For $V_{gb} < V_{fb}$,

$$Q_{DEPL,s} = -C_{ox} \cdot hyp3(V_{gb} - V_{fb}, -V_{bs} + V_{t1}, \epsilon_6), \tag{C.362}$$

$$Q_{DEPL,d} = -C_{ox} \cdot hyp3(V_{gb} - V_{fb}, -V_{bd} + V_{t1,d}, \epsilon_6), \tag{C.363}$$

$$\epsilon_6 = 0.03, \tag{C.364}$$

$$hyp3 = hyp2(x, x_o, \epsilon) - hyp2(0, x_o, \epsilon), \tag{C.365}$$

$$V_{t1,d} = VTO + \Delta V_{to,d}, \tag{C.366}$$

$$\Delta V_{to,d} = K \cdot \left\{ \left[hyp4(-V_{bd}, VSBX, \epsilon_2) + \left(\frac{K}{KO}\right)^2 \cdot (PHIB + VSBX) \right]^{\frac{1}{2}} \right.$$

$$\left. - \left(\frac{K}{KO}\right) \cdot (PHIB + VSBX)^{\frac{1}{2}} \right\}$$

$$+ KO \left\{ [h_1 - hyp4(-V_{bd}, VSBX, \epsilon_2)]^{\frac{1}{2}} - PHIB^{\frac{1}{2}} \right\}, \tag{C.367}$$

$$V_{bd} = -V_{ds} + V_{bs}. \tag{C.368}$$

For $V_{gb} \geq V_{fb}$,

$$Q_{DEPL,s} = -C_{ox} \cdot KO$$

$$\cdot \left\{ -\frac{KO}{2} + \left[\left(\frac{KO}{2}\right)^2 + hyp3(V_{gb} - V_{fb}, -V_{bs} + V_{t1} - V_{fb}, \epsilon_6) \right]^{\frac{1}{2}} \right\}, \tag{C.369}$$

$$Q_{DEPL,d} = -C_{ox} \cdot KO$$

$$\cdot \left\{ -\frac{KO}{2} + \left[\left(\frac{KO}{2}\right)^2 + hyp3(V_{gb} - V_{fb}, -V_{bd} + V_{t1,d} - V_{fb}, \epsilon_6) \right]^{\frac{1}{2}} \right\}. \tag{C.370}$$

The Inversion Charge

$$Q_{INV} = Q_S + Q_D. \tag{C.371}$$

$$Q_S = -C_{ox} \left[\frac{1}{2} \cdot V_{gt3} + \Delta_2 \cdot V_{ds2} \cdot \left(\frac{1}{12} \cdot F_j - \frac{1}{60} \cdot F_j^2 - \frac{1}{6} \right) \right]. \tag{C.372}$$

$$Q_D = -C_{ox} \left[\frac{1}{2} \cdot V_{gt3} + \Delta_2 \cdot V_{ds2} \cdot \left(\frac{1}{12} \cdot F_j + \frac{1}{60} \cdot F_j^2 - \frac{1}{3} \right) \right]. \tag{C.373}$$

$$V_{ds2} = hyp5(V_{ds}, V_{dsat,q}, \epsilon_7). \tag{C.374}$$

$$\epsilon_7 = \lambda_8 \cdot \frac{V_{dsat,q}}{1 + V_{dsat,q}}. \tag{C.375}$$

$$\lambda_8 = 0.1. \tag{C.376}$$

$$F_j = \frac{(1 + \delta_2) \cdot (1 + THE3 \cdot V_{ds2}) \cdot V_{ds2}}{2 \cdot V_{gt3} - (1 + \delta_2) \cdot V_{ds2}}. \tag{C.377}$$

$$V_{dsat,q} = \frac{V_{gt3}}{1 + \delta_2} \cdot \frac{2}{1 + \left(1 + \frac{2 \cdot THE3 \cdot V_{gt3}}{1 + \delta_2}\right)^{\frac{1}{2}}}. \tag{C.378}$$

$$\Delta_2 = \frac{\partial V_{gt3}}{\partial V_{bs}} + \frac{\partial V_{gt3}}{\partial V_{gs}} + \frac{\partial V_{gt3}}{\partial V_{ds}}.$$ (C.379)

The Gate Charge

$$Q_{GATE} = -(Q_{DEPL} - Q_{INV}).$$ (C.380)

The Zero Bias Gate Capacitance

$$C_{gd,z} = C_{gs,z} = W_{eff} \cdot \mathbf{COL}.$$ (C.381)

Appendix D
The Extracted HSPICE Level 28 Model

```
*- - - - - - - - - - - - - - - - - - - - - - - - - - -
.model nch.1 nmos capop=9
+ level=28 lmlt=1.0 wmlt=1.0
+ lref=  .70u wref= 1.30u
+ lmin=  .70u lmax=  .90u
+ wmin= 1.30u wmax= 2.00u
+ acm=3 ldif=0.0u ld=0.1e-6 xl=-0.04e-6 wd=0.1e-6 xw=-0.2e-6
+ rsh=0 rs=  .70E-3 rd=  .70E-3
+ delvto=0 tox=13.0e-9 vddm=5.0
+ pb=0.79 mj=0.45 cj=1e-15
+ php=0.75 mjsw=0.34 cjsw=1e-15
+ cgdo=3.50E-10 cgso=3.50E-10 cjgate=1e-15
+ bex=-1.5602 tcv=333.6655e-6 fex=-6.9636
+ vfb0= -.4464739E+00 lvfb=  .1536338E+00 wvfb= -.1839674E+00 pvfb= -.2749503E+00
+ phi0=  .7500265E+00 lphi=  .0000000E+00 wphi=  .0000000E+00 pphi=  .0000000E+00
+ k1=  .7266487E+00 lk1= -.2913040E+00 wk1=  .3465182E+00 pk1=  .3698169E+00
+ k2=  .8304690E-01 lk2= -.2234572E-01 wk2=  .4534395E-01 pk2=  .1004390E+00
+ eta0=  .5022600E-01 leta=  .5653012E-03 weta=  .1968274E-02 peta=  .2232636E-01
+ x2e=  .0000000E+00 lx2e=  .0000000E+00 wx2e=  .0000000E+00 px2e=  .0000000E+00
+ x3e= -.3778300E-02 lx3e=  .7487688E-02 wx3e= -.2420846E-02 px3e= -.1025853E-01
+ muz=  .4107173E+03 lmuz=  .1180198E+02 wmuz=  .5216151E+02 pmuz=  .5724173E+01
+ x2m=  .3672400E+01 lx2m=  .1757693E+01 wx2m=  .3721165E+01 px2m=  .1525617E+00
+ u00=  .2431330E+00 lu0=  .7771082E-01 wu0= -.1007811E+00 pu0= -.3744918E-01
+ x2u0= -.6333500E-02 lx2u0=  .2685950E-02 wx2u0=  .2945005E-02 px2u0=  .1168531E-02
+ u1=  .2059145E+00 lu1=  .9325850E-01 wu1=  .2971563E-01 pu1=  .1242478E-01
+ x2u1=  .9019700E-02 lx2u1= -.1934085E-02 wx2u1= -.1506035E-02 px2u1= -.4917700E-02
+ x3u1=  .5585100E-02 lx3u1= -.1191661E-01 wx3u1= -.3795431E-03 px3u1= -.3741048E-02
+ x3ms=  .1151740E+02 lx3ms= -.1276805E+02 wx3ms=  .5837966E+01 px3ms=  .6058126E+01
+ n0=  .1592700E+01 ln0= -.6481869E-01 wn0= -.3305828E+00 pn0= -.4615414E+00
+ nb0=  .0000000E+00 lnb= -.2065421E-01 wnb=  .0000000E+00 pnb= -.4248867E-01
+ nd0=  .0000000E+00 lnd=  .0000000E+00 wnd=  .0000000E+00 pnd=  .0000000E+00
```

```
+ x33m= -.6586550E-01 lx33m=  .1615279E+02 wx33m=  .1260822E-01 px33m= -.5173939E+02
+ b1=   .4719200E-02 lb1= -.6180537E-02 wb1=  .2094994E-02 pb1= -.2482505E-01
+ b2=   .5316446E+00 lb2=  .5354785E+00 wb2= -.3222726E+00 pb2=  .2074899E+01
+ wfac=  .1259180E+02 lwfac= -.3214062E+01 wwfac=  .4842514E+01 pwfac= -.6764485E+01
+ wfacu=  .0000000E+00 lwfacu= -.3358424E-02 wwfacu= -.2087177E+01 pwfacu= -.7032065E+00
*- - - - - - - - - - - - - - - - - - - - - - -
.model nch.2 nmos capop=9
+ level=28 lmlt=1.0 wmlt=1.0
+ lref=   .90u wref= 1.30u
+ lmin=   .90u lmax= 2.00u
+ wmin= 1.30u wmax= 2.00u
+ acm=3 ldif=0.0u ld=0.1e-6 xl=-0.04e-6 wd=0.1e-6 xw=-0.2e-6
+ rsh=0 rs=   .90E-3 rd=   .90E-3
+ delvto=0 tox=13.0e-9 vddm=5.0
+ pb=0.79 mj=0.45 cj=1e-15
+ php=0.75 mjsw=0.34 cjsw=1e-15
+ cgdo=3.50E-10 cgso=3.50E-10 cjgate=1e-15
+ bex=-1.5602 tcv=333.6655e-6 fex=-6.9636
+ vfb0= -.5476819E+00 lvfb=  .1835916E+00 wvfb= -.2840750E-02 pvfb=  .3276313E-01
+ phi0=  .7500265E+00 lphi=  .0000000E+00 wphi=  .0000000E+00 pphi=  .0000000E+00
+ k1=   .9185485E+00 lk1= -.2179071E+00 wk1=  .1028971E+00 pk1= -.3580128E-01
+ k2=   .9776740E-01 lk2= -.2991901E-01 wk2= -.2085430E-01 pk2= -.2778708E-01
+ eta0=  .4985360E-01 leta=  .4638205E-01 weta= -.1273947E-01 peta= -.1075114E-01
+ x2e=  .0000000E+00 lx2e=  .0000000E+00 wx2e=  .0000000E+00 px2e=  .0000000E+00
+ x3e= -.8710900E-02 lx3e= -.8239098E-02 wx3e=  .4337074E-02 px3e=  .3657842E-02
+ muz=  .4029426E+03 lmuz=  .2678175E+02 wmuz=  .4773189E+02 pmuz=  .4908455E+02
+ x2m=  .2514500E+01 lx2m= -.6088790E+01 wx2m=  .3620664E+01 px2m= -.4668267E+01
+ u00=  .1919401E+00 lu0=  .8287973E-01 wu0= -.7611100E-01 pu0= -.2521245E-01
+ x2u0= -.8102900E-02 lx2u0= -.5709264E-02 wx2u0=  .2175222E-02 px2u0= -.3202469E-02
+ u1=  .1444794E+00 lu1=  .1391748E+00 wu1=  .2153066E-01 pu1= -.1905820E-02
+ x2u1=  .1029380E-01 lx2u1=  .3654394E-02 wx2u1=  .1733556E-02 px2u1=  .4438533E-02
+ x3u1=  .1343530E-01 lx3u1=  .5016845E-02 wx3u1=  .2084916E-02 px3u1=  .5340487E-02
+ x3ms=  .1992850E+02 lx3ms=  .9474427E+00 wx3ms=  .1847108E+01 px3ms=  .4120066E+01
+ n0=  .1635400E+01 ln0=  .1415048E-01 wn0= -.2653713E-01 pn0= -.6625631E-01
+ nb0=  .1360620E-01 lnb=  .5295839E-02 wnb=  .2798990E-01 pnb=  .1579727E-01
+ nd0=  .0000000E+00 lnd=  .0000000E+00 wnd=  .0000000E+00 pnd=  .0000000E+00
+ x33m= -.1070670E+02 lx33m= -.4776435E+02 wx33m=  .3409652E+02 px33m=  .4375793E+02
+ b1=   .8790700E-02 lb1= -.5092339E+00 wb1=  .1844878E-01 pb1= -.3913268E-01
+ b2=   .1788920E+00 lb2=  .1136053E+00 wb2= -.1689136E+01 pb2= -.1773474E+01
+ wfac=  .1470910E+02 lwfac= -.6365780E+01 wwfac=  .9298697E+01 pwfac= -.3541109E-01
+ wfacu=  .2212400E-02 lwfacu= -.6302704E-01 wwfacu= -.1623932E+01 pwfacu= -.1611849E+01
*- - - - - - - - - - - - - - - - - - - - - - - - - - -
.model nch.3 nmos capop=9
+ level=28 lmlt=1.0 wmlt=1.0
+ lref= 2.00u wref= 1.30u
+ lmin= 2.00u lmax=20.00u
+ wmin= 1.30u wmax= 2.00u
+ acm=3 ldif=0.0u ld=0.1e-6 xl=-0.04e-6 wd=0.1e-6 xw=-0.2e-6
+ rsh=0 rs= 2.00E-3 rd= 2.00E-3
+ delvto=0 tox=13.0e-9 vddm=5.0
+ pb=0.79 mj=0.45 cj=1e-15
+ php=0.75 mjsw=0.34 cjsw=1e-15
+ cgdo=3.50E-10 cgso=3.50E-10 cjgate=1e-15
```

```
+ bex=-1.5602 tcv=333.6655e-6 fex=-6.9636
+ vfb0= -.7215376E+00 lvfb=   .2355670E+00 wvfb= -.3386646E-01 pvfb=   .7000168E-01
+ phi0=   .7500265E+00 lphi=   .0000000E+00 wphi=   .0000000E+00 pphi=   .0000000E+00
+ k1=   .1124900E+01 lk1= -.2565814E+00 wk1=   .1367999E+00 pk1=   .5166882E-02
+ k2=   .1260998E+00 lk2= -.5357896E-01 wk2=   .5459230E-02 pk2=   .3275869E-02
+ eta0=   .5931200E-02 leta=   .1644401E-02 weta= -.2558468E-02 peta=   .3660582E-03
+ x2e=   .0000000E+00 lx2e=   .0000000E+00 wx2e=   .0000000E+00 px2e=   .0000000E+00
+ x3e= -.9087235E-03 lx3e= -.1372422E-02 wx3e=   .8732088E-03 px3e= -.5204838E-02
+ muz=   .3775811E+03 lmuz= -.4871590E+02 wmuz=   .1250307E+01 pmuz=   .4976694E+02
+ x2m=   .8280400E+01 lx2m= -.6626292E+01 wx2m=   .8041371E+01 px2m=   .5118864E+01
+ u00=   .1134555E+00 lu0=   .9004770E-01 wu0= -.5223559E-01 pu0= -.4308886E-01
+ x2u0= -.2696400E-02 lx2u0= -.3414846E-02 wx2u0=   .5207863E-02 px2u0=   .5292805E-02
+ u1=   .1268510E-01 lu1=   .2450874E-01 wu1=   .2333540E-01 pu1=   .4508607E-01
+ x2u1=   .6833200E-02 lx2u1=   .2251270E-02 wx2u1= -.2469601E-02 px2u1= -.4619261E-02
+ x3u1=   .8684500E-02 lx3u1=   .3143566E-01 wx3u1= -.2972364E-02 px3u1= -.8078737E-02
+ x3ms=   .1903130E+02 lx3ms=   .2643658E+02 wx3ms= -.2054471E+01 px3ms=   .1139751E+02
+ n0=   .1622000E+01 ln0=   .1738869E-01 wn0=   .3620559E-01 pn0=   .1312407E+01
+ nb0=   .8591200E-02 lnb=   .1659896E-01 wnb=   .1303035E-01 pnb=   .2517580E-01
+ nd0=   .0000000E+00 lnd=   .0000000E+00 wnd=   .0000000E+00 pnd=   .0000000E+00
+ x33m=   .3452470E+02 lx33m=   .2222146E+03 wx33m= -.7340915E+01 px33m=   .3738533E+02
+ b1=   .4910198E+00 lb1= -.3842913E+00 wb1=   .5550621E-01 pb1= -.2636856E+01
+ b2=   .7131120E-01 lb2= -.3236173E-01 wb2= -.9709918E-02 pb2=   .3605817E+01
+ wfac=   .2073730E+02 lwfac= -.5532534E+01 wwfac=   .9332230E+01 pwfac= -.1964278E+02
+ wfacu=   .6189710E-01 lwfacu= -.1617488E-01 wwfacu= -.9755979E-01 pwfacu=   .2931266E+00
*- - - - - - - - - - - - - - - - - - - - - - - - - - -
.model nch.4 nmos capop=9
+ level=28 lmlt=1.0 wmlt=1.0
+ lref=20.00u wref= 1.30u
+ lmin=20.00u lmax=1.0m
+ wmin= 1.30u wmax= 2.00u
+ acm=3 ldif=0.0u ld=0.1e-6 xl=-0.04e-6 wd=0.1e-6 xw=-0.2e-6
+ rsh=0 rs=20.00E-3 rd=20.00E-3
+ delvto=0 tox=13.0e-9 vddm=5.0
+ pb=0.79 mj=0.45 cj=1e-15
+ php=0.75 mjsw=0.34 cjsw=1e-15
+ cgdo=3.50E-10 cgso=3.50E-10 cjgate=1e-15
+ bex=-1.5602 tcv=333.6655e-6 fex=-6.9636
+ vfb0= -.8434611E+00 lvfb=   .0000000E+00 wvfb= -.7009755E-01 pvfb=   .0000000E+00
+ phi0=   .7500265E+00 lphi=   .0000000E+00 wphi=   .0000000E+00 pphi=   .0000000E+00
+ k1=   .1257700E+01 lk1=   .0000000E+00 wk1=   .1341256E+00 pk1=   .0000000E+00
+ k2=   .1538309E+00 lk2=   .0000000E+00 wk2=   .1175850E-02 pk2=   .0000000E+00
+ eta0=   .5080100E-02 leta=   .0000000E+00 weta= -.2747931E-02 peta=   .0000000E+00
+ x2e=   .0000000E+00 lx2e=   .0000000E+00 wx2e=   .0000000E+00 px2e=   .0000000E+00
+ x3e= -.1983927E-03 lx3e=   .0000000E+00 wx3e=   .3567101E-02 px3e=   .0000000E+00
+ muz=   .4027952E+03 lmuz=   .0000000E+00 wmuz= -.2450780E+02 pmuz=   .0000000E+00
+ x2m=   .1171000E+02 lx2m=   .0000000E+00 wx2m=   .5391978E+01 px2m=   .0000000E+00
+ u00=   .6684910E-01 lu0=   .0000000E+00 wu0= -.2993390E-01 pu0=   .0000000E+00
+ x2u0= -.9289629E-03 lx2u0=   .0000000E+00 wx2u0=   .2468442E-02 px2u0=   .0000000E+00
+ u1=   .0000000E+00 lu1=   .0000000E+00 wu1=   .0000000E+00 pu1=   .0000000E+00
+ x2u1=   .5668000E-02 lx2u1=   .0000000E+00 wx2u1= -.7878893E-04 px2u1=   .0000000E+00
+ x3u1= -.7585800E-02 lx3u1=   .0000000E+00 wx3u1=   .1208984E-02 px3u1=   .0000000E+00
+ x3ms=   .5348400E+01 lx3ms=   .0000000E+00 wx3ms= -.7953530E+01 px3ms=   .0000000E+00
+ n0=   .1613000E+01 ln0=   .0000000E+00 wn0= -.6430628E+00 pn0=   .0000000E+00
```

```
+ nb0=   .0000000E+00 lnb=   .0000000E+00 wnb=   .0000000E+00 pnb=   .0000000E+00
+ nd0=   .0000000E+00 lnd=   .0000000E+00 wnd=   .0000000E+00 pnd=   .0000000E+00
+ x33m= -.8048790E+02 lx33m=  .0000000E+00 wx33m= -.2669061E+02 px33m=  .0000000E+00
+ b1=   .6899192E+00 lb1=   .0000000E+00 wb1=   .1420276E+01 pb1=   .0000000E+00
+ b2=   .8806080E-01 lb2=   .0000000E+00 wb2= -.1875989E+01 pb2=   .0000000E+00
+ wfac=   .2360080E+02 lwfac=   .0000000E+00 wwfac=   .1949883E+02 pwfac=   .0000000E+00
+ wfacu=   .7026880E-01 lwfacu=   .0000000E+00 wwfacu= -.2492747E+00 pwfacu=   .0000000E+00
*- - - - - - - - - - - - - - - - - - - - - - - -

.model nch.5 nmos capop=9
+ level=28 lmlt=1.0 wmlt=1.0
+ lref=   .70u wref= 2.00u
+ lmin=   .70u lmax=   .90u
+ wmin= 2.00u wmax=20.00u
+ acm=3 ldif=0.0u ld=0.1e-6 xl=-0.04e-6 wd=0.1e-6 xw=-0.2e-6
+ rsh=0 rs=   .70E-3 rd=   .70E-3
+ delvto=0 tox=13.0e-9 vddm=5.0
+ pb=0.79 mj=0.45 cj=1e-15
+ php=0.75 mjsw=0.34 cjsw=1e-15
+ cgdo=3.50E-10 cgso=3.50E-10 cjgate=1e-15
+ bex=-1.5602 tcv=333.6655e-6 fex=-6.9636
+ vfb0= -.3570453E+00 lvfb=   .2872902E+00 wvfb=   .2353502E+00 pvfb= -.6204770E-01
+ phi0=   .7500265E+00 lphi=   .0000000E+00 wphi=   .0000000E+00 pphi=   .0000000E+00
+ k1=   .5582023E+00 lk1= -.4710761E+00 wk1= -.3105996E+00 pk1= -.2725868E-01
+ k2=   .6100470E-01 lk2= -.7119452E-01 wk2= -.1079932E+00 pk2=   .2653264E-01
+ eta0=   .4926920E-01 leta= -.1028779E-01 weta= -.1637168E-02 peta=   .3119020E-01
+ x2e=   .0000000E+00 lx2e=   .0000000E+00 wx2e=   .0000000E+00 px2e=   .0000000E+00
+ x3e= -.2601500E-02 lx3e=   .1247447E-01 wx3e=   .1242902E-02 px3e= -.5142344E-02
+ muz=   .3853610E+03 lmuz=   .8533284E+01 wmuz= -.5982987E+01 pmuz=   .3936444E+02
+ x2m=   .1863500E+01 lx2m=   .1683531E+01 wx2m=   .3246631E+01 px2m=   .2933084E+01
+ u00=   .2921233E+00 lu0=   .9591528E-01 wu0= -.1450060E+00 pu0= -.3915100E-01
+ x2u0= -.7765100E-02 lx2u0=   .2117914E-02 wx2u0= -.4168266E-02 px2u0=   .4043472E-02
+ u1=   .1914694E+00 lu1=   .8721867E-01 wu1=   .1533684E+00 pu1=   .3891799E-01
+ x2u1=   .9751800E-02 lx2u1=   .4564638E-03 wx2u1=   .2937389E-03 px2u1=   .2999085E-02
+ x3u1=   .5769600E-02 lx3u1= -.1009804E-01 wx3u1= -.1591938E-01 px3u1=   .1427875E-02
+ x3ms=   .8679500E+01 lx3ms= -.1571297E+02 wx3ms=   .5382072E+01 px3ms=   .6654316E+01
+ n0=   .1753400E+01 ln0=   .1595417E+00 wn0= -.1202146E-01 pn0= -.6770437E-01
+ nb0=   .0000000E+00 lnb=   .0000000E+00 wnb= -.1159519E+00 pnb= -.1760150E+00
+ nd0=   .0000000E+00 lnd=   .0000000E+00 wnd=   .0000000E+00 pnd=   .0000000E+00
+ x33m= -.7199450E-01 lx33m=   .4130388E+02 wx33m= -.2009651E-02 px33m=   .2391212E+02
+ b1=   .3700800E-02 lb1=   .5887194E-02 wb1= -.3233739E-02 pb1=   .5072961E-01
+ b2=   .6883049E+00 lb2= -.4731532E+00 wb2=   .3890354E+00 pb2= -.1781677E+01
+ wfac=   .1023780E+02 lwfac=   .7422968E-01 wwfac=   .3169101E+01 pwfac=   .4635038E+02
+ wfacu=   .1014600E+01 lwfacu=   .3384781E+00 wwfacu=   .1767659E+01 pwfacu=   .3397311E+01
*- - - - - - - - - - - - - - - - - - - - - - - -

.model nch.6 nmos capop=9
+ level=28 lmlt=1.0 wmlt=1.0
+ lref=   .90u wref= 2.00u
+ lmin=   .90u lmax= 2.00u
+ wmin= 2.00u wmax=20.00u
+ acm=3 ldif=0.0u ld=0.1e-6 xl=-0.04e-6 wd=0.1e-6 xw=-0.2e-6
+ rsh=0 rs=   .90E-3 rd=   .90E-3
+ delvto=0 tox=13.0e-9 vddm=5.0
+ pb=0.79 mj=0.45 cj=1e-15
```

```
+ php=0.75 mjsw=0.34 cjsw=1e-15
+ cgdo=3.50E-10 cgso=3.50E-10 cjgate=1e-15
+ bex=-1.5602 tcv=333.6655e-6 fex=-6.9636
+ vfb0= -.5463010E+00 lvfb=  .1676650E+00 wvfb=  .2762248E+00 pvfb=  .3431878E+00
+ phi0=  .7500265E+00 lphi=  .0000000E+00 wphi=  .0000000E+00 pphi=  .0000000E+00
+ k1=  .8685291E+00 lk1= -.2005037E+00 wk1= -.2926426E+00 pk1= -.5057040E+00
+ k2=  .1079049E+00 lk2= -.1641140E-01 wk2= -.1254719E+00 pk2= -.1542735E+00
+ eta0=  .5604640E-01 leta=  .5160830E-01 weta= -.2218407E-01 peta= -.1537473E-01
+ x2e=  .0000000E+00 lx2e=  .0000000E+00 wx2e=  .0000000E+00 px2e=  .0000000E+00
+ x3e= -.1081920E-01 lx3e= -.1001722E-01 wx3e=  .4630478E-02 px3e=  .1784777E-02
+ muz=  .3797396E+03 lmuz=  .2921209E+01 wmuz= -.3191476E+02 pmuz=  .3277760E+02
+ x2m=  .7544550E+00 lx2m= -.3819494E+01 wx2m=  .1314428E+01 px2m= -.5658790E+01
+ u00=  .2289385E+00 lu0=  .9513578E-01 wu0= -.1192149E+00 pu0= -.6681754E-01
+ x2u0= -.9160300E-02 lx2u0= -.4152508E-02 wx2u0= -.6831950E-02 px2u0= -.9813053E-02
+ u1=  .1340131E+00 lu1=  .1401012E+00 wu1=  .1277307E+00 pu1=  .1324155E+00
+ x2u1=  .9451100E-02 lx2u1=  .1496774E-02 wx2u1= -.1681942E-02 px2u1=  .1088048E-02
+ x3u1=  .1242180E-01 lx3u1=  .2420774E-02 wx3u1= -.1686001E-01 px3u1= -.1830919E-01
+ x3ms=  .1903060E+02 lx3ms= -.1055367E+01 wx3ms=  .9984661E+00 px3ms=  .4427811E+01
+ n0=  .1648300E+01 ln0=  .4635841E-01 wn0=  .3257956E-01 pn0=  .2382522E+00
+ nb0=  .0000000E+00 lnb= -.2383392E-02 wnb=  .0000000E+00 pnb=  .7737461E-01
+ nd0=  .0000000E+00 lnd=  .0000000E+00 wnd=  .0000000E+00 pnd=  .0000000E+00
+ x33m= -.2728140E+02 lx33m= -.6903558E+02 wx33m= -.1575439E+02 px33m=  .3180365E+02
+ b1= -.1774561E-03 lb1= -.4902111E+00 wb1= -.3665245E-01 pb1=  .3694624E-01
+ b2=  .1000000E+01 lb2=  .9757109E+00 wb2=  .1562735E+01 pb2=  .1681439E+01
+ wfac=  .1018890E+02 lwfac= -.6348566E+01 wwfac= -.2736474E+02 pwfac= -.3343776E+02
+ wfacu=  .7916237E+00 lwfacu=  .7205105E+00 wwfacu= -.4703586E+00 pwfacu= -.5703746E+00
*- - - - - - - - - - - - - - - - - - - - - - - - - -
.model nch.7 nmos capop=9
+ level=28 lmlt=1.0 wmlt=1.0
+ lref= 2.00u wref= 2.00u
+ lmin= 2.00u lmax=20.00u
+ wmin= 2.00u wmax=20.00u
+ acm=3 ldif=0.0u ld=0.1e-6 xl=-0.04e-6 wd=0.1e-6 xw=-0.2e-6
+ rsh=0 rs= 2.00E-3 rd= 2.00E-3
+ delvto=0 tox=13.0e-9 vddm=5.0
+ pb=0.79 mj=0.45 cj=1e-15
+ php=0.75 mjsw=0.34 cjsw=1e-15
+ cgdo=3.50E-10 cgso=3.50E-10 cjgate=1e-15
+ bex=-1.5602 tcv=333.6655e-6 fex=-6.9636
+ vfb0= -.7050747E+00 lvfb=  .2015384E+00 wvfb= -.4876363E-01 pvfb= -.8990945E-02
+ phi0=  .7500265E+00 lphi=  .0000000E+00 wphi=  .0000000E+00 pphi=  .0000000E+00
+ k1=  .1058400E+01 lk1= -.2590930E+00 wk1=  .1862438E+00 pk1=  .9088589E-01
+ k2=  .1234460E+00 lk2= -.5760195E-01 wk2=  .2062042E-01 pk2=  .3281062E-01
+ eta0=  .7174900E-02 leta=  .1466455E-02 weta= -.7624662E-02 peta= -.3600856E-01
+ x2e=  .0000000E+00 lx2e=  .0000000E+00 wx2e=  .0000000E+00 px2e=  .0000000E+00
+ x3e= -.1333200E-02 lx3e=  .1157708E-02 wx3e=  .2940349E-02 px3e=  .1393889E-01
+ muz=  .3769733E+03 lmuz= -.7290817E+02 wmuz= -.6295415E+02 pmuz=  .8441577E+02
+ x2m=  .4371400E+01 lx2m= -.9114629E+01 wx2m=  .6673131E+01 px2m=  .7340655E+01
+ u00=  .1388478E+00 lu0=  .1109937E+00 wu0= -.5594067E-01 pu0= -.1884796E-01
+ x2u0= -.5228000E-02 lx2u0= -.5987737E-02 wx2u0=  .2460714E-02 px2u0=  .3714185E-02
+ u1=  .1341500E-02 lu1=  .2591897E-02 wu1=  .2337191E-02 pu1=  .4515661E-02
+ x2u1=  .8033700E-02 lx2u1=  .4496744E-02 wx2u1= -.2712291E-02 px2u1=  .5718715E-02
+ x3u1=  .1012940E-01 lx3u1=  .3536283E-01 wx3u1=  .4782398E-03 px3u1= -.4016430E-01
```

```
+ x3ms=  .2003000E+02 lx3ms=  .2089612E+02 wx3ms= -.3194537E+01 px3ms= -.3477488E+02
+ n0=  .1604400E+01 ln0= -.6205870E+00 wn0= -.1930381E+00 pn0= -.1286198E+01
+ nb0=  .2257000E-02 lnb=  .4360725E-02 wnb= -.7327142E-01 pnb= -.7251314E-01
+ nd0=  .0000000E+00 lnd=  .0000000E+00 wnd=  .0000000E+00 pnd=  .0000000E+00
+ x33m=  .3809320E+02 lx33m=  .2040411E+03 wx33m= -.4587149E+02 px33m= -.3862158E+03
+ b1=  .4640376E+00 lb1=  .8975139E+00 wb1= -.7163944E-01 pb1= -.1355040E+00
+ b2=  .7603130E-01 lb2= -.1785190E+01 wb2= -.2953642E-01 pb2= -.1015545E+02
+ wfac=  .1620080E+02 lwfac=  .4016040E+01 wwfac=  .4299805E+01 pwfac=  .2227267E+02
+ wfacu=  .1093220E+00 lwfacu= -.1586670E+00 wwfacu=  .6976885E-01 pwfacu= -.1098348E+00
*- - - - - - - - - - - - - - - - - - - - - - - - - -
.model nch.8 nmos capop=9
+ level=28 lmlt=1.0 wmlt=1.0
+ lref=20.00u wref= 2.00u
+ lmin=20.00u lmax=1.0m
+ wmin= 2.00u wmax=20.00u
+ acm=3 ldif=0.0u ld=0.1e-6 xl=-0.04e-6 wd=0.1e-6 xw=-0.2e-6
+ rsh=0 rs=20.00E-3 rd=20.00E-3
+ delvto=0 tox=13.0e-9 vddm=5.0
+ pb=0.79 mj=0.45 cj=1e-15
+ php=0.75 mjsw=0.34 cjsw=1e-15
+ cgdo=3.50E-10 cgso=3.50E-10 cjgate=1e-15
+ bex=-1.5602 tcv=333.6655e-6 fex=-6.9636
+ vfb0= -.8093859E+00 lvfb=  .0000000E+00 wvfb= -.4411015E-01 pvfb=  .0000000E+00
+ phi0=  .7500265E+00 lphi=  .0000000E+00 wphi=  .0000000E+00 pphi=  .0000000E+00
+ k1=  .1192500E+01 lk1=  .0000000E+00 wk1=  .1392036E+00 pk1=  .0000000E+00
+ k2=  .1532593E+00 lk2=  .0000000E+00 wk2=  .3638479E-02 pk2=  .0000000E+00
+ eta0=  .6415900E-02 leta=  .0000000E+00 weta=  .1101245E-01 peta=  .0000000E+00
+ x2e=  .0000000E+00 lx2e=  .0000000E+00 wx2e=  .0000000E+00 px2e=  .0000000E+00
+ x3e= -.1932400E-02 lx3e=  .0000000E+00 wx3e= -.4274067E-02 px3e=  .0000000E+00
+ muz=  .4147087E+03 lmuz=  .0000000E+00 wmuz= -.1066456E+03 pmuz=  .0000000E+00
+ x2m=  .9088900E+01 lx2m=  .0000000E+00 wx2m=  .2873795E+01 px2m=  .0000000E+00
+ u00=  .8140030E-01 lu0=  .0000000E+00 wu0= -.4618544E-01 pu0=  .0000000E+00
+ x2u0= -.2128900E-02 lx2u0=  .0000000E+00 wx2u0=  .5383467E-03 px2u0=  .0000000E+00
+ u1=  .0000000E+00 lu1=  .0000000E+00 wu1=  .0000000E+00 pu1=  .0000000E+00
+ x2u1=  .5706300E-02 lx2u1=  .0000000E+00 wx2u1= -.5672153E-02 px2u1=  .0000000E+00
+ x3u1= -.8173500E-02 lx3u1=  .0000000E+00 wx3u1=  .2126626E-01 px3u1=  .0000000E+00
+ x3ms=  .9214700E+01 lx3ms=  .0000000E+00 wx3ms=  .1480406E+02 px3ms=  .0000000E+00
+ n0=  .1925600E+01 ln0=  .0000000E+00 wn0=  .4726650E+00 pn0=  .0000000E+00
+ nb0=  .0000000E+00 lnb=  .0000000E+00 wnb= -.3574047E-01 pnb=  .0000000E+00
+ nd0=  .0000000E+00 lnd=  .0000000E+00 wnd=  .0000000E+00 pnd=  .0000000E+00
+ x33m= -.6751330E+02 lx33m=  .0000000E+00 wx33m=  .1540240E+03 px33m=  .0000000E+00
+ b1= -.4927163E-03 lb1=  .0000000E+00 wb1= -.1506009E-02 pb1=  .0000000E+00
+ b2=  .1000000E+01 lb2=  .0000000E+00 wb2=  .5226667E+01 pb2=  .0000000E+00
+ wfac=  .1412220E+02 lwfac=  .0000000E+00 wwfac= -.7227959E+01 pwfac=  .0000000E+00
+ wfacu=  .1914440E+00 lwfacu=  .0000000E+00 wwfacu=  .1266165E+00 pwfacu=  .0000000E+00
*- - - - - - - - - - - - - - - - - - - - - - - - - -
.model nch.9 nmos capop=9
+ level=28 lmlt=1.0 wmlt=1.0
+ lref= .70u wref=20.00u
+ lmin= .70u lmax= .90u
+ wmin=20.00u wmax=1.0m
+ acm=3 ldif=0.0u ld=0.1e-6 xl=-0.04e-6 wd=0.1e-6 xw=-0.2e-6
+ rsh=0 rs= .70E-3 rd= .70E-3
```

```
+ delvto=0 tox=13.0e-9 vddm=5.0
+ pb=0.79 mj=0.45 cj=1e-15
+ php=0.75 mjsw=0.34 cjsw=1e-15
+ cgdo=3.50E-10 cgso=3.50E-10 cjgate=1e-15
+ bex=-1.5602 tcv=333.6655e-6 fex=-6.9636
+ vfb0= -.4921315E+00 lvfb=  .3229043E+00 wvfb=  .0000000E+00 pvfb=  .0000000E+00
+ phi0=  .7500265E+00 lphi=  .0000000E+00 wphi=  .0000000E+00 pphi=  .0000000E+00
+ k1=  .7364802E+00 lk1= -.4554301E+00 wk1=  .0000000E+00 pk1=  .0000000E+00
+ k2=  .1229906E+00 lk2= -.8642370E-01 wk2=  .0000000E+00 pk2=  .0000000E+00
+ eta0=  .5020890E-01 leta= -.2819033E-01 weta=  .0000000E+00 peta=  .0000000E+00
+ x2e=  .0000000E+00 lx2e=  .0000000E+00 wx2e=  .0000000E+00 px2e=  .0000000E+00
+ x3e= -.3314900E-02 lx3e=  .1542607E-01 wx3e=  .0000000E+00 px3e=  .0000000E+00
+ muz=  .3887951E+03 lmuz= -.1406110E+02 wmuz=  .0000000E+00 pmuz=  .0000000E+00
+ x2m=  .0000000E+00 lx2m=  .0000000E+00 wx2m=  .0000000E+00 px2m=  .0000000E+00
+ u00=  .3753543E+00 lu0=  .1183872E+00 wu0=  .0000000E+00 pu0=  .0000000E+00
+ x2u0= -.5372600E-02 lx2u0= -.2029569E-03 wx2u0=  .0000000E+00 px2u0=  .0000000E+00
+ u1=  .1034391E+00 lu1=  .6488054E-01 wu1=  .0000000E+00 pu1=  .0000000E+00
+ x2u1=  .9583200E-02 lx2u1= -.1264949E-02 wx2u1=  .0000000E+00 px2u1=  .0000000E+00
+ x3u1=  .1490700E-01 lx3u1= -.1091761E-01 wx3u1=  .0000000E+00 px3u1=  .0000000E+00
+ x3ms=  .5590300E+01 lx3ms= -.1953241E+02 wx3ms=  .0000000E+00 px3ms=  .0000000E+00
+ n0=  .1760300E+01 ln0=  .1984026E+00 wn0=  .0000000E+00 pn0=  .0000000E+00
+ nb0=  .6655400E-01 lnb=  .1010290E+00 wnb=  .0000000E+00 pnb=  .0000000E+00
+ nd0=  .0000000E+00 lnd=  .0000000E+00 wnd=  .0000000E+00 pnd=  .0000000E+00
+ x33m= -.7084100E-01 lx33m=  .2757882E+02 wx33m=  .0000000E+00 px33m=  .0000000E+00
+ b1=  .5556900E-02 lb1= -.2323056E-01 wb1=  .0000000E+00 pb1=  .0000000E+00
+ b2=  .4650065E+00 lb2=  .5494927E+00 wb2=  .0000000E+00 pb2=  .0000000E+00
+ wfac=  .8418800E+01 lwfac= -.2652994E+02 wwfac=  .0000000E+00 pwfac=  .0000000E+00
+ wfacu=  .0000000E+00 lwfacu= -.1611509E+01 wwfacu=  .0000000E+00 pwfacu=  .0000000E+00
*- - - - - - - - - - - - - - - - - - - - - - - - - - - - - -

.model nch.10 nmos capop=9
+ level=28 lmlt=1.0 wmlt=1.0
+ lref=  .90u wref=20.00u
+ lmin=  .90u lmax= 2.00u
+ wmin=20.00u wmax=1.0m
+ acm=3 ldif=0.0u ld=0.1e-6 xl=-0.04e-6 wd=0.1e-6 xw=-0.2e-6
+ rsh=0 rs=  .90E-3 rd=  .90E-3
+ delvto=0 tox=13.0e-9 vddm=5.0
+ pb=0.79 mj=0.45 cj=1e-15
+ php=0.75 mjsw=0.34 cjsw=1e-15
+ cgdo=3.50E-10 cgso=3.50E-10 cjgate=1e-15
+ bex=-1.5602 tcv=333.6655e-6 fex=-6.9636
+ vfb0= -.7048484E+00 lvfb= -.2931773E-01 wvfb=  .0000000E+00 pvfb=  .0000000E+00
+ phi0=  .7500265E+00 lphi=  .0000000E+00 wphi=  .0000000E+00 pphi=  .0000000E+00
+ k1=  .1036500E+01 lk1=  .8976009E-01 wk1=  .0000000E+00 pk1=  .0000000E+00
+ k2=  .1799232E+00 lk2=  .7213842E-01 wk2=  .0000000E+00 pk2=  .0000000E+00
+ eta0=  .6877960E-01 leta=  .6043308E-01 weta=  .0000000E+00 peta=  .0000000E+00
+ x2e=  .0000000E+00 lx2e=  .0000000E+00 wx2e=  .0000000E+00 px2e=  .0000000E+00
+ x3e= -.1347700E-01 lx3e= -.1104164E-01 wx3e=  .0000000E+00 px3e=  .0000000E+00
+ muz=  .3980580E+03 lmuz= -.1589246E+02 wmuz=  .0000000E+00 pmuz=  .0000000E+00
+ x2m=  .0000000E+00 lx2m= -.5714638E+00 wx2m=  .0000000E+00 px2m=  .0000000E+00
+ u00=  .2973654E+00 lu0=  .1334877E+00 wu0=  .0000000E+00 pu0=  .0000000E+00
+ x2u0= -.5238900E-02 lx2u0=  .1479984E-02 wx2u0=  .0000000E+00 px2u0=  .0000000E+00
+ u1=  .6069830E-01 lu1=  .6409740E-01 wu1=  .0000000E+00 pu1=  .0000000E+00
```

```
+ x2u1=  .1041550E-01 lx2u1=  .8722562E-03 wx2u1=  .0000000E+00 px2u1=  .0000000E+00
+ x3u1=  .2209910E-01 lx3u1=  .1292987E-01 wx3u1=  .0000000E+00 px3u1=  .0000000E+00
+ x3ms=  .1845750E+02 lx3ms= -.3596841E+01 wx3ms=  .0000000E+00 px3ms=  .0000000E+00
+ n0=  .1629600E+01 ln0= -.9039348E-01 wn0=  .0000000E+00 pn0=  .0000000E+00
+ nb0=  .0000000E+00 lnb= -.4679484E-01 wnb=  .0000000E+00 pnb=  .0000000E+00
+ nd0=  .0000000E+00 lnd=  .0000000E+00 wnd=  .0000000E+00 pnd=  .0000000E+00
+ x33m= -.1823870E+02 lx33m= -.8729022E+02 wx33m=  .0000000E+00 px33m=  .0000000E+00
+ b1=  .2086030E-01 lb1= -.5114174E+00 wb1=  .0000000E+00 pb1=  .0000000E+00
+ b2=  .1030219E+00 lb2=  .1059938E-01 wb2=  .0000000E+00 pb2=  .0000000E+00
+ wfac=  .2589570E+02 lwfac=  .1284402E+02 wwfac=  .0000000E+00 pwfac=  .0000000E+00
+ wfacu=  .1061600E+01 lwfacu=  .1047894E+01 wwfacu=  .0000000E+00 pwfacu=  .0000000E+00
*- - - - - - - - - - - - - - - - - - - - - - - - - - - - -
.model nch.11 nmos capop=9
+ level=28 lmlt=1.0 wmlt=1.0
+ lref= 2.00u wref=20.00u
+ lmin= 2.00u lmax=20.00u
+ wmin=20.00u wmax=1.0m
+ acm=3 ldif=0.0u ld=0.1e-6 xl=-0.04e-6 wd=0.1e-6 xw=-0.2e-6
+ rsh=0 rs= 2.00E-3 rd= 2.00E-3
+ delvto=0 tox=13.0e-9 vddm=5.0
+ pb=0.79 mj=0.45 cj=1e-15
+ php=0.75 mjsw=0.34 cjsw=1e-15
+ cgdo=3.50E-10 cgso=3.50E-10 cjgate=1e-15
+ bex=-1.5602 tcv=333.6655e-6 fex=-6.9636
+ vfb0= -.6770854E+00 lvfb=  .2066991E+00 wvfb=  .0000000E+00 pvfb=  .0000000E+00
+ phi0=  .7500265E+00 lphi=  .0000000E+00 wphi=  .0000000E+00 pphi=  .0000000E+00
+ k1=  .9514999E+00 lk1= -.3112597E+00 wk1=  .0000000E+00 pk1=  .0000000E+00
+ k2=  .1116103E+00 lk2= -.7643458E-01 wk2=  .0000000E+00 pk2=  .0000000E+00
+ eta0=  .1155130E-01 leta=  .2213464E-01 weta=  .0000000E+00 peta=  .0000000E+00
+ x2e=  .0000000E+00 lx2e=  .0000000E+00 wx2e=  .0000000E+00 px2e=  .0000000E+00
+ x3e= -.3020900E-02 lx3e= -.6842931E-02 wx3e=  .0000000E+00 px3e=  .0000000E+00
+ muz=  .4131077E+03 lmuz= -.1213611E+03 wmuz=  .0000000E+00 pmuz=  .0000000E+00
+ x2m=  .5411589E+00 lx2m= -.1332802E+02 wx2m=  .0000000E+00 px2m=  .0000000E+00
+ u00=  .1709565E+00 lu0=  .1218120E+00 wu0=  .0000000E+00 pu0=  .0000000E+00
+ x2u0= -.6640400E-02 lx2u0= -.8119603E-02 wx2u0=  .0000000E+00 px2u0=  .0000000E+00
+ u1=  .0000000E+00 lu1=  .0000000E+00 wu1=  .0000000E+00 pu1=  .0000000E+00
+ x2u1=  .9590500E-02 lx2u1=  .1214318E-02 wx2u1=  .0000000E+00 px2u1=  .0000000E+00
+ x3u1=  .9854900E-02 lx3u1=  .5841632E-01 wx3u1=  .0000000E+00 px3u1=  .0000000E+00
+ x3ms=  .2186360E+02 lx3ms=  .4085619E+02 wx3ms=  .0000000E+00 px3ms=  .0000000E+00
+ n0=  .1715200E+01 ln0=  .1176642E+00 wn0=  .0000000E+00 pn0=  .0000000E+00
+ nb0=  .4431330E-01 lnb=  .4598178E-01 wnb=  .0000000E+00 pnb=  .0000000E+00
+ nd0=  .0000000E+00 lnd=  .0000000E+00 wnd=  .0000000E+00 pnd=  .0000000E+00
+ x33m=  .6442250E+02 lx33m=  .4257211E+03 wx33m=  .0000000E+00 px33m=  .0000000E+00
+ b1=  .5051572E+00 lb1=  .9752904E+00 wb1=  .0000000E+00 pb1=  .0000000E+00
+ b2=  .9298460E-01 lb2=  .4043832E+01 wb2=  .0000000E+00 pb2=  .0000000E+00
+ wfac=  .1373280E+02 lwfac= -.8768015E+01 wwfac=  .0000000E+00 pwfac=  .0000000E+00
+ wfacu=  .6927610E-01 lwfacu= -.9562410E-01 wwfacu=  .0000000E+00 pwfacu=  .0000000E+00
*- - - - - - - - - - - - - - - - - - - - - - - - - - - - -
.model nch.12 nmos capop=9
+ level=28 lmlt=1.0 wmlt=1.0
+ lref=20.00u wref=20.00u
+ lmin=20.00u lmax=1.0m
+ wmin=20.00u wmax=1.0m
```

```
+ acm=3 ldif=0.0u ld=0.1e-6 xl=-0.04e-6 wd=0.1e-6 xw=-0.2e-6
+ rsh=0 rs=20.00E-3 rd=20.00E-3
+ delvto=0 tox=13.0e-9 vddm=5.0
+ pb=0.79 mj=0.45 cj=1e-15
+ php=0.75 mjsw=0.34 cjsw=1e-15
+ cgdo=3.50E-10 cgso=3.50E-10 cjgate=1e-15
+ bex=-1.5602 tcv=333.6655e-6 fex=-6.9636
+ vfb0= -.7840676E+00 lvfb=   .0000000E+00 wvfb=   .0000000E+00 pvfb=   .0000000E+00
+ phi0=  .7500265E+00 lphi=   .0000000E+00 wphi=   .0000000E+00 pphi=   .0000000E+00
+ k1=   .1112600E+01 lk1=   .0000000E+00 wk1=   .0000000E+00 pk1=   .0000000E+00
+ k2=   .1511709E+00 lk2=   .0000000E+00 wk2=   .0000000E+00 pk2=   .0000000E+00
+ eta0=  .9497620E-04 leta=   .0000000E+00 weta=   .0000000E+00 peta=   .0000000E+00
+ x2e=   .0000000E+00 lx2e=   .0000000E+00 wx2e=   .0000000E+00 px2e=   .0000000E+00
+ x3e=   .5208268E-03 lx3e=   .0000000E+00 wx3e=   .0000000E+00 px3e=   .0000000E+00
+ muz=   .4759211E+03 lmuz=   .0000000E+00 wmuz=   .0000000E+00 pmuz=   .0000000E+00
+ x2m=   .7439400E+01 lx2m=   .0000000E+00 wx2m=   .0000000E+00 px2m=   .0000000E+00
+ u00=   .1079098E+00 lu0=   .0000000E+00 wu0=   .0000000E+00 pu0=   .0000000E+00
+ x2u0= -.2437900E-02 lx2u0=   .0000000E+00 wx2u0=   .0000000E+00 px2u0=   .0000000E+00
+ u1=   .0000000E+00 lu1=   .0000000E+00 wu1=   .0000000E+00 pu1=   .0000000E+00
+ x2u1=  .8962000E-02 lx2u1=   .0000000E+00 wx2u1=   .0000000E+00 px2u1=   .0000000E+00
+ x3u1= -.2037990E-01 lx3u1=   .0000000E+00 wx3u1=   .0000000E+00 px3u1=   .0000000E+00
+ x3ms=  .7174741E+00 lx3ms=   .0000000E+00 wx3ms=   .0000000E+00 px3ms=   .0000000E+00
+ n0=   .1654300E+01 ln0=   .0000000E+00 wn0=   .0000000E+00 pn0=   .0000000E+00
+ nb0=   .2051430E-01 lnb=   .0000000E+00 wnb=   .0000000E+00 pnb=   .0000000E+00
+ nd0=   .0000000E+00 lnd=   .0000000E+00 wnd=   .0000000E+00 pnd=   .0000000E+00
+ x33m= -.1559199E+03 lx33m=   .0000000E+00 wx33m=   .0000000E+00 px33m=   .0000000E+00
+ b1=   .3717020E-03 lb1=   .0000000E+00 wb1=   .0000000E+00 pb1=   .0000000E+00
+ b2= -.2000000E+01 lb2=   .0000000E+00 wb2=   .0000000E+00 pb2=   .0000000E+00
+ wfac=   .1827090E+02 lwfac=   .0000000E+00 wwfac=   .0000000E+00 pwfac=   .0000000E+00
+ wfacu=   .1187687E+00 lwfacu=   .0000000E+00 wwfacu=   .0000000E+00 pwfacu=   .0000000E+00
```

Appendix E
The Binned BSIM2 Model

```
*- - - - - - - - - - - - - - - - - - - - - - - - - -
.model nch.1 nmos capop=9
+ level=39 lmlt=1.0 wmlt=1.0
+ lmin=  .70u lmax=  .90u
+ wmin= 1.30u wmax= 2.00u
+ acm=3 ldif=0.0u ld=0.1e-6 xl=-0.04e-6 wd=0.1e-6 xw=-0.2e-6
+ delvto=0 tox=130 vdd=5.0
+ mu0=406.3019  lmu0=-14.7  wmu0=-33.3
+ mu0b=-8.7893  lmu0b=-0.695  wmu0b=6.35
+ mus0=600  lmus0=0  wmus0=0
+ musb=0  lmusb=0  wmusb=0
+ ua0=46.0971e-6  lua0=0.135  wua0=2.68e-4
+ uab=-0.0084854  luab=-7.18e-3  wuab=-5.41e-3
+ ub0=0.0094305  lub0=8.22e-3  wub0=-2.45e-3
+ ubb=66.1530e-6  lubb=6.14e-4  wubb=1.70e-3
+ vfb=-0.8227073  lvfb=0.188  wvfb=0
+ phi=0.7499925  lphi=0  wphi=0
+ k1=1.0895  lk1=-0.220  wk1=8.10e-2
+ k2=0.1379169  lk2=-0.0262  wk2=-1.17e-2
+ eta0=0.0055953  leta0=0.0219  weta0=5.22e-3
+ etab=0  letab=0  wetab=0
+ u10=0  lu10=0.122  wu10=0
+ u1b=0  lu1b=2.82e-3  wu1b=0
+ u1d=0  lu1d=0.892  wu1d=0
+ vof0=0.6947874  lvof0=0.177  wvof0=0.410
+ vofb=-0.0943277  lvofb=0.425  wvofb=-8.43e-2
+ vofd=-0.07  lvofd=0  wvofd=0
+ vglow=-0.0087477  lvglow=-0.0470  wvglow=-2.83e-2
+ vghigh=0.1933212  lvghigh=0.0170  wvghigh=4.68e-2
+ n0=1.624  ln0=0  wn0=9.00e-3
+ nb=0  lnb=0  wnb=0.363
+ nd=0  lnd=0  wnd=0
+ mu20=0.0239792  lmu20=-0.0110  wmu20=-2.16e-2
```

"""

```
+ mu2b=0  lmu2b=0  wmu2b=0
+ mu2g=0  lmu2g=0  wmu2g=0
+ mu30=0.8758548  lmu30=-0.378  wmu30=-0.788
+ mu3b=0  lmu3b=0  wmu3b=0
+ mu3g=1.0378  lmu3g=-9.20e-3  wmu3g=-0.936
+ mu40=0.4100637  lmu40=0.176  wmu40=-0.369
+ mu4b=0  lmu4b=0  wmu4b=0
+ mu4g=0  lmu4g=0  wmu4g=0
+ ai0=-0.2295403  lai0=0.0876  wai0=0.207
+ aib=0  laib=0  waib=0
+ bi0=0  lbi0=0  wbi0=0.180
+ bib=0  lbib=0  wbib=0
*- - - - - - - - - - - - - - - - - - - - - - - - -
.model nch.2 nmos capop=9
+ level=39 lmlt=1.0 wmlt=1.0
+ lmin=  .90u lmax= 2.00u
+ wmin= 1.30u wmax= 2.00u
+ acm=3 ldif=0.0u ld=0.1e-6 xl=-0.04e-6 wd=0.1e-6 xw=-0.2e-6
+ delvto=0 tox=130 vdd=5.0
+ mu0=406.3019  lmu0=-14.5  wmu0=-33.3
+ mu0b=-8.7893  lmu0b=-2.25  wmu0b=6.35
+ mus0=600  lmus0=0  wmus0=0
+ musb=0  lmusb=0  wmusb=0
+ ua0=46.0971e-6  lua0=0.131  wua0=2.68e-4
+ uab=-0.0084854  luab=-1.03e-2  wuab=-5.41e-3
+ ub0=0.0094305  lub0=1.18e-2  wub0=-2.45e-3
+ ubb=66.1530e-6  lubb=8.80e-4  wubb=1.70e-3
+ vfb=-0.8227073  lvfb=0.172  wvfb=0
+ phi=0.7499925  lphi=0  wphi=0
+ k1=1.0895  lk1=-0.179  wk1=8.10e-2
+ k2=0.1379169  lk2=-0.0191  wk2=-1.17e-2
+ eta0=0.0055953  leta0=0.0223  weta0=5.22e-3
+ etab=0  letab=0  wetab=0
+ u10=0  lu10=0.0964  wu10=0
+ u1b=0  lu1b=0.0150  wu1b=0
+ u1d=0  lu1d=1.47  wu1d=0
+ vof0=0.6947874  lvof0=0.0165  wvof0=0.410
+ vofb=-0.0943277  lvofb=0.147  wvofb=-8.43e-2
+ vofd=-0.07  lvofd=0  wvofd=0
+ vglow=-0.0087477  lvglow=-1.75e-3  wvglow=-2.83e-2
+ vghigh=0.1933212  lvghigh=1.98e-3  wvghigh=4.68e-2
+ n0=1.624  ln0=-1.32e-2  wn0=9.00e-3
+ nb=0  lnb=0.220  wnb=0.363
+ nd=0  lnd=0  wnd=0
+ mu20=0.0239792  lmu20=-0.0112  wmu20=-2.16e-2
+ mu2b=0  lmu2b=0  wmu2b=0
+ mu2g=0  lmu2g=0  wmu2g=0
+ mu30=0.8758548  lmu30=-0.578  wmu30=-0.788
+ mu3b=0  lmu3b=0  wmu3b=0
+ mu3g=1.0378  lmu3g=-0.589  wmu3g=-0.936
+ mu40=0.4100637  lmu40=-0.0343  wmu40=-0.369
+ mu4b=0  lmu4b=0  wmu4b=0
+ mu4g=0  lmu4g=0  wmu4g=0
```

```
+ ai0=-0.2295403  lai0=0.104  wai0=0.207
+ aib=0  laib=0  waib=0
+ bi0=0  lbi0=0  wbi0=0.180
+ bib=0  lbib=0  wbib=0
*- - - - - - - - - - - - - - - - - - - - - - - -
.model nch.3 nmos capop=9
+ level=39 lmlt=1.0 wmlt=1.0
+ lmin= 2.00u lmax= 20.0u
+ wmin= 1.30u wmax= 2.00u
+ acm=3 ldif=0.0u ld=0.1e-6 xl=-0.04e-6 wd=0.1e-6 xw=-0.2e-6
+ delvto=0 tox=130 vdd=5.0
+ mu0=406.3019  lmu0=-45.8  wmu0=-33.3
+ mu0b=-8.7893  lmu0b=-0.422  wmu0b=6.35
+ mus0=600  lmus0=0  wmus0=0
+ musb=0  lmusb=0  wmusb=0
+ ua0=46.0971e-6  lua0=0.132  wua0=2.68e-4
+ uab=-0.0084854  luab=-1.92e-3  wuab=-5.41e-3
+ ub0=0.0094305  lub0=1.39e-2  wub0=-2.45e-3
+ ubb=66.1530e-6  lubb=1.55e-4  wubb=1.70e-3
+ vfb=-0.8227073  lvfb=0.220  wvfb=0
+ phi=0.7499925  lphi=0  wphi=0
+ k1=1.0895  lk1=-0.245  wk1=8.10e-2
+ k2=0.1379169  lk2=-0.0510  wk2=-1.17e-2
+ eta0=0.0055953  leta0=0.0167  weta0=5.22e-3
+ etab=0  letab=0  wetab=0
+ u10=0  lu10=2.80e-3  wu10=0
+ u1b=0  lu1b=0.187  wu1b=0
+ u1d=0  lu1d=3.59  wu1d=0
+ vof0=0.6947874  lvof0=0.132  wvof0=0.410
+ vofb=-0.0943277  lvofb=0.279  wvofb=-8.43e-2
+ vofd=-0.07  lvofd=0  wvofd=0
+ vglow=-0.0087477  lvglow=-4.84e-3  wvglow=-2.83e-2
+ vghigh=0.1933212  lvghigh=0.0194  wvghigh=4.68e-2
+ n0=1.624  ln0=7.04e-2  wn0=9.00e-3
+ nb=0  lnb=0  wnb=0.363
+ nd=0  lnd=0  wnd=0
+ mu20=0.0239792  lmu20=-0.0257  wmu20=-2.16e-2
+ mu2b=0  lmu2b=0  wmu2b=0
+ mu2g=0  lmu2g=0  wmu2g=0
+ mu30=0.8758548  lmu30=-1.54  wmu30=-0.788
+ mu3b=0  lmu3b=0  wmu3b=0
+ mu3g=1.0378  lmu3g=-1.74  wmu3g=-0.936
+ mu40=0.4100637  lmu40=-0.583  wmu40=-0.369
+ mu4b=0  lmu4b=0  wmu4b=0
+ mu4g=0  lmu4g=0  wmu4g=0
+ ai0=-0.2295403  lai0=0.336  wai0=0.207
+ aib=0  laib=0  waib=0
+ bi0=0  lbi0=0  wbi0=0.180
+ bib=0  lbib=0  wbib=0
*- - - - - - - - - - - - - - - - - - - - - - - -
.model nch.4 nmos capop=9
+ level=39 lmlt=1.0 wmlt=1.0
+ lmin= 20.0u lmax= 1000u
```

```
+ wmin= 1.30u wmax= 2.00u
+ acm=3 ldif=0.0u ld=0.1e-6 xl=-0.04e-6 wd=0.1e-6 xw=-0.2ǝ-6
+ delvto=0 tox=130 vdd=5.0
+ mu0=406.3019  lmu0=0  wmu0=-33.3
+ mu0b=-8.7893  lmu0b=0  wmu0b=6.35
+ mus0=600  lmus0=0  wmus0=0
+ musb=0  lmusb=0  wmusb=0
+ ua0=46.0971e-6  lua0=0  wua0=2.68e-4
+ uab=-0.0084854  luab=0  wuab=-5.41e-3
+ ub0=0.0094305  lub0=0  wub0=-2.45e-3
+ ubb=66.1530e-6  lubb=0  wubb=1.70e-3
+ vfb=-0.8227073  lvfb=0  wvfb=0
+ phi=0.7499925  lphi=0  wphi=0
+ k1=1.0895  lk1=0  wk1=8.10e-2
+ k2=0.1379169  lk2=0  wk2=-1.17e-2
+ eta0=0.0055953  leta0=0  weta0=5.22e-3
+ etab=0  letab=0  wetab=0
+ u10=0  lu10=0  wu10=0
+ u1b=0  lu1b=0  wu1b=0
+ u1d=0  lu1d=0  wu1d=0
+ vof0=0.6947874  lvof0=0  wvof0=0.410
+ vofb=-0.0943277  lvofb=0  wvofb=-8.43e-2
+ vofd=-0.07  lvofd=0  wvofd=0
+ vglow=-0.0087477  lvglow=0  wvglow=-2.83e-2
+ vghigh=0.1933212  lvghigh=0  wvghigh=4.68e-2
+ n0=1.624  ln0=0  wn0=9.00e-3
+ nb=0  lnb=0  wnb=0.363
+ nd=0  lnd=0  wnd=0
+ mu20=0.0239792  lmu20=0  wmu20=-2.16e-2
+ mu2b=0  lmu2b=0  wmu2b=0
+ mu2g=0  lmu2g=0  wmu2g=0
+ mu30=0.8758548  lmu30=0  wmu30=-0.788
+ mu3b=0  lmu3b=0  wmu3b=0
+ mu3g=1.0378  lmu3g=0  wmu3g=-0.936
+ mu40=0.4100637  lmu40=0  wmu40=-0.369
+ mu4b=0  lmu4b=0  wmu4b=0
+ mu4g=0  lmu4g=0  wmu4g=0
+ ai0=-0.2295403  lai0=0  wai0=0.207
+ aib=0  laib=0  waib=0
+ bi0=0  lbi0=0  wbi0=0.180
+ bib=0  lbib=0  wbib=0
*- - - - - - - - - - - - - - - - - - - - - - - - - -
.model nch.5 nmos capop=9
+ level=39 lmlt=1.0 wmlt=1.0
+ lmin=  .70u lmax=  .90u
+ wmin= 2.00u wmax= 20.0u
+ acm=3 ldif=0.0u ld=0.1e-6 xl=-0.04e-6 wd=0.1e-6 xw=-0.2e-Є
+ delvto=0 tox=130 vdd=5.0
+ mu0=406.3019  lmu0=-14.7  wmu0=-57.6
+ mu0b=-8.7893  lmu0b=-0.695  wmu0b=6.91
+ mus0=600  lmus0=0  wmus0=0
+ musb=0  lmusb=0  wmusb=0
+ ua0=46.0971e-6  lua0=0.135  wua0=6.75e-4
```

```
+ uab=-0.0084854  luab=-7.18e-3  wuab=-4.97e-3
+ ub0=0.0094305  lub0=8.22e-3  wub0=-4.14e-3
+ ubb=66.1530e-6  lubb=6.14e-4  wubb=1.48e-3
+ vfb=-0.8227073  lvfb=0.188  wvfb=-3.20e-3
+ phi=0.7499925  lphi=0  wphi=0
+ k1=1.0895  lk1=-0.220  wk1=0.112
+ k2=0.1379169  lk2=-0.0262  wk2=-3.20e-3
+ eta0=0.0055953  leta0=0.0219  weta0=0.0163
+ etab=0  letab=0  wetab=0
+ u10=0  lu10=0.122  wu10=0
+ u1b=0  lu1b=2.82e-3  wu1b=0
+ u1d=0  lu1d=0.892  wu1d=0
+ vof0=0.6947874  lvof0=0.177  wvof0=0.256
+ vofb=-0.0943277  lvofb=0.425  wvofb=-0.102
+ vofd=-0.07  lvofd=0  wvofd=0
+ vglow=-0.0087477  lvglow=-0.0470  wvglow=-2.48e-2
+ vghigh=0.1933212  lvghigh=0.0170  wvghigh=4.96e-2
+ n0=1.624  ln0=0  wn0=0.384
+ nb=0  lnb=0  wnb=0
+ nd=0  lnd=0  wnd=0
+ mu20=0.0239792  lmu20=-0.0110  wmu20=-3.84e-2
+ mu2b=0  lmu2b=0  wmu2b=0
+ mu2g=0  lmu2g=0  wmu2g=0
+ mu30=0.8758548  lmu30=-0.378  wmu30=-1.41
+ mu3b=0  lmu3b=0  wmu3b=0
+ mu3g=1.0378  lmu3g=-9.20e-3  wmu3g=-1.66
+ mu40=0.4100637  lmu40=0.176  wmu40=-0.656
+ mu4b=0  lmu4b=0  wmu4b=0
+ mu4g=0  lmu4g=0  wmu4g=0
+ ai0=-0.2295403  lai0=0.0876  wai0=0.0336
+ aib=0  laib=0  waib=0
+ bi0=0  lbi0=0  wbi0=0
+ bib=0  lbib=0  wbib=0
*- - - - - - - - - - - - - - - - - - - - - - - - -
.model nch.6 nmos capop=9
+ level=39 lmlt=1.0 wmlt=1.0
+ lmin=  .90u lmax= 2.00u
+ wmin= 2.00u wmax= 20.0u
+ acm=3 ldif=0.0u ld=0.1e-6 xl=-0.04e-6 wd=0.1e-6 xw=-0.2e-6
+ delvtc=0 tox=130 vdd=5.0
+ mu0=406.3019  lmu0=-14.5  wmu0=-57.6
+ mu0b=-8.7893  lmu0b=-2.25  wmu0b=6.91
+ mus0=600  lmus0=0  wmus0=0
+ musb=0  lmusb=0  wmusb=0
+ ua0=46.0971e-6  lua0=0.131  wua0=6.75e-4
+ uab=-0.0084854  luab=-1.03e-2  wuab=-4.97e-3
+ ub0=0.0094305  lub0=1.18e-2  wub0=-4.14e-3
+ ubb=66.1530e-6  lubb=8.80e-4  wubb=1.48e-3
+ vfb=-0.8227073  lvfb=0.172  wvfb=-3.20e-3
+ phi=0.7499925  lphi=0  wphi=0
+ k1=1.0395  lk1=-0.179  wk1=0.112
+ k2=0.1379169  lk2=-0.0191  wk2=-3.20e-3
+ eta0=0.0055953  leta0=0.0223  weta0=0.0163
```

```
+ etab=0   letab=0   wetab=0
+ u10=0   lu10=0.0964   wu10=0
+ u1b=0   lu1b=0.0150   wu1b=0
+ u1d=0   lu1d=1.47   wu1d=0
+ vof0=0.6947874   lvof0=0.0165   wvof0=0.256
+ vofb=-0.0943277   lvofb=0.147   wvofb=-0.102
+ vofd=-0.07   lvofd=0   wvofd=0
+ vglow=-0.0087477   lvglow=-1.75e-3   wvglow=-2.48e-2
+ vghigh=0.1933212   lvghigh=1.98e-3   wvghigh=4.96e-2
+ n0=1.624   ln0=-1.32e-2   wn0=0.384
+ nb=0   lnb=0.220   wnb=0
+ nd=0   lnd=0   wnd=0
+ mu20=0.0239792   lmu20=-0.0112   wmu20=-3.84e-2
+ mu2b=0   lmu2b=0   wmu2b=0
+ mu2g=0   lmu2g=0   wmu2g=0
+ mu30=0.8758548   lmu30=-0.578   wmu30=-1.41
+ mu3b=0   lmu3b=0   wmu3b=0
+ mu3g=1.0378   lmu3g=-0.589   wmu3g=-1.66
+ mu40=0.4100637   lmu40=-0.0343   wmu40=-0.656
+ mu4b=0   lmu4b=0   wmu4b=0
+ mu4g=0   lmu4g=0   wmu4g=0
+ ai0=-0.2295403   lai0=0.104   wai0=0.0336
+ aib=0   laib=0   waib=0
+ bi0=0   lbi0=0   wbi0=0
+ bib=0   lbib=0   wbib=0
*- - - - - - - - - - - - - - - - - - - - - - - -
.model nch.7 nmos capop=9
+ level=39 lmlt=1.0 wmlt=1.0
+ lmin= 2.00u lmax= 20.0u
+ wmin= 2.00u wmax= 20.0u
+ acm=3 ldif=0.0u ld=0.1e-6 xl=-0.04e-6 wd=0.1e-6 xw=-0.2e-6
+ delvto=0 tox=130 vdd=5.0
+ mu0=406.3019   lmu0=-45.8   wmu0=-57.6
+ mu0b=-8.7893   lmu0b=-0.422   wmu0b=6.91
+ mus0=600   lmus0=0   wmus0=0
+ musb=0   lmusb=0   wmusb=0
+ ua0=46.0971e-6   lua0=0.132   wua0=6.75e-4
+ uab=-0.0084854   luab=-1.92e-3   wuab=-4.97e-3
+ ub0=0.0094305   lub0=1.39e-2   wub0=-4.14e-3
+ ubb=66.1530e-6   lubb=1.55e-4   wubb=1.48e-3
+ vfb=-0.8227073   lvfb=0.220   wvfb=-3.20e-3
+ phi=0.7499925   lphi=0   wphi=0
+ k1=1.0895   lk1=-0.245   wk1=0.112
+ k2=0.1379169   lk2=-0.0510   wk2=-3.20e-3
+ eta0=0.0055953   leta0=0.0167   weta0=0.0163
+ etab=0   letab=0   wetab=0
+ u10=0   lu10=2.80e-3   wu10=0
+ u1b=0   lu1b=0.187   wu1b=0
+ u1d=0   lu1d=3.59   wu1d=0
+ vof0=0.6947874   lvof0=0.132   wvof0=0.256
+ vofb=-0.0943277   lvofb=0.279   wvofb=-0.102
+ vofd=-0.07   lvofd=0   wvofd=0
+ vglow=-0.0087477   lvglow=-4.84e-3   wvglow=-2.48e-2
```

```
+ vghigh=0.1933212  lvghigh=0.0194  wvghigh=4.96e-2
+ n0=1.624  ln0=7.04e-2  wn0=0.384
+ nb=0  lnb=0  wnb=0
+ nd=0  lnd=0  wnd=0
+ mu20=0.0239792  lmu20=-0.0257  wmu20=-3.84e-2
+ mu2b=0  lmu2b=0  wmu2b=0
+ mu2g=0  lmu2g=0  wmu2g=0
+ mu30=0.8758548  lmu30=-1.54  wmu30=-1.41
+ mu3b=0  lmu3b=0  wmu3b=0
+ mu3g=1.0378  lmu3g=-1.74  wmu3g=-1.66
+ mu40=0.4100637  lmu40=-0.583  wmu40=-0.656
+ mu4b=0  lmu4b=0  wmu4b=0
+ mu4g=0  lmu4g=0  wmu4g=0
+ ai0=-0.2295403  lai0=0.336  wai0=0.0336
+ aib=0  laib=0  waib=0
+ bi0=0  lbi0=0  wbi0=0
+ bib=0  lbib=0  wbib=0
*- - - - - - - - - - - - - - - - - - - - - - - - - -
.model nch.8 nmos capop=9
+ level=39 lmlt=1.0 wmlt=1.0
+ lmin= 20.0u lmax= 1000u
+ wmin= 2.00u wmax= 20.0u
+ acm=3 ldif=0.0u ld=0.1e-6 xl=-0.04e-6 wd=0.1e-6 xw=-0.2e-6
+ delvto=0 tox=130 vdd=5.0
+ mu0=406.3019  lmu0=0  wmu0=-57.6
+ mu0b=-8.7893  lmu0b=0  wmu0b=6.91
+ mus0=600  lmus0=0  wmus0=0
+ musb=0  lmusb=0  wmusb=0
+ ua0=46.0971e-6  lua0=0  wua0=6.75e-4
+ uab=-0.0084854  luab=0  wuab=-4.97e-3
+ ub0=0.0094305  lub0=0  wub0=-4.14e-3
+ ubb=66.1530e-6  lubb=0  wubb=1.48e-3
+ vfb=-0.8227073  lvfb=0  wvfb=-3.20e-3
+ phi=0.7499925  lphi=0  wphi=0
+ k1=1.0895  lk1=0  wk1=0.112
+ k2=0.1379169  lk2=0  wk2=-3.20e-3
+ eta0=0.0055953  leta0=0  weta0=0.0163
+ etab=0  letab=0  wetab=0
+ u10=0  lu10=0  wu10=0
+ u1b=0  lu1b=0  wu1b=0
+ u1d=0  lu1d=0  wu1d=0
+ vof0=0.6947874  lvof0=0  wvof0=0.256
+ vofb=-0.0943277  lvofb=0  wvofb=-0.102
+ vofd=-0.07  lvofd=0  wvofd=0
+ vglow=-0.0087477  lvglow=0  wvglow=-2.48e-2
+ vghigh=0.1933212  lvghigh=0  wvghigh=4.96e-2
+ n0=1.624  ln0=0  wn0=0.384
+ nb=0  lnb=0  wnb=0
+ nd=0  lnd=0  wnd=0
+ mu20=0.0239792  lmu20=0  wmu20=-3.84e-2
+ mu2b=0  lmu2b=0  wmu2b=0
+ mu2g=0  lmu2g=0  wmu2g=0
+ mu30=0.8758548  lmu30=0  wmu30=-1.41
```

```
+ mu3b=0   lmu3b=0   wmu3b=0
+ mu3g=1.0378   lmu3g=0   wmu3g=-1.66
+ mu40=0.4100637   lmu40=0   wmu40=-0.656
+ mu4b=0   lmu4b=0   wmu4b=0
+ mu4g=0   lmu4g=0   wmu4g=0
+ ai0=-0.2295403   lai0=0   wai0=0.0336
+ aib=0   laib=0   waib=0
+ bi0=0   lbi0=0   wbi0=0
+ bib=0   lbib=0   wbib=0
*- - - - - - - - - - - - - - - - - - - - - - - -
.model nch.9 nmos capop=9
+ level=39 lmlt=1.0 wmlt=1.0
+ lmin=  .70u lmax=  .90u
+ wmin= 20.0u wmax= 1000u
+ acm=3 ldif=0.0u ld=0.1e-6 xl=-0.04e-6 wd=0.1e-6 xw=-0.2e-6
+ delvto=0 tox=130 vdd=5.0
+ mu0=406.3019   lmu0=-14.7   wmu0=0
+ mu0b=-8.7893   lmu0b=-0.695   wmu0b=0
+ mus0=600   lmus0=0   wmus0=0
+ musb=0   lmusb=0   wmusb=0
+ ua0=46.0971e-6   lua0=0.135   wua0=0
+ uab=-0.0084854   luab=-7.18e-3   wuab=0
+ ub0=0.0094305   lub0=8.22e-3   wub0=0
+ ubb=66.1530e-6   lubb=6.14e-4   wubb=0
+ vfb=-0.8227073   lvfb=0.188   wvfb=0
+ phi=0.7499925   lphi=0   wphi=0
+ k1=1.0895   lk1=-0.220   wk1=0
+ k2=0.1379169   lk2=-0.0262   wk2=0
+ eta0=0.0055953   leta0=0.0219   weta0=0
+ etab=0   letab=0   wetab=0
+ u10=0   lu10=0.122   wu10=0
+ u1b=0   lu1b=2.82e-3   wu1b=0
+ u1d=0   lu1d=0.892   wu1d=0
+ vof0=0.6947874   lvof0=0.177   wvof0=0
+ vofb=-0.0943277   lvofb=0.425   wvofb=0
+ vofd=-0.07   lvofd=0   wvofd=0
+ vglow=-0.0087477   lvglow=-0.0470   wvglow=0
+ vghigh=0.1933212   lvghigh=0.0170   wvghigh=0
+ n0=1.624   ln0=0   wn0=0
+ nb=0   lnb=0   wnb=0
+ nd=0   lnd=0   wnd=0
+ mu20=0.0239792   lmu20=-0.0110   wmu20=0
+ mu2b=0   lmu2b=0   wmu2b=0
+ mu2g=0   lmu2g=0   wmu2g=0
+ mu30=0.8758548   lmu30=-0.378   wmu30=0
+ mu3b=0   lmu3b=0   wmu3b=0
+ mu3g=1.0378   lmu3g=-9.20e-3   wmu3g=0
+ mu40=0.4100637   lmu40=0.176   wmu40=0
+ mu4b=0   lmu4b=0   wmu4b=0
+ mu4g=0   lmu4g=0   wmu4g=0
+ ai0=-0.2295403   lai0=0.0876   wai0=0
+ aib=0   laib=0   waib=0
+ bi0=0   lbi0=0   wbi0=0
```

```
+ bib=0   lbib=0   wbib=0
*- - - - - - - - - - - - - - - - - - - - - - - -
.model nch.10 nmos capop=9
+ level=39 lmlt=1.0 wmlt=1.0
+ lmin=  .90u lmax= 2.00u
+ wmin= 20.0u wmax= 1000u
+ acm=3 ldif=0.0u ld=0.1e-6 xl=-0.04e-6 wd=0.1e-6 xw=-0.2e-6
+ delvto=0 tox=130 vdd=5.0
+ mu0=406.3019  lmu0=-14.5  wmu0=0
+ mu0b=-8.7893  lmu0b=-2.25  wmu0b=0
+ mus0=600  lmus0=0  wmus0=0
+ musb=0  lmusb=0  wmusb=0
+ ua0=46.0971e-6  lua0=0.131  wua0=0
+ uab=-0.0084854  luab=-1.03e-2  wuab=0
+ ub0=0.0094305  lub0=1.18e-2  wub0=0
+ ubb=66.1530e-6  lubb=8.80e-4  wubb=0
+ vfb=-0.8227073  lvfb=0.172  wvfb=0
+ phi=0.7499925  lphi=0  wphi=0
+ k1=1.0895  lk1=-0.179  wk1=0
+ k2=0.1379169  lk2=-0.0191  wk2=0
+ eta0=0.0055953  leta0=0.0223  weta0=0
+ etab=0  letab=0  wetab=0
+ u10=0  lu10=0.0964  wu10=0
+ u1b=0  lu1b=0.0150  wu1b=0
+ u1d=0  lu1d=1.47  wu1d=0
+ vof0=0.6947874  lvof0=0.0165  wvof0=0
+ vofb=-0.0943277  lvofb=0.147  wvofb=0
+ vofd=-0.07  lvofd=0  wvofd=0
+ vglow=-0.0087477  lvglow=-1.75e-3  wvglow=0
+ vghigh=0.1933212  lvghigh=1.98e-3  wvghigh=0
+ n0=1.624  ln0=-1.32e-2  wn0=0
+ nb=0  lnb=0.220  wnb=0
+ nd=0  lnd=0  wnd=0
+ mu20=0.0239792  lmu20=-0.0112  wmu20=0
+ mu2b=0  lmu2b=0  wmu2b=0
+ mu2g=0  lmu2g=0  wmu2g=0
+ mu30=0.8758548  lmu30=-0.578  wmu30=0
+ mu3b=0  lmu3b=0  wmu3b=0
+ mu3g=1.0378  lmu3g=-0.589  wmu3g=0
+ mu40=0.4100637  lmu40=-0.0343  wmu40=0
+ mu4b=0  lmu4b=0  wmu4b=0
+ mu4g=0  lmu4g=0  wmu4g=0
+ ai0=-0.2295403  lai0=0.104  wai0=0
+ aib=0  laib=0  waib=0
+ bi0=0  lbi0=0  wbi0=0
+ bib=0  lbib=0  wbib=0
*- - - - - - - - - - - - - - - - - - - - - - - -
.model nch.11 nmos capop=9
+ level=39 lmlt=1.0 wmlt=1.0
+ lmin= 2.00u lmax= 20.0u
+ wmin= 20.0u wmax= 1000u
+ acm=3 ldif=0.0u ld=0.1e-6 xl=-0.04e-6 wd=0.1e-6 xw=-0.2e-6
+ delvto=0 tox=130 vdd=5.0
```

```
+ mu0=406.3019   lmu0=-45.8   wmu0=0
+ mu0b=-8.7893   lmu0b=-0.422   wmu0b=0
+ mus0=600   lmus0=0   wmus0=0
+ musb=0   lmusb=0   wmusb=0
+ ua0=46.0971e-6   lua0=0.132   wua0=0
+ uab=-0.0084854   luab=-1.92e-3   wuab=0
+ ub0=0.0094305   lub0=1.39e-2   wub0=0
+ ubb=66.1530e-6   lubb=1.55e-4   wubb=0
+ vfb=-0.8227073   lvfb=0.220   wvfb=0
+ phi=0.7499925   lphi=0   wphi=0
+ k1=1.0895   lk1=-0.245   wk1=0
+ k2=0.1379169   lk2=-0.0510   wk2=0
+ eta0=0.0055953   leta0=0.0167   weta0=0
+ etab=0   letab=0   wetab=0
+ u10=0   lu10=2.80e-3   wu10=0
+ u1b=0   lu1b=0.187   wu1b=0
+ u1d=0   lu1d=3.59   wu1d=0
+ vof0=0.6947874   lvof0=0.132   wvof0=0
+ vofb=-0.0943277   lvofb=0.279   wvofb=0
+ vofd=-0.07   lvofd=0   wvofd=0
+ vglow=-0.0087477   lvglow=-4.84e-3   wvglow=0
+ vghigh=0.1933212   lvghigh=0.0194   wvghigh=0
+ n0=1.624   ln0=7.04e-2   wn0=0
+ nb=0   lnb=0   wnb=0
+ nd=0   lnd=0   wnd=0
+ mu20=0.0239792   lmu20=-0.0257   wmu20=0
+ mu2b=0   lmu2b=0   wmu2b=0
+ mu2g=0   lmu2g=0   wmu2g=0
+ mu30=0.8758548   lmu30=-1.54   wmu30=0
+ mu3b=0   lmu3b=0   wmu3b=0
+ mu3g=1.0378   lmu3g=-1.74   wmu3g=0
+ mu40=0.4100637   lmu40=-0.583   wmu40=0
+ mu4b=0   lmu4b=0   wmu4b=0
+ mu4g=0   lmu4g=0   wmu4g=0
+ ai0=-0.2295403   lai0=0.336   wai0=0
+ aib=0   laib=0   waib=0
+ bi0=0   lbi0=0   wbi0=0
+ bib=0   lbib=0   wbib=0
*- - - - - - - - - - - - - - - - - - - - - - - - -
.model nch.12 nmos capop=9
+ level=39 lmlt=1.0 wmlt=1.0
+ lmin= 20.0u lmax= 1000u
+ wmin= 20.0u wmax= 1000u
+ acm=3 ldif=0.0u ld=0.1e-6 xl=-0.04e-6 wd=0.1e-6 xw=-0.2e-6
+ delvto=0 tox=130 vdd=5.0
+ mu0=406.3019   lmu0=0   wmu0=0
+ mu0b=-8.7893   lmu0b=0   wmu0b=0
+ mus0=600   lmus0=0   wmus0=0
+ musb=0   lmusb=0   wmusb=0
+ ua0=46.0971e-6   lua0=0   wua0=0
+ uab=-0.0084854   luab=0   wuab=0
+ ub0=0.0094305   lub0=0   wub0=0
+ ubb=66.1530e-6   lubb=0   wubb=0
```

```
+ vfb=-0.8227073   lvfb=0   wvfb=0
+ phi=0.7499925   lphi=0   wphi=0
+ k1=1.0895   lk1=0   wk1=0
+ k2=0.1379169   lk2=0   wk2=0
+ eta0=0.0055953   leta0=0   weta0=0
+ etab=0   letab=0   wetab=0
+ u10=0   lu10=0   wu10=0
+ u1b=0   lu1b=0   wu1b=0
+ u1d=0   lu1d=0   wu1d=0
+ vof0=0.6947874   lvof0=0   wvof0=0
+ vofb=-0.0943277   lvofb=0   wvofb=0
+ vofd=-0.07   lvofd=0   wvofd=0
+ vglow=-0.0087477   lvglow=0   wvglow=0
+ vghigh=0.1933212   lvghigh=0   wvghigh=0
+ n0=1.624   ln0=0   wn0=0
+ nb=0   lnb=0   wnb=0
+ nd=0   lnd=0   wnd=0
+ mu20=0.0239792   lmu20=0   wmu20=0
+ mu2b=0   lmu2b=0   wmu2b=0
+ mu2g=0   lmu2g=0   wmu2g=0
+ mu30=0.8758548   lmu30=0   wmu30=0
+ mu3b=0   lmu3b=0   wmu3b=0
+ mu3g=1.0378   lmu3g=0   wmu3g=0
+ mu40=0.4100637   lmu40=0   wmu40=0
+ mu4b=0   lmu4b=0   wmu4b=0
+ mu4g=0   lmu4g=0   wmu4g=0
+ ai0=-0.2295403   lai0=0   wai0=0
+ aib=0   laib=0   waib=0
+ bi0=0   lbi0=0   wbi0=0
+ bib=0   lbib=0   wbib=0
```

Index

M